Grundlehren der mathematischen Wissenschaften 327

A Series of Comprehensive Studies in Mathematics

Springer
Berlin
Heidelberg
New York
Hong Kong
London
Milan
Paris
Tokyo

Preface to the English Edition

This book discusses a beautiful and central problem in mathematics, which involves geometry, number theory, coding theory and group theory. The geometrical objects we consider are *lattices*, i.e. discrete subgroups of maximal rank in a Euclidean space. To such an object we attach its canonical *sphere packing*, namely the set of (non-overlapping) spheres centred at all points of the lattice whose common radius is half the minimal distance of two lattice points. Assuming some regularity conditions, a sphere packing has a density. The question of estimating the highest possible density of a sphere packing in a given dimension n is a fascinating and difficult problem: the answer is known only up to dimension 3, and the case of dimension 3 was settled very recently by Hales, who gave a positive answer to an old conjecture of Kepler. The case of lattice packings is slightly easier, though still highly non-trivial: in this case, the answer is known up to dimension 8, thanks to the difficult 1935 paper [Bl2] of Blichfeldt.

The book is centred on the study of *extreme lattices*, those on which the density of the canonical sphere packing attains a local maximum, and various related questions. This is based on the French version published by Masson in 1996 under the title "Les réseaux parfaits des espaces euclidiens". However, it is very far from being a word-for-word translation. Every chapter has been rewritten, some completely, For example, I found a much simpler proof of the classification of 5-dimensional perfect lattices after the French version was printed, and as a result Section 4 of Chapter 6 has been completely changed.

A more detailed list of the major changes can be found in the *footnotes* to the *Introduction* (which otherwise follows the French version). The reader who possesses the French Edition can find an erratum and a full list of the changes on the web page http://www.math.u-bordeaux.fr/~martinet/. A number of readers of the French version supplied lists of corrections and of course these have all been made in the new edition. I should like to thank more specially Anne-Marie Bergé, Philippe Calame and Maurice Mischler for their comments.

During the six years since the French version was published there has been important progress in some areas directly connected with the main topics of the book. For this reason an essentially descriptive appendix has been added, whose main objective is to discuss the work of B.B. Venkov and others connecting the local theory of lattices with *spherical designs.*

I must mention here the recent work of Bavard, who has incorporated the theory of Chapter 10 (which considers families of lattices which form orbits under the action of a Lie subgroup of the linear group) in a more general setting in the context of Riemannian geometry. This has provided unified definitions for some ad hoc notions that were introduced in order to obtain reasonable finiteness theorems or conditions to guarantee that certain lattices are algebraic. However, it would have taken too much space to include a full discussion of this work here, and I can only refer the reader to Bavard's papers mentioned in the bibliography.

Finally, the bibliography has been updated and the index has been greatly enlarged. There is an extensive list of symbols.

The following are the most important changes that have been made (besides the modifications to the appendices and the bibliography):

- Chapter 1: Section 1.9 contains further material on integral lattices.
- Chapter 4: Changes in Section 4.8, and six new exercises.
- Chapter 6: Section 6.4 has been completely rewritten.
- Chapter 8: Section 8.1 has been completely rewritten to take into account the modifications to the appendices.
- Chapter 9: There is a new Section 9.7 devoted to recent results by Batut and by Bavard.
- Chapter 14: The first section has been shortened, and a new theorem has been incorporated in Section 14.6.

Many improvements were suggested to me by Neil Sloane after he read in detail my manuscript. I thank him for his important contribution to the book. I would like to thank also Henri Cohen for his help in using Springer's Latex.

Finally, I express my heartfelt thanks to my wife Titou for her patience during the years I have been writing the two editions of this book.

Talence, March 28th, 2002.

Jacques Martinet

General Principles for the Notation

The following notation will be used throughout the book unless explicitly stated to the contrary. (For instance, it may happen that we consider a sequence $L_1, L_2, \ldots$ of lattices.)

1. E denotes a Euclidean space, whose dimension is denoted by n.
2. The notation $F \perp F' \perp \ldots$ is used for *orthogonal* direct sums.
3. In a Euclidean space F of dimension $n+1$ (resp. n), $(\varepsilon_0, \varepsilon_1, \ldots, \varepsilon_n)$ (resp. $(\varepsilon_1, \ldots, \varepsilon_n)$) denotes an orthonormal basis for F. Latin letters, usually $(e_1, \ldots, e_n)$, are used for a basis which is not *a priori* orthonormal.
4. The symbols Λ, Λ', L, L', M etc. denote lattices. A subscript such as Λ_m indicates the dimension.

The notation for duality deserves special comments. The symbol M^* means first that M is a lattice, and that we consider its dual *over* $\mathbb{Z}$, even if M is endowed with some other algebraic structure. If we need to consider the dual of M over some ring containing strictly $\mathbb{Z}$, we write $M^\sharp$. Finally the set of invertible elements in a ring R is denoted by $R^\times$.

The notation from linear algebra is essentially the standard one ($\mathrm{End}(V)$, $\mathrm{GL}(V)$, $\mathrm{SL}(V)$, $\mathrm{O}(V)$). As for matrices, we denote by $\mathcal{M}_{p,q}(R)$ the module of matrices with p rows and q columns over a ring R, and use the shorter notation $\mathcal{M}_n(R)$ for $\mathcal{M}_{n,n}(R)$.

Finally, we use two notions of norm. In a Euclidean space the norm $x \cdot x$ is denoted by $N(x)$. In an algebra L/K the determinant of the endomorphism $y \mapsto xy$ is denoted by $\mathrm{N}(x)$ (or $\mathrm{N}_{L/K}(x)$). Note that $N(x) = \|x\|^2$ is the square of the classical Euclidean norm.

Statements (theorems, propositions, definitions, tables, etc.) are numbered according to the scheme **Theorem a.b.c.**, **Table a.b.c.**, etc., where **a** is the number of the chapter, **b** is that of the section, and **c** denotes the **c**-th statement in Section **b** of Chapter **a**, independently of its nature. A reference to Theorem 5 of Section 3 in Chapter 4 is thus written in the form Theorem 4.3.5.

Introduction

This is based on the Introduction to the French version, the main differences being the addition of footnotes to indicate places where significant changes have been made in the English version.

Throughout this book, E denotes an n-dimensional *Euclidean space*, i.e. a finite-dimensional real vector space endowed with a positive definite symmetric bilinear form, namely the *scalar product on E*. A *lattice in E* is a discrete finitely generated subgroup of E of maximal rank, i.e. of rank n.

Let Λ be a lattice, and let d be the smallest distance to the origin of the other points of Λ; d is also the smallest distance between two points of Λ. Consequently, balls of radius $R = \frac{d}{2}$ centred at points of Λ do not overlap (their intersection contains at most one point). We say that they are *packed* by Λ. The *density* of this packing is an important invariant of Λ, sometimes called by abuse of language "the density of Λ". Another important invariant is the *kissing number of* Λ, the number of length d vectors in Λ (the *minimal vectors of* Λ), which we will usually denote by $2s$ (these vectors occur in pairs $\pm x$).

Lattices for which this density is a local maximum, i.e. such that this density does not increase when one performs a sufficiently small deformation, were called *extreme lattices* by Korkine and Zolotareff. Their goal, which they were able to carry out up to dimension 5, was to classify extreme lattices in a given dimension and then to extract from this classification a list of the absolutely densest ones, which they called the *absolutely extreme* lattices. However, we prefer to call these *critical lattices*, in conformity with the tradition in the geometry of numbers.

The notion of a *perfect lattice* is a less restrictive one, which can be expressed within the framework of linear algebra, whereas inequalities are required to characterize extreme lattices. These are the lattices with the property that the set of projections onto the lines containing the minimal vectors spans the space of symmetric endomorphisms of E. Less formal properties will show up later, for instance the following, which goes back to Voronoi: a lattice Λ is perfect if the image of Λ under any sufficiently small deformation which is not a similarity of E contains fewer minimal vectors than Λ itself. The formal definition shows that the inequality $s \geq \dfrac{n(n+1)}{2}$ holds for all perfect lattices. Let us also quote Voronoi's finiteness theorem: up to similarity, there exist only finitely many perfect lattices in any given dimension.

*

* *

In fact, the theory of perfect lattices and sphere packings was developed up to the 1960s as part of the theory of positive definite real quadratic forms. However, we prefer to work with the *Hermite invariant* of a lattice Λ rather than with its density. This is defined by

$$\gamma(\Lambda) = N(\Lambda)\ \det(\Lambda)^{-1/n}$$

where $N(\Lambda)$ (the *norm* or *minimum of* Λ) is the minimum on $\Lambda \smallsetminus \{0\}$ of the scalar products $N(x) = x.x$, and $\det(\Lambda)$, the *determinant of* Λ, is the determinant of the matrix of pairwise scalar products of the vectors of any $\mathbb{Z}$-basis $\mathcal{B}$ for Λ (the *Gram matrix of* $\mathcal{B}$). Then the density is proportional to $\gamma(\Lambda)^{n/2}$. Since the density is bounded from above (by 1), one may consider the upper bound of $\gamma(\Lambda)$ on the set of all n-dimensional lattices: this is the *Hermite constant for dimension* n, denoted by γ_n. (A more natural invariant would be γ^n, which was shown by Korkine and Zolotareff to take rational values on perfect lattices.)

The definitions above, which were expressed in terms of lattices, are easily translated into the language of *quadratic forms over* $\mathbb{R}^n$ (homogeneous degree 2 polynomials in n variables $x_1, \dots, x_n$): with a given a pair $(\Lambda, \mathcal{B})$ where Λ is a lattice and $\mathcal{B} = (e_1, \dots, e_n)$ is a basis for Λ over $\mathbb{Z}$, we associate the quadratic form

$$q(x_1, \dots, x_n) = N(x_1 e_1 + \cdots + x_n e_n)\,.$$

To replace $\mathcal{B}$ by another basis amounts to replacing q by an *equivalent form* (equivalent under a transformation of $\mathrm{GL}_n(\mathbb{Z})$). Passing to quotients, we establish a one-to-one correspondence between isometry classes of lattices and equivalence classes of quadratic forms. This dictionary relating lattices and quadratic forms is studied in detail in Chapter 1; notice that it induces a one-to-one correspondence between similarity classes of lattices, the natural object in the theory of perfect or extreme lattices, and classes of quadratic forms *up to proportionality*.

When working with quadratic forms, the existence of the constant γ_n is not evident. It was established by Hermite in a letter to Jacobi (dated August 6, 1845), although Lagrange (resp. Gauss) had already calculated it for dimension 2 (resp. 3).

After Hermite's work, the problem of determining γ_n beyond dimension 3 was considered. The case of dimensions 4 and 5 was solved by Korkine and Zolotareff in a series of three papers published in *Mathematische Annalen* between 1872 and 1877, in which they indeed determine *all* perfect forms in dimension $n \leq 5$ (although without using the word "perfect", which was introduced by Voronoi some thirty years later).

The translation in terms of lattices was done by Minkowski, who realized that *estimating the minima of the set of all real (positive definite) quadratic forms on* one *particular lattice* (for instance on $\mathbb{Z}^n$, which is the aim of the classical theory) is essentially the same problem as *estimating the minima on the set of all lattices in* $\mathbb{R}^n$ *on* one *particular form* (for instance, those of $x_1^2 + \cdots + x_n^2$, which defines the canonical Euclidean structure on $\mathbb{R}^n$).

This extremely original idea, of fundamental importance despite its simplicity, constitutes the birth certificate of the geometry of numbers, a new branch of mathematics whose autonomous existence can be reasonably dated to 1896, the year Minkowski's book *Geometrie der Zahlen* appeared.

As for the Hermite constant, Minkowski's geometrical methods prove the existence of upper bounds which are linear in n, whereas the arguments that Hermite used to prove existence give only exponential bounds.

Minkowski never considered perfect forms, and the next step was made by Voronoi; it consists of three articles published in Crelle's journal between 1907 and 1909. It is his first article which chiefly interests us: Voronoi proves here his finiteness theorem referred to above, characterizes the extreme forms among perfect forms (they must have the additional property of being *eutactic*), and develops an algorithm with which he recovers the classification results found thirty years earlier by Korkine and Zolotareff.

Unfortunately, Voronoi died in 1908, leaving his research unfinished. He was clearly working on the case of dimension 6, where he had found the first non-eutactic perfect form, but one had to wait half a century for the completion of the 6-dimensional classification, obtained by Barnes in 1957. Barnes also discovered many other perfect lattices, including the lattice K_{11}, and together with G.E. Wall also found Λ_{15} and Λ_{16}. These are the densest lattices known in dimensions 11, 15 and 16 and are widely believed to be the critical lattices in these dimensions.

We must mention two important results that were obtained in the period between the works of Voronoi and Barnes. The first is the determination by Blichfeldt in 1935 of the Hermite constant in dimensions 6, 7 and 8 (and the reduction by Mordell in 1944 of the calculation of γ_8 to that of γ_7); the other is Coxeter's 1951 paper on root lattices and their relatives, and the discovery in a joint work with Todd of the K_{12} lattice, which they showed to be extreme, and which is very likely the densest 12-dimensional lattice.

After Barnes, among various work devoted to local methods, I would like to mention the following:

• Watson's work, extending the methods of Korkine and Zolotareff in order to attempt to classify perfect lattices in dimensions 6 and 7.

• Kaye Stacey's work (1975), which obtained an essentially correct list of the 7-dimensional perfect lattices. However, she was unable to establish that her list was complete, for lack of an efficient identification algorithm; her results were confirmed by Jaquet in 1990. Barnes and Jaquet worked with the Voronoi algorithm, whereas Stacey used the of Korkine–Zolotareff–Watson methods.

• The construction of important lattices, by Leech (in particular, the fundamental 24-dimensional lattice which bears his name), by Conway and Sloane (for instance, the "laminated lattices" Λ_n), by Quebbemann in dimension 32, ... , all of which are perfect.

• The systematic study by Conway and Sloane of perfect lattices up to dimension 7.

• The definition by Bergé and Martinet of new notions of perfection and extremality restricted to special classes of lattices (e.g., lattices which are extreme with respect to a given automorphism group or with a given section, or as isodual lattices, or dual-extreme lattices – an analogue of the Hermite invariant for which a lattice and its dual play a joint rôle).

*

* *

We begin with a chapter on general properties of lattices. The second chapter on inequalities is more technical and is not used much before Chapter 6. The story really begins in Chapter 3,[1] where we introduce the notions of extremality, perfection and eutaxy of a lattice, and the analogous notions involving the pair consisting of a lattice and its dual. It is thus quite possible to start reading the book at the third chapter. This chapter is a particular case of a more general theory to be developed in Chapter 10, where we shall consider families of lattices which are orbits under the action of a closed subgroup of the general linear group. However, the basic techniques developed in Chapter 3 will play a fundamental rôle, and a detailed study of the classical situation may prove useful for understanding Chapter 10.

Chapter 4 is devoted to root lattices, whose rôle in numerous domains of mathematics (group theory, Lie algebras, etc.) is well-known. It is followed by a fifth chapter dealing with various lattices which are easily constructed as modifications of root lattices. Both these chapters contain many applications of the theory developed in Chapter 3. Moreover, the families of lattices that we construct in these two chapters contain almost all perfect lattices up to dimension 6. (The complete list is obtained by considering two extra families of lattices, which we construct in Chapter 8.)

Chapter 6 is devoted to the classification of perfect lattices. We give complete proofs up to dimension 5, but content ourselves with a description of the known results in dimensions 6 and 7 and a few indications for dimension 8, since classification is not known from $n = 8$ onwards. The proofs we give in this chapter originate from the work of Korkine and Zolotareff, although we must emphasize that they proved more than they stated: for instance, the

[1] We also introduce here the notion (not mentioned in the French edition) of a strongly eutactic lattice, which plays an important rôle in the study of spherical designs.

necessary condition $s(\Lambda) \geq \frac{n(n+1)}{2}$ suffices to ensure perfection (and even extremality) up to dimension 5, except for 5-dimensional lattices possessing a $\mathbb{D}_4$-section with the same norm, where one must assume the existence of five directions of minimal vectors outside the section.[2] We also describe all dual-extreme lattices up to dimension 4 (the classification is not known from $n = 5$ onwards), but we do not give the complicated proof of the classification in dimension 4.

Chapter 7 is devoted to the classical Voronoi algorithm. Now we drop the language of lattices for that of quadratic forms, which is here more convenient. (However, the lattice point of view shows up in certain places, and Voronoi himself sometimes used it without saying so.) We describe Voronoi's procedure, which allows one to attach to every perfect form (defined up to proportionality by giving it a fixed minimum on $\mathbb{Z}^n$) a polyhedral convex cone, its *Voronoi domain*, and to every facet of this cone a new perfect form, which is *contiguous* to the original one (through the given facet), giving in this way the set of equivalence classes of perfect forms a structure of a finite, connected graph. Following Voronoi, we use this method to recover the classification of perfect forms of dimension $n \leq 5$ that we established in the preceding chapter.

The reader *could* learn the Voronoi algorithm by directly reading Chapter 13, ignoring the first four sections of Chapter 7. However, the remark we made above about Chapters 3 and 10 applies to Chapters 7 and 13. Note also that reading Chapters 6 and 7 directly after Chapter 3 (in any order) is also possible for a reader having some acquaintance with the zoology of lattices constructed in Chapters 4 and 5.

Chapter 8 is a continuation of the constructions of lattices given in Chapters 4 and 5 that were interrupted by two chapters on classification problems. We consider here constructions of a more algebraic nature, making use of orders in semi-simple algebras, essentially in fields with complex multiplication or quaternion fields. This chapter contains various original results. The constructions of Barnes and Coxeter–Todd, which make use of the ring of Eisenstein integers, are described and generalized in Sections 8.4 and 8.5, after some analogous constructions making use of the Hurwitz order have been performed in Sections 8.2 and 8.3. This is followed by sections describing new constructions relying on the structure of some left ideals in quaternion skew-fields whose centres are no longer the field of rational numbers. Numerous classical lattices are endowed in this way with various algebraic structures, and new lattices, often unimodular, are obtained; a résumé of the original results of this chapter appeared in the proceedings of the Paris Number Theory Seminar of 1992–1993.

Chapter 9, which is based on results by A.-M. Bergé and myself, deals with classifications of lattices (or of pairs of a lattice and its dual) according to

[2] Here is one important difference from the French edition: following [Mar5], we have written much simpler proofs, relying on methods of Watson.

properties of their minimal vectors. As an application, we prove a finiteness theorem which contains results of Voronoi (on perfect lattices) and Avner Ash (on eutactic lattices), and also a dual version due to A.-M. Bergé of Voronoi's finiteness theorem. This subdivision of lattices into finitely many *minimal classes*, which contains the classification of perfect lattices (those for which the minimal class reduces to their similarity class), is known only for $n \leq 5$ (resp. only for $n \leq 3$ in case of pairs (Λ, Λ^*)). We present the proofs in dimensions $n \leq 4$ (resp. $n \leq 3$). Some results of Watson about the kissing number have a natural interpretation in terms of minimal classes; it would be interesting to extend at least partially these classification results beyond dimension 5. (Minimal classes are kind of "orbifold"; they correspond to the cell decomposition in the space of positive definite quadratic forms; this cell decomposition has been considered by the Russian school, notably by Štogrin, Baranovskii, and Ryshkov.)

Chapter 10 gives a broad generalization[3] of Voronoi's characterization of extreme lattices as those which are both perfect and eutactic. Notions of perfection and eutaxy for a subspace $\mathcal{T}$ of the space $\mathrm{End}^s(E)$ of all symmetric endomorphisms were defined in Chapter 3. We consider here families of lattices that constitute a homogeneous space under the action of a closed (possibly connected) subgroup $\mathcal{G}$ of $\mathrm{GL}(E)$, which we also assume to be invariant under transposition. Being a Lie subgroup of $\mathrm{GL}(E)$, $\mathcal{G}$ possesses a tangent space at the identity, and Voronoi's theory generalizes, working in the symmetrized set $\mathcal{T} \subset \mathrm{End}^s(E)$ of this tangent space.

The preceding theory is applied in Chapter 11 to two examples of a great practical importance.

The first one, studied in the first four sections, is that of G-lattices, the set of lattices in E whose automorphism group contain a given finite subgroup G of the orthogonal group $\mathrm{O}(E)$. The idea of considering such lattices originates in algebraic number theory, G being a Galois group acting on units modulo torsion.

Our second example, studied in the remainder of the chapter, is that of lattices endowed with a given isometry σ onto their duals. The case of symplectic lattices (i.e., such that $\sigma^2 = -\mathrm{Id}$) is of particular importance in the theory of complex Abelian varieties.

In Chapter 12, we consider the following two questions: How can one classify the sections of a given lattice? And how can one characterize the perfection and eutaxy properties of a lattice for which we know *a priori* a section?

These problems somewhat resemble those which were examined in Chapter 10, and it might well be possible to unify the two theories by considering

[3] Notions of perfection and eutaxy, which generalize those we consider in this book, have been developed recently by Bavard within the framework of Riemannian geometry.

lattices which are parametrized by submanifolds with boundary in the vector space of endomorphisms. No such theory is yet available.

In Chapter 13 – the last largely theoretical one – we consider extensions of the classical classical Voronoi algorithm that was studied in Chapter 7. The theory is applied relative to a subspace $\mathcal{T}$ of $\mathrm{End}^s(E)$, but is in practice useless unless the contiguity algorithm stabilizes an interesting class of lattices. This is the case with G-lattices in the sense of Chapter 11: $\mathcal{T}$ is the subspace consisting of elements of $\mathrm{End}^s(E)$ which commute with G and the operation of "$\mathcal{T}$-contiguity" transforms a G-lattice into a new G-lattice; this property also holds for the family of lattices possessing a given section with the same minimum. Unfortunately, we are able to generalize the Voronoi algorithm only to families which constitute a cone[4] inside an affine subspace of $\mathrm{End}^s(E)$. To discover an algorithm for families such as those of symplectic lattices would be of a great interest for the solution of classification problems.

We have collected in the final chapter, Chapter 14, some numerical data (Gram matrices of perfect or eutactic lattices, tables of invariants of some remarkable lattices). Numerical complements are also available on the WEB; see the Batut–Martinet Catalogue of Perfect Lattices

http://www.math.u-bordeaux.fr/~martinet/

and the Nebe–Sloane database

http://www.research.att.com/~njas/lattices/;

see also the tables in Chapter 6.

The book ends with two appendices, the first one of an algebraic nature written on request of certain colleagues who wished a guide to algebraic constructions of lattices. It is certainly not an accident if so many "beautiful" lattices, especially in even dimensions, and above all in dimensions divisible by 4, possess rich algebraic structures, and the search for algebraic constructions is indeed still an active domain in lattice theory. The second appendix is a short account of the connections which exist between the theory of lattices and that of spherical designs.[5]

*

* *

Every chapter except the first and the last contains numerous exercises, some of which are not in the French edition. Their numbers vary from chapter to chapter. In general they can be solved by hand, but in a few cases computer programs (especially those for finding minimal vectors) can be used to avoid

[4] Recently, Bavard has considered in [Bav3] a generalization which deals with symplectic lattices in the Siegel space parametrized by the Poincaré upper half-plane.

[5] This appendix is new; the first appendix is a contraction of the four appendices of the French version. This is partially compensated for by an expansion of Section 8.1.

tedious calculations. For the sake of simplicity, those chapters involving heavy computational methods contain only few exercises.

Each chapter except the last (and the appendices), ends with "notes", where we quote some results which could not be incorporated within the chapter. The notes also include a number of historical remarks. They have been extended to include results discovered after the French edition was written.

*

* *

This book is devoted to the study of *local methods* in lattice theory. I have chosen to centre it on the perfection property. I realized while writing it that eutaxy, which appears at first sight to be just a minor restriction, is just as important as perfection. Moreover, the discovery of the existence of an amazing number of 8-dimensional perfect lattices[6] in comparison with smaller dimensions (3, 7, and 33 in dimensions 5, 6, and 7), strongly suggests that the eutaxy property could be used for limiting the number of lattices that must be considered when classifying extreme lattices. For this reason, the eutaxy property also plays a major rôle in this book.

Venkov has recently discovered that the theory of modular forms can be used to prove *a priori* that certain lattices are extreme, or at least eutactic; the proofs rely on the theory of *spherical designs*. We shall not give a detailed study of this theory, about which nothing had been written at the time the French edition was printed. However, an appendix, of a purely descriptive nature, has been added to the English edition; the proofs can be read in the recent article [Ven3].

Another recent breakthrough was made independently by Elkies and Shioda, who constructed dense lattices using algebraic geometry. They obtain integral lattices for which they can determine the minimum and the determinant, and even the kissing number. However, up to now, this theory has not yielded results concerning perfection nor eutaxy.

*

* *

The notion of an integral lattice, although it appears here and there in this book (perfect lattices are proportional to integral lattices, as one knows since the time of Korkine, Zolotareff, and Voronoi), was not at the heart of our study. As a consequence, the content of this book is essentially disjoint from that of Conway and Sloane's "bible" (*Sphere Packings, Lattices and Groups* [C-S], [C-S'], [C-S"]), which remains a necessary tool for any one who wishes to study lattices.

---

[6] 10916 were known at the time this introduction was translated

Contents

1 General Properties of Lattices 1
- 1.1 Lattices in Real Vector Spaces 1
- 1.2 Lattices in Euclidean Spaces 4
- 1.3 Duality 7
- 1.4 Automorphism Groups 12
- 1.5 Bilinear and Quadratic Forms 17
- 1.6 Quadratic Forms 18
- 1.7 The Dictionary Relating Lattices and Quadratic Forms 20
- 1.8 Packings 24
- 1.9 More on Integral Lattices 26
- 1.10 Tensor Product and Exterior Powers 30
- 1.11 Notes on Chapter 1 33

2 Geometric Inequalities 37
- 2.1 The Hadamard Inequality 37
- 2.2 The Hermite Inequality 39
- 2.3 The Mordell Inequality 41
- 2.4 Mahler's Compactness Theorem 43
- 2.5 Lattice Constants 45
- 2.6 Extreme Lattices for an Open Star Body 47
- 2.7 The Lattice Constant for a Convex Star Body 51
- 2.8 Generalizations of the Hermite Invariant 55
- 2.9 The HKZ Reduction 59
- 2.10 Exercises for Chapter 2 62
- 2.11 Notes on Chapter 2 63

3 Perfection and Eutaxy 67
- 3.1 Symmetric Endomorphisms 67
- 3.2 Linear Forms on Spaces of Endomorphisms 73
- 3.3 Linear Inequalities 77
- 3.4 A Characterization of Extreme Lattices 78
- 3.5 Perfect Configurations 81
- 3.6 Eutactic Configurations and Extreme Lattices 84
- 3.7 The Lamination Process 92

3.8 Dual-Extreme Lattices ... 94
3.9 Exercises for Chapter 3 ... 100
3.10 Notes on Chapter 3. ... 105

4 Root Lattices ... 109
4.1 The $\mathbb{Z}^n$ Lattice ... 110
4.2 The $\mathbb{A}_n$ Lattice ... 110
4.3 The $\mathbb{D}_n$ Lattice ... 112
4.4 The $\mathbb{D}_n^+$ Packing and the $\mathbb{E}_8$ Lattice ... 114
4.5 The Lattices $\mathbb{E}_7$ and $\mathbb{E}_6$ ... 117
4.6 Graphs and Inclusions Between Root Lattices ... 120
4.7 Perfection and Eutaxy ... 124
4.8 Some Other Constructions for Root Lattices ... 125
4.9 Residual Quadratic Forms ... 129
4.10 Root Systems ... 131
4.11 Exercises for Chapter 4 ... 138
4.12 Notes on Chapter 4 ... 145

5 Lattices Related to Root Lattices ... 147
5.1 The Coxeter–Barnes Lattices $\mathbb{A}_n^r$ ... 147
5.2 The Coxeter Lattices $\mathbb{A}_n^r$ ... 153
5.3 Barnes's Lattices P_n ... 157
5.4 Craig's Difference Lattices ... 163
5.5 Lattices Related to the $\mathbb{D}_n$ Lattice ... 171
5.6 Unimodular Lattices ... 174
5.7 Around the Leech Lattice ... 177
5.8 Exercises for Chapter 5 ... 182
5.9 Notes on Chapter 5 ... 187

6 Low-Dimensional Perfect Lattices ... 189
6.1 A Combinatorial Characterization of the $\mathbb{A}_n$ Lattices ... 190
6.2 Perfect Lattices up to Dimension 4 ... 194
6.3 Dual-Extreme Lattices up to Dimension 4 ... 196
6.4 Perfect Lattices in Dimension 5 ... 200
6.5 Perfect Lattices in Dimensions 6 and 7 ... 208
6.6 Some Indications About 8-Dimensional Perfect Lattices ... 212
6.7 Exercises for Chapter 6 ... 219
6.8 Notes on Chapter 6 ... 223

7 The Voronoi Algorithm ... 227
7.1 Voronoi Domains ... 227
7.2 Contiguity ... 234
7.3 Finiteness Results ... 237
7.4 The Voronoi Graphs ... 238
7.5 Lattices Contiguous to $\mathbb{A}_n$ ... 241

7.6 The Voronoi Algorithm in Dimension 4 242
7.7 The Facets of $\mathbb{D}_n$ and the 5-Dimensional Perfect Lattices 244
7.8 Determination of the Contiguous Form 253
7.9 Perfect Forms in Dimensions 6 and 7 254
7.10 Exercises for Chapter 7 258
7.11 Notes on Chapter 7 260

8 Hermitian Lattices 263
8.1 Complex and Quaternionic Structures 263
8.2 Hurwitz Lattices: Enlargements of $\mathbb{D}_n$ 268
8.3 Hurwitz Lattices: Around Dimension 16 274
8.4 Eisenstein Lattices: A Construction of Barnes 280
8.5 Eisenstein Lattices: The Coxeter–Todd Lattice 284
8.6 A General Construction of Hermitian Lattices 292
8.7 Quadratic Hermitian Structures 298
8.8 Beyond Dimension 24 306
8.9 Exercises for Chapter 8 310
8.10 Notes on Chapter 8 316

9 The Configurations of Minimal Vectors 321
9.1 Minimal-Equivalent Lattices 321
9.2 Classes of Dimension $n \leq 3$ 329
9.3 Classification in Dimension 4 333
9.4 Weakly Eutactic Lattices in a Minimal Class 339
9.5 The Classification of Eutactic Lattices 343
9.6 Perfect Pairs of Lattices 348
9.7 Complements 354
9.8 Exercises for Chapter 9 357
9.9 Notes on Chapter 9 361

10 Extremal Properties of Families of Lattices 363
10.1 Some Elementary Results on Lie Groups 364
10.2 Perfection and Eutaxy 366
10.3 Extremality 368
10.4 Minimal Classes 373
10.5 Dual-Extreme Lattices 374
10.6 The Rankin Invariants 376
10.7 Exercises for Chapter 10 379
10.8 Notes on Chapter 10 380

11 Group Actions 383
11.1 Rational and Integral Representations 383
11.2 G-Lattices 385
11.3 G-extreme Lattices 388
11.4 Cyclotomic Lattices 393

11.5 Isodual Lattices, Modular Lattices, and Normal Lattices 400
11.6 Normal Lattices 403
11.7 Extreme Symmetric and Symplectic Lattices 407
11.8 Isodual Lattices: Examples and Classification Results 414
11.9 Rationality and Finiteness Questions 417
11.10 Exercises for Chapter 11 421
11.11 Notes on Chapter 11 425

12 Cross-Sections 427
12.1 Embedding a Lattice in a Larger One 427
12.2 X-Rays of Lattices 430
12.3 Lattices with a Fixed Cross-Section 433
12.4 A Characterization of Relatively Extreme Lattices 437
12.5 Patchwork Lattices 438
12.6 Exercises for Chapter 12 439
12.7 Notes on Chapter 12 440

13 Extensions of the Voronoi Algorithm 443
13.1 Contiguity Relative to a Space of Symmetric Matrices 444
13.2 The Voronoi Algorithm Relative to a Space of Symmetric Matrices 448
13.3 Perfect *G*-Lattices 451
13.4 Two-Dimensional Centralizers 454
13.5 Cyclotomic Lattices 458
13.6 Lattices with a Fixed Section and Patchwork Lattices 460
13.7 Examples 463
13.8 Exercises for Chapter 13 464
13.9 Notes on Chapter 13 465

14 Numerical Data 467
14.1 Low-Dimensional Perfect Lattices 467
14.2 Root Lattices 468
14.3 Eutactic Lattices up to Dimension 4 469
14.4 The Hermite Constant 472
14.5 Invariants Related to Duality 474
14.6 The Kissing Number 476

15 Appendix 1: Semi-Simple Algebras and Quaternions 479
15.1 Semi-Simple Algebras 479
15.2 Quaternion Algebras 481
15.3 Algebraic Lattices over Dedekind Domains 482
15.4 Arithmetic in Separable Algebras 484
15.5 Number Fields 485
15.6 Quaternions Again 486
15.7 Ideal Class Set 487

16 Appendix 2: Strongly Perfect Lattices 489
16.1 Spherical Designs 489
16.2 Strong Perfection 491
16.3 An Infinite Series 492
16.4 Modular Lattices 493
16.5 Group Theory 494
16.6 Designs on Grassmannian Varieties 495

References 497

List of Symbols 511

Index 517

1 General Properties of Lattices

Introduction

The aim of this chapter is to give an account of the basic results which are needed to work with *lattices*, i.e. *discrete subgroups of maximal rank* in *Euclidean vector spaces*. Throughout the book, E stands for a real vector space of finite dimension n. From Section 1.2 onwards, the space is assumed to be Euclidean.

1.1 Lattices in Real Vector Spaces

In this section, E denotes a real vector space of finite dimension n, endowed with its natural topology, which can be defined by an arbitrary norm.

Definition 1.1.1. *A* lattice in E *is a subgroup* Λ *of* E *which satisfies the following property: there exists a basis* $\mathcal{B} = (e_1, e_2, \dots, e_n)$ *for* E *such that* Λ *is the set of all* $\mathbb{Z}$*-linear combinations of the* e_i. *Then* $\mathcal{B}$ *is called a* basis for the lattice. *The set* $P = \{\sum_i x_i e_i \mid 0 \le x_i < 1\}$ *is the* fundamental parallelotope for $(\Lambda, \mathcal{B})$. *A lattice in a proper subspace of* E *is called a* relative lattice. *(To put emphasis on the dimension of* Λ*, we sometimes say* "full lattice" *instead of* "lattice".*) We denote by* $\mathcal{L}$ *the set of lattices in* E.

Clearly, two bases for E define the same lattice if and only if there exists an element of $\mathrm{GL}_n(\mathbb{Z})$ (i.e., an $n \times n$ integral invertible matrix) which maps one basis onto the other. Moreover, relative lattices are closed discrete subgroups of E, and a relative lattice Λ is a lattice if and only if E/Λ is compact (for it is then the image of the closure of some fundamental parallelotope of Λ under the canonical surjection $E \to E/\Lambda$).

Conversely, discrete subgroups of E are relative lattices; this is a consequence of the following more general result (see e.g. [Boul], Chapter VII):

Theorem 1.1.2. *A closed subgroup of* E *is of the form* $\Lambda \oplus F'$ *where* Λ *is a lattice in some subspace* F *of* E *and* F' *is a subspace of* E *such that the sum* $F + F'$ *is direct (i.e.* $F \cap F' = \{0\}$ *; we then write* $F + F' = F \oplus F'$*).* □

As a consequence, we see that (closed) discrete subgroups of E are of rank at most n. This result, sometimes referred to as the *Jacobi–Bravais theorem*, justifies the definition of a lattice that we gave in the introduction.

Let Λ be a lattice in E and let F be a subspace of E, of dimension m. Then $\Lambda' = \Lambda \cap F$ is a relative lattice in F and is a full lattice in F if and only if F contains m independent elements of Λ. In all cases, the quotient Λ/Λ' is a finitely generated torsion-free $\mathbb{Z}$-module, hence a free module, and it is possible to extend any basis for Λ' to a basis for Λ by lifting in Λ a basis for Λ/Λ'. Thus:

Proposition 1.1.3. *For any lattice Λ in E and any subspace F of E, $\Lambda \cap F$ is a relative lattice in F, any basis for which can be extended to a basis for Λ.* □

Let F' be a complementary subspace of F in E.

Proposition 1.1.4. *A necessary and sufficient condition for the projection of Λ onto F parallel to F' to be a discrete subgroup of F is that $\Lambda \cap F'$ be a full lattice in F'. This projection is then a full lattice in F.*

Proof. The rank of the projection of Λ is equal to $n - r_1$, where r_1 denotes the rank of the kernel of the projection map. This rank, which is the rank of $\Lambda \cap F'$, is at most $r = \dim F'$. Thus the projection is discrete if and only if $\operatorname{rk} \Lambda \cap F' = r$, i.e. if and only if $\Lambda \cap F'$ is a full lattice in F'. □

Let $\mathcal{B}_0$ be a basis for E. Given n vectors $e_1, e_2, \dots, e_n$ in E, we can define the *determinant in $\mathcal{B}_0$* of this system. This is a real number, which is nonzero if and only if these vectors are independent. Now, given two bases $\mathcal{B}$ and $\mathcal{B}'$ for a lattice, the base change matrix is an element of $\mathrm{GL}_n(\mathbb{Z})$. Hence, given two lattices Λ_1 and Λ_2 together with bases $\mathcal{B}_1$ and $\mathcal{B}_2$, the strictly positive real number

$$\chi(\Lambda_1, \Lambda_2) = |\det{}_{\mathcal{B}_1}(\mathcal{B}_2)|$$

depends solely on Λ_1 and Λ_2. We call it the *relative invariant of Λ_1 and Λ_2*. It has the following properties:

Proposition 1.1.5. *1.* $\chi(\Lambda, \Lambda) = 1$.
2. $\chi(\Lambda_1, \Lambda_2) = \chi(\Lambda_2, \Lambda_1)^{-1}$.
3. $\chi(\Lambda_1, \Lambda_2)\,\chi(\Lambda_2, \Lambda_3) = \chi(\Lambda_1, \Lambda_3)$.
4. If $\Lambda_1 \supset \Lambda_2$, $\chi(\Lambda_1, \Lambda_2) = [\Lambda_1 : \Lambda_2]$.

Proof. Properties (1), (2), (3) follow immediately from the well-known transformation formulae for determinants. To prove (4), we make use of the *elementary divisors theorem* in the theory of Principal Ideal Domains (see Bourbaki, Alg. VII), which asserts the existence of a basis $(e_1, e_2, \dots, e_n)$ for Λ_1 and of positive integers $a_1, a_2, \dots, a_n$ such that $\mathcal{B}_2 = (a_1 e_1, a_2 e_2, \dots, a_n e_n)$

is a basis for Λ_2. Obviously $\det_{\mathcal{B}_1}(\mathcal{B}_2) = \prod_i a_i$, and the isomorphism $\Lambda_1/\Lambda_2 \simeq \oplus\mathbb{Z}/a_i\mathbb{Z}$ shows that $[\Lambda_1;\Lambda_2]$ is also equal to $\prod_i a_i$. □

We apply the proposition above to make another index calculation:

Proposition 1.1.6. *Let Λ be a lattice in E and let E_1 and E_2 two complementary subspaces of E (i.e., $E = E_1 \oplus E_2$) such that $\Lambda_1 = \Lambda \cap E_1$ and $\Lambda_2 = \Lambda \cap E_2$ are lattices in E_1 and E_2 respectively. Let p_1 (resp. p_2) be the projection onto E_1 (resp. E_2) parallel to E_2 (resp. E_1), and let $\Lambda'_i = p_i(\Lambda)$. Then Λ'_i, $i = 1,2$ is a lattice in E_i containing Λ_i, and we have*

$$[\Lambda'_1 : \Lambda_1] = [\Lambda'_2 : \Lambda_2] = [\Lambda : \Lambda_1 \oplus \Lambda_2] = [\Lambda'_1 \oplus \Lambda'_2 : \Lambda]\,.$$

Proof. Because Λ_i is its own projection onto E_i, Λ'_i is a lattice which contains Λ_i. Let $(e'_1, e'_2, \dots, e'_r)$ be a basis for Λ'_1, let e_i, $1 \le i \le r$ be lifts in Λ of the e'_i, and let $(e_{r+1}, e_{r+2}, \dots, e_n)$ be a basis for Λ_2. To every $x \in \Lambda$, let us attach the components $a_1, a_2, \dots, a_r$ of its projection x' on Λ'_1 and the vector $y = a_1e_1 + a_2e_2 + \cdots + a_re_r \in \Lambda$. Then $x - y$ belongs to the kernel of p_1, hence to $E_2 \cap \Lambda = \Lambda_2$. Consequently, $(e_1, e_2, \dots, e_n)$ is a basis for Λ.

Let $(f_1, f_2, \dots, f_r)$ be a basis for Λ_1. Since the differences $e'_i - e_i$ belong to E_2, we have $\det_{(e'_1,\dots,e'_r,e_{r+1},\dots,e_n)}(e_1, \dots, e_r, e_{r+1}, \dots, e_n) = 1$, whence

$$\begin{aligned}[\Lambda'_1 : \Lambda_1] &= \det{}_{(e'_1,\dots,e'_r)}(f_1, \dots, f_r)\\ &= \det{}_{(e'_1,\dots,e'_r,e_{r+1},\dots,e_n)}(f_1, \dots, f_r, e_{r+1}, \dots, e_n)\\ &= \det{}_{(e_1,\dots,e_r,e_{r+1},\dots,e_n)}(f_1, \dots, f_r, e_{r+1}, \dots, e_n) = [\Lambda : \Lambda_1 \oplus \Lambda_2]\,.\end{aligned}$$

Similarly, we have $[\Lambda'_2 : \Lambda_2] = [\Lambda : \Lambda_1 \oplus \Lambda_2]$, and finally,

$$[\Lambda'_1 \oplus \Lambda'_2 : \Lambda] = \frac{[\Lambda'_1 \oplus \Lambda'_2 : \Lambda_1 \oplus \Lambda_2]}{[\Lambda : \Lambda_1 \oplus \Lambda_2]} = \frac{[\Lambda'_1 : \Lambda_1][\Lambda'_2 : \Lambda_2]}{[\Lambda : \Lambda_1 \oplus \Lambda_2]} = [\Lambda : \Lambda_1 \oplus \Lambda_2]\,. \ \square$$

The relative invariant may be viewed as the quotient of the volumes of fundamental parallelotopes relative to the two lattices, the unit volume being defined by some basis for E, for instance the basis $\mathcal{B}_0$ we used previously. In practice, the unit volume will be defined by an orthonormal basis, once a Euclidean structure has been chosen. This is what will be done in the forthcoming sections and chapters.

The image $u(\Lambda)$ of a lattice Λ under an element $u \in \mathrm{GL}(E)$ is again a lattice. Starting with a particular lattice Λ_0, we obtain all the lattices in E in this way, and the equality $v(\Lambda) = u(\Lambda)$ holds if and only if $v^{-1}u$ belongs to the *stabilizer* $\mathrm{GL}(\Lambda_0)$ *of* Λ_0. We can therefore identify the set $\mathcal{L} = \mathcal{L}(E)$ of lattices in E with the set of left cosets of $\mathrm{GL}(E)$ modulo $\mathrm{GL}(\Lambda_0)$ (the sets of the form $u\,\mathrm{GL}(\Lambda_0), u \in \mathrm{GL}(E)$). Choosing a basis $\mathcal{B}_0$ for Λ_0, we can identify $\mathrm{GL}(E)$ with $\mathrm{GL}_n(\mathbb{R})$ and $\mathrm{GL}(\Lambda_0)$ with $\mathrm{GL}_n(\mathbb{Z})$, and thus $\mathcal{L}$ with $\mathrm{GL}_n(\mathbb{R})/\mathrm{GL}_n(\mathbb{Z})$. This identification gives $\mathcal{L}$ a topology, for which a fundamental system of neighbourhoods of $\Lambda = u(\Lambda_0)$ is the set of images by u of a fundamental system of neighbourhoods of the unit matrix I_n in $\mathrm{GL}_n(\mathbb{R})$. Alternatively, we may consider a basis $\mathcal{B}_0 = (e_1, e_2, \dots, e_n)$ for E, a fundamental system of neighbourhoods $\mathcal{V}_i$ of each vector e_i in E and the sets of lattices which possess a basis $(e'_1, \dots, e'_n)$ with $e'_i \in \mathcal{V}_i$ for $i = 1, \dots, n$.

1.2 Lattices in Euclidean Spaces

From now on we suppose that E is endowed with a *Euclidean structure*, i.e. a positive definite bilinear form $(x, y) \mapsto x \cdot y$, called the *scalar product for E*. Thus we have $x \cdot x > 0$ for any nonzero $x \in E$. There is a corresponding norm on the vector space E, namely $\|x\| = \sqrt{x \cdot x}$, and $(x, y) \mapsto \|x - y\|$ is a *distance* on E. *However, the expression* "**norm of** x" *will denote its* **square** $\|x\|^2$, written $N(x)$. This has the advantage of getting rid of square roots in many formulae.

Definition 1.2.1. *For a lattice Λ, the* minimal norm of Λ, *or simply the* norm *or* minimum of Λ, *is*

$$N(\Lambda) = \min \Lambda = \inf_{x \in \Lambda, x \neq 0} N(x)\,,$$

and we set $\|\Lambda\| = \sqrt{N(\Lambda)}$. The sphere of Λ, *or the set of* minimal vectors of Λ, *is* $S(\Lambda) = \{x \in \Lambda \mid \|x\| = \|\Lambda\|\}$. *Finally,* $|S(\Lambda)|$ *is called the* kissing number[1] *of* Λ, *and we set* $s(\Lambda) = \frac{1}{2}|S(\Lambda)|$.

We recall some standard conventions about matrices. Given a basis $\mathcal{B} = (e_1, \dots, e_n)$ for E and a family $\mathcal{F}$ of m vectors $x_1, x_2, \dots, x_m$, the entries of the matrix M of $\mathcal{F}$ with respect to $\mathcal{B}$ are the components $m_{i,j}$ of x_j expressed in terms of the vectors e_i (i is the row index and j the column index of M); we thus have the formula $x_i = \sum_k m_{k,i} e_k$, in which the summation is performed on the *first* index.
[However, when M represents a linear map relative to some pair of bases, one computes the components of the image of a vector by a formula of the form $y_i = \sum_k m_{i,k} x_k$, with the summation on the *second* index.]

Recall also that a basis $\mathcal{B} = (e_1, \dots, e_n)$ is called *orthogonal* if $e_i \cdot e_j = 0$ for $j \neq i$ and *orthonormal* if moreover $N(e_i) = 1$ for all i. Unless otherwise stated, the notation ε_i will be used exclusively to denote elements of an orthonormal basis.

Definition 1.2.2. *The* Gram matrix *of a an ordered set* $\mathcal{F} = (x_1, \dots, x_s)$ *of vectors in* E *is the matrix* $\mathrm{Gram}(\mathcal{F})$ *of the scalar products* $x_i \cdot x_j$.

---

[1] N.J.A. Sloane supplied the following comment: "The terminology comes from billiards, where two balls are said to <u>kiss</u> if they are touching. I think the term was first introduced in the senior mathematical literature in Leech and Sloane ([Le-Sl], 1971)."

This is a symmetric $s \times s$ matrix, for which we shall give a nice interpretation in Section 1.3 in the case where $\mathcal{F}$ is a basis – see Lemma 1.3.6 below.

Lemma 1.2.3. *Given two families $\mathcal{F} = (x_i)$ and $\mathcal{F}' = (x'_i)$, with respective matrices M and M' with respect to a same orthonormal basis for E, the matrix of the scalar products $x_i \cdot x'_j$ is $A = {}^tMM'$. In particular, if $M = M'$, we have* $\det A = \det M^2$.
[The notation ${}^tM$ stands for the *transpose* of the matrix M.]

Proof. From $x_i = \sum_k m_{k,i}\,\varepsilon_k$ and $x'_j = \sum_\ell m_{\ell,j}\,\varepsilon_\ell$, we immediately deduce the equalities $x_i \cdot x'_j = \sum_{k,\ell} m_{k,i}\,m_{\ell,j}\,\delta_{k,\ell} = \sum_k m_{k,i}\,m_{k,j} = \sum_k {}^tm_{i,k}\,m_{k,j}$. □

As a consequence, we see that the determinant of an orthonormal basis with respect to another orthonormal basis is equal to ± 1. This remark justifies the following definition:

Definition 1.2.4. *The* discriminant of a family $\mathcal{F}$ of n vectors in E *is the absolute value of the determinant of $\mathcal{F}$ with respect to an orthonormal basis. The* discriminant of a lattice Λ *is the discriminant of any basis for it. The square of the discriminant of $\mathcal{F}$ (resp. of Λ) is the* determinant of $\mathcal{F}$ *(resp.* of Λ*).* Notation*: $\Delta(\mathcal{F})$, $\Delta(\Lambda)$ for the discriminants,* $\det(\mathcal{F})$, $\det(\Lambda)$ *for the determinants.*

Note that $\Delta(\Lambda) = \chi(\Lambda, \Lambda_0)$ (see Proposition 1.1.5) for any lattice Λ_0 generated by an orthonormal basis for E.

The use of the absolute value in the definition of the discriminant shows that the notion of a determinant is a more natural one. The lattices we shall consider will often have rational determinants but irrational discriminants.

The following result is an immediate consequence of Lemma 1.2.3:

Proposition 1.2.5. *The determinant of a lattice is the determinant of any one of its bases.* □

Let $\mathcal{B}$ and $\mathcal{B}'$ be two bases for the same lattice Λ, and let Q be the base change matrix. The matrices M and M' of $\mathcal{B}$ and of $\mathcal{B}'$ in the same orthonormal basis for E are are linked by the relation $M' = MQ$ (right multiplication); on the entries, the relation reads $m'_{k,i} = \sum_\ell m_{k,\ell}\,q_{\ell,i}$.

The connection between the corresponding Gram matrices A and A' is then (see Lemma 1.2.3) $A' = {}^tQAQ$ (whence again the invariance of the determinant). This relation shows that all Gram matrices for a lattice are integral provided one of them is (a remark which will be stated in the next section in terms of duality), and allows us to make the following definition:

Definition 1.2.6. *A lattice is* integral *if any one of its Gram matrices is integral (i.e., belongs to $\mathcal{M}_n(\mathbb{Z})$), and is* unimodular *if this Gram matrix is* *integral with determinant* 1. *An integral lattice is* primitive *if any one of its Gram matrices has coprime entries. A lattice is* rational *if it is proportional to an integral lattice.*

A calculation of scalar products immediately yields:

Proposition 1.2.7. *A lattice Λ is integral (resp. primitive) if and only if the scalar products of vectors in Λ are integers (resp. and generate $\mathbb{Z}$).* □

Integral lattices will play an important rôle in what follows, since perfect lattices (see Chapter 3) are proportional to integral lattices. The norm obviously takes integral values on an integral lattice. The following definition introduces an important property of the norms on an integral lattice:

Definition 1.2.8. *We say that a lattice Λ is* even *(resp.* odd*) if it is integral, and if the norm takes only even values (resp. at least one odd value) on Λ.*

The identity $x \cdot y = \frac{1}{2}(N(x+y) - N(x) - N(y))$ shows that a lattice whose vectors all have even norm is integral, and indeed even, and that the norms on a primitive lattice cannot have a common divisor $m > 2$. Moreover, a lattice is even if and only if one of its Gram matrices is integral with even entries on the diagonal.

On the other hand, an odd lattice Λ possesses a basis $(e_1, e_2, \dots, e_n)$ such that $N(e_i) = e_i \cdot e_i$ is odd for $i = 1$ and an even number for $i \geq 2$. Thus the set of vectors of even norm in Λ is an (even) sublattice of index 2 in Λ, canonically attached to Λ, and sometimes called the *even part* of Λ.

We now return to the Euclidean setting for studying projections, already mentioned in Section 1.1; the connection with duality will be considered in Section 1.3.

Proposition 1.2.9. *1. Let $e_1, e_2, \dots, e_n$ be n vectors in E, let F be an r-dimensional subspace of E containing $e_1, \dots, e_r$, and for $i > r$, let e'_i be the projection of e_i onto $F^\perp$. Then*

$$\Delta(e_1, e_2, \dots, e_n) = \Delta(e_1, \dots, e_r)\Delta(e'_{r+1}, \dots, e'_n)\,.$$

2. Let Λ be a lattice, let F be a subspace of E such that $\Lambda \cap F$ is a lattice in F, and let p be the orthogonal projection onto $F^\perp$. Then

$$\det(p(\Lambda)) = \det(\Lambda)\left(\det(\Lambda \cap F)^{-1}\right)\,.$$

Proof. (1) Choose an orthonormal basis $\mathcal{B}$ for E which is the union of bases $\mathcal{B}_1$ and $\mathcal{B}_2$ for F and $F^\perp$ and decompose each e_i for $i > r$ into a sum $e''_i + e'_i$ on F and $F^\perp$. Then the left–hand side appears as a sum of 2^{n-r} determinants, among which those which involve a vector e''_i are zero since $e''_i \in F$. Hence the left–hand side is equal to $\Delta(e_1, \dots, e_r, e'_{r+1}, \dots, e'_n)$. Since the components in $\mathcal{B}_1$ of the e'_i and in $\mathcal{B}_2$ of the e_j are zero, $\Delta(e_1,\dots,e_n)$ is equal to the product of the two determinants of the proposition.

(2) Let $r = \dim F$. Since $\Lambda/\Lambda \cap F$ is torsion-free, one can find a basis $(e_1,\dots,e_n)$ for Λ such that $(e_1,\dots,e_r)$ is a basis for $\Lambda \cap F$, and (2) follows immediately from (1). □

The previous proposition allows us to calculate the determinant of a projection. It is more difficult to find the minimal norm. We end this section with a useful inequality in that direction (we say that $x \in \Lambda$ is primitive if it is not of the form $m\,y$ for some integer $m > 1$ and $y \in \Lambda$).

Proposition 1.2.10. *Let Λ be a lattice in E, let e be a primitive nonzero vector in Λ, let F be the hyperplane $(\mathbb{R}e)^\perp$ of E, let p be the orthogonal projection onto F, and let $x \in \Lambda$, $x \notin \mathbb{Z}e$. Then the orthogonal projection of x satisfies the inequality*

$$N(p(x)) \geq N(x - ke) - \frac{N(e)}{4},$$

where k denotes the integer (or one of the two integers) for which the distance between $x - ke$ and F is minimal. In particular, we have

$$N(p(\Lambda)) \geq N(\Lambda) - \frac{N(e)}{4},$$

and equality is attained precisely on the projections of those vectors (if any) of $S(\Lambda)$ whose distance from F is equal to $\frac{1}{2}\|e\|$.

Proof. We have

$$p(x) = p(x - ke) \quad \text{and} \quad N(x - ke) = N(p(x)) + N(x - ke - p(x)).$$

By the definition of k, the distance from $x - ke - p(x)$ to F is bounded above by $\frac{1}{2}\|e\|$, whence the first inequality, which in turn implies the second one. Moreover, if x is such that $N(p(\Lambda)) = N(p(x)) = N(p(x - ke)) - \frac{1}{4}N(e)$, we necessarily have $N(x) = N(\Lambda)$ and $N(x) - N(p(x)) = \frac{1}{4}\,N(e) = \left(\frac{1}{2}\|e\|\right)^2$. □

1.3 Duality

Let us first recall a few general algebraic definitions. Given a *commutative* ring R and an R-module X, the *dual module of* X is $X^* = \mathrm{Hom}_R(X, R)$; a *bilinear form on* X is a map $B : X \times X \to R$ which is linear with respect to both variables. The set of bilinear forms on X carries a natural structure of an R-module, which is R-isomorphic to $\mathrm{Hom}_R(X, X^*)$ via the map $x \mapsto (y \mapsto B(x,y))$. We say that B is *non-degenerate* if this map is an isomorphism. In practice we only consider *symmetric* or *skew-symmetric* forms, in which case orthogonality between elements $x, y \in X$ (defined by $B(x, y) = 0$) is a symmetric relation.

Suppose now that X is a free R-module, and let $\mathcal{B} = (e_1, \ldots, e_n)$ be a basis for X. For every R-module Y, there exists a unique homomorphism $f : X \to Y$ which takes prescribed values on the elements of $\mathcal{B}$. In particular, there exists a unique family of elements $f_i \in X^*$ such that $f_i(e_j) = \delta_{ij}$ (the Kronecker symbol). Clearly $\mathcal{B}' = (f_1, \ldots, f_n)$ is a basis for X^*, the *dual basis of* $\mathcal{B}$. When X is endowed with a non-degenerate bilinear form B, the dual basis of $\mathcal{B}$ can be identified with a basis for X itself, denoted by $\mathcal{B}^* = (e_1^*, \ldots, e_n^*)$, which is characterized by the equalities $B(e_i, e_j^*) = \delta_{ij}$.

All this can be applied to the space E, viewed as an $\mathbb{R}$-module and endowed with the standard scalar product.

Definition 1.3.1. *Let Λ be a lattice in E. The* dual lattice of Λ *is*

$$\Lambda^* = \{x \in E \mid \forall y \in \Lambda,\, x \cdot y \in \mathbb{Z}\}.$$

Note that, for any basis $\mathcal{B}$ for Λ, the dual basis $\mathcal{B}^*$ for E is a $\mathbb{Z}$-basis for Λ^*. Hence Definition 1.3.1 is consistent with the algebraic considerations above.

Proposition 1.3.2. *1. The self-dual bases for E are the orthonormal bases.*

2. For any basis $\mathcal{B}$ for E, the matrix of $\mathcal{B}^*$ in an orthonormal basis is the transpose of the inverse of the matrix of $\mathcal{B}$ in the same basis; in particular, for any lattice $\Lambda \subset E$ and for any automorphism u of E, we have

$$u(\Lambda)^* = {}^t u^{-1}(\Lambda^*).$$

[The transpose ${}^t u$ of $u \in \mathrm{GL}(E)$ is defined by $\forall x, y \in E, x \cdot u(y) = {}^t u(x) \cdot y$.]

3. Given two bases $\mathcal{B}_1$ and $\mathcal{B}_2$ for E, the matrix in $\mathcal{B}_2^*$ of $\mathcal{B}_1^*$ is the transpose of the matrix in $\mathcal{B}_1$ of $\mathcal{B}_2$.

Proof. The first assertion is obvious. To prove the second, let $\mathcal{B}_0 = (\varepsilon_1, \ldots, \varepsilon_n)$ be an orthonormal basis for E, and let $\mathcal{B} = (e_1, \ldots, e_n)$. The matrix $A = \mathrm{Mat}_{\mathcal{B}_0}(\mathcal{B})$ has entries $a_{i,j}$ such that $e_i = \sum_k a_{k,i}\,\varepsilon_k$ for all i, hence $a_{j,i} = e_i \cdot \varepsilon_j$. We describe in the same way $B = \mathrm{Mat}_{\mathcal{B}_0}(\mathcal{B}^*)$. The relations $e_i = \sum_k a_{k,i}\,\varepsilon_k$ show that we have $\delta_{i,j} = e_i \cdot e_j^* = \sum_k a_{k,i}\varepsilon_k e_j^* = \sum_k a_{k,i} b_{k,j}$, i.e. ${}^t AB = I_n$, which is precisely (2). We easily deduce the last assertion from the second one (to which it reduces when $\mathcal{B}_2 = \mathcal{B}_0$) by making use of the transitivity relation $\mathrm{Mat}_{\mathcal{B}_1^*}(\mathcal{B}_2^*) = \mathrm{Mat}_{\mathcal{B}_1^*}(\mathcal{B}_0)\,\mathrm{Mat}_{\mathcal{B}_0}(\mathcal{B}_2^*)$. □

From assertion (2), we immediately deduce:

Corollary 1.3.3. *The determinant of a lattice and of its dual are mutually inverse.* □

We now return to the study of projections.

Proposition 1.3.4. *Let Λ be a lattice and let F be a subspace of E. Then $\Lambda \cap F$ is a lattice in F if and only if $\Lambda^* \cap F^\perp$ is a lattice in $F^\perp$. Moreover, if these conditions are satisfied, the dual of $\Lambda \cap F$ is the orthogonal projection onto F of the dual of Λ. In other words, the following formula holds:*

$$(\Lambda \cap F)^* = p_F(\Lambda^*)\,.$$

Proof. If $\Lambda\cap F$ is a lattice, one can find a basis $(e_1, \dots, e_n)$ for Λ whose first r vectors constitute a basis for $\Lambda\cap F$. Then the last $n-r$ vectors $(e_{r+1}^*, \dots, e_n^*)$ of the dual basis belong to $\Lambda^*\cap F^\perp$. Since $\dim F^\perp = n-r$, $\Lambda^*\cap F^\perp$ is a lattice in $F^\perp$, and the converse is also true, as one sees by exchanging Λ and Λ^*. It is now clear that the orthogonal projections on F of $e_1^*, \dots, e_r^*$ constitute the dual basis in F of the basis $(e_1, \dots, e_r)$, and thus generate $(\Lambda \cap F)^*$. □

Combining Propositions 1.3.4 and 1.2.9, we easily obtain:

Corollary 1.3.5. *Let $r \le n$ be an integer, let $\mathcal{B} = (e_1, \dots, e_n)$ be a basis for E, and let F be the subspace of E generated by $e_1, \dots, e_r$. Then*

$$\det(e_1, \dots, e_n) = \det(e_1, \dots, e_r)\det(e_{r+1}^*, \dots, e_n^*)^{-1}\,;$$

similarly, if Λ and $\Lambda \cap F$ are both lattices, we have

$$\det(\Lambda)\det(\Lambda^* \cap F^\perp) = \det(\Lambda \cap F)\,. \;\square$$

We now explain the translations in the language of duality of the integrality properties of lattices mentioned in Section 1.2. We begin with a few formulae involving a basis and its dual basis, together with an interpretation of Gram matrices. A *similarity of E with ratio* $\lambda \neq 0$ is an endomorphism σ of E which satisfies the identity: $\forall\, x,\, y \in E,\ \sigma(x)\cdot\sigma(y) = \lambda^2(x\cdot y)$.

Lemma 1.3.6. *Let $\mathcal{B} = (e_1, e_2, \dots, e_n)$ be a basis for E, let $A = (a_{i,j})$ be the Gram matrix of $(\mathcal{B})$, let $u \in \mathrm{GL}(E)$, and let $P = (p_{i,j})$ be the matrix of u with respect to the bases $\mathcal{B}$ and $\mathcal{B}^*$.*

1. *The Gram matrix of $\mathcal{B}$ is the matrix of the identity with respect to the bases $\mathcal{B}$ and $\mathcal{B}^*$:*
$$\mathrm{Gram}(\mathcal{B}) = \mathrm{Mat}(Id, \mathcal{B}, \mathcal{B}^*)\,.$$
2. *The matrices Q of u in $\mathcal{B}$ and P of u with respect to the bases $\mathcal{B}$ and $\mathcal{B}^*$ are linked by the relation $P = AQ$.*
3. *We have* $\mathrm{Gram}(u(\mathcal{B})) = {}^tQ\,\mathrm{Gram}(\mathcal{B})\,Q$.
4. *The matrix with respect to the bases $\mathcal{B}$ and $\mathcal{B}^*$ of the transpose of u is the transpose of P, i.e.*
$$\mathrm{Mat}({}^tu, \mathcal{B}, \mathcal{B}^*) = {}^t\mathrm{Mat}(u, \mathcal{B}, \mathcal{B}^*)\,.$$
5. *u is a similarity of ratio λ if and only if*
$${}^tP = \lambda^2 A P^{-1} A\,.$$

Proof. (1) If $e_j = \sum_k a_{k,j}\, e_k^*$, then $e_i \cdot e_j = a_{i,j}$.

(2) Put $v = \mathrm{Id}$ in the general base change relation

$$\mathrm{Mat}(v \circ u, \mathcal{B}, \mathcal{B}') = \mathrm{Mat}(v, \mathcal{B}'', \mathcal{B}')\,\mathrm{Mat}(u, \mathcal{B}, \mathcal{B}'')\,.$$

(3) Let $Q = (q_{i,j})$. Then we have $u(e_i) = \sum_k q_{k,i}\, e_k$ for all i, and thus

$$u(e_i) \cdot u(e_j) = \sum_{k,\ell} q_{k,i}\,(e_k \cdot e_\ell)\, q_{\ell,j}\,.$$

(4) We have $p_{i,j} = u(e_i) \cdot e_j$ for all i, j and thus ${}^t u(e_i) \cdot e_j = e_i \cdot u(e_j) = p_{j,i}$.

(5) We simply consider the relation ${}^t u = \lambda^2 u^{-1}$ with respect to the bases $\mathcal{B}$ and $\mathcal{B}^*$: the matrix on the left is ${}^t P$, whereas that on the right is $\lambda^2 A Q^{-1} = \lambda^2 A (P^{-1} A)$. □

Proposition 1.3.7. *Let Λ be a lattice.*

1. *Λ is integral if and only if it is contained in its dual.*
2. *When this condition is satisfied, the index $[\Lambda^* : \Lambda]$ is equal to the determinant of Λ.*
3. *Λ is unimodular if and only if Λ is equal to its dual Λ^*.*

Proof. Let $x \in \Lambda$. Since $x \cdot y$ is integral for all $y \in \Lambda$, x is an element of Λ^*, whence (1). Since the index $[\Lambda^* : \Lambda]$ is the relative invariant (Proposition 1.5) of Λ^* and Λ, we have

$$[\Lambda^* : \Lambda] = \Delta(\Lambda)\,\Delta(\Lambda^*)^{-1} = \Delta(\Lambda)^2 = \det(\Lambda)\,,$$

which proves (2). Finally, (3) follows easily from (1) and (2). □

We shall return to integrality in Section 1.9, and end this section with some applications of duality to inclusions of lattices. There will be some analogies with Proposition 1.3.4, but we replace linear forms by linear forms modulo certain integers.

Let Λ be a lattice and let M be a sublattice of Λ. As a sub-$\mathbb{Z}$-module of Λ, M can be defined as the intersection of the kernels of finitely many homomorphisms $\varphi_i : \Lambda \to \mathbb{Z}/m_i\mathbb{Z}$: we can for instance first choose an integer $m > 0$ with $m\Lambda \subset M$, then consider among the elements of $\mathrm{Hom}(\Lambda, \Lambda/m\Lambda) \simeq \mathrm{Hom}(\Lambda, \mathbb{Z}/m\mathbb{Z})^n$ those which annihilate M, and finally extract from this set a finite set of generators. (We can eventually choose for the m_i the elementary divisors of Λ/M.) Thanks to the Euclidean structure of E, it is possible to find for all i a vector f_i such that $\varphi_i = (x \mapsto f_i \cdot x)$.

Proposition 1.3.8. *Let Λ be a lattice, let $f_1, \dots, f_r$ be a finite set of vectors in Λ, let m_i, $1 \le i \le r$ be integers, and let $M \subset \Lambda$ be the intersection of the kernels of the homomorphisms $x \mapsto f_i \cdot x \bmod m_i$ (thus M is a sublattice of Λ). Then:*

1. *M^* is the lattice generated in E by Λ^* and the vectors $\frac{1}{m_i} f_i$.*
2. *The two $\mathbb{Z}$-modules M^*/Λ^* and Λ/M are isomorphic.*

Proof. Let P be the $\mathbb{Z}$-module generated in E by Λ^* and the r vectors $\frac{1}{m_i} f_i$. Then M^* contains Λ^* as well as the vectors $\frac{1}{m_i} f_i$. Hence it contains P. An induction argument reduces the proof of the opposite inequality to the case where M is defined by a single congruence $f \cdot x \equiv 0 \mod p$ with p prime. If f belongs to $p\Lambda^*$, then $M = \Lambda$ and $M^* = \Lambda^* = P$. Otherwise, we have $[\Lambda : M] = p$, hence $[M^* : \Lambda^*] = p = [P : \Lambda^*]$, whence again the equality $P = M^*$ in this case. We have thus proved (1). We now prove (2) by making use once more of the elementary divisors theorem. There exist a basis $(e_1, \dots, e_n)$ for Λ and integers $a_1, \dots, a_n$ such that $(a_1 e_1, \dots, a_n e_n)$ is a basis for M. Then $(\frac{1}{a_1} e_1, \dots, \frac{1}{a_n} e_n)$ is clearly a basis for M^*, and we conclude by using the isomorphisms

$$M^*/\Lambda^* \simeq \bigoplus_i \frac{1}{a_i} \mathbb{Z}/\mathbb{Z} \simeq \bigoplus_i \mathbb{Z}/a_i\mathbb{Z} \simeq \Lambda/M . \; \square$$

Example 1.3.9. *Let $(e_1, \dots, e_n)$ be a basis for Λ, let m be a positive integer, let $a_1, \dots, a_n$ be integers, and let*

$$M = \Big\{ x = \sum_i x_i e_i \in \Lambda \,\Big|\, \sum_i a_i x_i \equiv 0 \mod m \Big\} .$$

Then M^* is generated by Λ^* together with the vector $e = \frac{1}{m} \sum_i a_i e_i^*$.

Finally, we describe an application of duality to the determination of the even lattice Λ' attached to a given odd lattice Λ (see the comments in Section 1.2, after Definition 1.2.8). Since $[\Lambda : \Lambda'] = 2$, there exist vectors $c \in \Lambda^*$, well-defined modulo $2\Lambda^*$, such that

$$\Lambda' = \{ x \in \Lambda \mid c \cdot x \equiv 0 \mod 2 \} .$$

Definition 1.3.10. *A vector c as above is called a* characteristic *or a* parity vector in Λ.

[More generally, any surjective homomorphism $\varphi : \Lambda \to \mathbb{Z}$ (resp. $\varphi_m : \Lambda \to \mathbb{Z}/m\mathbb{Z}$) is of the form $x \mapsto e \cdot x$ (resp. $x \mapsto e \cdot x \mod m$) for some $e \in \Lambda^*$, which is unique modulo $m\Lambda^*$, and is thus a kind of *characteristic vector* for φ.]

The case where Λ has an odd determinant d yields an interesting simplification: the map $x \mapsto dx$ induces an isomorphism of $\Lambda^*/2\Lambda^*$ onto $\Lambda/2\Lambda$. Parity vectors may then be chosen in Λ, they are well-defined modulo 2Λ, and $c \cdot c \mod 8$ does not depend on the particular choice of c: if c, c' are two parity vectors, then $c' = c + 2x$ for some $x \in \Lambda$, which implies

$$c' \cdot c' = c \cdot c + 4(c \cdot x + x \cdot x) \equiv c \cdot c \mod 8$$

since $x \cdot x \equiv c \cdot x \mod 2$.

The considerations above trivially apply to even lattices, taking $c = 0$ in this case. In all cases, we have attached to every integral lattice of odd determinant an *invariant modulo* 8, which takes the value 0 on even lattices. [Parity vectors also exist in Λ when d is even; however, they no longer belong to a single class modulo 2Λ.]

1.4 Automorphism Groups

Two lattices in two real vector spaces V and V' of the same dimension are isomorphic, since for instance the second one is the image of the first one under a one-to-one element of $\mathcal{L}(V, V')$. When considering lattices in Euclidean spaces E and E', the natural notion of isomorphism does not allow arbitrary elements of $\mathcal{L}(E, E')$, but only isometries of E onto E'. We are thus led to the following definition:

Definition 1.4.1. *Let E and E' be two Euclidean spaces. An* isometry of a lattice $\Lambda \subset E$ onto a lattice $\Lambda' \subset E'$ *is an isometry of E onto E' which maps Λ onto Λ'. An* automorphism of Λ *is an isometry of Λ onto itself. The group of all automorphisms of Λ is denoted by* $\mathrm{Aut}(\Lambda)$. *We write $\Lambda \simeq \Lambda'$ (resp. $\Lambda \sim \Lambda'$) to indicate that Λ and Λ' are isometric (resp.* similar, *i.e. that they become isometric when rescaled to the same minimum).*
[We also use the symbol "$\simeq$" to indicate an algebraic isomorphism (between groups, rings, etc.).]

This definition shows that $\mathrm{Aut}(\Lambda) = \mathrm{GL}(\Lambda) \cap \mathrm{O}(E)$, where $\mathrm{O}(E)$ is the *orthogonal group of E*, i.e. the group of one-to-one isometries of E. Now $\mathrm{GL}(\Lambda)$ is a discrete subgroup of $\mathrm{GL}(E)$, whereas $\mathrm{O}(E)$ is compact. Hence:

Theorem 1.4.2. *The automorphism group of a lattice is finite.* □

The determinant (or the discriminant) of a lattice, as well as its number of minimal vectors, are clearly invariants of its isometry class. The same is true for the isomorphism class of its automorphism group, and in particular for the order of this group.

The proof of Theorem 1.4.2 given above, although short, has the disadvantage that it does not yield any procedure for calculating the automorphism group. The notion of the spectrum that we now introduce will allow us to make such a calculation, and also provides us with important new invariants of the isometry classes.

Definition 1.4.3. *Let Λ be a lattice, and let $\mathcal{F}$ be a finite family of nonzero vectors in Λ, invariant under the symmetry $x \mapsto -x$. The* spectrum (with respect to $\mathcal{F}$) *of a vector $x \in \Lambda$ is the sequence $(a_1, \alpha_1), \dots, (a_k, \alpha_k)$, $0 \le \alpha_1 < \cdots < \alpha_k)$ where the α_i are the absolute values of the scalar products*

$x \cdot y$, $y \in \mathcal{F}$, *and* a_i *is the number of pairs* $\pm y, y \in \mathcal{F}$ *with* $|x \cdot y| = \alpha_i$. *The* spectrum of $\mathcal{F}$ *is the sequence* (m_i, a_i, α_i) *where* m_i *is the number of pairs* $\pm x, x \in \mathcal{F}$ *with spectrum* (a_i, α_i). *The* minimal spectrum of Λ *(or simply the* spectrum of Λ*) is the spectrum of the set* $S(\Lambda)$ *of minimal vectors in* Λ.

We shall use the spectrum essentially for minimal vectors of integral lattices. In this case, following [C-S5], we shall give for a given $x \in S(\Lambda)$ the sequence consisting of the number of pairs $\pm y$ with $|x \cdot y| = 0, 1, \ldots, \lfloor \frac{N(\Lambda)}{2} \rfloor$, thus omitting the pair $\pm x$. For the spectrum of $S(\Lambda)$, we give similarly the number of vectors (up to sign) with a given spectrum.

It is clear that the minimal spectrum of a lattice is still an invariant of its isometry class, and that the spectrum of a vector $x \in \Lambda$ with respect to $S(\Lambda)$ solely depends on the orbit of x under $\operatorname{Aut}(\Lambda)$. The orbit decomposition of the vectors in $S(\Lambda)$ is also an important invariant of the isometry class of Λ.

Let $t = \frac{1}{2}|\mathcal{F}|$. The spectra of the vectors in $\mathcal{F}$ satisfy the relation $\sum_i a_i = t$, and we have $\sum_i m_i a_i = t^2$ for the spectrum of $\mathcal{F}$ itself.
[Warning: for the equalities above to hold, one must not use the convention of [C-S5].]
When $\operatorname{Aut}(\Lambda)$ acts transitively on $\mathcal{F}$, all elements of $\mathcal{F}$ have the same spectrum.

We now sketch a constructive proof of Theorem 1.4.2 relying on the spectrum. Indeed, we consider more generally the problem of deciding whether two given lattices are isometric; however, before trying to construct isometries, one should first compute some invariants of the lattices, e.g. determinant, number of minimal vectors, etc., or some invariants involving the dual lattice to be defined later.

Thus let Λ and Λ' be two lattices and let $\mathcal{F}_0$ be a finite generating family of Λ, invariant under $x \mapsto -x$; one can choose for $\mathcal{F}_0$ the union $\mathcal{B} \cup -\mathcal{B}$ for some basis $\mathcal{B}$ for Λ. Let $0 < M_1 < M_2 \cdots < M_k$ be the norms of the various vectors in $\mathcal{F}_0$, let $\mathcal{F}_i$ be the set of the vectors in Λ of norm M_i, and let $\mathcal{F} = \cup \mathcal{F}_i$. Finally, let $\mathcal{F}'_i = \{x' \in \Lambda' \mid N(x') = M_i\}$ and $\mathcal{F}' = \cup_i \mathcal{F}'_i$.

If there exists an isometry f of Λ onto Λ', we have the equality $f(\mathcal{F}_i) = \mathcal{F}'_i$ for all i, hence $\mathcal{F}_i$ and $\mathcal{F}'_i$ have the same spectrum for all i, and in particular the same cardinality.

Assume that the sets $\mathcal{F}_i$ and $\mathcal{F}'_i$ have the same cardinality, say c_i, and choose for all i a system $\mathcal{G}_i$ of representatives for the pairs $\pm x, x \in \mathcal{F}_i$, the elements of which we denote by $e_{i,j}$ $(1 \le j \le c_i)$. If an isometry f exists, it maps $\mathcal{G}_i$ into $\mathcal{F}'_i$ for all i, and is well defined by the set $\{f(e_{i,j})\}$.

Choose elements $e'_{i,j} \in \Lambda'$ such that, $\forall i, j$ and $\forall k \neq j$, $e'_{i,j} \in \mathcal{F}'_i$ and $e'_{i,j} \neq e'_{i,k}$. The existence of an isometry f such that $f(e_{i,j}) = e'_{i,j}$ requires that the following two conditions be satisfied:

1. $\sum_j \lambda_j e'_{i,j} = 0$ for all i and for every $\mathbb{Z}$-linear relation $\sum_j \lambda_j e_{i,j} = 0$ between the $e_{i,j}$;
2. $\forall i, j, k, \ell$, $e'_{i,j} \cdot e'_{k,\ell} = e_{i,j} \cdot e_{k,\ell}$.

(Condition (1) corresponds to the linearity of f.)

Conversely, when conditions (1) and (2) are satisfied, there exists a (unique) isometry f with $f(e_{i,j}) = f(e'_{i,j})$.

We have thus reduced the search for an isometry (or all isometries) between Λ and Λ' to a finite amount of verifications. Though there can be a great number of calculations to perform, our problem can effectively be solved in this way with the help of a computer when the dimension and kissing number are not too large. In practice, Λ and Λ' are defined up to isometry by the Gram matrices for some bases $\mathcal{B}$ and $\mathcal{B}'$, and the algorithm reduces to the determination of all vectors in Λ and Λ' whose norm is smaller than some known bound. This can be effectively done by means of the decompositions as sums of squares of the integral quadratic forms $x \mapsto N(x)$ defined by Λ and Λ'.

In the important case of lattices which are generated by their minimal vectors, we can take $\mathcal{F} = S(\Lambda)$. Note, however, that it might well be impossible to extract from S a basis for Λ: one cannot extract a basis for $\mathbb{Z}$ from the generating set $\{2, 3\}$, and a lattice generated by its minimal vectors having no basis of minimal vectors has recently been constructed by Conway and Sloane ([C-S10]).

A somewhat more general case, important in the theory of perfection, is that of lattices possessing n independent vectors (*well–rounded lattices* in the terminology of Ash and McConnell ([A-MC])):

(WR) $$S(\Lambda) \ \ \textit{spans} \ \ E\,.$$

Note than $S(\Lambda)$ may generate a strict sublattice of Λ. The following proposition reduces the calculation of the automorphism group to the case above.

Proposition 1.4.4. *Let Λ be a well–rounded lattice and let Λ' be the sublattice generated by its minimal vectors. Then Λ' is invariant under* $\mathrm{Aut}(\Lambda)$*, and the restriction to Λ' of automorphisms of Λ yields an injective homomorphism of* $\mathrm{Aut}(\Lambda)$ *into* $\mathrm{Aut}(\Lambda')$.

Proof. Since $\mathrm{Aut}(\Lambda)$ preserves the norm, it stabilizes $S(\Lambda)$. Thus the restriction map $\varphi' : \mathrm{Aut}(\Lambda) \to \mathrm{Aut}(\Lambda')$ is well defined. Any $\sigma \in \ker \varphi$ acts trivially on $S(\Lambda') = S(\Lambda)$. Since $S(\Lambda)$ spans E, σ is the identity of E.
[The restriction map is not always surjective; see Proposition 4.4.2.] □

There are two important groups related to the automorphism group of a lattice Λ: the subgroup $\mathrm{Aut}^+(\Lambda) = \{\sigma \in \Lambda \mid \det(\sigma) = +1\}$, of index 1 or 2, and the quotient $\overline{\mathrm{Aut}}(\Lambda) = \mathrm{Aut}(\Lambda)/\{\pm \mathrm{Id}\}$. For odd n, $\mathrm{Aut}(\Lambda)$ is isomorphic to the direct product $\{\pm \mathrm{Id}\} \times \mathrm{Aut}^+(\Lambda)$. For even n, examples of both split and non-split extensions exist (see Sections 4.2 and 4.3).

More generally, for $G \subset \mathrm{Aut}(\Lambda)$, one can define $\overline{G} = G/(G \cap \{\pm \mathrm{Id}\})$, and in particular a group $\overline{\mathrm{Aut}}(\Lambda)^+$. Many examples for which this last group is a simple group occur in the ATLAS. Other important normal subgroups are the *Weyl groups*, generated by hyperplane reflections of various kinds, discussed in Section 4.10.

We say that a lattice Λ is the *orthogonal sum* of lattices $\Lambda_1, \dots, \Lambda_r$ in subspaces $E_1, \dots, E_r$ of E if $\Lambda = \Lambda_1 + \dots + \Lambda_r$ and if any two subspaces E_i, E_j are orthogonal; *notation*: $\Lambda = \Lambda_1 \perp \dots \perp \Lambda_r$. Then both the sums $\perp_i E_i$ and $\perp_i \Lambda_i$ are direct. We also say that Λ is *irreducible* if it is nonzero and if it is not the orthogonal sum of two nonzero relative sublattices.

For an orthogonal decomposition $\Lambda = \Lambda_1 \perp \dots \perp \Lambda_r$, we have

$$N(\Lambda) = \min_i N(\Lambda_i) \quad \text{and} \quad S(\Lambda) = \bigcup_{N(\Lambda_i)=N(\Lambda)} S(\Lambda_i)\,.$$

In particular, Λ is well rounded if and only if all the Λ_i are themselves well rounded and have the same minimum.

Theorem 1.4.5. (Eichler, Kneser.) *A lattice Λ possesses an orthogonal decomposition $\Lambda = \Lambda_1 \perp \dots \perp \Lambda_r$ into irreducible components Λ_i, and this decomposition is unique up to a permutation of the Λ_i.*

Proof. The existence of a decomposition is easily proved by induction on $\dim \Lambda$: if Λ is not irreducible, it is of the form $\Lambda = \Lambda' \perp \Lambda''$ with nonzero Λ', Λ'', and we can then decompose both Λ' and Λ'' into irreducible components.

To prove that the decomposition is unique, we first need two definitions. We say that a nonzero vector $x \in \Lambda$ is *reduced* if it cannot be written as a sum of two nonzero shorter vectors, and that two reduced vectors x, y are *contiguous* if there exists a chain $x_0 = x, x_1, \dots, x_k = y$ of reduced vectors in Λ with $x_i \cdot x_{i+1} \neq 0$ for $1 \le i < k$.

These definitions imply the following three properties (an induction argument establishes the third one, taking into account the fact that the lengths of the vectors in Λ constitute a discrete set in $\mathbb{R}$):

1. A sum of two nonzero orthogonal vectors is not reduced.
2. Contiguity is an equivalence relation.
3. Every nonzero vector in E is a sum of reduced vectors.

Let $\mathcal{R}$ be a system of representatives for the contiguity classes of reduced vectors and, for $x \in \mathcal{R}$, let $\Lambda(x)$ be the subgroup of Λ generated by the reduced vectors contiguous to x. For distinct $x, y \in \mathcal{R}$, the sum $\Lambda(x) + \Lambda(y)$ is orthogonal: by (3), it suffices to prove the orthogonality of arbitrary reduced vectors $x' \in \Lambda(x)$ and $y' \in \Lambda(y)$, and this is plainly true, since x and y would be otherwise contiguous.

For $x_1, \dots, x_k \in \mathcal{R}$, we have $k \le \sum_{i \le k} \dim \Lambda(x_i) \le \dim \Lambda$ since an orthogonal sum is in particular a direct sum. Hence $\mathcal{R}$ is finite, and we have $\Lambda = \sum_{x \in \mathcal{R}} L(x)$ again by (3). This completes the proof. □

Say that Λ is *isotypic* if its irreducible components are isometric. By putting together isometric components, we obtain the *isotypic* decomposition of a lattice. This notion is an important one for the description of automorphism groups.

Theorem 1.4.6. *Let Λ be a lattice. Its automorphism group is the direct product of the automorphism groups of its isotypic components. Moreover, if Λ is isotypic, then* $\mathrm{Aut}(\Lambda)$ *is canonically isomorphic to a semi-direct product* $\mathrm{Aut}(\Lambda_0)^r \rtimes S_r$*, where we have identified Λ with a power Λ_0^r of an irreducible lattice Λ_0.*

Proof. An isometry $f : \Lambda \to \Lambda'$ between two lattices sends an irreducible component of Λ onto an isometric irreducible component of Λ', and thus maps an isotypic component of Λ onto an isometric isotypic component of Λ'. Applied to the case $\Lambda' = \Lambda$, this remark proves the first assertion of the theorem.

Suppose $\Lambda \sim \Lambda_0^r$ where Λ_0 is irreducible. The symmetric group S_r permutes the irreducible components of Λ, and we can identify $\mathrm{Aut}(\Lambda_0)^r \rtimes S_r$ with a subgroup of $\mathrm{Aut}(\Lambda)$. It is clear that we obtain the whole automorphism group in this way: any automorphism f of Λ defines a permutation σ of S_r which extends naturally to $\mathrm{Aut}(\Lambda)$, and $f \circ \sigma^{-1}$ then belongs to $\mathrm{Aut}(\Lambda_0)^r$. □

We end this section with some remarks on the reduction of automorphisms of a lattice Λ modulo an integer $m \geq 2$. We denote by $\mathrm{GL}(\Lambda/m\Lambda) \simeq \mathrm{GL}_n(\mathbb{Z}/m\mathbb{Z})$ the automorphism group of the $\mathbb{Z}$-module $\Lambda/m\Lambda$, without taking account of the additional structure induced on this quotient by the Euclidean structure of E.

More generally, let G be a subgroup of $\mathrm{Aut}(\Lambda)$ and let $\Lambda' \subset \Lambda$ be a sublattice invariant under G. The map which to $u \in G$ associates the residue map $x \mapsto u(x) \mod \Lambda'$ induces a homomorphism of G into $\mathrm{GL}(\Lambda/\Lambda')$. We consider the case where $\Lambda' = m\Lambda$.

Theorem 1.4.7. *Let $m \geq 2$ be a natural integer and let $\phi : \mathrm{Aut}(\Lambda) \to \mathrm{GL}(\Lambda/m\Lambda)$ be the reduction homomorphism modulo m.*

1. *If $m \geq 3$, ϕ is injective.*
2. *If $m = 2$, $\ker \phi$ contains $\{\pm \mathrm{Id}\}$, and reduces to $\{\pm \mathrm{Id}\}$ if and only if Λ is irreducible.*

Proof. Let $u \in \operatorname{Ker} \phi$, and let $v = u - \mathrm{Id} \in \mathrm{End}(E)$. To every $x \in \Lambda$ we attach the sequence (x_i) in E defined by $x_0 = x$ and $x_{i+1} = \frac{1}{m}(v(x_i))$ (hence $x_i = \frac{1}{m^i} v^i(x)$). We have $u(x) \equiv x \mod m\Lambda$ for every $x \in \Lambda$; hence v maps Λ into $m\Lambda$, so that (x_i) is a sequence in Λ. Using the triangle inequality, we have

$$\|x_{i+1}\| = \frac{1}{m}\|u(x_i) - x_i\| \leq \frac{1}{m}\|u(x_i)\| + \|x_i\| \leq \frac{2}{m}\|x_i\|,$$

and the equality $\|x_{i+1}\| = \|x_i\|$ holds if and only if $x_i = 0$ or $m = 2$ and $\frac{1}{2}\|x_i\| = \|x_i - u(x_i)\|$, i.e. $m = 2$ and $u(x_i) = -x_i$.

Since $\{\|x\| \mid x \in \Lambda\}$ is a discrete subset of $\mathbb{R}$, the decreasing sequence (x_i) is constant for i sufficiently large, and this constant is zero for $m \geq 3$, since the sequence is then strictly decreasing. Thus for $m \geq 3$ (resp. for

$m = 2$), for all $x \in E$ and for i sufficiently large, we have $(u - \mathrm{Id})^i(x) = 0$ (resp. $(u^2 - \mathrm{Id})^i(x) = 0$), i.e. $u - \mathrm{Id}$ (resp. $u^2 - \mathrm{Id}$) is nilpotent. Now the orthogonal transformations u and u^2 are *semi-simple* (i.e. possess a diagonal form over $\mathbb{C}$), and so are $u - \mathrm{Id}$ and $u^2 - \mathrm{Id}$. Since they are nilpotent, they are indeed zero.

Hence $\ker\phi$ is trivial for $m \geq 3$, and contains only orthogonal symmetries for $m = 2$. In this last case, if $u \neq \pm\mathrm{Id}$ is such a symmetry, let F be its invariant subspace, and let $r = \dim F$ $(0 < r < n)$. For $x \in E$ written in the form $x = y + z$, $y \in F$, $z \in F^\perp$, we have $u(x) = y - z = x - 2z$, and z belongs to Λ (because of the inclusion $(u - \mathrm{Id})(\Lambda) \subset 2\Lambda$). But $y = x - z$ also belongs to Λ. Hence Λ is the orthogonal sum of $\Lambda \cap F$ and $\Lambda \cap F^\perp$ (and the proof indeed yields a simple general description of $\ker\phi$). □

1.5 Bilinear and Quadratic Forms

The aim of this short section is to make precise some questions of vocabulary needed in the forthcoming two sections. Here, R is a commutative ring and M is an R-module. We denote by $\mathcal{B}^s(M)$ (resp. $\mathcal{B}^a(M)$) the R-module of symmetric (resp. alternating) bilinear forms on M. (Recall that B is *alternating* if $B(x, x) = 0$ identically on M.)

Alternating forms are skew-symmetric, because of the identity

$$B(x,y) + B(y,x) = B(x+y, x+y) - B(x,x) - B(y,y)\,.$$

The converse is true whenever the map $x \mapsto 2x$ from M to M is injective; this condition is certainly satisfied when 2 is invertible in R and M is torsion-free, e.g. when R is a field of characteristic $p \neq 2$. In the opposite direction, when $2 = 0$ in R, both notions of symmetric and skew-symmetric forms coincide.

Definition 1.5.1. A quadratic form on M *is a map* $Q : M \to M$ *such that:*

1. $\forall \lambda \in R$ *and* $\forall x \in M$, $Q(\lambda x) = \lambda^2 Q(x)$.
2. $B_Q : (x, y) \mapsto Q(x+y) - Q(x) - Q(y)$ *is a (symmetric) bilinear form.*

We denote by $\mathcal{Q}(M)$ *the* R*-module of quadratic forms on* M.

For all $x \in M$, we have $B_Q(x,x) = 2Q(x)$. Hence the homomorphism $Q \mapsto B_Q$ of $\mathcal{Q}(M)$ in $\mathcal{B}^s(M)$ is injective, surjective, or bijective according to whether $x \mapsto 2x$ is. In this last case, we instead attach to Q the form $B' = \frac{1}{2}B_Q$; we then have the formula $Q(x) = B'(x,x)$. This is what we did in Section 1.2 with $R = \mathbb{R}$, $M = E$, $Q(x) = N(x)$ and $B'(x,y) = x \cdot y$. (When $x \mapsto 2x$ is the null map in M, B_Q is alternating.)

The considerations above apply to a lattice Λ, taking $R = \mathbb{Z}$ and $M = \Lambda$. To define integral lattices, we choose the point of view of bilinear forms rather than that of quadratic forms, which might give rise to half-integral bilinear forms.

Conversely, let M be a rank-n, finitely generated, torsion-free $\mathbb{Z}$-module endowed with a symmetric bilinear form. This form extends to the $\mathbb{Q}$-vector space $V = \mathbb{Q}\bigotimes_{\mathbb{Z}} M$, and moreover we can embed V into $E = \mathbb{R}\bigotimes_{\mathbb{Q}} V$ and identify M with a lattice in E. When the initial form is positive definite, E is a Euclidean space, and we recover the situation of Section 1.2. It is customary to call a "lattice" any pair (M, B) consisting of a free $\mathbb{Z}$-module M of rank n and a positive definite symmetric bilinear form B, keeping in mind the possibility of embedding M into a Euclidean space E.

1.6 Quadratic Forms

In this section, V denotes a vector space of dimension n over a field K, Q a quadratic form over V, and B the corresponding bilinear form. We describe some invariants which will allow us in some cases to characterize the pair (V, Q) (we shall simply say "the form Q") up to isometry. We pay special attention to the case where K is a finite field, which occurs when we consider the reduction of a lattice modulo a prime.

Besides the dimension n of V, we consider the rank r of Q (or of B, i.e. the rank of a Gram matrix for B), and the codimension r' of the kernel $\ker Q$ of Q, the largest subspace of V on which Q is identically zero. The form Q defines a form on $V/\ker Q$, still denoted by Q. Replacing V by $V/\ker Q$ if necessary, we may and shall assume in the remainder of this section that $r' = n$. We then say that the form Q is *nonsingular*. We say that Q is *non-degenerate* if B is non-degenerate, a condition which is *a priori* more restrictive than the previous one.

Let m be the integer such that $r' = 2m$ or $r' = 2m + 1$.

If $\operatorname{Char} K \neq 2$, then $r' = r$. If $\operatorname{Char} K = 2$, r is even (since B is then an alternating form), but r' may be greater than r. It is easily verified that V is an orthogonal sum of a non-degenerate quadratic subspace and of lines on which the restriction of Q is of the form λX^2, $\lambda \in K^*$.

When K is perfect (for instance, when K is finite), the identity $\sum \lambda_i x_i^2 = (\sum \lambda_i^{1/2} x_i)^2$ shows that r' is at most $r + 1$.

Let us first consider the case where Q is non-degenerate. If $\operatorname{Char} K \neq 2$ (resp. if $\operatorname{Char} K = 2$), V can be decomposed into an orthogonal sum of lines (resp. of planes), so that in a suitable basis Q can be written in the form

$$Q(X) = \sum_{i \leq n} \lambda_i X_i^2, \ \lambda_i \in K^*$$

$$(\text{resp. } Q(X) = \sum_{i \leq m} a_i X_i^2 + X_i Y_i + b_i Y_i^2, \ a_i, \, b_i \in K).$$

Then $d_Q = \prod \lambda_i \in K^*/K^{*2}$ (resp. $\operatorname{Arf}_Q = \sum_i a_i b_i \in K/\wp(K)$) is an invariant of the isometry class of Q, the *discriminant* (resp. the *Arf* *invariant*) of Q, which is multiplicative (resp. additive) on orthogonal sums. [In characteristic $p > 0$, $\wp$ denotes the *Artin–Schreier map* $x \mapsto x^p - x$.]

Whatever Char K may be, a nonsingular quadratic form is an orthogonal sum of m planes and 0 or 1 line. The quadratic planes which represent 0 are called *hyperbolic planes*. On such a plane Q can be given the form $Q(X, Y) = XY$. They are characterized by the property $d_Q = -1 \mod K^{*2}$ (resp. $\mathrm{Arf}_Q = 0 \mod \wp(K)$).

From now on, we suppose that K is the finite field $\mathbb{F}_q$ and that r is positive (recall that $n = r' \geq 2$). Then K/K^{*2} and $K/\wp(K)$ are groups of order 2, a non-trivial element of which we denote by α.

For even n, there are up to proportionality two nonsingular (here, non-degenerate) quadratic forms, which are proportional to forms denoted by Q_n^+ and Q_n^-, and thus two orthogonal groups, denoted by $\mathrm{O}_n^+(\mathbb{F}_q)$ and $\mathrm{O}_n^-(\mathbb{F}_q)$, namely $Q_n^+ = X_1Y_1 + \cdots + X_mY_m$ (the neutral form) and $Q_n^- = Q' \perp Q_{m-1}^+$, where Q' is the form $X_1^2 - \alpha Y_1^2$ if q is odd, and the form $X_1^2 + X_1Y_1 + \alpha Y_1^2$ otherwise.

For odd n, there is a single form when q is even and two proportional forms otherwise, and thus one orthogonal group. We represent the forms by $Q_n^\circ = X_1^2 \perp Q_{n-1}^+$, and we denote by $\mathrm{O}_n(\mathbb{F}_q)$ the corresponding orthogonal group.

Assuming now $n \geq 3$, we explain how to calculate the order of the group $\mathrm{O}(Q)$ of a nonsingular quadratic form Q over a finite field $\mathbb{F}_q$. We first define in a non-traditional way a group $\mathrm{SO}(Q)$ of index 1 or 2 in $\mathrm{O}(Q)$. For odd q, $\mathrm{SO}(Q)$ is as usual the kernel of the determinant (which maps $\mathrm{O}(Q)$ onto $\{\pm 1\}$). For even q, there exists in the non-degenerate case an *additive determinant* related to the Arf invariant, and $\mathrm{SO}(Q)$ is its kernel, still of index 2 in $\mathrm{O}(Q)$. In the degenerate case, there is no Arf invariant, and we set $\mathrm{SO}(Q) = \mathrm{O}(Q)$.

For $Q = Q^+, Q^-, Q^\circ$, the orders N of $\mathrm{SO}(Q)$, denoted respectively by N^+, N^-, N°, are given by the following formulae, where $\epsilon \in \{\pm 1\}$ is identified with $\pm$:

$$N^\circ = q^{m^2}(q^{2m} - 1)(q^{2m-2} - 1) \cdots (q^2 - 1) \qquad \textbf{(1.6.1)}$$

and

$$N^\pm = q^{m(m-1)}(q^m - \epsilon)(q^{2m-2} - 1)(q^{2m-4} - 1) \ldots (q^2 - 1)\,. \qquad \textbf{(1.6.2)}$$

For odd q, we define the *spinor norm* of $\mathrm{SO}(q)$ onto $\mathbb{F}_q/\mathbb{F}_q^2$ by attaching to a product of reflections along vectors $x_1, x_2, \ldots, x_{2k}$ the element

$$Q(x_1)Q(x_2) \ldots Q(x_{2k}) \in \mathbb{F}_q/\mathbb{F}_q^2\,.$$

Its kernel is denoted by $\Omega_n(\mathbb{F}_q)$. This is a group whose centre is trivial if n is odd or if q is even, and is $\{\pm \mathrm{Id}\}$ otherwise. For $n \geq 5$, the corresponding projective group $P\Omega_n(Q)$ is the simple group attached to Q (except $P\Omega_5 \simeq \mathrm{Sp}_4(\mathbb{F}_2) \simeq S_6$). For all $n \geq 3$, its order is

$$|P\Omega_n^*(\mathbb{F}_q)| = N^*/d\,, \tag{1.6.3}$$

where

$$d = (2, q-1) \text{ if } * = \circ, (4, q^m - (\pm 1)) \text{ if } * = \pm\,,$$

and $*$ stands for $+, -$ or $\circ$. This is the group denoted by $\mathrm{O}_n(q)$ or $\mathrm{O}_n^{\pm}(q)$ in the ATLAS.

There exist in low dimensions isomorphisms between orthogonal groups on the one hand, and linear, *symplectic* or *unitary groups* on the other hand. Recall that the *symplectic group* Sp_n (n even) is the subgroup of GL_n which preserves the standard alternating form, and that the unitary group U_n preserves the standard Hermitian form. The following are the isomorphisms involving the groups $P\Omega$:

$$P\Omega_3(\mathbb{F}_q) \simeq \mathrm{PSL}_2(\mathbb{F}_q)\,, \quad P\Omega_4^+(\mathbb{F}_q) \simeq \mathrm{PSL}_2\,\mathbb{F}_q \times \mathrm{PSL}_2\,\mathbb{F}_q\,,$$
$$P\Omega_4^-(\mathbb{F}_q) \simeq \mathrm{PSL}_2(\mathbb{F}_{q^2})\,, \quad P\Omega_5(\mathbb{F}_q) \simeq \mathrm{PSp}_4(\mathbb{F}_q)\,,$$
$$P\Omega_6^+(\mathbb{F}_q) \simeq \mathrm{PSL}_4(\mathbb{F}_q)\,, \quad P\Omega_6^-(\mathbb{F}_q) \simeq \mathrm{PSU}_4(\mathbb{F}_q)\,.$$

There are also interesting isomorphisms involving symmetric and alternating groups. We shall encounter some of them when calculating automorphism groups of various classical lattices (Section 4.9).

1.7 The Dictionary Relating Lattices and Quadratic Forms

We still denote by E a Euclidean space of dimension n and by $\mathcal{L}$ the set of lattices in E. Let us choose a particular lattice $\Lambda_0 \in \mathcal{L}$. We saw in Section 1.1 that the map $u \mapsto u(\Lambda_0)$ from $\mathrm{GL}(E)$ to $\mathcal{L}$ induces a bijection from $\mathrm{GL}(E)/\mathrm{GL}(\Lambda_0)$ onto $\mathcal{L}$. Two lattices $u(\Lambda_0)$ and $v(\Lambda_0)$ are isometric (resp. similar) if and only if there is an equality $v = wus$ with appropriate elements $s \in \mathrm{GL}(\Lambda_0)$ and $w \in \mathrm{O}(E)$ (resp. $w \in \mathbb{R}^*\,\mathrm{O}(E)$). We can thus identify the sets of isometry classes (resp. of similarity classes) of lattices in E with

$$\mathrm{O}(E)\backslash \mathrm{GL}(E)/\mathrm{GL}(\Lambda_0) \quad (\text{resp. } \mathbb{R}^*\,\mathrm{O}(E)\backslash \mathrm{GL}(E)/\mathrm{GL}(\Lambda_0)).$$

For odd n, since $\mathrm{GL}(E) = \{\pm \mathrm{Id}\} \times \mathrm{GL}^+(E)$, the identifications above can be transformed into

$$\mathrm{SO}(E)\backslash \mathrm{GL}^+(E)/\mathrm{SL}(\Lambda_0) \quad (\text{resp. } \mathrm{SO}(E)\backslash \mathrm{SL}(E)/\mathrm{SL}(\Lambda_0)).$$

Choose for Λ_0 a lattice generated by an orthonormal basis $\mathcal{B}_0$ for E. We may then identify E with $\mathbb{R}^n$ and Λ_0 with $\mathbb{Z}^n$. Isometry (resp. similarity) classes are now in one-to-one correspondence with the sets

$$\mathrm{O}_n(\mathbb{R})\backslash \mathrm{GL}_n(\mathbb{R})/\mathrm{GL}_n(\mathbb{Z}) \quad (\text{resp. } \mathbb{R}^*\,\mathrm{O}_n(\mathbb{R})\backslash \mathrm{GL}_n(\mathbb{R})/\mathrm{GL}_n(\mathbb{Z})),$$

or for odd n with

$$\mathrm{SO}_n(\mathbb{R})\backslash \mathrm{GL}_n^+(\mathbb{R})/\,\mathrm{SL}_n(\mathbb{Z}) \quad (\text{resp. } \mathrm{SO}_n(\mathbb{R})\backslash \mathrm{SL}_n(\mathbb{R})/\,\mathrm{SL}_n(\mathbb{Z})).$$

We denote by $X, Y, \ldots$ the $n \times 1$ matrices (columns) which represent vectors $x, y, \ldots$ of E identified with $\mathbb{R}^n$.

With every quadratic form Q on E, we associate the symmetric bilinear form B such that $Q(x) = B(x,x)$ identically on E (thus $B = \frac{1}{2}B_Q$ in the notation of Section 5), and the symmetric matrix $A = (a_{i,j}) \in M_n(\mathbb{R})$ such that $Q(x) = {}^tXAX$ (then $B(x,y) = {}^tYAX$). We shall often identify the quadratic form Q with the bilinear form B and the corresponding symmetric matrix A.

We attach to any $u \in \mathrm{GL}(E)$ the quadratic form $Q \circ u$, as well as the corresponding bilinear form $B \circ u : (x,y) \mapsto B(u(x), u(y))$ and the symmetric matrix ${}^tPAP$, where $P = \mathrm{Mat}_{\mathcal{B}_0}(u)$.

We say that two quadratic forms Q and Q' are *equivalent* if $Q' = Q \circ u$ for some $u \in \mathrm{GL}(\Lambda_0)$. In terms of matrices, equivalence corresponds to the existence of $P \in \mathrm{GL}_n(\mathbb{Z})$ such that $A' = {}^tPAP$. (We consider equivalence modulo $\mathrm{GL}_n(\mathbb{Z})$, *not* modulo $\mathrm{SL}_n(\mathbb{Z})$.) We shall also make use of *equivalence up to proportionality*.

We restrict ourselves to positive definite forms, or equivalently to symmetric matrices whose eigenvalues, which of course are real, are indeed strictly positive. The set of these forms on E (resp. on $\mathbb{R}^n$) is denoted by $\mathcal{Q}^+(E)$ (resp. $\mathcal{Q}_n^+(\mathbb{R})$).

We now return to the Euclidean space E. The map $x \mapsto \|x\|^2 = N(x)$ is a canonical positive definite quadratic form, with corresponding bilinear form the scalar product $(x,y) \mapsto x \cdot y$. The map $u \mapsto (x \mapsto \|u(x)\|^2)$ of $\mathrm{End}(E)$ into $\mathcal{Q}^+(E)$ is surjective (by *Sylvester's law of inertia*: real quadratic forms are classified up to equivalence by their *signature*, i.e. the numbers of positive and negative squares in a decomposition as a sum of squares of independent linear forms), and the stabilizer of the canonical form is the orthogonal group $\mathrm{O}(E)$. Thus quadratic forms over E, and these same forms up to proportionality, are in one-to-one correspondence with the right equivalence classes

$$\mathrm{O}(E)\backslash \mathrm{GL}(E) \quad \text{and} \quad \mathbb{R}^* \mathrm{O}(E)\backslash \mathrm{GL}(E)\,,$$

and, for odd n, with the right equivalence classes

$$\mathrm{SO}(E)\backslash \mathrm{GL}^+(E) \quad \text{and} \quad \mathrm{SO}(E)\backslash \mathrm{SL}(E)\,.$$

As for the sets of equivalence classes and equivalence classes up to proportionality, they are in one-to-one correspondence with the double classes

$$\mathrm{O}(E)\backslash \mathrm{GL}(E)/\,\mathrm{GL}(\Lambda_0) \quad \text{and} \quad \mathbb{R}^* \mathrm{O}(E)\backslash \mathrm{GL}(E)/\,\mathrm{GL}(\Lambda_0)\,,$$

and, for odd n, with

$$\mathrm{SO}(E)\backslash\,\mathrm{GL}^+(E)/\,\mathrm{SL}(\Lambda_0) \quad \text{and} \quad \mathrm{SO}(E)\backslash\,\mathrm{SL}(E)/\,\mathrm{SL}(\Lambda_0)\,,$$

which take the following form after the identification of E with $\mathbb{R}^n$:

$$\mathrm{SO}_n(\mathbb{R})\backslash\,\mathrm{GL}_n^+(\mathbb{R})/\,\mathrm{SL}_n(\mathbb{Z}) \quad \text{and} \quad \mathrm{SO}_n(\mathbb{R})\backslash\,\mathrm{SL}_n(\mathbb{R})/\,\mathrm{SL}_n(\mathbb{Z})\,.$$

We observe that lattices and quadratic forms yield the same double coset decomposition. The direct transformation of isometry classes of lattices into equivalence classes of forms can be done in the following way: start with a pair $(\Lambda, \mathcal{B})$ of a lattice Λ and a basis $\mathcal{B}$ for Λ, and consider the quadratic form $Q(x) = \|x_1 e_1 + \cdots + x_n e_n\|^2$. We have $Q \circ u = Q$ for all $u \in \mathrm{O}(E)$. To change the basis for Λ amounts to replacing Q by an equivalent form. In this correspondence, we have $N(\Lambda) = m(Q)$, where $m(Q) = \inf_{x \in \mathbb{Z}^n \smallsetminus \{0\}} Q(x)$ is the *minimum of* Q, and to $S(\Lambda)$ corresponds

$$S(Q) = \{x \in \mathbb{Z}^n \mid Q(x) = m(Q)\}\,.$$

The group $\mathrm{Aut}(\Lambda)$ can be identified with the *automorphism group of* Q, defined by $\mathrm{Aut}(Q) = \{P \in \mathrm{GL}_n(\mathbb{Z}) \,|\, {}^tPAP = A\}$, where A is the matrix of Q, whose entries are $a_{i,j} = e_i \cdot e_j$.

Let us make these identifications more precise. Let $u \in \mathrm{GL}(E)$ with matrix $P = (p_{i,j})$ in $\mathcal{B}$. If $x \in E$ has components x_i in $\mathcal{B}$ and if $X \in \mathcal{M}_{n,1}(\mathbb{R})$ is the one-column matrix with entries $x_1, \dots, x_n$, $x' = u(x)$ satisfies the transformation formulae

$$x'_i = \sum_k p_{i,k}\, x_k \Longleftrightarrow X' = PX\,,$$

and the matrix of $Q' = Q \circ u$ is $A' = {}^tPAP$.

Note that the same formula arises when P is viewed as the base change matrix from $\mathcal{B}$ to a new basis $\mathcal{B}' = (e'_1, e'_2, \dots, e'_n)$, where we then have $e'_i = \sum_k p_{k,i}\, e_k$. Thus:

$$u(\Lambda) \subset \Lambda \Longleftrightarrow P \in \mathcal{M}_n(\mathbb{Z}),\ u(\Lambda) = \Lambda \Longleftrightarrow P \in \mathrm{GL}_n(\mathbb{Z})$$

and

$$u \in \mathrm{O}(E) \Longleftrightarrow {}^tPAP = A\,.$$

The schemes (see Lemma 1.3.6)

$$\mathcal{B} \xrightarrow[P]{u} \mathcal{B} \xrightarrow[A]{\mathrm{Id}} \mathcal{B} \quad \text{and} \quad \mathcal{B} \xrightarrow[P^{-1}]{u^{-1}} \mathcal{B} \xrightarrow[A]{\mathrm{Id}} \mathcal{B}$$

again show that u is orthogonal if and only if

$$AP^{-1} = {}^t(AP) \Longleftrightarrow P^{-1} = A^{-1}\,{}^tPA\,.$$

We shall return later to other forms of this correspondence, involving various notions (e.g., perfection, eutaxy, extremality, Voronoi domains) that will be defined later, and shall now consider the orthogonalization and decomposition into sums of squares of quadratic forms, a procedure to be considered again in the next chapter (Section 2.9) together with "Hermite reduction". [Roughly speaking, "reduction ", when referring to lattices, means a search for bases with nice properties; when referring to quadratic forms, it means the search for equivalent forms whose coefficients lie in certain not too large domains.]

Given a lattice Λ together with a basis $\mathcal{B} = (e_1, \ldots, e_n)$, we construct by the *Gram–Schmidt orthogonalization* procedure a new basis $\mathcal{B}' = (e'_1, \ldots, e'_n)$ for Λ defined in the following way: let E_r be the subspace of E spanned by $e_1, \ldots, e_r$; we first take $e'_1 = e_1$, and then assuming by induction that $e'_1, \ldots, e'_r \in \langle e_1, \ldots, e_r \rangle$ have already been constructed, we choose for e'_{r+1} the unique vector $x \equiv e_{r+1} \mod E_r$ which is orthogonal to E_r. A variant consists in replacing e'_i by $e''_i = \dfrac{1}{\|e'_i\|} e'_i$ for all i. The new basis $\mathcal{B}''$ is then *the* orthonormal basis for E such that, for all $r \leq n$, $(e''_1, \ldots, e''_r)$ is a basis for E_r which defines the same orientation as $(e_1, \ldots, e_r)$.

This Gram–Schmidt procedure has the following geometric interpretation: first define e''_1 as above, then project Λ onto the hyperplane F orthogonal to e_1, then play the same game with the projection of $\mathcal{B}$ onto $F, \ldots$

Given a positive definite quadratic form

$$Q(x) = \sum_i a_{i,i}\, x_i^2 + \sum_{i<j} 2a_{i,j}\, x_i x_j\,,$$

we define n *outer* coefficients A_i and $\frac{n(n-1)}{2}$ *inner* coefficients $\alpha_{i,j}$, with $1 \leq i < j \leq n$, using its classical *decomposition into squares*, an algorithm which is particularly easy to perform for definite forms. Since $a_{1,1}$ is nonzero, we may write Q in the form $Q(x) = \sum_{i=1}^n A_i(x_i + \sum_{j>i} \alpha_{i,j} x_j)^2$ by first using the identity

$$Q(x_1, \ldots, x_n) = a_{1,1}\Big(x_1 + \sum_{i=2}^n \frac{a_{1,i}}{a_{1,1}}\, x_i\Big)^2 + Q'(x_2, \ldots, x_n)\,,$$

which defines A_1 and $\alpha_{1,i}$, $1 \leq i \leq n$, and then continuing by induction.

In the dictionary, the orthogonal projection on the first r vectors of $\mathcal{B}$ corresponds to a *partial decomposition as a sum of squares*, in which Q is expressed as a sum of r squares and a form in the $n-r$ variables $x_{r+1}, \ldots, x_n$.

The dual lattice corresponds to the *inverse form*, defined by the inverse of its matrix, which may be calculated by dividing the *adjoint matrix* by the determinant. (Recall that the entries of the adjoint of $A = (a_{i,j})$ are the *cofactors* $(-1)^{i+j} d_{i,j}$ where $d_{i,j}$ is the determinant of the matrix obtained by omitting the i-th row and the j-th column from A.) The adjoint matrix defines the *adjoint form*.

Here is a summary of the dictionary:

Table 1.7.1. Dictionary relating lattices and quadratic forms.

Lattices	*Quadratic forms*
(space E, orthonormal basis)	(space $\mathbb{R}^n$, canonical basis)
(lattice, basis (e_i)) mod. isometries	form $x \mapsto \|\sum_i x_i e_i\|^2$
lattice up to isometry	class of form
lattice up to similarity	class of form up to proportionality
$N(\Lambda) = \|\Lambda\|^2$, $S(\Lambda)$	$m(Q)$, $S(Q)$
Gram–Schmidt construction	decomposition into squares
projection on $\langle e_1, \dots, e_r\rangle^\perp$	r squares + form in $x_{r+1}, \dots, x_n$
dual lattice	inverse form (with matrix A^{-1})

1.8 Packings

In this section, we denote by $\mathcal{P}$ a closed discrete subspace of the Euclidean space E. The set $\mathcal{P}$ is often a lattice, in which case it will be denoted by Λ.

Definition 1.8.1. *Let $A \subset E$. We say that $\mathcal{P}$* packs *A if any two sets among the $A + x, x \in \mathcal{P}$ intersect only at boundary points. We also say that $\{A + x \,|\, x \in \mathcal{P}\}$ is a packing. Let $x_0 \in E$, and let B_R be the ball of radius R centred at x_0 . The limit δ for $R \to \infty$ (if it exists) of*

$$\mathrm{vol}(B_R \cap (\cup_{x\in\mathcal{P}} A + x)) / \mathrm{vol}(B_R)$$

is the density of the packing.

In practice, we only consider convex, bounded sets. When A is a convex set, it has a well–defined volume, and moreover $\mathrm{vol}(A) = \mathrm{vol}(\overline{A})$. We also restrict ourselves to sufficiently regular sets P, so that the density is well defined; note that it is in any case possible to consider *upper* and *lower densities*.

The (upper) density is of course bounded above by 1. It is of great importance to estimate this density for balls having the greatest possible radius (indeed $\frac{1}{2}\|\Lambda\|$) which can be packed by a lattice Λ. This problem led Minkowski to create the "geometry of numbers"; see the notes on Chapter 2 (Section 2.11). The result (see Chapter 2) reads as follows (ω_n denotes the volume of the unit ball of E):

Proposition 1.8.2. *Suppose $\mathcal{P}$ is a lattice Λ, and A is a ball packed by Λ whose radius is as large as possible. Then the packing has density*

$$\delta = \frac{\omega_n \|\Lambda\|^n}{2^n \Delta(\Lambda)} = \frac{\omega_n}{2^n} \frac{N(\Lambda)^{n/2}}{\det(\Lambda)^{1/2}} \,. \;\square$$

We now consider a few other notions, which will be used for lattices, but which are more conveniently expressed in the general setting of packings.

Definition 1.8.3. *The* Voronoi cell of a point $x_0 \in \mathcal{P}$ *is*

$$V_{x_0}(\mathcal{P}) = \{x \in E \mid \forall y \in \mathcal{P} : \|x - x_0\| \leq \|x - y\|\}\,.$$

This is in fact a (convex) polytope in case $\mathcal{P}$ is a lattice, since for given y, x_0, the equation $\|x - y\| = \|x - x_0\|$ is that of a hyperplane in E. [The words "cell" or "polytope" are preferred to the word "domain", which we shall reserve for subsets of the $\frac{n(n+1)}{2}$-dimensional space attached to the lattice.]

The various Voronoi cells of a lattice are translates of the cell containing the origin, and constitute a *tessellation* of E. In particular, the volume of any Voronoi cell is the discriminant of Λ.

Let us return to general packings.

Definition 1.8.4. *The* distance of $x \in E$ to $\mathcal{P}$ *is* $d(x, \mathcal{P}) = \min_{y \in \mathcal{P}} \|y - x\|$. *A* hole in $\mathcal{P}$ *is a point $x \in E$ for which $d(x, \mathcal{P})$ attains a local maximum; according to the terminology of [C-S], a hole is called a* deep hole *if the maximum is absolute, and a* shallow hole *otherwise.*

The deep holes in a lattice Λ are of fundamental importance for the construction of dense lattices containing Λ as a hyperplane section with the same minimal norm as Λ – see [C-S1] for an important example.[2]

The holes in a lattice Λ are the images under translations by elements of Λ of those which lie in a given fundamental parallelotope. Thus there are only finitely many holes modulo Λ. The sets of deep holes and of shallow holes are invariant under $\mathrm{Aut}(\Lambda)$ (and also under the affine *group of symmetries of* Λ, the semi-direct product of $\mathrm{Aut}(\Lambda)$ and Λ). Note that the holes are the vertices of the Voronoi cells of Λ.

The sphere packings associated to a lattice are called *regular*. Besides these packings, which are the most interesting for us, it is useful to consider the *semi-regular sphere packings*, those which are of the form $\mathcal{E} = \cup_{i=1}^{r} x_i + \Lambda$ for some lattice Λ. For a recent example of such a packing, see [Var]. Without loss of generality, we may assume that $x_1 = 0$ and that the x_i are distinct modulo Λ, since for all $x \in E$, $x + \Lambda$ depends solely on x modulo Λ.

Such a semi-regular packing is regular if and only if the $x_i \mod \Lambda$ constitute a subgroup of E/Λ. For instance, if $r = 2$, say $\mathcal{E} = \Lambda \cup (x + \Lambda)$, the packing $\mathcal{E}$ is regular if and only if $2x \in \Lambda$.

Let us now look at lattices obtained by adjoining a single vector to a given lattice. We thus consider a lattice Λ equipped with a basis $e_1, \ldots, e_n$. The

[2] (Remark of N.J.A. Sloane.) The holes in a lattice, deep or not, are also of great importance in applications. In chemistry, for example, one often finds crystals with atoms of one type at lattice points and smaller atoms of different types at the holes.

lattices $\Lambda' \supset \Lambda$ such that Λ'/Λ is a cyclic group of a given order $r > 1$ are generated by Λ and a vector $e = \frac{1}{r}(a_1e_1 + \cdots + a_ne_n)$ where $a_1, \dots, a_n$ are integers whose g.c.d. is prime to r.

Suppose that $a_n = 1$. Then $e_1, \dots, e_{n-1}, e$ is a basis for Λ'. Let $x = \sum_{i=1}^n x_i e_i \in \Lambda$, and set $y_i = rx_i + a_iy$ for $i < n$ and $y_n = y$. Then $x \in \Lambda'$ if and only if rx is of the form $\sum_{i=1}^n y_i e_i$ where y_i is integral and satisfies the congruence $a_i y_i \equiv y_n \mod r$ for every $i < n$. Denoting by f the quadratic form $x \mapsto \|\sum x_ie_i\|^2$, we see that the condition $\|\Lambda'\| = \|\Lambda\|$ can be expressed in the following way:

$$y_n \not\equiv 0 \mod r \iff f(y_1, \dots, y_n) \geq r^2\|\Lambda\|^2 .$$

In practice, we are led to calculate the minimum of f on integers which satisfy one of the congruences $y_n \equiv a \mod r$ for $a = 1, 2, \dots, \lfloor \frac{r}{2} \rfloor$.

We end this section with a very simple example, that of the cubic lattice $\Lambda = \mathbb{Z}^n$ in $E = \mathbb{R}^n$. To find its holes, we reduce by translation and permutation of the coordinates to an exploration of the closed cube $C = \{x \in \mathbb{Z}^n \mid \forall i, 0 \leq x_i \leq \frac{1}{2}\}$.

For all $x \in C$, we have $d(x, \Lambda) = \|x\|$. Hence $\mathbb{Z}^n$ contains a unique hole modulo $\mathbb{Z}^n$, necessarily a deep one, namely the point $x \in C$ with all its coordinates equal to $\frac{1}{2}$. Semi-regular packings with $r = 2$ based on $\Lambda = \mathbb{Z}^n$ are of the form $\Lambda \cup (x + \Lambda)$ with $x \in C$ and $\|x\| \geq 1$. They exist only for $n \geq 4$. They are regular if and only if the components of x are half-integral, and are then isometric to one of the $n - 3$ lattices constructed with vectors x whose first $k \geq 4$ components are equal to $\frac{1}{2}$ and the following $n - k$ components are zero.

1.9 More on Integral Lattices

This section is mainly devoted to integral lattices, though we shall sometimes ignore this hypothesis. We first recall a few results about modules over a principal ideal domain R.

Let M be an R-module and let M' be a submodule of M, both having the same rank n. (When $R = \mathbb{Z}$, this amounts to saying that $[M : M'] < \infty$.) There then exists a basis $\mathcal{B} = (e_1, e_2, \dots, e_n)$ for M and nonzero elements $a_1, a_2, \dots, a_n$ of R such that $\mathcal{B}' = (a_1 e_1, a_2 e_2, \dots, a_n e_n)$ is a basis for M', and a_i divides a_{i-1} for $2 \leq i \leq n$. The a_i are called the *elementary divisors* or the *invariant factors* of the pair (M, M'), and the sequence $a_1, \dots, a_n$ is sometimes called the *Smith invariant of the pair* (M, M').

[More precisely, uniqueness holds not for the sequence a_i itself, but only for the principal ideals (a_i) they generate; for $R = \mathbb{Z}$, we choose for every i the positive generator of (a_i). Note that it is the use of ideals rather than numbers which can be generalized to more general rings such as Dedekind domains. Note also that the definition is of a local nature. Local methods will be used in the proof of Theorem 1.9.10 below; they also yield a direct definition of elementary divisors over Dedekind domains.]

To a matrix $P \in \mathcal{M}_{p,q}(R)$, we attach the elementary divisors of $M = R^q$ and its submodule M' generated by the columns of P.

Definition 1.9.1. *Let Λ be a lattice which is proportional to an integral lattice, and let λ be the smallest positive real number μ such that $\Lambda_1 = \sqrt{\mu}\,\Lambda$ is integral. The set of elementary divisors (over $\mathbb{Z}$) for (Λ_1^*, Λ_1) is called the* Smith invariant of Λ, *and denoted by* Smith(Λ).

In other words, λ is the smallest real number μ such that $\mu\,(x \cdot y)$ is integral for all $x, y \in \Lambda$, or such that the inclusion $\mu\Lambda \subset \Lambda^*$ holds; this is also the unique positive real number μ such that $\sqrt{\mu}\,\Lambda$ is primitive in the sense of Definition 1.2.6.

Given an integral lattice Λ, we sometimes say that λ as defined above is the *level of* Λ, though this word sometimes applies (for reasons related to modular forms) only to *even* lattices Λ, with μ such that $\mu\Lambda^*$ is moreover *even*.

Proposition 1.9.2. *The Smith invariant of a lattice solely depends on its similarity class.*

Proof. It suffices to prove that if Λ is any primitive lattice and u any isometry, then Smith(Λ) = Smith$(u(\Lambda))$. But we have $u(\Lambda)^* = u(\Lambda^*)$ (because $u \in \mathrm{O}(E)$). Hence u induces an isomorphism of Λ^* onto $u(\Lambda)^*$. □

Let Λ be an integral lattice, and let $\mathcal{B} = (e_1, \dots, e_n)$ be a basis for Λ. Then $P = \mathrm{Gram}(\mathcal{B})$ has integral entries, and thus elementary divisors, which may be compared with those of Λ.

Proposition 1.9.3. Smith(P) = Smith(Λ).

Proof. The isomorphism of $\mathbb{Z}^n$ onto Λ^* which maps the canonical basis $(\varepsilon_1, \dots, \varepsilon_n)$ of $\mathbb{Z}^n$ onto $\mathcal{B}^*$ maps the j-th column of P, i.e. $\sum_i (e_i \cdot e_j)\,\varepsilon_i$, onto the vector $\sum_i (e_i \cdot e_j)\,e_j^*$ of Λ^*. But this vector is equal to e_j, as one sees by calculating its scalar products with the vectors of $\mathcal{B}$. We have thus constructed an isomorphism of $\mathbb{Z}^n$ onto Λ^* which maps $P\mathbb{Z}^n$ onto Λ. □

Let a be a positive integer. When one replaces an integral lattice Λ by a "multiple" $\sqrt{a}\Lambda$, the elementary divisors of the pair (Λ^*, Λ) are then multiplied by a (and $\det(\Lambda)$ by a^n). However, since we mainly consider lattices only up to similarity, we choose to define Smith(Λ) so as to obtain an invariant of similarity classes.

Moreover, note that with our definition, $a_n = 1$: since a_n divides all the other elementary divisors, an integral lattice with $a_n > 1$ cannot be primitive.

The following proposition allows us to compute the Smith invariant of the dual lattice:

Proposition 1.9.4. *For any lattice Λ proportional to an integral one with Smith invariant $(a_1,\dots,a_n)$, we have*

$$\mathrm{Smith}(\Lambda^*) = \Big(\frac{a_1}{a_n} = a_1, \frac{a_1}{a_{n-1}}, \dots, \frac{a_1}{a_2}, \frac{a_1}{a_1} = 1\Big).$$

Proof. We may assume that Λ is primitive. Then

$$\mathrm{Smith}(\Lambda) = \mathrm{Smith}(\Lambda^*, \Lambda).$$

Since a_1 is the smallest of the positive integers m such that $m\Lambda^* \subset \Lambda$, $\mathrm{Smith}(\Lambda^*) = \mathrm{Smith}(\Lambda, a_1\Lambda^*)$, and we are reduced to the following lemma about principal ideal domains:

Lemma 1.9.5. *Let R be a P.I.D., let M a free R-module, let $M' \subset M$ be a free submodule with the same rank n and let* $\mathrm{Smith}(M, M') = (a_1, a_2, \dots, a_n)$. *(In particular, (a_1) is the annihilator of M/M'.) Then the elementary divisors (well defined up to products by invertible elements of R) of $(M', a_1 M)$ are* $\frac{a_1}{a_n}, \frac{a_1}{a_{n-1}}, \dots \frac{a_1}{a_2}, \frac{a_1}{a_1} = 1$.

Proof. Just choose a basis $(e_1, e_2, \dots, e_{n-1}, e_n)$ for M such that $(a_1\, e_1, a_2\, e_2, \dots, a_{n-1}\, e_{n-1}, a_n\, e_n)$ is a basis for M', and compare it to the basis $(a_1\, e_1, a_1\, e_2, \dots, a_1\, e_{n-1}, a_1\, e_n)$ for $a_1 M$. □

In the remainder of this section, Λ denotes an integral lattice in E and $M \subset \Lambda$ a relative lattice. Let $F = \mathbb{R}M$, and for two relative lattices X and Y, let $Y_X^\perp = X \cap \mathbb{R}Y^\perp$ be the *orthogonal complement of Y in X*.

Proposition 1.9.6. *$\Lambda \cap F^\perp$ is a lattice in $F^\perp$.*

Proof. Proposition 1.3.4 shows that it suffices to prove that $\Lambda^* \cap F$ is a lattice in F, and this is a consequence of the inclusion $\Lambda \subset \Lambda^*$. □
[Note that it suffices to suppose that Λ^* is proportional to a sublattice of Λ.]

Corollary 1.3.5 and Proposition 1.9.3 show that the following equalities hold:

$$\begin{aligned}\det(M) &= \det(\Lambda)\,\det(F^\perp \cap \Lambda^*)\\ &= \det(\Lambda)\,\frac{\det(M_\Lambda^\perp)}{[F^\perp \cap \Lambda^* : F^\perp \cap \Lambda]^2}\end{aligned} \qquad (*)$$

and

$$\det(M^\perp) = \det(\Lambda)\,\frac{\det(M)}{[F \cap \Lambda^* : F \cap \Lambda]^2}\,. \qquad (*')$$

Eliminating the determinants of M and $M^\perp$ in the equalities above, we obtain the following generalization of Proposition 1.3.7 (2):

Proposition 1.9.7. *For every subspace F of E whose intersection with Λ is a lattice in F, the following equality holds:*

$$\det(\Lambda) = [F \cap \Lambda^* : F \cap \Lambda]\,[F^\perp \cap \Lambda^* : F^\perp \cap \Lambda]\,. \square$$

We now turn to the special case of unimodular lattices.

Proposition 1.9.8. *Suppose that $M = \mathbb{R}M \cap \Lambda$. If Λ is unimodular,*

$$\det(M_\Lambda^\perp) = \det(M) \qquad \text{and} \qquad [\Lambda : M \perp M_\Lambda^\perp] = \det(M)\,.$$

Proof. When Λ is unimodular, $\det(\Lambda) = 1$ and $\Lambda^* = \Lambda$. This immediately establishes the first assertion, and the second is a consequence of the first, thanks to the equalities

$$\det(M \perp M') = \det(M)\,\det(M') = \det(\Lambda)\,[\Lambda : M \perp M']^2\,,$$

which hold because M and M' generate complementary subspaces of E. $\square$

The case where M itself is unimodular deserves special attention. Replacing $\det(M_\Lambda^\perp)$ in $(*)$ by its value calculated with the formula above yields the equality

$$\det(M) = [\Lambda : M \perp M_\Lambda^\perp]\,[F \cap \Lambda^* : F \cap \Lambda]\,, \qquad (**)$$

which for $\det(M) = 1$ implies that the indices in the right–hand side both take the value 1. Moreover, the condition $M = \mathbb{R}M \cap \Lambda$ is automatically satisfied for all relative unimodular lattices contained in Λ because of the index formula $\det(M) = \det(\mathbb{R}M \cap \Lambda)\,[\mathbb{R}M \cap \Lambda : M]^2$. We have thus proved:

Proposition 1.9.9. *Let Λ an integral lattice and let $M \subset \Lambda$ be a unimodular relative lattice. Then M is an orthogonal summand in Λ. More precisely, Λ decomposes into the orthogonal sum $\Lambda = M \perp M_\Lambda^\perp$, and in particular, $\det(M_\Lambda^\perp) = \det(\Lambda)$.* $\square$

[Direct proof: one first proves as above the equality $M = \mathbb{R}M \cap \Lambda$, which implies the existence of a basis $(e_1, \dots, e_n)$ for Λ such that $(e_1, \dots, e_r)$ is a basis for M. Let $x = \sum_{j=1}^n a_j\, e_j \in \Lambda$. We must find a $y \in M$ with $x - y \in M_\Lambda^\perp$. Let us write it in the form $y = \sum_{j=1}^r b_j e_j$. We are then reduced to finding integer solutions to the linear system

$$\sum_{j=1}^r (e_i \cdot e_j)\, b_j = c_i,\ i = 1, \dots, r\,,$$

and this has a unique solution (b_j) in $\mathbb{Z}^r$, since $\det(e_i \cdot e_j) = \det(M) = 1$.]

In general, it is not possible to predict the value of the determinant of $M_\Lambda^\perp$ just from the determinants of M and Λ. The following theorem is a useful partial result which contains as special cases the calculations of $\det(M_\Lambda^\perp)$ we have given in the preceding two propositions:

Theorem 1.9.10. *We assume as before that Λ/M is torsion-free. Then $\det(M^{\perp}_{\Lambda})$ divides the product $\det(M)\det(\Lambda)$, and the ratio $\dfrac{\det(M)\det(\Lambda)}{\det(M^{\perp}_{\Lambda})}$ is moreover a square whose prime factors divide g.c.d.$(\det(M),\det(\Lambda))$. In particular, the equality $\det(M^{\perp}_{\Lambda}) = \det(M)\det(\Lambda)$ holds whenever M and Λ have coprime determinants.*

Proof. $(*')$ shows that $\det(M^{\perp}_{\Lambda})$ divides the product $\det(M)\det(\Lambda)$ and that these two quantities have a square ratio.

To prove the second part of the theorem, we extend the results of Propositions 1.9.8 and 1.9.9 to the case when $\mathbb{Z}$ is replaced by its local ring at a prime p (i.e. by the subring $\mathbb{Z}_{(p)}$ of $\mathbb{Q}$ of those rational numbers whose denominator is prime to p). This is a P.I.D. in which (p) is the unique nonzero prime ideal. The hypotheses of Theorem 1.9.10 show that at least one of the lattices M, Λ is $\mathbb{Z}_{(p)}$-unimodular (i.e., its determinant is a unit in $\mathbb{Z}_{(p)}$), and the arguments used to prove 1.9.8 and 1.9.9 can now be applied to establish 1.9.10. □

A useful notion concerning integral lattices is that of a partial dual. Let Λ be integral, and let m be the annihilator of Λ^*/Λ. For each divisor a of m, let $\Lambda[a] = \Lambda + a\,\Lambda^*$. We have $\Lambda[1] = \Lambda^*$ and $\Lambda[m] = \Lambda$. It is easily verified that $\Lambda[a]^* = \Lambda[b]$ any time there is a decomposition $m = ab$ with coprime a, b. Such a lattice $\Lambda[a]$ is called a *partial dual to* Λ.

We end this section with the description of a transformation relying on local principles, due to Watson ([Wat1]; see also [Ne-Pl] and [Sch-SP]), which reduces (or preserves) the determinant of a given integral lattice Λ. For each prime divisor p of $\det(\Lambda)$, consider the sublattice $\sqrt{p}\,(\frac{1}{p}\,\Lambda\cap\Lambda^*)$, then replace the scalar product $x\cdot y$ by $\frac{1}{p}\,x\cdot y$ every time the congruence $x\cdot y\equiv 0 \mod p$ holds on $\Lambda\cap\Lambda^*$, and repeat the same procedure over and over until it stabilizes. We finally obtain a lattice which satisfies the following two properties:

1. The annihilator of Λ^*/Λ is square free;
2. For all p, $\dim_{\mathbb{Z}/p\mathbb{Z}}(\Lambda^*/\Lambda)_p \le \frac{n}{2}$.

1.10 Tensor Product and Exterior Powers

Again let R be a commutative ring. We can define ([Bou2], Chapter II, § 3) the *tensor product* $M\bigotimes_A N$ of two R-modules M and N. This is an R-module which is characterized up to a unique isomorphism by the existence of an R-bilinear map $(x,y)\mapsto x\otimes y$ from $M\times N$ into $M\bigotimes_A N$ such that any bilinear map from $M\times N$ into an R-module P factorizes through $M\bigotimes_A N$ in a unique linear map. Any element of $M\bigotimes_A N$ (also written $M\bigotimes N$) is a sum of *split tensors* $x\otimes y$, $x\in M, y\in N$.

In practice, one constructs homomorphisms from $M \bigotimes N$ by making use of the factorization property and then proves their uniqueness by evaluating them on split tensors.

For example, given two further R-modules M', N' and two homomorphisms $u : M \to M'$ and $v : N \to N'$, there exists a unique homomorphism $u \otimes v : M \bigotimes N \to M' \bigotimes N'$ satisfying the identity $(u \otimes v)(x \otimes y) = u(x) \otimes v(y)$.

Similarly, given bilinear forms b on M and c on N, there exists a unique bilinear form $b \otimes c$ on $M \bigotimes N$ satisfying the identity $(b \otimes c)(x \otimes y, x' \otimes y') = b(x, y) c(x', y')$.

Suppose now that M and N are free with respective ranks p and q. Then $M \bigotimes N$ is free with rank pq: if $\mathcal{B} = (e_1, e_2, \dots, e_p)$ and $\mathcal{B}' = (e'_1, e'_2, \dots, e'_q)$ are bases for M and N respectively, $\mathcal{B} \otimes \mathcal{B}' = (e_i \otimes e'_j)$ is a basis for $M \bigotimes N$. [This is a consequence of the following more general result: if N is free of rank $q > 0$, then any element of $M \bigotimes N$ can be written in a unique way as a sum $\sum_{j=1}^{q} x_j \otimes e'_j$ with $x_j \in M$.]

Given $u \in \operatorname{End}(M)$ and $v \in \operatorname{End}(N)$, the determinants of u, v, $u \otimes v$ are well defined and satisfy ([Bou2], Chapter III, § 8, n° 7)

$$\det(u \otimes v) = \det(u)^{\dim N} \det(v)^{\dim M}.$$

Let Λ_1 and Λ_2 be two lattices in Euclidean spaces E_1 and E_2, of respective dimensions p and q. Then $\Lambda_1 \bigotimes_{\mathbb{Z}} \Lambda_2$ embeds naturally in the Euclidean space $E = E_1 \bigotimes_{\mathbb{R}} E_2$, whose scalar products are well defined on split tensors by the formula $(x \otimes y) \cdot (x' \otimes y') = (x \cdot x') \times (y \cdot y')$. This makes $\Lambda = \Lambda_1 \bigotimes \Lambda_2$ a lattice in E.

Proposition 1.10.1. *We have*

$$(\Lambda_1 \bigotimes \Lambda_2)^* = \Lambda_1^* \bigotimes \Lambda_2^* \quad \text{and} \quad \det(\Lambda_1 \bigotimes \Lambda_2) = \det(\Lambda_1)^q \det(\Lambda_2)^p.$$

Proof. (1) The inclusion $(\Lambda_1 \bigotimes \Lambda_2)^* \subset \Lambda_1^* \bigotimes \Lambda_2^*$ is obvious. To prove the opposite inclusion, we consider a basis $\mathcal{B} = (e_1, \dots, e_q)$ for Λ_2. Let $x = \sum_{i=1}^{q} x_j \otimes e_j^* \in (\Lambda_1 \bigotimes \Lambda_2)^*$. Now $x \cdot (y \otimes e_i)$ is integral for all $y \in \Lambda_1$ and all $i \in [1, q]$. But $x \cdot (y \otimes e_i) = x \cdot x_i$. Hence $y \cdot x_i$ is integral for all i and all $y \in \Lambda_1$, x_j lies in Λ_1^* for all j, and x lies in $(\Lambda_1 \bigotimes \Lambda_2)^*$.

(2) Choose $u_1 \in \operatorname{End}(E_1)$ and $u_2 \in \operatorname{End}(E_2)$ with $u_i(\Lambda_i) = \Lambda_i^*$, and observe that the determinants of Λ_1, Λ_2, $\Lambda_1 \bigotimes \Lambda_2$ are the squares of those of u_1, u_2, $u_1 \otimes u_2$ respectively. □

The difficult problem is the determination of $S(\Lambda_1 \bigotimes \Lambda_2)$. Split tensors $x_1 \otimes x_2$, $x_1 \in S(\Lambda_1)$, $x_2 \in S(\Lambda_2)$ have norm $N(\Lambda_1)\, N(\Lambda_2)$, which proves the upper bound $N(\Lambda_1 \bigotimes \Lambda_2) \le N(\Lambda_1)\, N(\Lambda_2)$. But this inequality may be strict, and non-split minimal vectors may exist even when equality holds. We refer the reader to [Ki], Chapter 7 for numerous results on this problem; see also [M-H], Chapter II, § 9, and [Cou5].

The tensor product of modules is commutative and associative up to isomorphism; more precisely, one can identify $M\bigotimes N$ with $N\bigotimes M$ (resp. $(M\bigotimes N)\bigotimes P$ with $M\bigotimes(N\bigotimes P)$) by means of the isomorphism which sends $x\otimes y$ onto $y\otimes x$ (resp. $(x\otimes y)\otimes z$ onto $x\otimes(y\otimes z)$). We can thus consider the tensor product of $r \geq 0$ copies of a same module M, denoted by $\overset{r}{\bigotimes}M$, whose elements are sums of split tensors $x_1\otimes x_2\otimes\dots\otimes x_r$ which we write without parentheses. (Warning: however, $x\otimes y$ and $y\otimes x$ are generally distinct elements of $M\bigotimes M$!) We have $\overset{0}{\bigotimes}M = A$ and $\overset{1}{\bigotimes}M = M$.

Definition 1.10.2. *For any integer $r \geq 0$, the r-th* exterior power *of M, denoted by $\overset{r}{\bigwedge}M$, is the quotient of $\overset{r}{\bigotimes}M$ by its submodule generated by split tensors having two identical components; we denote by $x_1 \wedge x_2 \wedge \cdots \wedge x_r$ the image of $x_1\otimes x_2\otimes\dots\otimes x_r$ in $\overset{r}{\bigwedge}M$.*

The symmetric group S_r acts on $\overset{r}{\bigwedge}M$, and the transformation formula reads $x_{\sigma(1)} \wedge x_{\sigma(2)} \wedge \cdots \wedge x_{\sigma(r)} = \varepsilon(\sigma)\, x_1 \wedge x_2 \wedge \cdots \wedge x_r$. When M is free with rank n, $\overset{r}{\bigwedge}M$ is zero for $r > n$ and free with rank $\binom{n}{r}$ for $0 \leq r \leq n$; moreover we have $\overset{0}{\bigwedge}M = A$ and $\overset{1}{\bigwedge}M = M$, and $\overset{n}{\bigwedge}M$ is a rank-1 module.

Every $u \in \operatorname{End}(E)$ defines an endomorphism $\overset{r}{\wedge}u$ of $\overset{r}{\bigwedge}M$; for $r = n$, this is multiplication by $\det(u)$.

We refer to [Bou2], Alg. IX, § 2, Proposition 6 for a proof of the following result:

Theorem 1.10.3. *Suppose that M is equipped with a bilinear form $(x,y) \mapsto f(x,y)$. Then there exists a unique bilinear form $\overset{r}{\wedge}f$ on $\overset{r}{\bigwedge}M$ satisfying the identity*

$$(\overset{r}{\wedge}f)\,(x_1 \wedge \cdots \wedge x_r, y_1 \wedge \cdots \wedge y_r) = \det(f(x_i, y_j))\,. \;\square$$

Given a Euclidean space E and a lattice Λ in E, the theorem above produces a lattice $\overset{r}{\bigwedge}\Lambda$ in $\overset{r}{\bigwedge}E$. (The fact that exterior powers of positive definite forms are positive definite can be proved by making use of the existence of orthonormal bases for such forms, see [Bou2], Alg. IX, § 7, Proposition 1.)

Proposition 1.10.4. *Let M be a lattice in E.*

1. *For $1 \leq r \leq n$, $\det(\overset{r}{\bigwedge}M) = \det(M)^{\binom{n-1}{r-1}}$.*
2. *The dual of $\overset{r}{\bigwedge}M$ is $\overset{r}{\bigwedge}M^*$.*

Proof. The first assertion follows immediately from the formula $\det(\overset{r}{\wedge}u) = \det(u)^{\binom{n-1}{r-1}}$ giving the determinant of exterior powers of an endomorphism u

of a vector space V, which is easily proved in matrix form over an algebraic closure of $\mathbb{Q}$ by making use of a triangular reduction.

The second assertion results from the embedding $(\bigwedge^r M)^* \hookrightarrow \bigwedge^r M^*$, which must be onto since the two sides have equal determinants by (1). □

Let us fix an orientation of E by choosing an orthonormal basis $\mathcal{B}$ for E. It is immediate that the map

$$(x_1 \wedge x_2 \wedge \cdots \wedge x_{n-r}, y_1 \wedge y_2 \wedge \cdots \wedge y_r) \mapsto \det_{\mathcal{B}_0}(x_1, \ldots, x_{n-r}, y_1, \ldots, y_r)$$

is well–defined, and induces a canonical isometry $\sigma : \bigwedge^{n-r} E \to (\bigwedge^r E)^* \simeq (\bigwedge^r E)$.

Proposition 1.10.5. *For every lattice M in E, the map σ induces an isometry of $\bigwedge^{n-r} M$ onto $\sqrt{\det(M)}\bigwedge^r M^*$.*

Proof. The two sides have equal determinants, and $\det_{\mathcal{B}_0}(x_1, \ldots, x_{n-r}, y_1, \ldots, y_r)^2$ (for $x_i, y_j \in M$) belongs to the ideal $(\det(M))$ of $\mathbb{Z}$. □

We end this chapter with a statement about automorphism groups, whose proof will be left to the reader. Given a lattice M, an automorphism σ of M and an integer $r \in [1, n-1]$, one easily verifies that there exists a unique automorphism σ' of $\bigwedge^r M$ satisfying the identity $\sigma'(x_1 \wedge x_2 \wedge \cdots \wedge x_r) = \sigma(x_1) \wedge \sigma(x_2) \wedge \cdots \wedge \sigma(x_r)$.

Proposition 1.10.6. *The map $\sigma \mapsto \sigma'$ is a homomorphism of* $\mathrm{Aut}(M)$ *into* $\mathrm{Aut}(\bigwedge^r M)$ *which is injective for odd r and has kernel* $\{\pm \mathrm{Id}\}$ *for even r.* □

[Example: the lattice E_8 (Section 4.4) defines a representation of degree 8 of its automorphism group, whose exterior square induces a representation of degree 28 of the simple group $\mathrm{O}_8^+(2)$, the smallest possible degree for a faithful representation of this group, see Section 4.9 and [ATLAS], p. 86; numerous other examples (including that of the famous Leech lattice) can be found in the ATLAS.]

As tensor products and exterior powers do not generally preserve the perfection property, the notions introduced in Section 1.10 will not be used much in the rest of the book.

1.11 Notes on Chapter 1

The notion of a lattice was known at the beginning of the 19th century, in particular to Bravais, who produced a classification of three-dimensional lattices prompted by applications to crystallography, and to Jacobi, who required it for the theory of elliptic functions (doubly periodic meromorphic functions). The theorem according to which any discrete subgroup of $\mathbb{R}^n$ must have rank at most n is sometimes known as "the Jacobi–Bravais theorem".

However, the theory of real quadratic forms was developed by Lagrange and Gauss, and up to the end of the 19th century this was independent of lattice theory. The connection was made only in 1891, by Minkowski, when he created the geometry of numbers – see the notes on Chapter 2 (Section 2.11). Nevertheless, despite the importance of Minkowski's work, the theory of real quadratic forms went on growing aside from lattice theory. Lattices made an appearance in Coxeter's paper [Cox2] in the early 1950s (see the notes on Chapter 4, Section 4.12), but the systematic use of lattices in the theory of perfection dates really to the late 1960s. Note that some aspects of the theory (e.g., the Voronoi algorithm) still properly belong to the domain of quadratic forms.

Section 1.3. The dual lattice has been known for a long time to crystallographers. It is a standard comment that "X-ray diffraction patterns show the dual lattice". Among mathematicians, it has long been considered only in terms of adjoint forms, until its systematic use in the geometry of numbers, notably by Mordell and Mahler.

The denomination "parity vector" instead of "characteristic vector" is a recent suggestion of Conway and Sloane in [C-S"].

Section 1.4. Automorphisms were first studied in detail by crystallographers for whom they constitute (together with minimal vectors) the keystone for the classification of crystallographic systems.

The automorphism group of the $\mathbb{A}_n$ lattice was found by Korkine and Zolotareff in 1873 ([K-Z2]). Remarkable examples of lattices related to various simple groups can be found in the ATLAS.

Section 1.5, 1.6, 1.7. There is a large literature on quadratic and Hermitian forms, see in particular [Sch2]. A summary of the main properties of classical groups can be found in the ATLAS; for the proofs, see [Die].

Section 1.8. We do not develop the theory of non-regular sphere packings as we intend to concentrate on lattices. However, the best upper bounds in print for densities are in fact valid for any packing, regular or not. We refer the reader to Rogers [Rog]; see also the paper [Ka-L] by Kabatiansky and Levenshtein for more recent developments.

Section 1.9. The results of this section can be widely generalized to "lattices" in the sense of quadratic forms which are merely assumed to be non-degenerate, see e.g. [Mi-H] or [Sch2]. However, orthogonal decomposition must be handled with some care. For instance, the equality $E = F \perp F^\perp$ no longer holds if the restriction of the form to F is degenerate: in this case we have only the dimension formula $\dim E = \dim F + \dim F^\perp$.

I do not know to whom the notion of partial duality should be attributed. It is used for instance by Rains and Sloane in [R-S]. For an application to perfect lattices, see [Ven3], Section 19.

Section 10. Formal properties of tensor products and exterior powers can be found in [Bou2], from which the assertions of this section can be easily derived. Precise results on the minima of tensor products are established in the book [Ki] by Kitaoka. I made use of [Cou3] and [Cou4] for exterior powers.

2 Geometric Inequalities

Introduction

The aim of this chapter is to prove certain inequalities which will allow us to get lower bounds (and sometimes upper bounds) for lattice constants (defined in Section 2.5) for some important subsets of $\mathbb{R}^n$. Though our aim is to study balls (one says traditionally "spheres") centred at the origin, some more general bodies are dealt with, either because it seems useful to consider there more general questions, or in order to give applications of spheres to various questions in the geometry of numbers.

After three sections centred on questions related to spheres and a technical one of frequent use in the geometry of numbers, we turn to the general definition of lattice constants, and then restrict ourselves first to star bodies, then to convex bodies.

We return to spheres in Section 2.8, where we establish various inequalities involving a lattice and its dual that will be used in Chapter 6.

Notation: E is a Euclidean space of dimension n, and p_F denotes the orthogonal projection onto a subspace F of E.

2.1 The Hadamard Inequality

The aim of this section is to prove upper bounds for the determinant of a lattice in function of the determinants of a family of relative complementary sublattices.

Theorem 2.1.1. (Hadamard inequality.) *Let Λ be a lattice in E and let $e_1, e_2, \dots, e_n$ be n independent vectors in Λ. Then we have the inequality*

$$\det(\Lambda) \le N(e_1)N(e_2)\cdots N(e_n)\,,$$

and equality holds if and only if the vectors e_i are pairwise orthogonal and generate Λ.

Proof. Let Λ' be the lattice generated by $e_1, \dots, e_n$. We have

$$\det(\Lambda) = [\Lambda : \Lambda']^{-2} \det(\Lambda'),$$

and we are thus reduced to the case where $e_1, \dots, e_n$ constitute a basis for Λ. We now use an induction argument, starting with the obvious case of dimension 1.

Let F be the hyperplane of E generated by $e_1, \dots, e_{n-1}$. By Proposition 1.2.9, we have $\det(\Lambda) = \det(\Lambda \cap F)\, N(p_{F^\perp}(e_n))$. This immediately proves the required inequality, and shows that equality holds if and only if $e_i \cdot e_j = 0$ for $i < j < n$ and e_n is orthogonal to F. □

To generalize the inequality above, we must first have a closer look at orthogonal projections.

Proposition 2.1.2. *Let Λ be a relative lattice and let F be a subspace of E, both having same dimension r. Then we have $\det(p_F(\Lambda)) \leq \det(\Lambda)$, and equality holds if and only if Λ is contained in F.*

Proof. Let $G = \mathbb{R}\Lambda$. If $p = p_F$ maps G onto a strict subspace of F, $\det(p_F(\Lambda))$ is zero, and there is nothing to prove. We may thus assume that p_F induces an isomorphism of G onto F.

Let $\mathcal{B}$ be a basis for Λ, $\mathcal{B}_0$ an orthonormal basis for G, and $\mathcal{B}_0'$ an orthonormal basis for F. We have

$$\det(p(\Lambda)) = [\det{}_{\mathcal{B}_0'}(p(\Lambda))]^2 = [\det{}_{\mathcal{B}_0'}(p(\mathcal{B}_0)) \det{}_{p(\mathcal{B}_0)}(p(\Lambda))]^2,$$

where $|\det_{p(\mathcal{B}_0)}(p(\Lambda))| = |\det_{\mathcal{B}_0}(\Lambda)| = \Delta(\Lambda)$, and $|\det_{\mathcal{B}_0'}(p(\mathcal{B}_0))|$ is the discriminant of the projection of the cubic lattice with basis $\mathcal{B}_0 = (\varepsilon_1, \dots, \varepsilon_r)$.

Applying now the Hadamard inequality to this lattice, we obtain

$$(\det{}_{\mathcal{B}_0'}(p(\mathcal{B}_0)))^2 \leq N(p(\varepsilon_1)) \cdots N(p(\varepsilon_r)) \leq N(\varepsilon_1) \cdots N(\varepsilon_r) = 1,$$

with equality if and only if $N(p(\varepsilon_i)) = N(\varepsilon_i)$, i.e. $\forall i,\ \varepsilon_i \in F$, i.e. $\Lambda \subset F$. □

Theorem 2.1.3. (Generalized Hadamard inequality.) *Let Λ be a lattice which is the direct sum of relative lattices $\Lambda_1, \Lambda_2, \dots, \Lambda_k$. Then we have*

$$\det(\Lambda) \leq \det(\Lambda_1)\, \det(\Lambda_2) \cdots \det(\Lambda_k),$$

and equality holds if and only if the direct sum is orthogonal.

Proof. By induction, we are immediately reduced to the case where $k = 2$. Let Λ_2' be the orthogonal projection of Λ_2 onto $\mathbb{R}\Lambda_1^\perp$. We have $\det(\Lambda) = \det(\Lambda_1)\, \det(\Lambda_2')$, and we conclude the proof by applying Proposition 2.1.2 to the pair (Λ_2, Λ_2'). □

Remark 2.1.4. *One can state the Hadamard inequality without assuming the vectors to be independent. The result then still holds provided one adds to the equality cases the possibility that one of the vectors be zero.*

Inequalities 2.1.1 and 2.1.3 will serve very often to bound from above the index of a sublattice. This is done in the following way: let Λ be a well–rounded lattice and let Λ' be a sublattice of Λ generated by n independent minimal vectors; we have the relation $[\Lambda : \Lambda']^2 = \det(\Lambda')/\det(\Lambda)$. Lower bounds for the determinants of such lattices are known: we have $\det(\Lambda) \geq \Gamma_n^2 N(\Lambda)^n$ (see below, Corollary 2.5.6), which can also be written with the notation of Definition 2.2.6 below in the form $\det(\Lambda) \geq \gamma_n^{-n} N(\Lambda)^n$. Combining these lower bounds with the upper bound of $\det(\Lambda')$ given by Theorem 2.1.1, we obtain the inequality $[\Lambda : \Lambda'] \leq \gamma_n^{n/2}$.

Let us now give an example involving Theorem 2.1.3.

Suppose that n is even, say $n = 2m$, and that Λ possesses an automorphism σ with minimal polynomial $X^2 + X + 1$. Then for every minimal $x \in \Lambda'$, the lattice $\langle x, \sigma x\rangle$ is a hexagonal lattice, with minimal vectors $\pm x, \pm\sigma x, \pm\sigma^2 x$ and determinant $\frac{3}{4}N(\Lambda')$. By Theorem 2.1.3, we have the upper bound $\det(\Lambda') \leq \left(\frac{3}{4}\right)^m N(\Lambda')^n$ which yields the better inequality $[\Lambda : \Lambda'] \leq \left(\frac{3}{4}\right)^{n/4} \gamma_n^{n/2}$.

[Example in dimension 8: using the known value $\gamma_8 = 2$, we bound $[\Lambda : \Lambda']$ from above by 16 in general and by 9 in case there exists a σ as above; these two bounds are optimal, see Section 4.8.]

2.2 The Hermite Inequality

Theorem 2.2.1. (Hermite inequality.) *Every lattice Λ of E has a basis $(e_1, , \dots, e_n)$ which satisfies the inequality*

$$N(e_1)N(e_2)\cdots N(e_n) \leq \left(\frac{4}{3}\right)^{n(n-1)/2} \det(\Lambda)\,.$$

Applied to a basis vector having the shortest possible norm, Theorem 2.2.1 yields:

Corollary 2.2.2. *Every lattice Λ of E satisfies the inequality*

$$N(\Lambda) \leq \left(\frac{4}{3}\right)^{(n-1)/2} \det(\Lambda)^{1/n}\,. \ \square$$

Proof of 2.2.1. Let e_1 be a minimal vector in Λ, and let $H = e_1^\perp$. Denote by Λ' the orthogonal projection of Λ onto H and for every $x \in \Lambda$, let x' be the projection of x onto H. The proof relies on the following two lemmas:

Lemma 2.2.3. *Every $x' \in \Lambda'$ is the projection of some $x \in \Lambda$ with $N(x) \leq \frac{4}{3}N(x')$.*

Lemma 2.2.4. *Let $e_2', \dots, e_n'$ be a basis for Λ'. For all $i \geq 2$, let $e_i \in \Lambda$ have projection e_i'. Then $(e_1, e_2, \dots, e_n)$ is a basis for Λ.*

Taking for granted the two lemmas above, we prove Theorem 2.2.1 by induction, the case $n = 1$ being trivial. There exists a basis $e'_2, \dots, e'_n$ of Λ' such that $N(e'_2) \cdots N(e'_n) \le (\frac{4}{3})^{(n-1)(n-2)/2} \det(\Lambda')$. Lifting e'_i to $e_i \in \Lambda$ as in Lemma 2.2.3, we obtain using Lemma 2.2.4 a basis $e_1, \dots, e_n$ of Λ for which the inequality $N(e_i) \le \frac{4}{3} N(e'_i)$ holds for all $i \ge 2$.

By Proposition 1,2.9, we also have $\det(\Lambda) = \det(\Lambda') \, N(e_1)$, whence

$$N(e_1)N(e_2)\cdots N(e_n) \le \left(\frac{4}{3}\right)^{(n-1)(n-2)/2} \left(\frac{4}{3}\right)^{(n-1)} \det(\Lambda) = \left(\frac{4}{3}\right)^{n(n-1)/2} \det(\Lambda) \, . \quad \square$$

Proof of 2.2.3. We may suppose that x' is nonzero. Lift x' to some $x_0 \in \Lambda$. Then, for all $m \in \mathbb{Z}$, $x_m = x_0 + m e_1$ is also a lifting of x' in Λ. Choose m in order that its projection on $\mathbb{R}e_1$ be of length $\ell \le \frac{1}{2} \|e_1\|$. We then have $|\cos(x_m, e_1)| \le \frac{1}{2}$, whence $\frac{N(x')}{N(x_m)} = \sin^2(x_m, e_1) \ge \frac{3}{4}$. $\square$

Proof of 2.2.4. Let $x \in \Lambda$. Its projection x' onto H is a $\mathbb{Z}$-linear combination of the e'_i, say $x' = \sum_{i=2}^n a_i e'_i$. Then $y = \sum_{i=2}^n a_i e_i$ also projects onto x'. Hence $x - y \in \Lambda \cap \mathbb{R}e_1$. Since e_1 is minimal, it generates $\Lambda \cap \mathbb{R}e_1$. $\square$

Definition 2.2.5. The Hermite invariant of Λ *is*

$$\gamma(\Lambda) = N(\Lambda)/\det(\Lambda)^{1/n} \, ;$$

The centre density of Λ *is* $\delta(\Lambda) = 2^{-n}\gamma(\Lambda)^{n/2}$.

Note that the centre density of Λ is proportional to the density of the sphere packing canonically attached to Λ, and that both invariants solely depend on the similarity class of Λ.

We can now state Corollary 2.2.2 in the form $\gamma(\Lambda) \le (\frac{4}{3})^{(n-1)/2}$, which justifies the validity of the following definition:

Definition 2.2.6. The Hermite constant for dimension n *is* $\gamma_n = \sup_\Lambda \gamma(\Lambda)$, *where Λ runs over all lattices in E. A* critical lattice *is an n-dimensional lattice Λ such that $\gamma(\Lambda) = \gamma_n$.*

The upper bound of γ_n given by Corollary 2.2.2 is obviously optimal for $n = 1$, and also for $n = 2$, since a *hexagonal lattice* (a 2-dimensional lattice having a basis (x, y) of two minimal vectors such that $N(x) = N(y)$ and $x \cdot y = \frac{1}{2} N(x)$) has a Hermite invariant of precisely $\frac{2}{\sqrt{3}}$. The proof of Theorem 2.2.1 immediately shows that the equality $\gamma(\Lambda) = \gamma_2$ holds only on such lattices. We shall see later that the inequality in Corollary 2.2.2 is not the best possible from $n = 3$ onwards.

For further reference, we now state as a proposition the results of the calculations done at the end of Section 2.1:

Proposition 2.2.7. *If a lattice Λ has minimal norm m (and possesses n independent minimal vectors, but this hypothesis is not a necessary one) then* $\det(\Lambda) \geq \gamma_n^{-n} m^n$. □

To translate in terms of quadratic forms the results of this section is an easy exercise. We just do it for the Hermite inequality, which can be stated also for indefinite forms (it is not difficult to see that the general case follows from the positive definite case, see the end of Section 2.5):

Theorem 2.2.8. (Hermite.) *Let Q be a non-degenerate quadratic form in n variables, of discriminant D. There exists a basis $(e_1, \ldots, e_n)$ of $\mathbb{Z}^n$ such that*

$$|Q(e_1) \cdots Q(e_n)| \leq \left(\frac{4}{3}\right)^{(n-1)/2} |D|^{1/n}. \ \square$$

2.3 The Mordell Inequality

Theorem 2.3.1. (Mordell.) *For every $n \geq 3$, we have $\gamma_n \leq \gamma_{n-1}^{(n-1)/(n-2)}$, and equality holds if and only if for every n-dimensional critical lattice Λ, both Λ and Λ^* possess orthogonal minimal vectors whose orthogonal hyperplanes intersect Λ^* and Λ along $(n-1)$-dimensional critical lattices with the same norm as Λ^* and Λ respectively. Moreover, Λ^* is then also a critical lattice.*

Corollary 2.3.2. *For $n \geq m \geq 2$, we have*

$$\gamma_n \leq \gamma_m^{(n-1)/(m-1)}.$$

Proof. Fix m and use induction on $n - m$. □

Applying the preceding corollary with $m = 2$ and making use of the value $\gamma_2 = \frac{2}{\sqrt{3}}$ proved at the end of Section 2.2, we recover for $n \geq 2$ the bound $\gamma_n \leq (\frac{4}{3})^{(n-1)/2}$ of Corollary 2.2.2

Proof of 2.3.1. Since there is nothing to prove if $n = m$, we may assume that $n \geq 3$. Let Λ be lattice in E, let x be a minimal vector in Λ^*, let H be the hyperplane $(\mathbb{R}x)^\perp$ and let M be the lattice $\Lambda \cap H$ in H. Using the relation $\det(M) = \det(\Lambda)\, N(x) = \det(\Lambda)\, N(\Lambda^*)$, we deduce from the very definition of Hermite's invariant that

$$\gamma(M) = N(M) \left(\det(\Lambda) N(\Lambda^*)\right)^{-1/(n-1)}. \quad \textbf{(2.3.3)}$$

Replacing $N(\Lambda^*)$ in 2.3.3 by its value taken from the equalities

$$N(\Lambda^*) = \gamma(\Lambda^*)\ \det(\Lambda^*)^{1/n} = \gamma(\Lambda^*)\ \det(\Lambda)^{-1/n}\,,$$

we obtain

$$\gamma(M) = \frac{N(M)}{\gamma(\Lambda^*)^{1/(n-1)}\det(\Lambda)^{1/n}} = \frac{\gamma(\Lambda)}{\gamma(\Lambda^*)^{1/(n-1)}}\,\frac{N(M)}{N(\Lambda)}\,. \tag{2.3.4}$$

Similarly, let $M' = x'^{\perp} \cap \Lambda^*$. For any x' minimal in Λ, we have

$$\gamma(M') = \frac{\gamma(\Lambda^*)}{\gamma(\Lambda)^{1/(n-1)}}\,\frac{N(M')}{N(\Lambda^*)}\,. \tag{2.3.5}$$

Taking the $(n-1)$-th power of (2.3.4) and multiplying by (2.3.5), we obtain

$$\gamma(M)^{n-1}\gamma(M') = \gamma(\Lambda)^{n(n-2)/(n-1)}\,\frac{N(M)^{(n-1)}}{N(\Lambda)^{(n-1)}}\,\frac{N(M')}{N(\Lambda^*)}\,. \tag{2.3.6}$$

Since the last two terms in the right–hand side of 2.3.6 are at least 1, we have the inequality

$$\gamma(\Lambda)^{n(n-2)/(n-1)} \le \gamma(M)^{n-1}\gamma(M')\,. \tag{2.3.7}$$

Since both $\gamma(M)$ and $\gamma(M')$ are bounded from above by γ_{n-1}, we have

$$\gamma(M)^{n-1}\gamma(M') \le \gamma_{n-1}^{n}\,, \tag{2.3.8}$$

and the inequality of Theorem 2.3.1 now follows from (2.3.7).

For equality to hold in Theorem 2.3.1, it is first necessary that the second term in the right–hand side of (2.3.6) be equal to 1, i.e. that M and Λ have equal norms. This happens if and only if the minimal vectors $x \in \Lambda^*$ and $x' \in \Lambda$ are orthogonal. When this condition is satisfied, the third term in (2.3.6) is then also equal to 1. (We can even prove a slightly more precise result: for all $x \in S(\Lambda)$, there exists $x' \in S(\Lambda^*)$ orthogonal to x, and for all $x' \in S(\Lambda^*)$, there exists $x \in S(\Lambda)$ orthogonal to x'.) Note that the existence of a pair of orthogonal minimal vectors in Λ and Λ^* suffices to ensure the validity of both equalities $N(M) = N(\Lambda)$ and $N(M^*) = N(\Lambda^*)$.

Equality must also hold in (2.3.8), and this happens if and only if M and M' are critical for dimension $n-1$. But equality then holds in (2.3.7). Hence we have completely characterized lattices for which equality holds in Theorem 3.1.

Finally, the fact that Λ^* must also be a critical lattice follows from the comparison of (2.3.4) and (2.3.5). □

We shall return later in Section 2.8 to the Mordell inequality to study the function $\gamma' : \Lambda \mapsto \|\Lambda\|\,\|\Lambda^*\|$.

Equality holds in Mordell's inequality whenever Λ is unimodular and possesses a pair of orthogonal minimal vectors. It is known (resp. likely true, resp.

possible) that critical lattices in dimension 8 (resp. 24, resp. 48) are of this type, with minimal norm 2 (resp. 4, resp. 6). One knows since the publication of Korkine and Zolotareff's first paper in 1872 that equality holds in dimension 4. It is generally conjectured that the same holds true in dimensions 12 and 16 (see [C-S], Chapter 6, Section 2, p. 167) and maybe also in dimension 32.

These remarks are summarized in Table 2.3.9 below; definitions of the lattices which appear there can be found in Chapter 4, Chapter 8 and [C-S], Chapter 7, Section 5, p. 195; see also Table 14.4.1. Columns are organized in pairs of a lattice preceded by its densest hyperplane section. (In all cases, $\mathrm{Aut}(\Lambda)$ acts transitively on $S(\Lambda^*)$, so that hyperplane sections are well defined up to isometry.)

Table 2.3.9. A list of examples for which equality is expected in Th. 2.3.1.

n	3	4	7	8	11	12	15	16	23	24	47	48
γ_n^n	2	2^2	2^6	2^8	$\frac{2^{20}}{3^5}$	$\frac{2^{24}}{3^6}$	2^{21}	2^{24}	2^{44}	2^{48}	6^{46}	6^{48}
lattice	$\mathbb{A}_3$	$\mathbb{D}_4$	$\mathbb{E}_7$	$\mathbb{E}_8$	K_{11}	K_{12}	Λ_{15}	Λ_{16}	Λ_{23}	Λ_{24}	$*$	P_{48*}

Remark 2.3.10. In all cases, the lattice Λ is similar to its dual lattice. This is obvious for unimodular Λ, and in general related to the fact that there might well exist a single similarity class of critical lattices for relatively small dimensions.

2.4 Mahler's Compactness Theorem

Among various compactness theorems due to Mahler, we have selected the following statement, which is the most useful one in the applications to geometry of numbers. The topology on the set $\mathcal{L}$ of lattices has been defined in Section 1.1. Clearly both functions $\Lambda \mapsto N(\Lambda)$ and $\Lambda \mapsto \det(\Lambda)$ are continuous.

Theorem 2.4.1. (Mahler.) *Let $\mathcal{F}$ be a family of lattices which satisfies the following two conditions:*

1. *There exists $m > 0$ such that $N(\Lambda) \geq m$ for every $\Lambda \in \mathcal{F}$.*
2. *There exists $M > 0$ such that $\det(\Lambda) \leq M$ for every $\Lambda \in \mathcal{F}$.*

Then every sequence of lattices in $\mathcal{F}$ possesses a convergent subsequence. [In other words, $\mathcal{F}$ is a relatively compact set in $\mathcal{L}$, i.e. it has a compact closure.]

We first transform condition (2) into a more appropriate one:

Lemma 2.4.2. *The family $\mathcal{F}$ satisfies the following condition:*
2'. There exists $M' > 0$ such that every $\Lambda \in \mathcal{F}$ has a basis whose vectors are of norm at most M'.

Proof of 2.4.2. The result is obvious for $n = 1$. Let $n > 1$. By Hermite's inequality, every lattice $\Lambda \in \mathcal{F}$ has a basis $(e_1, e_2, \dots, e_n)$ which satisfies the inequality $N(e_1)N(e_2)\cdots N(e_n) \le (\frac{4}{3})^{n(n-1)/2} M$. We have $N(e_j) \ge m$ for all j, hence, for all i,

$$N(e_i) \le \left(\frac{4}{3}\right)^{n(n-1)/2} M \prod_{j \ne i} N(e_j)^{-1} \le \left(\frac{4}{3}\right)^{n(n-1)/2} M\, m^{-(n-1)}. \ \square$$

Proof of 2.4.1. We now prove the relative compactness of $\mathcal{F}$ under hypotheses (1) and (2'). Let $\Lambda^{(p)}$ be a sequence of lattices in $\mathcal{F}$, and for every p let $\mathcal{B}^{(p)} = (e_1^{(p)}, e_2^{(p)}, \dots, e_n^{(p)})$ be a basis for $\Lambda^{(p)}$ as in Lemma 2.4.2. By the compactness of the ball of radius M', we can extract from $(e_1^{(p)})$, $p \in \mathbb{N}$ a converging subsequence, i.e. find a subset I of $\mathbb{N}$ such that $(e_1^{(p)})$, $p \in I$ converges. We then repeat the argument for $(e_2^{(p)})$, $p \in I$ and so on, and we are finally reduced to the case where all sequences $(e_i^{(p)})$, $p \in \mathbb{N}$ $(1 \le i \le n)$ are convergent. Set $e_i = \lim_p e_i^{(p)}$, and let Λ be the subgroup of E generated by $e_1, \dots, e_n$. It suffices to prove that these vectors are linearly independent, for this implies that Λ is a lattice and that this is the limit of the sequence $(\Lambda^{(p)})$.

By Proposition 2.2.7, we have the inequality

$$\det(e_1^{(p)}, e_2^{(p)}, \dots, e_n^{(p)}) \ge \gamma_n^{-n} m^n .$$

Since the determinant is a continuous function, the same inequality holds for the system $e_1, \dots, e_n$. Hence these vectors have a nonzero determinant; in particular, they are independent. $\square$

Corollary 2.4.3. *The set $\mathcal{E}_m$ of well-rounded lattices in E having a given minimal norm $m > 0$ is compact.*

Proof. Let $\Lambda \in \mathcal{E}_m$ and let Λ' be the sublattice of Λ generated by $S(\Lambda)$. We have $\det(\Lambda') \le m^n$ by Hadamard's inequality and thus

$$\det(\Lambda) = \frac{\det(\Lambda')}{[\Lambda : \Lambda']^2} \le m^n .$$

The hypotheses of Theorem 2.4.1 are thus satisfied by the set $\mathcal{E}_m$, and we must now just prove that $\mathcal{E}_m$ is closed in $\mathcal{L}$.

So, let Λ be a lattice which is the limit of a sequence (Λ_p) of lattices in $\mathcal{E}_m$. Choose for each p a basis $\mathcal{B}_p$ for Λ_p. By the argument used in the proof of Theorem 2.4.1, we can extract from $\mathcal{B}_p$ a sequence whose limit is a basis $\mathcal{B}$ for Λ. The continuity of the norm shows that the vectors of $\mathcal{B}$ all have norm m. $\square$

2.5 Lattice Constants

Definition 2.5.1. *Let A be a subset of E. We say that a lattice Λ is* admissible *for A, or A*-admissible, *if $\Lambda \cap A$ only contains the origin of E (or is empty). The infimum of the discriminants of admissible lattices for A is the* lattice constant *of A, denoted by $\kappa(A)$. (We set $\kappa(A) = +\infty$ if no A-admissible lattice exists.) We say that Λ is* critical *for A if it is admissible for A and if $\Delta(\Lambda) = \kappa(A)$. More generally, we say that a lattice whose discriminant is a local minimum among those which are A-admissible is* extreme *for A. Critical lattices are sometimes called* absolutely extreme lattices.

We remark that any lattice which is admissible for some subset A of E is also admissible for $A \cup -A$. For this reason, we shall only consider subsets of E which are symmetric with respect to the origin.

Proposition 2.5.2. *1. If $A' \subset A$, then $\kappa(A') \leq \kappa(A)$.*
2. If A is bounded, then $\kappa(A) < \infty$.
3. If the origin belongs to the interior of A, then $\kappa(A)$ is strictly positive.

Proof. The first assertion is obvious. To prove (2), we consider an orthonormal basis $(\varepsilon_1, \ldots, \varepsilon_n)$ for E and for any $M > 0$ the product C_M of the intervals $[-M\varepsilon_i, +M\varepsilon_i]$. For M sufficiently large, the cube C_M contains A, and the lattice with basis $(M\varepsilon_1, \ldots, M\varepsilon_n)$, which is admissible for C_M, is also admissible for A.

We now prove (3). If $0 \in \operatorname{Int}(A)$, A contains a ball with radius $m > 0$. If L is admissible for A, we have $\|\Lambda\| \geq m$ and thus $\Delta(\Lambda) \geq \gamma_n^{-n/2} m^{n/2}$ by Proposition 2.2.7. □

In practice, we restrict ourselves to open sets which contain the origin. We then have the following existence theorem:

Theorem 2.5.3. *Let A be an open set containing the origin. Then, if admissible lattices exist for A (for instance, if A is bounded), critical lattices also exist for A.*

Proof. The definition of $\kappa(A)$ shows that there exists for every integer $p > 0$ an admissible lattice Λ_p with discriminant $\Delta_p \leq \kappa(A) + \frac{1}{p}$. By Mahler's compactness theorem, we can extract from (Λ_p) a subsequence converging to some lattice Λ. By the continuity of the discriminant, we have $\Delta(\Lambda) = \lim_{p \to \infty} \Delta(\Lambda_p) = \kappa(A)$. Any $x \in \Lambda \smallsetminus \{0\}$ is the limit of a sequence (x_p) of elements of $\Lambda_p \smallsetminus \{0\}$ (p running through the indices of the subsequence). Since the complementary set A' of A is closed, x belongs to A'. We have thus $\Lambda \cap A \subset \{0\}$. □

The principal aim of geometry of numbers is to find as sharp as possible lower bounds for $\kappa(A)$ for certain subsets A of E which play a role in various problems of an arithmetical nature. We can then obtain interesting theorems when we succeed in translating some properties we wish to prove into statements involving A-admissible lattices. Another interesting question is that of the description of critical or more generally of extreme lattices, and even more generally of all admissible lattices.

In practice, we essentially consider sets of the following type:

Definition 2.5.4. *We say that* $A \subset E$ *is a* star body (with respect to the origin) *if whenever* A *contains some* $x \in E$*, it then also contains the whole segment* $[0, x]$. *A* distance-function *is a continuous map* $f : E \to \mathbb{R}_{\geq 0}$ *which is "homogeneous" in the following sense: there exists* $d > 0$ *(the* degree of f*) such that the equality* $f(\lambda x) = |\lambda|^d f(x)$ *holds for all* $x \in E$ *and all* $\lambda \in \mathbb{R}$. *The* canonical open star body attached to f *is* $A = \{x \in E \mid f(x) < 1\}$. *For any lattice* Λ*, we set* $f(\Lambda) = \{\inf f(x) \mid x \in \Lambda, x \neq 0\}$ *and* $\gamma_{n,f}(\Lambda) = f(\Lambda)/\Delta(\Lambda)^{d/n}$. *Finally, let* $\gamma_{n,f} = \sup_\Lambda \gamma_{n,f}(\Lambda)$.

Note that the open set A attached to f is bounded if and only if f does not vanish outside the origin. Note also that conversely, given a positive real number d, an open symmetric star body is attached to a unique distance-function of degree d provided it satisfies a suitable regularity condition.

The invariant $\gamma_{n,f}$ depends on Λ solely up to proportionality, and is the Hermite invariant in the case where $f(x) = \|x\|^2$; the constant $\gamma_{n,f}$ is the infimum of the real numbers t such that any lattice Λ contains a nonzero vector x with $f(x) \leq t\Delta(\Lambda)^{d/n}$.

Proposition 2.5.5. *With the notation above, the following conditions on the positive real number* λ *are equivalent:*

1. *Every lattice* Λ *contains a nonzero vector* x *such that* $f(x) < \lambda\,\Delta(\Lambda)^{d/n}$.
2. *Every lattice* Λ *of discriminant* 1 *contains a nonzero vector* x *such that* $f(x) < \lambda$.
3. *We have* $\lambda > \kappa(A)^{-d/n}$.

Proof. From the homogeneity of f, we see that condition (1) is equivalent to the apparently more general following condition:

$$\forall t > 0, \forall \Lambda \in \mathcal{L}, \exists x \in t\Lambda, x \neq 0, \text{ such that } f(x) < \lambda\,\Delta(t\Lambda)^{d/n}.$$

Choosing t such that $\det(t\Lambda) = 1$, we immediately see that (1) and (2) are equivalent. Choosing then t such that $\lambda\,\Delta(t\Lambda)^{d/n} = 1$ shows that the inequality in (1) characterizes the condition $x \in A$. The existence of such a nonzero x is equivalent to the inequality $\Delta(t\Lambda) < \kappa(A)$, itself equivalent to $\lambda > \kappa(A)^{-d/n}$. □

Corollary 2.5.6. *We have* $\gamma_{n,f} = \kappa(A)^{-d/n}$*; in particular, the lattice constant* Γ_n *of the unit ball of* E *is related to the Hermite constant by the equality* $\Gamma_n = \gamma_n^{-n/2}$. □

For any $u \in \mathrm{GL}(E)$, set $f_u = f \circ u$. From the relation $f_u(u^{-1}(\Lambda)) = f(\Lambda)$, we immediately deduce the transformation formula

$$\gamma_{n,f_u} = \gamma_{n,f} / |\det(u)|^{d/n}\,.$$

Corollary 2.5.6 is an illustration of the dictionary which exists between distance-functions and star bodies. Here is the interpretation of Definition 2.5.1 in terms of distance-functions:

Definition 2.5.7. *We say that a lattice Λ is* extreme *(resp.* critical, *or sometimes* absolutely extreme*) for f if the function $\gamma_{n,f}$ attains a local (resp. global) maximum at Λ. When f is the Euclidean norm, we simply say* extreme *(resp.* critical, *or sometimes* absolutely extreme*).*

Let us now apply the preceding remarks to positive definite quadratic forms, taking for f the Euclidean norm $x \mapsto \|x\|^2 = N(x)$. The map $u \mapsto f_u$ is a one-to-one correspondence between $\mathrm{GL}(E)$ and the set of positive definite quadratic forms. Taking $E = \mathbb{R}^n$ and $N(x) = x_1^2 + \cdots + x_n^2$, a quadratic form Q may be written $Q(x) = {}^tXAX$ for a unique symmetric matrix A, whose determinant is the discriminant $D(Q)$ of Q. Changing X into $u(X)$ transforms $D(Q)$ into $D(Q)\det(u)^2$. By the definition of the Hermite constant, we then have:

Proposition 2.5.8. *Let Q be a positive definite quadratic form in n real variables. Then every lattice Λ contains a nonzero vector x such that*

$$Q(x) \le \gamma_n\, D(Q)^{1/n} \det(\Lambda)^{1/n}\,. \ \square$$

When u runs through $\mathrm{GL}_n(\mathbb{R})$, $u(\mathbb{Z}^n)$ runs through the set of all lattices in $\mathbb{R}^n$. The statement above if thus trivially equivalent to the following one, that we obtain by restricting ourselves to the cubic lattice $\mathbb{Z}^n$: for any positive definite quadratic form Q in n real variables, there exists a nonzero integral x such that $Q(x) \le \gamma_n D(q)^{1/n}$, and we recover Hermite's original statement by replacing γ_n by $(\frac{4}{3})^{n(n-1)/4}$ in the previous inequality.

[Actually, Hermite only assumes that Q is non-degenerate, but the replacement of $|x_1^2 + \cdots + x_r^2 - x_{r+1}^2 - \cdots - x_n^2|$ by the expression $x_1^2 + \cdots + x_n^2$ which takes larger values on $\mathbb{Z}^n$ reduces immediately the general case to the positive definite one.]

2.6 Extreme Lattices for an Open Star Body

In this section, f is a distance-function and $A = \{x \in E \mid f(x) < 1\}$ the open star body canonically attached to it.

Theorem 2.6.1. *If A is bounded (i.e. if f is nonzero outside the origin), every A-extreme lattice possesses n linearly independent points on the boundary ∂A of A.*

Proof. Suppose the contrary. Let Λ be an extreme lattice for A and let F be the subspace of E generated by $\Lambda \cap \mathrm{Fr}(A)$. Set $r = \dim F$, and suppose $r < n$. Since the inequalities $f(x) < M$ have only finitely many solutions in $\Lambda \smallsetminus \Lambda \cap F$ and at least one for M sufficiently large, the infimum m of $f(x)$ on $\Lambda \smallsetminus \Lambda \cap F$ is a minimum and is strictly greater than 1.

For $0 < \lambda \leq 1$, let $u_\lambda \in \mathrm{GL}(E)$ defined by $u_\lambda(x) = x$ for $x \in F$ and $u_\lambda(x) = \lambda x$ for $x \in F^\perp$. We have $\det(u_\lambda) = \lambda^{n-r} < 1$ for $\lambda < 1$. The function $(\lambda, x) \mapsto f(u_\lambda(x))f(x)^{-1}$ is continuous on $[\frac{1}{2}, 1] \times (E \smallsetminus \{0\})$ and invariant under the homothetic transformations on x; its values are those which are attained on the compact set $[\frac{1}{2}, 1] \times \partial A$. Consequently, as a function of x, f possesses a minimum m_λ which is itself a continuous function of λ and thus tends to 1 when λ tend to 1. Choose now $\mu < 1$ with $m_\mu m \geq 1$. For $x \in \Lambda \smallsetminus \{0\}$ and $\lambda \in [\mu, 1]$, we have $f(u_\lambda(x)) \geq 1$: for $x \in F$, because $f(u_\lambda(x)) = f(x)$; for $x \notin F$, because $f(u_\lambda(x)) \geq m_\mu m \geq 1$. Hence the lattices $u_\lambda(\Lambda)$ are A-admissible for $\mu \leq \lambda < 1$, a fact which contradicts the extremality of Λ since any neighbourhood of Λ contains such a lattice. □

We now return to arbitrary distance-functions.

Definition 2.6.2. *An* automorphism of f *(or of A) is an element u of* $\mathrm{GL}(E)$ *such that* $f \circ u = f$.

Example 2.6.3. *If f is the Euclidean norm,* $\mathrm{Aut}(f) = \mathrm{O}(E)$.

Proposition 2.6.4. *If A-admissible lattices exist (i.e., if $f(\Lambda) \neq 0$), the automorphisms of f have determinant ± 1.*

Proof. Let $u \in \mathrm{Aut}(f)$. If Λ is an A-admissible lattice, so are the lattices $u^p(A)$ for all integers $p \geq 0$. We thus have

$$|\det(u)|^p = \Delta(u^p(\Lambda))/\Delta(\Lambda) \geq \kappa(A)/\Delta(\Lambda)$$

for all $p \geq 0$, hence $|\det(u)| \geq 1$. Replacing u by u^{-1} yields the opposite inequality. □

Clearly, if Λ is admissible for A, so are all lattices of the form $\lambda u(\Lambda)$ with $u \in \mathrm{Aut}(A)$ and $\lambda \geq 1$. The fact that it may happen that lattices of this form are the only admissible lattices in a sufficiently small neighbourhood of Λ is an astonishing property of some functions, for which we give the following definition:

Definition 2.6.5. *A lattice Λ is* isolated for A *if there exists a neighbourhood of Λ which contains no admissible lattices except those of the form above. (In terms of f: we have $f(\Lambda') < f(\Lambda)$ on a sufficiently small neighbourhood of Λ except if $\Lambda' = \lambda u_\lambda(\Lambda)$ for some $\lambda \geq 1$ and $u \in \mathrm{Aut}(f)$.)*

Isolated lattices are extreme, but the converse is false, see Proposition 2.6.6 below.

Isolation phenomena often raise particularly delicate questions. They showed up in the study of some functions which play a crucial role in algebraic number theory or in Diophantine approximation. Here are a few examples: critical lattices are isolated when f is an indefinite quadratic form in two, three or four variables (by a theorem of Margulis, the canonical star body attached to f does not possess any admissible lattice for more than four variables), or when f is one of the two canonical cubic forms ($x_1x_2x_3$ and $x_1(x_2^2+x_3^2)$) associated to cubic fields. (Note that real quadratic fields correspond to indefinite quadratic forms in two variables; here, infinitely many *successive minima ("the Markov chain")* are known, see [Cas1], Chapter II.) We shall not investigate any more isolated lattices, contenting ourselves with the following negative result about spheres:

Proposition 2.6.6. *Isolated lattices for the Euclidean norm do not exist for* $n \geq 2$.

Proof. Observe that transformations of the form $x \mapsto \lambda u(x)$ ($\lambda \in \mathbb{R}^*$, $u \in \mathrm{Aut}(\|\cdot\|)$) are similarities. Let F be a hyperplane of E and let $u_\lambda \in \mathrm{GL}(E)$ be defined by $u_\lambda(x) = x$ if $x \in F$ and $u_\lambda(x) = \lambda x$ if $x \in F^\perp$. For $\lambda > 1$, the transformations u_λ do not lower the norm and are not similarities. This proves that a lattice which is admissible for some sphere S cannot be isolated for S. □

We now turn to a generalization of the notion of a lattice constant, involving several vectors instead of a single one.

Definition 2.6.7. *Let Λ be a lattice and let f be a distance-function which takes positive values outside the origin. (In other words, the canonical open set A attached to f is bounded.) The* successive minima *of f on Λ are the n real numbers $m_1, m_2, \dots, m_n$ defined in the following way: for $1 \leq r \leq n$, m_r is the infimum of the real numbers λ such that there exist r independent vectors $x \in \Lambda$ with $f(x) \leq \lambda$; we simply say* successive minima *when $f(x) = x \cdot x$ is the norm. For $1 \leq r \leq n$, let $f_r(\Lambda) = m_1 \cdot m_2 \cdots m_r \cdot \det(\Lambda)^{-r/n}$ and let $f_r = \sup_{\Lambda \subset E} f_r(\Lambda)$.*

Note that m_1 is the minimum of f on Λ, denoted by $f(\Lambda)$ in Section 2.5.

The greatest lower bounds used to define the m_i are actually attained (because Λ is a discrete group); this justifies the name "minima" we used for them. Moreover, if g is a distance-function which is greater than f, then the upper bound $f_r(\Lambda) \leq g_r(\Lambda)$ holds for every lattice Λ. An easy compactness argument shows that it is always possible to chose for g a function which is proportional to the Euclidean norm. Thanks to this remark, the results we shall prove below will show the existence of the least upper bound f_r for any function f.

We cannot hope to prove better upper bounds than $f_r \leq \gamma_{n,f}^r$, since critical lattices for A possess n independent points on ∂A (Theorem 2.6.1). An important theorem is that equality holds in the case where f is the Euclidean norm:

Theorem 2.6.8. (Minkowski.) *The successive minima for the Euclidean norm of an n-dimensional lattice Λ satisfy for all $r \leq n$ the inequality*

$$m_1 \cdot m_2 \cdots m_r \leq \gamma_n^r \det(\Lambda)^{r/n}.$$

We state as a corollary the following equivalent formulation of the theorem above:

Corollary 2.6.9. *Every n-dimensional lattice contains n independent vectors $e_1, \ldots, e_n$ which satisfy for all $r \leq n$ the inequality*

$$N(e_1) \cdot N(e_2) \cdots N(e_r) \leq \gamma_n^r \det(\Lambda)^{r/n}. \square$$

Proof of 2.6.8. The proof will be by descending induction on the smallest index $k \leq n$ such that $m_k < m_{k+1}$, the results being obvious for $k = n$, since all minima are then equal and thus bounded from above by the Hermite constant γ_n. The proof resembles that of Theorem 2.6.1.

For all $i \leq n$, let $e_i \in \Lambda$ with norm m_i, and let F be the subspace of E generated by $e_1, \ldots, e_k$. For any $\lambda > 0$, let $u_\lambda \in \mathrm{GL}(E)$ defined by $u_\lambda(x) = x$ for $x \in F$ and $u_\lambda(x) = \lambda x$ for $x \in F^\perp$. When $\lambda < 1$ is sufficiently close to 1, $u_\lambda(\Lambda)$ has minimum m_1 and the minima m_i' of $u_\lambda(\Lambda)$ satisfy the conditions $m_1 = m_1' = \cdots = m_k' < m_{k+1}' \leq \cdots \leq m_n'$. Replacing Λ by $u_\lambda(\Lambda)$ multiplies the determinant by $\lambda^{2(n-k)}$ without changing the first k minima.

Choose for every index $i > k$ a vector x_i' with norm m_i' and set $x_i = u_\lambda^{-1}(x_i')$. If $x = y + z$, $y \in F, z \in F^\perp$, we have $\frac{N(u_\lambda(x))}{N(x)} \geq \lambda^2$, and this inequality is strict for $x \notin F^\perp$. For a suitable permutation of the indices $i > k$, we have $N(x_j) > m_j$ if $j > k$. We thus have the inequalities

$$\prod_{i>k} m_i' > \lambda^{2(n-k)} \prod_{i>k} N(x_i) \geq \lambda^{2(n-k)} \prod_{i>k} m_i;$$

with the previous notation, we have $f_n(u_\lambda(\Lambda)) > f_n(\Lambda)$.

Let μ be the greatest lower bound of those λ for which the inequality $N(u_\lambda(\Lambda)) \geq m_1$ holds. When λ tends to μ, u_λ tends to u_μ and $u_\lambda(\Lambda)$ tends to $u_\mu(\Lambda)$. By the definition of μ, we have $N(u_\lambda(\Lambda)) < N(u_\mu(\Lambda))$ for all $\lambda < \mu$. This shows that the first $k+1$ minima of $u_\mu(\Lambda)$ are equal, otherwise one could apply the construction above to the lattice $u_\mu(\Lambda)$. $\square$

From Corollary 2.6.9, we easily obtain an upper bound for the index of a sublattice which reduces when all minima coincide to the one we sketched at the end of Section 2.1:

Corollary 2.6.10. *Let Λ be a lattice and let Λ' be a sublattice of Λ generated by vectors which represent the minima of Λ. We then have $[\Lambda : \Lambda'] \leq \gamma_n^{n/2}$.*

Proof. Let $(e_1, \ldots, e_n)$ be a basis for Λ' with $N(e_i) = m_i$. Then $[\Lambda : \Lambda'] = \Delta(\Lambda') \cdot \Delta(\Lambda)^{-1}$. The Hadamard inequality yields for the first discriminant the upper bound

$$\Delta(\Lambda') \leq \|e_1\| \cdot \|e_2\| \cdots \|e_n\|$$

whereas Theorem 2.6.8 yields for the second discriminant the lower bound

$$\Delta(\Lambda) \geq \|e_1\| \cdot \|e_2\| \cdots \|e_n\| \cdot \gamma(\Lambda)^{-n/2} \,. \ \square$$

For $n \leq 3$, we have $\gamma_n^{n/2} < 2$ by Hermite's inequality, a bound which shows that lattices of dimension $n \leq 3$ are generated by representatives for their minima. We shall see in Chapter 6 that the result is true when $n = 4$ if one excludes lattices similar to the *centred cubic lattice* C_4 (C_n is generated by the vectors $\varepsilon_1, \ldots, \varepsilon_n$ of an orthonormal basis for $\mathbb{R}^n$ together with $\varepsilon = \frac{\varepsilon_1 + \cdots + \varepsilon_n}{2}$) for which both indices 1 and 2 are possible, depending on the chosen representatives for the successive minima.

For all $n \geq 5$, the lattice C_n provides an example of a lattice which cannot be generated by representatives for its successive minima: such vectors much be chosen among $\{\pm\varepsilon_1, \pm\varepsilon_2, \ldots, \pm\varepsilon_n\}$, which generate the lattice $\mathbb{Z}^n$, of index 2 in C_n.

From dimension 5 onwards, the problem of estimating the product of the norms of n vectors of a *basis* for a lattice is not equivalent to the same question for vectors only assumed to be independent. We shall contend ourselves with the following theorem that we sate without a proof (see [vdW], Section 6, Theorem 7), which improves on Hermite's inequality 2.2.1 and is optimal for $n \leq 5$:

Theorem 2.6.11. (van der Waerden.) *Let $\delta_0, \delta_1, \ldots, \delta_p, \ldots$ be the sequence of rational numbers defined inductively by the formulae*

$$\delta_0 = 1 \qquad \textit{and} \qquad \delta_p = \max\left(1, \frac{1}{4}(\delta_0 + \delta_1 + \cdots + \delta_{p-1})\right) .$$

Then a lattice of dimension n with successive minima $m_1, m_2, \ldots, m_n$ has a basis $(e_1, e_2, \ldots, e_n)$ which satisfies the inequalities $N(e_i) \leq \delta_i m_i$ for $1 \leq i \leq n$. $\square$

2.7 The Lattice Constant for a Convex Star Body

Among symmetric star bodies, convex sets play an important special rôle, due to the possibility discovered by Minkowski of obtaining good lower bounds of their lattice constants by applying simple remarks from measure theory.

We use the notation $\mathrm{vol}(A)$ to denote the volume (Lebesgues measure) of a subset A of E. (All the sets we consider, in particular convex sets, are measurable, even in the more elementary sense of Riemann and Jordan.) Here is Minkowski's well-known result:

Theorem 2.7.1. (Minkowski's theorem on convex bodies.) *Let $C \subset E$ be a convex, symmetrical set (with respect to the origin). Then its lattice constant satisfies the inequality*

$$\kappa(C) \geq 2^{-n}\,\mathrm{vol}(C)\,.$$

For the proof, we need the following lemma which gives an upper bound for the volume of a convex set packed by a lattice:

Lemma 2.7.2. *Let Λ be a lattice, and let A be a measurable subset of E, with measure $\mathrm{vol}(A) > \Delta(\Lambda)$. Then A contains two distinct vectors whose difference belongs to Λ.*

Proof of 2.7.2. It amounts to saying that there exists a nonzero element $x \in \Lambda$ such that $A \cap (A + x)$ is not empty. Suppose, conversely, that the sets $(A+x), x \in \Lambda$ are disjoint. Let P be a fundamental parallelotope of Λ. Since the sets $P-x, x \in \Lambda$ constitute a partition of E, we have $A = \cup_{x\in\Lambda} A \cap (P-x)$, whence

$$\begin{aligned}\mathrm{vol}(A) &= \mathrm{vol}\left(\cup_{x\in\Lambda} A \cap (P-x)\right) = \sum_{x\in\Lambda} \mathrm{vol}\left(A \cap (P-x)\right) \\ &= \mathrm{vol}\left(\cup_{x\in\Lambda} (A+x) \cap P\right) \leq \mathrm{vol}(P) = \Delta(\Lambda)\,. \ \square\end{aligned}$$

Proof of 2.7.1. We want to prove that a lattice of discriminant $\Delta < 2^{-n}\,\mathrm{vol}(C)$ possesses a point other than 0 in C. Applying Lemma 2.7.2 to $A = \frac{1}{2}C$, we see that there exists $x \in \Lambda, x \neq 0$ such that $A \cap (A + x) \neq \emptyset$. Let $y \in A \cap (A + x)$. Then $y' = y - x$ is an element of A, and so is $y'' = -y'$ by the symmetry of A. Since A is a convex set, $z = \frac{y+y''}{2}$ is also an element of A. Hence $x = 2z$ belongs to $2A = C$. □

As an example, we now apply Theorem 2.7.1 to the unit ball of E. To this end, we first calculate its volume.

Proposition 2.7.3. *The volume ω_n of the unit ball of $\mathbb{R}^n$ is given for all $n \geq 0$ by the following two formulae, depending on the parity of n:*

$$\omega_{2m} = \frac{\pi^m}{m!} \quad \textit{and} \quad \omega_{2m+1} = \frac{2^{m+1}\pi^m}{1\cdot 3\cdot 5\ldots(2m+1)}\,.$$

Proof. The formulae are obvious for $n = 0$ and $n = 1$; for $n \geq 2$, we prove them by induction, using a shift from n to $n-2$. We must calculate the n-tuple integral

$$\int \cdots \int_{x_1^2+\cdots+x_n^2 \leq 1} dx_1 \ldots dx_n \,;$$

we express it as a double integral in the disc $x_1^2 + x_2^2 \leq 1$ of the value of the $(n-2)$-tuple integral

$$\int \cdots \int_{x_3^2+\cdots+x_n^2 \leq 1-x_1^2-x_2^2} dx_3 \ldots dx_n \,.$$

We recognize here the volume of the ball of radius $\sqrt{1-x_1^2-x_2^2}$ in dimension $n-2$, for which the induction hypothesis gives the value

$$\omega_n = \iint \omega_{n-2}(1-x_1^2-x_2^2)^{(n-2)/2} \, dx_1 dx_2 \,.$$

Using polar coordinates, we transform the integral above into

$$\omega_n = \omega_{n-2} \iint (1-\rho^2)^{n/2} \rho \, d\rho \, d\theta = 2\pi\omega_{n-2} \int_0^1 (1-\rho^2)^{(n-2)/2} \rho \, d\rho \,,$$

which we evaluate by means of the change of coordinates $u = 1 - \rho$. This yields the equality

$$\omega_n = 2\pi\omega_{n-2} \int_0^1 u^{(n-2)/2} \frac{du}{2} = \frac{2\pi}{n} \omega_{n-2}$$

which completes the proof. □

Theorem 2.7.4. (Minkowski.) *The Hermite constant for dimension n satisfies the inequality $\gamma_n \leq 4\omega_n^{-2/n}$.*

Proof. Just use Corollary 2.5.6 and and Theorem 2.7.1. □

Remark 2.7.5. *The upper bounds of γ_n which can be derived from the inequalities of Hermite or Mordell are of exponential type ("there exists a constant $K > 1$ such that $\gamma_n \leq K^n$"), whereas the inequality of Minkowski is linear in n: using the asymptotic estimation $4\omega^{-2/n} \sim \frac{2n}{\pi e}$, we see that it suffices to consider the case of even dimensions, for which we can apply the* Stirling formula *$m! \sim m^m e^{-m}\sqrt{2\pi m}$. A more detailed study (see [M-H]) yields the explicit upper bound $\gamma_n < 1 + \frac{n}{4}$ which is valid for for all n.*

The end of this section is devoted to applications of the Hermite constant to number fields, and will not be used in the sequel. We shall essentially state results with only sketchy proofs, except when they have something to do with spheres.

Definition 2.7.6. *Let $n \geq 1$ and $r_1, r_2 \geq 0$ be integers such that $n = r_1 + 2r_2$. Let $E = \mathbb{R}^{r_1} \times \mathbb{C}^{r_2} \simeq \mathbb{R}^n$ where $\mathbb{C}$ is endowed with the Euclidean structure which makes $(1, i)$ an orthonormal basis for $\mathbb{C}$. We denote by A_{r_1,r_2} the open star body of E defined by the inequality*

$$|x_1 \cdot x_2 \cdots x_{r_1} z_1^2 \cdot z_2^2 \cdots z_{r_2}^2| \leq 2^{r_2},$$

and by κ_{r_1,r_2} its lattice constant.
[We have used on $\mathbb{C}$ half the natural scalar product $\frac{1}{2}\operatorname{Tr}_{\mathbb{C}/\mathbb{R}}(z\overline{z})$; this accounts for the occurrence of 2^{r_2} in the right–hand side of the formula displayed above.]

Recall that the *signature of a number field K* is the pair (r_1, r_2) such that among the $n = [K : \mathbb{Q}]$ embeddings of K in $\mathbb{C}$, r_1 are real and $2r_2$ are non-real. The following theorem is the key for the classical inequality of Minkowski on ideal classes of number fields. To a *lattice M in K* (i.e., a free rank-n sub-$\mathbb{Z}$-module of K), we attach a lattice Λ in the usual sense by embedding M in E with the $r_1 + r_2$ embeddings up to conjugacy of K into $\mathbb{C}$. Recall also that the *discriminant $d_K(M)$ of M* is the determinant $\det \operatorname{Tr}(e_i e_j)$ of the trace form on any $\mathbb{Z}$-basis $(e_1, \ldots, e_n)$ for M. When M is the ring $\mathbb{Z}_K$ of (algebraic) integers of K, $d_K(M)$ is the *discriminant d_K of K*.

We now easily prove the following theorem, in which $N_{K/\mathbb{Q}}(\cdot)$ is the usual norm of algebraic number theory (notice the difference between "Λ" and "N"):

Theorem 2.7.7. *For any lattice M in a number field with signature (r_1, r_2), there exists a nonzero $x \in M$ such that*

$$|N_{K/\mathbb{Q}}(x)| \leq \kappa_{r_1,r_2}^{-1} \sqrt{|d_K(M)|}. \;\square$$

[The inequality would be more conveniently written in the form

$$|N_{K/\mathbb{Q}}(x)| \leq \sqrt{\frac{|d_K(M)|}{\kappa'_{r_1,r_2}}},$$

where $\kappa' = \kappa^2$, for instance $N(x) \leq \sqrt{d_K/3}$ for K imaginary quadratic.]

The discriminants of a fractional ideal $\mathfrak{a}$ of K and of K itself are linked by the formula $d_K(\mathfrak{a}) = d_k N_{K/\mathbb{Q}}(\mathfrak{a})^2$. Applying the preceding theorem to an ideal in the inverse class of the ideal class we wish to consider, one proves the following fundamental result:

Theorem 2.7.8. *Every ideal class of K (i.e. of $\mathbb{Z}_K$) contains a nonzero integral ideal $\mathfrak{a}$ with norm $N(\mathfrak{a}) \leq \kappa_{r_1,r_2}^{-1} \sqrt{|d_K(M)|}$.* $\square$

In view of the theorem above, to prove sharp lower bounds for κ_{r_1,r_2} is an important problem in algebraic number theory. Minkowski's classical theorem consists in applying Theorem 2.7.1 to a suitably chosen convex subset of A_{r_1,r_2}. Minkowski's choice is $n\,2^{r_2/n}B_{r_1,r_2}$ where

$$B_{r_1,r_2} = \left\{x \in \mathbb{E} \,\middle|\, |x_1| + \cdots + |x_{r_1}| + 2|z_1| + \cdots + 2|z_{r_2}| < 1\right\},$$

proving the inclusion $n2^{r_2/n}B_{r_1,r_2} \subset A_{r_1,r_2}$ by means of the arithmetic-geometric mean inequality. The evaluation of the volume of B_{r_1,r_2} is then done by induction on r_2: for $r_2 = 0$, the result is well-known (and is easily proved by induction of r_1), since $B_{r_1,0}$ is a simplex; the reduction from r_2 to r_2-1 is then of the same kind that the one we used to prove Proposition 2.7.3. The result is $\mathrm{vol}(B_{r_1,r_2}) = \left(\frac{\pi}{4}\right)^{r_2}\frac{2^{n-r_2}}{n!}$, from which we immediately deduce:

Theorem 2.7.9. (Minkowski.) $\kappa_{r_1,r_2} \geq \left(\frac{\pi}{4}\right)^{r_2}\frac{n^n}{n!}$. □

The constants κ_{r_1,r_2} are known only (but with some interesting refinements) up to degree 3. Curiously, for small values of n and of r_2, embedding a sphere in A_{r_1,r_2} sometimes sometimes yields the best known lower bounds for κ_{r_1,r_2}. The result depends solely on the degree of the field.

Theorem 2.7.10. (Minkowski.) $\kappa_{r_1,r_2} \geq \left(\frac{n}{\gamma_n}\right)^{n/2}$.

Proof. For $1 \leq j \leq r_2$, set $z'_j = \frac{z_j}{\sqrt{2}}$. With this notation, A_{r_1,r_2} is defined by the inequality $\prod_{i=1}^{r_1} x_i^2 \prod_{j=1}^{r_2} |z'_j|^4 < 1$. Applying now the arithmetic-geometric mean inequality to the $n = r_1 + 2r_2$ positive real numbers $x_i^2, |z'_j|^2$ ($i \leq r_1, j \leq r_2$, with repetition of the z'_j), we see that A_{r_1,r_2} contains the open ball of radius $\sqrt{n}$, whence the lower bound

$$\kappa_{r_1,r_2} \geq n^{n/2}\Gamma_n = \left(\frac{n}{\gamma_n}\right)^{n/2}. \ \square$$

2.8 Generalizations of the Hermite Invariant

We still denote by E a Euclidean space of dimension n.

Definition 2.8.1. *The* dual Hermite invariant of a lattice Λ *is*

$$\gamma'(\Lambda) = \|\Lambda\| \cdot \|\Lambda^*\| = \left(N(\Lambda)\,N(\Lambda')\right)^{1/2},$$

and the dual Hermite constant for dimension n *is* $\gamma'_n = \sup_\Lambda \gamma'(\Lambda)$. *We say that a lattice* Λ *is* dual-extreme *(resp.* dual-critical*) if the invariant* γ' *attains on it a local (resp. absolute) maximum.*

A lattice and its dual lattice play symmetric rôles in this definition, and the invariant γ' can thus be defined for pairs (Λ, Λ^*). Also, since $\det(\Lambda^*) = \det(\Lambda)^{-1}$, the definition of the Hermite invariant shows that

$$\gamma'(\Lambda)^2 = N(\Lambda)\,N(\Lambda^*) = \gamma(\Lambda)\,\gamma(\Lambda^*) \leq \gamma_n^2\,,$$

from which we deduce for $\gamma'(\Lambda)$ the upper bound $\gamma'(\Lambda) \leq \gamma_n$ which justifies the existence of the supremum used in the definition of γ'_n.

Proposition 2.8.2. *1. The dual invariant $\gamma'(\Lambda)$ of a lattice Λ depends solely on its similarity class.*

2. Dual-critical lattices exist in every dimension.

3. We have $\gamma'_n \leq \gamma_n$, and equality holds if and only if there exists a critical lattice whose dual is also critical (and all dual-critical lattices are then critical).

Proof. (1) The equality $u(\Lambda)^* = {}^t u^{-1}(\Lambda^*)$ holds for all $u \in \mathrm{GL}(E)$. When u is a homothetic transformation of ratio λ (resp. an isometry), we have $u(\Lambda)^* = \lambda^{-1}\Lambda^*$ (resp. $u(\Lambda)^* = u(\Lambda^*)$), whence the invariance of the product $N(\Lambda)N(\Lambda^*)$.

To prove (2), we consider a sequence Λ_p of lattices such that $\gamma'_n = \lim_{p\to\infty} \gamma'(\Lambda_p)$. Replacing if necessary Λ_p by a proportional lattice, we may assume that Λ_p has norm 1 for every p. We have

$$\begin{aligned}\det(\Lambda_p^*) &= N(\Lambda_p^*)^n \gamma(\Lambda_p^*)^{-1} = \gamma'(\Lambda_p^*)^n \gamma(\Lambda_p^*)^{-1}\\ &\geq \gamma'(\Lambda_p^*)^n \gamma_n^{-1} \to_{p\to\infty} (\gamma'_n)^n \gamma_n^{-1} > 0\,.\end{aligned}$$

Hence the determinants of the Λ_p are bounded from above, and Mahler's compactness theorem shows that Λ_p possesses a subsequence which converges to some lattice Λ. Since the map $u \mapsto {}^t u^{-1}$ is continuous on $\mathrm{GL}(E)$, the map $\Lambda \mapsto \Lambda^*$ is continuous on $\mathcal{L}$. Consequently, $\gamma'(\Lambda_p^*)$ tends to $\gamma'(\Lambda)$ on the subsequence, whence the equality $\gamma'(\Lambda) = \gamma'_n$.

The inequality $\gamma'_n \leq \gamma_n$ in (3) follows from the upper bound $\gamma'(\Lambda) \leq \gamma_n$ mentioned after Definition 2.8.1, and the equality $\gamma'(\Lambda)^2 = \gamma(\Lambda)\cdot\gamma(\Lambda^*)$ shows that $\gamma'_n = \gamma_n$ holds whenever Λ and Λ^* are both critical. Conversely let Λ be a dual-critical lattice. If $\gamma'_n = \gamma_n$, we have

$$\gamma(\Lambda)\cdot\gamma(\Lambda^*) = \gamma'(\Lambda)^2 = \gamma'^2_n = \gamma_n^2\,,$$

and this shows that the inequalities $\gamma(\Lambda) \leq \gamma_n$ and $\gamma(\Lambda^*) \leq \gamma_n$ are indeed both equalities, i.e. that both Λ and Λ^* are critical. □

We now turn to generalizations of the constants γ_n and γ'_n involving more than a single minimal vector.

Definition 2.8.3. *Let Λ be a lattice. For $0 \leq k \leq n$, let $d_k(\Lambda)$ be the minimum of the determinants of k independent vectors in Λ.*

1. *(Rankin.) Let* $\gamma_{n,k}(\Lambda) = d_k(\Lambda)\det(\Lambda)^{-k/n}$*, and let* $\gamma_{n,k} = \sup_\Lambda \gamma_{n,k}(\Lambda)$ *(the* Rankin invariant *and* constant (of order k of L)*).*
2. *Let* $\gamma'_{n,k}(\Lambda) = \sqrt{d_k(\Lambda)\, d_k(\Lambda^*)}$*, and let* $\gamma'_{n,k} = \sup_\Lambda \gamma'_{n,k}(\Lambda)$*.*

We shall exclude the trivial cases $k = 0$ and $k = n$, for which we have $\gamma_{n,k}(\Lambda) = \gamma'_{n,k}(\Lambda) = 1$. For $k = 1$, we recover the previous invariants $\gamma(\Lambda)$ and $\gamma'(\Lambda)$.

The existence of $d_k(\Lambda)$ is proved in the following way: let $d > 0$; using γ_k, we see that there exists $M > 0$ such that the solutions in Λ to the inequality $\det(x_1, \dots, x_k) \le d$ (if any) can be found among vectors of norm at most M; thus this inequality can be solved on a finite set, whence the existence of the minimum $d_k(\Lambda)$.

Using the Hadamard inequality, we see that $d_k(\Lambda)$ is bounded from above by the product of the first k minima of Λ. This proves the existence of the constants $\gamma_{n,k}$, from which the existence of the constants $\gamma'_{n,k}$ follows immediately.

The proposition below is just an explicit restatement of some inequalities we used to prove the existence statements above.

Proposition 2.8.4. *Let Λ be a lattice. For $0 < k < n$, we have:*

1. $\gamma_{n,k}(\Lambda) \le \gamma(\Lambda)^k$ *and* $\gamma'_{n,k}(\Lambda)^2 = \gamma_{n,k}(\Lambda)\,\gamma_{n,k}(\Lambda^*)$*;*
2. $\gamma'_{n,k} \le \gamma_{n,k} \le (\gamma_n)^k$*.* □

One also easily proves that the constants $\gamma_{n,k}$ and $\gamma'_{n,k}$ are attained (the proofs are similar to the one we gave previously for $k = 1$).

The following proposition makes more precise the links which exist between the invariants of a lattice and of its dual:

Proposition 2.8.5. *Let $k \in [1, n-1]$ be an integer.*

1. *For every lattice Λ, we have* $\gamma_{n,k}(\Lambda) = \gamma_{n,n-k}(\Lambda^*)$ *and* $\gamma'_{n,k}(\Lambda) = \gamma'_{n,n-k}(\Lambda) = \gamma'_{n,k}(\Lambda^*) = \gamma'_{n,n-k}(\Lambda^*)$*.*
2. *We have the symmetry relations* $\gamma_{n,k} = \gamma_{n,n-k}$ *and* $\gamma'_{n,k} = \gamma'_{n,n-k}$*.*

Proof. The first assertion follows immediately from the equality (Corollary 1.3.5)

$$\det(e_1, \dots, e_k) = \det(e_1, \dots, e_n)\, \det(e^*_{k+1}, \dots, e^*_n)$$

applied to a basis $(e_1, \dots, e_n)$ of E and to its dual basis, which implies that every lattice Λ satisfies the equality

$$d_{n-k}(\Lambda^*) = d_k(\Lambda^*)\, \det(\Lambda)^{-1}.$$

The second assertion is then an immediate consequence of the first one. □

The following proposition, when applied inductively, improves on the upper bounds of the constants $\gamma_{n,k}$ given by Proposition 2.8.4 (2):

Theorem 2.8.6. (Rankin.) *For all integers k, h with $0 \leq k \leq h \leq n$, we have $\gamma_{n,k} \leq \gamma_{h,k} \left(\gamma_{n,h}\right)^{k/h}$.*

Proof. Let Λ be a lattice in E, let $(e_1, \ldots, e_h)$ be a family of vectors in Λ on which $d_h(\Lambda)$ is attained, and let M be the lattice they generate in the subspace F of E they span. We have $d_k(\Lambda) \leq d_k(M) \leq \det(e_1, \ldots, e_k)$ and $d_h(\Lambda) \leq \det(e_1, \ldots, e_k, e_{k+1}, \ldots, e_h)$, hence

$$\gamma_{n,k}(\Lambda) \leq \frac{\det(e_1, \ldots, e_k)}{\det(e_1, \ldots, e_n)^{k/n}} = \frac{\det(e_1, \ldots, e_k)}{\det(e_1, \ldots, e_h)^{k/n}} \cdot \left(\frac{\det(e_1, \ldots, e_h)}{\det(e_1, \ldots, e_n)^{h/n}}\right)^{k/h}$$
$$\leq \gamma_{h,k}(M) \cdot \gamma_{n,h}(\Lambda)^{k/h},$$

from which we deduce the inequality we want by taking the supremum of each side. □

We now make use of the constants $\gamma'_{n,k}$ to obtain new upper bounds for the Rankin constants:

Theorem 2.8.7. *Let k be an integer with $0 \leq k \leq n/2$. The following two inequalities then hold:*

1. $(\gamma_{n,k})^n \leq (\gamma_{n-k,k})^{n-k} \, (\gamma'_{n,k})^{2k}$.
2. $\gamma'_{n,2k} \leq (\gamma'_{n-k,k})^2$.

When n is even, applying assertion (1) with $k = \frac{n}{2}$ yields the inequality $\gamma_{n,k} \leq \gamma'_{n,k}$. Since the opposite inequality has been proved in Proposition 2.8.4, we have:

Corollary 2.8.8. *If n is even, then $\gamma'_{n,n/2} = \gamma_{n,n/2}$.* □

Another consequence of assertion (1) in Theorem 2.8.7, which also relies on the inequality $\gamma'_{n,k} \leq \gamma_{n,k}$, is the following generalization of Mordell's inequality, which we recover by taking $k = 1$:

Corollary 2.8.9. *For all k with $0 \leq k \leq n$, we have*

$$(\gamma_{n,k})^{n-2k} \leq (\gamma_{n-k,k})^{n-k}. \ \square$$

As for the dual constants, assertion (2) in Theorem 2.8.7 yields by taking $n = 2k+1$ the following upper bound:

Corollary 2.8.10. *For all positive k, we have*

$$\gamma'_{2k+1} \leq \gamma'^2_{k+1} \cdot \ \square$$

Proof of 8.7(1). Let Λ be a lattice in E and let $M \subset \Lambda$ be a relative lattice of dimension $n-k$ and determinant $d_{n-k}(\Lambda)$. Using the definition of the invariant $\gamma'_{n,n-k}$, we can express the determinant of M in the forms

$$\det(M) = \gamma'_{n,n-k}(\Lambda)^2\, d_{n-k}(\Lambda^*) = \gamma'_{n,n-k}(\Lambda)^2 \det(\Lambda)\, d_k(\Lambda)^{-1}.$$

From the definition of the invariant $\gamma_{n-k,k}$, we deduce the equality

$$d_k(M)^{n-k} = \gamma_{n-k,k}(M)^{n-k} \det(M)^k.$$

The obvious upper bound $d_k(\Lambda) \leq d_k(M)$ together with the expression of $\det(M)$ given by the first equality yields the new upper bound

$$d_k(\Lambda)^{n-k} \leq \gamma_{n-k,k}(M)^{n-k}\, \gamma'_{n,n-k}(\Lambda)^2 \det(\Lambda)^k\, d_k(\Lambda)^{-k}$$

from which we deduce what we need by taking the supremum of each side and making use of the equality $\gamma'_{n,n-k} = \gamma'_{n,k}$. □

To prove the second assertion in Theorem 2.8.7, we need a lemma:

Lemma 2.8.11. *Let Λ be a lattice in E and let k and h be integers such that $0 \leq k \leq h \leq n$. Then $d_h(\Lambda) \leq (\gamma'_{n-k,h-k})^2\, d_k(\Lambda)\, d_{h-k}(\Lambda^*)^{-1}$.*

Proof of 2.8.11. Let $M \subset \Lambda$ be a relative lattice of dimension k and determinant $d_k(\Lambda)$, let F be the subspace of E spanned by M, let p be the orthogonal projection from E onto $F^\perp$, and let $N_0 \subset p(\Lambda)$ be an $(h-k)$-dimensional lattice with determinant $d_{h-k}(p(\Lambda))$. Let, moreover, $N \subset \Lambda$ be the relative $(h-k)$-dimensional lattice in Λ obtained by lifting a basis for N_0. We have $d_h(\Lambda) \leq \det(M \oplus N) = \det(M)\det(p(N)) = d_k(\Lambda)\det(N_0)$, and also, since $p(\Lambda) = \Lambda^* \cap F^\perp$ (Proposition 1.3.4)

$$\begin{aligned}\det(N_0) &= d_{h-k}\left((\Lambda^* \cap F^\perp)^*\right)\\ &= \gamma'_{n-k,h-k}(\Lambda^* \cap F^\perp)\, d_{h-k}(\Lambda^* \cap F^\perp)^{-1}\\ &\leq \gamma'_{n-k,h-k}\, d_{h-k}(\Lambda^*)^{-1}. \ \square\end{aligned}$$

Proof of 8.7(2). Multiplying together the two equations obtained by applying Lemma 2.8.11 to Λ and to Λ^*, we get the new inequality

$$\gamma'_{n,h}(\Lambda) \leq (\gamma'_{n-k,h-k})^2\, \gamma'_{n,k}(\Lambda)\, \gamma'_{n,h-k}(\Lambda)^{-1}$$

whose last two terms collapse when h is given the value $2k$, yielding then precisely the inequality we wanted to prove. □

2.9 The HKZ Reduction

Traditionally, by "reduction of quadratic forms" (real, non-degenerate, and here even positive definite) is meant a procedure which transforms a given quadratic form into an equivalent one whose coefficients satisfy certain inequalities, depending on the reduction, which ensure that there are only finitely many reduced integral quadratic forms with a bounded discriminant (whence Hermite's theorem, which asserts the finiteness of the class number of quadratic forms with a given discriminant).

Reduction for lattices consists in finding bases whose elements have scalar products satisfying suitable inequalities, as one can see through the dictionary relating lattices and quadratic forms of Table 1.7.1.

The *Hermite reduction*, sometimes called *Korkine–Zolotareff reduction*, and that we prefer to call *HKZ reduction*, relies on the method we used in Section 2.2 to prove Theorem 2.2.1. We first explain it in terms of lattices.

We thus consider a lattice Λ in a Euclidean space E of dimension $n \geq 2$ (there is nothing to do for $n \leq 1$.) Let us *choose* a minimal vector e_1 of Λ, and let Λ' be the orthogonal projection of Λ onto $F = \mathbb{R}e_1^\perp$; Λ' is a lattice in F. We may assume by induction that we know a reduced basis $(e'_2, \dots, e'_n)$ for Λ'. For all $i \geq 2$, e'_i is the projection of some $e_i \in \Lambda$ which is well defined only modulo $\mathbb{Z}e_1$. It is thus possible to choose e_i in such a way that its projection on $\mathbb{R}e_1$ be of length at most half the length of e_1, i.e. that the inequalities $|e_1 \cdot e_i| \leq \frac{1}{2}N(e_1)$ be satisfied. We obtain in this way a basis $(e_1, e_2, \dots, e_n)$ for Λ, an explicit construction of which being obtained (once an e_1 has been found) by choosing e'_2 minimal in Λ', then projecting Λ' onto the subspace of F orthogonal to e'_2, etc.

[However (remark of Gindraux), this procedure does not allow further restrictions on the sign of some scalar products $e_i \cdot e_j$, as was erroneously written in the French edition. Such restrictions are possible for the *Minkowski reduction*, see [Borel].]

To translate the HKZ reduction into the more traditional language of quadratic forms, one must interpret the matrix A of a given positive definite form $Q(X) = {}^tXAX$ as a Gram matrix of some lattice Λ.

Replacing if need be Q by an equivalent form Q_1, we may assume that $(1, 0, \dots, 0)$ is a minimal vector for Q_1. (Note that such a transformation relies on the *choice* of some minimal vector for Q.) The orthogonal projection previously used corresponds now to a partial decomposition as a sum of squares, namely

$$Q_1(X_1, X_2, \dots, X_n) = A_1(X_1 + \alpha_{1,2}X_2 + \cdots + \alpha_{1,n}X_n)^2 + Q'(X_2, \dots, X_n)$$

in which the form Q' corresponds to the lattice previously called Λ'.

Applying the same procedure to Q' yields after $n-1$ steps a decomposition as a sum of squares of a form equivalent to Q:

$$\begin{aligned} Q \sim\, & A_1(X_1 + \alpha_{1,2}X_2 + \cdots + \alpha_{1,n}X_n)^2 \\ & + A_2(X_2 + \alpha_{2,3}X_3 + \cdots + \alpha_{2,n}X_n)^2 + \cdots \\ & + A_{n-1}(X_{n-1} + \alpha_{n-1,n}X_n)^2 + A_nX_n^2\,, \end{aligned}$$

in which the n *outer coefficients* $A_1, A_2, \dots, A_n$ are the minimal norms of the successive projections of Λ. As for the $\frac{n(n-1)}{2}$ *inner coefficients* $\alpha_{i,j}$,

2.7.3. Show that the inclusion $A_{r_1-2,r_2+1} \subset A_{r_1,r_2}$ holds for all $r_1 \geq 2$. Use this to deduce the inequality $\kappa_{r_1,r_2} \leq \kappa_{r_1-2,r_2+1}$. [For $n \geq 3$, the only known admissible lattices for A_{r_1,r_2} are of the form $\lambda\, u(\Lambda), \lambda \geq 1, u \in \mathrm{Aut}(A_{r_1,r_2})$ where Λ is attached to some number field of degree n with a larger r_1.]

2.7.4. 1. Show that the distance-function f attached to the open set A_{r_1,r_2} is of *compact type* in the following sense: there exists a compact set C in E such that, for all $x \in E$ with $f(x) = 1$, there exists $u \in \mathrm{Aut}(f)$ which sends x in C.

2. Show that for any function of compact type, if admissible lattices exist, there exist critical lattices with $f(x) = 1$.

3. Show that there exist for $\mathbb{A}_{r_1,r_2}$ critical lattices containing the vector $(1, 1, \ldots, 1)$ in $\mathbb{R}^{r_1} \times \mathbb{C}^{r_2}$.

2.11 Notes on Chapter 2

§ 2.1. Hadamard published his inequality in 1893, cf. [Be-B], § 11. It is surprising that this result had not appeared earlier.

§ 2.2. Charles Hermite (born in 1822) initiated in 1843 ([He3], pp. 10–37) a correspondence with the celebrated mathematician Carl Jacobi. Hermite proved the existence of his constant in his third letter, dated August 6, 1845 ([He3], pp. 100–121). The letters which follow this are devoted to what became the *Hermite reduction of quadratic forms*, described in Section 9.

§ 2.3. The Mordell inequality dates back to 1944 ([Mor] ; cf. [Cas2], Chapter X, Section 3). However, the transition from dimension 3 to dimension 4 had been done a long time ago (in [K-Z1], dated 1872) by Korkine and Zolotareff. They did not notice that the same thing could (conjecturally) be done between dimensions 7 and 8 (for which they had guessed the critical forms); nor did Blichfeldt, whose difficult determination of $\gamma_6, \gamma_7, \gamma_8$ (in [Bl2], 1935, whose results were previously announced in [Bl1] and later verified by Watson in [Wat2]) would have been somewhat shortened. Curiously, Mordell quotes
[K-Z3] but not [K-Z1]. Probably he did not notice the possible interpretation (which is actually not so easy) of [K-Z1] in terms of lattices.

§ 2.4. Kurt Mahler published his compactness theorem in 1946. For more details, see [Cas2], Chapter V, §§ 3 and 4. Cassels gives a proof due to Chabauty which does use any result from the geometry of numbers.

§§ 2.5, 2.6. These two sections are devoted to the basic questions of the geometry of numbers, developing the necessary material for various translations, notably between quadratic forms and lattices. Theorem 2.6.1 for spheres goes back to Korkine and Zolotareff ([K-Z3], 1877).

§ 2.7. We have explained the basic work of Minkowski which is at the origin of the foundation of the geometry of numbers. The chief idea (see Proposition 2.5.5) is that finding upper bounds on $\mathbb{Z}^n$ for a distance-function f is the same as finding lower bounds on the discriminants of lattices which are admissible for the canonical open set attached to f. The fact that a lattice is admissible for the unit ball is equivalent to the fact that this lattice packs the ball of radius $\frac{1}{2}$, which yields immediately the upper bound $\gamma_n \leq 4\omega_n^{-2/n}$, since a packing has density at most 1. This simple and extremely original argument appeared in Minkowski's paper *Über die positiven quadratischen Formen und über kettenbruchähnliche Algorithmen* (Crelle 107 (1891), 278–297, = [Min1], VIII, 243–260). Minkowski then noticed that his argument works for any symmetric convex set, and developed this idea in his book *Geometrie der Zahlen* ([Min2], 1896).

The application to ideal classes is explained in a letter that Minkowski wrote to Hermite in 1991 (after a letter to Hilbert at the end of the preceding year), cf. [Min1], IX, 261–263. In particular, this theorem of Minkowski proved a ten-year-old conjecture of Kronecker, namely that the absolute value of the discriminant of a number field of degree at least 2 is greater than 1. At the end of this letter, Minkowski remarks that for non-real cubic fields, one obtains a slightly better inequality by embedding a sphere (despite its smaller volume) in $A_{1,1}$ rather than $B_{1,1}$; this remark is our Theorem 2.7.9. The reason for this success (especially useful for totally complex quartic and sextic fields) is that much work has been done on spheres, whose lattice constants are known up to dimension 8, a fairly high dimension for the geometry of numbers.

§ 2.8. The invariant γ' and the constant γ'_n (the "Bergé–Martinet constant" in [C-S9]) were introduced by Anne-Marie Bergé and myself in [B-M1] (1989). The motivation was the analytical paper [Zi] by Zimmert (1981) in which the author studied a variant of Minkowski's theorem on ideal classes consisting in looking at upper bounds for the product of the norms of a class and a "twin" class (defined using duality for the trace form), product of this class by the class of the codifferent. A useful reference is Oesterlé's paper [Oe1], which gives an interpretation of Zimmert's paper in terms of "explicit formulae". The geometric translation of Zimmert's inequality involves the product of the minima of the distance-function f attached to the set A_{r_1,r_2} on a lattice and on its dual. Replacing f by the Euclidean norm yields the invariant γ'. We considered at the same time the invariants and constants $\gamma_{n,k}$ and $\gamma'_{n,k}$ as we wanted to analyse Mordell's inequality in some detail, and we realized later that the first had been defined some 35 years earlier by Rankin in [Ran1].

§ 2.9. The HKZ reduction appears in letters from Hermite to Jacobi beginning in 1845, and then in Korkine and Zolotareff's paper [K-Z2] where the new inequalities of Theorems 2.9.3 and 2.9.4 are proved. A short proof of the inequality in Theorem 2.9.3 was given by Blichfeldt ([Bl2], Section 3), but without the characterization of the forms for which equality holds. (In the same way, Blichfeldt did not determine critical forms for the constants $\gamma_6, \gamma_7, \gamma_8$ he determined, and until recently the only reference for this question was the long posthumous paper [Vet] by Vetchinkin.)

Blichfeldt's paper contains numerous inequalities involving the *inner coefficients* of the HKZ reduction, as well as this very interesting comment ([Bl2], Section 2):

Certain relations among the outer coefficients were established by Korkine and Zolotareff:

$$A_{i+1} \geq \frac{3}{4} A_i\,, \qquad A_{i+2} \geq \frac{2}{3}\, A_i\,; \tag{2.2}$$

$$A_{i+1}\, A_{i+2}\, A_{i+3}\, A_{i+4} \geq \frac{1}{8}\, A_i^4\,. \tag{2.3}$$

These relations are sufficient to give the precise minima for $n \leq 5$. The relation (2.3) follows from the theorem of these authors in Math. Ann. **11** (1877), p. 192. This same relation, as well as the precise minima for $6, 7, 8$ variables to be established in this paper, would immediately follow from (2.2) and from a certain inequality that we would reasonably expect to be true, namely $A_{i+4} \geq \frac{1}{2}\, A_i$. But such an inequality is not always true, as is seen from the form

$$\begin{aligned} &\Big(x_1 - \frac{1}{2}(x_2 + x_3 + x_4 + x_5)\Big)^2 + \frac{3}{4}\,\Big(x_2 - \frac{1}{3}(x_3 + x_4 - x_5)\Big)^2 \\ &+\frac{2}{3}\,\Big(x_3 - \frac{1}{2}x_4 - \frac{1}{4}x_5\Big)^2 + \frac{1}{2}\,(1+r)\Big(x_4 - \frac{1}{2}x_5\Big)^2 + \frac{1}{2}\,\Big(1 - \frac{r}{4}\Big)x_5^2 \end{aligned} \tag{2.4}$$

where r is an arbitrary positive number $\leq \frac{1}{4}$.

It seems likely that Korkine and Zolotareff tried to prove the inequality $\frac{A_1}{A_5} \leq 2$ and that there failure to do so led them to restrict themselves to dimensions ≤ 5 and to develop there the local methods that we shall explain in Chapter 6. This may explain the comparatively long time which separates the publication of [K-Z2] and [K-Z3].

After Batut and Bergé, Blichfeldt's counter-example, though quite correct, can be bypassed by carefully choosing the minimal vectors at each step of the HKZ reduction procedure; this yields natural definitions of invariants $\gamma''^{\,\max} = \gamma''$ and $\gamma''^{\,\min}$, for which a detailed analysis was recently made by Gindraux ([Gi]).

The article [K-Z2] ends with a proof of the *strict* inequality $\frac{A_1}{A_5} < \frac{9}{4}$. (The non-strict inequality is a consequence of (2.9.2) and Theorem 2.9.3.) This allows them to show *a priori* that the successive minima of a 5-dimensional lattice generate a sublattice of index 1 or 2, whereas it would not have been possible to exclude index 3 using only the non-strict inequality. We shall return to this index problem in Chapter 6.

3 Perfection and Eutaxy

Introduction

The aim of this chapter is to give a characterization of extreme lattices, those lattices on which the Hermite invariant attains a local maximum. To this end, we introduce in Section 3.2 the two notions of perfection and eutaxy, and prove in Section 3.4 that they characterize extreme lattices; in particular, we obtain an algorithm which allows us to decide whether a given lattice is extreme, whose justification relies on the technical material in Sections 3.1 and 3.3. Then we give in Sections 3.5 and 3.6 a detailed analysis of the properties of perfection and eutaxy.

In Section 3.7 we show that the "laminating" process of Conway and Sloane produces perfect – but not necessarily extreme – lattices. Section 3.8 is devoted to another kind of extremal property involving both a lattice and its dual.

As usual E denotes an n-dimensional Euclidean vector space.

3.1 Symmetric Endomorphisms

Definition 3.1.1. *Let* $u \in \mathrm{End}(E)$*. The* transpose *of* u *is the endomorphism* ${}^t u$ *such that* $u(x) \cdot y = x \cdot {}^t u(y)$ *for all* $x, y \in E$*. We say that* u *is* symmetric *(resp.* skew-symmetric *or* alternating*) if* ${}^t u = u$ *(resp.* ${}^t u = -u$*). We denote by* $\mathrm{End}^s(E)$ *(resp.* $\mathrm{End}^a(E)$*) the real vector space of symmetric (resp. skew-symmetric) endomorphisms of* E*, and by* Sym_n *the space of* $n \times n$ *symmetric matrices.*

One easily verifies that the transpose of a given $u \in \mathrm{End}(E)$ exists and is unique, and that $u \mapsto {}^t u$ is an *involution* of $\mathrm{End}(E)$, i.e., that it satisfies the rule ${}^t(v \circ u) = {}^t u\, {}^t v$.

Proposition 3.1.2. *Let* $\mathcal{B}$ *be a basis for* E*, let* $\mathcal{B}^*$ *be its dual basis, let* $\mathcal{B}_0$ *be an orthogonal basis for* E*, and let* $u, v \in \mathrm{End}(E)$*. The following three conditions are equivalent:*

1. *Each of the matrices of* u *and of* v *with respect to the bases* $\mathcal{B}$ *and* $\mathcal{B}^*$ *is the transpose of the other.*

2. *Each of the matrices of u and of v in the base $\mathcal{B}_0$ is the transpose of the other.*
3. $v = {}^t u$.

Proof. The matrix $(m_{i,j})$ of an element $u \in \mathrm{End}(E)$ with respect to the bases $\mathcal{B}$ and $\mathcal{B}'$ is characterized by the relations $u(e_j) = \sum_k m_{k,j}\, e'_k$, which can also be written in the form $m_{i,j} = e'^*_i \cdot u(e_j)$. Taking $\mathcal{B}' = \mathcal{B}^*$, we find for u and ${}^t u$ matrices $(m_{i,j})$ and $(m'_{i,j})$ such that $m_{i,j} = e_i \cdot u(e_j)$ and $m'_{i,j} = e_i \cdot {}^t u(e_j) = u(e_i) \cdot e_j$. This first shows the equivalence of (1) and (2), and then that of (2) and (3) because $\mathcal{B}_0^* = \mathcal{B}_0$. □

Let $u \in \mathrm{End}^s(E)$. The definition of the transpose of u shows that the orthogonal of any subspace of E stable under u is also stable under ${}^t u$. Moreover, we know that u has real eigenvalues (extend for instance u to a Hermitian endomorphism of the complex space $\mathbb{C} \bigotimes E$). Combining these two remarks, we easily prove that the matrix of u is a diagonal matrix in some orthonormal basis for E. [Matrix interpretation: for any real symmetric matrix $A \in \mathcal{M}_n(\mathbb{R})$, there exists an orthogonal matrix $P \in \mathcal{M}_n(\mathbb{R})$ (i.e., we have $P^{-1} = {}^t P$) such that ${}^t PAP$ is a diagonal matrix.]

Definition 3.1.3. *We say that $u \in \mathrm{End}^s(E)$ is* positive *if its eigenvalues are non-negative,* definite *if they are nonzero, and thus* positive definite *if they are strictly positive. We denote by $\mathrm{End}^{s+}(E)$ (resp. $\mathrm{End}^{s++}(E)$) the set of positive (resp. positive definite) symmetric endomorphisms.*

We recover the usual definitions of the theory of real quadratic forms by applying the definitions above to the form $x \mapsto x \cdot u(x)$. The notion of a positive, or definite, or positive definite matrix is defined in the same way.

Example 3.1.4. *For any $u \in \mathrm{End}(E)$, ${}^t uu$ is a (symmetric) positive endomorphism.*

Indeed, if x is any eigenvector for ${}^t uu$ with corresponding eigenvalue λ, we have $\lambda\,(x \cdot x) = {}^t uu(x) \cdot x = u(x) \cdot u(x)$, whence $\lambda = \frac{\|u(x)\|^2}{\|x\|^2} \geq 0$.

Lemma 3.1.5. *Let u be a symmetric endomorphism and let m be a positive integer.*

1. *The centralizer of u in $\mathrm{End}(E)$ is the set of the endomorphisms which stabilize all the eigenspaces of u.*
2. *If u is positive or if m is odd, u and u^m have the same centralizer.*

Proof. Let F be an eigenspace of E with corresponding eigenvalue λ, and let $v \in \mathrm{End}(E)$. If u and v commute, we have $u(v(x)) = v(u(x)) = v(\lambda x) = \lambda v(x)$ for all $x \in F$, whence the inclusion $v(F) \subset F$.

Conversely, let $v \in \mathrm{End}(E)$ which stabilizes the eigenspaces $E_1, \dots, E_k$ of u, and for all i, let v_i be the restriction of v to E_i. Since E is the direct sum of the E_i and since the restrictions u_i of u to E_i are homothetic transformations, we have $uv(x) = u_i v_i(x) = v_i u_i(x) = vu(x)$ for all $x \in E_i$, whence (1).

Denote by λ_i the eigenvalue attached to the eigenspace E_i of u. For $x \in E_i$, we have $u^m(x) = \lambda_i^m x$, which shows that λ_i^m is for all i an eigenvalue of u^m whose corresponding eigenspace E'_i contains E_i. But the hypotheses we have made on u and m show that the map $i \mapsto \lambda_i^m$ is injective. This implies that the sum $\sum_i E_i$ is a direct one, and shows the equality $\sum \dim E'_i \geq \sum \dim E_i = n$, whence $\sum \dim E'_i = n$ and $\dim E'_i = \dim E_i$ for all i (because of the lower bound $\dim E'_i \geq \dim E_i$). Finally, the subspaces E'_i and E_i coincide for all i, and (2) is now an immediate consequence of (1). □

Theorem 3.1.6. *A positive symmetric endomorphism possesses a unique positive square root.*

Proof. Let $\mathcal{B}$ be an orthonormal basis for E in which the matrix of u is a diagonal matrix D. The endomorphism v defined in the basis $\mathcal{B}$ by the diagonal matrix whose diagonal terms are the square roots of the diagonal terms of D is a positive square root of u. By Lemma 3.1.5 (2), v and $u = v^2$ have the same centralizer, and Lemma 3.1.5 (1) then shows that v is characterized by the equations $v(x) = \sqrt{\lambda}\, x$ for every eigenvalue λ of u and every vector x in the corresponding eigenspace. This proves that v is unique. □

In the sequel, the unique positive square root of a positive symmetric endomorphism u will be called the *square root of* u, and the same convention applies to real positive symmetric matrices. In both cases, we shall use the usual notation for a square root. The equality $\sqrt{uv} = \sqrt{u}\,\sqrt{v}$ holds whenever u and v commute, and we also have $\sqrt{u^{-1}} = (\sqrt{u})^{-1}$ whenever u is invertible.

We now come to a theorem which will play a prominent role in the sequel: the decomposition of an endomorphism into symmetric and orthogonal components.

Theorem 3.1.7. *Any $u \in \mathrm{GL}(E)$ possesses unique decompositions into each of the forms $u = v\,w$ and $u = w'\,v'$ with positive symmetric v, v' and orthogonal w, w'.*

Proof. From $u = v\,w$ (resp. $u = w'\,v'$), we deduce the relation $u\,{}^t\!u = v^2$ (resp. ${}^t\!u\,u = v'^2$). The uniqueness of v and of v' then follows from Theorem 3.1.6, and that of w and of w' from the equalities $w = v^{-1}u$ and $w' = u\,v'^{-1}$.

To prove the existence of these decompositions, we apply Example 3.1.4 and Theorem 3.1.6: let v (resp. v') be the square root of $u\,{}^t\!u$ (resp. of ${}^t\!u\,u$), and let $w = v^{-1}u$ (resp. $w' = u\,v'^{-1}$). We then easily verify that ${}^t\!w\,w = {}^t\!w'\,w' = \mathrm{Id}$, and thus that w and w' are orthogonal. □

We now study how the endomorphisms constructed in Theorems 3.1.6 and 3.1.7 vary with u.

Proposition 3.1.8. *1. The map $u \mapsto \sqrt{u}$ from* $\mathrm{End}^{s++}(E)$ *onto itself is a diffeomorphism.*

2. The maps from $\mathrm{GL}(E)$ *to* $\mathrm{End}^s(E)$ *(resp. to* $\mathrm{O}(E)$*) which associate with an automorphism of E its left or right symmetric (resp. orthogonal) components are differentiable.*

Proof. The way the symmetric orthogonal components have been constructed in the proof of Theorem 3.1.7 shows that the assertions in (2) are immediate consequences of assertion (1), which we now prove.

Let $u \in \mathrm{End}^{s++}(E)$ and let $h \in \mathrm{End}^s(E)$. The identity $(u+h)^2 = u^2 + hu + uh + h^2$ shows that the tangent map to $u \mapsto u^2$ (the differential map) is $h \mapsto hu + uh$. If h lies in its kernel, we have $uh = -hu$, hence $u^2 h = hu^2$, and thus $uh = hu$ by Lemma 3.1.5 (2). We therefore have $uh = 0$, hence also $h = 0$. We now know that the map $u \mapsto u^2$ is one-to-one and that its differential map is invertible at every point of $\mathrm{End}^{s++}(E)$. The reciprocal map then possesses the same properties. □

The uniqueness assertion of Theorem 3.1.7 allows the extension of the decomposition to some subspaces of E. Here is an example that we shall need later to study families of lattices possessing a prescribed automorphism group:

Proposition 3.1.9. *Let G be a subgroup of* $\mathrm{O}(E)$ *and let u be an endomorphism of E which commutes with G. Then the symmetric and orthogonal components (on each side) of u also commute with G.*

Proof. With the notation of Theorem 3.7.1, let $g \in G$. We have

$$u = gug^{-1} = (gvg^{-1})\,(gwg^{-1})\,.$$

Then gwg^{-1} is obviously orthogonal and gvg^{-1} is symmetric, for $g^{-1} = {}^t g$. We thus have $gvg^{-1} = v$ and $gwg^{-1} = w$, and similar equalities hold for the other decomposition. □

To obtain suitable geometrical interpretations of some of the notions we are going to introduce in the next section, we shall need to work sometimes with the dual space of $\mathrm{End}^s(E)$. To this end, we introduce a Euclidean structure on this space of endomorphisms.

Proposition 3.1.10. *The map $u \mapsto \mathrm{Tr}(u^2)$ is a positive definite quadratic form on* $\mathrm{End}^s(E)$ *with corresponding bilinear form $(u,v) \mapsto \mathrm{Tr}(uv)$, and the resulting identification of* $\mathrm{End}^s(E)$ *with* $\mathrm{End}^s(E)^*$ *transforms* $\mathrm{Id} \in \mathrm{End}^s(E)$ *into* $\mathrm{Tr} \in \mathrm{End}^s(E)^*$.

We denote by $\langle u,v \rangle$ the scalar product $\mathrm{Tr}(uv)$, *and call it the* Voronoi scalar product. *We use the same notation on the space* Sym_n *of real symmetric matrices.*

Proof. The map $u \mapsto \operatorname{Tr}(u^2)$ is clearly a quadratic form with corresponding bilinear form $(u, v) \mapsto \operatorname{Tr}(uv)$. Let $u \in \operatorname{End}^s(E)$. Since its eigenvalues are real, those of u^2 are non-negative. This shows that the form $u \mapsto \operatorname{Tr}(u^2)$ is positive. Moreover, if $\operatorname{Tr}(u^2) = 0$, the eigenvalues of u are all zero, since their sum is zero, and u itself is then zero, since it possesses a diagonal form. Hence $u \mapsto \operatorname{Tr}(u^2)$ is a definite form. Finally, the last assertion is a reformulation of the equalities $\langle \operatorname{Id}, u \rangle = \operatorname{Tr}(\operatorname{Id} \circ u) = \operatorname{Tr}(u)$. □

Among the various symmetric endomorphisms of E, the orthogonal projections onto lines in E will play a crucial rôle. Recall (Proposition 1.3.4) that p_D denotes the orthogonal projection onto a line D; given $x \neq 0$ in E, we set $p_x = p_{\mathbb{R}\,x}$. We have

$$p_x(y) = \frac{x \cdot y}{N(x)}\, x\,.$$

Proposition 3.1.11. *1. For all* $u \in \operatorname{End}^s(E)$, *we have*

$$\operatorname{Tr}(u \circ p_x) = \operatorname{Tr}(p_x \circ u) = \frac{u(x) \cdot x}{N(x)}\,.$$

2. The (orthogonal) projections onto the various lines of E *span* $\operatorname{End}^s(E)$.

Proof. (1) Consider a basis $\mathcal{B} = (f_1, \ldots, f_n)$ for E with $f_1 = x$ and $f_1 \cdot f_i = 0$ for all $i > 1$. We have $p_x(f_i) = 0$ for $i > 1$. The map $u \circ p_x$ is thus zero on $\mathbb{R}x^{\perp}$, so that the trace of $u \circ p_x$ is the coefficient of $u(f_1)$ on f_1. If $u(f_1) = \sum_{i=1}^{n} x_i\, f_i$, we have $u(f_1) \cdot f_1 = x_1\,(f_1 \cdot f_1)$, whence $x_1 = \frac{u(x) \cdot x}{N(x)}$. Taking into account the symmetry of the bilinear trace form, this completes the proof of (1).

(2) It suffices to show that an element $u \in \operatorname{End}^s(E)$ which is orthogonal to all the projections is the null map. By (1), we have $\langle u, p_x \rangle = \frac{u(x) \cdot x}{N(x)}$. Hence the quadratic form $x \mapsto u(x) \cdot x$ is identically zero on E, so that all scalar products $u(x) \cdot y$ are zero. For every $x \in E$, $u(x)$ is thus orthogonal to all vectors of E, whence $u = 0$. □

We now explain a way to calculate with projections by means of matrices. We still consider a pair of a basis $\mathcal{B}$ and its dual basis $\mathcal{B}^*$, but we express the (orthogonal) projections with respect to the bases $\mathcal{B}^*$ and $\mathcal{B}$, for exchanging $\mathcal{B}$ and $\mathcal{B}^*$ simplifies some formulae. We have $p_x(e_j^*) = \frac{x \cdot e_j^*}{N(x)} x$ whence (using the components x_i of x in $\mathcal{B}$)

$$N(x)\, p_x(e_j^*) = \Big(\sum_k x_k\, e_k \cdot e_j^*\Big) \Big(\sum_i x_i\, e_i\Big) = \sum_i x_i x_j\, e_i\,.$$

We are thus led to the following definition:

Definition 3.1.12. *Let x be a nonzero vector in E, whose components $x_1, \dots, x_n$ in some basis $\mathcal{B}$ are represented by the column-matrix X. We denote by P_x or by P_X the matrix $X\,{}^tX$* (a symmetric matrix of order n).

Explicitly, we have

$$P_X = \begin{pmatrix} x_1 \\ x_2 \\ \vdots \\ x_n \end{pmatrix} \begin{pmatrix} x_1 & x_2 & \dots & x_n \end{pmatrix} = \begin{pmatrix} x_1^2 & x_1x_2 & \dots & x_1x_n \\ x_2x_1 & x_2^2 & \dots & x_2x_n \\ \vdots & \vdots & \ddots & \vdots \\ x_nx_1 & x_nx_2 & \dots & x_n^2 \end{pmatrix}.$$

The proof of the following proposition results immediately from the calculation we have done before Definition 3.1.12:

Proposition 3.1.13. *Let $\mathcal{B}$ be a basis for E and let $x \in E$ be represented by the column-matrix X. Then the matrix of $N(x)\, p_x$ with respect to the bases $\mathcal{B}^*$ and $\mathcal{B}$ is $P_X = X\,{}^tX$.* [Warning: note the inversion between $\mathcal{B}$ and $\mathcal{B}^*$ with respect to the usual ordering.] □

Let x be a nonzero element in E. The matrices P_X are of rank 1 and the corresponding quadratic forms are

$$Y \mapsto {}^tY\,(X{}^tX)\,Y = {}^t({}^tXY)\,({}^tXY)\,,$$

or denoting by $y_1, y_2, \dots, y_n$ the components of y,

$$(y_1, y_2, \dots, y_n) \mapsto (x_1y_1 + x_2y_2 + \dots + x_ny_n)^2.$$

When x runs through $E \smallsetminus \{0\}$, we find in this way all rank-1 positive quadratic forms, and two such forms are equal if and only if they correspond to equal or opposite vectors.

To study the Hermite invariant we shall use either of the following two convexity results with which we end Section 3.1.

Proposition 3.1.14. *Let v be a nonzero symmetric endomorphism, let I be an interval such that the function $t \mapsto 1 + \lambda t$ is positive for all $t \in I$ and for all eigenvalues λ of v, and for $t \in I$, let $u_t = \mathrm{Id} + t\,v$. Then u_t is a positive definite symmetric endomorphism, the map $t \mapsto \det(u_t)$ is strictly logarithmically concave, and the map $t \mapsto \dfrac{1}{\det(u_t)}$ is strictly convex.*

Proof. Since both functions we consider are C^∞ on I, we just have to show that the second derivative of the logarithm of the first function (resp. of the second function) is strictly negative (resp. positive) on I. As a function of the eigenvalues $\lambda_1, \dots, \lambda_n$ of v, we have $\det(u_t) = \prod_i (1 + \lambda_i t)$.

For the first function, we find

$$\frac{d^2}{dt^2} \log \prod_i (1 + \lambda_i t) = -\sum \frac{\lambda_i^2}{(1 + \lambda_i t)^2}\,,$$

and this expression is strictly negative, since at least one of the eigenvalues is nonzero.

The result for the second function follows from the previous one by applying the following identity, valid for any strictly positive twice differentiable function:

$$\left(\frac{1}{f}\right)'' = -\frac{1}{f}(\log f)'' + \frac{f'^2}{f^3}\,. \ \square$$

3.2 Linear Forms on Spaces of Endomorphisms

In this section, we denote by $\mathcal{T}$ a subspace of $\mathrm{End}^s(E)$; $\mathrm{Tr}(u)$ stands for the trace of the endomorphism u.

Definition 3.2.1. *For all $x \in E$, let φ_x be the linear form $u \mapsto u(x) \cdot x$ on $\mathrm{End}^s(E)$; for every line $D \subset E$, let $\varphi_D = \varphi_x$ for any unit vector $x \in D$. When there is no risk of confusion, we use the same notation to denote the restrictions to the subspace $\mathcal{T}$ of the linear forms above; this convention applies in particular to Definition 3.2.2 below.*

The equality $\varphi_{\lambda x} = \lambda^2 \varphi_x$ justifies the notation φ_D. In practice, we consider φ_x only up to a positive factor, so that the choice of any nonzero $x \in D$ would have been suitable.
[*However, for the notion of strong eutaxy defined below, one must be careful and avoid repeated lines or vectors, or non-constant numbers of vectors defining the same line.*]

Definition 3.2.2. *Let $\mathcal{F}$ be a finite family of nonzero vectors in E and let $\mathcal{D}$ be a finite family of lines in E.*

1. *We say that $\mathcal{F}$ (resp. $\mathcal{D}$) is $\mathcal{T}$-*perfect *if the forms φ_x, $x \in \mathcal{F}$ (resp. the forms φ_D, $D \in \mathcal{D}$) span the dual space $\mathcal{T}^*$ of $\mathcal{T}$. More generally, the rank r of the system φ_x (or φ_D) is called the $\mathcal{T}$-*perfection rank *of the family, and the codimension* $\dim \mathcal{T} - r$ *its $\mathcal{T}$-*perfection corank.
2. *We say that real numbers $\rho_x, x \in \mathcal{F}$ (resp. $\rho_D, D \in \mathcal{D}$) are $\mathcal{T}$-*eutaxy coefficients *for $\mathcal{F}$ (resp.* for *$\mathcal{D}$) if there exists a relation* $\mathrm{Tr} = \sum_{x\in\mathcal{F}} \rho_x \varphi_x$ *(resp.* $\mathrm{Tr} = \sum_{D\in\mathcal{D}} \rho_D \varphi_D$*) between the restrictions to $\mathcal{T}$ of the forms* Tr, φ_x, φ_D.
3. *We say that the family $\mathcal{F}$ (resp. $\mathcal{D}$) is $\mathcal{T}$-*weakly eutactic *if it possesses $\mathcal{T}$-eutaxy coefficients, that it is $\mathcal{T}$-*semi-eutactic *(resp. $\mathcal{T}$-*eutactic, *resp.* strongly eutactic*) if it is weakly eutactic with positive (resp. strictly positive, resp. equal) eutaxy coefficients.*

   [Warning: For *strong eutaxy*, we must consider either distinct lines, or vectors of equal norm such that no two of them are collinear.]

When $\mathcal{T} = \mathrm{End}^s(E)$, one simply says perfect, weakly eutactic, semi-eutactic, eutactic, *and* strongly eutactic.

To say that $\mathcal{F}$ (or $\mathcal{D}$) is $\mathcal{T}$-perfect amounts to the same thing as saying that the intersection of the kernels of the restriction to $\mathcal{T}$ of the forms φ_x (or φ_D) is reduced to $\{0\}$.

Note that perfect families are those which have corank 0, and that they are always weakly eutactic; note also that strongly eutactic families are actually eutactic: let $s = |\mathcal{F}|$; applying a given eutaxy relation $\mathrm{Tr} = \sum_{x \in \mathcal{F}} \rho_x \varphi_x$ to the identity, we obtain the new relation

$$\sum_{x \in \mathcal{F}} N(x)\, \rho_x = n \qquad \mathbf{(3.2.2')}$$

from which we deduce that equal eutaxy coefficients have the common positive value

$$\rho = \frac{1}{N(x)}\, \frac{n}{s}\,. \qquad \mathbf{(3.2.2'')}$$

The notions of a perfect or of a eutactic family $\mathcal{F}$ of vectors solely depend on the set of lines which contain them; it is possible to suppress a vector in $\mathcal{F}$ whenever $\mathcal{F}$ contains another one which is proportional to it, and hence to restrict oneself to families of vectors with a given norm. The natural notions are thus those of a family (one also says *configuration*) of perfect or eutactic lines. However, we shall have to work with vectors rather than with lines, and shall essentially apply Definition 3.2.2 to families of vectors.

The following trivial proposition is nevertheless very useful:

Proposition 3.2.3. *1. If $\mathcal{T}'$ is a subspace of $\mathcal{T}$, any $\mathcal{T}$-perfect (resp. $\mathcal{T}$-eutactic) configuration is $\mathcal{T}'$-perfect (resp. $\mathcal{T}'$-eutactic).*
2. Any $\mathcal{T}$-perfect configuration has cardinality at least $\dim \mathcal{T}$. □

We now consider more closely the case where $\mathcal{T}$ is the whole space $\mathrm{End}^s(E)$.

Lemma 3.2.4. *For any nonzero $x \in E$, the duality defined by* $\mathrm{Tr}(u^2)$ *(see Proposition 3.1.10) transforms $N(x)\, p_x$ into φ_x.*

Proof. This is just a reformulation of Proposition 3.1.11. □

Theorem 3.2.5. *Let $\mathcal{F}$ be a finite family of nonzero vectors in E, let $\mathcal{B} = (e_1, \dots, e_n)$ be a basis for E, with Gram matrix $A = (a_{i,j})$ (i.e., we have $a_{i,j} = e_i \cdot e_j$), and let $(a^*_{i,j})$ be the Gram matrix of $\mathcal{B}^*$ (i.e., we have $(a^*_{i,j}) = A^{-1}$). For all $x \in E$, denote by $c_1(x), \dots, c_n(x)$ the components of x in $\mathcal{B}$. The following conditions relative to a family ρ_x, $x \in \mathcal{F}$ of real numbers are then equivalent:*
1. The ρ_x are eutaxy coefficients for $\mathcal{F}$.
2. We have the identity $\mathrm{Id}_E = \sum_{x \in \mathcal{F}} \rho_x N(x)\, p_x$.
3. For all $y \in E$, $N(y) = \sum_{x \in \mathcal{F}} \rho_x\, (y \cdot x)^2$.

3′. For all $y, z \in E$, $y \cdot z = \sum_{x\in\mathcal{F}} \rho_x\,(y\cdot x)\,(z\cdot x)$.
4. For all i *and all* j, $a^*_{i,j} = \sum_{x\in\mathcal{F}} \rho_x\,c_i(x)\,c_j(x)$.
5. We have the identity $A^{-1} = \sum_{x\in\mathcal{F}} \rho_x\,P_x$.

Proof. The equivalence of (1) and (2) is an immediate consequence of Lemma 3.2.4, that of (3) and (3′) results from the correspondence between quadratic and bilinear forms, and that of (4) and (5) is obvious, as (4) is just an explicit formulation of (5). The remainder of the proof is organized along the logical scheme (2) ⇒ (3′) ⇒ (4) ⇒ (2).

If (2) is satisfied, we have for all $y \in E$ the relation

$$y = \sum_{x\in\mathcal{F}} \rho_x\,N(x)\,p_x(y) = \sum_{x\in\mathcal{F}} \rho_x\,(x\cdot y)\,x$$

and (3′) follows, as one sees by performing the scalar product of both sides with z.

One obtains (4) from (3′) by putting $y = e_i$ and $z = e_j$.

Finally, if (4) is satisfied, we recover the preceding formula by setting $y = \sum \lambda_i e_i^*$ and $z = \sum \mu_j e_j^*$: we actually have

$$y\cdot z = \sum_{i,j} \lambda_i\mu_j \sum_x \rho_x\,(x\cdot e_i^*)\,(x\cdot e_j^*) = \sum_x \rho_x \Big(x\cdot\sum_i \lambda_i\,e_i^*\Big)\Big(x\cdot\sum_j \mu_j\,e_j^*\Big)$$
$$= \sum_x \rho_x\,(x\cdot y)\,(x\cdot z)$$

and the relation $y = \sum_x \rho_x\,(x\cdot y)$ follows, since both sides have the same scalar product with any $z \in E$. □

Corollary 3.2.6. *Any weakly eutactic family spans* E*; in particular any perfect family spans* E.

Proof. Let $\mathcal{F}$ be a weakly eutactic family, and let $y \in E$ be orthogonal to all vectors in $\mathcal{F}$. Then, by condition (2) above, we have

$$y = \sum \rho_x\,N(x)\,(x\cdot y)\,y = 0\,. \;\square$$

Remark 3.2.7. *When* $\mathcal{T}$ *is no longer the whole space* $\mathrm{End}^s(E)$*, Assertion (2) above still holds, with the* p_x *replaced by their orthogonal projections* ω_x *onto* $\mathcal{T}$ *(in* $\mathrm{End}^s(E)$*, for the scalar product* $\langle .\,,.\rangle$*).*

We now return to the general case of an arbitrary subspace $\mathcal{T}$ of $\mathrm{End}^s(E)$ and give the definitions we need for lattices.

Definition 3.2.8. *We say that a lattice is* $\mathcal{T}$-perfect *(resp.* weakly $\mathcal{T}$-eutactic, *resp.* $\mathcal{T}$-semi-eutactic, *resp.* $\mathcal{T}$-eutactic, *resp.* strongly $\mathcal{T}$-eutactic*) if the set of its minimal vectors constitutes a* $\mathcal{T}$*-perfect (resp. a weakly* $\mathcal{T}$*-eutactic,* *resp. a $\mathcal{T}$-semi-eutactic, resp. a $\mathcal{T}$-eutactic, resp. a strongly $\mathcal{T}$-eutactic) family. When $\mathcal{T}$ is the whole space* $\mathrm{End}^s(E)$, *one simply says* perfect, weakly eutactic, semi-eutactic, eutactic *and* strongly eutactic. *We define as in Definition 3.2.2 the* ($\mathcal{T}$)-perfection rank *and* corank *of a lattice.*

Let Λ be a lattice. To a basis $\mathcal{B}$ for Λ, we can attach a symmetric matrix A, namely the Gram matrix of $\mathcal{B}$, whence also a quadratic form Q on $\mathbb{R}^n$. By Theorem 2.5, the properties of perfection or eutaxy of Λ are equivalent to the corresponding properties for Q (or for A) as defined below. (For the sake of simplicity, we ignore the case of a general subspace $\mathcal{T}$ of $\mathrm{End}^s(E)$.)

Definition 3.2.9. *Let Q be a positive definite quadratic form on $\mathbb{R}^n$ with matrix $A \in \mathrm{Sym}_n$. We say that Q (or A) is* perfect *if the matrices P_X, $X \in S(Q)$ span* Sym_n, *and that it is* weakly eutactic *if there exists a relation*

$$A^{-1} = \sum_{X \in S(Q)} \rho_X P_X\,.$$

We define similarly semi-eutaxy, eutaxy and strong eutaxy *for forms and symmetric matrices.* [Warning: the coefficients ρ_X coincide with those of Theorem 3.2.5 only up to the factor $N(X) = {}^tXX$.]

More generally, when going from lattices to matrices (or to quadratic forms), we attach to $\mathcal{T}$ a subspace $\mathcal{T}_{\mathrm{mat}}$ (or $\mathcal{T}_{\mathrm{quad}}$) of Sym_n (or of the space $\mathcal{Q}_n$ of quadratic forms on $\mathbb{R}^n$), which we shall often simply denote by $\mathcal{T}$. Definition 3.2.9 applies to this situation, provided that Sym_n should be replaced by $\mathcal{T}$ and the P_X by their projections Ω_X onto $\mathcal{T}$.

The following statement, which only involves the usual notion of perfection, is an easy consequence of the mere definition of perfection. Due to its great importance, we state it as a theorem, which actually characterizes perfect lattices or forms:

Theorem 3.2.10. (Korkine and Zolotareff.)

1. *A perfect lattice with given norm m is well defined up to isometry by the components in some basis of its minimal vectors.*
2. *A perfect quadratic form with given minimum m is uniquely defined by the set of its minimal vectors in $\mathbb{Z}^n$.*

Proof. The dictionary quadratic forms–lattices shows that the two statements above are equivalent. Let us consider the case of quadratic forms.

Let Q_1 and Q_2 be two perfect quadratic forms with the same set $S \subset \mathbb{Z}^n$ of minimal vectors, and let Q be the form $Q_1 - Q_2$. For all $X \in S$, we have $Q_1(X) = Q_2(X) = m$, hence $Q(X) = 0$. The matrix A of Q is thus such that $\forall X \in S$, ${}^tXAX = 0$, a condition equivalent to $\forall X \in S$, $\langle A, P_X \rangle = 0$. Since the P_X, $X \in S$ generate Sym_n, this implies $A = 0$. □

From the theorem above, we now deduce another result which also goes back to Korkine and Zolotareff, and which accounts for the crucial rôle that integral lattices play in the theory of the Hermite constant:

Proposition 3.2.11. *A perfect lattice is proportional to an integral lattice.*

Proof. Let Λ be a perfect lattice. We may rescale Λ so as to give it the norm 1. Let A be the Gram matrix of some basis for Λ. The corresponding quadratic form is determined by its minimal vectors, and these vectors have, moreover, integral components. Thus the entries of A are the solution of a linear system with integer coefficients, from which we can extract a Cramer system (because A is uniquely determined by this system). These entries are thus rational numbers. Consequently, $\sqrt{m}\,\Lambda$ is an integral lattice for any m such that $m\,A$ has entries in $\mathbb{Z}$. □

3.3 Linear Inequalities

We prove here a theorem which presently belongs to the so-called linear programming theory. We shall make use of it in the next section to characterize extreme lattices.

Theorem 3.3.1. (Stiemke, [Sti].) *Let V be a real vector space and let $\varphi_1, \dots, \varphi_r$ be linear forms on V. The following conditions are equivalent:*

1. *Every $x \in V$ which is a solution of the system of linear inequalities $\varphi_i(x) \geq 0$ for $i = 1, 2, \dots, r$ is a solution of the linear system $\varphi_i(x) = 0$ for $i = 1, 2, \dots, r$.*
2. *There exist strictly positive real numbers $\rho_1, \rho_2, \dots, \rho_r$ such that $\rho_1\varphi_1 + \rho_2\varphi_2 + \cdots + \rho_r\varphi_r = 0$.*

Proof. The implication (2) ⇒ (1) is obvious, and we now prove the other one (the only one which will be useful). There is nothing to prove if $V = \{0\}$ or if $r \leq 1$. We thus suppose $V \neq \{0\}$ and $r \geq 2$, and prove the theorem by induction on r. Since Assertions (1) and (2) are invariant under replacement of V by the intersection $W = \cap_i \ker\varphi_i$, we may and shall assume that the linear forms φ_i generate $V^* = \mathcal{L}(V, \mathbb{R})$.

Let $m \leq r$ be the largest integer such that there are m forms among the φ_i with the following property: there exists $x \in V$ on which these forms take values which are positive but not all zero. After permutation of the indices, we may assume that these forms are the first m forms.

Let us first show that $\varphi_1, \dots, \varphi_m$ span V^*. If it were not the case, we could find a nonzero element $y \in \cap_{1\leq i\leq m} \ker\varphi_i$ and an index $k > m$ with $\varphi_k(y) \neq 0$. Since such a k would contradict the definition of m, we certainly have $m = r$. But this contradicts the fact that V^* is spanned by the r linear forms φ_i.

We now distinguish two cases.

Case 1: $r > m+1$. The fact that m is as large as possible allows us to apply the induction hypothesis to the $r-1 > m$ forms $\varphi_1, \dots, \varphi_{r-1}$: there exists a relation $\rho'_1\varphi_1 + \dots + \rho'_{r-1}\varphi_{r-1} = 0$ with strictly positive coefficients ρ'_i. We can write φ_r as a linear combination $\lambda_1\varphi_1 + \dots + \lambda_{r-1}\varphi_{r-1}$. Choose $\rho_r > 0$ and sufficiently small in order that $\rho_i = \rho'_i - \rho_r\lambda_i$ be positive for all $i < r$. We then have the equality $\sum_{1\le i\le r} \rho_i\,\varphi_i = 0$, which completes the proof in this case.

Case 2: $r = m+1$. We apply the induction hypothesis to $H = \operatorname{Ker}\varphi_r$. Let $x \in H$. Since $\varphi_r(x) = 0$, the inequalities $\varphi_i(x) \ge 0$ for all $i < r$ imply that we have $\varphi_i(x) = 0$ for all $i < r$.

On the one hand, there exist $\rho_1, \dots, \rho_{r-1} > 0$ such that the linear form $\sum_{i<r} \rho_i\,\varphi_i$ is zero on H. On the other hand, the definition of m shows that there exists $v \in V$ with $\varphi_r(v) < 0$ and $\varphi_i(v) \ge 0$ for all $i < r$, with at least one of the $v_i \ne 0$. The inequalities $\sum_{i<r} \rho_i\,\varphi_i(v) > 0$ and $\varphi_r(v) < 0$ show that there exists $\rho_r > 0$ such that the linear form $\sum_{1\le i\le r} \rho_i\,\varphi_i$ is zero at v. Since this form is also zero on H, it is zero on the whole space V. □

Here is an equivalent statement which shows up the convexity property which lies behind Stiemke's theorem:

Theorem 3.3.2. *Let K be the convex polytope defined by the inequalities $\varphi_i(x) \ge 0$. Then K reduces to the intersection of the kernels of the φ_i if and only if there exist strictly positive real numbers ρ_i such that $\sum_i \rho_i\varphi_i = 0$.* □

3.4 A Characterization of Extreme Lattices

We keep in this section the notation of Sections 3.1 and 3.2, but the results of Section 3.2 will be used only for $\mathcal{T} = \operatorname{End}^s(E)$. Given a lattice $\Lambda \subset E$, recall (Chapter 1, Definition 1.2.1 and Chapter 2, Definition 2.2.5) that $S = S(\Lambda)$ is the set of minimal vectors in Λ and $\gamma(\Lambda) = N(\Lambda)\det(\Lambda)^{-1/n}$ is its Hermite invariant.

Definition 3.4.1. *We say that Λ is* extreme *if the Hermite invariant attains a local maximum on Λ (for the topology on the set $\mathcal{L}$ of lattices in E defined in Section 1.1 of Chapter 1), and that it is* critical *(or* absolutely extreme*) if the Hermite invariant attains the absolute maximum on Λ.*

Equivalently, the density of the sphere packing attached to Λ (see Definition 1.8.1) is a local maximum. Clearly, the notion of extremality solely depends on the similarity class of Λ.

To test whether Λ is extreme, we study the family $u(\Lambda)$ when u runs through some neighbourhood of the identity of $\mathrm{GL}(E)$. The topology on

$\mathrm{End}(E)$ is defined by an arbitrary norm (all norms are equivalent). A classical choice is $\|u\| = \sup_{x \in E \smallsetminus \{0\}} \frac{\|u(x)\|}{\|x\|}$; one can as well use the norm $\sqrt{\mathrm{Tr}({}^t\!uu)}$.

Thanks to Theorem 3.1.7, we could restrict ourselves to symmetric u, and even assume that u preserves the norm of Λ, since extremality is invariant under similarity. The way the determinant of Λ transforms is obvious: we have $\det(u(\Lambda)) = \det(\Lambda)\det(u)^2$. To evaluate the norm of Λ, we shall need the following lemma:

Lemma 3.4.2. *There exists a neighbourhood $\mathcal{V}$ of the identity in* $\mathrm{GL}(E)$ *such that, for $u \in \mathcal{V}$, the minimal vectors in $u(\Lambda)$ are the images of some minimal vectors in Λ; in other words, we have $S(u(\Lambda)) \subset u(S(\Lambda))$ for all $u \in \mathcal{V}$.*

Proof. Set $N_1 = N(\Lambda)$ and $N_2 = \min_{x \in \Lambda, N(x) > N(\Lambda)} N(x)$, and let $\mathcal{V} = \left\{u \in \mathrm{GL}(E) \mid \|u^{-1}\| < \sqrt{N_2/N_1}\right\}$. We clearly have $N(u(y)) > N(u(x))$ for all $u \in \mathcal{V}$ whenever x is minimal and $N(y)$ is larger than N_2. □

To see how $u(\Lambda)$ varies with u, we shall make use of a series expansion of $\gamma(u(\Lambda))$ in an appropriate neighbourhood of the identity. However, the calculations are greatly simplified by making use of ${}^t\!uu$ rather than u. Writing ${}^t\!uu = \mathrm{Id} + v$, we transform by $v \mapsto {}^t\!uu$ a fundamental system of neighbourhoods of 0 in $\mathrm{End}^s(E)$ into a fundamental system of neighbourhoods of Id in $\mathrm{GL}(E)$. Recall (Definition 3.2.1) that $\varphi_x(v) = v(x) \cdot x$ for $x \in E$ and $v \in \mathrm{End}^s(E)$.

Lemma 3.4.3. *There exists a neighbourhood $\mathcal{V}$ of* 0 *in* $\mathrm{End}^s(E)$ *such that $N(u(\Lambda)) = N(\Lambda)$ if and only if* $\min_{x \in S} \varphi_x(v) = 0$.

Proof. By Lemma 3.4.2, provided $\mathcal{V}$ is sufficiently small, we have

$$N(u(\Lambda)) = \min_{x \in S} N(u(x)),$$

and the equalities

$$N(u(x)) = u(x) \cdot u(x) = {}^t\!uu(x) \cdot x = \varphi_x(\mathrm{Id}) + \varphi_x(v) = N(x) + \varphi_x(v)$$

yield

$$N(u(\Lambda)) = N(\Lambda) + \min_{x \in S} \varphi_x(v). \ \square$$

We now look at the determinant of u.

Lemma 3.4.4. 1. *There exists a neighbourhood $\mathcal{V}$ of* 0 *in* $\mathrm{End}^s(E)$ *such that, for $v \in \mathcal{V}$ with $\mathrm{Tr}(v) \le 0$ and $u \in \mathrm{GL}(E)$ such that ${}^t\!uu = \mathrm{Id} + v$, we have $u \in \mathrm{O}(E)$ or $\det(u) < 1$.*

2. *Let C be a closed cone in* $\mathrm{End}^s(E)$ *such that $\mathrm{Tr}(v)$ is strictly positive for every nonzero $v \in C$. There exists then $\alpha > 0$ such that*

$$v \in C \ \textit{and} \ 0 < \|v\| < \alpha \Longrightarrow \det(\mathrm{Id} + v) > 1.$$

Proof. (1) Let $\lambda_1, \lambda_2, \ldots, \lambda_n$ be the (real) eigenvalues of v; those of $\mathrm{Id} + v$ are $1+\lambda_1, 1+\lambda_2, \ldots, 1+\lambda_n$; they are strictly positive on any sufficiently small neighbourhood $\mathcal{V}$ of 0. We can thus consider on the interval $[0,1]$ the two functions $t \mapsto \varphi_v(t) = \det(\mathrm{Id} + tv)$ and $\psi_v = \log \varphi_v$. The derivative of the second one is $\sum_{i=1}^n \dfrac{\lambda_i}{1+\lambda_i t}$, and we thus have

$$\psi_v'(0) = -\sum_{i=1}^n \lambda_i = -\operatorname{Tr}(v) \leq 0\,.$$

If $v = 0$, u is an isometry.

If $v \neq 0$, ψ_v is strictly concave by Proposition 3.1.14, and this implies the inequality $\psi_v(t) < 0$ on $[0,1]$. In particular, we have $\psi_v(1) < 0$, hence $\det(u) = (\varphi_v(1))^{1/2} < 1$.

(2) Let $\Sigma = \{w \in \mathrm{End}^s(E) \mid \|w\| = 1\}$ be the unit sphere of $\mathrm{End}^s(E)$ and let $w \in \mathcal{C} \cap \Sigma$. The function ψ_w is defined in any sufficiently small neighbourhood of 0, and we have by our hypothesis $\operatorname{Tr}(w) > 0$, hence $\varphi_w'(0) > 0$; there thus exists $t_w > 0$ such that $\psi_w(t_w) > 0$. Since the function $w' \mapsto \psi_{w'}(t_w)$ is continuous on $\mathcal{C} \cap \Sigma$, there exists an open neighbourhood $\mathcal{V}(w)$ of w in $\mathcal{C} \cap \Sigma$ such that $\psi_{w'}$ is positive at t_w for all $w' \in \mathcal{V}(w)$, and also by convexity on the whole interval $(0, t_w]$.

From the covering $\bigcup_{w \in \mathcal{C} \cap \Sigma} \mathcal{V}(w)$ of the compact set $\mathcal{C} \cap \Sigma$, we can extract a finite covering $\bigcup_{1 \leq i \leq r} \mathcal{V}(w_i)$. Let $\alpha = \min(t_{w_1} \cdots t_{w_r})$, let $v \in \mathcal{C}$ such that $0 < \|v\| < \alpha$, and let $w = \frac{v}{\|v\|} \in \Sigma$. There exists i, $1 \leq i \leq r$, such that $w \in \mathcal{V}(w_i)$. Hence $f_w(t)$ is strictly positive on the interval $(0,\alpha) \subset (0, t_{w_i}]$, and in particular for $t = \|v\|$. This is equivalent to the inequality $\det(\mathrm{Id} + v) > 1$. □

The calculation of the Hermite invariant by means of Lemmas 3.4.3 and 3.4.4, with $\mathcal{C} = \{v \in \mathrm{End}^s(E) \mid \forall x \in S(\Lambda),\, \varphi_x(v) \geq 0\}$ in 3.4.4 (2), immediately yields:

Theorem 3.4.5. (Korkine and Zolotareff) *A lattice Λ is extreme if and only if the following implication holds:*

$$v \in \mathrm{End}^s(E)\,, \quad \min_{x \in S(\Lambda)} \varphi_x(v) = 0 \quad \textit{and} \quad \operatorname{Tr}(v) \leq 0 \Longrightarrow v = 0\,.$$

Moreover if Λ is extreme, there exists a neighbourhood of Λ on which the Hermite invariant of any lattice which is not similar to Λ is strictly smaller than that of Λ. □

We have previously defined (Definition 3.2.8) the notions of a perfect and of a eutactic lattice. They allow the following characterization of extreme lattices, which is the central result of this section:

Theorem 3.4.6. (Voronoi.) *A lattice is extreme if and only if it is both perfect and eutactic.*

Proof. The proof relies on the characterization of extreme lattices we have given in Theorem 3.4.5.

Let us first show that a eutactic and perfect lattice Λ is extreme. Let $v \in \mathrm{End}^s(E)$ such that $\min_{x \in S} \varphi_x(v) = 0$ and $\mathrm{Tr}(v) \leq 0$. In particular, we have $\varphi_x(v) \geq 0$ and $\mathrm{Tr}(v) \leq 0$, so that the definition of eutaxy implies that $\varphi_x(v)$ is zero for all $x \in S$. Since Λ is perfect, v must be zero.

Conversely, let Λ be an extreme lattice. Let us first show that Λ is perfect. Consider an element $v \in \mathrm{End}^s(E)$ such that $\varphi_x(v) = 0$ for all $x \in S$. We must verify that $v = 0$. Replacing if need be v by $-v$, we may assume that $\mathrm{Tr}(v) \leq 0$. Since $\min_{x \in S} \varphi_x(v) = 0$, Theorem 3.4.5 shows that v is zero.

To prove that Λ is also eutactic, we apply Stiemke's Theorem 3.3.1 to the set of linear forms φ_x, $x \in S$ and $-\mathrm{Tr}$ on $V = \mathrm{End}^s(E)$. We are reduced to prove that the inequalities $\varphi_x(v) \geq 0$ and $\mathrm{Tr}(v) \leq 0$ imply the equalities $\varphi_x(v) = 0$ and $\mathrm{Tr}(v) = 0$. For $k \in \mathbb{R}$, let $v' = v - k\,\mathrm{Id}$. We have $\mathrm{Tr}(v') = \mathrm{Tr}(v) - kn$ and $\varphi_x(v') = \varphi_x(v) - kN(\Lambda)$, hence $\min_{x \in S} \varphi_x(v') = \min_{x \in S} \varphi_x(v) - kN(\Lambda)$. Choose k in order that this minimum be zero. We clearly have $k \geq 0$ and $\mathrm{Tr}(v') \leq \mathrm{Tr}(v) \leq 0$. Theorem 3.4.5 shows that v' is zero, hence that v is multiplication by k. From $\varphi_x(v) = -kN(\Lambda) \leq 0$ and $\mathrm{Tr}(v) = kn \geq 0$, we deduce that $k = v = 0$, and this proves in particular that $\mathrm{Tr}(v) = 0$ and that the equalities $\varphi_x(v) = 0$ hold for all $x \in S$. □

Corollary 3.4.7. (Korkine and Zolotareff.) *The n-th power of the Hermite constant is a rational number.*

Proof. A lattice whose Hermite invariant is maximum is extreme, hence in particular perfect, and we may apply Proposition 3.2.11. □

3.5 Perfect Configurations

We keep the notation of the previous sections, and denote by S a configuration of vectors in E (S is thus a finite set of nonzero vectors in E). In practice, S is the set of minimal vectors of some lattice Λ; this is the reason why we work with vectors rather than with the family of lines that they define, though this point of view would have been more natural.

We study the perfection of S in the usual sense, disregarding the possible extensions of this notion to various subspaces $\mathcal{T}$ of $\mathrm{End}^s(E)$ other than $\mathrm{End}^s(E)$ itself, even when this does not cause any difficulty, as is the case for the following trivial (but useful) result:

Proposition 3.5.1. *Any configuration which contains a perfect configuration is perfect.* □

Theorem 3.5.2. *1. A necessary and sufficient condition for a configuration to be perfect is that it should not be contained in any quadratic cone.*
2. A perfect configuration is not contained in the union of two hyperplanes; in particular, a perfect configuration cannot be the union of two orthogonal strict subsets.
3. The vectors of a perfect configuration span E.

Proof. (2) results from (1) applied to the degenerated cones which are the union of two hyperplanes and (3) is the particular case of (2) in which the two hyperplanes collapse to a single one. We now prove (1) in terms of matrices.

To say that a configuration $S \subset \mathbb{R}^n$ is not perfect amounts to saying that the matrices X^tX, $X \in S$ are contained in some hyperplane H inside Sym_n. Since H is the subspace of Sym_n orthogonal to some nonzero symmetric matrix A for the Voronoi scalar product $\langle U, V\rangle = \mathrm{Tr}(UV)$ of Proposition 3.1.10, we have for every $X \in S$ the equivalences

$$X^tX \in H \Longleftrightarrow \langle A, X^tX\rangle = 0 \Longleftrightarrow \mathrm{Tr}(A^tXX) = 0$$
$$\Longleftrightarrow \mathrm{Tr}({}^tXAX) = 0 \Longleftrightarrow {}^tXAX = 0\,,$$

and the last equality means that the elements of S belong to the cone defined by the equation ${}^tXAX = 0$. □

The forthcoming perfection criterion, though far from being a general one, is nevertheless very often useful, see e.g. the case of root lattices (Chapter 4, Section 4.7). This is a kind of converse to the second assertion of the preceding theorem, as it shows that when S is not contained in the union of two hyperplanes and possesses a perfect hyperplane section, it is then itself perfect.

Proposition 3.5.3. *1. Let H be a hyperplane of E such that $S \cap H$ is a perfect configuration in H. Then S is perfect if and only if the vectors in S which lie outside H span E.*
2. If there exist three distinct hyperplanes H_1, H_2, H_3 of E such that $S \cap H_i$ is for every i a perfect configuration in H_i, S is then perfect.
3. In order that a lattice Λ in E possessing a perfect hyperplane section with the same norm as Λ be perfect, it is necessary and sufficient that $S(\Lambda)$ should contain n independent vectors outside this hyperplane section.
4. A lattice Λ possessing three distinct hyperplane perfect sections having the same norm as Λ is perfect.

Proof. (1) The necessity of the condition follows immediately from Theorem 3.5.2 (the fact that $H \cap S$ be perfect is irrelevant here). To show the sufficiency of the condition, we consider the space $\mathcal{P}_H = \{p_H \circ u \mid u \in \mathrm{End}^s(E)\}$ (p_H stands for the orthogonal projection onto H), which has codimension n in $\mathrm{End}^s(E)$. Choose a vector $e \neq 0$ in $H^\perp$ and a basis $\mathcal{B}$ for E consisting of n

vectors $e_1, \dots, e_n$ which lie in S but not in H. It suffices to show that there cannot exist a relation $\sum_i \lambda_i p_{e_i} \in \mathcal{P}_H$ with real not all zero λ_i. We now have the equivalences

$$\sum_i \lambda_i p_{e_i} \in \mathcal{P}_H \Longleftrightarrow \sum_i \lambda_i p_e \circ p_{e_i} = 0 \Longleftrightarrow \forall y \in E, \sum_i \lambda_i (e_i \cdot y)(e_i \cdot e) = 0\,.$$

When one takes $y = e_j^*$, the last equality reduces to $\lambda_j (e_j \cdot e) = 0$, which implies $\lambda_j = 0$ since e_j does not lie in H.

(2). Since it is perfect, $S \cap H_2$ contains $n-1$ independent vectors outside $H_1 \cap H_2$. Since these vectors do not lie in H_1, (1) shows that it suffices to prove that there exists in $S \cap H_3$ a vector which does not lie in $H_1 \cup H_2$, a result which follows from Theorem 3.5.2 (2) applied to the hyperplanes $H_1 \cap H_3$ and $H_2 \cap H_3$ of H_3.

Finally, (3) and (4) are immediate consequences of (1) and (2). □

We shall return to the question of relative perfection in Chapter 12, contenting ourselves in this section with the case of codimension 1 just dealt with above, for which the result can be expressed using only geometric properties of vectors in E. We now prove an important finiteness theorem:

Theorem 3.5.4. (Voronoi.) *For a given dimension n, the number of similarity classes of perfect lattices and the number equivalence classes up to proportionality of perfect quadratic forms are finite.*

Proof. The two statements are clearly equivalent. We shall prove the first one by showing the finiteness assertion up to isometry for norm 1 lattices.

Let thus Λ be a perfect lattice of norm 1. Since the minimal vectors in Λ span E, we have $\det(\Lambda) \leq 1$ by the Hadamard inequality (Theorem 2.1.1). The Hermite inequality (Theorem 2.2.1) then shows that there exists a basis $\mathcal{B} = (e_1, \dots, e_n)$ for Λ for which the norms of the basis vectors are bounded from above by some constant c_n which solely depends on the dimension.

For $x = \sum_i a_i e_i \in \Lambda$, we have

$$|a_i|^2 = \frac{\det(\langle e_1, \dots, e_{i-1}, x, e_{i+1}, \dots, e_n \rangle)}{\det(\langle e_1, \dots, e_{i-1}, e_i, e_{i+1}, \dots, e_n \rangle)}\,.$$

Replacing the numerator by the upper bound given by the Hadamard inequality and the denominator by the lower bound given by the Hermite inequality, we obtain the upper bound

$$|a_i|^2 \leq \gamma_n^n c_n^{n-1} N(x)\,,$$

which bounds $|a_i|^2$ for $1 \leq i \leq n$ and all $x \in \Lambda$ by $C_n N(x)$, where the constant C_n depends only on n.

The components in $\mathcal{B}$ of any $x \in S(\Lambda)$ are thus bounded from above by $\sqrt{C_n}$. Hence the set of the components of the minimal vectors of a perfect norm 1 lattice in a basis which satisfies the Hermite inequality is a subset of a finite subset of $\mathbb{Z}^n$, and our claim now follows from Theorem 3.2.10. □

We end this section with applications of the notion of perfection to the study of the kissing number.

Theorem 3.5.5. *An n-dimensional lattice for which the kissing number is maximum among lattices of dimension n is perfect.*

Proof. Let Λ be such a lattice. If it is not perfect, there exists a nonzero $v \in \mathrm{End}^s(E)$ such that $\varphi_x(v) = 0$ for all $x \in S(\Lambda)$ (see Definitions 3.2.1 and 3.2.2). For $\lambda \geq 0$ and sufficiently small for the square root to exist, let $u_\lambda = \sqrt{\mathrm{Id} - \lambda v}$. Lemma 3.4.3 shows that the lattice $u_\lambda(\Lambda)$ has the same norm as Λ for sufficiently small λ and satisfies the property $S(u_\lambda(\Lambda)) = u_\lambda(S(\Lambda))$. Since $\det(u_\lambda(\Lambda)) = \det(u_\lambda)\,\det(\Lambda)$ strictly decreases and tends to 0 on the interval of definition of u_λ, the norm of $u_\lambda(\Lambda)$ becomes smaller than that of Λ for all sufficiently large λ . Let μ be the supremum of the λ for which $N(u_\lambda(\Lambda)) = N(\Lambda)$ and $S(u_\lambda(\Lambda)) = u_\lambda(S(\Lambda))$. The lattice $u_\mu(\Lambda)$ has clearly the same norm as Λ but at least one pair of minimal vectors outside $u_\lambda(S(\Lambda))$. We thus have $s(u_\mu(\Lambda) > s(\Lambda)$. □

Finer results have been obtained for dimensions 8 and 24 for the proof of which we refer the reader to [C-S], Chapter 14; the lattice $\mathbb{E}_8$ (resp. the Leech lattice Λ_{24}) is defined in Section 4.4 (resp. in Sections 5.7 and 8.7):

Theorem 3.5.6. (Bannai and Sloane.) *In dimension* 8 *(resp.* 24*), the configuration of the centres of* $240 = 120 \times 2$ *(resp.* $196560 = 98280 \times 2$*) spheres of radius* 1 *in contact with the unit sphere of E is similar to that of the minimal vectors in* $\mathbb{E}_8$ *(resp. in* Λ_{24}*).* □

From this theorem, we first deduce that the maximum of s for dimension 8 (resp. 24) is equal to 120 (resp. to 98280), a result first proved by Odlyzko and Sloane; see [C-S], Chapter 13, then that it is attained only on $\mathbb{E}_8$ (resp. on Λ_{24}), and finally that these two lattices are perfect. Actually, for a given dimension n, if an inequality $s(\Lambda) \geq s_0$ has only finitely many solutions up to similarity, then all lattices with $s(\Lambda) \geq s_0$ are perfect.

Some other examples of lattices for which a characterization can be obtained via their kissing numbers will be seen in Chapter 6.

3.6 Eutactic Configurations and Extreme Lattices

We keep the notation of the preceding sections. The aim of this section is to obtain methods which will allow us to prove that some configurations are eutactic without finding explicitly the eutaxy coefficients. We shall also give a theoretical characterization of eutaxy. For the sake of simplicity, we do not consider $\mathcal{T}$-eutaxy for $\mathcal{T} \neq \mathrm{End}^s(E)$.

Proposition 3.6.1. *A union of weakly eutactic (resp. semi-eutactic, resp. eutactic) configurations is weakly eutactic (resp. semi-eutactic, resp. eutactic); a union of* disjoint *strongly eutactic configurations is strongly eutactic.*

Proof. We consider line configurations. An induction argument shows that it suffices to prove the proposition for a union $S = S_1 \cup S_2$ of two configurations. By weak eutaxy, there exist two eutaxy relations $\mathrm{Id} = \sum_{x \in S_1} \rho_D \varphi_D$ and $\mathrm{Id} = \sum_{D \in S_2} \pi_D \varphi_D$.

Let $\alpha, \beta > 0$ with $\alpha + \beta = 1$. Then the eutaxy relation

$$\mathrm{Id} = \sum_{D \in S_1} \alpha \rho_D + \sum_{\Delta \in S_2} \beta \rho_\Delta$$

holds for S. This proves the proposition for (weak, semi-) eutaxy. In the case where strong eutaxy holds for S_1 and S_2, the eutaxy relations for S_1 and S_2 take the forms $\mathrm{Id} = \rho \sum_{D \in S_1} \varphi_D$ and $\mathrm{Id} = \pi \sum_{D \in S_2} \varphi_D$ and yield a eutaxy relation for S with the two coefficients $\alpha\rho$ and $\beta\pi$, which are equal when choosing $\alpha = \frac{\pi}{\rho+\pi}$ and $\beta = \frac{\rho}{\rho+\pi}$. [Variant for strong eutaxy: since the partial sums $\sum p_x$ are proportional to the identity, so is the global sum.] □

Recall (Corollary 3.2.6) that the vectors of a weakly eutactic configuration span E.

Theorem 3.6.2. *A configuration which contains a perfect and eutactic configuration is itself perfect and eutactic.* [However, without the perfection hypothesis, only weak eutaxy is invariant by extension; see Exercise 3.6.1.]

Proof. Let S be a perfect and eutactic configuration, and let S' be a configuration which contains S. By induction on $|S'| - |S|$, we restrict ourselves to the case where S' is of the form $S \cup \{D'\}$. By Proposition 3.5.1, S' is perfect; let us now prove that it is also eutactic. By hypothesis, there exists a relation $\mathrm{Id} = \sum_{D \in S} \rho_D \, p_D$ with strictly positive coefficients ρ_D. Since S is perfect, we can express $p_{D'}$ as a linear combination of the p_x, $x \in S$: there exists a relation $p_{D'} = \sum_{D \in S} \lambda_D \, p_D$ with real coefficients λ_D. This yields for S' the new eutaxy relations

$$\mathrm{Id} = \mu p_{D'} + \sum_{D \in S} (\rho_D - \mu \lambda_D) p_D$$

with arbitrary $\mu \in \mathbb{R}$. For sufficiently small strictly positive $\mu > 0$, this is a eutaxy relation with strictly positive coefficients. □

We now easily deduce from Theorem 3.6.2 an extremality criterion for lattices:

Theorem 3.6.3. *Let Λ be a lattice. The following conditions are equivalent:*

1. *Λ is extreme.*
2. *The lattice Λ_S generated by the minimal vectors in Λ is extreme.*
3. *Λ contains an extreme lattice Λ' with the same norm.*

Proof. (1) $\Rightarrow$ (2). Let Λ_S be the subgroup of Λ generated by $S(\Lambda)$. Theorem 3.5.2 shows that Λ_S is a sublattice of Λ. Since $S(\Lambda_S) = S(\Lambda)$, Λ_S is both perfect and eutactic, hence extreme.

(2) $\Rightarrow$ (3). Take $\Lambda' = \Lambda_S$.

(3) $\Rightarrow$ (1). Since Λ' and Λ have the same norm, we have the inclusion $S(\Lambda') \subset S(\Lambda)$. Since Λ' is extreme, the configuration $S(\Lambda')$ is both perfect and eutactic. By Theorem 3.6.2, the same properties hold for $S(\Lambda)$, and Λ is thus extreme, since it is both perfect and eutactic. □

We now look at connections which exist between the eutaxy property and automorphisms. We denote by S a configuration of vectors in E. As usual, the definitions and results which follow can easily be translated in terms of line configurations.

Definition 3.6.4. An automorphism of S *is an orthogonal transformation which preserves S.*

When S generates E, $\mathrm{Aut}(S)$ is a finite group. The orthogonal group $\mathrm{O}(E)$ acts on $\mathrm{End}^s(E)$ by the rule $s \cdot u = s \circ u \circ s^{-1}$.

For all nonzero $x \in E$ and for all $s \in \mathrm{O}(E)$, we have $s \circ p_x \circ s^{-1} = p_{sx}$, because of the following equalities, valid for all $y \in E$:

$$(s \circ p_x \circ s^{-1})(y) = \frac{x \cdot s^{-1}(y)}{x \cdot x}\, sx = \frac{sx \cdot y}{x \cdot x}\, sx = p_{sx}(y)\,.$$

The linear forms φ_x behave in a similar way under the action of the orthogonal group: we indeed have

$$\varphi_x(s \cdot u) = (s^{-1} \circ u \circ s)(x) \cdot x = u(sx) \cdot sx = \varphi_{sx}(u)$$

for all $x \in E$, $u \in \mathrm{End}^s(E)$, and $s \in \mathrm{O}(E)$.

Given a finite subgroup G of $\mathrm{O}(E)$ and an orbit ω of G on S, we define $p_\omega \in \mathrm{End}^s(E)$ by $p_\omega = \frac{1}{|G|} \sum_{x \in \omega} p_x$ and $\varphi_\omega \in (\mathrm{End}^s(E))^*$ by $\varphi_\omega = \sum_{x \in \omega} \varphi_x$.

Proposition 3.6.5. *Let $\mathcal{O}$ be the set of orbits of G acting on S. The following conditions are equivalent:*

1. *The configuration S is weakly eutactic (resp. eutactic).*
2. *There exists a relation $\mathrm{Tr} = \sum_{\omega \in \mathcal{O}} \rho_\omega \varphi_\omega$ with real (resp. with strictly positive) coefficients ρ_ω.*

Proof. Suppose that S is weakly eutactic, and let

$$\mathrm{Tr} = \sum_{x \in S} \lambda_x \, \varphi_x$$

be a eutaxy relation with real coefficients λ_x. For all $u \in \mathrm{End}^s(E)$ and all $s \in G$, we have

$$\mathrm{Tr}(u) = \mathrm{Tr}(s^{-1} \circ u \circ s) = \sum_{x \in S} \lambda_x \, \varphi_{sx}(u) \Longleftrightarrow \mathrm{Tr} = \sum_{x \in S} \lambda_x \, \varphi_{sx} \,,$$

whence by summation on the elements of G:

$$|G| \; \mathrm{Tr} = \sum_{x \in S} \lambda_x \sum_{s \in G} \varphi_{sx} \,.$$

Taking $\rho_\omega = \sum_{x \in \omega} \lambda_x$, we obtain a relation of the form we want, and the coefficients ρ_ω are plainly strictly positive when the λ_x are.

The converse statements are obvious. □

To test whether a given lattice Λ is eutactic, it is thus interesting to find a subgroup of $\mathrm{Aut}(\Lambda)$ which is as large as possible. The following two theorems list examples for which we can directly prove the eutaxy property, for a lattice or for its dual. (The complete statement of the second theorem anticipates Section 3.8.)

Theorem 3.6.6. *Let G be a subgroup of* $\mathrm{Aut}(S)$. *Suppose that one of the following conditions holds:*

1. *The trace form belongs to the span in* $\mathrm{End}^s(E)$ *of the φ_x, $x \in S$, and G acts transitively on S.*
2. *G acts irreducibly on E.*

Then S is a strongly eutactic configuration.

[One says that G acts *irreducibly* or is *irreducible* on E if the only invariant subspaces of E are E itself and $\{0\}$; this means that the representation of G afforded by the embedding $\mathrm{O}(E) \subset \mathrm{GL}(E)$ is irreducible *over* $\mathbb{R}$.]

Theorem 3.6.7. *Let Λ be a perfect lattice. Suppose that one of the following conditions holds:*

1. *G acts transitively on $S(\Lambda)$.*
2. *G acts irreducibly on E.*

Then Λ is extreme, and both Λ and Λ^* are strongly eutactic (and dual-extreme in the sense of Definition 3.8.1 below).

Proof of 3.6.6 under hypothesis (1). The existence of a eutaxy relation with a single eutaxy coefficient is an immediate consequence of Proposition 3.6.5 (2). (This unique coefficient is equal to $\frac{s}{n}$, see Formula (2) in Theorem 3.2.5.)

The proof under hypothesis (2) will rely on the following lemma:

Lemma 3.6.8. *If G is irreducible, the set of symmetric endomorphisms of E which commute with G is the 1-dimensional subspace of E generated by the identity.*

Proof of 3.6.6 under hypothesis (2). Let us take provisionally for granted the lemma above, and consider the orbit ω of some element $x \in S$. The endomorphism $p_\omega = \sum_{s\in G} p_{sx}$ commutes with G, and is thus of the form $\lambda\,\mathrm{Id}$ for an appropriate $\lambda \in \mathbb{R}$. This shows that the orbit ω is a strongly eutactic configuration, and so is S by Proposition 3.6.1, since it is a disjoint union of orbits. □

Proof of 3.6.8. Let $u \in \mathrm{End}^s(E)$ which commutes with G. Let x be an eigenvector for u, with corresponding eigenvalue λ (which exists in $\mathbb{R}$ since u is symmetric). For all $s \in G$, we have

$$u(sx) = u \circ s(x) = s \circ u(x) = s(\lambda\, x) = \lambda\,(sx)\,.$$

Since G is irreducible, the sx, $s \in G$ span E. Hence u is the homothetic transformation $x \mapsto \lambda\, x$. □

Proof of 3.6.7. By Theorem 3.4.6 (and Corollary 3.8.6 below), it suffices to prove that both Λ and Λ^* are strongly eutactic. Under assumption (2), this is a direct consequence of Theorem 3.6.6, which also applies to Λ^*, since $\mathrm{Aut}(\Lambda) = \mathrm{Aut}(\Lambda^*)$. Moreover, since Λ is perfect, assumption (1) in Theorem 3.6.6 is satisfied, which shows that Λ is strongly eutactic.

To prove that Λ^* is strongly eutactic under assumption (1), consider the sum $u \in \mathrm{End}^s(E)$ of all orthogonal projections onto the directions of minimal vectors in Λ^*. By the perfection of Λ, u is a linear combination of orthogonal projections onto the directions of minimal vectors in Λ. Taking the average under G of both sides of this equality, we obtain a relation of the form

$$\sum_{x\in S(\Lambda)/\{\pm 1\}} p_x = \lambda \sum_{y\in S(\Lambda^*)/\{\pm 1\}} p_y\,,$$

$\lambda \in \mathbb{R}$. Since Λ is strongly eutactic, the left–hand side is proportional to the identity. The same result thus holds for the right–hand side. □

Remark 3.6.9. *One can generalize Theorem 3.6.6 to $\mathcal{T}$-eutactic configurations such that $\mathcal{T}$ is invariant under G.*

Following Coxeter and Hadwiger, we shall now prove a geometrical characterization of the eutaxy condition. We consider line configurations with s elements, with which we associate one of the vector configurations, denoted by S, obtained by choosing arbitrarily a unit vector on each line.

Theorem 3.6.10. *1. If S is weakly eutactic, then for all subspaces F of E, the lines in S which do not lie inside F span a subspace of E which contains the orthogonal of F.*

2. Any weakly eutactic configuration contains at least n lines which span E.

3. Weakly eutactic configurations which contain exactly n lines are the configurations of n orthogonal lines, and they are strongly eutactic.

Proof. (1) A eutaxy relation for S can be written in the form

$$\mathrm{Id} = \sum_{x \in S \cap F} \lambda_x\, p_x + \sum_{y \in S \smallsetminus F} \lambda_y\, p_y\,.$$

Applying both sides to a vector $e \in F^\perp$, we obtain $e = \sum_{y \in S \smallsetminus F} \lambda_y p_y(e)$.

(2) This result, previously proved as Corollary 3.2.6, follows immediately from (1).

(3) By induction on n, we see that among the configurations of n lines, only those which constitute an orthogonal system may be weakly eutactic. That these configurations are actually strongly eutactic results from the identity $\mathrm{Id} = \sum_{i=1}^n p_{\varepsilon_i}$ which holds for any orthogonal basis $(\varepsilon_1, \varepsilon_2, \dots, \varepsilon_n)$ for E. □

Proposition 3.6.11. *Any orthogonal projection of a (weakly, semi-) eutactic configuration is a (weakly, semi-) eutactic configuration.*

[For this statement to make sense, one must of course remove the lines or vectors which project to $\{0\}$.]

Proof. Let S be a (weakly) eutactic configuration of unit vectors in E and let F be a subspace of E. Since for $E \supset F \supset G$, the orthogonal projection onto G can be obtained by performing first the orthogonal projection onto F and then the orthogonal projection *in* F onto G, we may assume by induction that F is a hyperplane of E.

Let e be a unit vector in $F^\perp$ and let S' be the projection of S onto F (after having removed the vector 0 when e or $-e$ belongs to S). Any vector $x \in S$ can be written in a unique way as a sum $x = x' + \lambda\, e$ with $x' \in S' \cup \{0\}$ and $\lambda \in \mathbb{R}$. There exist coefficients $\rho_x \in \mathbb{R}$ such that

$$\forall\, y \in E,\ y = \sum_{x \in S} \rho_x\, (x \cdot y)\, x\,.$$

For $y \in F$, this also reads

$$y = \sum_{x \in S} \rho_x\, (x' \cdot y)\, x' + \sum_{x \in S} \rho_x \lambda_x\, (x' \cdot y)\, e\,.$$

Since the left–hand side and the first sum in the equality above belong to F, the second sum is zero, and we obtain (after removing the vector 0 if need be) a eutaxy relation for S', whose coefficients $\rho'_x = \rho_x\,(x' \cdot y)$ are (strictly) positive whenever the ρ_x are. □

Theorem 3.6.12. (Hadwiger.) *For a configuration S of s lines in E to be eutactic, it is necessary and sufficient that it should be the orthogonal projection on E of an orthogonal configuration of s lines in an s-dimensional Euclidean space F containing E.*

Proof. We know by Theorem 3.6.10 and Proposition 3.6.11 that we have $s \geq n$ and that our claim is true if $s = n$. Suppose now $s > n$. By induction on $s - n$, we are reduced to show that S is the projection onto E of some eutactic configuration lying in an $(n+1)$-dimensional space F containing E.

We thus embed E into Euclidean space F of dimension $n+1$ which we decompose as an orthogonal sum $F = E \perp \mathbb{R}e$ for some unit vector $e \in F$. We then lift all vectors $x \in S$ to vectors $x' = x + \lambda_x\, e \in F$, where the λ_x are real numbers to be chosen later. We obtain in this way a set S' of s vectors in F.

Let ρ_x, $x \in S$ be the eutaxy coefficients for S. We shall now show that for a suitable choice of the λ_x, we obtain a decomposition the identity of F of the form $\mathrm{Id} = \sum_{x \in S} \rho_x N(x')\, p_{x'}$. The existence of such an equality is equivalent to the condition

$$\forall y \in F\,,\ y = \sum_{x \in S} \rho_x (x' \cdot y)\, x + \sum_{x \in S} \rho_x \lambda_x (x' \cdot y)\, e \qquad (*)$$

which we are going to check for $y \in E$ and for $y = e$.

If $y \in E$, we have $x' \cdot y = x \cdot y$ and thus

$$\sum_{x \in S} \rho_x (x' \cdot y)\, x = \sum_{x \in S} \rho_x (x \cdot y)\, x = y\,,$$

so that condition $(*)$ is equivalent to the equality $\sum_{x \in S} \rho_x \lambda_x\,(x \cdot y) = 0$. Since $s > n$, there exists between the vectors in S a non-trivial relation $\sum_{x \in S} \alpha_x\, x = 0$. We then satisfy condition $(*)$ in this case by taking $\lambda_x = \lambda\,\frac{\alpha_x}{\rho_x}$ where $\lambda \in \mathbb{R}$ can be arbitrarily chosen.

If $y = e$, we have

$$\sum_{x \in S} \rho_x\,(x' \cdot y)\, x = \sum_{x \in S} \rho_x \lambda_x\, x = \lambda \sum_{x \in S} \alpha_x\, x = 0\,,$$

and condition $(*)$ reduces to

$$\lambda = \left(\sum_{x \in S} \frac{\alpha_x^2}{\rho_x} \right)^{-1/2},$$

a condition which it is possible to ensure since the various terms in the sum above are positive or zero, but not all zero. □

The eutaxy condition for a configuration S of vectors in E can be expressed in terms of *eutactic stars*: these are vector configurations T for which the identity $N(y) = \sum_{x' \in T}(x' \cdot y)^2$ holds on E. Starting from a eutactic configuration S and making use of condition (3) in Theorem 3.2.5, we transform S into a eutactic star S' by multiplying each vector in S by its eutaxy coefficient.

Among the various sufficient conditions for eutaxy to hold, we quote the following one, which is particularly useful to handle dual-extreme lattices, a notion to be defined in Section 3.8.

Theorem 3.6.13. *Suppose that E is an orthogonal direct sum of subspaces $E_1, \dots, E_r$. For all i, let S_i be a configuration (say, of vectors) in E_i and let Λ_i be a lattice in E_i. Let S be the union $S = \cup_i S_i$ and let Λ be the (orthogonal) sum $\Lambda = \Lambda_1 + \cdots + \Lambda_r$. Then:*

1. *S is (weakly, semi-) eutactic if and only if all S_i are (weakly, semi-) eutactic.*
2. *S is strongly eutactic if and only if all S_i are strongly eutactic and the ratio $\dfrac{|S_i|}{\dim E_i}$ is independent of i.*
3. *Λ is (weakly, semi-) eutactic if and only if Λ_i is (weakly, semi-) eutactic for all i and all Λ_i have the same norm.*
4. *Λ is strongly eutactic if and only if Λ_i is strongly eutactic for all i, all Λ_i have the same norm, and the ratio $\dfrac{s(\Lambda_i)}{\dim \Lambda_i}$ is independent of i.*

Proof. (1) For any i and any nonzero $x \in E_i$, we have $p_{E_i} \circ p_x = p_x$. Thus S is eutactic in E, S_i is eutactic in E_i for all i. Conversely, a eutaxy relation for S_i can be written in the form $p_{E_i} = \sum_{x \in S_i} \rho_x \, p_x$, and the equality $\mathrm{Id} = \sum_i p_{E_i}$ transforms the set of eutaxy relations for the S_i into a eutaxy relation for S, with appropriate signs of the coefficients.

(2) From a strong eutaxy relation $\mathrm{Id} = \rho \sum_{x \in S} p_x$, we deduce the equalities $p_{E_i} = \rho \sum_{x \in S_i} p_x$ which are strong eutaxy relations for the S_i in which moreover $\dfrac{|S_i|}{\dim E_i} = \rho$ is constant. Conversely, the eutaxy relation obtained using the identity $\mathrm{Id} = \sum_i p_{E_i}$ is a relation of strong eutaxy whenever the coefficients of strong eutaxy for the S_i do not depend on i.

(3) and (4) We have $N(\Lambda) = \min_i N(\Lambda_i)$. Hence $S(\Lambda)$ does not span E if the lattices Λ_i do not have the same norm. Conversely, if all Λ_i have the same norm, then $S(\Lambda) = \cup_i S(\Lambda_i)$, and (3) and (4) are direct consequences of (1) and (2) respectively. □

3.7 The Lamination Process

Let $n_0 \geq 0$ be an integer, let Λ_0 be a lattice in a Euclidean space E_0 of dimension n_0, and let $N_0 = N(\Lambda_0)$. Embed E_0 as a hyperplane in a Euclidean space E_1 of dimension $n_0 + 1$. We are interested in lattices $\Lambda_1 \subset E_1$ with norm N_0 and such that $\Lambda_1 \cap E_0 = \Lambda_0$.

Definition 3.7.1. *We say that $\Lambda_1 \subset E_1$ is* extreme relative to Λ_0 *(resp.* critical relative to Λ_0*) if its Hermite constant is a local (resp. an absolute) maximum on the set of lattices in E_1 with norm N_0 whose intersection with E_0 is Λ_0.*

Proposition 3.7.2. *Let Λ_0 be a lattice in E_0 and let $\Lambda_1 \subset E_1$ be extreme relative to Λ_0. Then the set of minimal vectors in Λ_1 lying outside E_0 spans E_1.*

Proof. Otherwise, the minimal vectors in E_1 lie in some hyperplane $H \neq E_0$ in E_1. Let $\alpha \in (0, \frac{\pi}{2})$ be the angle of E_0 and H (non-oriented angle of two unit vectors e_0 and e orthogonal to E_0 and to H respectively). For $0 < \theta < \frac{\pi}{2}$, let e_θ be the vector of the plane $\langle e_0, e\rangle$ whose angle with e_0 (resp. e) is $\alpha - \theta$ (resp. θ), and let H_θ be the hyperplane of E_1 which is orthogonal to e_θ. Consider the rabatment u_θ which maps H onto H_θ (u_θ is the identity on E_0 and transforms e into $(\sin\theta)\, e + (\cos\theta)\, e_0$).

For all $x \in H$, we have $N(u_\theta(x)) = N(x)$. Since the vectors in $S(\Lambda_1)$ lie in $E_0 \cup H$, the image under u_θ of $S(\Lambda_1)$ is made of vectors of norm N_0. Since u_θ is continuous and $u_0 = \mathrm{Id}$, the images under u_θ of vectors in Λ_1 which do not lie in $E_0 \cup H$ all have for sufficiently small θ a norm $N > N_0$. The equalities

$$\det(u_\theta(\Lambda_1)) = \det(u_\theta)^2 \cdot \det(\Lambda_1) = (\cos^2\theta)\,\det(\Lambda_1)$$

show that the determinant of $u_\theta(L_1)$ is strictly smaller than that of Λ_1 for $\theta > 0$, and this contradicts the hypothesis that the Hermite invariant of Λ_1 is a local maximum. □

Applying Proposition 3.5.3, we obtain:

Corollary 3.7.3. *A lattice which is extreme relative to a perfect lattice is perfect.* □

Remark 3.7.4. *We shall prove in Chapter 12 (in any codimension) a characterization* à la Voronoi *of relatively extreme lattices as relatively perfect and relatively eutactic lattices, and also show that a lattice which is perfect relative to a perfect lattice is perfect. In this section, we have restricted ourselves to the case of codimension* 1 *(for which the statement above reduces to Proposition 3.5.3).*
However, there exist non-extreme lattices which are extreme relative to an

extreme lattice: it may actually happen that a deformation of Λ_1 does not preserve the isometry class of Λ_0, and examples in codimension 1 have been given by Barnes ([Bar5], I, p. 64); an explicit example is provided by the lattices $L_{2r}^r \subset L_{2r+1}^r$ of Definition 8.4.4.

Definition 3.7.5. *Let Λ_0 be a lattice of dimension n_0, and let $n \geq n_0$ be an integer.*

1. *We say that an n-dimensional lattice Λ is a* weakly laminated lattice above Λ_0 *if there exists a sequence $L_0 = \Lambda_0, L_1, \ldots, L_{n-n_0} = \Lambda$ of lattices of dimensions $n_0, n_0+1, \ldots, n$, each term of which is critical relative to the preceding one.*
2. *We say that Λ is* (strongly) laminated above L_0 *if its Hermite invariant is maximal among all n-dimensional weakly laminated lattices.*

Laminated lattices above a given lattice Λ_0 have the same norm as Λ_0, and those which are strongly laminated moreover share the same value for their determinant. By Corollary 3.7.3, laminated lattices (strongly or weakly) above a perfect lattice are perfect. We say *laminated lattices* without any other precision for the particular case considered by Conway and Sloane:

Definition 3.7.6. *An n-*dimensional laminated lattice *is an n-dimensional strongly laminated lattice above the trivial lattice $\{0\}$ to which is given the norm 4. We denote by Λ_n any n-dimensional laminated lattice, with possibly a superscript when uniqueness does not hold in dimension n.*

For $n \geq 1$, they are laminated above $2\mathbb{Z}$, but to consider dimension 0 has some importance to take advantage of various symmetries. Conway and Sloane ([C-S], Chapter 6) have determined the determinants of the laminated lattices up to dimension 48 and found all laminated lattices up to dimension 25. These determinants can be calculated by simple symmetry rules once they are known in dimensions 1 to 4. For the proofs, see [C-S], Chapter 6.

Theorem 3.7.7. (Conway and Sloane.) *The values λ_n of the determinants of the laminated lattices Λ_n in dimensions $n \leq 48$ satisfy the following rules, which determine them:*

1. $\lambda_0 = 1$, $\lambda_1 = 4$, $\lambda_2 = 12$, $\lambda_3 = 32$, $\lambda_4 = 64$.
2. $\lambda_n = 2^{2n-8}\lambda_{8-n}$ *for* $0 \leq n \leq 8$.
3. $\lambda_n = 2^{16-n}\lambda_{n-8}$ *for* $8 \leq n \leq 16$.
4. $\lambda_n = \lambda_{24-n}$ *for* $0 \leq n \leq 24$.
5. $\lambda_n = 2^{24-n}\lambda_{n-24} = 2^{24-n}\lambda_{48-n}$ *for* $24 \leq n \leq 48$. □

Laminated lattices are integral up to dimension 24, and are unique except in dimensions 11, 12 and 13 where there exist respectively 2, 3 and 3 isometry classes, which have different kissing numbers and which we distinguish by superscripts min, mid, max, according to the following table:

Table 3.7.7′. Laminated lattices in dimensions 11-13

$n = 11$	$s(\Lambda_{11}^{\mathrm{min}}) = 216$		$s(\Lambda_{11}^{\mathrm{max}}) = 219$
$n = 12$	$s(\Lambda_{12}^{\mathrm{min}}) = 312$	$s(\Lambda_{12}^{\mathrm{mid}}) = 316$	$s(\Lambda_{12}^{\mathrm{max}}) = 324$
$n = 13$	$s(\Lambda_{13}^{\mathrm{min}}) = 444$	$s(\Lambda_{13}^{\mathrm{mid}}) = 445$	$s(\Lambda_{13}^{\mathrm{max}}) = 453$

[We shall construct in Chapter 8, Sections 8.6 to 8.8, all laminated lattices up to dimension 24 as well as a "principal series" between dimensions 25 and 48, whose members become integral when rescaled to norm 8 (the smallest possible norm for $25 \le n \le 48$, see Exercise 3.7.2).]

It should be noticed that among strongly laminated lattices above a given lattice, there may exist dead-ends. This is the case for $\Lambda_{13}^{\mathrm{mid}}$: laminated lattices above $\Lambda_{13}^{\mathrm{mid}}$ have a determinant larger than that of Λ_{14}. This is the only example among the usual laminated lattices of dimension up to 24.

Modifications of the lamination process, involving algebraic structures (given automorphism groups, module structures over various orders of number fields or quaternion skew-fields) have been considered. Note, however, that lattices which can be constructed in this way are not *a priori* perfect, at least for the notion of perfection defined in this chapter.

Besides the previous laminations that they name "geometrical laminations", Plesken and Pohst ([Pl-P1], [Pl-P2]), have defined "arithmetical laminations", in both a weak and a strong sense. One starts with an integral lattice Λ_0 of dimension n_0 and norm m_0, and then consider integral lattices of norm m_0 which contain Λ_0 as a codimension 1 section and which, moreover, has a basis obtained by adjoining a new minimal vector to a basis for Λ_0. These are the *weak arithmetical laminations over* Λ_0, and one obtains the *strong arithmetical laminations over* Λ_0 by keeping only those lattices with the largest possible Hermite invariant. (As previously, a discrepancy may appear from dimension $n_0 + 2$ onwards.) There is no reason for perfection to be preserved. It seems likely that the weak laminated lattices in the geometrical sense above $\Lambda_{13}^{\mathrm{mid}}$ lie among the arithmetical ones, which in particular allows the embedding of $\Lambda_{13}^{\mathrm{mid}}$ into Λ_{17} (the embedding in Λ_{16} is not possible; see Chapter 8).

3.8 Dual-Extreme Lattices

We still consider a Euclidean vector space of dimension n. Recall (Chapter 2, Section 2.8) that the "dual Hermite invariant" of a lattice $\Lambda \subset E$ is

$$\gamma'(\Lambda) = \|\Lambda\| \cdot \|\Lambda^*\| = (N(\Lambda)\, N(\Lambda^*))^{1/2} = (\gamma(\Lambda)\, \gamma(\Lambda^*))^{1/2}.$$

We have $\gamma'(\Lambda) = \gamma'(\Lambda^*) = (\gamma(\Lambda)\, \gamma(\Lambda^*))^{1/2} \le \gamma_n$, which proves the existence of $\gamma'_n = \sup_\Lambda \gamma'(\Lambda)$.

Definition 3.8.1. *We say that Λ is* dual-extreme *if the invariant γ' attains a local maximum on Λ.*

Notice that Λ and Λ^* play symmetric rôles in this definition, so that a lattice is dual-extreme if and only if its dual is.

We now prove a first necessary condition for a lattice to be dual-extreme, which is of a great practical importance, and which constitutes a fundamental lemma for the proof of the characterization of dual-extreme lattices given in Theorem 3.8.4 below.

Proposition 3.8.2. *Each of the sets of minimal vectors of a dual-extreme lattice and of its dual spans E.*

Proof. Let Λ be a lattice which does not satisfy the condition above, and let $\mathcal{V}$ be a neighbourhood of Λ in the space $\mathcal{L}$ of lattices. We shall show that $\mathcal{V}$ contains a lattice whose invariant γ' is strictly larger than that of Λ. Exchanging if need be Λ and Λ^*, we may assume that $S(\Lambda)$ spans a subspace $F \ne E$ of E. We shall use the following obvious generalization of Lemma 3.4.2: there exists in $\mathrm{End}(E)$ a neighbourhood the identity on which all minimal vectors in $u(\Lambda)$ and in $u(\Lambda)^*$ are images of minimal vectors in Λ and Λ^* respectively.

For nonzero $\lambda \in \mathbb{R}$, let $u_\lambda \in \mathrm{GL}(E)$ be the map which is the identity on F and multiplication by λ on $F^\perp$, and let $\Lambda' = u_\lambda(\Lambda)$.

Since $S(\Lambda)$ is contained in F, it is invariant under u_λ. For λ sufficiently close to 1, Λ' belongs to $\mathcal{V}$, and the norms of the vectors in Λ' which do not belong to F remain strictly larger than that of Λ. The equalities $\Lambda'^* = {}^t u_\lambda^{-1}(\Lambda^*) = u_{\lambda^{-1}}(\Lambda^*)$ show that u_λ strictly increases the norms of the elements of (Λ^*) which do not belong to F. If $S(\Lambda^*) \cap F = \emptyset$, we have $N(\Lambda'^*) > N(\Lambda^*)$, and Λ is not dual-extreme.

Suppose now that $S(\Lambda^*) \cap F$ is not empty. The minimal vectors in Λ' and in Λ'^* then all belong to F. Let w be an endomorphism of E which is zero on F and which maps $F^\perp$ into F. Its transpose ${}^t w$ is zero on $F^\perp$ and maps F into $F^\perp$. Let us choose w in such a way that the kernel of ${}^t w$ does not contain any vector in $S(\Lambda'^*)$. For any $\mu \in \mathbb{R}$, let $v_\mu = \mathrm{Id} + \mu\, w$. We have $({}^t w)^2 = w^2 = 0$, hence ${}^t v_\mu^{-1} = \mathrm{Id} - \mu\, {}^t w$. For μ sufficiently close to 0, $v_\mu(\Lambda')$ belongs to $\mathcal{V}$, and the norms of the elements in $v_\mu(\Lambda')$ (resp. in $v_\mu(\Lambda'^*)$) which do not belong to F are strictly greater that $N(\Lambda')$ (resp. than $N(\Lambda'^*)$). Since the restriction of v_μ to F is the identity, we have $S(v_\mu(\Lambda')) = S(\Lambda') = S(\Lambda)$, whence $N(v_\mu(\Lambda')) = N(\Lambda)$. For $x \in F$, ${}^t v_\mu$ transforms x into $x + x_\mu$ where $x_\mu = \mu {}^t w(x)$ is orthogonal to x. We thus have $N(v_\mu(x)) = N(x) + N(x_\mu) > N(x)$ for $x \notin \mathrm{Ker}\, w$, which shows that the norm of the dual of $v_\mu(\Lambda')$ is strictly larger than that of Λ'^*.

To summarize, for the choices above of u_λ and of v_μ, the lattice $v_\mu \circ u_\lambda(\Lambda)$ belongs to $\mathcal{V}$ and has the same norm as Λ, but the norm of its dual is strictly larger than that of Λ^*. Hence the lattice Λ is not dual-extreme. □

To characterize dual-extreme lattices, we need analogues to the notions of perfection and eutaxy. These are provided by the following definition:

Definition 3.8.3. *We say that a pair (S, S') of configurations of lines (or of vectors) in E is* dual-perfect *if $S \cup S'$ is a perfect configuration, i.e. if the projections p_x, $x \in S \cup S'$ span* $\mathrm{End}^s(E)$, *and that it is* dual-eutactic *if there are two families ρ_x, $x \in S$ and ρ'_y, $y \in S'$ of strictly positive real numbers such that $\sum_{x\in S} \rho_x\, p_x = \sum_{y \in S'} \rho'_y\, p_y$. We say that a lattice Λ is* dual-perfect *(resp.* dual-eutactic*) if the pair $(S(\Lambda), S(\Lambda^*))$ is dual-perfect (resp. dual-eutactic).*

Given a basis $\mathcal{B}$ for Λ with Gram matrix A, expressing the projections onto minimal vectors with respect to the pair $(\mathcal{B}, \mathcal{B}^*)$ of bases for E shows that Λ is dual eutactic if and only if there exists a relation

$$\sum_{\pm X} \rho_X\, X^t X = A^{-1} \left(\sum_{\pm Y} \rho'_Y Y {}^t Y \right) A^{-1} \qquad (\mathbf{3.8.3'})$$

with strictly positive coefficients ρ_X, ρ'_Y, where X (resp. Y) runs through the column-matrices of the components in $\mathcal{B}$ (resp. in $\mathcal{B}^*$) of the vectors in $S(\Lambda)$ (resp. in $S(\Lambda^*)$).

Definition 3.8.3 is clearly invariant under the exchange of S and S', and Definitions 3.8.1 and 3.8.3 are invariant under the exchange of Λ and Λ^*.

Definition 3.8.3 possesses the following dual version, which makes use as usual of linear forms φ_x on $\mathrm{End}^s(E)$, that we give in the case of a lattice: Λ is dual-perfect if and only if the forms φ_x, $x \in S(\Lambda) \cup S(\Lambda^*)$ span $\mathrm{End}^s(E)^*$, and dual-eutactic if and only if there exists a relation of the form $\sum_{x \in S(\Lambda)} \rho_x\, \varphi_x = \sum_{y \in S(\Lambda^*)} \rho'_y\, \varphi_y$ with strictly positive coefficients ρ_x, ρ'_y.

The following theorem is an analogue to Korkine and Zolotareff's characterization of extreme lattices (Theorem 3.4.5):

Theorem 3.8.4. *A necessary and sufficient condition for a lattice Λ to be dual-extreme is that the system of inequalities*

$$\varphi_x(v) \ge 0\,, x \in S(\Lambda) \quad \textit{and} \quad \varphi_y(v) \le 0\,, y \in S(\Lambda^*)$$

only have the solution $v = 0$ in $\mathrm{End}^s(E)$. *Moreover, when these conditions are satisfied, there exists a neighbourhood of Λ on which any lattice Λ' such that $\gamma'(\Lambda') \ge \gamma'(\Lambda)$ is similar to Λ (and we thus have $\gamma'(\Lambda') = \gamma'(\Lambda)$).*

Proof. Suppose first that there exists a nonzero $v \in \mathrm{End}^s(E)$ which satisfies the system of inequalities above. Let $u \in \mathrm{End}(E)$ such that ${}^t u u = \mathrm{Id} + \varepsilon\, v$

where $\varepsilon > 0$ has been chosen in such a way that the minimal vectors in $u(\Lambda)$ (resp. in $u(\Lambda)^* = {}^t u^{-1}(\Lambda^*)$) come from $S(\Lambda)$ (resp. from $S(\Lambda^*)$); see Lemma 3.4.2. Consider the formula $N(u(\Lambda)) = N(\Lambda) + \min_{x \in S(\Lambda)} \varphi_x(v)$ (see the proof of Lemma 3.4.3) and its analogue $N(u(\Lambda)^*) = N(\Lambda^*) + \min_{y \in S(\Lambda^*)} \varphi_y(w)$ for Λ^* in which w is defined by $\mathrm{Id} + w = (\mathrm{Id} + v)^{-1}$. Replacing if need be $\mathcal{V}$ by a smaller neighbourhood, we may suppose that the series expansion of $(\mathrm{Id} + v)^{-1}$ converges, and this yields for w the expansion

$$w = -v(\mathrm{Id} - v + v^2 + \cdots + (-1)^{m-1} v^m + \ldots).$$

The fact that v is a solution to the inequalities of Theorem 3.8.4 shows that, provided that v is sufficiently close to 0, both inequalities $N(u(\Lambda)) \geq N(\Lambda)$ and $N(u(\Lambda)^*) \geq N(\Lambda^*)$ hold, whence $\gamma'(u(\Lambda)) \geq \gamma'(\Lambda)$. [One can make use of a convexity lemma similar to Lemma 3.4.4; we shall not give the details here, as a more general result will be proved in Chapter 10.]

If equality holds, we have in particular $N(u(\Lambda)) = N(\Lambda)$ which implies by Lemma 3.4.3 that $S(u(\Lambda)) = \{x \in S(\Lambda) \mid v(x) = 0\}$ is contained in some hyperplane of E. Proposition 3.8.2 then shows that every neighbourhood of $u(\Lambda)$ contains a lattice Λ_1 such that $\gamma'(\Lambda_1) > \gamma'(u(\Lambda))$, and thus that Λ is not dual-extreme.

Conversely, assuming that the system of inequalities $\forall\, x \in S(\Lambda), \varphi_x(v) \geq 0$ and $\forall\, y \in S(\Lambda^*)$, $\varphi_y(v) \leq 0$ only has the solution $v = 0$ in $\mathrm{End}^s(E)$, we show that every lattice Λ' sufficiently close to Λ, such that $\gamma'(\Lambda') \geq \gamma'(\Lambda)$, is similar to Λ.

Let $u \in \mathrm{End}(E)$ such that $\Lambda' = u(\Lambda)$; composing u with a suitable similarity, we may assume that it is symmetric and that $N(u(\Lambda^*)) = N(\Lambda^*)$, and we now prove that $u = \mathrm{Id}$. Otherwise, we could write $u^2 = \mathrm{Id} + v$ for some $v \neq 0$. For u sufficiently close to the identity, we would have $N(u(\Lambda)) \geq N(\Lambda)$, whence $v(x) \cdot x = \varphi_x(v) \geq 0$ for all $x \in S(\Lambda)$ and similarly $\varphi_y(v) \leq 0$ for all $y \in S(\Lambda^*)$, and this would contradict the hypothesis $u \neq \mathrm{Id}$. □

Following [B-M1], we are now able to give for dual-extreme lattices a characterization *à la Voronoi*:

Theorem 3.8.5. (Bergé–Martinet.) *A lattice is dual-extreme if and only if it is both dual-perfect and dual-eutactic.*

Proof. Let $\mathcal{F} = \{\varphi_x, x \in S(\Lambda)\} \cup \{-\varphi_y,\, y \in S(\Lambda^*)\}$. Let us first show that a lattice Λ which is dual-perfect and dual-eutactic is dual-extreme. Let $v \in \mathrm{End}^s(E)$ such that $f(v) \geq 0$ for all $f \in \mathcal{F}$. Since Λ is dual-eutactic, there are coefficients $\rho_f > 0$ such that $\sum_{f \in \mathcal{F}} \rho_f\, f = 0$, from which we deduce $\sum_{f \in \mathcal{F}} \rho_f\, f(v) = 0$. We thus have $f(v) = 0$ for all $f \in \mathcal{F}$. Since Λ is dual-perfect, $\mathcal{F}$ generates the dual space of $\mathrm{End}^s(E)$. Consequently, v is zero, and Λ is dual-extreme by Theorem 3.8.4.

Conversely, let Λ be a dual-extreme lattice. Any element v of $\mathrm{End}^s(E)$ such that $f(v) = 0$ for all $f \in \mathcal{F}$ must be zero by Theorem 3.8.4. Hence

Λ is dual-perfect. To prove the eutaxy condition, we make use of Stiemke's theorem: if $v \in \mathrm{End}^s(E)$ is such that $f(v)$ is ≥ 0 for all $f \in \mathcal{F}$, Theorem 3.8.4 shows that v is zero, thus in particular that $f(v) = 0$ on $\mathcal{F}$, and we conclude by Theorem 3.3.1. □

Corollary 3.8.6. *An extreme lattice whose dual is eutactic is dual-extreme. In particular, an extreme lattice is dual-extreme any time its automorphism group acts transitively on the set of its minimal vectors or irreducibly on E.* [The condition that the dual-lattice should be eutactic can be weakened: semi-eutaxy suffices; see Exercise 3.8.8.]

Proof. Let Λ be such a lattice. Since it is perfect, it is in particular dual-perfect. Moreover, since both Λ and Λ^* are eutactic, there are strictly positive coefficients ρ_x, $x \in S(\Lambda)$ and ρ'_y, $y \in S(\Lambda^*)$ such that

$$\mathrm{Id} = \sum_{x \in S(\Lambda)} \rho_x \, p_x = \sum_{y \in S(\Lambda^*)} \rho'_y \, p_y \,,$$

and eliminating Id in these relations yields a relation of dual-eutaxy for Λ. The last assertions follow from Theorem 3.6.7. □

Corollary 3.8.7. *The inequality $s(\Lambda) + s(\Lambda^*) \geq \frac{n(n+1)}{2} + 1$ holds for any dual-extreme lattice Λ.*

Proof. The existence of a non-trivial linear relation between the p_x, $x \in S(\Lambda) \cup S(\Lambda^*)$ shows the lower bound $s + s^* > \dim \mathrm{End}^s(E)$. □

Some of the results we proved in Sections 3.5 and 3.6 generalize to dual-perfect or dual-eutactic configurations For instance, it is easily checked that dual-perfect lattices are irreducible. Similarly, Proposition 3.6.5. possesses the following counterpart, which is proved in the same way:

Proposition 3.8.8. *A lattice Λ is dual-eutactic if and only if there exists a relation $\sum_{x \in S(\Lambda)} \rho_x \, p_x = \sum_{x' \in S(\Lambda^*)} \rho_{x'} \, p_{x'}$ with coefficients ρ_x (resp. $\rho_{x'}$) which are strictly positive and constant on the orbits of $S(\Lambda)$ (resp. of $S(\Lambda^*)$) under* $\mathrm{Aut}(\Lambda)$. □

The following proposition does not rely on perfection and eutaxy properties, and will be proved directly by means of suitable deformations:

Proposition 3.8.9. *Let Λ be a dual-extreme lattice possessing a hyperplane extreme section of the same norm as Λ which is orthogonal to a minimal vector in Λ^*. Then Λ is extreme.*

Proof. Let $x' \in S(\Lambda^*)$ such that $H = (\mathbb{R}x')^\perp$ is the hyperplane defining the section above of Λ. On a sufficiently small neighbourhood of the identity of GL(E), we can construct a continuous map $u \mapsto \sigma$ in O(E) such that σ maps $u(x')$ into the line $\mathbb{R}x'$ (consider for instance a 2-dimensional rotation). Thus, replacing u by $\sigma \circ u$, we may assume that u preserves H, whence $u(\Lambda) \cap H = u(\Lambda \cap H)$. By Corollary 1.3.5, the determinants of Λ and $\Lambda \cap H$ satisfy the relation

$$\det(\Lambda \cap H) = \det(\Lambda)\, N(\Lambda^*)\,.$$

Since $\gamma(\Lambda)^n = N(\Lambda)^n \det(\Lambda)^{-1}$, $\gamma(\Lambda \cap H)^{n-1} = N(\Lambda)^{n-1} \det(\Lambda \cap H)^{-1}$, and $\gamma'(\Lambda)^2 = N(\Lambda)\, N(\Lambda^*)$, we have the further relation

$$\gamma(\Lambda)^n = \gamma(\Lambda \cap H)^{n-1}\, \gamma'(\Lambda)^2.$$

For any $u \in$ GL(E) sufficiently close to the identity, we have $\gamma'(u(\Lambda)) \leq \gamma'(\Lambda)$ and $\gamma(u(\Lambda \cap H)) \leq \gamma(\Lambda \cap H)$ since Λ is dual-extreme and $\Lambda \cap H$ is extreme, hence also $\gamma(u(\Lambda)) \leq \gamma(\Lambda)$. □

Corollary 3.8.10. *A dual-extreme lattice possessing a critical hyperplane section with the same norm is extreme.*

Proof. The determinant of a critical section is in particular minimal among all hyperplane sections (of any norm) of Λ. Its orthogonal in Λ^* has thus also the smallest possible determinant, i.e. the smallest possible norm, since it is a 1-dimensional lattice. □

It is worth noticing that, in contrast to the case of usual perfection, there are generally infinitely many similarity classes of dual-perfect lattices. Exercise 3.8.3 below shows the existence of a continuous one-parameter family in dimension 2, and further examples exist in larger dimensions (see Chapter 6, Section 6.3 and the corresponding exercises). A more restrictive notion of dual-perfection to be defined later yields a reasonable analogue to Voronoi's Theorem 3.5.4 (see Theorem 9.6.1). We can also prove that the set of similarity classes of dual-eutactic lattices in a given dimension whose minimal vectors span E is finite (see Proposition 10.5.3). This will justify the first assertion in following result, which we only state in this section:

Theorem 3.8.11. (A.-M. Bergé, [Ber1].)

1. *There are only finitely many similarity classes of dual-extreme lattices in a given dimension.*
2. *Dual-extreme lattices are proportional to algebraic lattices; in particular, the values of the invariant γ'_n on dual-extreme lattices are algebraic.* □

These remarks suggest the following questions, reproduced from the French edition:

Questions 3.8.12. *1. Do there exist dual-extreme lattices with irrational invariant γ'^2 ?*

2. *For what dimensions do there exist dual-extreme lattices which do not satisfy Corollary 3.8.6?*
3. *Can one improve for $n > 1$ on the inequality $s(\Lambda) + s(\Lambda^*) \geq \frac{n(n+1)}{2} + 1$ of Corollary 3.8.7?*

Since the publication of the French edition of this book, partial answers were found.

A 5-dimensional example – the smallest possible dimension – of an irrational dual-extreme lattice has been produced by A.-M. Bergé ([Ber4]). Her example shares with those of Martinet ([Mar2], which concern even dimensions $n \geq 8$) the following properties: none of the lattices Λ, Λ^* is perfect and at most one is eutactic, so that they cannot be proved to be eutactic using Corollary 3.8.6, and they satisfy the relation $s + s^* = \frac{n(n+1)}{2} + n$, giving the sum $s + s^*$ a smaller value that the smallest one previously known for $n \geq 2$, namely $\frac{n(n+1)}{2} + n + 1$, attained on $(\mathbb{A}_n, \mathbb{A}_n^*)$).

Examples of dual-extreme lattices for which $s + s^* = \frac{n(n+1)}{2} + 1$ have recently been obtained in [B-M6] for all $n \geq 8$ even, as cross-sections of some Coxeter's lattices $\mathbb{A}_n^r$ (defined in Section 5.2).

These partial results, however, are far from giving a complete answer to Questions 3.8.12 from dimension 5 onwards.

3.9 Exercises for Chapter 3

3.1.1. Let $u \in \mathrm{End}^s(E)^{++}$. Show that u possesses 2^n square roots if its eigenvalues are all different, and infinitely many otherwise.

3.1.2. Show that any endomorphism possesses left and right decompositions into positive symmetric and orthogonal components, but that only the symmetric ones are unique if u is not invertible.

3.1.3. 1. Show that $u \mapsto \mathrm{Tr}({}^t u u)$ is a positive definite quadratic form on $\mathrm{End}(E)$, with polar form $(u, v) \mapsto \frac{1}{2}\,\mathrm{Tr}({}^t u v + {}^t v u)$.

2. Show that $\mathrm{End}(E)$ is the orthogonal sum of $\mathrm{End}^s(E)$ and $\mathrm{End}^a(E)$ for each of the quadratic forms $u \mapsto \mathrm{Tr}(u^2)$ and $u \mapsto \mathrm{Tr}({}^t u u)$.

3. Show that $\mathrm{Tr}(u^2)$ has signature $(\frac{n(n+1)}{2}, \frac{n(n-1)}{2})$ on $\mathrm{End}(E)$.

3.1.4. Let $f = \sum_{p \geq 0} a_p T^p$ an entire complex series, let R be its convergence radius, let V be a complex vector space of finite dimension n and let $u \in \mathrm{End}(V)$. Show that the series $\sum a_p u^p$ converges if the eigenvalues of u have a modulus $< R$, and diverges if at least one modulus is strictly larger than 1. [Reduce first to the case of triangular matrices, then use the decomposition of such a matrix as a sum of a diagonal matrix D and an upper triangular matrix T such that $DT = TD$ and $T^n = 0$.]

3.1.5. For any $u \in \mathrm{End}(E)$, let $\exp(u) = \sum_{p=0}^{+\infty} \frac{u^p}{p!}$ be exponential map.

1. Verify that the series above converges everywhere in $\mathbb{C}$.
2. Show that the exponential map is a diffeomorphism of $\mathrm{End}^s(E)$ onto $\mathrm{End}^{s++}(E)$.
3. Define u^m on $\mathrm{End}^{s++}(E)$ for every real m.

3.1.6. Let $e, f \neq 0$ be two vectors in E. Prove the formula $\operatorname{Tr} p_f \circ p_e = \frac{(e \cdot f)^2}{(e \cdot e)(f \cdot f)}$.

3.2.1. Let Λ be a weakly eutactic lattice. Show that there exists a eutaxy relation $\operatorname{Id} = \sum_{x \in S(\Lambda)/\pm} \rho_x p_x$ with rational coefficients ρ_x if and only if Λ is rational, i.e. proportional to an integral lattice.
[To prove the "if" part, rescale Λ to a rational norm and use Theorem 3.2.5 (5); to prove the "only if" part, extract from S a basis for the span of the p_x, $x \in S$.]

3.2.2. (Venkov's index formula.) Preliminary questions:

(a) Show that the ranks of a finite system of vectors in E and of the corresponding Gram matrix are equal.

(b) Show that for an $s \times s$ (real) matrix with eigenvalues $\lambda_1, \dots, \lambda_s$, the eigenvalues of $\bigwedge^k M$ are the products $\lambda_{i_1} \dots \lambda_{i_k}$ with $i_1 < \dots < i_k$.

1. Let Λ be a well-rounded lattice with set of minimal vectors $S = \{\pm x_1, \dots, \pm x_s\}$, let M be the Gram matrix of $S' = (x_1, \dots, x_s)$ and let $\lambda_1 \geq \dots \geq \lambda_n > \lambda_{n+1} = \dots = \lambda_s = 0$ be its eigenvalues. Show that $\operatorname{Tr}(\bigwedge^n M) = \lambda_1 \dots \lambda_n$.

2. Show that $\operatorname{Tr}(\bigwedge^n M) = \det(\Lambda) \sum_X [\Lambda : \Lambda_X]^2$ where X runs through the subsets of n independent vectors in S' and Λ_X stands for the sublattice of Λ with basis X.

From now on, we suppose that Λ is strongly eutactic.

3. Show that $M^2 = \frac{s}{n} N(\Lambda)\, M$; show that the nonzero eigenvalues of M are equal to $\frac{s}{n} N(\Lambda)$.

4. Prove that any strongly eutactic lattice satisfies the *Venkov index formula*

$$\sum_X [\Lambda : \Lambda_X]^2 = \frac{s^n}{n^n} \gamma(\Lambda)^n.$$

3.3.1. Let φ_i, $0 \leq i \leq k$ be linear forms on a real vector space V. Consider the following three statements:
(1) $\varphi_i(x) \geq 0,\ 0 \leq i \leq k \Rightarrow x = 0$.
(2) $\varphi_i(x) \geq 0,\ 0 \leq i \leq k \Rightarrow \varphi_i(x) = 0,\ 0 \leq i \leq k$.
(3) $\varphi_i(x) = 0,\ 1 \leq i \leq k \Rightarrow x = 0$.
Prove the equivalence $(1) \Longleftrightarrow (2) + (3)$.
[To prove $(1) \Rightarrow (3)$, use the transformation $x \mapsto -x$.]

3.5.1. Prove directly that the set of minimal vectors of an extreme lattice is not contained in the union of two hyperplanes.
[Use a rabatment of one of the hyperplanes onto the other one.]

3.5.2. Show that three distinct lines in a plane constitute a perfect configuration.

3.5.3. For any subspace F of E, let p_F be the orthogonal projection onto F and let $\mathcal{P}_F$ be the image in $\operatorname{End}^s(E)$ of $u \mapsto p_F \circ u$.

1. Let F and F' be two subspaces of E and let $F'' = F \cap F'$. Prove the equality $\mathcal{P}_{F''} = \mathcal{P}_F \cap \mathcal{P}_{F'}$.

2. Let m, m', m'' be the respective dimensions of F, F', F''. Show that the dimension of $\mathcal{P}_F + \mathcal{P}_{F'}$ is $M = \frac{m(m+1)}{2} + \frac{m'(m'+1)}{2} - \frac{m''(m''+1)}{2}$.

3. Take for F and F' two hyperplanes of E. Show that if they are distinct (resp. if $F' = F$), $\mathcal{P}_F + \mathcal{P}_{F'}$ has codimension 1 (resp. n) in $\operatorname{End}^s(E)$.

4. Deduce from 3 a direct proof of statements (2) and (3) in Theorem 3.5.2.

5. Let Λ be a lattice possessing two perfect hyperplane sections with the same norm as Λ. Show that Λ is perfect if and only if it possesses a minimal vector which belongs to none of the two hyperplanes. [Other method: use Proposition 5.3.]

3.5.4. Let E_1 and E_2 be two Euclidean spaces of dimensions n_1 and n_2, and let $S_1 \subset E_1$ and $S_2 \subset E_2$ be two (finite) configurations.

1. Show that $\mathrm{End}^s(E_1) \bigotimes \mathrm{End}^s(E_2)$ has codimension $\frac{n_1 n_2 (n_1-1)(n_2-1)}{4}$ in $\mathrm{End}^s(E_1 \bigotimes E_2)$.

2. Prove the equality $p_{x \bigotimes y} = p_x \bigotimes p_y$ for all nonzero $x \in E_1$ and $y \in E_2$. [Consider bases (e_i) for E_1 and (f_j) for E_2 with $e_1 = x$, $f_1 = y$ and $e_i \cdot x = f_i \cdot y = 0$ otherwise.]

3. Deduce from 2 that the perfection rank of $S_1 \bigotimes S_2 = \{x_1 \otimes x_2 \mid x_i \in S_i\}$ is the product of the perfection ranks of S_1 and of S_2.

4. Show that the tensor product of two lattices of dimensions greater than 1 is not perfect whenever its minimal vectors are split tensors.

3.6.1. Let S be the 2-dimensional eutactic configuration consisting of two orthogonal lines. Show that a configuration S' of at least four lines containing S is eutactic if and only if the lines in $S' \smallsetminus S$ do not all lie in the same angular domain defined by the lines in S.

3.6.2. Let S be a 2-dimensional configuration consisting of three distinct lines.

1. Show that it is possible to choose an orientation of the plane and three unit vectors v_0, v_1, v_2 on these lines in such a way that the angles $\theta_1 = \widehat{v_1, v_0}$ and $\theta_2 = \widehat{v_0, v_2}$ both belong to the interval $(0, \frac{\pi}{2})$.

2. Show that the eutaxy coefficients ρ_0, ρ_1, ρ_2 are given by the formulae

$$\rho_1 = \frac{\cos\theta_2}{\sin\theta_1 \sin(\theta_1+\theta_2)}, \quad \rho_2 = \frac{\cos\theta_1}{\sin\theta_2 \sin(\theta_1+\theta_2)} \quad \text{and} \quad \rho_0 = 2 - \rho_1 - \rho_2.$$

3. Transform the condition $\rho_1 + \rho_2 < 2$ into
(1) $\sin 2\theta_1 + \sin 2\theta_2 < 4 \sin\theta_1 \sin\theta_2 \sin(\theta_1+\theta_2)$;
(2) and then into $\cos(\theta_1+\theta_2) < 0$.

4. Use the last two questions to deduce that three lines in a Euclidean plane constitute a eutactic configuration if and only if, once chosen an indexing D_1, D_2, D_3 of the lines such that the (non-oriented) angles $\widehat{D_1, D_2}$ and $\widehat{D_2, D_3}$ are acute and that D_2 lies between D_1 and D_3, the angle $\widehat{D_1, D_3}$ is then obtuse.

3.6.3. Write down a proof of Theorem 3.6.4 which does not rely on the Voronoi theorem. [Use Exercise 5.1 and observe that for a given pair $\Lambda' \subset \Lambda$ of lattices, the index $[u(\Lambda) : u(\Lambda')]$ does not depend on $u \in \mathrm{GL}(E)$.]

3.6.4. Let S be a configuration of lines containing a eutactic configuration S' such that the orthogonal projections on the lines of S' and on those of S generate the same subspace of $\mathrm{End}^s(E)$. Show that S is eutactic. [Argue as in the proof of Theorem 3.6.2.]

3.6.5. Let S be a configuration of lines in E such that the orthogonal projections on the lines in S are independent in $\mathrm{End}^s(E)$, and let S' be a configuration strictly contained in S.

1. Show that if S' is eutactic, then S is not.

2. Let $\Lambda \simeq \mathbb{Z}^n \subset \mathbb{R}^n$ where $n = m^2$, $m \geq 3$, let $(\varepsilon_1, \varepsilon_2, \ldots, \varepsilon_n)$ be its canonical orthonormal basis, let $e = \frac{1}{m}(\varepsilon_1 + \varepsilon_2 + \cdots + \varepsilon_n)$, and let Λ' be the lattice generated by Λ and e. Show that $S(\Lambda')$ contains $S(\Lambda)$, but that Λ' is not eutactic.

3.6.6. Show that the tensor product of two (weakly, semi-, strongly) eutactic configurations is a (weakly, semi-, strongly) eutactic configuration.

3.7.1. Let Λ be a laminated lattice (in particular, Λ has norm 4), and let $n = \dim \Lambda$.

1. Show that Λ is irreducible.

2. Use the question above to prove that Λ is not an integral lattice for any $n > 24$. [Use Proposition 1.9.9.]

3. Prove that $\det(\Lambda)$ is nevertheless an integer for $n \leq 32$, by showing that for $24 \leq n \leq 32$, they coincide with those of the lattices $\frac{1}{\sqrt{2}}\Lambda_n$, $0 \leq n \leq 8$, for which the determinants are $1, 2, 3, 4, 4, 4, 3, 2, 1$.

4. Show that unimodular lattices of dimension 32 and norm 4 are not similar to laminated lattices, although they have the same Hermite invariant.
[Examples can be found in Section 6.7 and in [K-V].]

5. Let Λ' be an odd unimodular lattice of norm 3 and dimension n equal to 27, 28 or 29. Show that its sublattice of even norm vectors (of index 2 in Λ') has the same Hermite invariant as Λ_n, but is not a laminated lattice.
[These lattices have been classified for $n = 27, 28$ by Bacher and Venkov, see [Bc-V2] and also [Bc-V1]; one of them is $\Lambda' = \bigwedge^2 \mathbb{E}_8$, see Section 1.10.]

3.7.2. Let Λ be an integral lattice of dimension n, determinant d and norm m, similar to a laminated lattice Λ_n.

1. Prove that $m^n = 4^n d \lambda_n^{-1}$; deduce from this that m is divisible by 4 for $24 \leq n \leq 48$.

2. Show that m is divisible by 8 for $33 \leq n \leq 48$.

3. Prove the inequality $m \geq 8$ for $25 \leq n \leq 48$.

3.8.1. Let E be a Euclidean plane and let $\mathcal{B} = (e_1, e_2)$ be a basis for E whose Gram matrix has determinant 1. Let $\mathcal{B}^* = (e_1^*, e_2^*)$ its dual basis.

1. Prove that there exists an orthonormal basis $(\varepsilon_1, \varepsilon_2)$ for E in which $e_1 = a\varepsilon_1$ and $e_2 = c\varepsilon_1 + a^{-1}\varepsilon_2$ for some $a, c \in \mathbb{R}$ with $a > 0$ and $c \geq 0$.

2. Prove that $e_1^* = a^{-1}\varepsilon_1 - c\varepsilon_2$ and $e_2^* = a\varepsilon_2$.

3. Prove that a rotation of angle $\pm\frac{\pi}{2}$ maps $\mathcal{B}$ onto $(e_2^*, -e_1^*)$ or onto $(-e_2^*, e_1^*)$.

4. Deduce from this that any 2-dimensional lattice of determinant 1 is directly isometric to its dual.

5. Show that any 2-dimensional lattice lattice is directly similar to its dual.

6. Show that $s(\Lambda) = s(\Lambda^*) \in \{1, 2, 3\}$ for any 2-dimensional lattice Λ,

3.8.2. Let Λ be a 2-dimensional lattice. Prove that if (e_1, e_2) is a basis for Λ which represents its two minima (i.e, e_1 is minimal and e_2 is minimal among those vectors in Λ which are not proportional to e_1), then e_2^*, e_1^* are the two minima of Λ^*. [Use Question 5 in Exercise 8.1.]

3.8.3. The aim of this exercise is to classify 2-dimensional dual-perfect, dual-eutactic and dual-extreme lattices. Let Λ be a norm 1 lattice in $\mathbb{R}^2$, endowed with its canonical basis and the corresponding orientation.

1. Suppose that $s(\Lambda) \geq 2$. Show that Λ is the image under a rotation of a lattice with basis (e_1, e_2) such that $e_1 = \varepsilon_1$, $N(e_2) = 1$ and $\frac{\pi}{3} \leq \theta \leq \frac{\pi}{2}$, where $\theta = \widehat{e_1, e_2}$.

2. We consider in what follows a lattice Λ together with a basis (e_1, e_2) as above. Show that its dual basis (e_1^*, e_2^*) is defined by the equations $\|e_1^*\| = \|e_2^*\| = \frac{1}{\sin\theta}$ and $\widehat{e_1^*, e_1} = \widehat{e_2, e_2^*} = \frac{\pi}{2} - \theta$.

3. Show that the only dual-extreme 2-dimensional lattices are the hexagonal lattices; prove that $\gamma'^2 = \frac{4}{3}$.

4. Show that Λ is dual-perfect except if $\theta = \frac{\pi}{2}$. [Use Exercise 5.2.]

5. Show that Λ is dual-eutactic if and only if $\theta = \frac{\pi}{3}$ or $\theta = \frac{\pi}{2}$. [Notice that for $\frac{\pi}{3} < \theta < \frac{\pi}{2}$, the linear combinations $\lambda\, p_{e_1} + \mu\, p_{e_2}$ in $\mathrm{End}(E)$ with $\lambda, \mu > 0$ map e_1 inside the sector bounded by the half-lines generated by e_1 and e_2.]; use this to recover the result of Question 3.

6. Show that a 2-dimensional lattice Λ with $s(\Lambda) = 1$ cannot be dual-eutactic. [Use Exercise 8.2.]

3.8.4. Let S and S' be two finite families of nonzero vectors in E.

1. Show that for nonzero vectors $x, y \in E$, we have $\mathrm{Tr}(p_y \circ p_x) = \cos^2(\widehat{x, y})$.

2. Show that a vector $e \in E$ such that there exists a relation with strictly positive coefficients $\sum_{x \in S} \rho_x\, p_x(e) = 0$ is orthogonal to all vectors in S.

3. Show that if (S, S') is dual-eutactic, every hyperplane which contains S' also contains S.

4. Use this to deduce that if (S, S') is a dual-eutactic pair, then S and S' generate the same subspace of E.

3.8.5. Let $t \in (\frac{3}{4}, 1)$ and let Λ_t be the lattice $t\mathbb{A}_1 \perp \mathbb{A}_2$, whose Gram matrix in an appropriate basis (e_1, e_2, e_3) is $\begin{pmatrix} 2t & 0 & 0 \\ 0 & 2 & -1 \\ 0 & -1 & 2 \end{pmatrix}$.

1. Show that $S(\Lambda) = \{e_1\}$ and $S(\Lambda^*) = \{e_1^*\}$.

2. Show that the lattices Λ_t are dual-eutactic.

3. Show that the equality $S(u(\Lambda)) = \{u(e_1)\}$ and $S(u(\Lambda^*)) = \{u(e_1)^*\}$ still holds for every $u \in \mathrm{GL}(E)$ sufficiently close to the identity, but that the lattices $u(\Lambda_t)$ are no longer dual-eutactic unless $u(e_1)$ is orthogonal to the plane $\langle u(e_2), (e_3) \rangle$.

3.8.6. Let $\mathcal{B} = (e_1, e_2, \ldots, e_n)$ be a basis for E. Denote by p_i (resp. p'_i) the orthogonal projection onto e_i (resp. e_i^*). Let $\sum_j \lambda_j\, p_j = \sum_k \mu_k\, p'_k$ be a relation of dual-eutaxy with nonzero coefficients λ_j, μ_k.

1. Prove the equalities $\frac{1}{(e_i \cdot e_i)}\, \lambda_i e_i = \sum_k \mu_k\, \frac{e_i^* \cdot e_k^*}{e_k^* \cdot e_k^*}\, e_k^*$ $(i = 1, 2, \ldots, n)$. [Use projections onto the e_i^*.]

2. Deduce from 1 that the relations $\frac{e_i \cdot e_j}{e_i \cdot e_i}\, \lambda_i = \frac{e_i^* \cdot e_j^*}{e_j^* \cdot e_j^*}\, \mu_j$ hold for all i and all j.

3. Prove the equalities $\mu_i = \lambda_i$ and $\frac{(e_i \cdot e_j)^2}{(e_i \cdot e_i)(e_j \cdot e_j)} = \frac{(e_i^* \cdot e_j^*)^2}{(e_i^* \cdot e_i^*)(e_j^* \cdot e_j^*)}$.

4. Show that if the vectors of $\mathcal{B}$ and of $\mathcal{B}^*$ all have the same norm, then the Gram matrices of $\mathcal{B}$ and of $\mathcal{B}^*$ are proportional.

5. Show that if Λ is a dual-eutactic lattice such that $S(\Lambda)$ and $S(\Lambda^*)$ are up to their signs the vectors of a basis and of its dual basis, Λ is similar to $\mathbb{Z}^n$.

3.8.7. Show that the maximum of $s + s^*$ in a given dimension n is attained on dual-perfect lattices. (Compare Theorem 3.5.5.)

3.8.8. Let Λ be a perfect lattice.

1. Show that if there exists a relation $\sum_{x \in S(\Lambda)} a_x\, p_x = \sum_{y \in T} b_y\, p_y$ for some $T \subset S(\Lambda^*)$ with strictly positive coefficients a_x, b_y, then Λ is dual-eutactic.
[Argue as in the proof of Theorem 3.6.2.]
2. Show that if Λ is extreme and if Λ^* is semi-eutactic, then Λ is dual-extreme.

3.10 Notes on Chapter 3.

The notion of an extreme quadratic form appears for the first time in Korkine and Zolotareff's 1873 article [K-Z2], where they moreover give a list of extreme forms, but proofs appear only in the 1877 paper [K-Z3] (Mathematische Annalen, volume **11**). The aim of this last article is to classify extreme forms up to dimension 5 and to derive from this classification the precise value of γ_5.

In [K-Z3], Korkine and Zolotareff give a characterization of extremality by means of inequalities which are equivalent to those of Theorem 3.4.5, they prove that extreme forms are perfect (without giving a name to perfection; the word "perfect" was used for the first time by Voronoi some thirty years later), and that perfect forms are well defined by the components in the canonical basis for $\mathbb{Z}^n$ of the set of their minimal vectors, a result stated in the form

> Toute forme extrême a au moins $\frac{n(n+1)}{2}$ représentations de son minimum qui déterminent complètement cette forme, en supposant que son minimum soit donné.

([K-Z3], p. 252). From this result, they deduce the remarkable fact that γ_n^n is a rational number. They however do not prove the finiteness theorem for perfect forms, but the way they wrote up their results makes it likely that they considered this theorem as true.

The lower bound $s \geq \frac{n(n+1)}{2}$ for the number of pairs of minimal vectors in any perfect lattice was extended in 1953 by Swinnerton-Dyer to lattices which are extreme for a bounded convex set (cf. [Cas2], Chapter V, Theorem VIII).

After [K-Z3], the theory of extreme forms falls asleep for thirty years until the publication in 1908 by Voronoi of his paper [Vo1]. The chief aim of Voronoi is the description of an algorithm which (at least theoretically) allows the classification of perfect forms in any given dimension (see the notes on Chapter 7). But Voronoi's paper also contains two results which we have proved in this chapter: the characterization of extreme forms as those which are perfect and eutactic (the notion of a eutactic form is due to Voronoi, who gives no name to it; the word "eutaxy" was used for the first time in this setting by Coxeter in 1951 in [Cox2]), and the proof of the finiteness theorem for perfect forms.

The theory of perfect forms again falls asleep until Coxeter's 1951 paper quoted above, except for a paper [Hof] by Hofreiter dating back to 1933, which states an erroneous classification of 6-dimensional perfect forms.

Coxeter studies the notions of perfection and eutaxy for their own interest, and replaces (partially) for the first time forms by lattices. He met eutactic configurations in previous works by Schläfli and by Hadwiger on regular polytopes. His paper contains a description of new extreme lattices related to root lattices, to which we shall return later in Chapter 5, the characterization of eutactic configurations in terms of eutactic stars, and various conjectures, sometimes not well inspired (see Conway and Sloane's comments at the beginning of [C-S5]).

The study of regular polytopes is also the motivation of the joint paper [B-C] of 1940 with Richard Brauer (slightly generalized in [Cox2], no. 4.3). The proof involves a lot of calculations, and it is not clear that representations to be irreducible only on $\mathbb{R}$, not on $\mathbb{C}$. The short proof of Section 3.6 has been given to me by A.-M. Bergé.

A great deal of work on perfect forms took place in England from 1955 onwards. This yielded the discovery of numerous new perfect forms as well as classification theorems. As for what concerns the subject of this chapter, we must in particular quote Barnes's 1957 paper [Bar2] in which he brought to light Stiemke's theorem. The proof of Theorem 3.4.6 we have given essentially follows Barnes's. It should be noticed that Voronoi's proof is of the same nature, except that he had to prove Stiemke's theorem in the particular case of linear forms on spaces of symmetric matrices. At this stage, one must point out a difficulty in the proof of Theorem 3.4.5, related to uniformity problems in power series expansions in several variables. Thus many proofs (including one in which I was involved) cannot be considered as completely correct; a similar remark applies to Theorems 3.8.4 and 3.8.5. Kneser ([Kn1], 1955) was the first to publish a completely correct proof, in which he made use of convexity. Such a convexity argument is used in Conway and Sloane's [C-S5] (quite correct up to a forgotten logarithm). The proof I gave here is modelled on their proof. I would like to thank A.-M. Bergé and R. Coulangeon for having drawn my attention on the difficulty above, which we shall again consider in Chapter 10, where convexity underlies in the use of the exponential map which goes from the tangent space at the origin of a Lie group to the group itself.

The method of Proposition 3.5.3, which yields a characterization of perfect lattices possessing a perfect hyperplane section, was used by Barnes in [Bar5], I, pp. 64–67, where he proves the perfection of a certain family Q_n of forms (see Section 5.3) by making use of the perfection of its family P_{n-1}, previously proved in [Bar3].

Theorem 3.5.5 was pointed out to me by Oesterlé; Watson considers in [Wat5] that it was known to Voronoi.

Proposition 3.6.5 is Scott's Theorem 2.1 in [Sco1].

Conway and Sloane published the important paper *Laminated Lattices* in *Annals of Mathematics* (vol. **116**, 1982, pp. 593–620); it is reproduced as the basis for Chapter 6 of [C-S]. For the arithmetical laminations of Plesken and Pohst, see [Pl-P1], [Pl-P2] and the appendix to Chapter 6 of [C-S].

I know of no published proof of the extremality of laminated lattice at least up to dimension 24, for which it actually holds. (There is no reason for extremality to hold in any dimension.) These lattices are expected to be the (only) densest lattices up to dimension 25 except in dimensions 11, 12, 13. Extremality is proved explicitly for dimensions $n \le 8$ in Chapter 4, and for $n \le 24$ and $n \equiv 0$ or $1 \mod 4$ in Chapter 8 (where it is shown that the set of minimal vectors contains a configuration similar to $S(\mathbb{D}_n)$); Theorem 3.6.6 applies to some other dimensions, e.g. $23, 24$.

Curiously, for $n \le 24$, laminated lattices are the densest known lattices, except for the three dimensions for which they are not unique, were the densest lattices known are K_{11}, K_{12} and K_{13}. Lattices in the K_n series no longer represent the densest known lattices in dimensions $n \le 10$ and $n \ge 14$, and it is precisely for dimensions 10 and 14 that a bifurcation to the lattices called K'_n in Chapter 8 appears!

The notion of a dual-extreme lattice was introduced in [B-M1], published in 1989. Most of the results of Section 3.8 come from this paper or from Bergé's [Ber1]. The motivation for [B-M1] was a paper by Zimmert on "twin classes" in number fields ([Zi], 1981; see also Oesterlé, [Oe1], where Zimmert's theory in explained in the setting of "Weil's explicit formulae"). The geometrical interpretation of Zimmert's inequalities (obtained via partial zeta functions) involves the product of the minima on a lattice and on its dual of a quadratic form Q depending on the signature of the field. This product reduces to $N(\Lambda)\,N(\Lambda^*)$ when one replaces Q by the Euclidean norm in $\mathbb{R}^n$.

Besides the various extremality problems which have been dealt with in this chapter or which we are going to handle later, the Humbert problem, that we shall not consider anywhere else in this book, deserves a special mention. It consists in estimating a Hermite like invariant $\gamma_{(K)}$ relative to a number field K (and a dimension n) and taking into account the $r_1 + r_2$ embeddings $\sigma_i : K \to \mathbb{C}$, with the usual notation, namely that the σ_i are real for $i \le r_1$, and non-real, pairwise non-conjugates for $i > r_1$. Set $d_i = 1$ if $i \le r_1$ and $d_i = 2$ if $i > r_1$.

Explicitly, a *Humbert form* is a system $\mathcal{A} = (A_1, \dots, A_{r_1+r_2})$ of positive definite matrices, symmetric real for $i \le r_1$, Hermitian complex for $i > r_1$. Given a column-vector $X \in \mathbb{Z}_K^n$, set $\mathcal{A}[X] = \prod_i ({}^t\sigma_i(X) A_i \sigma_i(X))^{d_i}$, and define the *minimum* and the *determinant* of the Humbert form by

$$\mu(\mathcal{A}) = \min\{\mathcal{A}[X] \mid X \in \mathbb{Z}_K^n \smallsetminus \{0\}\} \quad \text{and} \quad \det(\mathcal{A}) \prod_i \det(A_i)^{d_i}.$$

The *Hermite–Humbert invariant* is then

$$\gamma_{(K)}(\mathcal{A}) = \frac{\mu(\mathcal{A})}{\det(A)^{1/n}}\,.$$

(Hence $\gamma_{(\mathbb{Q})}$ is the usual Hermite invariant.)

The theory was founded by Humbert in his papers [Hmb1] and [Hmb2]. Convenient notions of perfection and eutaxy have been given by Coulangeon after work by Baeza and Icaza ([B-I], [I]), with which he was able to prove an analogue of Voronoi's Theorem 4.6 ([Cou6]). Some explicit computation have been done in [B-C-I-O] for quadratic fields. The Humbert problem has also been considered in the setting of linear algebraic groups by Watanabe; see [O-W], [Wata1], [Wata2].

4 Root Lattices

Introduction

Root lattices are orthogonal sums of irreducible lattices which are either the lattice $\mathbb{Z}$ or the lattices of norm 2 consisting of two infinite families $\mathbb{A}_n$, $\mathbb{D}_n$ and three *exceptional* lattices $\mathbb{E}_6$, $\mathbb{E}_7$, $\mathbb{E}_8$. They play a crucial rôle in Coxeter's classification of groups generated by reflections (the roots are then vectors which are orthogonal to the hyperplanes defining the reflections, whence the name given to these lattices). Those which are irreducible are extreme. Moreover, they are involved in numerous constructions of extreme lattices. Finally, the Hermite constant is attained on root lattices in dimensions up to 8 (but in no higher dimension).

However, Witt's classification theorem is certainly the main reason for the ubiquity of root lattices: an integral lattice generated by vectors of norm 1 or 2 is isometric to a root lattice, and is thus an orthogonal sum of lattices isometric to $\mathbb{Z}$, $\mathbb{A}_n$, $\mathbb{D}_n$ or $\mathbb{E}_n$. No result of this kind is known (or likely to exist) for vectors having a larger norm.

We study root lattices individually and independently of the classification of root systems that we state without proof in Section 4.10. All statements in Sections 4.1 to 4.9 are proved, with one exception which is postponed to Chapter 5. In each case we determine various invariants: the minimal vectors of the lattice and of its dual, the determinant, the automorphism group, the *Weyl group* (the subgroup generated by the orthogonal reflections in the minimal vectors), and we exhibit special bases constituted by norm 2 vectors, which we call Korkine and Zolotareff bases (resp. Coxeter bases), for which scalar products are 0 or 1 (resp. 0 or -1).

Although reducible lattices are not perfect, we begin with the lattice $\mathbb{Z}^n$, which plays a key rôle in the construction of other root lattices.

As usual E will denote a Euclidean space of dimension n, which will be either the space $\mathbb{R}^n$ or the hyperplane in $\mathbb{R}^{n+1}$ defined by $\sum_{i=0}^n x_i = 0$; the canonical basis for $\mathbb{R}^n$ (resp. for $\mathbb{R}^{n+1}$) is denoted by (ε_i) with $1 \le i \le n$ (resp. with $0 \le i \le n$).

4.1 The $\mathbb{Z}^n$ Lattice

We consider the lattice $\mathbb{Z}^n$ of integral points $\mathbb{R}^n$. This trivial lattice has norm 1, determinant 1, minimal vectors $\pm\varepsilon_i$ and thus $s = n$.

The determination of the automorphism group of an orthogonal sum of irreducible lattices (Theorem 1.4.6) immediately yields the following statement:

Theorem 4.1.1. *The group* $\operatorname{Aut}(\mathbb{Z}^n)$ *is isomorphic to the semi-direct product* $\{\pm 1\}^n \rtimes S_n$*, where the symmetric group* S_n *acts on* $\{\pm 1\}^n$ *by permuting the* n *components.* □

This group has order $2^n\, n!$ and its centre $\{\pm\operatorname{Id}\}$ is a direct factor for even n but not for odd n. It can be identified with the symmetry group of a cube centred at the origin. It acts transitively on the set of minimal vectors. Since S_n is k-fold transitive for all $k \le n$, $\operatorname{Aut}(\mathbb{Z}^n)$ acts transitively for all $k \le n$ on the subspaces generated by k independent minimal vectors. For $n = 1$ (resp. 2 ; resp. 3), $\operatorname{Aut}(\mathbb{Z}^n)$ is isomorphic to the cyclic group C_2 of order 2 (resp. to the dihedral group D_4 of order 8; resp. to the direct product $S_4 \times C_2$).

The norm 1 (resp. norm 2) vectors in $\mathbb{Z}^n$ are the $2n$ vectors $\pm\varepsilon_i$ (resp. the $2n(n-1)$ vectors $\pm\varepsilon_i \pm \varepsilon_j$, $i < j$). Denoting by s_x the orthogonal reflection along x (i.e., with respect to the hyperplane $\mathbb{R}x^\perp$), we see immediately that $s_{\varepsilon_i - \varepsilon_j}$ (resp. $s_{\varepsilon_i + \varepsilon_j}$) maps $(\varepsilon_i, \varepsilon_j)$ onto $(\varepsilon_j, \varepsilon_i)$ (resp. $(-\varepsilon_j, -\varepsilon_i)$) and leaves the other vectors ε_k invariant. Hence:

Proposition 4.1.2. *The orthogonal reflections defined by the norm 2 (resp. by the norm* 1 *or* 2*) vectors in* $\mathbb{Z}^n$ *generate the subgroup of index* 2 *of* $\operatorname{Aut}(\mathbb{Z}^n)$ *consisting of those transformations which induce an even number of changes of signs on the vectors* ε_i *(resp. the whole group* $\operatorname{Aut}(\mathbb{Z}^n)$*).* □

4.2 The $\mathbb{A}_n$ Lattice

For all $n \ge 1$, let $\mathcal{B}_0 = (\varepsilon_0, \varepsilon_1, \dots, \varepsilon_n)$ be the canonical basis for $\mathbb{R}^{n-1}$, and let $\mathbb{Z}^{n+1}$ be the lattice in $\mathbb{R}^{n+1}$ with basis $\mathcal{B}_0$. Let H be the hyperplane defined by the equation $\sum_{i=0}^n x_i = 0$, and for $1 \le i \le n$, let $e_i = \varepsilon_i - \varepsilon_0$ and $e'_i = \varepsilon_i - \varepsilon_{i-1}$. Clearly $\mathcal{B} = (e_1, e_2, \dots, e_n)$ and $\mathcal{B}' = (e'_1, e'_2, \dots, e'_n)$ are bases for H.

Definition 4.2.1. *The* lattice $\mathbb{A}_n$ *is the intersection* $\mathbb{Z}^{n+1} \cap H$.

This is actually a lattice in $E = H$, for which $\mathcal{B}$ and $\mathcal{B}'$ are bases that we call *Korkine–Zolotareff* and *Coxeter bases* respectively. The Gram matrices for these two bases have entries $a_{i,j}$ and $a'_{i,j}$ given by $a_{i,i} = a'_{i,i} = 2$, $a_{i,j} = 1$

for $j \neq i$, $a'_{i,j} = -1$ for $|j-i| = 1$ and $a'_{i,j} = 0$ for $|j-i| \geq 2$. The corresponding quadratic forms are respectively $2U_n$ and $2U'_n$, where

$$U_n = \sum_{1\leq i\leq n} x_1^2 + \sum_{1\leq i<j\leq n} x_1 x_j \quad \text{and} \quad U'_n = \sum_{1\leq i\leq n} x_i^2 - \sum_{1\leq i\leq n-1} x_i x_{i+1}\,.$$

The form U_n was given by Korkine and Zolotareff; $\mathbb{A}_1$ is similar to $\mathbb{Z}$.

Proposition 4.2.2. *The lattice $\mathbb{A}_n$ has the following properties:*

1. $\det(\mathbb{A}_n) = n+1$.
2. *$\mathbb{A}_n$ has norm 2, and $s(\mathbb{A}_n) = \frac{n(n+1)}{2}$.*
3. *For $n \geq 2$, the restriction to H of the automorphisms of $\mathbb{Z}^{n+1}$ induces an isomorphism of $\{\pm \mathrm{Id}\} \times S_{n+1}$ onto $\mathrm{Aut}(\mathbb{A}_n)$; the Weyl group $W(\mathbb{A}_n)$ has index 2 in $\mathrm{Aut}(\mathbb{A}_n)$. It can be identified with S_{n+1} acting on the vectors $\varepsilon_0, \varepsilon_1, \dots, \varepsilon_n$ and is transitive on $S(\mathbb{A}_n)$.*
4. *The vectors in $S(\mathbb{A}_n)$ have spectrum $(\frac{(n-1)(n-2)}{2}, 0)$, $(2(n-1), 1)$.*

Proof. (1) The lattice $\mathbb{Z}^{n+1}$ is generated by ε_0 and $\mathbb{A}_n$, and $H^\perp$ is the line directed by the vector $e = \sum \varepsilon_i$. We have $N(e) = n+1$ and $e \cdot \varepsilon_0 = 1$.

(2) A nonzero vector in $\mathbb{A}_n$ has at least two nonzero components on $\mathcal{B}_0$. This immediately shows that $\mathbb{A}_n$ has norm 2 and that $S(\mathbb{A}_n)$ is the set of the $n(n+1)$ vectors $(\varepsilon_i - \varepsilon_j)$, $j \neq i$.

(3) The automorphisms described in (3) preserve both $\mathbb{Z}^{n+1}$ and H. Hence the group $\{\pm \mathrm{Id}\} \times S_{n+1}$ acts on $\mathbb{A}_n$, and this action is faithful thanks to the hypothesis $n \geq 2$. Conversely, an automorphism of $\mathbb{A}_n$ possesses a unique extension to $\mathbb{R}^{n+1}$ which stabilizes the vector e. Since the ε_i are linear combinations of e and of elements of $S(\mathbb{A}_n)$, this extension preserves $S(\mathbb{Z}^{n+1})$. It is then readily verified that the only elements of $\mathrm{Aut}(\mathbb{Z}^{n+1})$ which preserve H are those of (3).

(4) This results from an easy calculation. □

We now turn to the dual lattice $\mathbb{A}_n^*$ of $\mathbb{A}_n$. Notice that $\mathbb{A}_n$ is a sublattice of $\mathbb{A}_n^*$, since $\mathbb{A}_n$ is integral.

Proposition 4.2.3. *We have $N(\mathbb{A}_n^*) = \dfrac{n}{(n+1)}$, $s(\mathbb{A}_n^*) = n+1$ if $n \geq 2$, and the quotient $\mathbb{A}_n^*/\mathbb{A}_n$ is cyclic of order $n+1$. In particular, $\sqrt{(n+1)}$ is the smallest multiplier which makes $\mathbb{A}_n^*$ an integral lattice.*

Proof. Since $\mathbb{A}_n = \mathbb{Z}^{n+1} \cap H$, $\mathbb{A}_n^*$ is the orthogonal projection onto H of $\mathbb{Z}^{n+1}$. Now the projections $\tilde{\varepsilon}_i$ on H of the ε_i are the vectors

$$\varepsilon_i - \frac{e \cdot \varepsilon_i}{e \cdot e}\, e = \frac{1}{n+1}\Big(n\,\varepsilon_i - \sum_{k\neq i} \varepsilon_k\Big),$$

of norm $\frac{n}{(n+1)}$. We have $N(\tilde{\varepsilon}_i) = \frac{n}{n+1}$ and $\tilde{\varepsilon}_i \cdot \tilde{\varepsilon}_j = -\frac{1}{n+1}$ for $j \neq i$.

In the basis $(\tilde{\varepsilon}_i)$, $1 \le i \le n$ of $\mathbb{A}_n^*$, the norm form q relative to $\mathbb{A}_n^*$ can be written after multiplication by $(n+1)$ as a sum of the two quadratic forms $q_1(x) = \sum_{1\le i<j\le n}(x_j - x_i)^2$ and $q_2(x) = \sum_{1\le i\le n} x_i^2$. If the x_i are distinct, we have $q_1(x) \ge n-1$, whence $(n+1)q(x) \ge n$, with equality exactly for the vectors $\pm\tilde{\varepsilon}_i$, $i \ge 1$. Otherwise, we have $(n+1)q(x) = q_2(x) \ge \sum x_i^2 \ge n$ with equality on $\pm\sum_{i\ge 1}\tilde{\varepsilon}_i = \mp\tilde{\varepsilon}_0$. This proves the first two assertions of the proposition.

Let m be a positive integer. It is clear that $m\,\tilde{\varepsilon}_i$ cannot be a $\mathbb{Z}$-linear combination of the ε_i unless m is a multiple of $n+1$, and also that the inclusion $(n+1)\mathbb{A}_n^* \subset \mathbb{A}_n$ holds. The annihilator of $\mathbb{A}_n^*/\mathbb{A}_n$ is thus $n+1$. Since $\det(A_n) = n+1$, $\mathbb{A}_n^*/A_n$ is cyclic. □

From the description of $S(\mathbb{A}_n^*)$, it results that the canonical isomorphism between the automorphism group of a lattice and its dual identifies $\mathrm{Aut}(\mathbb{A}_n)$ with the product $\{\pm\,\mathrm{Id}\}\times S_{n+1}$, the group S_{n+1} acting on the $n+1$ directions of minimal vectors of $\mathbb{A}_n^*$.

We end this section with the special case of dimension 2, where we shall construct $\mathbb{A}_2$ inside $\mathbb{C}$ endowed with its scalar product $\lambda\cdot\mu = \mathrm{Tr}(\lambda\overline{\mu})$. Let $\omega = \frac{-1+i\sqrt{3}}{2}$ and let $\mathfrak{O}$ be the subring $\mathbb{Z}[\omega]$ of $\mathbb{C}$. The proof of the following proposition is immediate.

Proposition 4.2.4. *The lattice $\mathfrak{O} \subset \mathbb{C}$ is isometric to $\mathbb{A}_2$ and multiplication by $(1-\omega)^{-1}$ is a similarity of ratio $\frac{1}{\sqrt{3}}$ and angle $-\frac{\pi}{6}$ of $\mathfrak{O}$ onto its dual. The group of unitary automorphisms of $\mathbb{A}_2$ is the cyclic group W' whose elements are the 6 rotations $z \mapsto \pm z, z \mapsto \pm\omega^{\pm 1}z$ of $\mathbb{C}$.* □

[All 2-dimensional lattices are similar to their dual (see Exercise 3.8.1); for a cyclotomic construction of the lattices $\mathbb{A}_{p-1}$, p an odd prime, see Section 5.4.]

4.3 The $\mathbb{D}_n$ Lattice

Let $n \ge 2$ be an integer and let $E = \mathbb{R}^n$.

Definition 4.3.1. *We denote by $\mathbb{D}_n$ the set of vectors in $\mathbb{Z}^n$ whose coordinates have an even sum.*

Thus, in the language of Chapter 1, Definitions 1.2.8 and 1.3.10, $\mathbb{D}_n$ is the even lattice attached to $\mathbb{Z}^n$ (or the even part of $\mathbb{Z}^n$), and $\varepsilon_1+\cdots+\varepsilon_n$ is a parity (or characteristic) vector for $\mathbb{Z}^n$.

The following two systems of n vectors in $\mathbb{D}_n$ constitute bases for $\mathbb{D}_n$ that we call *Korkine–Zolotareff* and *Coxeter bases* respectively:

$$e_1 = \varepsilon_1+\varepsilon_2,\ e_2 = \varepsilon_1-\varepsilon_2,\ e_3 = \varepsilon_1-\varepsilon_3, \dots, e_n = \varepsilon_1-\varepsilon_n;$$
$$e_1' = \varepsilon_1+\varepsilon_2,\ e_2' = \varepsilon_2-\varepsilon_1,\ e_3' = \varepsilon_3-\varepsilon_2, \dots, e_n' = \varepsilon_n-\varepsilon_{n-1}.$$

Gram matrices for these bases have the following entries $a_{i,j}$ and $a_{i,j}'$:

$a_{i,i} = a'_{i,i} = 2$, $a_{1,2} = a'_{1,2} = a_{2,1} = a'_{2,1} = 0$, and, for $j \neq i$ and i or $j \neq 1$, $a_{i,j} = 1$, $a'_{i,j} = -1$ if $|j-i| = 1$ and $a'_{i,j} = 0$ if $|j-i| \geq 2$.

The corresponding quadratic forms are respectively $2V_n$ and $2V'_n$, where (with the notation of Section 4.2)

$$V_n = U_n - x_1x_2 \quad \text{and} \quad V'_n = U'_n + x_1x_2 - x_1x_3\,.$$

The form V_n is Korkine and Zolotareff's.

The $\mathbb{D}_n$ lattice can also be constructed in the following way: one starts with the lattice $2\,\mathbb{Z}^n$ together with its basis $(2\,\varepsilon_1, \ldots, 2\,\varepsilon_n)$, and replace then $2\,\varepsilon_i$ by $\varepsilon_1 + \varepsilon_i$ for $i \geq 2$, adding thus to $2\,\mathbb{Z}^n$ the centres of the 2-dimensional faces of its fundamental cube, whence the name of *face–centred cubic lattice* often given to $\mathbb{D}_n$.

Note that Definition 4.3.1 makes also sense for $n = 1$, where $D_1 = 2\,\mathbb{Z}$ has norm 4. For $n = 2$, we have $\mathbb{D}_2 \simeq \mathbb{A}_1 \perp \mathbb{A}_1$. The case of dimension 3 is also somewhat special:

Proposition 4.3.2. *$\mathbb{D}_3$ is isometric to $\mathbb{A}_3$.*

Proof. The basis $(\varepsilon_1 + \varepsilon_2, \varepsilon_2 + \varepsilon_3, \varepsilon_3 + \varepsilon_1)$ for $\mathbb{D}_3$ defines the form U_3. □

[We have thus interpreted $\mathbb{A}_3$ as the familiar face–centred cubic lattice of crystallographers.]

We now turn to the determination of the standard invariants of $\mathbb{D}_n$:

Theorem 4.3.3. *Let $n \geq 3$ be an integer.*

1. *$\mathbb{D}_n$ has norm 2 and determinant 4.*
2. *We have $S(\mathbb{D}_n) = \{\pm\varepsilon_i \pm \varepsilon_j \mid 1 \leq i < j \leq n\}$, and in particular $s(\mathbb{D}_n) = n(n-1)$.*
3. *The dual lattice of $\mathbb{D}_n$ is $\mathbb{D}_n^* = \mathbb{Z}^n \cup (e + \mathbb{Z}^n)$ where $e = \frac{1}{2}\sum_{i=1}^n \varepsilon_i$.*
4. *The quotient $\mathbb{D}_n^*/\mathbb{D}_n$ is a group of order 4, cyclic when n is odd and of type $(2,2)$ when n is even.*
5. *The restriction to $\mathbb{D}_n$ of the automorphisms of $\mathbb{Z}^n$ induces an isomorphism of $\mathrm{Aut}(\mathbb{Z}^n) \simeq \{\pm 1\}^n \rtimes S_n$ onto $\mathrm{Aut}(\mathbb{D}_n)$, except for $n = 4$ where the image has index 3: in all cases, the homomorphism above induces an isomorphism of the subgroup of index 2 of $\mathrm{Aut}(\mathbb{Z}^n)$ of those elements which induce an even number of changes of signs of the ε_i onto the Weyl group $W(\mathbb{D}_n)$ of $\mathbb{D}_n$, and this group acts transitively on $S(\mathbb{D}_n)$.*
6. *The spectrum of the vectors in $S(\mathbb{D}_n)$ is $(n^2 - 5n + 7, 0)$, $(4(n-2), 1)$.*

Proof. The statements (1), (2), and (6) are easy, and (3) is a consequence of the fact that $\mathbb{D}_n$ is the kernel of the map $x \mapsto 2\,e \cdot x \mod 2$ of $\mathbb{Z}^n$ onto $\mathbb{Z}/2\mathbb{Z}$. We now use (3) to prove (5).

By duality, the restriction to $\mathbb{D}_n$ of $\mathrm{Aut}(\mathbb{Z}^n)$ induces an injective homomorphism of $\mathrm{Aut}((\mathbb{Z}^n)^*)$ into $\mathrm{Aut}(\mathbb{D}_n^*)$. For $n \neq 4$, the norm 1 vectors in $\mathbb{D}_n^*$ are those of $(\mathbb{Z}^n)^*$, whence the result in this case (and the equality $s(\mathbb{D}_n^*) = n$). For $n = 4$, $S(\mathbb{D}_4^*) = S(\mathbb{Z}^4) \cup S(e + \mathbb{Z}^4)$ has 2×12 elements.

Let F be the hexagonal plane lattice $\langle \varepsilon_1, e\rangle$, whose orthogonal is $\langle \varepsilon_2-\varepsilon_3, \varepsilon_3-\varepsilon_4\rangle$. The consideration of the plane rotation $\varepsilon_1 \mapsto e$, $e \mapsto e-\varepsilon_1$ of plane F shows that the cardinality of the orbit of ε_1 under the action of $\mathrm{Aut}(\mathbb{D}_4^*)$ is divisible by 3, hence the result for $n = 4$. (With the notation of Proposition 4.3.4 below, a special automorphism of order 3 is given by (right) multiplication by $(1-\omega)$.) The statement about the Weyl group results immediately from Proposition 4.1.2.

It remains to prove (4). We know that $\mathbb{D}_n^*/\mathbb{D}_n$ has order $\det(\mathbb{D}_n) = 4$. This quotient is obviously cyclic for odd n (because $2e$ does not belong to $\mathbb{D}_n$), and non–cyclic for even n, with generators $e, \varepsilon_1 \mod \mathbb{D}_n$. □

We now look at the special case of dimension 4; for the notation, we refer to Appendix 1, Section 15.2. We endow the skew–field $\mathbb{H}$ of the Hamilton quaternions with the Hermitian structure for which $\lambda\cdot\mu = \frac{1}{2}\,\mathrm{Trd}(\lambda\cdot\overline{\mu})$ (mind the factor $\frac{1}{2}$!), so that $(1,i,j,k)$ is an orthonormal basis for the $\mathbb{R}$-vector space $\mathbb{H}$, which we identify in this way with $\mathbb{R}^4$, and we denote by $\mathfrak{M}$ the Hurwitz order, a $\mathbb{Z}$-basis for which is $(1,i,j,\omega)$ where $\omega = \frac{-1+i+j+k}{2}$; we view it as a left module over itself. We then have the following proposition, whose proof is immediate and that we thus leave to the reader:

Proposition 4.3.4. *The ring $\mathfrak{M}$ is equal to the lattice $\mathbb{D}_4^*$, the two-sided ideal $\mathfrak{A} = \mathfrak{M}(1+i) = \{x+yi+zj+tk \mid x,y,z,t \in \mathbb{Z},\, x+y+z+t \equiv 0 \mod 2\}$ is equal to the lattice $\mathbb{D}_4$, and the map $q \mapsto q(1+i)$ is a similarity of ratio $\sqrt{2}$ which maps $\mathbb{D}_4^*$ onto $\mathbb{D}_4$. The subgroup W' of unitary automorphisms of* $\mathrm{Aut}(\mathbb{D}_4)$ *is the set of right multiplications by the* 24 *units*

$$\left\{\pm 1, \pm i, \pm j, \pm k, \frac{\pm 1 \pm i \pm j \pm k}{2}\right\}$$

of $\mathfrak{M}$. □

4.4 The $\mathbb{D}_n^+$ Packing and the $\mathbb{E}_8$ Lattice

Let $n \geq 8$ be an integer, and let $e = \frac{1}{2}\sum_{i=1}^n \varepsilon_i$.

Definition 4.4.1. *We denote by $\mathbb{D}_n^+$ the packing* $\mathbb{D}_n \cup (e+\mathbb{D}_n)$.

The elements of $e+\mathbb{D}_n$ are the vectors of the form $x = \frac{1}{2}\sum_i x_i\,\varepsilon_i$ where the x_i are odd integers whose sum is congruent to n modulo 4. We thus have $N(x) \geq \frac{n}{4}$ on $e+\mathbb{D}_n$, and equality holds exactly on the 2^{n-1} vectors $\frac{1}{2}\sum \pm\varepsilon_i$ with an *even* number of minus signs. Since the distance between two distinct vectors in $\mathbb{D}_n$ is at least $\sqrt{2}$, $\mathbb{D}_n^+$ is like $\mathbb{D}_n$ a packing of spheres of radius $\frac{1}{2}\sqrt{2}$, but it is twice denser.

We have the equivalences

$$\mathbb{D}_n^+ \text{ is a lattice} \iff 2e \in \mathbb{D}_n \iff n \equiv 0 \mod 2\,.$$

From now on, we suppose that n is even (and still at least 8).

The lattice $\mathbb{D}_n^+$ is clearly a half-integral lattice of determinant 1 which is integral if and only if n is divisible by 4. This is then even or odd according to whether n is congruent to 0 or to 4 modulo 8. In all cases, $\mathbb{D}_n^+$ is similar to its dual, see Exercise 4.4.2.

Proposition 4.4.2. *For even $n \geq 10$, we have $S(\mathbb{D}_n^+) = S(\mathbb{D}_n)$, $s(\mathbb{D}_n^+) = n(n-1)$, and $\mathrm{Aut}(\mathbb{D}_n^+)$ is isomorphic by restriction to $W(\mathbb{D}_n)$, of order $2^{n-1}\,n!$*

Proof. We have $N(x) \geq \frac{n}{4} > 2$ on $e+\mathbb{D}_n$. This shows that $S(\mathbb{D}_n^+) = S(\mathbb{D}_n)$ and $s(\mathbb{D}_n^+) = n(n-1)$, and also that the subgroup of $\mathrm{Aut}(\mathbb{D}_n)$ which stabilizes $\mathbb{D}_n^+$ is precisely $W(\mathbb{D}_n)$. The inclusion $W(\mathbb{D}_n) \subset \mathrm{Aut}(\mathbb{D}_n^+)$ is now an equality, since norm 2 vectors in $\mathbb{D}_n^+$ generate $\mathbb{D}_n$. □

When $n = 8$, the argument above still identifies $W(\mathbb{D}_8)$ with a subgroup of $\mathrm{Aut}(\mathbb{D}_8^+)$. The remainder of this section is devoted to the case $n = 8$.

Definition 4.4.3. *We denote by $\mathbb{E}_8$ the lattice $\mathbb{D}_8^+$.*

Let $e' = \frac{1}{2}(\varepsilon_1 + \varepsilon_8) - \frac{1}{2}(\varepsilon_2 + \varepsilon_3 + \varepsilon_4 + \varepsilon_5 + \varepsilon_6 + \varepsilon_7)$. Among the numerous reasonable bases for $\mathbb{E}_8$, we distinguish the following two bases:

(i) (*Korkine and Zolotareff basis*)

$$e_1 = e',\ e_2 = \varepsilon_1 + \varepsilon_2,\ e_3 = \varepsilon_1 - \varepsilon_2,\ e_4 = \varepsilon_1 - \varepsilon_3, \dots, e_8 = \varepsilon_1 - \varepsilon_7;$$

(ii) (*Coxeter basis*)

$$e_1' = e',\ e_2' = \varepsilon_1 + \varepsilon_2,\ e_3' = \varepsilon_2 - \varepsilon_1, e_4' = \varepsilon_3 - \varepsilon_2, \dots, e_8' = \varepsilon_7 - \varepsilon_6.$$

The corresponding quadratic forms are respectively

$$W_8 = V_8 - x_2x_3 = U_8 - x_2(x_1 + x_3)$$

and

$$W_8' = x_1^2 + x_2^2 + \cdots + x_8^2 - x_1x_3 - x_4(x_2 + x_3 + x_5) - x_5x_6 - x_6x_7 - x_7x_8.$$

Up to the exchange of x_3 and x_8, the form W_8 is Korkine and Zolotareff's.

Theorem 4.4.4. *1. The elements of $S(\mathbb{E}_8)$ are the $56 \cdot 2 = 112$ vectors $\pm\varepsilon_i \pm \varepsilon_j$, $(1 \leq i < j \leq 8)$ of $S(\mathbb{D}_8)$ together with the $64 \cdot 2 = 128$ vectors $\frac{1}{2}\sum_{i=1}^{n} \pm\varepsilon_i$ with an even number of minus signs; we have $s(\mathbb{E}_8) = 120$.*

2. The group $\mathrm{Aut}(\mathbb{E}_8)$ has order $2^{14} \cdot 3^5 \cdot 5^2 \cdot 7 = 696729600$, and is equal to the Weyl group $W(\mathbb{E}_8)$.

3. $\mathrm{Aut}(\mathbb{E}_8)$ acts transitively on $S(\mathbb{E}_8)$ and on the set of pairs of minimal vectors with a given angle. In particular, it acts transitively on the sets of square and hexagonal lattices of norm 2.

4. The spectrum of minimal vectors in $\mathbb{E}_8$ is $(63, 0)$, $(56, 1)$.

Proof. The statements (1) and (4) are trivial. To prove (2), we exhibit various elements of the Weyl group and study the stabilizer of the two vectors $\varepsilon_7 - \varepsilon_8$ and $\varepsilon_7 + \varepsilon_8$.

Let $e'' = \frac{1}{2}\sum_{k=1}^{8} a_k\,\varepsilon_k$ and $x = \frac{1}{2}\sum_{k=1}^{8} u_k\,\varepsilon_k$ be two vectors in $S(e+\mathbb{D}_8)$. One verifies immediately the following formulae (recall that s_x stands for the reflection along $x \in E$):

(a) $s_x(\varepsilon_i) = \frac{1}{4}\,(3\,\varepsilon_i - \sum_{k\neq i} u_i u_k\,\varepsilon_k)$, whence for $\eta = \pm 1$,
(b) $s_x(\varepsilon_i + \eta\,\varepsilon_j) = \frac{3-\eta\,u_i u_j}{4}\,(\varepsilon_i + \eta\,\varepsilon_j) - \frac{1}{4}\sum_{k\neq i,j} u_k(u_i + \eta\,u_j)\,\varepsilon_k$, as well as
(c) $s_x(e'') = \frac{1}{8}\sum_i(3a_i\,\varepsilon_i - a_i u_i \sum_{k\neq i} u_k\,\varepsilon_k)$.

Formula (b) shows that one can map $\varepsilon_1 + \varepsilon_2$ (resp. $\varepsilon_1 - \varepsilon_2$) onto any element of $S(e+\mathbb{D}_8)$ such that $(\varepsilon_i, \varepsilon_j)$ have signs $(+,+)$ (resp. $(+,-)$). Since $W(\mathbb{D}_8) \subset W(\mathbb{E}_8)$ acts transitively on $S(\mathbb{D}_8)$, $W(\mathbb{E}_8)$ acts transitively on $S(\mathbb{E}_8)$.

Let G be the stabilizer of $\varepsilon_7 + \varepsilon_8$ in $W(\mathbb{E}_8)$. The vectors in $S(\mathbb{E}_8)$ which are orthogonal to $\varepsilon_7 + \varepsilon_8$ are:

(i) the two vectors $\pm(\varepsilon_7 - \varepsilon_8)$;
(ii) the 60 vectors $\pm\varepsilon_i \pm \varepsilon_j$, $1 \le i < j \le 6$;
(iii) the 64 vectors $\pm\frac{1}{2}(\varepsilon_7 - \varepsilon_8 + \sum_{k=1}^{6} u_k\,\varepsilon_k)$ such that $\prod_k u_k = -1$.

We are going to prove that G permutes transitively these 126 vectors. We remark that $\varepsilon_7 + \varepsilon_8$ is invariant under $s_{\varepsilon_i \pm \varepsilon_j}$ for $i,\,j \le 6$, under $s_{\varepsilon_7 - \varepsilon_8}$, and under the reflections s_x whenever $u_8 = -u_7$. One transforms $\varepsilon_7 - \varepsilon_8$ into $-(\varepsilon_7 - \varepsilon_8)$ by $s_{\varepsilon_7 - \varepsilon_8}$ and then $\pm(\varepsilon_7 - \varepsilon_8)$ into any of the 32 vectors $\pm\frac{1}{2}(\varepsilon_7 - \varepsilon_8 + \sum_{k\le 6} v_k \varepsilon_k)$ by suitable symmetries s_x which fix $\varepsilon_7 + \varepsilon_8$. If x is such that $u_1 = u_2 = -1$ and $u_k = 1$ for $k \ge 3$, we have $s_x(e) = \varepsilon_1 + \varepsilon_2$; we then obtain the 60 vectors $\pm\varepsilon_i \pm \varepsilon_j$, $i,\,j \le 6$ by using transformations $s_{\varepsilon_{i'} \pm \varepsilon_{j'}}$.

The calculation of $|W(\mathbb{E}_8)|$ is now easy. Let G' be the stabilizer in G of $\varepsilon_7 - \varepsilon_8$. We have $|W(\mathbb{E}_8)| = 240 \cdot 126 \cdot |G'|$. But the orthogonal in $\mathbb{E}_8$ of $\langle \varepsilon_7 + \varepsilon_8,\, \varepsilon_7 - \varepsilon_8 \rangle = \langle \varepsilon_7,\, \varepsilon_8 \rangle$ is the lattice $\mathbb{D}_6$ (as a matter of fact, ε_n has orthogonal $\mathbb{D}_{n-1}$ in both $\mathbb{D}_n^+$ and $\mathbb{D}_n$). Since G' contains $W(\mathbb{D}_6)$ (which is generated by the $s_{\varepsilon_i \pm \varepsilon_j}$ for $i,\,j \le 6$), we have

$$\begin{aligned} |\operatorname{Aut}(\mathbb{E}_8)| \ge |W(\mathbb{E}_8)| \ge 240 \cdot 126 \cdot |W(\mathbb{D}_6)| &= (2^4 \cdot 3 \cdot 5) \cdot (2 \cdot 3^2 \cdot 7) \cdot (2^5 \cdot 6!) \\ &= 2^{14} \cdot 3^5 \cdot 5^2 \cdot 7. \end{aligned}$$

The parity rule for vectors in $S(e + \mathbb{E}_8)$ shows that the restriction to $\mathbb{D}_6$ of its stabilizer in $\operatorname{Aut}(\mathbb{E}_8)$ is contained in $W(\mathbb{D}_6)$. This proves the opposite inequality for $|\operatorname{Aut}(\mathbb{E}_8)|$.

The transitivity of $\operatorname{Aut}(\mathbb{E}_8)$ on $S(\mathbb{E}_8)$ and on pairs of orthogonal minimal vectors has been proved with (2). To complete the proof of (3), we just have to study the action of G on the set of pairs of minimal vectors with scalar product 1 (the same proof works for the scalar product -1). Let $T = \{x \in S(\mathbb{E}_8) \mid x \cdot (\varepsilon_7 + \varepsilon_8) = 1\}$. The elements of T are:

(i) $\pm\varepsilon_i + \varepsilon_7$, $i \le 6$ (12 vectors);
(ii) $\pm\varepsilon_i + \varepsilon_8$, $i \le 6$ (12 vectors);
(iii) $\frac{1}{2}(\varepsilon_7 + \varepsilon_8 + \sum_{k\le 6} u_k\,\varepsilon_k)$, $u_k \in \{\pm 1\}$, $\prod_k u_k = +1$ (32 vectors).
Hence $|T| = 56$. We now transform $\varepsilon_6 + \varepsilon_7$ into $\pm\varepsilon_i + \varepsilon_7$ by $s_{\varepsilon_i \pm \varepsilon_6}$, then $\pm\varepsilon_i + \varepsilon_7$ into $\pm\varepsilon_i + \varepsilon_8$ by $s_{\varepsilon_7 - \varepsilon_8}$, and finally the 32 remaining vectors are images of $\varepsilon_i + \varepsilon_7$ by reflections s_x with x such that $u_7 = u_8 = 1$. □

Other transitivity properties of $\mathrm{Aut}(\Lambda)$ (for $\Lambda = \mathbb{E}_8$ as well as for $\Lambda = A_n$ or $\Lambda = \mathbb{D}_n$) can be proved, concerning vectors of norm $N > 2$ or sublattices of dimension $m > 2$; see Sections 4.5 and 4.6, and Exercises). The case of dimensions 1 and 2 which have been dealt with above are, however, of a special importance for the next section.

Remark 4.4.5. *Another way for constructing $\mathbb{D}_n^+$ is to use double congruences, in analogy with constructions due to Barnes and Wall, see Exercise 4.4.3 and Sections 8.3 and 8.5.*

4.5 The Lattices $\mathbb{E}_7$ and $\mathbb{E}_6$

Definition 4.5.1. *We denote by $\mathbb{E}_7$ (resp. $\mathbb{E}_6$) the orthogonal in $\mathbb{E}_8$ of the relative lattice generated by $f_8 = \varepsilon_7 + \varepsilon_8$ (resp. by f_8 and $f_7 = \varepsilon_6 + \varepsilon_8$).*

We obtain in this way lattices of respective dimensions 7 and 6. Since f_8 (resp. f_8 and f_7) generate a lattice isometric to $\mathbb{A}_1$ (resp. to $\mathbb{A}_2$), we have $\det(\mathbb{E}_7) = 2$ (resp. $\det(\mathbb{E}_6) = 3$). These two lattices clearly have norm 2 and are generated by norm 2 vectors. Actually, we obtain Korkine and Zolotareff as well as Coxeter bases for them by suppressing the last or the last two vectors in the corresponding bases for $\mathbb{E}_8$. The corresponding quadratic forms $2W_n$ and $2W'_n$ ($n = 7, 6$) are then obtained by suppressing in W_8 and W'_8 the terms containing x_8 or x_8, x_7. The forms W_7 and W_6 are Korkine and Zolorareff's forms Y and X, up to the exchange of x_3 and x_7 or of x_3 and x_6.

We now describe the principal invariants of $\mathbb{E}_7$ and of $\mathbb{E}_6$.

Theorem 4.5.2. *1. The minimal vectors in $\mathbb{E}_7$ are the* 60 *vectors $\pm\varepsilon_i \pm \varepsilon_j \in S(\mathbb{D}_6)$, $(1 \le i < j \le 6)$, the* 2 *vectors $\pm(\varepsilon_7 - \varepsilon_8)$, and the* 64 *vectors $\pm\frac{1}{2}(\varepsilon_8 - \varepsilon_7 + \sum_{k\le 6} u_k\,\varepsilon_k)$, $u_k \in \{\pm 1\}$, $\prod_k u_k = +1$; we have $s(\mathbb{E}_7) = 63$.*
2. The group $\mathrm{Aut}(\mathbb{E}_7)$ has order $2^{10} \cdot 3^4 \cdot 5 \cdot 7 = 2903040$; it coincides with $W(\mathbb{E}_7)$.
3. $\mathrm{Aut}(\mathbb{E}_7)$ acts transitively on $S(\mathbb{E}_7)$.
4. The minimal vectors in $\mathbb{E}_7$ have spectrum $(30,0)$, $(32,1)$.

Theorem 4.5.3. *1. The minimal vectors in $\mathbb{E}_6$ are the* 40 *vectors $\pm\varepsilon_i \pm \varepsilon_j \in S(\mathbb{D}_5)$ $(1 \le i < j \le 5)$ and the* 32 *vectors $\pm\frac{1}{2}(\varepsilon_8 - \varepsilon_7 - \varepsilon_6 + \sum_{k\le 5} u_k\,\varepsilon_k)$, $u_k \in \{\pm 1\}$, $\prod_k u_k = +1$; we have $s(\mathbb{E}_6) = 36$.*

2. *The group* $W(\mathbb{E}_6)$ *has order* $2^7 \cdot 3^4 \cdot 5 = 51840$, *and* $\mathrm{Aut}(\mathbb{E}_6)$ *is the direct product* $\{\pm \mathrm{Id}\} \times W(\mathbb{E}_6)$.
3. $\mathrm{Aut}(\mathbb{E}_6)$ *acts transitively on* $S(\mathbb{E}_6)$.
4. *The minimal vectors in* $\mathbb{E}_6$ *have spectrum* $(15,0)$, $(20,1)$.

Proof of 4.5.2. The minimal vectors in E_8 which are orthogonal to $\varepsilon_7+\varepsilon_8$ have been listed in the proof of Theorem 4.4.4; this proves (1) and (4). In the course of this proof, we have proved the transitivity of $W(\mathbb{E}_8)$ on $S(\mathbb{E}_8) \cap \langle \varepsilon_7 + \varepsilon_8 \rangle^\perp$ by an argument relying solely on the consideration of orthogonal reflections. This proves that $|W(\mathbb{E}_7)|$ and $|\mathrm{Aut}(\mathbb{E}_7)|$ are both of order $|\mathrm{Aut}(\mathbb{E}_8)|/240$ and completes the proof of (3). □

Proof of 5.3. Statement (1) follows immediately from Theorem 4.5.2 (1) by extracting from $S(\mathbb{E}_7)$ the vectors which are orthogonal to $\varepsilon_6 + \varepsilon_8$.

To prove (2), we first remark that $W(\mathbb{E}_7)$ acts transitively on the set of vectors in $S(\mathbb{E}_7)$ with scalar product 1 with $\varepsilon_7 + \varepsilon_8$ (this results from the proof of Theorem 4.5.2 (2)). Consequently, we see that $W(\mathbb{E}_6)$ has order at least $|W(\mathbb{E}_7)|/56 = 2^7 \cdot 3^4 \cdot 5$. We then show that equality holds by identifying the orthogonal in $\mathbb{E}_6$ of $\varepsilon_5 + \varepsilon_6$ with $\mathbb{D}_5$, and noticing that $|W(\mathbb{D}_5)| = 2^4 \cdot 5!$ and that the orbit under $W(\mathbb{E}_6)$ of $\varepsilon_5 + \varepsilon_6$ contains 27 elements, the number of elements in $S(\mathbb{E}_7)$ having scalar product 1 with $\varepsilon_6 - \varepsilon_7$. (List: $\pm\varepsilon_i + \varepsilon_6$ and $\pm\varepsilon_i - \varepsilon_7$, $i \le 5$; $-\varepsilon_7 - \varepsilon_8$; 16 vectors of the form $\frac{1}{2}(\varepsilon_8 - \varepsilon_7 - \varepsilon_6 + \sum_{k\le 5} u_k\, \varepsilon_k)$.)

This argument moreover shows that $-\mathrm{Id}$ does not belong to $W(\mathbb{E}_6)$, because the same remark applies to $W(\mathbb{D}_5)$. Hence $\mathrm{Aut}(\mathbb{E}_6)$ contains the direct product $\{\pm \mathrm{Id}\} \times W(\mathbb{E}_6)$. Since $|\mathrm{Aut}(\mathbb{D}_5)| = 2|W(\mathbb{D}_5)|$, we have the bound $|\mathrm{Aut}(\mathbb{E}_6)| \le 2|W(\mathbb{E}_6)$, which completes the proof of (2). Finally, (3) is proved along the same line as the analogous statements concerning $\mathbb{E}_7$ and $\mathbb{E}_8$. □

We now turn to the study of the lattices dual to $\mathbb{E}_6$ and $\mathbb{E}_7$. We identify with $\mathbb{A}_1$ (resp. to $\mathbb{A}_2$) the relative lattice generated inside $\mathbb{E}_8$ by f_8 (resp. by f_8 and f_7).

With the convention above, $\mathbb{E}_7 \perp \mathbb{A}_1$ is the kernel of the map $x \mapsto x \cdot f_8$ mod 2. This yields the identifications

$$\mathbb{E}_7^* \perp \mathbb{A}_1^* = (\mathbb{E}_7 \perp \mathbb{A}_1)^* = \mathbb{E}_8 \cup \left(\frac{1}{2} f_8 + \mathbb{E}_8\right),$$

which allow us to obtain $\mathbb{E}_7^*$ as $\mathbb{E}_7^* = \mathbb{E}_7 \cup (v + \mathbb{E}_7)$, where v denotes a vector in $(\frac{1}{2} f_8 + \mathbb{E}_8)$ orthogonal to f_8, for instance $\frac{1}{2}\sum_1^6 \varepsilon_k$ or one of the vectors $\varepsilon_i \pm \frac{1}{2}(\varepsilon_7 - \varepsilon_8)$. The following proposition is now clear:

Proposition 4.5.4. *1. The lattice* $\Lambda = \mathbb{E}_7^*$ *possesses the following invariants:* $N(\Lambda) = \frac{3}{2}$, $\det(\Lambda) = \frac{1}{2}$, $s(\Lambda) = 28$; *its minimal vectors are the 24 vectors* $\pm\varepsilon_i \pm \frac{1}{2}(\varepsilon_7 - \varepsilon_8)$, $i \le 6$ *and the 32 vectors* $\frac{1}{2}\sum_{k\le 6} \pm\varepsilon_k$ *with an odd number of minus signs.*

2. *The lattice $\mathbb{E}_7^*$ is half-integral, and $\sqrt{2}\,\mathbb{E}_7^*$ is an integral lattice of norm* 3 *and determinant* 2^6. □

Similarly, $\mathbb{E}_6 \perp \mathbb{A}_2$ is the kernel of the map $x \mapsto x \cdot (f_7 + f_8) \mod 2$, since $x - \alpha f_7 - \beta f_8$ is orthogonal to f_7 and f_8 for $\alpha = \frac{1}{3}(2x \cdot f_7 - x \cdot f_8)$ and $\beta = \frac{1}{3}(-x \cdot f_7 + 2x \cdot f_8)$. We thus have the identifications

$$\mathbb{E}_6^* \perp \mathbb{A}_2^* = (\mathbb{E}_6 \perp \mathbb{A}_2)^* = \mathbb{E}_8 \cup \Big(\frac{1}{3}(f_7 + f_8\Big) + \mathbb{E}_8) \cup \Big(-\frac{1}{3}(f_7 + f_8) + \mathbb{E}_8 \Big),$$

which show that $\mathbb{E}_6^*$ is the union $\mathbb{E}_6^* = \mathbb{E}_6 \cup (w + \mathbb{E}_6) \cup (-w + \mathbb{E}_6)$, where w denotes a vector in $\pm\frac{1}{3}(f_7 + f_8) + \mathbb{E}_8$ orthogonal to f_7 and f_8, for instance $w = \frac{2}{3}(\varepsilon_8 - \varepsilon_7 - \varepsilon_6)$, of norm $\frac{4}{3}$.

The coefficients on the basis $(\varepsilon_1, \ldots, \varepsilon_8)$ for $\mathbb{R}^8$ of an element x of $\mathbb{E}_6^*$ have denominators 1 or 2 if $x \in \mathbb{E}_6$, and 3 or 6 if $x \in \pm w + \mathbb{E}_6$.

Proposition 4.5.5. *1. The lattice $\Lambda = \mathbb{E}_6^*$ possesses the following invariants: $N(\Lambda) = \frac{4}{3}$, $\det(\Lambda) = \frac{1}{3}$, $s(\Lambda) = 27$; its minimal vectors are the two vectors $\pm w$, the* 20 *vectors $\pm\varepsilon_i \pm \frac{1}{3}(\varepsilon_8 - \varepsilon_7 - \varepsilon_6)$, $i \le 5$ and the* 32 *vectors $\pm\big(\frac{1}{6}(\varepsilon_8 - \varepsilon_7 - \varepsilon_6) + \frac{1}{2}\sum_{k\le 5} \pm\varepsilon_k\big)$ with an odd number of minus signs in the inner sum.*

2. *The lattice $\sqrt{3}\,\mathbb{E}_6^*$ is integral of norm* 4 *and determinant* 3^5.

Proof. We have $N(w) = \frac{4}{3}$. This proves (2), and shows moreover that the minimal vectors in $\mathbb{E}_6$ have denominators 3 or 6. The denominator 3 corresponds to vectors in $\pm w + \mathbb{D}_8$. To preserve the norm $\frac{4}{3}$ when introducing nonzero components coming from $\mathbb{D}_8$, we must subtract from w a vector of the form $\varepsilon_8 - \varepsilon_7 - \varepsilon_6 \pm \varepsilon_i$. We get 10 pairs $\pm x$ in this way. On the other hand, the vector $w - e = \frac{1}{6}(\varepsilon_8 - \varepsilon_7 - \varepsilon_6) + \frac{1}{2}(-\varepsilon_1 + \varepsilon_2 + \varepsilon_3 + \varepsilon_4 + \varepsilon_5)$ still has norm $\frac{4}{3}$, and this norm keeps the same value when we add to $w - e$ a vector in $\mathbb{D}_8 \cap \mathbb{E}_6$. We get in this way the list of vectors which lie in $S(\mathbb{E}_6^*)$ but not in $w + \mathbb{D}_8$. We have thus completed the proof of (1). □

Remark 4.5.6. *It can be proved that among perfect lattices of dimension n between* 2 *and* 7, *only $\mathbb{E}_7^*$ is an odd lattice in the normalization which renders it integral and primitive. Moreover, the* 28 *lines defined by its minimal vectors constitute a maximal equiangular family. There is an analogous example in dimension* 23, *discovered by Venkov, of a perfect integral lattice of minimum* 5 *with $s = \frac{n(n+1)}{2} = 276$. For more details, see [Le-Se], [Ven3], and Appendix 2.*

We end this section with a study of the action of the automorphism groups of $\mathbb{E}_6$ and $\mathbb{E}_7$ on the minimal vectors in their dual lattices.

Proposition 4.5.7. *The automorphism groups of $\mathbb{E}_6$ and $\mathbb{E}_7$ act transitively on $S(\mathbb{E}_6^*)$ and $S(\mathbb{E}_7^*)$ respectively.*

Proof. Let us begin with $\mathbb{E}_7^*$. It is clear that the Weyl group of $\mathbb{D}_6 \subset \mathbb{E}_7$ acts transitively on each of the two sets $\pm\varepsilon_i \pm \frac{1}{2}(\varepsilon_7 - \varepsilon_8)$ and $\frac{1}{2}(\sum \pm\varepsilon_i)$, since $W(\mathbb{E}_7) = \mathrm{Aut}(\mathbb{E}_7)$ permutes transitively the vectors $\pm\varepsilon_i$ and induces all system of an even number of sign changes on the $\pm\varepsilon_i$. Finally, the orthogonal symmetry along e' exchanges $\frac{1}{2}\sum_{i=1}^{6} \varepsilon_i$ and $\varepsilon_1 + \frac{1}{2}(\varepsilon_7 - \varepsilon_8)$, since these vectors have difference e'.

Let us now turn to the case of $\mathbb{E}_6^*$. One verifies that the Weyl group $W(\mathbb{E}_6)$ acts transitively on the directions of minimal vectors in $\mathbb{E}_6^*$. (The Weyl group however is not transitive on $S(E_6^*)$.) An argument similar to that we used for $\mathbb{E}_7^*$ shows that $W(\mathbb{E}_6)$ is transitive on each of the three families (with respectively 2, 20 and 32 vectors) described in Proposition 4.5.5 (1).

Let $x = \varepsilon_8 - \varepsilon_7 - \varepsilon_6$ and $y = \frac{1}{2}(\varepsilon_1 + \varepsilon_2 + \varepsilon_3 + \varepsilon_4 + \varepsilon_5)$. We must find elements in $W(\mathbb{E}_6)$ which permute $w = \frac{2}{3}x$, $w' = \frac{1}{3}x + \varepsilon_1$ and $w'' = -\frac{1}{6}x + y$. Since $w' - w'' = e'$, $s_{e'}$ exchanges w' and w'', and, since $w + w'' \in S(\mathbb{E}_6)$, we can transform the orbit of w into that of w'' by making use of $s_{w+w''}$. □

4.6 Graphs and Inclusions Between Root Lattices

The graphs we shall have to handle are finite and *simple* (two vertices are connected by at most one edge). To a graph Γ with ordered set of vertices $(1, 2, \dots, n)$, we attach the matrix $A \in \mathcal{M}_n(\mathbb{R})$ defined by $a_{i,i} = 2$, $a_{i,j} = -1$ if $\{i, j\}$ is an edge of Γ, and $a_{i,j} = 0$ otherwise.

Definition 4.6.1. *Let Γ be a graph as above. The* quadratic form attached *to Γ is the form Q_Γ whose matrix in the canonical basis $(e_1, e_2, \dots, e_n)$ for $E = \mathbb{R}^n$ is the matrix A above. When Q_Γ is positive definite, we endow E with the Euclidean structure defined by Q_Γ, and the* lattice attached *to Γ is the lattice in E with basis $(e_1, e_2, \dots, e_n)$.*

It is clear that Q_Γ is the orthogonal sum of the quadratic forms attached to the various connected components of Γ.

Definition 4.6.2. *We denote by A_n, D_n, E_n the graphs in Figure 4.6.2' below, for which the index n is the number of vertices.*

Theorem 4.6.3. *The graphs in Definition 6.2 define lattices isometric to the lattices denoted by the same letter in the cases of A_n , D_n, E_6, E_7 and E_8. The lattices defined by the graphs E_5, E_4 (and E_3) are isometric to $\mathbb{D}_5$, $\mathbb{A}_4$ (and $\mathbb{A}_1 \perp \mathbb{A}_2$) respectively. (We shall sometimes denote these lattices by $\mathbb{E}_5, \mathbb{E}_4, \mathbb{E}_3$.) The graphs corresponding to these lattices are called* Dynkin diagrams *(or* Coxeter–Dynkin diagrams*).*

Proof. The verification of the statements for $\mathbb{A}_n$, $\mathbb{D}_n$, $\mathbb{E}_6$, $\mathbb{E}_7$ and $\mathbb{E}_8$ is immediate on the Coxeter bases previously defined. □

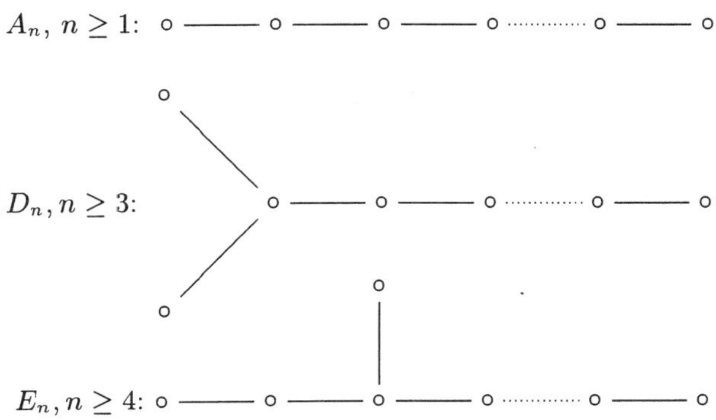

Figure 4.6.2′. The graphs of Definition 4.6.2.

Remark 4.6.4. *The form* $Q_{\mathbb{E}_9}$ *is positive of rank* 8 *; for* $n \geq 10$*, the form* $Q_{\mathbb{E}_n}$ *is a* Lorentz form, *i.e. an indefinite quadratic form with signature* $(n-1, 1)$. *This property of* E_{10} *can be used to construct* $\mathbb{E}_8$ *as the intersection of a lattice in a* 10*-dimensional Lorentz space with the orthogonal of a hyperbolic plane.*

The graphs above show various properties of root lattices, e.g. the existence of sections which still are root lattices. In particular, one verifies the existence of hyperplane sections $\mathbb{A}_{n-1}$ of $\mathbb{A}_n$, $n \geq 2$, $\mathbb{A}_{n-1}$ and $\mathbb{D}_{n-1}$ of $\mathbb{D}_n$, $n \geq 4$, $\mathbb{A}_{n-1}$, $\mathbb{D}_{n-1}$ and $\mathbb{E}_{n-1}$ of $\mathbb{E}_n$, $n \geq 4$, as well as "disconnected sections", e.g. $\mathbb{A}_k \perp \mathbb{A}_{n-k-1}$ of $\mathbb{A}_n$ or $\mathbb{D}_n$ (or even of $\mathbb{E}_n$), or $\mathbb{A}_1 \perp \mathbb{A}_1 \perp \mathbb{A}_{n-3}$ of $\mathbb{D}_n$, or $\mathbb{A}_1 \perp \mathbb{A}_2 \perp \mathbb{A}_{n-4}$ of $\mathbb{E}_n$ $(n \geq 5)$, ...

These graphs also show up relations between automorphism groups and Weyl groups:

Theorem 4.6.5. *In a root lattice, the Weyl group is a normal subgroup of the automorphism group, and the quotient group is isomorphic to the automorphism group of the corresponding graph.*

Proof. That the Weyl group is normal is a straightforward consequence of the formula $\sigma s_x \sigma^{-1} = s_{\sigma x}$, which holds for all $\sigma \in \mathrm{O}(E)$. Now the estimation of the quotient reduces immediately to the case of a connected graph, and we conclude by looking separately to the various cases: we have $W(\Lambda) = \mathrm{Aut}(\Lambda)$ if $\Lambda \simeq \mathbb{A}_1$, $\mathbb{E}_7$ or $\mathbb{E}_8$, $[\mathrm{Aut}(\Lambda) : W(\Lambda)] = 2$ if $\Lambda \simeq \mathbb{A}_n$, $n \geq 2$, if $\Lambda \simeq \mathbb{D}_n$, $n \geq 5$, or if $\Lambda \simeq \mathbb{E}_6$, and finally $\mathrm{Aut}(\Lambda)/W(\Lambda) \simeq S_3$ if $\Lambda \simeq \mathbb{D}_4$. □

However, one cannot read on the graph the inclusions between root lattices which have the same rank. We are now going to find the inclusions $\Lambda \supset \Lambda'$ between root lattices, but we shall limit ourselves to the case where Λ is irreducible: one can easily reduce the general case to this particular case by splitting Λ into its irreducible components; we give complete results in the case where Λ' itself is irreducible in Theorem 4.6.7 below.

We first prove a lemma which goes back to Korkine and Zolotareff:

Lemma 4.6.6. *The sublattice of $\mathbb{A}_n$ generated by n independent vectors $e_1, \dots, e_n$ of norm 2 of $\mathbb{A}_n$ is the whole lattice $\mathbb{A}_n$.*

Proof. By performing suitable permutations of the ε_i and the e_j, we may assume that $e_1 = \varepsilon_0 - \varepsilon_1$. Suppose we are reduced by permutations of the ε_i and base changes of the sublattice generated by the e_j to the case where $e_1 = \varepsilon_0 - \varepsilon_1, \dots, e_k = \varepsilon_0 - \varepsilon_k$ for some index $k < n$. Since the hyperplane generated by $\mathbb{A}_n$ in $\mathbb{R}^{n+1}$ intersects $\langle \varepsilon_0, \dots, \varepsilon_k \rangle^\perp$ along a strict subspace, there exist indices i and j with $i \le k$ and $j > k$ for which e_j is of the form $\pm(\varepsilon_i - \varepsilon_{i'})$. By replacing e_i by $e_i + e_j$ for positive i and then permuting $\varepsilon_{i'}$ and ε_{k+1} for $i' > k+1$ and also e_j and e_{k+1} for $j > k+1$, we reduce to the case where $e_{k+1} = \varepsilon_0 - \varepsilon_{k+1}$. We can now conclude by induction on k. □

We now consider inclusions $\Lambda \supset \Lambda'$ for which both Λ and Λ' are irreducible.

Theorem 4.6.7. *Let Λ and $\Lambda' \subset \Lambda$, $\Lambda' \neq \Lambda$ be two irreducible root lattices of the same dimension. Then one of the following statements holds:*

1. $\Lambda \simeq \mathbb{E}_7$, $\Lambda' \simeq \mathbb{A}_7$, $[\Lambda : \Lambda'] = 2$.
2. $\Lambda \simeq \mathbb{E}_8$, $\Lambda' \simeq \mathbb{D}_8$, $[\Lambda : \Lambda'] = 2$.
3. $\Lambda \simeq \mathbb{E}_8$, $\Lambda' \simeq \mathbb{A}_8$, $[\Lambda : \Lambda'] = 3$.

Proof. Taking into account the relation $\det(\Lambda') = \det(\Lambda)\,[\Lambda : \Lambda']^2$, we can exclude the possibilities $\Lambda \simeq \mathbb{A}_n$ and $\Lambda \simeq \mathbb{E}_6$ (the first of which is also a consequence of Lemma 4.6.6), and we also see that we must have $\Lambda' \simeq \mathbb{A}_n$ if $\Lambda \simeq \mathbb{D}_n$. In this last case, we must find n vectors among the $\pm\varepsilon_i \pm \varepsilon_j$, any two of them having scalar product 1. It can be checked immediately that at most $n-1$ vectors in $S(\mathbb{D}_n)$ may have mutual scalar products all equal to 1 except for $n = 3$, where there exists the system $(\varepsilon_1 + \varepsilon_2,\ \varepsilon_2 + \varepsilon_3,\ \varepsilon_3 + \varepsilon_1)$ which yields the isometry $\mathbb{D}_3 \simeq \mathbb{A}_3$.

We are thus left with the three possibilities stated in the theorem, the existence of which we must now prove.

For $\Lambda \simeq \mathbb{E}_8$ and $\Lambda' \simeq \mathbb{D}_8$, this is the mere definition of $\mathbb{E}_8$.

The lattice generated by the eight vectors $e, \varepsilon_1 + \varepsilon_2, \varepsilon_1 + \varepsilon_3, \dots, \varepsilon_1 + \varepsilon_8 \in \mathbb{E}_8$ is easily seen to be isometric to $\mathbb{A}_8$, whence the existence of Λ' with $\Lambda' \simeq \mathbb{A}_8$.

The seven vectors $\varepsilon_1 - \varepsilon_2, \varepsilon_1 - \varepsilon_3, \dots, \varepsilon_1 - \varepsilon_8$ generate in $\mathbb{E}_8$ a relative sublattice Λ' isometric to $\mathbb{A}_7$ and orthogonal to e, a vector orthogonal to a relative sublattice Λ which is isometric to $\mathbb{E}_7$. This proves the existence of the first possibility in Theorem 4.6.7. □

Remark 4.6.8. *The three inclusions in Theorem 4.6.7 have a simple interpretation in terms of duality. By Proposition 4.2.3, for any divisor r*

of $n+1$, there exists a unique sublattice of $\mathbb{A}_n^*$ containing $\mathbb{A}_n$ to index r. For $(n,r)=(7,2)$ (resp. $(8,3)$), one finds a lattice isometric to $\mathbb{E}_7$ (resp. to E_8), see Section 5.2 of the next chapter. We refer to Exercise 4.4.2 for the inclusion $\mathbb{D}_n \subset \mathbb{D}_n^*$.

Let us now give some examples (to be used in Section 4.9) of inclusions $\Lambda \supset \Lambda'$ in which only Λ is irreducible.

Lemma 4.6.9. *The orthogonal in E_6 of a norm 2 vector is isometric to $\mathbb{A}_5$.*

Proof. Consider the Coxeter basis $e_1' = e', e_2' = \varepsilon_2+\varepsilon_1$, $e_3' = \varepsilon_2-\varepsilon_1$, $e_4' = \varepsilon_3-\varepsilon_2$, $e_5' = \varepsilon_4-\varepsilon_3$, $e_6' = \varepsilon_5-\varepsilon_4$ for $\mathbb{E}_6$. Since $\mathrm{Aut}(\mathbb{E}_6)$ acts transitively on $S(\mathbb{E}_6)$, it suffices to determine ${e_2'}^{\perp}$; but we easily see that e_1, e_3, e_5, e_6 and $e_2+2e_4 = \varepsilon_1-\varepsilon_2+2\varepsilon_3$ constitute a basis for this lattice and that we can moreover replace e_2+2e_4 by $e_2'' = e_2'+2e_4'+e_3'+e_5' = \varepsilon_3+\varepsilon_4$. We then see that $(e_5', e_6', e_2'', e_1', e_3')$ is a Coxeter basis for a lattice isometric to $\mathbb{A}_5$. □

Theorem 4.6.10. *1. $\mathbb{E}_6$ contains to index 2 a lattice isometric to $\mathbb{A}_1 \perp \mathbb{A}_5$.*
2. $\mathbb{E}_7$ contains to index 3 a lattice isometric to $\mathbb{A}_2 \perp \mathbb{A}_5$.
3. $\mathbb{E}_6$ contains to index 3 a lattice isometric to $\mathbb{A}_2 \perp \mathbb{A}_2 \perp \mathbb{A}_2$.

Proof. The first statement is a consequence of Lemma 4.6.9 and of the equality $\det(\mathbb{A}_1 \perp \mathbb{A}_5) = 3\cdot 2^2$. Since $\mathbb{E}_6$ is the orthogonal of any $\mathbb{A}_2 \subset \mathbb{E}_8$, we can embed in $\mathbb{E}_8$ a lattice isometric to $\mathbb{A}_1 \perp \mathbb{A}_1 \perp \mathbb{A}_5$. Since $\mathbb{E}_7$ is the orthogonal of any $\mathbb{A}_1 \subset \mathbb{E}_8$, it contains a lattice isometric to $\mathbb{A}_2 \perp \mathbb{A}_5$, whence (2). Finally, with the notation of Lemma 4.6.9, we verify the equality $\langle e_5', e_6'\rangle^{\perp} = \langle e_1', e_3'\rangle \perp \langle e_2', e_4''\rangle$, where $e_4'' = \frac{1}{2}(\varepsilon_3+\varepsilon_4+\varepsilon_5+\varepsilon_8-\varepsilon_1-\varepsilon_2-\varepsilon_6-\varepsilon_7)$, which proves (3). □

Remark 4.6.11. *We shall give in Section 4.8 another proof of Statement (3) above, relying on Coxeter's construction of $\mathbb{E}_6$, which will immediately show that $\mathrm{Aut}(\mathbb{E}_6)$ acts transitively on its norm 2 hexagonal lattices.*

To find all inclusions $\Lambda' \subset \Lambda$ between root lattices with Λ irreducible amounts to finding all relative lattices $M \subset \Lambda$ such that both M and its orthogonal $M^{\perp}$ in Λ are generated by norm 2 vectors. It suffices to consider the case where M is irreducible and has rank at most half the rank of Λ. We are certain to find all maximal root sublattices in Λ among the lattices $M \perp M^{\perp}$, and we obtain the others by applying the same procedure to M and to $M^{\perp}$. We obtain the following result (for the proof, see Theorem 4.6.10 and Exercises 4.6.2 to 4.6.5):

Theorem 4.6.12. *The list of all possible inclusions $\Lambda' \subset \Lambda$ where Λ and Λ' are root lattices, Λ is irreducible and Λ' is maximal in Λ, ordered by the number of irreducible components of the sublattice, is as follows:*

1. $\mathbb{E}_7, \mathbb{D}_7$; $\mathbb{E}_8, \mathbb{D}_8$; $\mathbb{E}_8, \mathbb{A}_8$.

2. $\mathbb{E}_6, A_1 \perp \mathbb{A}_5$; $\mathbb{E}_7, \mathbb{A}_1 \perp \mathbb{D}_6$; $\mathbb{E}_7, \mathbb{A}_2 \perp \mathbb{A}_5$; $\mathbb{E}_8, \mathbb{A}_1 \perp \mathbb{E}_7$; $\mathbb{E}_8, \mathbb{A}_2 \perp \mathbb{E}_6$; $\mathbb{E}_8, \mathbb{A}_3 \perp \mathbb{D}_5$; $\mathbb{E}_8, \mathbb{A}_4 \perp \mathbb{A}_4$; $\mathbb{E}_8, \mathbb{D}_4 \perp \mathbb{D}_4$; $\mathbb{D}_n, \mathbb{D}_m \perp \mathbb{D}_{n-m}$ $(n \geq 5, m \leq n-3)$.
3. $\mathbb{E}_6, \mathbb{A}_2 \perp (\mathbb{A}_2 \perp \mathbb{A}_2)$; $\mathbb{D}_n, \mathbb{A}_1 \perp (\mathbb{A}_1 \perp \mathbb{D}_{n-2})$ $(n \geq 5)$.
4. $\mathbb{D}_4, \mathbb{A}_1 \perp \mathbb{A}_1 \perp \mathbb{A}_1 \perp \mathbb{A}_1$. □

4.7 Perfection and Eutaxy

We now study irreducible root lattices and the corresponding dual lattices from the point of view of perfection and eutaxy; however, we shall postpone the proofs for $\mathbb{E}_6^*$ and $\mathbb{E}_7^*$, to Section 5.2 of the next chapter, where we prove more general results. The proposition below, which plays an important rôle in the study of eutaxy, has been proved in each of the cases considered in Sections 4.2 to 4.5:

Proposition 4.7.1. *The automorphism group of an irreducible root lattice acts transitively on its set of minimal vectors and of the set of minimal vectors in its dual lattice.* □

Theorem 4.7.2. *The irreducible root lattices are extreme and strongly eutactic; their dual lattices are strongly eutactic, but they are not perfect, except for the following six exceptions (which are thus extreme lattices):* $\mathbb{E}_8^*$, *equal to* $\mathbb{E}_8$; $\mathbb{A}_1^*$, *proportional (but not equal) to* $\mathbb{A}_1$; $\mathbb{A}_2^*$ *and* $\mathbb{D}_4^*$, *similar but not proportional to* $\mathbb{A}_2$ *and* $\mathbb{D}_4$ *respectively;* $\mathbb{E}_6^*$ *and* $\mathbb{E}_7^*$, *which are not similar to root lattices.*

Corollary 4.7.3. *All irreducible root lattices are dual-extreme.*

For an extreme lattice whose dual is eutactic is dual-extreme. □

Proof of 4.7.2. We shall first show that irreducible root lattices are perfect; Proposition 4.7.1 together with Theorem 3.6.6 will then show that they are strongly eutactic, hence extreme by Theorem 3.4.6. (Variant: the automorphism groups of irreducible root lattices are $\mathbb{R}$-irreducible; this argument also shows that their dual lattices are also strongly eutactic.)

We prove the perfection of $\mathbb{A}_n$ by induction on n, starting with the trivial case of dimension 1. The $n-1$ vectors $\varepsilon_0 - \varepsilon_1, \dots, \varepsilon_0 - \varepsilon_{n-1}$ generate a hyperplane section of $\mathbb{A}_n$ which we can assume to be perfect since it is isometric to $\mathbb{A}_{n-1}$, and the n vectors $\varepsilon_i - \varepsilon_n$, $0 \leq i \leq n-1$, of $S(\mathbb{A}_n)$ are independent and all lie outside the previous hyperplane. Proposition 3.5.3 then shows that $\mathbb{A}_n$ is perfect.

The case of $\mathbb{D}_n$ is dealt with by an argument of the same kind starting with dimension 3 where the perfection results from the isometry $\mathbb{A}_3 \simeq \mathbb{D}_3$. For $n \geq 4$, the $n-1$ vectors $\varepsilon_1 - \varepsilon_2, \dots, \varepsilon_1 - \varepsilon_n$ generate a hyperplane section of

$\mathbb{D}_n$ which is isometric to $\mathbb{A}_{n-1}$, hence perfect, and the n vectors $\varepsilon_1+\varepsilon_i$, $i>2$, and $\varepsilon_2+\varepsilon_3$ are independent and lie outside the previous hyperplane.

We finally verify that the three exceptional lattices are perfect by making use of the ascending chain $\mathbb{D}_5 \subset \mathbb{E}_6 \subset \mathbb{E}_7 \subset \mathbb{E}_8$. □

[The arguments above also show that $\mathbb{D}_n^+$ is perfect for any $n \geq 8$, a result which yields a direct proof of the $\mathbb{E}_8$ case.]

We now turn to the dual lattices of irreducible root lattices. To prove that they are strongly eutactic, if we do not want to use the irreducibility over $\mathbb{R}$ of the automorphism group (which we have not proved), we can use Proposition 4.7.1 and the first assertion of Theorem 3.6.6.

The statements about $\mathbb{A}_1^*$, $\mathbb{A}_2^*$, $\mathbb{D}_4^*$ and $\mathbb{E}_8^*$ have also been seen in the course of previous sections. That $\mathbb{A}_n^*$ $(n \geq 3)$ and $\mathbb{D}_n^*$ $(n \geq 5)$ are not perfect follows from the equalities $s(\mathbb{A}_n) = n+1$ and $s(\mathbb{D}_n) = n$. That $\mathbb{E}_6^*$ and $\mathbb{E}_7^*$ are not root lattices is also easy to verify, since the primitive integral lattices to which they are proportional have respective norms 4 and 3.

We are thus left with the question of the perfection of the two lattices $\mathbb{E}_6^*$ and $\mathbb{E}_7^*$, and this will not be completely settled in the present chapter. For $\mathbb{E}_7^*$, it will result from its identification with an extreme lattice named $\mathbb{A}_7^4$. For $\mathbb{E}_6^*$, we are going to reduce the problem to dimension 5.

Let $x = \varepsilon_8-\varepsilon_7-\varepsilon_6$ and let y be the vector $\frac{1}{2}(x+\sum_{i=1}^{5}\varepsilon_i)$ of norm 2 in $\mathbb{E}_6$. Its orthogonal in $\mathbb{E}_6^*$ is a 5-dimensional lattice Λ with the same norm $(\frac{4}{3})$ as $\mathbb{E}_6^*$ and of determinant $\frac{2}{3}$, and moreover with $s(\Lambda) = 15$. It is easily verified that a half-system of minimal vectors in Λ is provided by vectors in the orbit under the group S_5 of $\frac{1}{3}x-\varepsilon_1$ (5 vectors) and of $\frac{1}{6}x+\frac{1}{2}(\varepsilon_1+\varepsilon_2-\varepsilon_3-\varepsilon_4-\varepsilon_5)$ (10 vectors), and that $S(\mathbb{E}_6^*)$ contains six independent vectors outside the hyperplane of Λ. We have thus reduced the proof of the perfection of $\mathbb{E}_6^*$ to that of Λ, which we shall identify in the next chapter with an extreme lattice named $\mathbb{A}_5^3$. We could also prove directly the perfection of Λ via the Gram matrix A of the vectors $\sigma^i(x)$, where σ denotes the cycle $(1,2,3,4,5) \in S_5$. (We easily find that the entries $a_{i,j}$ of A are $\frac{4}{3}$ if $|j-i|=0$, $\frac{1}{3}$ if $|j-i|=1$ or 4 and $-\frac{2}{3}$ if $|j-i|=2$ or 3.) □

[One can also extract from the work of Korkine and Zolotareff (see Chapter 6) that a 5-dimensional lattice with $s \geq 15$ and no $\mathbb{D}_4$-section with the same norm is perfect; this applies to Λ, see Chapter 6, Section 6.4.]

4.8 Some Other Constructions for Root Lattices

In this section, we make alternative constructions for root lattices which give them module structures over orders in imaginary quadratic fields or totally definite quaternion algebras with centre $\mathbb{Q}$. We shall return later (notably in Chapter 8) to this question, using there other algebras and constructing other kind of lattices. We also give some indications on constructions related to coding theory, which consist in "filling holes" in orthogonal sums of lattices. We refer to Appendix 1 for the general algebraic results that we shall need.

We denote by i (resp. ω) a root of unity of order 4 (resp. 3) in some skew-field. We embed $\mathbb{Z}[i]$ and $\mathbb{Z}[\omega]$ into $\mathbb{C}$, the last ring by $\omega \mapsto (-1+i\sqrt{-3})/2$. Given a prime number ℓ, we denote by H_ℓ the quaternion skew-field with centre $\mathbb{Q}$ which is ramified exactly at ℓ and at the infinite prime. For $\ell = 2$ (resp. $\ell = 3$), $\mathbb{H}_\ell$ is defined by a basis $(1, i, j, k)$ such that $i^2 = -1$, $j^2 = -1$ (resp. $j^2 = -3$), $ij = -ji = k$ (and consequently $k^2 = j^2$). We denote by $\mathfrak{M}_\ell$ a maximal order (see Appendix 1, Section 15.6) of H_ℓ which contains i and j ; $\mathfrak{M}_2$ is unique, equal to the Hurwitz order, with basis $(1, i, j, \omega)$ where $\omega = \frac{1}{2}(-1+i+j+k)$, and one takes for $\mathfrak{M}_3$ the order with basis $(1, i, \omega, i\omega)$ where $\omega = \frac{1}{2}(-1+j)$.

The bilinear form $(x, y) \mapsto \mathrm{Trd}(x\bar{y})$ is in all cases a scalar product, for which we have $N(u) = 2$ for any root of unity u. As the minimal vectors in the n-dimensional lattices Λ we are going to construct are associated with roots of unity, we shall consider only the four algebras referred to above, which are the only ones for which roots of unity constitute a system of rank n.

We shall adapt the constructions of $\mathbb{D}_n$ and $\mathbb{D}_n^*$ given in Section 4.3, replacing $\mathbb{Z}$ by a base ring A which is one of the four orders above and the ideal (2) by a suitable two-sided ideal $\mathfrak{a}$.

Together with $\mathbb{D}_n = \{x \in \mathbb{Z}^n \mid \sum_t x_t \equiv 0 \mod 2\}$, we consider the lattice $\mathbb{D}'_n = \{x \in \mathbb{Z}^n \mid x_t \equiv x_1 \mod 2\}$, isometric to $2\mathbb{D}_n^*$. The lattice A^m has dimension $n = 2m$ (resp. $n = 4m$) in the complex (resp. in the quaternionic) case, and determinant $\det(A)^m$, where $\det(A)$ takes one of the values 3, 4, 4, 9 according to whether A is $\mathbb{Z}[\omega]$, $\mathbb{Z}[i]$, $\mathfrak{M}_2$ or $\mathfrak{M}_3$. (*Warning*: in Section 4.3, we used the form $\frac{1}{2}$ Trd.)

Using one congruence modulo $\mathfrak{a}$ multiplies the determinant by the square of the norm of $\mathfrak{a}$ (hence by the fourth power of its reduced norm in the quaternion case). We denote by U the unit group of A, and by U_1 its subgroup defined by the congruence $u \equiv 1 \mod \mathfrak{a}$.

For each choice of the pair $(A, \mathfrak{a})$, we obtain lattices Λ_m and Λ'_m ($\mathbb{D}_n$ and $\mathbb{D}'_n$ if $A = \mathbb{Z}$ and $\mathfrak{a} = 2\mathbb{Z}$) which are dual to each other up to a similarity. We restrict ourselves here to the case where one obtains in this way a root lattice or the dual of such a lattice. Proofs (which are easy, especially when one can make use of Witt's theorem stated in Section 4.10 below) are generally left to the reader. We denote by $(\varepsilon_1, \varepsilon_2, \ldots, \varepsilon_m)$ the canonical basis for K^m, where K is one of the real algebras $\mathbb{R}$, $\mathbb{C}$ or $\mathbb{H}$.

Any minimal vector in Λ_m (resp. of Λ'_m) belongs to one of the following two types:

1. $\alpha\varepsilon_i$, $\alpha \in \mathfrak{a}$;
2. $\zeta\varepsilon_t + \zeta'\varepsilon_{t'}$, $\zeta, \zeta' \in U$, $\zeta + \zeta' \in \mathfrak{a}$
   (resp. $\zeta_1\varepsilon_1 + \cdots + \zeta_m\varepsilon_m$, $\zeta_1, \ldots, \zeta_m \in U$, $\zeta_i - \zeta_j \in \mathfrak{a}$).

Suppose that $\mathfrak{a}$ is a principal ideal, say $\mathfrak{a} = A\alpha$. Then type (1) (resp. type (2)) yields $m\,|U|$ vectors of norm $2\,N_{K/\mathbb{Q}}(\mathfrak{a})$ (resp. $\frac{m(m-1)}{2}\,|U|^2$ vectors of norm 4 in the Λ_m case and $|U|\,|U_1|^{m-1}$ vectors of norm $2m$ in the Λ'_m case).

First, we consider the case where $A = \mathbb{Z}[i]$ and $\mathfrak{a} = (1+i)$. We then have $N_{K/\mathbb{Q}}(\mathfrak{a}) = 2$, and $|U| = |U_1| = 4$. The Λ_m construction applied with the scaling factor $\frac{1}{2}$ yields an integral lattice with determinant 4, whose minimal vectors are the $8m(m-1)$ vectors $\zeta\varepsilon_t + \zeta'\varepsilon_{t'}$ and the $4m$ vectors $\zeta(1+i)\,\varepsilon_t$ ($\zeta^4 = \zeta'^4 = 1$). This lattice is obviously isometric to $\mathbb{D}_{2m}$. Thus $\mathbb{D}_n$ carries for even n a $\mathbb{Z}[i]$-module structure, and one recovers $\mathbb{D}_n^+$ (and in particular $\mathbb{E}_8$) by adjoining to $\mathbb{D}_n$ the vector $\frac{1}{1+i}(\varepsilon_1 + \cdots + \varepsilon_m)$.

Next, consider the case where $A = \mathbb{Z}[\omega]$ and $\mathfrak{a} = (1-\omega)$. We then have $N_{K/\mathbb{Q}}(\mathfrak{a}) = 3$, $|U| = 6$ and $|U_1| = 3$. The Λ'_m construction yields for $m \geq 3$ a norm 6 lattice of determinant $3^m \cdot 3^{2(m-1)} = 3^{3m-2}$, with $6m$ (resp. $6 \cdot 3 + 6 \cdot 3^2 = 72$) minimal vectors if m is ≥ 4 (resp. if $m = 3$). For $m = 3$, $\frac{1}{\sqrt{3}}\Lambda'_3$ is an integral lattice of norm 2 and determinant 3. We have thus found a construction (due to Coxeter) of $\mathbb{E}_6$ over the ring of Eisenstein integers. We prove similarly that Λ_m is an integral norm 4 lattice of determinant 3^{m+2} possessing 3^m pairs of minimal vectors; for $m = 3$, it is isometric to $\sqrt{3}\,\mathbb{E}_6^*$.

The order $\mathfrak{M}_3$ does not yield irreducible root lattices. We shall thus in this section simply notice that $\mathfrak{M}_3$ is isometric to $\mathbb{A}_2 \perp \mathbb{A}_2$ and that the analogue over $\mathfrak{M}_3$ of Coxeter's construction of $\mathbb{E}_6$ endows $\mathbb{E}_6 \perp \mathbb{E}_6$ with a $\mathfrak{M}_3$-module structure.

We finally consider the particularly interesting case where $A = \mathfrak{M}_2$ and $\mathfrak{a} = \mathfrak{M}_2(1+i)$. Let $A' = \mathbb{Z}[i, j, k]$. We have

$$\mathfrak{a} = \{x + yi + zj + tk \in A' \mid x + y + z + t \equiv 0 \mod 2\},$$

whence $\Lambda'_1 \sim \mathfrak{a} \sim \mathbb{D}_4$. We also have $\mathrm{Nrd}_{H_2/\mathbb{Q}}(\mathfrak{a}) = 2$, $|U| = 24$ and $|U_1| = 8$. The Λ'_m construction with the scaling factor $\frac{1}{2}$ yields a lattice of dimension $4m$, norm 2 and determinant 4^{m-2}, with $s = 24 \cdot 8^{m-1}$ for $m > 2$ and $s = 24 \cdot 8 + 24 \cdot 2 = 240$ for $m = 2$. We recognize the $\mathbb{E}_8$ lattice, which is in this way a (left) rank-2 free module over the Hurwitz order.

[Those who do not fear non-associative algebras can also consider the construction of $\mathbb{E}_8$ as a *maximal Cayley order*, which directly interprets the 240 minimal vectors in $\mathbb{E}_8$ as the 240 units in such an order; see [Cox1].]

Curiously, one can also (M. Laïhem) construct $\mathbb{E}_8$ as a rank-2 free module over A', whereas the mixed module structure $A \oplus A'$ can be used to construct $\mathbb{D}_8$.

We now turn to the second part of this section, in which we shall indicate how to obtain certain root lattices by filling holes in orthogonal sums of lattices of small dimensions. Let us start with the lattice $\mathbb{Z}^n$, and first of all with an even dimension $n = 2m \geq 4$. By adjoining to it all $\mathbb{Z}$-linear combinations of the vectors $\frac{1}{2}(\varepsilon_{2k+1} + \varepsilon_{2k+2} + \varepsilon_{2k+3} + \varepsilon_{2k+4})$ for $k \leq m-2$, we obtain a norm 1 lattice of determinant $4^{-(m-1)}$, which, once rescaled to the norm 2, becomes isometric to $\mathbb{D}_n$. (Such a construction could also have been done on $\mathbb{Z}[i]$.) We thus find $\mathbb{D}_4$ by adjunction of $\frac{1}{2}(1,1,1,1)$, then $\mathbb{D}_6$ by adjunction of $e = \frac{1}{2}(1,1,1,1,0,0)$ and $e' = \frac{1}{2}(0,0,1,1,1,1)$ ($e + e'$ is congruent to $\frac{1}{2}(1,1,0,0,1,1)$ modulo $\mathbb{Z}^6$). By embedding $\mathbb{Z}^6$ into $\mathbb{Z}^7$, we see that there is enough room for $\frac{1}{2}(1,0,1,0,1,0,1)$. We find in this way a lattice L_7' of determinant 4^{-3}, which is such that $\sqrt{2}\,L_7' \simeq \mathbb{E}_7$. Similarly, we see that the adjunction of $\frac{1}{2}(1,0,1,0,1,0,1,0)$ to L_4 yields a lattice L_8' similar to $\mathbb{E}_8$.

What we have just done is a standard construction in coding theory: we have first considered a vector subspace C of $\mathbb{F}_2^n$ (one says a *(linear, binary) code of length* n and *dimension* $k = \dim_{\mathbb{F}_2} C$, whose *weight* ω (the minimum number of nonzero components of a nonzero element of C) is at least 4, and then the lattice generated by $\mathbb{Z}^n$ and the lifting in $\mathbb{R}^n$ of C embedded in $\frac{1}{2}\mathbb{Z}^n/\mathbb{Z}^n$.

Note that a more traditional procedure consists in attaching to C the lattice Λ_C generated by $2\mathbb{Z}^n$ together with the pull back of C in $\mathbb{Z}^n$ ("construction A" in [C-S], Chapter 5, Section 2); the lattices D_{2m}, D_{2m}^+, $\mathbb{E}_8$, $\mathbb{E}_7$ could have been also found in this way, directly with their natural norm. It is also interesting to consider weights $\omega \neq 4$.

The pull back in $\mathbb{Z}^n$ of the code generated by the single vector $(1,1,\dots,1)$ (resp. by vectors whose components add to 0) is isometric to $\mathbb{D}_n^*$ (resp. to $\mathbb{D}_n$).

Let us now consider the *Hamming code of length* 7. This is the subspace H of $\mathbb{F}_2^7$ of dimension $k = 4$ whose elements are the vectors of the form $(a,b,c,d,b+c+d,c+d+a,d+a+b)$. We easily see that this code has weight 3, so that its pull back is a lattice Λ_H of norm 3. Its determinant is $2^{2(n-k)} = 64$. We shall see that Λ_H is isometric to $\sqrt{2}\,\mathbb{E}_7^*$.

Recall that the *dual code* of a code C of length n is its orthogonal in $\mathbb{F}_2^n$ for the scalar product $\sum x_i y_i$; for any code C, $\Lambda_{C^\perp}$ is isometric to $2\Lambda_C^*$, as one sees from the formula $\frac{1}{\sqrt{2}}\Lambda_{C^\perp} = (\frac{1}{\sqrt{2}}\Lambda_C)^*$.

Let $H^\perp$ be the dual of H. Extending then H to length 8 by adjoining the sum of its components as an 8th coordinate (the *parity check construction*), one obtains the *extended Hamming code H' of length* 8, generated by the matrix

$$\begin{pmatrix} 1&0&0&0&0&1&1&1\\ 0&1&0&0&1&0&1&1\\ 0&0&1&0&1&1&0&1\\ 0&0&0&1&1&1&1&0 \end{pmatrix}.$$

One easily sees that H' is self-dual. This implies that $\Lambda_{H'}$ has norm 4 and determinant 2^8, and that the norms of all vectors in $\Lambda_{H'}$ are divisible by 4, and shows that $\frac{1}{\sqrt{2}}\Lambda_{H'}$ is an integral even lattice of norm 2 and determinant 1. This last lattice must be isometric to E_8. As a consequence, we have similarities $\Lambda_{H^\perp} \sim \mathbb{E}_7$, whence also $\Lambda_H \sim \mathbb{E}_7^*$.

To fill holes in $\mathbb{Z}^n$ is an important method to construct new lattices with large Hermite invariants and/or large kissing numbers. There is no reason to restrict ourselves to the ring $\mathbb{Z}$ and the lattice $\mathbb{Z}^n$. One can as well handle $\mathbb{Z}[i]^n$ with binary codes via the isomorphism $\mathbb{Z}[i]/(1+i)\mathbb{Z}[i] \simeq \mathbb{F}_2$. Taking $A = \mathbb{Z}[\omega]$ and using *ternary codes* via the isomorphism $A/(1-\omega)A \simeq \mathbb{F}_3$ yields an interpretation of Coxeter's construction of $\mathbb{E}_6$: we recover $\mathbb{E}_6$ by adjoining the vector $\dfrac{1}{1-\omega}(1,1,1)$ to A^3. Replacing the trivial code $(1,1,1)$

by the *ternary tetracode* with generating matrix

$$\begin{pmatrix} 1 & 1 & 1 & 0 \\ 0 & 1 & -1 & 1 \end{pmatrix},$$

we obtain a construction of $\mathbb{E}_8$ over the ring of Eisenstein integers.

One can also give an interpretation of a previous construction of $\mathbb{E}_8$ over the Hurwitz order in terms of *quaternary codes* via the isomorphism $\mathfrak{M}_2/(1+i)\mathfrak{M}_2 \simeq \mathbb{F}_4$.

4.9 Residual Quadratic Forms

Reducing an even lattice Λ modulo a prime number p yields a vector space $\bar{\Lambda} = \Lambda/p\Lambda$ over the finite field $\mathbb{F}_p$ endowed with a quadratic form, namely $\overline{Q} = (x \mapsto \frac{1}{2}N(x))$. We are going to determine the structure of this quadratic space in case Λ is an irreducible root lattice. giving the invariants of $\bar{\Lambda}$ and/or its quotient by the kernel of $\overline{Q}$; we refer to Section 1.6 for the description of quadratic forms over a finite field.

When p is odd, it amounts to considering $\overline{Q}$ or the bilinear space defined by the scalar product modulo p. Then, when $\det(\Lambda)$ is not divisible by p, the residual form is non-degenerate, and is well defined up to proportionality by the quadratic character $\det(\Lambda) \mod p$, which fixes the type ($+$ or $-$) of $\overline{Q}$. We are thus left with one of the following two possibilities: $\Lambda = \mathbb{A}_n$ and $n \equiv -1 \mod p$, or $\Lambda = \mathbb{E}_6$, and $p = 3$.

Now $\mathbb{A}_n$ (resp. $\mathbb{E}_6$) possesses a hyperplane section isometric to $\mathbb{A}_{n-1}$ (resp. $\mathbb{D}_5$), whose determinant is not divisible by p. Hence the kernel of $\overline{Q}$ has dimension 1, and modulo its kernel, the form $\overline{Q}$ has rank $n-1$ (resp. 5) and discriminant $n \mod p$ (resp. is the $+$ form).

We now consider the case where $p = 2$. The form $\overline{Q}$ has then even rank, say $2m$, and its kernel has codimension $2m$ or $2m+1$ (over a field of characteristic 2, a nonsingular form may degenerate). We shall see that the behaviour of $\overline{Q}$ depends on its dimension *modulo* 8.

Theorem 4.9.1. *The form $\overline{Q}$ (of dimension $n = \dim \Lambda$) attached to an irreducible root lattice Λ is characterized up to isometry (and maybe also up to scale) by the following invariants:*

1. *$\Lambda = \mathbb{A}_n$, n even: $\overline{Q}$ has rank n, type $+$ (resp. $-$) if $n \equiv 0, 6$ (resp. $n \equiv 2, 4$) $\mod 8$.*
2. *$\Lambda = \mathbb{A}_n$, n odd: $\overline{Q}$ has rank $n-1$, is nonsingular if $n \equiv 1 \mod 4$, singular otherwise, and has type $+$ for $n \equiv -1 \mod 8$ and $-$ for $n \equiv 3 \mod 8$.*
3. *$\Lambda = \mathbb{D}_n$, n odd: $\overline{Q}$ is singular of rank $n-1$, and is of type $+$ (resp. $-$) for $n \equiv \pm 1 \mod 8$ (resp. for $n \equiv \pm 3 \mod 8$).*
4. *$\Lambda = \mathbb{D}_n$, n even: $\overline{Q}$ has rank $n-2$; its kernel has codimension 1 for $n \equiv 2 \mod 4$ ($\overline{Q}$ is then nonsingular modulo its kernel), and has codimension 2 otherwise, with type $+$ (resp. $-$) modulo its kernel for $n \equiv 0$ (resp. for $n \equiv 4$) $\mod 8$.*

5. *For* $\Lambda = \mathbb{E}_6$ *(resp.* $\mathbb{E}_8$*),* $\overline{Q}$ *is non-degenerate of type* $-$ *(resp.* $+$*).*
6. *For* $\Lambda = \mathbb{E}_7$*,* $\overline{Q}$ *is nonsingular.*

Proof. We prove (1) and (2) by direct calculation for $n \leq 4$ and by induction on n for $n \geq 5$. With the notation of Section 4.2, we have $\overline{Q} = U_n$ mod 2 where

$$U_n(X_1, X_2, \ldots, X_n) = X_1^2 + X_2^2 + \cdots + X_n^2 + X_1X_2 + X_2X_3 + \cdots + X_{n-1}X_n\,.$$

For $n \leq 4$, $\overline{Q}$ (denoted here by $\overline{Q}_n$) takes the forms $\overline{Q}_1 = X_1^2$, $\overline{Q}_2 = X_1^2 + X_1X_2 + X_2^2$, $\overline{Q}_3 = X_2^2 + X_2(X_1+X_3) + (X_1+X_3)^2$, and $\overline{Q}_4 = \overline{Q}_3 + X_4(X_3+X_4)$. The equality

$$\overline{Q}_n = \overline{Q}_3 + X_4(X_3 + X_4 + X_5) + \overline{Q}_{n-4}(X_5, X_6, \ldots, X_n)$$

then proves the result for $n \geq 5$.

We prove (3) and (4) again by direct calculation for $n \leq 5$ and then use (1) and (2) to calculate $\overline{Q}$ (denoted here by $\overline{Q}'_n$) for $n \geq 6$. With the notation of Section 4.3, we have $\overline{Q}'_n = V_n \mod 2$, where $V_n = U_n - X_1X_2$, whence $\overline{Q}'_2 = (X_1 + X_2)^2$, $\overline{Q}'_3 \sim \overline{Q}_3$ (see Proposition 4.3.2), $\overline{Q}'_4 = X_3^2 + X_3(X_1 + X_2 + X_4) + (X_1 + X_2 + X_4)^2$ and $\overline{Q}'_5 = \overline{Q}'_4 + X_5(X_4 + X_5)$. The result for $n \geq 6$ is then a consequence of the equality

$$\overline{Q}'_n = \overline{Q}'_5 + \overline{Q}_{n-5}(X_6, X_7, \ldots, X_n)\,.$$

Finally, we prove (5) and (6) by making use of the fact that $\mathbb{E}_6$, $\mathbb{E}_7$ and $\mathbb{E}_8$ contain to the odd index 3 lattices isometric to $\mathbb{A}_2 \perp \mathbb{A}_2 \perp \mathbb{A}_2$, to $\mathbb{A}_2 \perp \mathbb{A}_5$, and to $\mathbb{A}_8$ respectively (use Theorems 4.6.7 and 4.6.10), and by applying (1) to these sublattices. □

By reduction modulo p, we obtain an homomorphism of $\mathrm{Aut}(\Lambda)$ into the orthogonal group of a vector space over a finite field. The result modulo 2 is particularly interesting in the case of exceptional lattices:

Theorem 4.9.2. *Reduction modulo* 2 *induces isomorphisms*

$$\mathrm{Aut}(\mathbb{E}_6)/\{\pm\,\mathrm{Id}\} \simeq \mathrm{O}_6^-(\mathbb{F}_2),\ \mathrm{Aut}(\mathbb{E}_7)/\{\pm\,\mathrm{Id}\} \simeq \mathrm{O}_7(\mathbb{F}_2),$$

and

$$\mathrm{Aut}(\mathbb{E}_8)/\{\pm\,\mathrm{Id}\} \simeq \mathrm{O}_8^+(\mathbb{F}_2).$$

Proof. Since these three lattices are irreducible, we know by Theorem 1.4.7 of Chapter 1 that the kernel of the reduction modulo 2 is $\{\pm\,\mathrm{Id}\}$. Using Theorem 4.9.1 (5) and (6), we obtain the existence and the injectivity of the three homomorphisms of the theorem. The surjectivity then follows by verifying that the orders of the orthogonal groups are in each case four times (n even) or twice (n odd) the orders of the corresponding automorphism groups obtained in Theorems 4.5.3, 4.5.2 and 4.4.4.

[Since $\mathbb{E}_6$ and $\mathbb{E}_8$ have odd determinants, consideration of the order of the previous groups yields a proof of Theorem 4.9.1 (5) which does not depend on the results of Section 4.6.]

□

The known simplicity theorems for classical groups immediately yield:

Corollary 4.9.3. *The groups* $W(\mathbb{E}_6)^+$, $W(\mathbb{E}_7)^+$ *and* $W(\mathbb{E}_8)^+/\{\pm \operatorname{Id}\}$ *are simple non-Abelian groups.* □

In the same way, combining Proposition 4.2.2 (3) and Theorem 4.9.1 (1) and (2), we construct an isomorphism of the symmetric group S_{n+1} onto a subgroup of $\mathrm{O}_n^+(\mathbb{F}_2)$ if $n \equiv 0, 6 \mod 8$, of $\mathrm{O}_n^-(\mathbb{F}_2)$ if $n \equiv 2, 4 \mod 8$ and of $\mathrm{O}_n(\mathbb{F}_2)$ if $n \equiv 1 \mod 4$. This yields the "exceptional" isomorphisms $S_5 \simeq \mathrm{O}_4^-(\mathbb{F}_2)$ ($\simeq PSL_2(\mathbb{F}_4)$) and $S_6 \simeq \mathrm{O}_5(\mathbb{F}_2)$ ($\simeq \mathrm{PSp}_4(\mathbb{F}_2)$).

Singular forms can also be useful. For instance, reduction modulo 3 of $\mathbb{E}_6$ yields an isomorphism of $W(\mathbb{E}_6)$ onto $\mathrm{O}_5(\mathbb{F}_3)$ ($\simeq \mathrm{PSp}_4(\mathbb{F}_3)$), and hence an isomorphism between the two simple groups $\mathrm{O}_6^-(2)$ and $\mathrm{O}_5(3)$.

4.10 Root Systems

We still denote by E a Euclidean vector space of dimension n. Recall that for $x \in E \smallsetminus \{0\}$, s_x denotes the orthogonal reflection along x (i.e., the orthogonal reflection with respect to the hyperplane $(\mathbb{R}x)^\perp$), characterized by the formula $s_x(y) = y - 2\,\frac{x \cdot y}{x \cdot x}\,x$.

Definition 4.10.1. A root system in E *is a finite subset* R *of* E *which satisfies the following properties:*

1. *R is finite, it does not contain 0, and it generates E.*
2. *For all $x \in R$, s_x stabilizes R.*
3. *For all $x, y \in R$, $2\,\dfrac{x \cdot y}{x \cdot x}$ is an integer.*

The elements of R are called the roots of R. *The subgroup of $\mathrm{O}(E)$ generated by the reflections s_x, $x \in R$ is called the* Weyl group of R, *denoted by $W(R)$. We say that R is* irreducible *if it is not the union of non-empty root systems in orthogonal subspaces of E, and that it is* reduced *if all its roots are primitive (i.e., indivisible).*

There is a natural notion of an isomorphism between root systems; in particular, a root system possesses an *automorphism group*, which is a finite group containing its Weyl group as a normal subgroup.

Root systems are clearly symmetric with respect to the origin of E, and the image by any similarity of a root system is still a root system.

Bourbaki's definition of a root system ([Bou3], Lie VI, p. 142), does not make use of any Euclidean structure, the orthogonal reflections being replaced by reflections parallel to the roots with respect to kernel of nonzero linear forms attached to the roots. It is actually easy to define an appropriate Euclidean structure which transforms such a root system into a root system in the sense of Definition 4.10.1. This is done by an averaging argument, to be used further in Chapter 11: the average of the transforms under the Weyl group of an arbitrary positive definite quadratic form is again a positive definite quadratic form, which is now invariant under $W(R)$; applying Lemma 2 of [Bou3], p. 143 yields what we want.

We shall often refer to Bourbaki's Lie VI, leaving to the reader the task of translating Bourbaki's statements into Euclidean ones. Although the non–Euclidean definition is more intrinsic, the Euclidean definition we have chosen is more suitable for lattice theory. Other references on reflection groups and root systems: [G-B], [Hmp1], [Hmp2], [Kan].

The following proposition is a direct consequence of the definition of root systems:

Proposition 4.10.2. *Let Λ be a lattice in E, and let x be an indivisible vector in Λ. Then s_x stabilizes Λ if and only if $2\,\frac{x\cdot y}{x\cdot x}$ is an integer for all $y \in \Lambda$. In particular, if R is a root system whose elements are in Λ, Λ is invariant under the orthogonal reflections defined by the elements of R.* □

We shall now study a converse to the proposition above.

Definition 4.10.3. *Let Λ be a lattice. A* root *in Λ is a primitive vector r of Λ such that the orthogonal reflection along r stabilizes Λ.*

Theorem 4.10.4. *Let Λ be a lattice in E.*

1. *Any orthogonal reflection which stabilizes Λ is of the form s_r for some root r in Λ.*
2. *The set R of roots in Λ is a root system in the subspace of E it generates.*
3. *If Λ is an integral lattice, R contains all vectors in Λ of norm 1 or 2, and the norms of the various vectors in R all divide $2\,\mathrm{Ann}(\Lambda^*/\Lambda)$, and in particular $2\,\det(\Lambda)$.*
4. *Still assuming that Λ is integral, the scalar product of a root r with a minimal vector in Λ is equal to 0 if r has odd norm, and to 0 or $\pm\frac{1}{2}N(r)$ otherwise. In particular, if x and y are two minimal vectors and if one of them is a root, they generate a lattice similar to $\mathbb{Z}^2$ or to $\mathbb{A}_2$.*

Proof. Let σ be an orthogonal reflection which stabilizes Λ, and let H be the hyperplane of σ. For all $x \in \Lambda$, $x \notin H$, $y = x - \sigma x$ is a nonzero vector in $H^\perp \cap \Lambda$. This is of the form $k\,r$ for some primitive vector r in Λ and some integer k (both defined up to their sign by σ), and we thus have $\sigma = s_r$, whence (1).

The map $r \mapsto s_r$ defines an injection of the set of pairs $\{\pm r\}$ of elements of R into the finite group $\mathrm{Aut}(\Lambda)$. This shows that R satisfies the first condition in Definition 4.10.1.

Let $r \in R$. For all $x \in \Lambda$, $x - 2\,\frac{r\cdot x}{r\cdot r}\,r$ belongs to Λ, and so does $2\,\frac{r\cdot x}{r\cdot r}\,r$. Since r is primitive, this shows that $2\,\frac{r\cdot x}{r\cdot r}$ is an integer for all $x \in \Lambda$, and in particular for all $x \in R$, which is condition (3) of Definition 4.10.1.

To prove (2), there remains to show that s_r stabilizes R. Let $r, r' \in R$, and let $r'' = s_r(r')$. We have $r'' \cdot r'' = r' \cdot r'$, hence for all $x \in \Lambda$,

$$2\,\frac{x \cdot r''}{r'' \cdot r''} = \frac{2}{r' \cdot r'}\left(r' - 2\,\frac{r \cdot r'}{r \cdot r}\,r\right) \cdot x = 2\,\frac{r' \cdot x}{r' \cdot r'} - \left(2\,\frac{r' \cdot r}{r' \cdot r'}\right)\left(2\,\frac{r \cdot x}{r \cdot r}\right) \in \mathbb{Z}\,.$$

These equalities show that $s_{r''}$ stabilizes Λ. Since $s_r(r')$ is primitive (for $r'' = ky$ implies $r' = ks_r(y)$), we have $s_r(r') \in R$, which proves (2).

The fact that R contains all vectors of norm 1 or 2 in Λ is a consequence of Proposition 4.10.2 (these vectors are of course primitive). To prove (3), there just remains to show the assertion about the norms of the roots. Let r be a root in Λ. Since r is primitive, there exists a basis $\mathcal{B} = (e_1, e_2, \ldots, e_n)$ for Λ such that $e_1 = r$. Let d be the l.c.m. of the scalar products $r \cdot e_i$, $2 \le i \le n$. Since $2\,\frac{r\cdot e_i}{N(r)}$ is an integer, $N(r)$ divides $2d$. Set now $d' = N(r)$ if $N(r)$ is odd, and $d' = \frac{1}{2}\,N(r)$ otherwise. The first row of the Gram matrix of $\mathcal{B}$ is $(r \cdot r, r \cdot e_2, \ldots, r \cdot e_n)$; its elements are all divisible by d'. The last assertion of (3) now follows from the interpretation in terms of Gram matrices of the invariant factors we have given in Section 1.9.

Let us finally prove (4). We know that $N(r)$ is a divisor of $2\,r \cdot x$. If x is minimal, we have $N(r \pm x) \ge N(x)$, an inequality equivalent to

$$N(r) \pm 2r \cdot x \ge 0\,, \text{ i.e. } \frac{2\,|r \cdot x|}{N(r)} \le 1\,.$$

Hence this ratio must take one of the values 0 or ± 1, whence (4). □

We now state the classification of *reduced* root systems; we refer to Bourbaki, Lie VI for the proofs, which occupy a large part of Bourbaki's Lie VI. To this end, we first define four infinite series $\boldsymbol{A}_n$, $n \ge 1$, $\boldsymbol{B}_n$, $n \ge 2$, $\boldsymbol{C}_n$, $n \ge 3$, $\boldsymbol{D}_n$, $n \ge 4$, and five *exceptional systems* $\boldsymbol{E}_6$, $\boldsymbol{E}_7$, $\boldsymbol{E}_8$, $\boldsymbol{F}_4$, $\boldsymbol{G}_2$ of roots, indexed by their rank n. (The three lattices $\mathbb{E}_6$, $\mathbb{E}_7$, $\mathbb{E}_8$ are called the *exceptional root lattices*.) In each case, we give the number $2s$ of roots together with a definition in terms of lattices in a Euclidean space E, equal to $\mathbb{R}^n$ except for $\boldsymbol{A}_n$ and $\boldsymbol{G}_2$, where E is the hyperplane $\sum_0^n x_i = 0$ in $\mathbb{R}^{n+1}$. We shall identify $\mathbb{A}_2$ with the ring $\mathbb{Z}[\omega]$, $\omega^2 + \omega + 1 = 0$ of Eisenstein integers (Proposition 4.2.4).

$\boldsymbol{A}_n = S(\mathbb{A}_n)$ $2s = n(n+1)$ $(n \ge 1)$;
$\boldsymbol{B}_n = \{\pm\varepsilon_i\} \cup \{\pm\varepsilon_i \pm \varepsilon_j\} = S(\mathbb{Z}^n) \cup S(\mathbb{D}_n)$ $2s = 2n^2$ $(n \ge 2)$;
$\boldsymbol{C}_n = \{\pm 2\varepsilon_i\} \cup \{\pm\varepsilon_i \pm \varepsilon_j\} = S(2\mathbb{Z}^n) \cup S(\mathbb{D}_n)$ $2s = 2n^2$ $(n \ge 3)$;
$\boldsymbol{D}_n = S(\mathbb{D}_n)$ $2s = 2n(n-1)$ $(n \ge 4)$;
$\boldsymbol{E}_n = S(\mathbb{E}_n)$ $2s = 72,\ 126,\ 240$ $(n = 6, 7, 8)$;
$\boldsymbol{F}_4 = \{\pm\varepsilon_i\} \cup \{\pm\varepsilon_i \pm \varepsilon_j\} \cup \{\frac{1}{2}(\pm\varepsilon_1 \pm \varepsilon_2 \pm \varepsilon_3 \pm \varepsilon_4)\} = S(\mathbb{D}_4) \cup S(\mathbb{D}_4^*)$
$2s = 48$ $(n = 4)$;

$$G_2 = \{\pm(\varepsilon_i - \varepsilon_j)\} \cup \{\pm(2\varepsilon_i - \varepsilon_j - \varepsilon_k)\} = S(\mathbb{A}_2) \cup S((1-\omega)\mathbb{A}_2)$$
$$2s = 12 \quad (n = 2).$$

[By making use of a similarity which transforms $\mathbb{D}_4^*$ into $\mathbb{D}_4$, we obtain $\boldsymbol{F}_4$ as the set of vectors of norm 2 or 4 in $\mathbb{D}_4$; such an interpretation with duality also works for $\boldsymbol{G}_2$: we have $\boldsymbol{G}_2 = S(\mathbb{A}_2) \cup S(3\mathbb{A}_2^*)$.]

These root systems are irreducible and reduced, no two of them are similar, and the $\mathbb{Z}$-modules they generate are isometric to $\mathbb{A}_n$, $\mathbb{Z}^n$, $\mathbb{D}_n$, $\mathbb{D}_n$, $\mathbb{E}_n$, $\mathbb{D}_4^* \sim \mathbb{D}_4$, and $\mathbb{A}_2$ respectively. Those whose roots all have the same norm, namely $\boldsymbol{A}_n$, $\boldsymbol{D}_n$, and $\boldsymbol{E}_n$, are called *simple root systems*.

We can now state the classification theorem:

Theorem 4.10.5. *An irreducible and reduced root system is similar to one of the systems $\boldsymbol{A}_n, \dots, \boldsymbol{G}_2$ of the list above.*
[The non-reduced systems are similar to $\boldsymbol{B}_n \cup \boldsymbol{C}_n$.] □

The classification of root systems yields easily the fundamental result of this section, namely Witt's theorem, that we stated in the introduction, and which justifies the name "root lattice" given to the lattices $\mathbb{A}_n$, $\mathbb{D}_n$, $\mathbb{E}_n$:

Theorem 4.10.6. *An integral lattice which is generated by vectors of norm 1 or 2 is an orthogonal sum of lattices isometric to one of the lattices*

$$\mathbb{Z},\ \mathbb{A}_n,\ n \geq 1,\ \mathbb{D}_n,\ n \geq 4,\ \mathbb{E}_6,\ \mathbb{E}_7,\ \textit{or}\ \mathbb{E}_8\ .$$

Proof. We first get rid of norm 1 vectors by means of the following proposition:

Proposition 4.10.7. *In an integral lattice, norm 1 vectors generate a relative lattice isometric to a lattice $\mathbb{Z}^m$, which is an orthogonal summand.*

Proof of 4.10.7. Let Λ be an integral lattice, and let $\{\pm e_1, \pm e_2, \dots, \pm e_m\}$ be the set of its norm 1 vectors. For $j \neq i$, we have $|e_i \cdot e_j| < 1$, hence $e_i \cdot e_j = 0$, and the $\pm e_i$ thus generate a lattice Λ' isometric to $\mathbb{Z}^m$. Since Λ' is unimodular, we have the decomposition $\Lambda \simeq \Lambda' \perp \Lambda'^{\perp}$.
[Direct proof: to $x \in \Lambda$, associate $x' = \sum_1^m (x \cdot e_i) e_i$ and $x'' = x - x'$. It is then clear that $x = x' + x''$ with $x' \in \Lambda'$ and $x'' \in \Lambda'^{\perp}$.] □

Proof of 4.10.6. Applying now Proposition 4.10.7, we restrict ourselves to the case of an integral lattice Λ of norm 2 which is generated by its set R of norm 2 vectors; by Proposition 4.10.2, R is a root system, which is moreover obviously reduced. This shows that Λ is the orthogonal sum of the lattices generated by the irreducible components of R. We now just have to consider the case where R is irreducible and reduced. Since all roots have the same norm, the systems $\boldsymbol{B}_n$, $\boldsymbol{C}_n$, $\boldsymbol{F}_4$ and $\boldsymbol{G}_2$ cannot occur, and the only possible systems are thus $\boldsymbol{A}_n$ $\boldsymbol{D}_n$ $\boldsymbol{E}_6$, $\boldsymbol{E}_7$ and $\boldsymbol{E}_8$. □

It would have been more natural to call "root lattice" a lattice generated by its roots according to Definition 4.10.3. It results indeed from Proposition 4.10.4 and Theorem 4.10.6 that a lattice generated by its roots is an orthogonal sum of integral lattices generated by their norm 2 vectors, but we then cannot control the norm, since the various irreducible components may have different norms (a problem which does not arise with Bourbaki's definition of roots). For this reason, it is better to limit ourselves to lattices generated by norm 1 and 2 vectors (and in practice only by norm 2 vectors).

Thanks to the classification of root systems, it is often possible to prove theorems on root systems or root lattices by simply looking at all possible systems. The following theorem, which is not *a priori* evident, since it is not always possible to extract a basis from a given generating subset of a $\mathbb{Z}$-module (even if it is free and finitely generated) is such an example:

Corollary 4.10.8. *An integral lattice generated by vectors of norm* 1 *or* 2 *(resp. of norm* 2*) has a basis constituted by vectors of norm* 1 *or* 2 *(resp. of norm* 2*).*

Proof. It suffices to consider the irreducible case, which is easily dealt with by using Korkine and Zolotareff or Coxeter bases as defined in Sections 4.2 to 4.5. □

It is not possible to enter into the details of the proof of the classification theorem. The basic idea is to attach to each root system a generalized graph, its *Dynkin* or *Coxeter–Dynkin diagram.* In the case of simple root systems, this is a simple graph. A short proof of the classification theorem for simple systems (the most important case in the applications to lattice theory) can be read in Ebeling's book [Eb].

This yields for instance the following proposition, whose proof by examination of all systems is somewhat tedious:

Proposition 4.10.9. *The Coxeter bases for a root lattice are the images of one of them by an automorphism; in particular, the number of scalar products equal to* -1 *in a Coxeter basis* $(e_1, \dots, e_n)$ *for a root lattice with scalar products* $e_i.e_j \in \{0, -1\}$ *for* $i < j$ *is an invariant of the lattice.* □

Theorem 4.10.10. *Let* Λ *be an irreducible root lattice and let* G *be the subgroup of* $\mathrm{Aut}(\Lambda)$ *generated by the reflections which stabilize* Λ*. Then* G *is the whole group* $\mathrm{Aut}(\Lambda)$ *if* Λ *is isometric to* $\mathbb{A}_1 \sim \mathbb{Z}$, $\mathbb{A}_2$, $\mathbb{A}_3$, $\mathbb{D}_n$ $(n \geq 4)$, $\mathbb{E}_7$, *or* $\mathbb{E}_8$, *and is equal to the Weyl group of* Λ *(of index* 2 *in* $\mathrm{Aut}(\Lambda)$*) if* Λ *is isometric to* $\mathbb{A}_n$ $(n \geq 4)$ *or to* $\mathbb{E}_6$.

Proof. There is nothing to prove if Λ is isometric to $\mathbb{A}_1$, $\mathbb{E}_7$ or $\mathbb{E}_8$, since $W(\Lambda) = \mathrm{Aut}(\Lambda)$ in these cases. If $n = 3$ or $n \geq 5$, the collection of vectors in $\mathbb{D}_n$ having norm 2 or 4 constitute a $\boldsymbol{C}_n$ root system whose automorphism group contains strictly $W(\mathbb{D}_n)$. Since $[\mathrm{Aut}(\mathbb{D}_n) : W(\mathbb{D}_n)] = 2$ for $n \neq 4$, we have $\mathrm{Aut}(\mathbb{D}_n) = \mathrm{Aut}(\boldsymbol{C}_n)$ in these cases. This solves the $\mathbb{A}_3$ case because of the isometry $\mathbb{A}_3 \simeq \mathbb{D}_3$, and the cases of $\mathbb{A}_2$ and $\mathbb{D}_4$ are solved by the inclusions $\boldsymbol{G}_2 \subset \mathbb{A}_2$ and $\boldsymbol{F}_4 \subset \mathbb{D}_4$. We are thus left with the cases $\Lambda = \mathbb{A}_n$ $(n \geq 4)$ and $\Lambda = \mathbb{E}_6$. The classification of root systems shows that G is then reduced to $W(\Lambda)$, a group which we have calculated in Sections 4.2 and 4.5. [The classification theorem has been used only for the lattices isometric to $\mathbb{A}_n$, $n \geq 4$ and $\mathbb{E}_6$; it could have been avoided by a direct calculation.] □

In practice, we especially consider integral lattices of norm $N \geq 2$. Since no classification is known for lattices of norm $N \geq 3$, the classification of the root systems, applied to a related lattice, is often the only tool at our disposal to distinguish such lattices. This is a crucial tool in the classification of unimodular lattices up to dimension 24 (see Section 5.6). The classification theorem also shows that the adjunction of a norm 2 vector to a root lattice again yields a root lattice (i.e. basically nothing new); we recover in this way the inclusion problem that we have solved in Section 4.6.

We end this section with a short discussion of *Coxeter groups*. A general definition is the following one: this is a group G defined by a generating set S and the only relations $s^2 = 1$ for $s \in S$ and $(ss')^{m(ss')} = 1$ for $s, s' \in S$, where $m(t)$ stands for the order of an element $t \in G$ (cf. Bourbaki, Lie IV, Definition 3, p. 11). We focus on those groups which are finite, a condition which first requires that S be finite. We then consider the space $\mathbb{R}^S$ together with its canonical basis e_s, $(s \in S)$, which we endow with the quadratic form defined by $e_s \cdot e_{s'} = -\cos \frac{\pi}{m(ss')}$. It can then be proved that G is finite if and only if this form is positive definite (Lie V, Th. 2, p. 98). If this condition is satisfied, we obtain on $\mathbb{R}^S$ a Euclidean structure, and it can be verified that this identifies the elements of G with suitable orthogonal reflections of $\mathbb{R}^S$, interpreting in this way finite Coxeter groups as finite subgroups generated by reflections in orthogonal groups of Euclidean spaces. Finite Coxeter groups (or more precisely pairs (G, S)) have been classified. They are actually direct products of *irreducible* groups (Lie VI, Theorem 1, p. 193) belonging to the following list: we first have the Weyl groups of the systems $\boldsymbol{A}_n$, $\boldsymbol{C}_n$, $\boldsymbol{D}_n$, $\boldsymbol{E}_6$, $\boldsymbol{E}_7$, $\boldsymbol{E}_8$, $\boldsymbol{F}_4$, $\boldsymbol{G}_2$ (we have suppressed the system $\boldsymbol{B}_n$ which clearly defines the same group as the system $\boldsymbol{C}_n$), then some other groups associated to "exotic" regular polytopes in dimensions 2, 3 and 4, and which do not appear inside automorphism groups of lattices with the same dimension: the infinite family ("system $\boldsymbol{I}_n$") of dihedral groups D_n of order $2n$ ($n = 5$ or $n \geq 7$; for $n \leq 4$ or $n = 6$ the groups D_n appear as Weyl groups), which are isomorphic to the automorphism group of a regular polygon with n sides, and two exceptional groups ("systems $\boldsymbol{H}_3$ and $\boldsymbol{H}_4$"), the group of the regular icosahedron of order 120 (isomorphic to $A_5 \times C_2$) and a group of order $14400 = 120^2$ associated to a 4-dimensional polytope, which can be constructed in the following way: by lifting $A_5 \subset \mathrm{SO}_3(\mathbb{R})$ into the sphere $S^{(3)}$

embedded in $\mathbb{R}^4 \simeq \mathbb{H}$ (see Appendix 1, Section 15.7), we obtain a group $\widehat{A}_5$ (or $\widetilde{A}_5$) of order 120 whose elements are the vertices of a regular polytope $P \subset \mathbb{H}$. The maps $x \mapsto hxk$ ($h, k \in \widehat{A}_5$) constitute a subgroup of order $\frac{120^2}{2}$ of $\mathrm{Aut}(P)$, which we obtain by adjoining the quaternionic conjugation to the previous group.

Let (G, S) be a finite Coxeter group. Up to conjugacy, the product of the elements of S does not depend on the order in which they are taken, and one can even show that the element obtained in this way does not depend on the choice of $S \subset G$ defining the same group G as a Coxeter group. This element, well defined up to conjugacy, is called "the" *Coxeter transformation*; its order, which is an invariant of (G, S), is called the *Coxeter number of G*, denoted by h (see [Bou3], Lie VI, § 1.11).

Proposition 4.10.11. *The values of the Coxeter numbers of irreducible and reduced root systems are as follows:* $h(\boldsymbol{A}_n) = n+1$, $h(\boldsymbol{B}_n) = h(\boldsymbol{C}_n) = 2n$, $h(\boldsymbol{D}_n) = 2n-2$, $h(\boldsymbol{E}_6) = 12$, $h(\boldsymbol{E}_7) = 18$, $h(\boldsymbol{E}_8) = 30$, $h(\boldsymbol{F}_4) = 12$, $h(\boldsymbol{G}_2) = 6$; *for the roots of irreducible root lattices, we have* $h = \frac{2s}{n}$. □

The Coxeter number plays an important rôle in various formulae; here are two useful instances. The first one can be found in [Bou3], Lie V, Section 6.2, and the second one (which only applies to irreducible systems whose roots have the same length) can be proved by making use of the spectra which we have determined in Sections 4.2 to 4.6:

Proposition 4.10.12. *1. Let R be an irreducible and reduced root system. Then, for all $x \in \mathbb{Q}R$, we have*

$$\sum_{y \in R} \frac{(x \cdot y)^2}{(y \cdot y)} = h\,(x \cdot x)\,.$$

2. For any irreducible root lattice, the number of minimal vectors having a nonzero scalar product with a given minimal vector is equal to $4h-6$. □

Table 4.10.13 below contains an invariant which has not been defined beforehand, namely the *deficiency* $d = n - k$, which is the difference between the dimension n and the maximal number k of orthogonal pairs of roots of minimal length. (Deficiency is the denomination used in [Ker]; it is called *Defekt* in [K-V] and *défaut* in [Mar].) Its calculation is an easy exercise.

Table 4.10.13. Main invariants of root lattices.

Name	dim	det	s	h	Smith	N^*	s^*	d	γ	γ'
$\mathbb{Z}$	1	1	1	1	1	1	1	0	1	1
$\mathbb{A}_n$	$n \geq 2$	$n+1$	$\frac{n(n+1)}{2}$	$n+1$	$n+1$	n	$n+1$	$\lfloor \frac{n}{2} \rfloor$	$\frac{2}{(n+1)^{1/n}}$	$(\frac{2n}{n+1})^{1/2}$
$\mathbb{D}_4$	4	4	12	6	2^2	2	12	0	$2^{1/2}$	$2^{1/2}$
$\mathbb{D}_{2m+1}$	$n \geq 5$	4	$n(n-1)$	$2n-2$	4	4	n	1	$2^{(n-2)/n}$	$2^{1/2}$
$\mathbb{D}_{2m}$	$n \geq 6$	4	$n(n-1)$	$2n-2$	2^2	2	n	0	$2^{(n-2)/n}$	$2^{1/2}$
$\mathbb{E}_6$	6	3	36	12	3	4	27	2	$\frac{2}{3^{1/6}}$	$(\frac{8}{3})^{1/2}$
$\mathbb{E}_7$	7	2	63	18	2	3	28	0	$2^{6/7}$	$3^{1/2}$
$\mathbb{E}_8$	8	1	120	30	1	2	120	0	2	2

[As usual, given an integral lattice Λ, $s^* = s(\Lambda^*)$, and N^* is the norm of the primitive integral scaled copy of Λ^*.]

4.11 Exercises for Chapter 4

In the following exercises, the notation $S_m(\Lambda) = S_m$ will be used to denote the set of primitive vectors of norm m; we set $s_m = \frac{1}{2}|S_m|$. Note that for any even lattice, vectors of norm $2, 4, 6$ are always primitive, whereas $2S_2$ (which may be empty) is the set of divisible norm 8 vectors. For a general viewpoint over exercises on representatives for $\Lambda \bmod 2\Lambda$ or $\Lambda \bmod 3\Lambda$, see [Mar8].

4.2.1. 1. Show that $\mathrm{Aut}(\mathbb{A}_n)$ acts transitively on the set of relative lattices isometric to $\mathbb{A}_m$, $m < n$ contained in $\mathbb{A}_n$.

2. What are the sequences $m_1, \dots, m_k$ such that $\mathbb{A}_n$ contains a relative lattice isometric to $\mathbb{A}_{m_1} \perp \cdots \perp \mathbb{A}_{m_k}$? Show that $\mathrm{Aut}(\mathbb{A}_n)$ acts transitively on the set of sublattices having a given sequence $m_1, \dots, m_k$.

3. Show that $\mathbb{A}_n$ $(n \geq 3)$ contains $\frac{(n+1)n(n-1)}{6}$ hexagonal lattices of norm 2, which are in one-to-one correspondence with the triplets $\{i, j, k\}$.
[Consider for each triple $\{i, j, k\}$ the plane $\langle \varepsilon_i - \varepsilon_j, \varepsilon_i - \varepsilon_k \rangle$.]

4.2.2. The notation $A_i, \alpha_{i,j}$ is that of Section 1.7.

1. Show that in the decomposition as a sum of squares of the form $2U_n$ corresponding to Korkine and Zolotareff bases for $\mathbb{A}_n$, we have $A_i = 1 + \frac{1}{i}$ and $\alpha_{i,j} = \frac{1}{i+1}$.

2. Show that for Coxeter bases (form $2U'_n$), we have $A_i = 1 + \frac{1}{i}$, $\alpha_{i,i+1} = -\frac{i}{i+1}$ and $\alpha_{i,j} = 0$ for $j \geq i+2$.

3. Show that for the form $n \sum_i x_i^2 - 2 \sum_{i<j} x_i x_j$ used in the proof of Proposition 4.2.3, we have $A_i = (n+1)\frac{n+1-i}{n+2-i}$ and $\alpha_{i,j} = -\frac{1}{n+1-i}$.

4.2.3. 1. Show that the norms of nonzero vectors in $\mathbb{A}_2$ are the integers of the form $2m$ such that prime numbers congruent to $-1 \mod 3$ divide m to an even exponent. [Use the law of decomposition of primes in the ring of Eisenstein integers.]

2. Show that for every $n \geq 3$, any even positive integer is the norm of some vector in $\mathbb{A}_n$.

4.2.4. Let $n \geq 3$ be an integer.

1. Show that $\mathbb{A}_n$ contains vectors of both norms 4 and 6.

2. Show that under the action of $\mathrm{Aut}(\mathbb{A}_n)$, norm 4 vectors of $\mathbb{A}_n$ constitute a single orbit.

3. Show that under the action of $\mathrm{Aut}(\mathbb{A}_n)$, norm 6 vectors of $\mathbb{A}_n$ constitute one orbit if $n = 3$ or $n = 4$, and two orbits for $n \geq 5$.

4.2.5. The notation is that of Proposition 4.2.3. Let $\Lambda = \sqrt{n+1}\,\mathbb{A}_n^*$, endowed with a basis $(e_1, \dots, e_n)$ of minimal vectors such that $S(\Lambda) = \{\pm e_1, \dots, \pm e_{n+1}\}$ and $e_{n+1} = -(e_1 + \cdots + e_n)$.

1. Let $S' = \{\pm(e_i + e_j) \mid 1 \leq i < j \leq n+1\}$. Show that the $2\,\frac{n(n+1)}{2}$ vectors in S_2 have norm $N_2 = 2(n-1)$.

2. Show that $N(x) > 2(n-1)$ on Λ except if $x = 0$, if $x \in S$ ($n+1$ pairs of vectors), and if $x \in S'$ ($\frac{n(n+1)}{2}$ pairs of vectors).

3. Suppose that n is odd. Let Λ_p be the even lattice attached to Λ. Show that $N(\Lambda_p) = 2(n-1)$ and $s(\Lambda_p) = \frac{n(n+1)}{2}$.

4.2.6. This exercise is devoted to the hexagonal lattice.

1. Show that the three elements of $\mathbb{A}_2/2\mathbb{A}_2 \smallsetminus \{0\}$ can be represented by three minimal vectors (one for each pair $\pm x$).

2. Show that $\mathrm{Aut}(\mathbb{A}_2)$ acts transitively on S_6.

3. Show that the vectors in S_6 with a given image modulo 3 constitute an $\mathbb{A}_2$ configuration, and that the eight elements of $\mathbb{A}_2/3\mathbb{A}_2 \smallsetminus \{0\}$ can be represented by the six minimal vectors in $\mathbb{A}_2$ together with one pair $\pm x$ with $N(x) = 6$, according to the formula $\frac{3^2-1}{2} = s_2 + \frac{1}{3} s_6$.

4.3.1. Let $n \geq 5$ be an integer.

1. Show that relative lattices isometric to $\mathbb{A}_1 \perp \mathbb{A}_1$ contained in $\mathbb{D}_n$ share out among two orbits under the action of $\mathrm{Aut}(\mathbb{D}_n)$, with orthogonal lattices isometric to $\mathbb{D}_{n-2}$ (of determinant 4; representative: $\langle \varepsilon_1 + \varepsilon_2, \varepsilon_1 - \varepsilon_2 \rangle$ in one case, and to $\mathbb{A}_1 \perp \mathbb{A}_1 \perp \mathbb{D}_{n-4}$ (of determinant 16; representative: $\langle \varepsilon_1 + \varepsilon_2, \varepsilon_3 + \varepsilon_4 \rangle$) in the other case.

2. Show similarly that relative lattices isometric to $\mathbb{A}_3$ also share out among two orbits, with orthogonal lattices isometric to $\mathbb{D}_{n-3}$ in one case and to $2\mathbb{Z} \perp \mathbb{D}_{n-4}$ in the other case. [Representatives: $\langle \varepsilon_1 - \varepsilon_2, \varepsilon_1 - \varepsilon_3, -\varepsilon_2 - \varepsilon_3 \rangle$ and $\langle \varepsilon_1 - \varepsilon_2, \varepsilon_1 - \varepsilon_3, \varepsilon_1 - \varepsilon_4 \rangle$.]

4.3.2. 1. Show that in the decomposition as a sum of squares of the form $2V_n$ attached to Korkine and Zolotareff bases for $\mathbb{D}_n$, we have $A_1 = A_2 = 2$, $A_i = 1$ for $i \geq 3$, $\alpha_{1,j} = \alpha_{2,j} = \frac{1}{2}$ for $j \geq 3$ and $\alpha_{i,j} = 0$ otherwise.

2. Show that for Coxeter bases, the result is $A_1 = A_2 = 2$, $A_i = 1$ for $i \geq 3$, $\alpha_{1,3} = \alpha_{2,3} = -\frac{1}{2}$, $\alpha_{i,i+1} = -1$ for $i \geq 3$ and $\alpha_{i,j} = 0$ otherwise.

3. Show that for D_n^* equipped with the basis $(\varepsilon_1, \dots, \varepsilon_{n-1}, \frac{1}{2}\sum_{i=1}^n \varepsilon_i)$, the result is $A_1 = \cdots = A_{n-1} = 1$, $A_n = \frac{1}{4}$, $\alpha_{i,n} = \frac{1}{4}$ and $\alpha_{i,j} = 0$ otherwise.

4.3.3. Let $n \geq 3$ be an integer.

1. Show that $\mathbb{D}_n$ contains vectors of norm $2m$ for all $m \geq 1$.

2. Show that under the action of $\mathrm{Aut}(\mathbb{D}_n)$, the norm 4 vectors in $\mathbb{D}_n$ constitute one orbit for $n = 3$ and $n = 4$, and two orbits for $n \geq 5$.

3. Show that under the action of $\mathrm{Aut}(\mathbb{D}_n)$, the norm 6 vectors in $\mathbb{D}_n$ constitute one orbit for $n = 3, 4$ and 5, and two orbits for $n \geq 6$.

4.3.4. Let $n \geq 3$ be an integer. Show that $\mathbb{D}_n$ contains $\frac{2n(n-1)(n-2)}{3}$ hexagonal lattices of norm 2. [Associate with any triplet $\{i, j, k\}$ the four planes $\langle \varepsilon_i + \varepsilon_j, \varepsilon_i + \varepsilon_k \rangle$, $\langle \varepsilon_i + \varepsilon_j, \varepsilon_i - \varepsilon_k \rangle$, $\langle \varepsilon_i - \varepsilon_j, \varepsilon_i + \varepsilon_k \rangle$, and $\langle \varepsilon_i - \varepsilon_j, \varepsilon_i - \varepsilon_k \rangle$.]

4.3.5. 1. Show that $\mathrm{Aut}(\mathbb{D}_4)$ acts transitively on each of the sets S_2, S_4, S_6, and that $s_2 = s_4 = 12$, $s_6 = 48$.

2. Show that the nonzero classes of $\mathbb{D}_4/2\mathbb{D}_4$ are represented by twelve pairs of vectors in S_2, and three systems of vectors in S_4 with configuration $S(\mathbb{Z}^4)$, according to the formula $2^4 - 1 = s_2 + \frac{1}{4}s_3$.

3. Show that the pairs of nonzero classes of $\mathbb{D}_4/3\mathbb{D}_4$ are represented by 12 vectors in S_2, 12 vectors in S_4, and 16 systems of vectors in S_6 with configuration $S(\mathbb{A}_2) \sim S(\mathbb{A}_2^*)$, according to the formula $\frac{3^4-1}{2} = s_2 + s_4 + \frac{1}{3}S_6$.

4.4.1. Show that in the Coxeter basis $(e'_1, \dots, e'_8)$, the decomposition as a sum of squares of the quadratic form defined by $\mathbb{E}_8$ is $2(x_1 - \frac{1}{2}x_3)^2 + 2(x_2 - \frac{1}{2}x_4)^2 + \frac{3}{2}(x_3 - \frac{2}{3}x_4)^2 + \frac{5}{6}(x_4 - \frac{6}{5}x_5)^2 + \frac{4}{5}(x_5 - \frac{5}{4}x_6)^2 + \frac{3}{4}(x_6 - \frac{4}{3}x_7)^2 + \frac{2}{3}(x_7 - \frac{3}{2}x_8)^2 + \frac{1}{2}x_8^2$.

4.4.2. The aim of this exercise is to describe for $n \geq 3$ the set of lattices Λ lying between $\mathbb{D}_n$ and $\mathbb{D}_n^*$.

1. Show that for odd n, Λ is unique and is equal to $\mathbb{Z}^n$.

2. Show that when n is even, there are three such lattices, which are $\mathbb{Z}^n$, $\mathbb{D}_n^+$ and $\mathbb{D}_n^- = \langle \mathbb{D}_n, \frac{1}{2}(-\varepsilon_1 + \varepsilon_2 + \cdots + \varepsilon_n) \rangle$.

3. Show that $\mathbb{D}_n^-$ is the image of $\mathbb{D}_n^+$ under a suitable reflection.

4. Show that $(\mathbb{D}_n^+)^* = \mathbb{D}_n^-$ (resp. $= \mathbb{D}_n^+$) if $n \equiv 2$ (resp. $\equiv 0$) $\mod 4$.

5. Show that $\mathrm{Smith}(\sqrt{2}\,\mathbb{D}_n^+) = \mathrm{Smith}(\sqrt{2}\,\mathbb{D}_n^-) = (4, 2^{n-2})$ if $n \equiv 2 \mod 4$.

6. Show that $\mathbb{D}_4^+$ is isometric to $\mathbb{Z}^4$ and that $\sqrt{2}\,\mathbb{D}_6^+$ has minimum 3.

4.4.3. We consider on $\mathbb{Z}^n$ the congruences
(C1) $x_1 \equiv x_2 \equiv \cdots \equiv x_n \mod 2$ and (C2) $x_1 + x_2 + \cdots + x_n \equiv 0 \mod 4$.
We denote by Λ (resp. Λ') the lattice defined by (C1) (resp. by (C1) and (C2)).

1. Show that $\frac{1}{2}\Lambda$ is isometric to $\mathbb{D}_n^*$.

2. Show that $\frac{1}{2}\Lambda'$ is isometric to $\mathbb{D}_n^+$.

4.4.4. Show that for $n \equiv 4 \mod 8$ (resp. for $n \equiv 2 \mod 4$), the lattice of even vectors in $\mathbb{D}_n^+$ (resp. in $\sqrt{2}\,\mathbb{D}_n^+$) is isometric to $\mathbb{D}_n$ (resp. to $\sqrt{2}\,\mathbb{D}_n$).

4.4.5. Consider the following two quadratic form of Korkine and Zolotareff ([K-Z2]):
$W_n = \sum_{i=1}^n x_i^2 + \sum_{1 \leq j < k \leq n} x_j x_k - x_1 x_2 - x_2 x_n + \frac{n-8}{8} x_n^2$ ($n \geq 8$ even) and $T_n = \sum_{i=1}^n x_i^2 + \sum_{1 \leq j < k \leq n} x_j x_k - x_1 x_2 - x_2 x_n - \frac{1}{2} x_{n-1} x_n + \frac{n-9}{8} x_n^2$ ($n \geq 9$ odd).

1. Show that $2\,W_n$ represents the lattice $\mathbb{D}_n^+$.
[Consider the basis $(\varepsilon_1 + \varepsilon_2, \varepsilon_1 - \varepsilon_2, \varepsilon_1 + \varepsilon_3, \dots, \varepsilon_1 + \varepsilon_{n-1}, \frac{\varepsilon_1 + \varepsilon_2 + \cdots + \varepsilon_n}{2})$ for E.]

2. Show that $2\,T_n$ represents the perfect lattice $\mathbb{D}_{n,n-1}$ defined in Section 5.5.

[Consider the basis $(\varepsilon_1+\varepsilon_2,\varepsilon_1-\varepsilon_2,\varepsilon_1+\varepsilon_3,\dots,\varepsilon_1+\varepsilon_{n-1},\frac{\varepsilon_1+\varepsilon_2+\cdots+\varepsilon_{n-2}+\varepsilon_n}{2})$ for E.]

3. Show that this lattice possesses a hyperplane section isometric to the preceding one. [Suppress the $(n-1)$-th vector in the preceding basis.]

4.4.6. Show that $\mathbb{E}_8$ contains 1120 hexagonal lattices of norm 2.

4.4.7. Let Λ be an (integral) odd lattice with psl odd determinant d, and let Λ' be its sublattice of vectors of even norms. We identify the automorphism group G of Λ with a subgroup of $G' = \mathrm{Aut}(\Lambda')$. Let $c \in \Lambda$ be a parity vector for Λ (Definition 1.3.10).

1. Show that there is in Λ'^* a unique lattice Λ'' containing Λ' and of index d in Λ'^*; show that Λ'' contains Λ, and that $[\Lambda'':\Lambda]=[\Lambda:\Lambda']=2$.

2. Show that c belongs to $\Lambda\smallsetminus\Lambda'$ if $\Lambda''/\Lambda' \simeq \mathbb{Z}/4\mathbb{Z}$, and that c belongs to Λ' if $\Lambda''/\Lambda' \simeq \mathbb{Z}/2\mathbb{Z}\times\mathbb{Z}/2\mathbb{Z}$.

3. Show that the set of values taken by the scalar products on Λ'' is $\frac{1}{4}\,\mathbb{Z}$ if Λ''/Λ' is cyclic, and $\frac{1}{2}\,\mathbb{Z}$ otherwise.

4. Suppose that Λ''/Λ' is cyclic. Show that Λ is the only lattice M such that $[\Lambda'':M]=[M:\Lambda']=2$, and that $\mathrm{Aut}(\Lambda)=\mathrm{Aut}(\Lambda')$.

5. Suppose from now on that Λ''/Λ' is not cyclic. Show that there are exactly three lattices M (say $M_1=\Lambda$, M_2, M_3) such that $[\Lambda'':M]=[M:\Lambda']=2$, and prove that $[\mathrm{Aut}(\Lambda'):\mathrm{Aut}(\Lambda)]$ divides 6.

6. Show that M_2 and M_3 are both half-integral or both integral, and that the inequality $[\mathrm{Aut}(\Lambda'):\mathrm{Aut}(\Lambda)]\ge 3$ cannot hold unless M_2 and M_3 are both odd.

4.4.8. Work out the details of Exercise 4.4.7 for $\Lambda=\mathbb{Z}^n$ and $\Lambda'=\mathbb{D}_n$; consider separately the cases $n\equiv 2 \mod 4$, $n\ge 6$; $n=4$; $n\equiv 4 \mod 8$, $n\ge 12$; $n\equiv 0 \mod 8$. [Use Exercise 4.4.2 above.]

4.4.9. 1. Show that the cardinalities of the orbits of $W(\mathbb{D}_8)\subset \mathrm{Aut}(E_8)$ are as follows on *primitive* vectors of norm $N\le 8$ of $\mathbb{E}_8$:

Norm 2: $s_2=120=56+64$;
Norm 4: $s_4=1080=8+560+512$;
Norm 6: $s_6=3360=672+896+1792$;
Norm 8: $s_8=8640=4480+64+3584+512$.

2. Show that $\mathrm{Aut}(\mathbb{E}_8)$ acts transitively on each of the sets S_2,S_4,S_6,S_8.

3. Show that the nonzero classes of $\mathbb{E}_8/2\mathbb{E}_8$ are made of 120 classes of pairs of minimal vectors and of $\frac{1}{8}1080=135$ classes represented by systems of 16 vectors of norm 4 with configuration $S(\mathbb{Z}^8)$, according to the formula $2^8-1=s_2+\frac{1}{8}s_4$.

4. Show that the nonzero pairs $\pm c$ of classes of $\mathbb{E}_8/3\mathbb{E}_8$ are are made of 120 pairs of classes of vectors in S_2, 1080 pairs of classes of vectors in S_4, $\frac{1}{3}3360=1120$ pairs of classes represented by systems of six vectors in S_6 with configuration $S(\mathbb{A}_2)\sim S(\mathbb{A}_2^*)$, and by $\frac{1}{9}8640=960$ classes represented by systems of 18 vectors with configuration $S(\mathbb{A}_8^*)$, according to the formula $\frac{3^8-1}{2}=s_2+s_4+\frac{1}{3}s_6+\frac{1}{9}s_8$.

4.5.1. Show that the decomposition as a sum of squares of the quadratic form attached to a Coxeter basis for E_7 (resp. of E_6) can be deduced from that of $\mathbb{E}_8$ by replacing x_8 (resp. x_8 and x_7) by 0 in the formula proved in Exercise 4.4.1.

4.5.2. Show that the even part of $\sqrt{2}\,\mathbb{E}_7^*$ is isometric to $\sqrt{2}\,\mathbb{E}_7$.

4.5.3. Show that $\mathbb{E}_7$ (resp. $\mathbb{E}_6$) contains 336 (resp. 120) hexagonal lattices of norm 2.

4.6.1. Show that one obtains the same lattices by replacing $a_{i,j} = -1$ by $a_{i,j} = +1$ in Definition 4.6.1 wherever $\{i, j\}$ is an edge of the Dynkin diagram.

4.6.2. Let Λ and Λ' be two irreducible root lattices with $m = \dim \Lambda' < n = \dim \Lambda$.

1. Show that Λ' embeds into Λ if and only if Λ' is smaller than Λ for the ordering $\mathbb{A} \prec \mathbb{D} \prec \mathbb{E}$.

2. Show that the embedding $\mathbb{A}_m \subset \Lambda$ is unique up to an automorphism of Λ if $\Lambda = \mathbb{A}_n$ or if $\Lambda = \mathbb{D}_m$, except if $m = 3$, $\Lambda = \mathbb{D}_n$, and $n \geq 5$, where there exist two orbits of $\mathbb{A}_3$ under the action of $\mathrm{Aut}(\mathbb{D}_n)$. More precisely, verify that one can take the last m vectors as a basis $(e_1, \ldots, e_m)$ for $\mathbb{A}_m$, except if $m = 3$ and $\Lambda = \mathbb{D}_m$, where one can also take $e_3 = -\varepsilon_n - \varepsilon_{n-1}$. [Establish classifications first in $\mathbb{A}_n$ and then in $\mathbb{D}_n$ of those sequences $e_1, \ldots, e_m$ of norm 2 vectors in Λ for which $e_i \cdot e_j = -1$ if $|j - i| = 1$ and $e_i \cdot e_j = 0$ otherwise.]

3. Use the result above to deduce that the embeddings $\mathbb{D}_m \subset \mathbb{D}_n$ are unique up to an automorphism of $\mathbb{D}_n$ for all $m \geq 4$, and can be obtained by the first $m - 1$ vertices of the Dynkin diagram of $\mathbb{D}_n$.

4. List all norm 2 vectors in $\Lambda'^{\perp}$ and show that this lattice is not a root lattice in case $\Lambda = \mathbb{A}_n$ and $\Lambda' = \mathbb{A}_m$.

4.6.3. With the notation of Section 4.3 and Exercise 4.6.2, let $\Lambda = \mathbb{D}_n$, $n \geq 4$, and for $m < n$, let moreover $\Lambda_m = \langle e'_n, e'_{n-1}, \ldots, e'_{n-m+1} \rangle$ and $M_m = \langle e'_1, e'_2, \ldots, e'_m \rangle$, where (e'_i) is a basis for Λ which defines its Dynkin diagram.

1. Show that if $m = 1$, $e'_1, \ldots, e'_{n-1}$ and $e''_n = \varepsilon_{n-1} + \varepsilon_n$ constitute a basis for $\Lambda_m^{\perp}$; deduce from this that $\mathbb{A}_1^{\perp}$ is isometric to $\mathbb{A}_1 \perp \mathbb{D}_{n-2}$ (i.e., to $\mathbb{A}_1 \perp \mathbb{A}_3$ if $n = 5$ and to $\mathbb{A}_1 \perp \mathbb{A}_1 \perp \mathbb{A}_1$ if $n = 4$).

2. Show that for all $m > 1$, $\Lambda_m^{\perp}$ is not a root lattice, except maybe if $m = 3$ and Λ_m belongs to the second orbit of $\mathbb{A}_3$ lattices described in Exercise 4.6.1 2. [One may verify that the norm 2 vectors in $\Lambda_m^{\perp}$ are contained in a subspace of codimension $m - 1$.]

3. Show that in an embedding $\mathbb{A}_3 \hookrightarrow \mathbb{D}_n$ via the identification of $\mathbb{A}_3$ with $\langle \varepsilon_n - \varepsilon_{n-1}, \varepsilon_{n-1} - \varepsilon_{n-2}, -\varepsilon_n - \varepsilon_{n-1} \rangle$, then $\mathbb{A}_3^{\perp} = \langle e'_1, \ldots, e'_{n-3} \rangle \simeq \mathbb{D}_{n-3}$ (with the conventions $\mathbb{D}_2 = \mathbb{A}_1 \perp \mathbb{A}_1$ and $\mathbb{D}_1 = 2\mathbb{Z}$); hence, in $\mathbb{D}_4$, $\mathbb{A}_3^{\perp}$ is not a root lattice.

4. Show that $e''_{m+1} = \varepsilon_{m+2} + \varepsilon_{m+1}, e'_{m+2}, \ldots, e'_n$ is a basis for $M_m^{\perp}$. Deduce from this that $M_m^{\perp}$ is isometric to $\mathbb{D}_{n-m}$ (with the conventions above for $n - m \leq 2$).

4.6.4. In this exercise, we use duality to establish some complements to the preceding exercises.

1. Show that the hyperplane sections of $\mathbb{A}_n$ $(n \geq 2)$ which are isometric to $\mathbb{A}_{n-1}$ constitute a unique orbit under the action of $\mathrm{Aut}(\mathbb{A}_n)$, whose elements are the $n + 1$ relative lattices orthogonal to one of the vectors $\varepsilon_i - \frac{\varepsilon_0 + \varepsilon_1 + \cdots + \varepsilon_n}{n+1} \in \mathbb{A}_n^*$.

2. Show that the sections of $\mathbb{D}_4$ isometric to $\mathbb{A}_3 \simeq \mathbb{D}_3$ constitute one orbit with 12 elements under the action of $\mathrm{Aut}(\mathbb{D}_4)$,

3. We now consider an integer $n \geq 5$. Show that the sections of $\mathbb{D}_n$ which are isometric to $\mathbb{D}_{n-1}$ constitute one orbit with n elements under the action of both $\mathrm{Aut}(\mathbb{D}_n)$ and $W(\mathbb{D}_n)$.

4. Let $r > 1$ be an integer and let $v = a_1\varepsilon_1 + \cdots + a_n\varepsilon_n$ be a vector of $\mathbb{Z}^n \subset \mathbb{D}_n$ such that a_i is odd for $1 \leq i \leq r$ and even for $r + 1 \leq i \leq n$. Show that a vector $x = \pm\varepsilon_i \pm \varepsilon_j$ of norm 2 of $\mathbb{D}_n$ satisfies the congruence $v \cdot x \equiv 0 \mod 2$ if and only if

i, j are both smaller or both larger than r. Deduce from this that the orthogonal of v in $\mathbb{D}_n$ is not an irreducible root lattice except perhaps if $r = n$.

5. Show that the sections of $\mathbb{D}_n$ which are isometric to $\mathbb{A}_{n-1}$ constitute one orbit with 2^{n-1} elements under the action of $\mathrm{Aut}(\mathbb{D}_n)$, whose elements are the relative lattices which are orthogonal to the vectors $\frac{\varepsilon_1 \pm \varepsilon_2 \pm \cdots \pm \varepsilon_n}{2}$ of $\mathbb{D}_n^*$, and that there exist two orbits under the action of the Weyl group.

4.6.5. Show by calculating a determinant that in $\mathbb{E}_6$, $\mathbb{A}_3^\perp$ is not a root lattice.

4.6.6. Let $\Lambda = \mathbb{E}_7$, let $m \leq 3$ and let Λ_m be the relative lattice generated by the vectors $e'_{8-i}, i \leq 3$.

1. Show that $(e'_1, e'_2, e'_3, e'_4, e'_5$ and $e''_6 = \varepsilon_5 + \varepsilon_6)$ is a basis for $\Lambda_1^\perp$, corresponding to the graph D_6 (the edges are 6–1–3–4–2 and 4–5).

2. Show that $(e'_1, e'_2, e'_3, e'_4$ and $e''_5 = \frac{1}{2}(-\varepsilon_1 - \varepsilon_2 - \varepsilon_3 + \varepsilon_4 + \varepsilon_5 + \varepsilon_6 - \varepsilon_7 + \varepsilon_8)$ is a basis for $\Lambda_2^\perp$, corresponding to the graph A_5 (the edges are 1–3–4–2–5).

3. Show that $(e'_1, e'_2, e'_3$ and $e''_4 = \frac{1}{2}(\varepsilon_1 - \varepsilon_2 + \varepsilon_3 + \varepsilon_4 + \varepsilon_5 + \varepsilon_6 - \varepsilon_7 + \varepsilon_8)$ is a basis for $\Lambda_3^\perp$, corresponding to the graph $A_3 \cup A_1$ (the edges are 1–3–4).

4.6.7. 1. Show that the two orbits of $\mathbb{A}_1 \perp \mathbb{A}_1$ and of $\mathbb{A}_3$ in $\mathbb{D}_8$ both collapse to a single one in $\mathbb{E}_8$.

2. Prove the existence inside $\mathbb{E}_8$ of isometries $\mathbb{A}_1^\perp \simeq \mathbb{E}_7$, $\mathbb{A}_2^\perp \simeq \mathbb{E}_6$, $\mathbb{A}_3^\perp \simeq \mathbb{D}_5$, $\mathbb{A}_4^\perp \simeq \mathbb{A}_4$, and $\mathbb{D}_4^\perp \simeq \mathbb{D}_4$.

4.6.8. (Shioda, [Sh2].) Let r be one of the integers $6, 7, 8$. Let $d = 9 - r$, and denote by L_r a lattice endowed with a basis $(u_1, u_2, \ldots, u_r)$ for which $u_i \cdot u_j = \delta_{i,j} + \frac{1}{d}$. In $\mathbb{Q}L_r$, let $v_0 = \frac{1}{3}\sum_{i=1}^r u_i$, and let $\tilde{L}_r = L_r + \mathbb{Z}v_0$.

1. Show that the determinants of L_r and of $\tilde{L}_r$ are $\frac{9}{d}$ and $\frac{1}{d}$ respectively.

2. The letters i, j, k denoting distinct indices, let $\alpha_{i,j} = u_i - u_j$, $\beta_{i,j,k} = v_0 - (u_i + u_j + u_k)$, $\gamma_{i,j} = v_0 - (u_i + u_j)$, $\alpha_i = \alpha_{i,i+1}$, and $\beta = \beta_{1,2,3}$. Show that $L_r^0 = \langle \alpha_1, \ldots, \alpha_{r-1}, \beta \rangle$ is isometric to $\mathbb{E}_r$.
[Consider the Dynkin diagram defined by β and the α_i.]

3. Show that $\tilde{L}_r$ is isometric to $\mathbb{E}_r^*$.

4. Show that for $r = 6$ (resp. $r = 7$; resp. $r = 8$), the pairs of minimal vectors in L_r are represented by 15 vectors $\alpha_{i,j}$, 20 vectors $b_{i,j,k}$, and the vector b_0 (resp. 21 vectors $\alpha_{i,j}$, 35 vectors $\beta_{i,j,k}$, and seven vectors u_i; resp. 28 vectors $\alpha_{i,j}$, 56 vectors $\beta_{i,j,k}$, 28 vectors $\gamma_{i,j}$, and eight vectors u_i).

5. Show that for $r = 6$ (resp. $r = 7$), the pairs of minimal vectors in $\tilde{L}_r$ are represented by six vectors u_i, six vectors u'_i, and 15 vectors $\gamma_{i,j}$ (resp. seven vectors u_i and 21 vectors $\gamma_{i,j}$).

4.8.1. Let $K = \mathbb{Q}(\zeta)$ $(\zeta = \zeta_9)$, let $A = \mathbb{Z}[\zeta]$ be its ring of integers, and let $\mathfrak{p}$ be the prime ideal of A lying above 3. (Thus $\mathfrak{p} = (1 - \zeta)$ and $(3) = \mathfrak{p}^6$.)

1. Show that K has discriminant 3^9 and different $\mathcal{D} = \mathfrak{p}^9$.

2. Show that $\mathrm{Tr}(x\overline{y})$ is an integral $\mathbb{Q}$-bilinear form on $\mathfrak{p}^{-4}$; use this to deduce that $\Lambda = (\mathfrak{p}^2, \frac{1}{9}\mathrm{Tr}(x\overline{y}))$ is an integral lattice of determinant 3.

3. Show that Λ is even and that its six elements $(1 - \zeta)^2(1 + \zeta)\zeta^i, 0 \leq i \leq 5$ are norm 2 vectors which constitute a basis for Λ.

4. Conclude that Λ is isometric to $\mathbb{E}_6$. [One can use results of Section 4.10.]

5. Let G the semi-direct product $\mathrm{Gal}(\mathbb{Q}(\zeta)/\mathbb{Q}) \rtimes \langle\zeta\rangle$. Show that the set of pairs of minimal vectors in $\mathbb{E}_6$ is the union of two G-orbits, that of $(1-\zeta)^2(1+\zeta)$ and that of $(1-\zeta^3)$, with 27 and 9 elements respectively.

6. Show that $\Lambda' = (\mathfrak{p}^3, \frac{1}{9}\,\mathrm{Tr}(x\overline{y}))$ is isometric to $\mathbb{A}_2 \perp \mathbb{A}_2 \perp \mathbb{A}_2$.

7. Show that the dual of Λ' is represented by the ideal $\mathfrak{p}$ and that $\mathfrak{p}^4$ represents an integral lattice of norm 4 similar to $\mathbb{E}_6^*$.

4.8.2. (On the quaternion algebra ramified at 3 and ∞ over $\mathbb{Q}(\cos(2\pi/9))$.)

1. Let $K = \mathbb{Q}(\zeta_9 + \overline{\zeta}_9)$. Show that $\alpha = (1-\zeta_9)(1-\overline{\zeta}_9)$ is a totally positive element of K which generates a prime ideal $\mathfrak{p}$ in $\mathbb{Z}_K$ such that $(3) = \mathfrak{p}^3$, and that K has discriminant 3^4.

2. Show that the quaternion algebra H with centre K ramified at $\mathfrak{p}$ and at the 3 infinite places of K has discriminant 3^{18}.

3. Show that any maximal order $\mathfrak{M}$ of H endowed with the $\mathbb{Z}$-bilinear form $\frac{1}{3}\,\mathrm{Trd}(x\overline{y})$ is a lattice isometric à $(\mathbb{A}_2)^6$.

4. Show that replacing the previous form by $\frac{1}{3}\,\mathrm{Trd}(\frac{1}{\alpha}x\overline{y})$ yields a lattice isometric to $\mathbb{E}_6 \perp \mathbb{E}_6$. [To solve Questions 3 and 4, use the exercise above and embed $\mathbb{Q}(\zeta_9)$ as a maximal commutative subfield of H.]

4.8.3. Show that the sublattice of $\mathfrak{M}_3^3$ defined by the congruences $x_1 \equiv x_2 \equiv x_3$ modulo the ideal of reduced norm 3 of $\mathfrak{M}_3$ is isometric to $\sqrt{3}\,(\mathbb{E}_6 \perp \mathbb{E}_6)$. [Use the fact that one obtains $\sqrt{3}\,\mathbb{E}_6$ by the analogous congruences on the ring of Eisenstein integers.]

4.8.4. Let H be the skew-field of the usual quaternions over $k = \mathbb{Q}(\sqrt{2})$.

1. Show that $(1, \frac{1+i}{\sqrt{2}}, \frac{1+j}{\sqrt{2}}, \omega = \frac{-1+i+j+k}{2})$ is a basis over $A = \mathbb{Z}[\sqrt{2}]$ for a maximal order $\mathfrak{M}$ of H which contains the Hurwitz order $\mathfrak{M}_2$.

2. Show that the discriminant of this basis for the reduced trace over k is equal to 1; deduce from this that the discriminant of $\mathfrak{M}$ over $\mathbb{Z}$ for the reduced trace over $\mathbb{Q}$ is 2^{12}.

3. Show that $\Lambda = (\mathfrak{M}, \frac{1}{4}\,\mathrm{Trd}_{H/\mathbb{Q}}(\frac{1}{2-\sqrt{2}}\,x\overline{y}))$ is an even unimodular lattice of norm 2.

4. Conclude that $\Lambda \simeq \mathbb{E}_8$.

5. Prove that $\mathfrak{M}$ contains 48 units of reduced norm 1 over the centre, which are the 24 units of the Hurwitz order, the 12 units $\frac{\pm i \pm j}{\sqrt{2}}$, $\frac{\pm i \pm k}{\sqrt{2}}$, and $\frac{\pm j \pm k}{\sqrt{2}}$, of order 4, and the 12 units $\frac{\pm 1 \pm i}{\sqrt{2}}$, $\frac{\pm 1 \pm j}{\sqrt{2}}$, and $\frac{\pm 1 \pm k}{\sqrt{2}}$, of order 8.

6. Show that $S(\Lambda) = \bigcup (\mathfrak{M}_2^\times u)$ for u running through the subset $\{1,\, v = \frac{1+i}{\sqrt{2}},\, 1-\sqrt{2},\, (1-\sqrt{2})\,v,\, \frac{i+j}{\sqrt{2}} - j\}$ of $\mathfrak{M}$.

4.8.5. Let H be the skew-field of the usual quaternions over $k = \mathbb{Q}(\sqrt{5})$. Set $\tau = \frac{1+\sqrt{5}}{2}$ and $\tau' = \frac{1-\sqrt{5}}{2}$.

1. Show that $(1, i, \omega, \frac{i+\tau j+\tau' k}{2})$ is a basis over $\mathbb{Z}[\tau]$ for a maximal order $\mathfrak{M}$ of H which contains the Hurwitz order $\mathfrak{M}_2$.

2. Show that the set U of units of $\mathfrak{M}$ with reduced norm 1 over the centre is made of the 120 elements $\mathfrak{M}_2^\times\, u$ where $u = 1$ or $u = \frac{i \pm \tau j \pm \tau' k}{2}$.

3. Show that $\Lambda = (\mathfrak{M}, \frac{1}{2}\,\mathrm{Trd}_{H/\mathbb{Q}}(\frac{1}{5-\sqrt{5}}\,x\overline{y}))$ is an even unimodular lattice of norm 2.

4. Conclude that $\Lambda \simeq \mathbb{E}_8$.
5. Show that $S(\Lambda) = U \cup \{\frac{1-\sqrt{5}}{2} U\}$.

4.10.1.1. Show that if two roots r and $r' = k\,r$ are proportional, then k takes one of the values $\pm\frac{1}{2}$, ± 1, or ± 2. (Hence the reduced root systems are those which do not contain any pair $(r, 2r)$ of roots.)

2. Show that for all $n \geq 1$, $\boldsymbol{B}_n \cup \boldsymbol{C}_n$ is a non-reduced root system. (This is the only non-reduced rank-n system, cf. [Bou3], Lie VI, p. 222.)

4.12 Notes on Chapter 4

The root lattices $\mathbb{A}_n$, $\mathbb{D}_n$, $\mathbb{E}_6$, $\mathbb{E}_7$ and $\mathbb{E}_8$ were described for the first time as a whole by Korkine and Zolotareff in [K-Z2], dated 1873. They define them in terms of the corresponding quadratic forms, as it was always the case until the fifties. Their paper contains the definitions of forms U_n, V_n, X, Y, W_n, ($n \geq 8$ even), and T_n, ($n \geq 9$ odd); U_n, V_n, X, Y, W_8 correspond respectively to the lattices $\mathbb{A}_n$, $\mathbb{D}_n$, $\mathbb{E}_6$, $\mathbb{E}_7$, and $\mathbb{E}_8$, and a form Z introduced in 1877 corresponds to the 5-dimensional lattice we used to prove the perfection of $\mathbb{E}_6^*$ (the lattice $\mathbb{A}_5^3$ in Section 5.2 of next chapter).

Beforehand, the only known quadratic forms from the list above were the forms U_n (known to Hermite, who believed wrongly that γ_n could be attained on them for all n, a result known to Gauss for $n \leq 3$), and the form V_4, which appeared in [K-Z1] (dated 1872) in which they prove that this is the unique critical form in dimension 4. They prove in [K-Z3] (dated 1877) that U_n, V_n (and also Z) are extreme, leaving to the reader the proofs of the extremality of the other forms, which they need not consider to classify extreme forms of dimension $n \leq 5$.

Korkine and Zolotareff study only partially (in terms of adjoint forms) the duals of root lattices. They focus on the forms U_n, and find in this way their automorphism groups. The first complete study is Coxeter's ([Cox2], 1951), where lattices are considered together with quadratic forms. Coxeter shows that E_6^* and E_7^* are extreme, carries out a detailed study of the lattices which lie between $\mathbb{A}_n$ and $\mathbb{A}_n^*$ (see Section 5.2), and considers the lattices $\mathbb{D}_n^+$ (called $\mathbb{D}_n^2$; he does not, however, notice that they correspond to the forms W_n). It is also in this paper that the classification of root systems and Coxeter groups is applied to the study of extreme forms.

We refer the reader to Bourbaki's historical note to Lie IV-V-VI ([Bou3]) for an analysis of the development of these theories, which originate from the classification of semi-simple complex Lie algebras.

5 Lattices Related to Root Lattices

Introduction

The aim of this chapter is to give several constructions which often yield infinite series of extreme lattices depending on one or two discrete parameters. The lattices $\mathbb{A}_n$ are those which offer the greatest number of applications. After an initial section devoted to a general family found by Barnes of lattices containing $\mathbb{A}_n$, we specialize in Section 5.2 to a particular case previously considered by Coxeter, namely that of lattices which lie between $\mathbb{A}_n$ and $\mathbb{A}_n^*$. Section 5.3 is devoted to another series derived from the $\mathbb{A}_n$ series discovered by Barnes in relation with the classification of 6-dimensional perfect lattices. This series, which he named P_n, was later generalized by Craig, who interpreted $\mathbb{A}_n$ and P_n as the first two terms of a two-parameter family. This family yields the densest known lattices for some dimensions around a few thousand. After a short fifth section devoted to some lattices related to $\mathbb{D}_n$, we end the chapter with two sections devoted to unimodular lattices, which we include here because of the close relations which exist between root systems and unimodular lattices in dimensions up to about 24.

As usual, E denotes an n-dimensional Euclidean vector space.

5.1 The Coxeter–Barnes Lattices $\mathbb{A}_n^r$

Recall (Section 4.2) that the lattice $\mathbb{A}_n$ is the intersection of $\mathbb{Z}^{n+1} \subset \mathbb{R}^{n+1}$ with the hyperplane $E = (\varepsilon_0 + \cdots + \varepsilon_n)^\perp$. A basis for it is provided by the vectors $e_i = \varepsilon_0 - \varepsilon_i$, $1 \le i \le n$.

Definition 5.1.1. *Let $e = e_1 + \cdots + e_n$. For any positive rational number r, we denote by $\mathbb{A}_n^r$ the lattice in E with basis $(e_1, e_2, \dots, e_{n-1}, \frac{1}{r}e)$. Moreover we set $q = \frac{n+1}{r}$.*

Since the vectors $e_1, \dots, e_{n-1}, e$ are independent, $\mathbb{A}_n^r$ is a full lattice. The Gram matrix of the basis above is given by the equations $e_i \cdot e_i = 2$, $e_i \cdot e_j = 1$ for $j \ne i$, $\frac{1}{r}\, e \cdot e_i = \frac{n+1}{r}$ and $\frac{1}{r} e \cdot \frac{1}{r} e = \frac{n(n+1)}{r^2}$. Expressed in terms of the basis (ε_i) of $\mathbb{R}^{n+1}$, we have $e_i = \varepsilon_0 - \varepsilon_i$ and $e = n\varepsilon_0 - \sum_1^n \varepsilon_i$.

Proposition 5.1.2. *1. The dual basis of* $\mathcal{B} = \big(e_1, \dots, e_{n-1}, \frac{1}{r}e = \frac{1}{r}(n\varepsilon_0 - \varepsilon_1 - \dots - \varepsilon_n)\big)$ *is*

$$\mathcal{B}^* = \big(e_1 - e_n, \dots, e_{n-1} - e_n, -r(e_n - \tfrac{1}{n+1}e)\big)$$
$$= \big(e_1 - e_n, \dots, e_{n-1} - e_n, \tfrac{1}{q}(n\varepsilon_n - \varepsilon_0 - \dots - \varepsilon_{n-1})\big).$$

2. The dual of $\mathbb{A}_n^r$ *is isometric to* $\mathbb{A}_n^q$.

3. We have $(\mathbb{A}_n^r)^* = \mathbb{A}_n^q$ *if and only if* r *is an integer which divides* $n + 1$.

Proof. (1) is proved by an easy calculation, and we deduce (2) from (1) by noticing that one transforms $\mathcal{B}^*$ into the natural basis for $\mathbb{A}_n^q$ simply by exchanging ε_0 and ε_n and negating the last vector. To prove (3), it now suffices to determine when $\mathbb{A}_n^q \subset (\mathbb{A}_n^r)^*$. In view of the equations $\varepsilon_0 - \varepsilon_i = (\varepsilon_n - \varepsilon_i) - (\varepsilon_n - \varepsilon_0)$, this inclusion holds if and only if the following two conditions

$$\varepsilon_n - \varepsilon_0 \in (\mathbb{A}_n^r)^* \quad \text{and} \quad \frac{n\varepsilon_n - \varepsilon_0 - \dots - \varepsilon_{n-1}}{q} \in (\mathbb{A}_n^r)^*$$

are satisfied. Writing $\varepsilon_n - \varepsilon_0$ as a linear combination of the vectors in $\mathcal{B}^*$ with integral coefficients λ_i and comparing the components in both sides on $\varepsilon_0, \dots, \varepsilon_{n-1}$, we obtain

$$\lambda_n = -q \quad \text{and} \quad \lambda_i + \frac{\lambda_n}{q} = 0 \, (i = 1, \dots, n-1)$$

which show that the first condition holds if and only if q is an integer. A similar calculation taking into account the second condition yields the system

$$\frac{\mu_n}{q} = -\frac{n}{q} \quad \text{and} \quad \mu_i + \frac{\mu_n}{q} = \frac{1}{q} \, (i = 1, \dots, n-1)$$

with integral coefficients μ_i whose unique solution is

$$\mu_n = -n \quad \text{and} \quad \mu_1 = \dots = \mu_{n-1} = \frac{n+1}{q}.$$

This proves that the second condition is equivalent to the fact that $\frac{n+1}{q}$ is also an integer. Since $rq = n + 1$, we see that we obtain equivalent conditions by replacing q by r. □

We have $\mathbb{A}_n^1 = \mathbb{A}_n$, hence $\mathbb{A}_n^* = \mathbb{A}_n^{n+1}$. We also have the equivalences

$$\mathbb{A}_n^r \subset \mathbb{A}_n^{r'} \iff \frac{1}{r}e \in \Big\langle e_1, \dots, e_{n-1}, \frac{1}{r'}e \Big\rangle \iff \frac{1}{r}e \in \Big\langle \frac{1}{r'}e \Big\rangle$$
$$\iff \frac{r'}{r} \text{ is an integer,}$$

and then $\frac{r'}{r} = [\mathbb{A}_n^{r'} : \mathbb{A}_n^r]$, from which we deduce:

Proposition 5.1.3. *The lattice* $\mathbb{A}_n^r$ *has determinant* $\det(A_n^r) = \frac{n+1}{r^2}$. □

In particular, $\mathbb{A}_n^r \supset \mathbb{A}_n \iff r \in \mathbb{N}$, and when it contains $\mathbb{A}_n$, $\mathbb{A}_n^r$ is integral if and only if $r^2 \mid n+1$ and unimodular if and only if $r^2 = n+1$. Moreover, we see that the conditions $r \in \mathbb{N}$ and $r \mid n+1$ characterize the lattices which satisfy the double inclusion $\mathbb{A}_n \subset \mathbb{A}_n^r \subset \mathbb{A}_n^*$. We know (Proposition 4.2.3) that $\mathbb{A}_n^*/\mathbb{A}_n$ is a cyclic group of order $n+1$. We have thus proved:

Proposition 5.1.4. *The lattices Λ such that $\mathbb{A}_n \subset \Lambda \subset \mathbb{A}_n^*$ are the lattices $\mathbb{A}_n^r$ for which r is an integer which divides $n+1$. When these conditions are satisfied, we have $(\mathbb{A}_n^r)^* = \mathbb{A}_n^q$.* □

Definition 5.1.5. *The lattices $\mathbb{A}_n^r$ are called* Coxeter–Barnes lattices, *and* Coxeter lattices *when r divides $n+1$.*

We have seen in Section 4.2 that the group $\mathrm{Aut}(\mathbb{A}_n)$ is canonically isomorphic to $\{\pm \mathrm{Id}\} \times S_{n+1}$, S_{n+1} being identified with the permutation group on the ε_i, $0 \le i \le n$. The definition of $\mathbb{A}_n^r$ shows that the subgroup $\{\pm \mathrm{Id}\} \times S_n$ of $\mathrm{Aut}(\mathbb{A}_n)$, S_n acting on $\{\varepsilon_1, \dots, \varepsilon_n\}$ (or equivalently on $\{e_1, \dots, e_n\}$) stabilizes $\mathbb{A}_n^r$. We can prove more when r is an integer:

Proposition 5.1.6. *When r is an integer, the subgroup of* $\mathrm{Aut}(\mathbb{A}_n)$ *which stabilizes $\mathbb{A}_n^r$ contains $\{\pm \mathrm{Id}\} \times S_n$, S_n permuting $e_1, \dots, e_n$, and equality holds except if r divides $n+1$, in which case this subgroup is the whole group* $\mathrm{Aut}(\mathbb{A}_n)$*. In particular, we then have* $\mathrm{Aut}(\mathbb{A}_n^r) = \mathrm{Aut}(\mathbb{A}_n)$ *if $\mathbb{A}_n^r$ does not contain norm 2 vectors outside $\mathbb{A}_n$.*

Proof. The equality $e_n = r(\frac{1}{r}e) - \sum_{i<n} e_i$ shows that $\mathbb{A}_n^r$ is invariant under the action of S_n whenever r is an integer. It is well-known and easy to prove that a permutation group of degree $n+1$ which strictly contains S_n is the whole group S_{n+1}. Thus if $\mathrm{Aut}(\mathbb{A}_n^r)$ strictly contains S_n, it contains the transposition which exchanges ε_0 and ε_1. Since this transposition transforms e into $e - \frac{n+1}{r} e_1$, r must divide $n+1$, and the converse is clear, since $\mathbb{A}_n^r$ is then *the* sublattice of $\mathbb{A}_n^*$ which contains $\mathbb{A}_n$ with the index r.
[For non-integral r, $\mathrm{Aut}(\mathbb{A}_n^r)$ may contain strictly $\{\pm \mathrm{Id}\} \times S_n$; see Exercise 5.1.1.] □

We are now going to characterize those lattices $\mathbb{A}_n^r$ with integral r which are still of norm 2 (we have of course $N(\mathbb{A}_n^r) \le N(\mathbb{A}_n)$):

Theorem 5.1.7. (Barnes.) *Let r be a positive integer. Then $\mathbb{A}_n^r$ has norm 2 if and only if $r = 1$, or $r = 2$ and $n \ge 7$, or $r \ge 3$ and $2r \le n+1-\sqrt{n+1}$; the lattice $\mathbb{A}_n^r$ is then extreme. Moreover, it has the same set of minimal vectors as $\mathbb{A}_n$ (i.e., we have $s(\mathbb{A}_n^r) = \frac{n(n+1)}{2}$) whenever it has norm 2 except for the following exceptions: $s(\mathbb{A}_7^2) = 63$, $s(\mathbb{A}_8^2) = 71$, $s(\mathbb{A}_8^3) = 120$, $s(\mathbb{A}_9^3) = 129$, and $s(\mathbb{A}_n^r) = n^2$ for $n \ge 15$ and $n+1-\sqrt{n+1} = 2r$ (i.e. when $n = m^2-1$ and $r = \frac{m^2-m}{2}$ for some integer $m \ge 4$).*

Proof. For $r = 1$, $\mathbb{A}_n^r$ is the root lattice $\mathbb{A}_n$. This proves that the lattices $\mathbb{A}_n^r$ are extreme when they are of norm 2. We now suppose $r > 1$, and use the method sketched in Section 1.8.

Let $x = \sum_{i=1}^{n-1} x_i e_i + y\,\frac{1}{r}e$, $x_i, y \in \mathbb{Z}$ be a nonzero vector $\mathbb{A}_n^r$. We have

$$rx = (r(x_1 + \cdots + x_{n-1}) + ny)\varepsilon_0 - y\varepsilon_n - \sum_{i=1}^{n-1}(rx_i + y)\varepsilon_i\,.$$

Set $y_i = rx_i + y$ for $i < n$ and $y_n = y$. Then

$$N(rx) = \sum_{i=1}^{n} y_i^2 + \Big(\sum_{i=1}^{n} y_i\Big)^2.$$

For the n-tuple $(y_1, \dots, y_n)$ of integers to correspond to a vector $x \in \mathbb{A}_n^r$, it is necessary and sufficient that the y_i should be congruent modulo r to a same element $a \in \mathbb{Z}$. Replacing if need be x by $-x$, we may assume that we have $0 \le a \le \frac{r}{2}$, and x then belongs to $\mathbb{A}_n$ if and only if $a = 0$.

Lemma 5.1.8. *If $a \ne 0$, the shortest vectors, if their norm is not larger than 2, are those for which $a = 1$, and $y_i = 1$ or $1 - r$ for all i.*

Proof of 5.1.8. Let $z = \sum_{k=1}^{n} y_k$. By changing one of the y_i into $y_i + rh$, $h \in \mathbb{Z}$, we replace $N(rx)$ by $N(rx) + 2rh(rh + y_i + z)$. Since the difference must be non-negative, the inequalities $-r \le y_i + z \le +r$ must hold for all i. Summing on i yields the inequalities $-nr \le (n+1)z \le nr$, i.e. $-r < z < +r$, which imply for y_i the bound $-2r < y_i < 2r$. We must thus have $y_i = a - 2r$, $a - r$, a, or $a + r$. But we have $|y_i - y_j| \le 2r$ and $y_i - y_j \equiv 0 \mod r$. Hence the inequality $|y_i - y_j| > r$ may hold only if $|y_i - y_j| = 2r$.

Suppose that $y_i = a + r$ for some index i. We then have $z < 0$, which implies that there exists an index j such that y_j is negative, hence equal to $a - r$, and this is possible only if $z = -a$, since we must have both the inequalities $z \le y_j + r$ and $z \ge -y_i - r$. Let u, v, w be the number of indices such that y_i is equal to $a + r$, a, and $a - r$. We have $z = na + (u - w)r$, hence $w - u = qa$ $(q = \frac{n+1}{r})$, and $N(rx) = u(a + r)^2 + (v + 1)a^2 + w(r - a)^2$. The inequality $N(rx) \le 2r^2$ implies $u = 1$, and then because $r - a \ge \frac{r}{2}$: $w \le 3$, hence $q \le 2$, i.e. $r \ge \frac{n+1}{2}$, which contradicts the hypotheses of the lemma. Similarly, y_i can never be equal to $a - 2r$.

We have thus proved that the y_i are equal to a or to $a - r$, and we must now prove that a is indeed equal to 1.

Let k be the number of indices for which $y_i = a - r$ (we thus have $y_i = a$ for $n - k$ values of i). The norm of rx is the function

$$f_k(a) = k(r - a)^2 + (n - k)a^2 + (na - kr)^2,$$

that we must consider for the values of k which make it minimum. Introducing q as above, we see that

$$f_k(a) = r^2\left(\left(k+\frac{1}{2}-qa\right)^2+\left(qa-\frac{1}{4}\right)\right)-rqa^2,$$

an expression which shows that f_k is minimal for $k = \lfloor aq \rfloor$. Giving k this value, we obtain the new function

$$f(a) = r^2\left(\left(\lfloor aq\rfloor+\frac{1}{2}-qa\right)^2+\left(qa-\frac{1}{4}\right)\right)-rqa^2. \qquad (*)$$

Denoting by m the smallest positive quadratic residue of $a(n+1)$ modulo r, we obtain the new expression

$$f(a) = a(n+1)(r-a)-m(r-m) \geq a(n+1)(r-a)-\frac{r^2}{4}. \qquad (**)$$

We must now show that $f(a)$ is larger than $f(1)$ for $2 \leq a \leq \frac{r}{2}$ whenever we have $f(1) \geq 2r^2$.

First remark that r is small relatively to n: indeed, for $r > \frac{n}{2}$, we have

$$f_1(1,\dots,1,1-r) = (n-1)+(r-1)^2+(n-r)^2 < (2r-1)+(r-1)^2+r^2 = 2r^2;$$

we may thus assume that $r \leq \frac{n}{2}$.

For $a \geq 3$, we have $r \geq 6$ and

$$\begin{aligned} f(a) &\geq \left(aq-\frac{1}{4}\right)r^2 - a^2qr = (n+1)(r-a)a-\frac{r^2}{4} \\ &> 3(2r+1)\frac{r}{2}-\frac{r^2}{4} = 2r^2+\frac{3}{4}r(r-2) > 2r^2. \end{aligned}$$

For $a = 2$, we have $r \geq 4$, $n \geq 2r \geq 8$, and

$$f(2) \geq 2(n+1)(r-2)-\frac{r^2}{4} > 4r(r-2)-\frac{r^2}{4},$$

hence $f(r) \geq 2r^2$ if $r \geq 5$. We are thus left with the case $r = 4$. We then have $f(2) = 4n$, but, for $a = 1$, $|\lfloor aq\rfloor + \frac{1}{2} - q| \leq \frac{1}{4}$, whence $f(1) \leq 12q = 3(n+1)$, and consequently $f(1) < f(2)$ since $n \geq 8$. □

To prove Theorem 5.1.7, it suffices on the one hand to calculate the number s_1 of pairs $\pm(y_1,\dots,y_n)$ with $y_i \equiv 1 \mod r$ which give $N(rx)$ the value $f(1)$, and on the other hand to compare $f(1)$ with $2r^2$, since $N(\mathbb{A}_n^r) = \frac{f(1)}{r^2}$ and $s = s_1$ when $f(1) < r^2$, and $N(\mathbb{A}_n^r) = 2$ when $f(1) \geq r^2$, with $s = s_1 + \frac{n(n+1)}{2}$ (resp. $s = s_1$) if $f(1) = 2r^2$ (resp. if $f(1) > 2r^2$).

Lemma 5.1.9. *We have* $s_1 = \binom{n}{\lfloor \frac{a}{q}\rfloor}$ *if* $r \nmid n+1$, $s_1 = \binom{n+1}{q}$ *if* $r \mid n+1$ *and* $r > 2$, *and* $s_1 = \binom{n}{q}$ *if* $r \mid n+1$ *and* $r = 2$.

Proof of 5.1.9. If $r \nmid n+1$, q is not an integer and the equality $f_k(1) = f(1)$ holds only when $k = \lfloor q \rfloor$. Otherwise, q is an integer, $|k + \frac{1}{2} - q|$ is minimum for $k = q$ and $k = q - 1$, and the corresponding vectors coincide up to their sign if $r = 2$ but not if $r > 2$. We thus have $s_1 = \binom{n}{q}$ if $r = 2$ and $s_1 = \binom{n}{q} + \binom{n}{q-1} = \binom{n+1}{q}$ if $r > 2$. □

To complete the proof of Theorem 5.1.7, we finally compare $f(1)$ with $2r^2$ by considering separately the case of a small r ($r = 2, 3, 4$) and of a large r ($r > \frac{n+1}{3}$, i.e. $q < 3$).

Formula (**) shows that for $r = 2$ (resp. 3, resp. 4), we have $f(1) = n$ or $n + 1$ (resp. $f(1) = 2n$ or $2(n + 1)$, resp. $f(1) = 3n - 1$, $3n$ or $3(n + 1)$), the value of f, easily found, depending on n modulo r. We easily verify the assertions in Theorem 5.1.7 in these cases, with two exceptional values of s for $r = 2, 3$ and none for $r = 4$.

In the case where $q < 3$, we more precisely have $2 < q < 3$, whence

$$f(1) = r^2(\frac{5}{2} - q)^2 + r^2(q - \frac{1}{4}) - rq$$

by (*), a formula which can also be written $f(1) = (n+1-2r)^2 - (n+1) + 2r^2$ and which implies $f(1) \geq 2r^2$ if and only if we have $2r < n+1-\sqrt{n+1}$, and from which we moreover deduce that $s = \frac{n(n+1)}{2}$ if the inequality is strict and $s = \frac{n(n+1)}{2} + \binom{n}{2} = n^2$ otherwise.

We may now assume that we have $q \geq 3$ and $r \geq 5$. Formula (*) implies the inequalities $f(1) \geq r^2(q - \frac{1}{4}) - rq \geq \frac{11}{4}r^2 - 3r = 2r^2 + \frac{3}{4}r(r - 4) > 2r^2$ (because of $r \geq 5$), and moreover we have

$$n + 1 - \sqrt{n+1} \geq 3r - \sqrt{3r} = 2r + (r - \sqrt{3r}) > 2r$$

provided r is at least 4. □

The calculations which follow Definition 5.1.1 show that $\mathbb{A}_n^r$ is an integral lattice if and only if both $\frac{n+1}{r}$ and $\frac{n(n+1)}{r^2}$ are integers. In practice, it is better to rescale $\mathbb{A}_n^r$ to a primitive (integral) lattice:

Definition 5.1.10. *For any rational number r, we denote by $a_{n,r}$ (or simply by a) the smallest of the positive integers a such that $\sqrt{a}\,\mathbb{A}_n^r$ is integral, and by $\widetilde{\mathbb{A}}_n^r$ the integral lattice we obtain in this way.*

The mutual scalar products of the n vectors $e_i = \varepsilon_0 - \varepsilon_i$ ($1 \leq i \leq n-1$) and $e' = \frac{1}{r}(n\varepsilon_0 - \varepsilon_1 - \cdots - \varepsilon_n)$ are $e_i \cdot e_i = 2$, $e_i \cdot e_j = 1$ ($i \neq j$), $e_i \cdot e' = \frac{n+1}{r}$, and $e' \cdot e' = \frac{n(n+1)}{r^2}$. Hence $a_{n,r}$ is the smallest of the positive integers p which satisfy the two conditions

$$r \mid p(n+1) \qquad \text{and} \qquad r^2 \mid pn(n+1)\,.$$

In the particular case where r divides $n + 1$, which we shall consider in the next section, only the second condition is needed.

5.2 The Coxeter Lattices $\mathbb{A}_n^r$

We keep the notation of Section 5.1, but we restrict ourselves to lattices $\mathbb{A}_n^r$ for which r is an integer which divides $n+1$. Recall that $\mathbb{A}_n^r$ is then the unique sublattice of $\mathbb{A}_n^*$ which contains $\mathbb{A}_n$ to index r.

Theorem 5.2.1. (Coxeter.) *Let r be a divisor of $n+1$. Denote by N the norm of $\mathbb{A}_n^r$.*

1. *The lattices $\mathbb{A}_n^r$ are strongly eutactic; they are moreover perfect (and thus extreme), except if $r = n+1, n \geq 3$, or $r = 2$, $n = 3, 5$.*
2. *Except if $r = n+1$, if $r = \frac{n+1}{2}$ or if $r = 2$ and $n = 5$, $\mathbb{A}_n^r$ has norm 2, and we even have $S(\mathbb{A}_n^r) = S(\mathbb{A}_n)$ to within the two exceptions of $\mathbb{A}_7^2 \simeq \mathbb{E}_7$ and $\mathbb{A}_8^3 \simeq \mathbb{E}_8$.*
3. *If N is smaller than 2, the invariants N and s are given by the following formulae: $N = \frac{n}{n+1}$ and $s = n+1$ if $r = n+1$; $N = \frac{2(n-1)}{n+1}$ and $s = \frac{n(n+1)}{2}$ if n is odd and $r = \frac{n+1}{2}$, except for $n = 3$, where $\mathbb{A}_3^2 \simeq \mathbb{Z}^3$; $N = \frac{3}{2}$ and $s = 10$ if $n = 5$ and $r = 2$.*
4. *The lattices $\mathbb{A}_n^r$ are dual extreme to within the unique exception of $\mathbb{A}_3^2$.*
5. *The invariant $(\gamma_n')^2$ is as follows:*
   $\gamma_n'^2(\mathbb{A}_n^r) = 4$ *for* $3 \leq r \leq \frac{n+1}{3}$;
   $\gamma_n'^2(\mathbb{A}_n^2) = \gamma_n'^2\left(\mathbb{A}_n^{(n+1)/2)}\right) = \frac{4(n-1)}{n+1}$ *for* $n \geq 7$ *(odd)*;
   $\gamma_5'^2(\mathbb{A}_5^2) = \gamma_5'^2(\mathbb{A}_5^3) = 2$;
   $\gamma_n'^2(\mathbb{A}_n) = \gamma_n'^2(\mathbb{A}_n^{n+1}) = \frac{2n}{n+1}$ *for all* $n \geq 1$.

Proof. By Proposition 5.1.6, $\mathrm{Aut}(\mathbb{A}_n^r)$ contains the group $W(\mathbb{A}_n) \simeq S_{n+1}$, which acts irreducibly on E. This proves that $\mathbb{A}_n^r$ is strongly eutactic, and we now turn to the perfection property. When $r = 1$ (resp. $r = n+1$), $\mathbb{A}_n^r$ is isometric to $\mathbb{A}_n$ (resp. $\mathbb{A}_n^*$), and the assertions in Theorem 5.2.1 have been proved in Sections 5.2 and 5.6. We may now assume that $2 \leq r \leq \frac{n+1}{2}$. For $r \leq \frac{n+1}{3}$, we have $n+1-\sqrt{n+1} \geq 3r - \sqrt{3r} = 2r + r(1-\frac{\sqrt{3}}{\sqrt{r}}) > 2r$. The results we want to prove are then consequences of Theorem 5.1.7, except if $r = \frac{n+1}{2}$ or if we are in one of the particular cases listed in Theorem 5.1.7, for which $r = 2$ or 3. Moreover, for $(n,r) = (7,2)$ or $(8,3)$, since $\mathbb{A}_n^r$ is the unique integral lattice containing $\mathbb{A}_n$, the existence of isometries $\mathbb{A}_7^2 \simeq \mathbb{E}_7$ and $\mathbb{A}_8^3 \simeq \mathbb{E}_8$ results from Remark 4.6.8. It now suffices to study the lattices $\mathbb{A}_3^2$, $\mathbb{A}_5^2$, and $\mathbb{A}_n^{(n+1)/2}$, $n \geq 5$ odd.

The calculations we have done to prove Theorem 5.1.7 in the case $r = 2$ show that the minimal vectors in $\mathbb{A}_n^2$, $n = 3, 5$ are the vectors $\frac{1}{2}(\sum_0^n \pm\varepsilon_i)$ for which the sum contains $\frac{n+1}{2}$ minus signs. We then immediately see that $\mathbb{A}_3^2$ is isometric to $\mathbb{Z}^3$ and that $\mathbb{A}_5^2$ possesses only 10 pairs of minimal vectors. Hence these two lattices are not perfect.

To complete the proof of assertions (1) to (3) in Theorem 5.1.7, we just have to deal with the lattices $\mathbb{A}_n^{(n+1)/2}$ for odd $n \geq 5$. With the notation of

Lemma 5.1.8, we have $f(a) > 2r^2$ for $a \geq 3$. Then Formula (**) implies that $f(1) = (n+1)(r-1)$ and $f(2) = 2(n+1)(r-2)$ for $r \geq 4$, from which we deduce the lower bound $f(2) - f(1) = (n+1)(r-3) = \frac{(n+1)(n-5)}{2} \geq 0$, in which the inequality is strict for $n > 5$. This proves that these lattices have norm $\frac{f(1)}{r^2} = \frac{2(n-1)}{n+1}$, and Lemma 5.1.9 then shows that $s = \frac{n(n+1)}{2}$ and that their minimal vectors are the $\pm\frac{2}{n+1}\, e_{i,j}$, $i < j$ with

$$e_{i,j} = \frac{n-1}{2}(\varepsilon_j + \varepsilon_i) - \sum_{k \neq i,j} \varepsilon_k\,;$$

a direct verification shows that this still holds true for $n = 5$.

To get rid of cumbersome denominators, we consider the family S' of vectors deduced from the minimal vectors above by suppressing the factor $\frac{2}{n+1}$ and keeping only the + sign, and we denote by $p'_{k,\ell} \in \operatorname{End}^s(E)$ the composition of the projection onto $e_{k,\ell}$ followed by multiplication by $N(e_{k,\ell})$. We easily check that $p'_{k,\ell}(\varepsilon_i - \varepsilon_j) = 0$ if $(k,\ell) = (i,j)$ or if $k,\ell \notin \{i,j\}$, $r\, e_{i,\ell}$ if $k = i$, $\ell \neq j$, and $-re_{j,\ell}$ if $k \neq i$, $\ell = j$ ($r = \frac{n+1}{2}$). A relation $\sum_{k<\ell} \alpha_{k,\ell}\, p'_{k,\ell} = 0$ thus implies that $\sum_{k \neq i,j} \alpha_{i,k}\, e_{i,k} - \alpha_{j,k}\, e_{j,k} = 0$ for all i, j. Equating the components on ε_i yields the equation $\sum_{k \neq i,j} r\alpha_{i,k} + \alpha_{j,k} = 0$. By making use of the analogous relation obtained by exchanging i and j, we find that the equality $\sum_{k \neq i,j} \alpha_{k,i} = 0$ holds for all i and all j, which proves first that all $\alpha_{i,j}$ are equal, and then that they are zero.

Since the fifth assertion is an easy consequence of the first four, we just have now to prove assertion (4), i.e. to check that the lattices $\mathbb{A}_n^r$ are dual-extreme except for $(n,r) = (3,2)$. Since they are eutactic by (1), it suffices to prove that at least one of the two lattices $\mathbb{A}_n^r$, $\mathbb{A}_n^q$ ($q = \frac{n+1}{r}$) is extreme. But this follows from the previously proved assertions (1) and (2). □

Remark 5.2.2. *The proof above together with results proved in the first section shows that the automorphism group of a Coxeter lattice acts transitively on its set of minimal vectors.*

We now focus on the special case of lattices $\mathbb{A}_n^{(n+1)/2}$; we thus assume that $r = \frac{n+1}{2}$ and $q = 2$.

Proposition 5.2.3. *Let $n \geq 5$, and let*

$$x = \frac{1}{r}\left((r-1)(\varepsilon_0 + \varepsilon_1) - \sum_2^n \varepsilon_i\right) = \varepsilon_0 + \varepsilon_1 - \frac{1}{r}\sum_0^n \varepsilon_i\,.$$

1. *The orbit of x under S_{n+1} constitutes a half-system of $\frac{n(n+1)}{2}$ minimal vectors in $\mathbb{A}_n^{(n+1)/2}$.*
2. *For $n \equiv 1 \mod 4$ (resp. $n \equiv -1 \mod 4$), the multiplier $a_{n,(n+1)/2}$ which renders $\mathbb{A}_n^{(n+1)/2}$ integral (see Definition 5.1.10) is equal to $\frac{n+1}{2}$ (resp. to*

$\frac{n+1}{4}$), *and the rescaled lattice* $\widetilde{\mathbb{A}}_n^{(n+1)/2}$ *has norm* $n-1$ (*resp.* $\frac{n-1}{2}$) *and determinant* $2(\frac{n+1}{2})^{n-1}$ (*resp.* $(\frac{n+1}{4})^{n-1}$).

3. *The spectrum of minimal vectors in* $\widetilde{\mathbb{A}}_n^{(n+1)/2}$ *is given by the following rules:*
   $(8,1)$, $(6,2)$ *if* $n=5$;
   $(27,1)$ *if* $n=7$;
   $(\frac{(n-1)(n-2)}{2},2)$, $(2(n-1),\frac{n-3}{2})$ *if* $n \equiv 1 \mod 4$ *and* $n \geq 9$;
   $(\frac{(n-1)(n-2)}{2},1)$, $(2(n-1),\frac{n-3}{4})$ *if* $n \equiv -1 \mod 4$ *and* $n \geq 11$.
4. *The images of* x *under the operation of the cycle* $\sigma = (\varepsilon_0, \varepsilon_1, \dots, \varepsilon_{n-1})$ (*of order* n) *constitute a basis for* $\mathbb{A}_n^{(n+1)/2}$.
5. *Set* $k=1$ *if* $n \equiv 1 \mod 4$ *and* $k=2$ *if* $n \equiv -1 \mod 4$. *Then the Gram matrix of the basis* $\sigma^i x$ *for* $\widetilde{\mathbb{A}}_n^{(n+1)/2}$ *has entries* $a_{i,j} = \frac{b_{i,j}}{k}$ *where the circulating entries* $b_{i,j}$ *are defined as follows:*
   $b_{i,i} = n-1$, $b_{i,i+1} = b_{1,n} = \frac{n-3}{2}$ *and* $b_{i,j} = -2$ *for* $1 < |j-i| < n-1$.
6. *For* $n \neq 7$, $\mathrm{Aut}(\mathbb{A}_n^{(n+1)/2})$ *is isomorphic to* $\mathrm{Aut}(\mathbb{A}_n) \simeq \{\pm \mathrm{Id}\} \times S_{n+1}$; *it acts transitively on* $S(\mathbb{A}_n^{(n+1)/2})$.

Proof. Assertion (1) follows by inspection of the list of minimal vectors we gave at the end of the proof of Theorem 5.2.1.

The calculations of scalar products we have made after Definition 5.1.1 show that the scaling factor of the Coxeter lattices is the smallest of the integers a such that r divides aq, a condition which reads here $\frac{n+1}{2} \mid 2a$. This proves assertion (2).

Among the minimal vectors of the half-system considered in (1), there are $2(n-1)$ of them with coefficient $r-1$ in front of exactly one of the vectors ε_0, ε_1, and $\frac{(n-1)(n-2)}{2}$ with coefficient -1 in front of both ε_0, ε_1. The corresponding scalar products are

$$x \cdot \sigma x = \frac{1}{r^2}\left((r-1)^2 - 2(r-1) + n - 2\right) = \frac{1}{r^2}\left(r^2 - 4r + qr\right) = \frac{r-2}{r}$$

and

$$x \cdot \sigma^2 x = \frac{1}{r^2}\left(-4(r-1) + n - 3\right) = \frac{1}{r^2}\left(-4r + rq\right) = -\frac{2}{r}$$

respectively. This proves (3), and (5) easily follows from (3).

We now prove (4) by calculating a determinant. The Gram matrix of the vectors $\sqrt{r}\,\sigma^i x$ has entries $a_{i,j}$ which solely depend on $|j-i|$: $2(r-1)$ if $j-i=0$, $r-2$ if $|j-i|=1$ or $n-1$, and -2 otherwise. The invariance of its first row under the permutation $i \mapsto n-i$ shows that this is a circulating matrix. By performing on the rows of the matrix the combination $1, \omega, \dots, \omega^{n-1}$ for ω running through the set of n-th roots of unity, we obtain the formula

$$\det(a_{i,j}) = \prod_{\omega^n=1}\left(2(r-1) + (r-2)(\omega + \omega^{n-1}) - 2\sum_{i=2}^{n-2}\omega^i\right).$$

The term which corresponds to $\omega = 1$ is equal to 2, and the others can be written in the form

$$2(r-1) + r(\omega + \omega^{-1}) - 2\sum_{i=1}^{n-1} \omega^i = r(2 + \omega + \omega^{-1}).$$

Since ω is of odd order $m > 1$, $1 + \omega$ is a unit in the field of m-th roots of unity, with norm $2 + \omega + \omega^{-1}$ with respect to the maximal real subfield. This proves that our Gram matrix has determinant $\frac{1}{r^n}\,2r^{n-1} = \frac{2}{r} = \frac{n+1}{r^2}$, whence (4).

Finally, we prove (6). By Proposition 5.1.6, $\mathbb{A}_n^{(n+1)/2}$ is invariant under $\mathrm{Aut}(\mathbb{A}_n)$ (which proves the transitivity assertion) and we obtain in this way all automorphisms whenever the only norm 2 vectors in $\mathbb{A}_n^{(n+1)/2}$ or in its dual $\mathbb{A}_n^2$ are those of $\mathbb{A}_n$. This last condition is satisfied by the lattice $\mathbb{A}_n^2$ for all $n \geq 9$ (Theorem 5.2.1 (2)) and is easily verified for $\mathbb{A}_5^3$ by checking that $\widetilde{\mathbb{A}}_5^3$ possesses exactly 15 pairs of norm 6 vectors.
[Dimension 7 is an exception: we have $\mathrm{Aut}(\mathbb{A}_7^4) = \mathrm{Aut}(\mathbb{E}_7) = \{\pm\,\mathrm{Id}\} \times \mathrm{Aut}(\mathbb{E}_7)^+$, and S_8 has index 36 in $\mathrm{Aut}(\mathbb{E}_7)^+$.] □

We are now able to complete the proof of Theorem 4.7.2:

Proposition 5.2.4. *The lattices $\mathbb{E}_6^*$ and $\mathbb{E}_7^*$ are extreme.*

Proof. The case of $\mathbb{E}_7^*$ is immediate: we have $\mathbb{E}_7 \simeq \mathbb{A}_7^2$ by Theorem 5.2.1 (2), hence $\mathbb{E}_7^* \simeq \mathbb{A}_7^4$ by Proposition 5.1.2, and we conclude by Theorem 5.2.1 (1).

In the previous chapter, at the end of the proof of Theorem 4.7.2, we constructed a hyperplane section Λ of $\mathbb{E}_6^*$ with respect to which $\mathbb{E}_6^*$ is relatively perfect, and also produced a Gram matrix of Λ, which is immediately proved to coincide with the Gram matrix of $\widetilde{\mathbb{A}}_5^3$ we have given in Proposition 5.2.3. As a consequence $\mathbb{E}_6^*$ is perfect, hence extreme, as was noticed in the proof of Theorem 4.7.2. □

An interesting question about perfect lattices, to which we shall return in Chapter 12, is the classification of the perfect sections of a given lattice Λ which have the same norm as Λ, and in particular of the maximal ones, since they contain the other ones. The case of root lattices was proposed as exercises; the maximal sections are as follows:

- $\mathbb{A}_{n-1} \subset \mathbb{A}_n$ ($n \geq 2$).
- $\mathbb{A}_{n-1} \subset \mathbb{D}_n$ ($n \geq 4$), $\mathbb{D}_{n-1} \subset \mathbb{D}_n$ ($n \geq 5$).
- $\mathbb{A}_{n-1}, \mathbb{D}_{n-1} \subset \mathbb{E}_n$ ($n = 6, 7, 8$), $\mathbb{E}_{n-1} \subset \mathbb{E}_n$ ($n = 7, 8$).

The list of inclusions between $\mathbb{A}_n^r$ and root lattices is not known. We shall give below a few examples.

When r is an integer and $\mathbb{A}_n^r$ has norm 2 (see Theorem 5.1.7), $\mathbb{A}_n^r$ contains the lattice $\langle e_1, \dots, e_{n-1}\rangle = \mathbb{A}_{n-1}$ (the orthogonal of $\mathbb{Z}\varepsilon_n$ in $\mathbb{A}_n^r$), and we can only find lattices which contain $\mathbb{A}_{n-1}$ and which have the same set of norm 2 vectors than $\mathbb{A}_n$. Here is another result:

Proposition 5.2.5. *For any integer r dividing n, the orthogonal of $\mathbb{Z}\varepsilon_0$ in $\mathbb{A}_n^r$ is a hyperplane section of $\mathbb{A}_n^r$ isometric to the Coxeter lattice $\mathbb{A}_{n-1}^r$.* [Example: $\mathbb{A}_8^2 \supset \mathbb{A}_7^2 \simeq \mathbb{E}_7$.]

Proof. Consider the basis

$$\left(e_1 = \varepsilon_0 - \varepsilon_1, \dots, e_{n-1} = \varepsilon_0 - \varepsilon_{n-1}, e' = \frac{1}{r}\, e = \frac{1}{r}\,(n\,\varepsilon_0 - \varepsilon_1 - \dots - \varepsilon_n)\right)$$

for $\mathbb{A}_n^r$. Replacing e_i by $e_i' = \varepsilon_1 - \varepsilon_i$ for $2 \le i \le n-1$ and e' by $e'' = e' - \frac{n}{r}\, e_1 = \frac{1}{r}\,((n-1)\,\varepsilon_1 - \varepsilon_2 - \dots - \varepsilon_n)$, we obtain a new basis for $\mathbb{A}_n^r$, and it is then clear that $\Lambda = \langle e_2', \dots, e_{n-1}', e'' \rangle$ is isometric to $\mathbb{A}_{n-1}^r$. □

One cannot forecast perfect sections if $N(\mathbb{A}_n^r) < 2$. The case of lattices $\mathbb{A}_n^{(n+1)/2}$ is particularly interesting. By Proposition 5.2.3, they give for all $m \ge 3$ examples of primitive lattices of norm m. Since their spectrum does not contain any zero scalar product, they do not contain any minimal section similar to a root lattice of dimension $k \ge 3$.

The classification of perfect lattices (known up to dimension 7, see Chapter 6) shows that $\widetilde{\mathbb{A}}_5^2$, of norm 4, does not contain any norm 4 perfect section of dimension 4 or 3 (since they should be similar to root lattices) and that $\widetilde{\mathbb{A}}_7^4 \sim \mathbb{E}_7^*$, of norm 3, does not contain any norm 3 perfect section of dimension greater than 1 (since $\mathbb{Z}$ and $\widetilde{\mathbb{A}}_7^4$ are the only odd integral perfect lattices up to dimension 7). We conjecture that this is general, namely that Coxeter's lattices have no perfect k-dimensional sections with the same norm in the range $3 \le k \le n$.

5.3 Barnes's Lattices P_n

We describe in this section a family of lattices discovered by Barnes ([Bar3]), which will appear in the next section as the case $r = 2$ of the more general family of Craig's lattices $\mathbb{A}_n^{(r)}$ ($\mathbb{A}_n^{(1)}$ is $\mathbb{A}_n$). The Barnes lattices are extreme from dimension 6 onwards. However, we shall not give the complicated proof, but rather focus on properties of their dual lattices, which (as far as I know) cannot be found in the literature.

We keep the notation of Section 5.1. From Proposition 5.3.5 onwards, we suppose $n \ge 6$.

Definition 5.3.1. *We denote by P_n the sublattice of $\mathbb{A}_n$ defined by the congruence $\sum_{i=0}^n i x_i \equiv 0 \mod n+1$. We consider that the coordinates x_i as well as the vectors $\varepsilon_i \in \mathbb{R}^{n+1}$ are indexed modulo $n+1$.*

Consider the Coxeter basis $e'_i = \varepsilon_i - \varepsilon_{i-1}$ $(1 \le i \le n)$ for $\mathbb{A}_n$. The equality

$$\sum_{i=0}^{n} x_i \varepsilon_i = \sum_{i=1}^{n} y_i\, e'_i \text{ (with } \sum_{i=0}^{n} x_i = 0)$$

amounts to the relations

$$y_i = -\sum_{j=0}^{i} x_j \quad \text{and} \quad x_0 = -y_1,\, x_n = y_n\,, x_i = y_i - y_{i+1}\,(1 \le i \le n-1)\,,$$

and the congruence which defines P_n reduces to $\sum_{i=1}^{n} y_i \equiv 0 \mod n+1$.

It is easily checked that $P_1 = 2\mathbb{A}_1 \simeq 2\sqrt{2}\,\mathbb{Z}$ and $P_2 \simeq \sqrt{3}\,\mathbb{A}_2$ (thus $N(P_1) = 8$ and $N(P_2) = 6$).

Proposition 5.3.2. *For $n \ge 3$, P_n is an even lattice of norm 4, determinant $(n+1)^3$ and Hermite invariant $\gamma(P_n) = 4(n+1)^{-3/n}$. Moreover, we have $s(P_n) = \frac{(n-2)n(n+1)}{8}$ if n is even and $s(P_n) = \frac{(n-1)^2(n+1)}{8}$ if n is odd.*

Proof. Definition 5.3.1 shows that $[\mathbb{A}_n : P_n] = n+1$ for all $n \ge 1$, which proves that $\det(P_n) = (n+1)^3$ for all n, and that P_n has norm $N > 2$, since the vectors in $S(\mathbb{A}_n)$, namely the $\varepsilon_i - \varepsilon_j$, $j \ne i$, do not satisfy the basic congruence in 5.3.1. As a sublattice of the even lattice $\mathbb{A}_n$, P_n is itself even, and we thus have $N(P_n) = 4$ for all $n \ge 3$, since $\varepsilon_0 - \varepsilon_1 - \varepsilon_2 + \varepsilon_3$ is then a norm 4 vector, whence $\gamma(P_n) = 4(n+1)^{-3/n}$.

To describe $S(P_n)$, we first observe that the cycle $\sigma \in \mathrm{Aut}(\mathbb{A}_n)$ defined by $\sigma(\varepsilon_i) = \varepsilon_{i+1}$ and the symmetry with respect to the origin stabilize P_n, which allows us to focus on vectors of the form $x = \varepsilon_0 \pm \varepsilon_i \pm \varepsilon_j \pm \varepsilon_k$, $0 < i < j < k$ with exactly two + signs. Since the combination $+-+-$ may not occur, we may assume that there is a + sign in front of ε_k. Because of the congruence $\pm i \pm j + k \equiv 0 \mod n+1$, we must have $i + j = k$, an equation which possesses $\lfloor \frac{k-1}{2} \rfloor$ solutions (i,j) for each value of k. The number of minimal vectors of this form is thus equal to $\sum_{k=3}^{n} \lfloor \frac{k-1}{2} \rfloor$, an expression which is equal to $2(1+2+\cdots+\frac{n-2}{2})$ or $2(1+2+\cdots+\frac{n-3}{2}) + \frac{n-1}{2}$, according to whether n is even or odd. Since the orbits of σ contain $n+1$ vectors among which two are of the form $\varepsilon_0 - \varepsilon_i - \varepsilon_j + \varepsilon_k$ with $0 < i < j < k \le n$, the values of s are twice the numbers above. □

The Hermite invariant of P_n exceeds that of $\mathbb{A}_n$ from $n = 6$ and that of $\mathbb{D}_n$ from $n = 8$ onwards; we have $\gamma(P_7) = \gamma(D_7) = 2^{5/7}$.

We now come to the construction of bases for P_n. Recall that an element of $\mathbb{A}_n$ can be uniquely written as a sum $x = \sum_{i=1}^{n} a_i\,(\varepsilon_{i-1} - \varepsilon_i)$, and that $x \in P_n \Longleftrightarrow \sum a_i \equiv 0 \mod n+1$.

Lemma 5.3.3. *Let Λ be a lattice, and let $\mathcal{B} = (e_1, \dots, e_n)$ be a basis for Λ. Set $e_{n+1} = -\sum_{i=1}^{n} e_i$ (thus $\sum_{i=1}^{n+1} e_i = 0$). Then the vectors $e'_i = e_i - e_{i+1}$, $1 \le i \le n$ constitute a basis for a lattice Λ' of index $n+1$ in Λ.*

Proof. We obtain the same subgroup Λ' of Λ by replacing e'_i by $e''_i = e_1 - e_i$ for $i < n$. Since $e'_n = 2e_2 + \sum_{i<n} e_i$, we have $(n+1)e_n = e'_n - \sum_{i<n} e''_i$, whence the result. □

Applying the preceding lemma to $\mathbb{A}_n$ and its basis $(\varepsilon_{i-1} - \varepsilon_i)$, we obtain:

Proposition 5.3.4. *The vectors $e'_i = \varepsilon_{i-1} - 2\varepsilon_i + \varepsilon_{i+1}$, $1 \le i \le n$, which have norm 6 for $n \ge 2$, constitute a basis for P_n.* □

With the preceding notation, let us consider for $2 \le i \le n$ the $n-1$ vectors $e_i = e'_{i-1} + e'_i = \varepsilon_{i-2} - \varepsilon_{i-1} - \varepsilon_i + \varepsilon_{i+1}$. They all have norm 4, and it is clear that they constitute together with e'_n a basis for P_n. Let $e' = \varepsilon_0 - \varepsilon_n - \varepsilon_{n-2} + \varepsilon_{n-3}$. Since $e'_n = e' - e_{n-1}$, we see that P_n possesses for all $n \ge 4$ a basis made of norm 4 vectors.

Adding successively for $k = 1, \dots, \lfloor \frac{n-2}{2} \rfloor$ the vectors $\varepsilon_{n-2k} - \varepsilon_{n-2k-1} - \varepsilon_{n-2k-2} + \varepsilon_{n-2k-3}$ to e', we obtain for even n a basis whose elements are the vectors e_i for $1 \le i \le n$ (or for $2 \le i \le n+1$, this amounts to the same thing); they are the images of the first one by the powers of σ. Denote by G the cyclic group of order $n+1$ generated by σ. We have thus found for the G-lattice P_n (in the sense of Chapter 11) a circulating basis. Notice that E defines the augmentation representation of G.

For odd n, the e_i are no longer independent (they satisfy the relation $e_1 + e_3 + \cdots + e_n = 0$); we shall then consider the basis $(e_2, \dots, e_n, e)$ where $e = \varepsilon_0 - \varepsilon_n - \varepsilon_3 + \varepsilon_2$. It is easy to write down in each case the corresponding Gram matrix, which yields the following proposition:

Proposition 5.3.5. *Suppose $n \ge 6$. Let $e_i = \varepsilon_{i-2} - \varepsilon_{i-1} - \varepsilon_i + \varepsilon_{i+1}$ ($i \bmod n+1$), and let $e = \varepsilon_0 - \varepsilon_n - \varepsilon_3 + \varepsilon_2$. Then, for even n (resp. for odd n), the vectors $e_1, e_2, \dots, e_n$ (resp. $e_2, \dots, e_n, e$) constitute a basis of minimal vectors for P_n, and the entries $a_{i,j}$ of the corresponding Gram matrix are as follows:*

1. *n even: $a_{i,i} = 4, a_{1,2} = -1, a_{1,3} = -2, a_{1,4} = 1, a_{1,n-1} = 1, a_{1,n} = -2$, $a_{1,j} = 0$ for $5 \le j \le n-2$, and $a_{i,j} = a_{1,j-i+1}$ for $1 < i < j \le n$;*
2. *n odd: same formulae for $1 \le i \le j \le n-1$, then $a_{1,n} = -1$, $a_{2,n} = 0$, $a_{3,n} = 2$, $a_{4,n} = a_{n-2,n} = -1$, $a_{n-1,n} = 2$, and $a_{i,n} = 0$ for $5 \le i \le n-3$.*

□

We now turn to the description of some automorphisms of P_n. We have considered in the proof of Proposition 5.3.2 the cycle $\sigma : i \mapsto i+1 \mod n+1$ of $\operatorname{Aut}(\mathbb{A}_n)$, which stabilizes P_n. For $a \in (\mathbb{Z}/(n+1)\mathbb{Z})^\times$, the permutation $i \mapsto a\,i \mod n+1$ is an element of $\operatorname{Aut}(\mathbb{A}_n)$ which again stabilizes P_n. We have thus constructed a subgroup H of $\operatorname{Aut}(P_n)$, of order $\varphi(n)(n+1)$, which is canonically isomorphic to the semi-direct product

$$\mathbb{Z}/(n+1)\mathbb{Z} \rtimes \operatorname{Aut}(\mathbb{Z}/(n+1)\mathbb{Z}) \simeq \mathbb{Z}/(n+1)\mathbb{Z} \rtimes (\mathbb{Z}/(n+1)\mathbb{Z})^\times .$$

Proposition 5.3.6. *The subgroup of* $\mathrm{Aut}(\mathbb{A}_n)$ *which stabilizes* P_n *is the direct product* $G_n = \{\pm \mathrm{Id}\} \times H$*, of order* $2(n+1)\,\varphi(n)$*,*

Proof. The group $\mathrm{Aut}(\mathbb{A}_n)$ acts on the hyperplane of $\mathbb{R}^{n+1}$ generated by P_n and via this action, the group we want to determine can be identified with a subgroup of $\mathrm{O}_{n+1}(\mathbb{R})$, hence with a subgroup of $\mathrm{Aut}(\mathbb{Z}^{n+1}) \simeq \{\pm \mathrm{Id}\} \times S_{n+1}$, where S_{n+1} acts as a permutation group on the ε_i. Let $s \in \mathrm{Aut}(P_n)$. By composing s with a suitable automorphism of the form $\pm\sigma^k$ of P_n, we obtain an element $s' \in \mathrm{Aut}(P_n)$ which fixes ε_0. Let $a \in \mathbb{Z}$ such that $\varepsilon_a = s'(\varepsilon_1)$. Then ${s'}^i$ maps ε_1 onto ε_{a^i}. Since s' is an automorphism of P_n, a is prime to $n+1$. By composing s' with the inverse of the automorphism $\varepsilon_i \mapsto \varepsilon_{a^i}$, we obtain an element $s'' \in \mathrm{Aut}(P_n)$ which fixes both ε_0 and ε_1. But $(i-1)\,\varepsilon_0 - i\,\varepsilon_1 + \varepsilon_i$ must then be fixed by s'' for all i. This implies that s'' fixes ε_i for all i, and finally that $s'' = \mathrm{Id}$. □

To find the full automorphism group of P_n, we must see whether there may exist automorphisms which do not stabilize $\mathbb{A}_n$. The study of the dual of P_n (see below) will show that P_n has no extra automorphisms from $n = 11$ onwards. (But extra automorphisms do exist in the range $6 \le n \le 10$.)

Theorem 5.3.7. *Let* $n \ge 6$ *be an integer.*

1. *For* $n = 10$, $n = 12$, *and* $n \ge 14$, *we have* $N(P_n^*) = N(\mathbb{A}_n^*) = \frac{n}{n+1}$, *and even* $S(P_n^*) = S(\mathbb{A}_n^*)$ *if* $n > 10$.
2. *The invariants* $N(P_n^*)$, $s(P_n^*)$, *and* $r(P_n^*)$ *(the rank of* $S(P_n^*)$*) are as follows:*

   $n = 6$: $N = \frac{4}{7}$; $s = 21$; $r = 6$. $n = 7$: $N = \frac{1}{2}$; $s = 1$; $r = 1$.
   $n = 8$: $N = \frac{2}{3}$; $s = 3$; $r = 2$. $n = 9$: $N = \frac{5}{8}$; $s = 1$; $r = 1$.
   $n = 10$: $N = \frac{10}{11}$; $s = 66$; $r = 10$. $n = 11$: $N = \frac{3}{4}$; $s = 1$; $r = 1$.
   $n = 12$: $N = \frac{12}{13}$; $s = 13$; $r = 12$. $n = 13$: $N = \frac{7}{8}$; $s = 1$; $r = 1$.

   (For $n \ge 14$, *we have* $N = \frac{n}{n+1}$, $s = n+1$, *and* $r = n$.*)*
3. *The vectors in* P_n *whose norm is equal to* $N(\mathbb{A}_n^*)$ *belong to* $\mathbb{A}_n^*$ *for all* $n \ge 11$. *For* $n = 6, 7, 8, 9, 10$, *the number of pairs of such vectors is equal to* $21, 16, 27, 30, 66$ *respectively (and to* $n+1$ *for* $n \ge 11$*).*

Proof. It is not difficult to deduce the statements in (2) and (3) from the lower bounds we shall establish to prove (1), and this work can moreover be done easily with the help of a computer program. For these reasons, we only prove assertion (1).

Let σ be as above the cycle $(0, 1, \dots, n)$ of order $n+1$, and let $e_0 = n\varepsilon_0 - \sum_{i=1}^n \varepsilon_i$. The lattice $\mathbb{A}_n^*$ is generated by the vectors $\frac{e_i}{n+1}$ where $e_i = \sigma^i e_0$, and the lattice P_n^* by $\mathbb{A}_n^*$ and one vector $\frac{e}{n+1}$ for some primitive $e \in \mathbb{A}_n^*$. The lattice P_n is the sublattice of those elements x of $\mathbb{A}_n$ which satisfy the congruence $u \cdot x \equiv 0 \mod n+1$ where $u = \sum_{i=0}^n i\,\varepsilon_i$. Since $\varepsilon_0 + \varepsilon_1 + \cdots + \varepsilon_n \in \mathbb{A}_n^\perp$, one can replace u by $v = u - \frac{n}{2}\,(\varepsilon_0 + \varepsilon_1 + \cdots + \varepsilon_n)$.

Let us first consider the case where n is even, say $n = 2m$. We then have

$$v = -m\,\varepsilon_0 + (-m+1)\,\varepsilon_1 + \cdots + (m-1)\,\varepsilon_{n-1} + m\,\varepsilon_n \in \mathbb{A}_n\,,$$

and we can take $e = v$. We are reduced to comparing $N(e_i) = n+1$ with the norms of the vectors

$$x = z_0\,v + \sum_0^n x_i\,\varepsilon_i\,,\ (z_0,\,x_0,\,x_1,\ldots,x_n \in \mathbb{Z},\ z_0 \neq 0)\,.$$

The components of x on the ε_i are

$$\begin{aligned} y_0 &= -m\,z_0 + n\,x_0 - x_1 - \cdots - x_n \\ &= -m\,z_0 + (n+1)\,x_0 - (x_0 + x_1 + \cdots + x_n) \\ &\equiv -m\,z_0 \mod n+1 \end{aligned}$$

and

$$y_i = y_0 + (n+1)\,(x_i - x_0) + i\,z_0 \equiv (i-m)\,z_0 \mod n+1\,.$$

Set $d = (z_0, n+1)$, and define n' and m' by $d(n'+1) = n+1$ and $2m' = n'$. If $d = n+1$, we have $y_i \equiv 0 \mod n+1$ for all i, thus $N(x) \equiv 0 \mod (n+1)^2$, whence $N(x) \geq (n+1)^2 \geq n^2 + n$.

Suppose now $1 \leq d < n+1$. The y_i share out among sets of d elements divisible by d and pairwise non-congruent modulo $n+1$, which implies the following optimal lower bound

$$N(x) \geq 2d^3 \sum_{i=0}^{m'} i^2 = 2d^3\,\frac{m'(m'+1)(2m'+1)}{6} = \frac{(n+1)\big((n+1)^2 - d^2\big)}{12}\,.$$

When $d = 1$, the case which always holds for $n+1$ a prime, we have $n' = n$, and the right–hand side of the inequality above is the norm of $\frac{e}{n+1}$, which is greater than or equal to the norm of $\mathbb{A}_n^*$ for all $n \geq 10$, and even strictly greater for $n \geq 12$. For $n = 8$, we obtain the norm of P_n^* by choosing $d = 3$. For $n \geq 14$, we immediately get the lower bound $N(x) \geq n(n+1)$ by bounding d from above by $\frac{n+1}{3}$. This completes for even n all the assertions about $N(P_n)$.

We now give some indications concerning the case where n is odd. The vector v then no longer belongs to $\mathbb{A}_n$, but it can nevertheless be used in the same way. We actually have for any $x = \sum x_i\,\varepsilon_i \in \mathbb{A}_n$ the equations

$$x \cdot v = \sum_{i=0}^n \Big(-\frac{n}{2} + i\Big)\,x_i = -\frac{n}{2}\sum_{i=0}^n x_i + \sum_{i=0}^n i x_i = \sum_{i=0}^n i x_i\,,$$

which first show that v is a primitive vector in $\mathbb{A}_n^*$ (because $v \cdot (\varepsilon_i - \varepsilon_0) = i$), and then that $P_n^* = \langle \mathbb{A}_n^*, e\rangle$ with $e = \frac{v}{n+1}$ as in the even–dimensional case. We can perform analogous calculations for a vector $x = z_0 + \sum_{i=0}^n x_i\,\varepsilon_i$.

The components y_i of x on the vectors ε_i are such that

$$y_0 = -\frac{n}{2}\, z_0 + (n+1)\, x_0 \text{ and } y_i \equiv y_0 \mod (n+1)\mathbb{Z}\,.$$

We define similarly d and n', and then m' by $n' = 2m' + 1$. If $d = n + 1$, we have $z_0 \equiv 0 \mod n+1$; if y_0 is an integer, we have $N(x) \geq (n+1)^2$ as in the even case; otherwise, the y_i all are nonzero, which implies the lower bound $N(x) \geq (n+1)(\frac{n+1}{2})^2 > n^2 + n$. If $1 \leq d < N + 1$, since the y_i may be only half-integers, the lower bound we obtain is slightly worse than for even n, namely

$$N(x) \geq d^3 \sum_{i=-m'}^{i=+m'} \left(\frac{1}{2} - i\right)^2,$$

but it can be dealt with in a similar way, and it implies the general inequality only for $n \geq 15$ instead of $n \geq 12$. □

The following theorem, of which we shall prove only a small part as was announced at the beginning of the section, lists all that concerns the extremality properties of the family P_n:

Theorem 5.3.8. *Let $n \geq 6$ be an integer.*

1. *The lattice P_n is extreme.*
2. *The lattice P_n^* is extreme for $n = 6$ and $n = 10$, eutactic but not extreme for $n = 12$ and for $n \geq 14$, but is neither perfect nor eutactic for $n =$ 7, 8, 9, 11, 13.*
3. *The lattice P_n is dual-extreme for $n = 6$, $n = 10$, $n = 12$, and $n \geq 14$ but for no other value of $n \geq 6$.*
4. *We have ${\gamma'_6}^2(P_6) = \frac{16}{7}$ and ${\gamma'_n}^2(P_n) = \frac{4n}{n+1}$ for the other values of n (10, 12 or ≥ 14) for which P_n is dual-extreme.*

Proof. We refer to Barnes's original paper [Bar3] for (1). The fact that P_n^* is eutactic for $n = 12$ and $n \geq 14$ but for no other value of $n \geq 6$ except perhaps for $n = 6$ or $n = 10$ is a consequence of Theorem 5.3.7, as well as the fact that P_n and P_n^* are dual-extreme for $n = 12$ and $n \geq 14$, but that they are not for $n = 7, 8, 9, 11, 13$; the values of γ'_n also result from Theorem 5.3.7. We are now left with the two cases of dimensions 6 and 10. The explicit calculation is not too complicated, and can moreover be done on a computer, starting with the Gram matrices given in Proposition 5.3.5. □

Dimension 6 is special; we just quote here the following result, that we shall prove in the next section:

Theorem 5.3.9. *The lattices P_6 and P_6^* are similar.* □

We end this section with the description of another family of perfect lattices, also discovered by Barnes, denoted by Q_n. The basic idea is to replace the hyperplane section of R^{n+1} which we used to define the P_n lattices by a congruence in $\mathbb{R}^n$, in the same way we defined $\mathbb{D}_n$ after $\mathbb{A}_n$.

Definition 5.3.10. *Let Q_n be the sublattice of $\mathbb{Z}^n$ defined by the two congruences*

$$\sum_{i=1}^{n} x_i \equiv 0 \mod 4 \qquad \textit{and} \qquad \sum_{i=1}^{n} i\, x_i \equiv 0 \mod n\,.$$

The calculation of the determinant is immediate, and perfection is proved for $n \geq 8$ by showing that Q_n is then perfect relatively to P_{n-1}. We leave to the reader the determination of those minimal vectors which lie outside P_{n-1} (which moreover shows that Q_7 is not perfect). The following theorem lists the main properties of the lattices Q_n:

Theorem 5.3.11. *The lattices Q_n are integral lattices of norm 4 and determinant $16\,n^2$. They are perfect for all $n \geq 8$, and their kissing number is given by the following formulae:*

$$\begin{aligned} n \equiv \pm 1 \mod 4 : s &= \frac{1}{12}\,(n-1)(n-3)(2n-1) \\ n \equiv \ \ 2 \mod 4 : s &= \frac{1}{12}\,(n-1)(n-2)(2n-3) \\ n \equiv \ \ 0 \mod 4 : s &= \frac{1}{12}\,(n-3)(2n^2-3n+4)\,. \quad \square \end{aligned}$$

5.4 Craig's Difference Lattices

Consider a Euclidean space E of dimension n and a system $\mathcal{S}$ of $n+1$ vectors $e_0, e_1, \ldots, e_n$ of rank n which add to 0. Denote by Λ the lattice they span. Then n arbitrary vectors in $\mathcal{S}$ constitute a basis for Λ and for E. We consider that the e_i are indexed modulo $n+1$.

Definition 5.4.1. *The* difference lattice *attached to $\mathcal{S}$ is the lattice Λ' generated by the differences of any two vectors in $\mathcal{S}$.* Notation*: $\Delta(\Lambda, \mathcal{S})$, or simply $\Delta(\Lambda)$ if no ambiguity may arise.*

This is actually a lattice, as shown by the proposition below:

Proposition 5.4.2. *1. The map $x = \sum_{i=0}^{n} x_i\, e_i \in \Lambda \mapsto \sum_{i=0}^{n} x_i \mod n{+}1$ does not depend on the set of the x_i which define x, and is a surjective homomorphism of Λ onto $\mathbb{Z}/(n+1)\mathbb{Z}$ with kernel Λ'.*
2. The quotient Λ/Λ' is cyclic of order $n+1$.
3. We have $\det(\Lambda') = (n+1)^2 \det(\Lambda)$.

Proof. (2) is an immediate consequence of (1), and (3) results from (2) together with the formula $\det(\Lambda') = [\Lambda : \Lambda']^2 \det(\Lambda)$. Let us prove (1).

If $x = \sum_{i=0}^{n} x_i\, e_i = \sum_{i=0}^{n} y_i\, e_i$, we have $\sum_{i=0}^{n} (y_i - x_i)\, e_i = 0$. Since the dependence relations between vectors in $\mathcal{S}$ are proportional to the relation

$\sum e_i = 0$, the differences $y_i - x_i$ do not depend on i. Setting $\lambda = y_i - x_i$, we obtain

$$\sum_i y_i = \sum_i x_i + (n+1)\lambda \equiv \sum_i x_i \mod n+1\,.$$

This shows that the map $x \mapsto \sum_i x_i \mod n+1$ is well defined, and is moreover a surjective homomorphism of Λ onto $\mathbb{Z}/(n+1)\mathbb{Z}$. Its kernel clearly contains Λ', and (1) now follows from the equality $[\Lambda : \Lambda'] = n+1$, that we prove by a calculation of determinants: it is clear that n out of the $n+1$ vectors $e'_i = e_0 - e_i$, $1 \le i \le n+1$ constitute a basis for Λ', whose determinant on the basis $(e_1, e_2, \dots, e_n)$ for Λ is

$$\Delta = \begin{vmatrix} 1 & 0 & 0 & \dots & 0 & 1 \\ -1 & 1 & 0 & \dots & 0 & 1 \\ 0 & -1 & 1 & \dots & 0 & 1 \\ 0 & 0 & -1 & \dots & 0 & 1 \\ \vdots & \vdots & \vdots & & & \\ & & & & 0 & 1 \\ 0 & 0 & 0 & \dots & 1 & 1 \\ 0 & 0 & 0 & \dots & -1 & 2 \end{vmatrix} = \begin{vmatrix} 0 & 0 & 0 & \dots & 0 & n+1 \\ -1 & 1 & 0 & \dots & 0 & 1 \\ 0 & -1 & 1 & \dots & 0 & 1 \\ 0 & 0 & -1 & \dots & 0 & 1 \\ \vdots & \vdots & \vdots & & & \\ & & & & 0 & 1 \\ 0 & 0 & 0 & \dots & 1 & 1 \\ 0 & 0 & 0 & \dots & -1 & 2 \end{vmatrix} = \pm(n+1)\,. \quad \square$$

The $n+1$ vectors e'_i, $i \mod n+1$ add to 0. Hence we may consider the lattice $\Delta(\Lambda', \mathcal{S}')$, and then define by induction a sequence $\Lambda^{(1)} = \Lambda$, $\Lambda^{(2)} = \Lambda'$, $\Lambda^{(3)} = \Delta(\Lambda^{(2)})$, ... of lattices, whose determinants are

$$\det(\Lambda^{(r)}) = (n+1)^{2r-2}\det(\Lambda)\,;$$

the lattice $\Lambda^{(r)}$ is also denoted by $\Delta^{(r)}(\Lambda)$.

The difference procedure above, which transforms $\mathcal{S}^{(0)} = \{e_i, \mid i \mod n+1\}$
into $\mathcal{S}^{(1)} = \{e_{i+1} - e_i \mid i \mod n+1\}$ then transforms $S^{(1)}$ into

$$S^{(2)} = \{e_{i+2} - 2e_{i+1} + e_i \mid i \mod n+1\}\,,$$

and we obtain by induction

$$S^{(r)} = \{(e_{i+1} - e_i)^{(r)} \mid i \mod n+1\},$$

where

$$(e_{i+1} - e_i)^{(r)} = \sum_{j=0}^{j=r} (-1)^j \binom{r}{j} e_{i+j}.$$

It is useful to apply this construction in a slightly more general context: we consider a pair $(\Lambda_0, \mathcal{B})$ of a lattice $\Lambda^{(0)}$ in some $(n+1)$-dimensional Euclidean space F and of a basis $(e_0, e_1, \dots, e_n)$ for $\Lambda^{(0)}$. Then $\Lambda^{(1)} = \Delta(\Lambda^{(0)})$ is a lattice in the hyperplane F of E defined in $\mathcal{B}$ by the equation $\sum_{i=0}^n x_i = 0$, and we endow $\Lambda^{(0)}$ with the generating system $\mathcal{S} = \{e_{i+1} - e_i,\ i \mod n+1\}$, to which we can apply successively the preceding n-dimensional constructions. We shall more particularly consider the case where $F = \mathbb{R}^{n+1}$ and $\mathcal{B}$ is its canonical basis $(\varepsilon_0, \varepsilon_1, \dots, \varepsilon_n)$. We then have $\Lambda^{(1)} = \mathbb{A}_n$.

Definition 5.4.3. *The* r-th Craig lattice, *denoted by* $\mathbb{A}_n^{(r)}$, *is the* r*-th difference lattice constructed on the root lattice* $\mathbb{A}_n$.

Proposition 5.4.4. *We have* $\mathbb{A}_n^{(0)} = \mathbb{Z}^{n+1}$, $\mathbb{A}_n^{(1)} = \mathbb{A}_n$, *and* $\mathbb{A}_n^{(2)} = P_n$.

Proof. The result is evident for $r = 0$ and $r = 1$, and the case $r = 2$ is a consequence of Proposition 5.3.4, where we described Barnes's lattice P_n in terms of the second differences $S^{(2)} = \varepsilon_{i+2} - 2\varepsilon_{i+1} + \varepsilon_i$. □

We are now going to give an algebraic interpretation for the Craig lattices in terms of group rings. Recall that given a commutative ring R and a finite group G of order m, the *group ring* $R[G]$ *of* G *over* R is the free R-module with basis the elements of G, where the product extends that of G. We thus have

$$\Big(\sum_{s\in G} a_s s\Big)\cdot\Big(\sum_{t\in G} b_t t\Big) = \sum_{s,t\in G} a_s b_t\, st = \sum_{\sigma\in G}\Big(\sum_{s\in G} a_s b_{s^{-1}\sigma}\Big)\sigma\,.$$

The map $u : \sum a_s s \mapsto \sum a_s$ (the "unit character of G") is a homomorphism of $R[G]$ onto R, also called the *augmentation homomorphism*, and its kernel, denoted by I_G, is the *augmentation ideal*. Let $T = \sum_{s\in G} s$. The equality $(\sum_s a_s\, s)\cdot\sum_t t = (\sum_s a_s)\cdot\sum_t t$ shows that $R[G]\,T = T\,R[G] = RT$ (RT is thus a two-sided ideal of $R[G]$), and that $I_G \cap R[G]\,T = \{0\}$ provided m is not a divisor of zero in R, which is the only case we shall consider. Under this condition, the injection $I_G \to R[G]$ induced by composition with the canonical surjection $R[G] \to R[G]/RT$ is an injection of I_G into $R[G]/RT$.

The group ring $R[G]$ possesses an involution which extends the anti-automorphism $s \mapsto s^{-1}$ of G, and which transforms $\lambda = \sum_s a_s s$ into $\bar\lambda = \sum_s a_s\, s^{-1} = \sum_s a_{s^{-1}}\, s$. Now there is a *trace form on* $R[G]$, namely the trace of the endomorphism "multiplication by λ". Since the permutation matrix defined on the canonical basis for $R[G]$ by multiplication (on the right, say) by a given element $s \in G$ has only zeros on its main diagonal except for $s = 1_G$ where this is the unit matrix, we have $\mathrm{Tr}(\sum a_s s) = m\, a_1$.

We now endow $R[G]$ with the R-bilinear form $(\lambda,\mu) \mapsto \frac{1}{m}\,\mathrm{Tr}(\lambda\bar\mu)$. The elements of G constitute an orthonormal basis for $R[G]$ over R. In particular, the $\mathbb{R}$-algebra $\mathbb{R}[G]$ is then given a Euclidean structure together with a canonical orthonormal basis, in which its subring $\mathbb{Z}[G]$ is a lattice. When $m = n+1$, $\mathbb{Z}[G]$ is isometric to $\mathbb{Z}^{n+1}$ and I_G to $\mathbb{A}_n$. (From now on, the notation I_G only denotes the augmentation ideal of $\mathbb{Z}[G]$.) Since m is invertible in $\mathbb{R}$, we have a direct product decomposition $\mathbb{R}[G] = \mathbb{R}T \times \mathbb{R}I_G$ which yields the identification

$$\left(I_G, \frac{1}{n+1}\,\mathrm{Tr}(\lambda\bar\mu)\right) \simeq \left(I_G(\mathbb{Z}[G]/\mathbb{Z}T), \frac{1}{n+1}\,\mathrm{Tr}_{(\mathbb{Z}[G]/\mathbb{Z}T)/\mathbb{Z}}(\lambda\bar\mu)\right).$$

In the remainder of this section, we limit ourselves to the case when G is cyclic of order $n+1$, and we denote by σ a generator of G, so that we can identify $\mathbb{Z}[G]$ with $\mathbb{Z}^{n+1}$ via the map $(\sigma^i, i \mod n+1) \mapsto \varepsilon_i$. The augmentation ideal I_G is then the principal ideal $(1-\sigma)$, and the identity $\sum_{j=0}^{j=r}(-1)^j \binom{r}{j} \sigma^{i+j} = \sigma^i(1-\sigma)^r$ then shows:

Proposition 5.4.5. *For all $r \geq 1$, the ideal $I_G^r \subset \mathbb{R}[G]$ is isometric to the Craig lattice $\mathbb{A}_n^{(r)}$.* □

The group $\mathrm{Aut}(G)$ is cyclic of order $\varphi(n+1)$ (its elements are the maps $\tau_i : s \mapsto s^i$, i prime to $n+1$). Since the semi-direct product $\widetilde{G} = G \rtimes \mathrm{Aut}(G)$ stabilizes I_G, we can for all r identify $\widetilde{G}$ with a subgroup of $\mathrm{Aut}(\mathbb{A}_n^{(r)})$. That this produces the whole group $\mathrm{Aut}(\mathbb{A}_n^{(r)})$ depends on the pair (n, r).

Our knowledge of the lattices $\mathbb{A}_n^{(r)}$ is much better when $n+1$ is a prime number $p \geq 3$ (we use now the notation p rather than by m). We first have:

Proposition 5.4.6. *The lattices $\mathbb{A}_{p-1}^{(r)}$ and their duals are strongly eutactic.*

Proof. It suffices to show that its operation on E defines an $\mathbb{R}$-irreducible representation of $\widetilde{G}$, a result we prove by showing that its character χ satisfies the relation $\sum_{s \in \widetilde{G}} |\chi(s)|^2 = |\widetilde{G}|$. Now $1+\chi$ is the character of the permutation representation of $\widetilde{G}$ acting on G in the following way: σ acts by multiplication and τ_i transforms σ into σ^i, which yields the rule $(\tau_i\sigma^j) \cdot (\sigma^k) = \sigma^{ij+k}$. For all $s \in \widetilde{G}$, $1+\chi(s)$ is the number of fixed points of s.

We have $\chi(1) = p-1$, $\chi(\sigma^j) = -1$ for $1 \leq j \leq p-1$ and $\chi(\tau_i\sigma^j) = 0$, whence $\sum_s \chi(s)\,\bar{\chi}(s) = 1 \cdot (p-1)^2 + (p-1)(-1)^2 = p(p-1)$.

[The group $\widetilde{G}$ is isomorphic to the automorphism group of the affine line with p elements; the $p(p-2)$ homothetic transformations have one fixed point, and the non-trivial translations have none.] □

To go further on the case where $p = n+1$ is a prime, we identify I_G via $I_G \hookrightarrow R[G]/RT \simeq \mathbb{Z}[\zeta]$ with the prime ideal $\mathfrak{P} = (1-\zeta)$ lying above p in the ring $\mathbb{Z}[\zeta]$ of integers of the p-th cyclotomic field, where ζ stands for a primitive p-th root of unity; $\mathfrak{P}$ is the unique prime ideal in $\mathbb{Z}[\zeta]$ which lies above p, and we have $(p) = \mathfrak{P}^{p-1}$. This identifies $\mathbb{A}_n^{(r)}$ with the ideal $\mathfrak{P}^r$, and the scalar product is defined by $\frac{1}{p}\,\mathrm{Tr}_{\mathbb{Q}(\zeta)/\mathbb{Q}}(\lambda\overline{\mu})$. Thanks to this interpretation of $\mathbb{A}_n^{(r)}$ as the ideal $\mathfrak{P}^r$, we may now define $\mathbb{A}_n^{(r)}$ for any $r \in \mathbb{Z}$.

The following proposition allows us to compare for a given prime p the various lattices $\mathbb{A}_{p-1}^{(r)}$ and their duals. In particular, it proves Proposition 5.3.9 whose proof was postponed until this section. To state it, we denote by π a square root of $(-1)^{(p-1)/2}p$ in $Q(\zeta)$.

[By Galois theory, $\mathbb{Q}(\zeta)$ possesses a unique quadratic subfield K; since K must be unramified outside p, it must be the field $\mathbb{Q}(\sqrt{(-1)^{(p-1)/2}p})$, whence the existence of π; explicitly, the two *Gauss sums* (or *Lagrange resolvents*) $\sum_a \zeta^a$ calculated for $\left(\frac{a}{p}\right) = +1$

and for $\left(\frac{a}{p}\right) = -1$ are the two roots of the polynomial $X^2+X+\frac{p+1}{4}$ or $X^2+X-\frac{p-1}{4}$, according to whether p is congruent to -1 or to $+1$ modulo 4.]

Proposition 5.4.7. *For all $r \in \mathbb{Z}$, let*

$$\mathbb{A}_{p-1}^{(r)} = \left(\mathfrak{P}^r;\ (\lambda,\mu) \mapsto \frac{1}{p}\operatorname{Tr}_{\mathbb{Q}(\zeta)/\mathbb{Q}}(\lambda\overline{\mu})\right).$$

Then multiplication by π in $\mathbb{Q}[\zeta]$ is a similarity of ratio $\sqrt{p}$ which maps $\mathbb{A}_{p-1}^{(r)}$ onto $\mathbb{A}_{p-1}^{(r+(p-1)/2)}$ and ${\mathbb{A}_{p-1}^{(r)}}^*$ onto $\mathbb{A}_{p-1}^{(-r+(p+1)/2)}$. In particular, for $p \equiv -1$ mod 4, *the lattice $\mathbb{A}_{p-1}^{((p+1)/4)}$ is similar to its dual.*

Proof. We have $(p) = \mathfrak{P}^{p-1}$, hence $(\pi) = \mathfrak{P}^{(p-1)/2}$. This makes clear that multiplication by π is a similarity of ratio $\sqrt{p}$ and that it maps $\mathfrak{P}^r$ onto $\mathfrak{P}^{r+(p-1)/2}$.

The dual lattice with respect to the trace form to an ideal $\mathfrak{A}$ is $\mathfrak{A}^* = \mathcal{D}^{-1}\mathfrak{A}^{-1}$ where $\mathcal{D}$ is the different of $\mathbb{Q}(\zeta)$, equal to $\mathfrak{P}^{p-2}$. With our form $\frac{1}{p}\operatorname{Tr}$, we thus have

$$\mathfrak{A}^* = (p)\,\mathcal{D}^{-1}\mathfrak{A}^{-1} = \mathfrak{P}\,\mathfrak{A}^{-1},$$

which implies, taking $\mathfrak{A} = \mathfrak{P}^r$, the equality ${\mathbb{A}_n^{(r)}}^* = \mathfrak{P}^{1-r}$, and consequently $\pi{\mathbb{A}_n^{(r)}}^* = \mathfrak{P}^{1-r+(p-1)/2}$. □

The norm of Craig's lattices is generally not known: its determination is equivalent to a combinatorial problem known as the *Tarry–Escott* or *Prouhet and Tarry problem*; see [H-W], Chapter XXI. When $n+1$ is a prime, there exists an important general lower bound:

Theorem 5.4.8. *Let $n = p-1$, p an odd prime.*

1. (Craig.) *If $r < \frac{n}{2}$, then $N(\mathbb{A}_n^{(r)}) \geq 2r$.*
2. (Bachoc and Batut.) *If r is a strict divisor of $n = p-1$, then $N(\mathbb{A}_n^{(r)}) = 2r$.*

Proof. (1) (After Conway and Sloane.) Let $x \in \mathbb{A}_n^{(r)}$ be represented by an element $\alpha = u_0 + u_1\zeta + \cdots + u_{p-2}\zeta^{p-2}$ in the ideal $\mathfrak{P}^r$. We must show that at least $2r$ coefficients u_i are nonzero. Let $u(X)$ be the polynomial $u_0 + u_1X + \cdots + u_{p-2}X^{p-2}$. By identifying $\mathbb{Z}[\zeta]$ with the quotient ring $\mathbb{Z}[X]/(X^{p-1} + X^{p-2} + \cdots + X + 1)$, we see that there exist polynomials $a, b \in \mathbb{Z}[X]$ such that we have the identity

$$u(X) = a(X)(X-1)^r + b(X)(X^{p-1} + X^{p-2} + \cdots + X + 1). \qquad (*)$$

If $N(x) < 2r$, one can write $u(X)$ as a difference $\sum_{s\in S} X^s - \sum_{t\in T} X^t$ where S and T denote two (not necessarily disjoint) subsets of $\{0, 1, \ldots, n\}$ identified with $\mathbb{F}_p$ having at most $r-1$ elements, and where we allow repetitions in the sums above.

The identity (*) shows that the derivatives of u of order up to $r-1$ take values divisible by p at $X=1$. We thus have

$$\sum_{s\in S} s^i \equiv \sum_{t\in T} t^i \mod p \quad \text{for } i=0,1,\dots,r-1,$$

which implies that the elementary symmetric functions of S and of T coincide on $\mathbb{F}_p$, since they can be expressed as sums of powers by Newton's formulae, and hence that the polynomials $\prod_{s\in S}(X-s)$ and $\prod_{t\in T}(X-t)$ take the same values on $\mathbb{F}_p$. Since their degrees are at most $p-1$, they are equal (as elements in $\mathbb{F}_p[X]$). We thus have $S=T$, and consequently $u=0$.

(2) We have seen that an element $x=\sum_{a\in S}\zeta^a-\sum_{b\in T}\zeta^b\in\mathfrak{P}^r$ satisfies the congruences $x=\sum_{a\in S}a^i\equiv\sum_{b\in T}b^i \mod p$ for $0\le i\le r-1$. The converse is true, as one sees by noticing that these congruences imply that all derivatives of order at most $r-1$ of the polynomial $\sum_{a\in S}X^a-\sum_{b\in T}X^b\in\mathbb{F}_p[X]$ are zero at $X=1$, and hence that it is divisible by $(X-1)^r$.

Let $S=\{a\in\mathbb{F}_p\mid a^r=1\}$. Since r divides $p-1$, S possesses r elements. Let $c\in\mathbb{F}_p$, $c\notin S\, c\neq 0$, and let $T=cS$. This is a subset of $\mathbb{F}_p$ which is disjoint from S. Set $x=\sum_{a\in S}\zeta^a-\sum_{b\in T}\zeta^b$. This is an element of $\mathbb{A}_n$ of norm $2r$. We have

$$\sum_{b\in T} b^i = c^i\sum_{a\in S} a^i = 0$$

for all $i\le r-1$ in $\mathbb{F}_p$, since the sums of roots of unity whose order divides a given integer greater than 1 are zero. Hence x belongs to $\mathbb{A}_n^{(r)}$. □

The lattices $\mathbb{A}_{p-1}^{(r)}$ have been studied by Bachoc and Batut for $p\le 23$ in [Bac-Bt]; Table 5.4.9 below, in which each row corresponds to a prime number $p\in\{5,7,11,13,17,19,23\}$ and each column to a positive integer $r\le\frac{p-1}{2}$, is extracted from their paper.

For each pair (p,r), we indicate successively the following three invariants for $\mathbb{A}_{p-1}^{(r)}$: its norm, its half kissing number, and its perfection by "yes" or "no".

Problem 5.4.10. Find for $1\le r\le\frac{p+1}{4}$ the norm of $\mathbb{A}_{p-1}^{(r)}$, and decide whether this lattice is perfect.

The equality $N(\mathbb{A}_{p-1}^{(r)})=2r$ holds for $r=1$ (Proposition 4.2.2), for $r=2$ (Proposition 5.3.2), for $r=3$ (Exercise 5.4.2), and also for $r=\frac{p+1}{4}$, $p\equiv -1$ mod 4, thanks to the following result due to Elkies and quoted by Gross in [Gro]:

For $p\equiv 3 \mod 4$ and $r=\frac{p+1}{4}$, $\mathbb{A}_{p-1}^{(r)}$ has norm $2r$ and possesses $\frac{p(p-1)}{2}=\frac{n(n+1)}{2}$ pairs of minimal vectors.

Table 5.4.9. Norm and perfection of the lattices $\mathbb{A}_{p-1}^{(r)}$.

	1	**2**	**3**	**4**	**5**	**6**	**7**	**8**	**9**	**10**	**11**
5	2 10 yes	4 5 no									
7	2 21 yes	4 21 yes	6 7 no								
11	2 55 yes	4 110 yes	6 55 yes	10 66 yes	10 11 no						
13	2 78 yes	4 195 yes	6 156 yes	8 39 no	12 13 no	12 13 no					
17	2 136 yes	4 476 yes	6 544 yes	8 238 yes	12 408 no	16 153 no	16 17 no	16 17 no			
19	2 171 yes	4 684 yes	6 969 yes	8 684 yes	10 171 yes	12 57 no	18 19 no	18 19 no	18 19 no		
23	2 253 yes	4 1265 yes	6 2277 yes	8 2024 yes	10 506 yes	12 253 yes	18 506 yes	22 276 yes	22 23 no	22 23 no	22 23 no

This is a consequence of Elkies and Shioda's theory in which lattices are produced using the Néron–Tate height on some Abelian varieties defined over suitable function fields, or (which apparently amounts to the same thing), intersection forms on suitable surfaces which are fibred by elliptic curves (*Mordell–Weil lattices*). For details on this highly interesting theory, which yields numerous examples of lattices with a large Hermite invariant, we refer the reader to the series of papers by Shioda (see [Sh1] and its list of references) and to Oesterlé's talk ([Oe2] at Bourbaki's seminar. After Shioda, these methods do not seem to apply to other cases of Problem 5.4.10; apparently, neither is suitable to prove perfection results.

On the basis of Theorem 5.4.8, Table 5.4.9 and of the result of Elkies above, it was stated as a conjecture in [Bac-Bt] that the Craig lattices should be perfect lattices of norm $2r$ for $1 \leq r \leq \frac{p+1}{4}$, a conjecture stated as Conjecture 4.10 in the French edition of this book. Actually, the equality $N(\mathbb{A}_{p-1}^{(r)}) = 2r$ holds for all $p \leq 37$ on the interval $[1, \frac{p+1}{4}]$. However, a counter-example has been found by Sigrist for the next value of p: we have

$N(\mathbb{A}_{40}^{(9)}) = 20 > 2r = 18$, and other examples exist for larger values of p, e.g. $p = 43$.

A variant of Craig lattices attached to cyclotomic fields consists in making use of a twist by an ideal. Given a fractional ideal $\mathfrak{A}$ of $\mathbb{Q}[\zeta]$ and a totally positive element α in its maximal real subfield $\mathbb{Q}(\zeta+\zeta^{-1})$ satisfying the extra condition $\alpha\,\mathfrak{A}\overline{\mathfrak{A}} = \mathbb{Z}[\zeta+\zeta^{-1}]$, one considers the lattice

$$(\mathfrak{P}^r\,\mathfrak{A},\ \frac{1}{p}\ \mathrm{Tr}_{\mathbb{Q}(\zeta)/\mathbb{Q}}(\alpha x\overline{x})).$$

This is still an even lattice of rank $p-1$ and determinant p^{2r-1}. Quebbemann ([Q1]) has proved that such a lattice is isometric to $\mathbb{A}_{p-1}^{(r)}$ if and only if the ideal $\mathfrak{A}$ is principal and generated by an element a of $\mathbb{Q}(\zeta)$ for which we have an equality of the form $\alpha a\overline{a} = \epsilon\overline{\epsilon}$ for some unit ϵ of $\mathbb{Z}(\zeta)$. These lattices satisfy the properties listed in Proposition 5.4.7.

The smallest possible value of p for which new lattices occur is $p = 23$. In $\mathbb{Q}(\zeta_{23})$, we have the decomposition $2\,\mathbb{Z}[\zeta_{23}] = \mathfrak{Q}\,\overline{\mathfrak{Q}}$. The ideal $\mathfrak{Q}$ is non-principal and generated by 2 and $\frac{-1+\sqrt{-23}}{2}$. This case has been considered in [Bac-Bt]; to obtain a basis for this lattice, they made use of an algorithm computing the Hermite normal form (*HNF*) of a matrix which allows a basis to be extracted efficiently from a generating system of a lattice, that they applied to the $2\cdot 22$ elements $2(1-\zeta)^r\zeta^i$, $\frac{1+\sqrt{-23}}{2}(1-\zeta)^r\zeta^i$ of $\mathbb{Q}(\zeta)$ for $0 \le i \le 21$. (A consequence of results of [Q1] and [Bay1] is that for $p = 23$, there does not exist any other twisted Craig lattice than the one we have just described.)

The results are displayed in Table 5.4.11 below, with the notation of Table 5.4.9:

Table 5.4.11. Norm and perfection of the twisted lattices $\mathbb{A}_{22}^{(r)}$.

	1	**2**	**3**	**4**	**5**	**6**	**7**	**8**	**9**	**10**	**11**
23	4 17963 yes	4 759 yes	6 2024 yes	8 1771 yes	10 506 yes	12 506 yes	18 506 yes	24 506 yes	30 253 yes	36 253 yes	44 23 no

Table 5.4.11 shows that twisted Craig lattices have norm $2r$ for $p = 23$ and $2 \le r \le 8$. Note that these lattices were previously considered by Feit ([Fe1], Section 14) in connection with the Mathieu group M_{23}.

When $n+1$ is not a prime, it seems likely that the inequality $N(\mathbb{A}_n^{(r)}) \ge 2r$ holds only for $r = 1$ or $r = 2$. We have verified this property for all pairs (n, r) with $n \le 40$ and $r = 3, 4, 5$.

Thanks to a program of Batut testing perfection modulo a prime, it has been possible to extend the numerical verifications of the perfection property to other values of (p, r) satisfying the inequality $r \le \frac{p+1}{4}$.

The following table gives the invariant s of some more Craig lattices of norm $2r$ with $r \le \lfloor \frac{p+1}{4} \rfloor$, or their norm N when it exceeds $2r$.

Table 5.4.12. Lattices $\mathbb{A}_{p-1}^{(r)}$ with $29 \le p \le 47$ and $3 \le r \le (p+1)/4$.

	3	**4**	**5**	**6**	**7**	**8**	**9**	**10**	**11**	**12**
29	6496	7917	5684	1421	580					
31	8835	11625	7905	5735	930	465				
37	19092	31968	29304	14430	13320	1665	1776			
41	29520	57605	61008	36490	31980	4920	$N = 20$	1066		
43	36421	74046	86688	63812	25929	28896	2709	$N = 22$	903	
47	52969	122153	154583	141611	63779	69184	4324	$N = 22$	$N = 24$	1081

When $p = n + 1$ is prime, experimentation shows a great regularity provided that p is large relative to r, as in the case $r = 2$, for which we observed in Section 5.3 that the equality $N(\mathbb{A}_n^{(2)*}) = N(\mathbb{A}_n^*)$ (resp. $S(\mathbb{A}_n^{(2)*}) = S(\mathbb{A}_n^*)$) holds for all $n \ge 10$ (resp. $n \ge 14$). There probably exist for all r bounds n_0, n_1 such that the equality $N(\mathbb{A}_n^{(r)*}) = N(\mathbb{A}_n^*)$ (resp. $S(\mathbb{A}_n^{(r)*}) = S(\mathbb{A}_n^*)$) holds for all $p \ge n_0$ (resp. for all $p \ge n_1$). Here are the facts that we have observed:

Table 5.4.13. Lattices dual to $\mathbb{A}_{p-1}^{(r)}$.

	$r = 2$	$r = 3$	$r = 4$	$r = 5$
$S(\mathbb{A}_{p-1}^{(r)\,*}) \supset S(\mathbb{A}_{p-1}^*)$	$p \ge 11$	$p \ge 17$	$p \ge 23$	$p \ge 29$
$S(\mathbb{A}_{p-1}^{(r)\,*}) = S(\mathbb{A}_{p-1}^*)$	$p \ge 13$	$p \ge 19$	$p \ge 29$	$p \ge 31$

5.5 Lattices Related to the $\mathbb{D}_n$ Lattice

In this section, E is the Euclidean space $\mathbb{R}^n$ with $n \ge 3$. Recall that $\mathbb{D}_n$ is the lattice $\{\sum_1^n x_i \varepsilon_i \mid x_i \in \mathbb{Z}, \sum x_i \equiv 0 \mod 2\}$. We first classify lattices which contain $\mathbb{D}_n$ to index 2.

More generally, let Λ be a lattice, let $m > 1$ be an integer and let Λ' be a lattice containing Λ to index m and such that Λ'/Λ is a cyclic group. We have $\Lambda' = \langle \Lambda, \frac{1}{m}\, e\rangle$ for a suitably chosen vector $e \in \Lambda$. Conversely, we obtain in this way a lattice which satisfies the desired properties provided that e does not belong to any lattice $d\Lambda$ with $d > 1$ dividing m. We have $\langle \Lambda, \frac{1}{m}e\rangle = \langle \Lambda, \frac{1}{m}e'\rangle$ if $e' - ae \in m\Lambda$ for some integer a prime to m. Moreover, for all $\sigma \in \mathrm{Aut}(\Lambda)$, $\langle \Lambda, \frac{1}{m}e\rangle$ is isometric to $\langle \Lambda, \frac{1}{m}\sigma(e)\rangle$. Given a basis $(e_1, \dots, e_n)$ for Λ, it suffices to search e among the m^n vectors $\sum_1^n a_i\, e_i$, $a_i \mod m$ to find all lattices Λ' such that Λ'/Λ is cyclic of order m. For instance, taking $\Lambda = \mathbb{A}_n$ with its basis $(\varepsilon_0 - \varepsilon_i)$ and setting $m = r$, we see that $\mathbb{A}_n^r$ corresponds to the choice $a_i = 1$ for all i, which does not exhaust all possibilities with $\Lambda'/\mathbb{A}_n \simeq \mathbb{Z}/r\mathbb{Z}$.

Let us now turn to the case $\Lambda = \mathbb{D}_n$ and $m = 2$, considered in Section 4.4 to construct $\mathbb{D}_n^+$. Write $e = \sum_1^n x_i\, \varepsilon_i$, and consider the integer $t \in [0, n]$ (which must be even) such that x_i is odd for t values of i and even otherwise. By adding to e suitable multiples of vectors of the form $4\,\varepsilon_i$ or $2\,(\varepsilon_i + \varepsilon_j)$, we restrict ourselves to the case where the x_i belong to $\{-1, 0, 1, 2\}$ and the even values are zero except perhaps one of them. If $t = 0$, we have $\Lambda' = \mathbb{Z}^n$. Otherwise, if $x_j = 2$ for one index j, we get rid of the even term $x_j\,\varepsilon_j$ by adding to $\frac{e}{2}$ a vector $\pm\varepsilon_i - \varepsilon_j$ where i is such that x_i is odd.

Since the group $\mathrm{Aut}(\mathbb{D}_n)$ allows all changes of signs of the ε_i and any permutation of the indices i, we can, up to an isometry, choose $e = \sum_1^t \varepsilon_i$.

Proposition 5.5.1. *Let $t > 0$ be even, let $e = \sum_1^t \varepsilon_i \in \mathbb{D}_n$, and let $\mathbb{D}_{n,t}$ be the lattice generated by $\mathbb{D}_n$ and $\frac{e}{2}$. Then $\mathbb{D}_{n,t}$ is an extreme lattice if and only if $t \geq 8$, and any perfect lattice containing $\mathbb{D}_n$ to index 2 is isometric to a unique lattice $\mathbb{D}_{n,t}$, where t is one of the $\lfloor \frac{n-8}{2} \rfloor$ even integers in $[8, n]$. We have* $\det(\mathbb{D}_{n,t}) = 1$, $s(\mathbb{D}_{n,t}) = n(n-1)$ *for $t > 8$ and* $s(\mathbb{D}_{n,8}) = n(n-1) + 64$. *For $t = n$ (n is then even), $\mathbb{D}_{n,t}$ is the lattice $\mathbb{D}_n^+$. The lattices $\mathbb{D}_{n,t}$ are half-integral, and are integral if and only if $t = n \equiv 0 \mod 4$, i.e. in the $\mathbb{D}_{4m}^+$ case.*

Proof. We have seen that a lattice which contains $\mathbb{D}_n$ to index 2 must be isometric to $\mathbb{Z}^n$ (which is not perfect) or to a lattice $\mathbb{D}_{n,t}$ where t is even and strictly positive. The elements of $e + \mathbb{D}_n$ are of the form $\frac{1}{2}(\sum x_i\,\varepsilon_i)$ with $\sum x_i \equiv 0 \mod 4$, and those of $S(e + \mathbb{D}_n)$ are the vectors $\frac{1}{2}(\sum \pm\varepsilon_i)$ with an even number of minus signs in the sum. We have $N(\mathbb{D}_{n,t}) \geq 2$, and indeed $N(\mathbb{D}_{n,t}) = 2$, if and only if $t \geq 8$. When $t \geq 8$, $\mathbb{D}_{n,t}$ is extreme since its set of minimal vectors contains $S(\mathbb{D}_n)$. The calculation of s is obvious, and an enumeration of the norm $\frac{t}{4}$ vectors shows that the $\mathbb{D}_{n,t}$ lattices are pairwise non-isometric. Finally, the integrality assertions follow from the two equalities $\frac{e}{2} \cdot (\varepsilon_1 + \varepsilon_n) = \frac{1}{2}$ for $t < n$ and $\frac{e}{2} \cdot \frac{e}{2} = \frac{t}{4}$. □

The lattice $\widetilde{\mathbb{D}}_{9,8} = \sqrt{2}\,\mathbb{D}_{9,8}$ is an integral lattice of norm 4 and determinant $2^9 = 512$, first considered by Korkine and Zolotareff in [K-Z2] as the quadratic form T_9. Chaundy ([Chn], 1946) proved that it has the greatest possible Hermite invariant *among the lattices which possess a norm 4 hyperplane section similar to* $\mathbb{E}_8$. This is thus the laminated lattice Λ_9 (see Sections 3.9 and 8.2, where it appears as a section of the lattice $\Lambda_{12}^{\max}$ also constructed below).

The classification of lattices $\Lambda \supset \mathbb{D}_n$ with $\Lambda/\mathbb{D}_n$ non-cyclic can be done in the same way. Disregarding the uninteresting case $\Lambda \supset \mathbb{Z}^n$, one easily proves:

Proposition 5.5.2. *The lattices of norm* 2 *containing* $\mathbb{D}_n$ *to index* 4 *and with a non-cyclic quotient are classified up to isometry by three parameters* t, t'_1, t'_2 *which satisfy the following conditions:* t *and* $t' = t'_1 + t'_2$ *even,* $t, t' \geq 8$, $2t'_1 \leq t$, *and* $t'_1 \leq t'_2$ *if* $t' = t$; *the lattice corresponding to* (t, t'_1, t'_2) *is then* $\Lambda = \langle \mathbb{D}_n, \frac{e}{2}, \frac{e'}{2} \rangle$, *where*

$$e = \sum_{i=1}^{t} \varepsilon_i \quad \textit{and} \quad e' = \sum_{j=1}^{t'_1} \varepsilon_j + \sum_{k=t+1}^{t+t'_2} \varepsilon_k \,. \quad \square$$

These lattices, which exist from dimension 12 onwards, have determinant $\frac{1}{4}$, and become integral (with determinant 2^{n-4}) when rescaled to norm 4. For $n = 12$, there is a unique lattice, with $s = 324$ ($12 \cdot 11 = 132$ pairs of vectors belonging to $\mathbb{D}_{12}$ and $3 \cdot 64 = 192$ pairs of vectors obtained from $\frac{1}{2}e, \frac{1}{2}e'$, and $\frac{1}{2}(e - e')$ by performing even numbers of sign changes of the ε_i). In dimension 13, we find three lattices, corresponding to $(t, t'_1, t'_2) = (8,4,4), (8,3,5), (8,4,5)$.

The 12-dimensional lattice above of is similar to the laminated lattice $\Lambda_{12}^{\max}$. We shall again construct it in Chapter 8 as a module over the Hurwitz quaternion order, together with other lattices containing $\mathbb{D}_n$, with $n \equiv 0 \mod 4$.

Other perfect lattices related to $\mathbb{D}_n$ have been constructed by Barnes. We just state one example, and refer to [Bar5] for the proof.

Theorem 5.5.3. *For every integer* $t \geq 2$, *denote by* $\mathbb{D}_n^t$ *the sublattice of* $\mathbb{D}_n$ *defined in the basis* $(\varepsilon_1, \dots, \varepsilon_n)$ *for* $\mathbb{Z}^n$ *by the congruences*

$$x_1 \equiv x_2 \equiv \cdots \equiv x_{n-1} \mod t \quad \textit{and} \quad x_1 + x_2 + \cdots + x_n \equiv 0 \mod 2t \,.$$

Let $k \in [0, t]$ *be the integer such that* $n \equiv \pm k \mod 2t$.

1. *For all* $t \geq 2$, $\mathbb{D}_n^t$ *is an integral lattice, with* $N(\mathbb{D}_n^t) \leq 2t^2$ *and* $\det(\mathbb{D}_n^t) = 4t^{2(n-1)}$.
2. *For* $t \geq 3$, $\mathbb{D}_n^t$ *is perfect with norm* $2t^2$ *if and only if one of the following two conditions (a), (b) holds:*

$$\begin{array}{lll} \text{(a)}\ n \geq 2t^2 - k^2 & \text{and} & 0 \leq k \leq t-1; \\ \text{(b)}\ n \geq t(t+2) & \text{and} & k = t. \end{array}$$

3. The number of pairs of minimal vectors is $s = n(n-1)$ except in the following three cases:

$$\begin{array}{llll} \text{(a)} & n = t(t+2), & t \text{ odd} & : s = n^2 - 1; \\ \text{(b)} & n = t(t+2) - 1, & t \text{ odd} & : s = n^2; \\ \text{(c)} & n = 2t^2 - k^2, & & \\ & 0 \le k \le t-2,\ k^2 \pm k \equiv 0 \mod 2t & & : s = n(n-1) + 1. \end{array} \square$$

5.6 Unimodular Lattices

We still denote by E a Euclidean space of dimension n. Recall that a lattice Λ is *unimodular* if it is equal to its dual lattice; equivalent definition: Λ is integral and $\det(\Lambda) = 1$.

Theorem 5.6.1. *A perfect unimodular lattice which contains vectors of norm 1 or 2 is extreme, and isometric to one of the following pairwise non-isometric lattices: $\mathbb{Z}$, $\mathbb{E}_8$, $\mathbb{A}_{r^2-1}^r$ ($r \ge 4$ an integer), and $\mathbb{D}_n^+$ ($n \ge 12$ and divisible by 4).*

Proof. Let Λ be such a lattice, let Λ' be the lattice generated by $S(\Lambda)$, and let $r = [\Lambda : \Lambda']$. Since Λ is perfect, S is irreducible and spans E. Since Λ is unimodular, we have $\det(\Lambda') = r^2$.

If Λ contains a vector of norm 1, this vector generates a lattice isometric to $\mathbb{Z}$ which is moreover a direct orthogonal factor in Λ'. Hence we must have $n = 1$ and $\Lambda = \Lambda' \simeq \mathbb{Z}$ in this case.

Otherwise, Λ' is an irreducible root lattice of norm 2, hence isometric to one of the lattices $\mathbb{A}_n, \mathbb{D}_n, \mathbb{E}_6, \mathbb{E}_7, \mathbb{E}_8$. This first shows that Λ' and thus also Λ is extreme. Moreover, since Λ' has square determinant, it must be isometric to $\mathbb{A}_n$, and then $n+1 = r^2$, or to $\mathbb{D}_n$, and then $r = 2$, or to $\mathbb{E}_8$, and then $r = 1$ and $\Lambda = \Lambda' \simeq \mathbb{E}_8$. The $\mathbb{A}_n$ case yields the Coxeter lattice $\mathbb{A}_{r^2-1}^r$, and we must exclude the cases $r = 2$ and $r = 3$ because of the isometries $\mathbb{A}_3^2 \simeq \mathbb{Z}^3$ and $\mathbb{A}_8^3 \simeq \mathbb{E}_8$. As for $\mathbb{D}_n$, we know from Proposition 5.5.1, that the only possibility is $n \equiv 0 \mod 4$ and $n \ge 8$, and then $\Lambda \simeq \mathbb{D}_n^+$. We must exclude $\mathbb{D}_8^+ \simeq \mathbb{E}_8$.

Conversely, $\mathbb{A}_{r^2-1}^r$ ($r \ge 4$) and $\mathbb{D}_n^+$ ($n \ge 12$ and divisible by 4) are unimodular irreducible lattices, and two of them cannot be isometric, since they have different root systems. $\square$

The other perfect unimodular lattices must be of norm at least 3, i e. they must have an empty root system. Now, given an integer $m \ge 3$, there is no classification of integral lattices generated by vectors of norm m available, and all we can do is to consider small-dimensional examples.

However, there are nevertheless some restrictions on the possible dimensions of even unimodular lattices. A look at the lattices $\mathbb{D}_n^+$ and $\mathbb{A}_{r^2-1}^r$

considered above shows that they are even when their dimension is divisible by 8 and odd otherwise. This is related to the following general result:

Theorem 5.6.2. (Schoeneberg.) *The rank of an even unimodular lattice is divisible by* 8. □

We shall not prove this theorem, but simply give some indications on a possible strategy which yields a more general result (not on the original proof, which relied on the theory of modulor forms, [Scho], p. 520; see also [Hecke], no. 41 (1940), p. 867). The basic idea is to consider more generally *unimodular symmetric bilinear modules*, i.e. free $\mathbb{Z}$-modules M endowed with an integral symmetric bilinear form of discriminant ± 1. As a real form, this form has a *signature* (r_1, r_2).

Theorem 5.6.2 is a particular case of the following one:

Theorem 5.6.3. *The signature* (r_1, r_2) *of an even unimodular symmetric bilinear module satisfies the congruence* $r_2 \equiv r_1 \mod 8$. □

To state the next theorem, we introduce the following notation: given an integral bilinear symmetric module (M, Q) and an integer p, we denote by $p\,M$ the orthogonal sum of $|p|$ copies of (M, Q) if $p \geq 0$ and of $(M, -Q)$ if $p < 0$. We moreover denote by U the module $\mathbb{Z}^2$ endowed with the quadratic form $2x_1x_2$ (whose corresponding bilinear form is $x_1y_2 + x_2y_1$); it has determinant -1. For a rank-n lattice M, $M \perp U$ has signature $(n+1, 1)$, determinant $-\det(M)$ and same parity (even or odd) as M.

We have thus reduced the proof of Theorem 5.6.3 to the indefinite case, for which there exists a classification, both for odd and even forms:

Theorem 5.6.4. *An indefinite unimodular symmetric bilinear* $\mathbb{Z}$*-module which is odd (resp. even) is isometric to* $p\mathbb{Z} \perp (-q)\mathbb{Z}$, $p, q \in \mathbb{Z}$, $p, q > 0$ *(resp. to* $p\,U \perp q\,\mathbb{E}_8$, $p, q \in \mathbb{Z}, p > 0$*).* □

A proof of this theorem can be read in [M-H] or in [Se2] (or [Se'2]). It makes use of the invariant modulo 8 defined for lattices in Chapter 1, after Definition 5.3.10, which extends to indefinite forms. A more precise statement of Theorem 5.6.3 is that *for any unimodular module of signature* (r_1, r_2)*, this invariant is equal to* $r_1 - r_2 \mod 8$.

We shall now consider succinctly the classification of small-dimensional unimodular lattices. We begin with the even case (for which n is divisible by 8).

In dimension 8 (resp. 16), we only have $\mathbb{E}_8$ (resp. $\mathbb{E}_8 \perp \mathbb{E}_8$ and $\mathbb{D}_{16}^+$).

In dimension 24, there are 24 even lattices, among which 23 have norm 2 and one has norm 4, the celebrated *Leech lattice*, a construction of which will be given below. The complete list has been established by Niemeier ([Nie]; for another point of view, see Venkov's Chapter 18 in [C-S]) a few years after Conway's proof of the uniqueness of the Leech lattice (Invent. Mat. 7

(1969), 137–142, = [C-S], Chapter 12). One usually calls *Niemeier lattices* those which are of norm 2. Curiously, up to dimension 24, even unimodular lattices are defined by their root systems, which are of maximal rank if not empty. Besides $\mathbb{D}_{24}^+$, $\mathbb{A}_{24}^5$, and the Leech lattice quoted above, there are the obvious lattices $\mathbb{E}_8 \perp \mathbb{E}_8 \perp \mathbb{E}_8$ and $\mathbb{E}_8 \perp \mathbb{D}_{16}^+$, and 19 more lattices whose root system is reducible. Those which are perfect (and indeed strongly eutactic, hence extreme) are $\mathbb{E}_8, \mathbb{D}_{16}^+, \mathbb{D}_{24}^+, \mathbb{A}_{24}^5$ and the Leech lattice. The perfection of the Leech lattice results from its identification with the 24th laminated lattice Λ_{24}, or from the Bannai–Sloane theorem (stated as Theorem 3.5.6); this lattice is also strongly eutactic because its automorphism group is $\mathbb{R}$-irreducible, or because it acts transitively on its set of minimal vectors (see [ATLAS]).

As for odd lattices, it suffices to consider those of norm at least 2 (from which the other ones can be deduced by orthogonal sum with a power of $\mathbb{Z}$). One first finds $\mathbb{D}_{12}^+$, then a lattice with $n = 14$ and root system $2\boldsymbol{E}_7$, then the lattice $\mathbb{A}_{15}^4$, then a lattice with $n = 16$ and root system $2\boldsymbol{D}_8$, ... The classification of these lattices has been done by Conway and Sloane up to dimension 23 and by Borcherds in dimension 24 (see [C-S], Chapters 16 and 17); they are also characterized by their root systems, whose rank, however, need not be 0 or n. Further results concerning lattices of small determinant in dimensions 25 to 28 can be found in [Bor1] and [Bc-V2].

Empty root systems appear in dimensions 23 and 24 (a single lattice in each case, of norm 3, to be constructed in the next section).

The remainder of this section is devoted to the description of a construction of the Leech lattice Λ_{24}, due to Eva Bayer ([Bay1]).

Let $\zeta_{35} = \zeta$ be a root of unity of order 35 in $\mathbb{C}$. We consider the cyclotomic field $K = \mathbb{Q}(\zeta)$, of degree $\varphi(35) = 24$, and its maximal real subfield $F = \mathbb{Q}(\zeta + \zeta^{-1})$. By using a few results on cyclotomic fields (see e.g. the tables in [Was]), one can show that the different $\mathcal{D}$ of $F/\mathbb{Q}$ is *principal in the narrow sense*, i.e. possesses a totally positive generator. Let α be such a generator. Since K/F is unramified outside the infinite places (observe that $K = F(\zeta_5) = F(\zeta_7)$), the form $(x,y) \mapsto \mathrm{Tr}_{K/\mathbb{Q}}(\alpha^{-1}xy)$ has discriminant 1. Its Hermitian twist $h : (x,y) \mapsto \mathrm{Tr}_{K/\mathbb{Q}}(\alpha^{-1}x\overline{y})$ also has determinant 1, and is moreover even, because $h(x,x) = 2\,\mathrm{Tr}_{F/\mathbb{Q}}\left(\alpha^{-1}N_{K/F}(x)\right)$. Thus, endowed with h, $\mathbb{Z}_K$ is an even unimodular lattice in $E = \mathbb{R} \otimes K$, of dimension $n = [K : \mathbb{Q}] = 24$.

Now $\mathbb{Z}_K$ has norm at least 2 (because it is an even lattice), and its norm 2 vectors constitute a root system, which must be an orthogonal sum of systems isometric to $\boldsymbol{A}_n, \boldsymbol{D}_n, \boldsymbol{E}_6, \boldsymbol{E}_7$ or $\boldsymbol{E}_8$. The automorphisms of the lattice $\mathbb{Z}_K$, and in particular the automorphism $\theta \mapsto \zeta\theta$, with characteristic polynomial ϕ_{35}, preserve this root system. But it is easily checked that the non-empty root systems of rank $r \leq 24$ do not possess such an automorphism. By Conway's theorem, $\mathbb{Z}_K$ is isometric to the Leech lattice.

The construction above has been made explicit by Bachoc and Batut ([Bac-Bt]). Using Bayer's results on the unit signatures, they have determined a totally positive generator α for $\mathcal{D}_{F/\mathbb{Q}}$, and then constructed a Gram matrix, by means of which one can verify directly the properties of perfection and strong eutaxy. The theorem below lists some important properties of this lattice:

Theorem 5.6.5. *Let K be the cyclotomic field $\mathbb{Q}(\zeta_{35})$, with maximal real subfield $F = \mathbb{Q}(\zeta_{35} + \zeta_{35}^{-1})$.*

1. *There exists a totally positive generator α of the different $\mathcal{D}_{F/\mathbb{Q}}$ such that the Gram matrix $A = (a_{i,j})$ of the lattice $\Lambda = (\mathbb{Z}_K, \mathrm{Tr}_{K/\mathbb{Q}}(\alpha^{-1}x\overline{y})) \subset \mathbb{R} \otimes K$ in the basis $(1, \zeta, \dots, \zeta^{23})$ is defined by the equations $a_{i,j} = b_{|j-i|}$, where b is the vector* $(4, 1, -1, 0, 0, 0, 1, -1, -2, -1, -1, -1, 1, 1, -1, -1, 1, 2, 2, 1, -1, -1, 1, 1)$.
2. *Λ is isometric to the Leech lattice Λ_{24}.*
3. *Λ possesses bases of minimal vectors.*
4. *Λ is extreme and strongly eutactic, and contains $s = 98280 = 2^3 \cdot 3^3 \cdot 5 \cdot 7 \cdot 13$ pairs of minimal vectors acted on transitively by* $\mathrm{Aut}(\Lambda)$, *all of which have spectrum* $(46575, 0)$, $(47104, 1)$, $(4600, 2)$. □

[The reason why the entries of the Gram matrix above solely depend on the difference $|j - i|$ is that our $\mathbb{Z}$-basis comes from a $\mathbb{Z}[\zeta_{35}]$-basis for Λ. Actually, it even suffices to know the first 12 components of b, the last 12 being obtainable by using the relation $x \cdot \zeta^i x = \zeta^{-i}x \cdot x = \frac{1}{2}((\zeta^i + \zeta^{-i})x \cdot x)$ and calculating inductively $\zeta^i + \zeta^{-i}$ for $i \geq 12$ by means of the cyclotomic polynomial ϕ_{35}. This would not happen with Craig's construction of Λ_{24} over $\mathbb{Z}[\zeta_{39}]$ ([Cra3]), which makes use of a non-principal ideal.]

The short construction above of Λ_{24}, most certainly simple, throws no light on the minimal vectors nor on the automorphism group, except for the small subgroup $C_2 \times (\mathbb{C}_{35} \rtimes C_{24})$ of order 1680 generated by multiplications by roots of unity and Galois automorphisms. (The automorphism group of Λ_{24} has order $2^{22} \cdot 3^9 \cdot 5^4 \cdot 7^2 \cdot 11 \cdot 13 \cdot 23 \,\#\, 1680 \times 5 \cdot 10^{15}$.) We shall give more appropriate constructions in Sections 8.7 and 8.8.

The quotient $\mathrm{Aut}(\Lambda_{24})/\{\pm \mathrm{Id}\}$ is Conway's first simple group Co_1. The group $\mathrm{Aut}(\Lambda_{24})$ acts transitively on the sets of vectors of norm $4, 6, 8, 10, 14$ (the stabilizers of the vectors of norm 4 and 6 are Conway's simple groups Co_2 and Co_3); the vectors of norm 12 share out among two orbits, cf. [ATLAS], groups Co_1, Co_2, and Co_3.

5.7 Around the Leech Lattice

We construct in this section some lattices related to the Leech lattice Λ_{24}, including unimodular lattices of norm 3 in dimensions $n = 23$ and $n = 24$. In particular, we describe a neighbouring procedure of Kneser, a variant of which will be used in Chapter 8. As usual, E denotes an n-dimensional Euclidean space.

Let Λ be an integral lattice and let e be a nonzero vector in Λ^* such that the scalar product $e \cdot x$ takes odd values on Λ.

Definition 5.7.1. *The* half-lattice of Λ and e *is*

$$\Lambda_e = \{x \in \Lambda \mid e \cdot x \equiv 0 \mod 2\}\,.$$

The lattice Λ_e is integral, of index 2 in Λ, hence $\det(\Lambda_e) = 4\det(\Lambda)$. Two lattices Λ_e and $\Lambda_{e'}$ are isometric when there exists $\sigma \in \mathrm{Aut}(\Lambda)$ such that $e' \equiv \sigma(e) \mod 2\Lambda^*$. Moreover, as an integral lattice, Λ is a sublattice of Λ^*. We may thus apply the preceding definition to vectors $e \in \Lambda$.

Let $e \in \Lambda$, let $m \in \mathbb{Z}$ be the integer such that $N(e) = 2m$ or $2m+1$, and let $(n_0, n_1, \ldots, n_m)$ be the spectrum of Λ with respect to e: for $0 \le h \le m$, there are n_h pairs of minimal vectors in Λ such that $|e \cdot x| = h$, and we have $s(\Lambda) = 1 + \sum_{i-0}^{m} n_i$ or $s(\Lambda) = \sum_{i-0}^{m} n_i$ according to whether e belongs or does not belong to $S(\Lambda)$. (For $x \in S(\Lambda)$, the inequality $|e \cdot x| \le m$ comes from the lower bound $N(e \pm x) \ge N(x)$.) The following proposition is easy:

Proposition 5.7.2. *Let Λ be an integral lattice and let $e \in \Lambda$ have an odd scalar product with some minimal vector in Λ. Then we have*

$$s(\Lambda_e) = \sum_{i \equiv 0 \mod 2} n_i\,,$$

except if e is minimal and has even norm, where

$$s(\Lambda_e) = 1 + \sum_{i \equiv 0 \mod 2} n_i\,. \quad \square$$

We now describe a first construction making use of the lattice Λ_e.

Definition 5.7.3. *Let Λ be an integral lattice and let e be a nonzero vector in Λ^*. We denote by $P_e\Lambda$ the orthogonal projection of Λ_e (Definition 5.7.1) onto the hyperplane of E orthogonal to e.*

Since we perform a projection parallel to e and since some multiple of e belongs to Λ_e (if $m\Lambda^* \subset \Lambda$, then $2me \in \Lambda_e$), $P_e\Lambda$ is a full lattice in $\mathbb{R}e^\perp$.

Theorem 5.7.4. *Let Λ be an integral lattice and let $e \in \Lambda^*$ be a vector such that there exists $x \in \Lambda$ with $e \cdot x = 1$. Then:*

1. *$P_e\Lambda$ is a lattice of determinant $\dfrac{\det(\Lambda)}{N(f)}$ if $N(f)$ is odd and $\dfrac{4\det(\Lambda)}{N(f)}$ if $N(f)$ is even, where f is the smallest rational multiple of e which belongs to Λ, and the lattices $P_e\Lambda$ and $P_{\sigma(e')}\Lambda$ are isometric for any $\sigma \in \mathrm{Aut}(\Lambda)$ and $e' \equiv e \mod 2\Lambda^*$.*
2. *If $e \in \Lambda$, $P_e\Lambda$ is integral if and only if $N(e)$ divides 4.*

3. If Λ is unimodular, $P_e\Lambda$ is unimodular if and only if $N(e) = 4$.
4. If Λ is even, $P_e\Lambda$ is even if $N(e) \equiv 2 \mod 4$ and odd if $N(e) \equiv 0 \mod 4$.
5. If Λ is of norm 4 and if e is a minimal vector in Λ, $P_e\Lambda$ is integral of norm ≥ 3, unimodular if Λ is unimodular, and its vectors of norm 3 are the projections of the minimal vectors in Λ whose scalar product with e is equal to ± 2. Moreover, $s(P_e\Lambda)$ is half the number of such vectors.

Proof. By definition, f is a primitive vector in Λ. Consequently, the lattice $\mathbb{R}e \cap \Lambda_e$ is generated by $2f$ when $N(f)$ is odd and by f when $N(f)$ is even, which implies (1), taking into account the value of $\det(\Lambda_e)$ provided by Definition 5.7.1.

The orthogonal projection p onto $\mathbb{R}e^\perp$ is given by the formula $p(x) = x - \frac{e \cdot x}{e \cdot e} e$, which implies for the scalar products of the projections the formulae

$$p(x) \cdot p(y) = x \cdot y - \frac{(e \cdot x)(e \cdot y)}{N(e)} \quad \text{and} \quad N(p(x)) = N(x) - \frac{(e \cdot x)^2}{N(e)}\,.$$

There exists a vector $x_1 \in \Lambda_e$ with $e \cdot x_1 = 2$. The calculation of $N(p(x_1))$ shows that $P_e\Lambda$ cannot be integral unless $N(e)$ divides 4, and the calculation of the scalar product of the projections shows that this condition suffices. We thus have proved (2) and (4), and (3) is an immediate consequence of (1) and (2).

We are now left with (5). The lower bound $N(p(\Lambda_e)) \geq 3$ is a consequence of Proposition 1.2.10, which asserts that the inequality $N(p(\Lambda_e) \geq N(\Lambda_e) - \frac{N(e)}{4}$ holds, and proves that all minimal vectors in $P_e\Lambda$ are projections of those vectors of norm 4 of Λ whose distance to the projection hyperplane are equal to $\frac{\|e\|}{2}$, a condition which can be written $\|\frac{x \cdot e}{e \cdot e}\, e\| = \frac{\|e\|}{2}$, i.e. $|x \cdot e| = 2$. Since these vectors arise in pairs $x\,, e - x$, their contribution to $s(P_e\Lambda)$ is only $\frac{n_2}{2}$. □

The second construction we are going to describe is due to Kneser ([Kn2]). It makes use of the lattice Λ_e introduced in Definition 5.7.1 to construct a new integral lattice with the original determinant.

Definition 5.7.5. *Let Λ be an integral lattice. We say that an integral lattice Λ' is a* (Kneser-) neighbour of *Λ if the intersection $\Lambda \cap \Lambda'$ is of index 2 in both of them.*

This intersection is some lattice Λ_e in the sense of Definition 5.1.1, and we have $\Lambda' = \langle \Lambda_e, \frac{1}{2}f \rangle$ for some $f \in \Lambda_e$. Actually, when $\det(\Lambda)$ is odd, it is usually possible to chose e so as to have $\Lambda' = \langle \Lambda_e, \frac{1}{2}e \rangle$ (see Exercise 5.7.1). We then use the notation

$$\Lambda' = K_e\Lambda \;\text{ or }\; \Lambda' = \Lambda^e \qquad \textbf{(5.7.5')}$$

even if the lattice Λ' defined by this formula is not integral. Such an expression is possible if and only if

$$[\Lambda' : \Lambda_e] = 2 \Longleftrightarrow e \in \Lambda_e \Longleftrightarrow N(e) \equiv 0 \mod 2\,.$$

The proof of the following proposition is immediate:

Proposition 5.7.6. *Let Λ be an integral lattice and let $e \in \Lambda$ have even norm. Then the lattice $K_e\Lambda$ is a half-integral lattice with the same determinant as Λ, which is integral if and only if $N(e)$ is divisible by 4, odd if $N(e) \equiv 4 \mod 8$ and even if Λ_e is even and $N(e)$ is divisible by 8.*
[For some more details on Kneser neighbouring, see Exercise 5.7.1.] □

Note that Λ_e is even if and only if Λ is even or e is a characteristic vector for Λ; an example is provided by $\Lambda = \mathbb{Z}^n$, $n \equiv 0 \mod 8$ and $e = \varepsilon_1 + \varepsilon_n$, where $K_e\Lambda = \mathbb{E}_8$. When $N(e) = 4$, we recover the first construction of this section:

Proposition 5.7.7. *Let Λ be an integral lattice and let $e \in \Lambda$.*

1. *If $N(e) = 4$, $K_e\Lambda$ is isometric to $\mathbb{Z} \perp P_e\Lambda$.*
2. *If Λ is even and if $N(e) = 8$, $K_e\Lambda$ is even and has norm 2.*
3. *If Λ is even of norm 4 and if $N(e) = 12$, $K_e\Lambda$ is odd and has norm 1 or 3. The first case occurs if and only if Λ_e contains a minimal vector $x \in S(\Lambda)$ with $e \cdot x = -6$.*

Proof. (1) Since $\frac{1}{2}\,e$ has norm 1, the lattice it generates is isometric to $\mathbb{Z}$ and is thus an orthogonal component of $K_e\Lambda$. This yields an orthogonal decomposition $K_e\Lambda = \mathbb{Z}\frac{e}{2} \perp \Lambda'$ where $\Lambda' = \mathbb{R}e^\perp \cap \Lambda_e$ is clearly the orthogonal projection of $K_e\Lambda$ onto $\mathbb{R}e^\perp$. Since $\frac{1}{2}e + x$ and x have the same orthogonal projection onto $\mathbb{R}e^\perp$, Λ' is also the projection of Λ_e.

(2) The equality $N(\frac{1}{2}e + x) = 2 + e \cdot x + x \cdot x$ shows that $K_e\Lambda$ is even and possesses a norm 2 vector ($\frac{1}{2}\,e$ for instance).

(3) We now have $N(\frac{1}{2}\,e + x) = 3 + e \cdot x + x \cdot x \equiv 1 \mod 2$ for all $x \in \Lambda_e$ and $N(K_e\Lambda) \le N(\frac{1}{2}\,e) = 3$. Since $N(x) \ge 4$ on Λ_e, we must have $N(K_e\Lambda) = 1$, which proves the existence of an $x \in \Lambda'_e$ with $N(\frac{1}{2}\,e + x) = 1$. We thus have $e \cdot x + x \cdot x = -2$, whence $N(e + x) = 12 + 2\,e \cdot x + x \cdot x = 8 - x \cdot x$. But we also have $N(e + x) \ge 4$, hence $x \cdot x \le 4$, which implies $x \cdot x = 4$ and $e \cdot x = -6$. □

Remark 5.7.8. *With the lattice $\Lambda = \sqrt{2}\,\mathbb{D}_4$ (the centred cubic lattice with edges e_1, e_2, e_3, e_4 of length 2), taking $e = e_1 + e_2 + e_3$ and $x = -\frac{1}{2}(e_1 + e_2 + e_3 + e_4)$, we obtain an example for which $N(K_e\Lambda) = 1$.*

From now on, we take for Λ the Leech lattice Λ_{24}, and choose for it a basis $(e_1, e_2, \dots, e_{24})$ with the Gram matrix we constructed in Section 5.6 by means of the 35th roots of unity.

The possibility quoted in Remark 5.7.8 effectively occurs, since Λ contains $\sqrt{2}\,\mathbb{D}_4$. But there are also vectors v which are a sum of three pairwise orthogonal minimal vectors for which $K_v\Lambda$ has norm 3; the two possibilities (1 and

3) for $N(K_v\Lambda)$ correspond to the existence (pointed out in Section 5.6) of two orbits of norm 12 vectors in Λ (see [ATLAS], p. 180).

One finds in [C-S] (Chapter 12, Theorem 2) a description of $\Lambda/2\Lambda$ ($= \Lambda^*/2\Lambda^*$); representatives for the classes are:

1. $\{0\}$;
2. One norm 4 vector for each pair $\pm x$;
3. One norm 6 vector for each pair $\pm x$;
4. One norm 8 vector for each system of 24 pairs $\pm x_1, \dots, \pm x_{24}$ of pairwise orthogonal vectors

so that, denoting by N_m the number of pairs of norm m vectors in Λ, we have the relation

$$N_4 + N_6 + \frac{1}{24} N_8 = 2^{24} - 1\,.$$

A consequence of this description and of the transitivity properties of $\mathrm{Aut}(\Lambda)$ is that a half-lattice attached to the Leech lattice is isometric to one of the three lattices Λ_e constructed with a vector e of norm 4, 6 or 8. Let us denote by $\Lambda(4), \Lambda(6), \Lambda(8)$ such lattices, which could be canonically defined by taking for instance $e = e_1$, $e = e_1 - e_2$ and $e = e_1 + e_4$ respectively. It is more appropriate for explicit calculations to make another choice. It is clear that for all i, we obtain a basis for $\Lambda_{e_i^*}$ by replacing e_i by $2e_i$ in the basis $(e_1, e_2, \dots, e_{24})$ for Λ. It is possible to construct the lattices $\Lambda(4)$, $\Lambda(6)$, $\Lambda(8)$ as Λ_e-lattices by choosing $e = e_2^*$, $e = e_1^*$, and $e = e_{12}^*$, whose norms are 4, 14, and 8 respectively; checking that these three lattices are pairwise non-isometric shows that $\Lambda_{e_1^*}$ is isometric to $\Lambda(6)$. [There are nine lattices related to classes modulo 3; see [Mar8].]

Theorem 5.7.4 (or Proposition 5.7.7) implies that for any minimal vector $e \in \Lambda$, $P_e\Lambda$ is a unimodular lattice of norm 3. Hence the isometry class of $P_e\Lambda$ does not depend on the particular choice of $e \in S(\Lambda)$. We *denote by* O_{23} the lattice corresponding to $e = e_2^*$, the *shorter Leech lattice* in [C-S].

Consider $v = e_1 + e_4 + e_2^*$; this is a norm 12 vector in Λ. We *denote by* O_{24} the lattice $K_v\Lambda$, the *odd Leech lattice* in [C-S]. Proposition 5.7.7 shows that this is an odd unimodular lattice of norm 1 or 3. A more detailed analysis of the Kneser neighbouring procedure shows that its norm is actually equal to 3.

We have in this way constructed in dimensions 23 and 24 the unique odd unimodular lattices of norm ≥ 3 whose existence and uniqueness were asserted in Section 5.6. It is easy to exhibit Gram matrices for the lattices $\Lambda(4)$, $\Lambda(6)$, $\Lambda(8)$, O_{23}, and O_{24} using the known Gram matrix of the basis $(e_1, \dots, e_{24})$ for Λ_{24}.

Using a program which computes, for a given positive definite symmetric matrix A, the rank of the set of matrices ${}^tXX$ and compares with A^{-1} the sum of the ${}^tXX$ for $X \in S(A)$, we can test perfection and strong eutaxy of a lattice with Gram matrix A.

We find the maximal possible perfection rank for $\Lambda(4)$, $\Lambda(6)$, $\Lambda(8)$, and O_{23}, but only 277 (instead of 300) for O_{24}, and we verify that a unique eutaxy coefficient suffices for $\Lambda(8)$, O_{23} and O_{24}, but not for the other two lattices. Using the fact (see [ATLAS]) that the stabilizer in $\mathrm{Aut}(\Lambda)$ of a vector e of norm 4 or 6 acts transitively on the sets of vectors of $S(\Lambda)$ having a given scalar product with e, we know that there are at most two coefficients for $\Lambda(4)$ and $\Lambda(6)$, which are now easily computed. We obtain the following eutaxy relations with the Gram matrices A_4 of $\Lambda(4)$ and A_6 of $\Lambda(6)$:

$$A_4^{-1} = \frac{1}{9315} \sum_{X \in S_0} X\,{}^tX + \frac{1}{4600} \sum_{X \in S_2} X\,{}^tX$$

and

$$A_6^{-1} = \frac{243}{2049300} \sum_{X \in S_0} X\,{}^tX + \frac{275}{2049300} \sum_{X \in S_2} X\,{}^tX\,,$$

where $S_m = \{x \in S(\Lambda) \mid x \cdot e = m\}$. We also compute simultaneously the kissing number of each of these lattices. Curiously, it has the same value for $\Lambda(6)$ and $\Lambda(8)$. (One could also use the theta function of Λ to calculate the kissing numbers of its three lattices of index 2, and then obtain it for O_{23} and O_{24}.)

The following theorem lists the main properties of the five lattices above:

Theorem 5.7.9. (Bachoc and Batut, [Bac-Bt].) *The invariants of the lattices $\Lambda(4)$, $\Lambda(6)$, $\Lambda(8)$, O_{23}, and O_{24} are as follows:*

Lattice	*det.*	*norm*	s	*perfection*	*eutaxy*
$\Lambda(4)$	4	4	51176	*perfect*	*eutactic*
$\Lambda(6)$	4	4	49128	*perfect*	*eutactic*
$\Lambda(8)$	4	4	49128	*perfect*	*strongly eut.*
O_{23}	1	3	2048	*perfect*	*strongly eut.*
O_{24}	1	3	2300	*non-perfect*	*strongly eut.*

□

5.8 Exercises for Chapter 5

5.1.1. Show that $\mathrm{Aut}(\mathbb{A}_n^r)$ contains a subgroup isomorphic to $\{\pm\,\mathrm{Id}\} \times S_n$ whenever $\frac{n+1}{r}$ and $(n+1)r$ are integers.

5.1.2. Let $n \geq 7$ be an integer.

1. Suppose that n is odd, and let $m = \frac{n-1}{2}$. Show that the vectors in $\mathbb{A}_n^2$ whose norm is minimal among those which do not belong to $\mathbb{A}_n$ are obtained from $\frac{1}{2}(\varepsilon_0 + \cdots + \varepsilon_m -$

$\varepsilon_{m+1} - \cdots - \varepsilon_n)$ by a permutation of the indices $0, 1, 2, \ldots, n$.

2. Use this to deduce that there are $\binom{n}{m}$ pairs of such vectors, and justify the equality $s(\mathbb{A}_7^2) = 63$ stated in Theorem 5.1.7.

3. Suppose now that n is even, say $n = 2m$. Show that the vectors of $\mathbb{A}_n^2$ whose norm is minimal among those which do not belong to $\mathbb{A}_n$ are obtained from $\pm\frac{1}{2}(\varepsilon_1 + \cdots + \varepsilon_m - \varepsilon_{m+1} - \cdots - \varepsilon_n)$ by a permutation of the indices $1, 2, \ldots, n$.

4. Use this to deduce that there are $\frac{1}{2}\binom{n}{m}$ pairs of such vectors, and justify the equality $s(\mathbb{A}_8^2) = 71$ stated in Theorem 5.1.7.

5.1.3. 1. Show that the minimal vectors in $\mathbb{A}_8^3$ which do not belong to $\mathbb{A}_8$ are obtained by permutation of the indices $0, 1, \ldots, n$ from the $\binom{9}{3}$ pairs $\pm\frac{1}{3}(2(\varepsilon_0 + \varepsilon_1 + \varepsilon_2) - (\varepsilon_3 + \varepsilon_4 + \varepsilon_5 + \varepsilon_6 + \varepsilon_7 + \varepsilon_8))$, and verify the equality $s(\mathbb{A}_8^3) = 120$.

2. Show in the same way that the vectors in $\mathbb{A}_9^3$ which do not belong to $\mathbb{A}_9$ are obtained by permutation of the indices $0, 1, \ldots, n$ from the $\binom{9}{3}$ pairs $\pm\frac{1}{2}(2(\varepsilon_1 + \varepsilon_2 + \varepsilon_3) - (\varepsilon_4 + \varepsilon_5 + \varepsilon_6 + \varepsilon_7 + \varepsilon_8 + \varepsilon_9))$, and verify the equality $s(\mathbb{A}_9^3) = 129$.

5.1.4. Show that the lattices $\mathbb{A}_7^2$, $\mathbb{A}_8^2$, $\mathbb{A}_8^3$, and $\mathbb{A}_9^3$ have bases of minimal vectors.

5.1.5. 1. Show that the lattices $\mathbb{A}_n^2$ are even for $n \equiv -1 \mod 8$, odd for $n \equiv 3 \mod 8$ and half-integral but not integral otherwise.

2. Show that their elementary divisors (Smith invariants) are: $n \equiv -1 \mod 4$: $\frac{n+1}{4}$; $n \equiv 1 \mod 4$: $(n+1)2^{n-2}$; $n \equiv 0 \mod 2$: $(2n+2) \cdot 2^{n-3}$.

5.1.6. Show that $\mathbb{A}_8^{3/2}$ is similar to $\mathbb{D}_8$.

5.1.7. Show that a lattice containing $\Lambda = \mathbb{A}_n$ to index 2 is isometric to $\Lambda'_{n,t} = \langle \mathbb{A}_n, \mathbb{A}_t^2 \rangle$ with $t \le n$ odd; prove that $N(\Lambda'_{n,t}) = 2 \Longleftrightarrow t \ge 7$, and that $S(\Lambda'_{n,t}) = S(\mathbb{A}_n) \Longleftrightarrow t \ge 9$.

5.1.8. Let Λ be a lattice containing $\Lambda' = \mathbb{A}_n$ and such that Λ/Λ' is 2-elementary of order 4.

1. Prove that Λ is of the form $\Lambda = \langle \mathbb{A}_n, \mathbb{A}_t^2, \mathbb{A}_{t'}^2 \rangle$. [See Exercise 5.1.7.]

2. Prove that if $N(\Lambda) = 2$, then $n \ge 11$, and that if $N(\Lambda) = 2$ and $n = 11$, Λ is isometric to a uniquely defined lattice with $s = 171$.

5.2.1. With the notation of Proposition 5.2.3, suppose that $n \equiv 3 \mod 4$, and denote by Λ the lattice $\widetilde{\mathbb{A}}_n^{((n+1)/2)}$.

1. Show that Λ is odd of norm $\frac{n-1}{2}$, and that its even part Λ' is a lattice of norm $\frac{n+1}{2}$ whose minimal vectors come from those of a scaled copy of $\mathbb{A}_n$.

2. Show that if $n = 15$, $\Lambda' \simeq 2\,\mathbb{A}_{15}^4$.

5.2.2. Let n be an odd integer. Show that the even part of $\sqrt{n+1}\,\mathbb{A}_n^*$ is proportional to $\mathbb{A}_n^2$. [Compare Exercise 4.2.5.]

5.3.1. Show that P_4 is similar to $\mathbb{A}_4^*$.

5.3.2. 1. Let G be a group acting on a set, and let H be a normal subgroup of G. Show that G (and thus also G/H) acts on the set of the orbits of H.

2. We now take $G = \operatorname{Aut}(P_n)/\{\pm \operatorname{Id}\}$ and $H = \langle \pm\sigma \rangle/\{\pm \operatorname{Id}\}$, and we suppose that $n \ge 4$. Show that the orbits of $S(P_n)$ under H have $n+1$ elements, except for odd n, where the $\frac{n-1}{2}$ orbits of $\varepsilon_0 - \varepsilon_i - \varepsilon_{\frac{n+1}{2}} + \varepsilon_{\frac{n+1}{2}+i}$, $1 \le i \le \frac{n-1}{2}$ have $\frac{n+1}{2}$ elements.

3. Show that $-1 \in (\mathbb{Z}/(n+1)\mathbb{Z})^\times$ stabilizes the orbits of H.

4. Show that $\mathrm{Aut}(P_n)$ acts transitively on $S(P_n)$ for $n = 4, 5, 6$ and for no other value of $n \geq 4$.

5.3.3. Let n be even, and let $e = e_1 + \cdots + e_n = n\varepsilon_0 - \varepsilon_1 - \cdots - \varepsilon_n$ [see Definition 5.1.1].

1. Show that e belongs to $P_n = \mathbb{A}_n^{(2)}$.

2. Show that a vector in E of the form $\frac{e}{2} + x$, $x \in \mathbb{A}_n$ possesses at least n odd components on the vectors ε_i.

3. Deduce from this that the minimum of $N(\frac{e}{2} + x)$, $x \in \mathbb{A}_n$ is at least $\frac{n}{4}$ and that it is attained on $\frac{e}{2} + P_n$ for all $n \geq 4$. [Consider the differences $e - (2\varepsilon_0 - \varepsilon_i - \varepsilon_{-i})$ and $e - (\varepsilon_0 - \varepsilon_j - \varepsilon_k + \varepsilon_{j+k})$ for suitable integers i, j, k.]

4. Show that the lattice $\langle P_n, \frac{e}{2}\rangle$ is of norm at least 4 for all $n \geq 16$, and that its minimal vectors are those of P_n for $n \geq 18$.

5.3.4. Let Λ be the $\mathbb{A}_n$ lattice, endowed with a Korkine–Zolotareff basis $\mathcal{B}$ (Section 4.2), and let G be a finite Abelian group, of order m. For a homomorphism $\varphi : \Lambda \to G$, set $\Lambda_\varphi = \ker \varphi$.

1. Show that Λ_φ has norm $N \geq 4$ if and only if φ takes distinct, nonzero values on $\mathcal{B}$.

2. Show that there exists φ with $N(\Lambda_\varphi) \geq 4$ if and only if $m \geq n+1$.

Assume from now on that $N(\Lambda_\varphi) \geq 4$.

3. Show that if G is cyclic of order $n+1$, then Λ_φ is isometric to P_n.

4. Show that if $n = 3$ and G is elementary of order 4, then $\Lambda_\varphi \simeq 2\,\mathbb{Z}^3$ is not isometric to P_3.

5. Show that if $n = 7$ and G is elementary of order 8, then $\Lambda_\varphi \simeq \sqrt{2}\,\mathbb{D}_7$ is not isometric to P_7. [Verify that $\frac{1}{\sqrt{2}}\Lambda_\varphi$ is a root lattice with $s = 42$.]

5.3.5. Determine the minimal vectors in Q_n which do not lie in P_{n-1}, and prove Theorem 5.3.11. [See Barnes, [Bar5], pp. 64–67.]

The exercises relating to Section 5.4 have been written with the collaboration of Laurent Habsieger.

5.4.1. Consider a lattice $\mathbb{A}_n^{(r)}$ with $n \geq 2r$.

1. Show that $\mathbb{A}_n^{(r)} = \{x \in \mathbb{A}_n^{(r-1)} \mid \sum_{i=0}^n \binom{i}{r-1} x_i \equiv 0 \mod n+1\}$.

2. Show that its Smith invariant (Section 1.9) is equal to $(n+1)^{2r-1}$ whenever $n+1$ is a prime.

5.4.2. Let $n \geq 5$ be an integer. The aim of this exercise is to calculate the norm of the lattices $\mathbb{A}_n^{(3)}$.

1. Show that the orbits of norm 4 vectors in $\mathbb{A}_n^{(2)}$ under the cycles $(0, 1, 2, \ldots, n)$ are represented by vectors of the form $\varepsilon_0 - \varepsilon_i - \varepsilon_j + \varepsilon_{i+j}$ with $1 \leq i < j < i - j \leq n$. [See the proof of Proposition 5.3.2.]

2. Show that such a vector belongs to $\mathbb{A}_n^{(3)}$ if and only if ij is divisible by $n+1$.

3. Verify that $\varepsilon_0 - \varepsilon_1 - \varepsilon_2 + \varepsilon_4 + \varepsilon_5 - \varepsilon_6$ is a norm 6 vector in $\mathbb{A}_n^{(3)}$.

4. Use this to deduce that $\mathbb{A}_n^{(3)}$ has norm 6 if $n+1$ is prime or if $n = 8$, and norm 4 otherwise.

5.4.3. We keep the notation of Exercise 5.4.2.

1. Show that $\varepsilon_0 - \varepsilon_i - \varepsilon_j + \varepsilon_{i+j}$ belongs to $\mathbb{A}_n^{(4)}$ if and only if $ij \equiv \frac{ij(i+j)}{2} \equiv 0 \mod n+1$.

2. Show that $\mathbb{A}_n^{(4)}$ has norm 4 for all $n \geq 9$ such that $n+1$ is not a prime (and that $N(\mathbb{A}_8^{(4)}) = 6$).

5.4.4. The aim of this exercise is to describe the norm 6 vectors of $\mathbb{A}_n^{(3)}$ when $n+1$ is prime p.

1. Show that the norm 6 vectors in $\mathbb{A}_n^{(2)} = P_n$ possess (on the basis $(\varepsilon_0, \varepsilon_1, \dots, \varepsilon_n)$) six components equal to ± 1 or are conjugate by an automorphism to a vector of the form $\varepsilon_0 - 2\varepsilon_i + \varepsilon_{2i}$, $0 \leq 1 \leq n$.

2. Show that the vectors above do not belong to $\mathbb{A}_n^{(3)}$.

3. We suppose from now on that $p = n+1$ is prime. Show that, up to conjugacy by an automorphism, these vectors are of the form $\varepsilon_0 - \varepsilon_1 + \varepsilon_a + \varepsilon_b - \varepsilon_x - \varepsilon_y$, where $0, 1, a, b, x, y$ are distinct indices in $[0, n]$.

4. Show that such a vector belongs to $\mathbb{A}_n^{(3)}$ if and only if the congruences $a + b \equiv x + y + 1 \mod p$ and $a^2 + b^2 - x^2 - y^2 \equiv -1 \mod p$ hold.

5. Interpreting a and b as the roots of a polynomial of degree two with coefficients depending on x and y, show that $(x - y - 1)^2 - 4y$ must be a square modulo p.

6. Show that a, b, x, y can be expressed modulo p as functions of two parameters u and v by the formulae

$$y = \frac{u^2 - v^2}{4}\,;\; x = \frac{(u+2)^2 - v^2}{4}, \text{ and } a, b = \frac{x + y + 1 \pm v}{2}.$$

5.5.1. Let $n \geq t \geq 8$ (t even) be integers.

1. Show that $\mathbb{D}_{n,t}$ is similar to the sublattice of $\mathbb{Z}^n = \{\sum_1^n x_i\,\varepsilon_i\}$ such that $x_i \equiv x_1 \mod 4$ if $i \leq t$, $x_i \equiv 0 \mod 4$ if $i > t$, and $\sum x_i \equiv t x_1 \mod 4$.

2. Determine its dual lattice.

3. Show that for $t < n$, we have $S(\mathbb{D}_{n,t}^*) \simeq S(Z^{n-t})$.

4. Suppose now that $t = n$ (thus $\mathbb{D}_{n,t} = \mathbb{D}_n^+$). Show that for $n \equiv 0$ (resp. 2) modulo 4, $(\mathbb{D}_n^+)^*$ is equal to $\mathbb{D}_n^+$ (resp. to the sublattice $\mathbb{D}_n^- \neq \mathbb{D}_n^+$ of index 2 in $\mathbb{D}_n^*$ which is isometric to $\mathbb{D}_n^+$, see Remark 4.6.8).

5. Show that $\mathbb{D}_{n,t}$ is dual-extreme if and only if $t = n$.

5.5.2. Let $n > t \geq 8$ (t even) be integers. We consider inside $\mathrm{Aut}(\mathbb{D}_n)$ the following subgroups: $G_1 = W(\mathbb{D}_t)$ acting on $F_1 = \langle \varepsilon_1, \dots, \varepsilon_t \rangle \subset \mathbb{R}^n$, $G_2 = \mathrm{Aut}(\mathbb{Z}^{n-t}) \simeq \{\pm 1\}^n \rtimes S_{n-t}$ acting on $F_2 = \langle \varepsilon_{t+1} \dots, \varepsilon_n \rangle \subset \mathbb{R}^n$, and $G = G_1 \times G_2$.

1. Show that G is a subgroup of $\mathrm{Aut}(\mathbb{D}_{n,t})$.

2. Show that there are at most four orbits of G acting on $S(\mathbb{D}_{n,t})$, with representatives for and contributions to $s = s(\mathbb{D}_{n,t})$ as follows:

type I: $\varepsilon_1 + \varepsilon_2$, $s_1 = t(t-1)$;
type II ($t = 8$): $\frac{1}{2}e = \frac{\varepsilon_1 + \dots + \varepsilon_t}{2}$, $s_2 = 0 (t > 8)$ or 64;
type III: $\varepsilon_1 + \varepsilon_n$, $s_3 = 2t(n-t)$;
type IV ($n \geq t+2$): $\varepsilon_{n-1} + \varepsilon_n$, $s_4 = (n-t)(n-t-1)$.

3. Suppose that $t = 8$. Show that the numbers of pairs of minimal vectors in $\mathbb{D}_{n,t}$ with scalar product 0 with a given minimal vector x of type I, II, III, IV (if any) is respectively $n^2 - 5n + 39$, $n^2 - 17n + 132$, $n^2 - 5n + 7$, and $n^2 - 5n + 71$.

4. Show that G is the full automorphism group of $\mathbb{D}_{n,t}$.
[Use Question 3 if $t = 8$ and Proposition 1.4.4 if $t \geq 9$.]

5. Show that F_1 and F_2 are the irreducible components of the real representation afforded by the action of $\mathrm{Aut}(\mathbb{D}_{n,t})$ on $\mathbb{R}^n$.

6. Show that $\mathbb{D}_{n,t}$ is strongly eutactic if and only if $t \geq 10$.

5.7.1. The aim of this exercise is to give a few indications on the Kneser neighbouring procedure. Let $\ell \geq 2$ be an integer.

1. Let Λ be any lattice. For any $e \in \Lambda^*$, let $\Lambda_e = \{x \in \Lambda \mid e \cdot x \equiv 0 \mod \ell\}$. Show that the map $e \mapsto \Lambda_e$ induces a one-to-one correspondence between $\Lambda^*/\Lambda^{*\ell}$ modulo the action of $(\mathbb{Z}/\ell\mathbb{Z})^\times$ and the set of sublattices of Λ whose index divides ℓ; show that $\Lambda_e = \Lambda \Leftrightarrow e \in \ell\Lambda^*$.

2. Show that if Λ is integral and if $\det(\Lambda)$ is prime to ℓ, then the restriction from Λ^* to Λ induces an isomorphism of $\Lambda^*/\Lambda^{*\ell}$ onto Λ/Λ^ℓ.

3. Say that two lattices Λ and Λ' are *weak (ℓ)-neighbours* if the following three conditions are satisfied:

1. $M = \Lambda \cap \Lambda'$ is integral;
2. $\Lambda'/M \simeq \Lambda/M \simeq \mathbb{Z}/\ell\mathbb{Z}$;
3. $\forall x \in \Lambda, \forall x' \in \Lambda', \forall z \in M$, $x \cdot y$ and $x' \cdot y$ are integral.

Show that if Λ and Λ' are weak neighbours, there exists $e \in L$ such that

$$M = \{x \in \Lambda \mid e \cdot x \equiv 0 \mod \ell\} \text{ and } \Lambda' = \langle M, \tfrac{e}{\ell} \rangle.$$

Show moreover that for any $f_0 \in \Lambda \smallsetminus M$, $f = \ell f_0 \in M$ and $\Lambda = \langle M, \frac{f}{\ell} \rangle$.

4. Assume from now on that $\ell = 2$, and let Λ be an integral lattice. Show that the set of weak neighbours of Λ consists of two lattices Λ' and Λ'', that any of the three lattices Λ, Λ', Λ'' is a neighbour of the others, and that one or three of these lattices are integral.

5. Prove that none or two of the three lattices Λ, Λ', Λ'' are even.

6. Work out the case where $\Lambda = \mathbb{Z}^n$ and $M = \mathbb{D}_n$, and prove that all the possibilities listed in Questions 4 and 5 may indeed occur, depending on $n \mod 8$.

5.7.2. Show that the even part of an odd integral primitive lattice Λ is imprimitive if and only if all components of $\mathrm{Smith}(\Lambda)$ but one are even.
[Example: $\Lambda = (\sqrt{2}\,\mathbb{E}_7^*)$ is odd, $\mathrm{Smith}(\Lambda) = 2^6$, and $(\sqrt{2}\,\mathbb{E}_7^*)_{\text{even}} \simeq \sqrt{2}\,\mathbb{E}_8$.]

5.7.3. The aim of this exercise is to describe a remarkable 16-dimensional sublattice of O_{23}.

1. Show that the even part of O_{23} is isometric to the laminated lattice Λ_{23}. [Notice that Λ_{23} is the section of Λ_{24} by the hyperplane orthogonal to one of its minimal vectors.]

2. Construct $\sqrt{2}\,\mathbb{E}_7^*$ as a lattice inside O_{23}. [Use Exercise 5.7.2.]

3. Prove the existence inside O_{23} of a unique lattice O_{16} whose corresponding even lattice is isometric to Λ_{16}.
[It can be shown that the "anti–laminations" of O_{23} are unique to within an automorphism of O_{23} in dimensions $n \geq 16$, and that these lattices are all perfect; see "A Catalogue of Perfect Lattices", http://www.math.u-bordeaux.fr/~martinet/.]

5.9 Notes on Chapter 5

In his basic paper *Extreme forms* ([Cox2], 1951), H.S.M. Coxeter considers lattices Λ which satisfy the double inclusion $L \subset \Lambda \subset L^*$ for some root lattice L. Besides $\mathbb{Z}^n$, there are two lattices in the case where $L = \mathbb{D}_n$, n even, which are isometric to the lattice denoted by $\mathbb{D}_n^+$ in Chapter 4, and a family of lattices indexed by the divisors of $n+1$ in the case where $L = \mathbb{A}_n$, namely the family $\mathbb{A}_n^r$ (denoted by $\mathbb{A}_n^{+r}$ in [C-S5]); this family is studied in Section 5.2. (In the exceptional cases, there are only L and L^*.)

A few years later, Barnes noticed ([Bar5], 1958) that Coxeter's definition makes sense for all integers r, and defined an interesting enlargement of the previous family whenever r is sufficiently small relatively to n. Since a great deal of Coxeter's results is a particular case of those of Barnes, we began with Barnes's theory. For another point of view, see [Cra1].

A very interesting illustration of Barnes's generalization is provided by the extreme lattice $\mathbb{A}_8^2$, which I conjecture to be the 8-dimensional perfect lattice other than $\mathbb{E}_8$ on which both γ and s attain their maximum.

The family P_n was also discovered by Barnes as a byproduct of his classification of 6-dimensional perfect forms: his new lattice became in [Bar3] the first term P_6 of the series P_n of extreme lattices. Barnes wrote down his proof of the perfection only in the case (which he asserts to be the easiest one) of even dimensions. We have not proved the perfection of the P_n lattices, preferring to concentrate on unpublished properties of their duals.

The two-parameter family $\mathbb{A}_n^{(r)}$ which generalizes the families $\mathbb{A}_n$ and P_n corresponding to $r = 1, 2$ was discovered by Craig. Conway and Sloane ([C-S], Chapter 8) refer to a preprint (*Automorphisms of prime cyclotomic lattices*, reference [Cra5] in [C-S]) which I have never seen. Only a finite number of lattices $\mathbb{A}_n^{(r)}$ among those with $r \geq 3$ are known to be perfect. When $p = n+1$ is prime, each orbit of minimal vectors under the action of the group $C_2 \times (C_p \rtimes C_{p-1})$ contains $p(p-1) = 2 \cdot \frac{n(n+1)}{2}$ minimal vectors, a fact which might suggest that a single orbit could suffice to ensure perfection. After Bachoc, this is not general, even in the case $r = 2$ which was dealt with by Barnes. As for automorphisms, it is worth noticing that for some special values of the pair (n, r) (not only those of the form $(p-1, \frac{p+1}{4})$), the automorphism group of $\mathbb{A}_n^{(r)}$ is much larger that $2p(p-1)$.

Other families of perfect lattices depending on several parameters and related to the root lattices $\mathbb{A}_n$ (resp. $\mathbb{D}_n$) have been found by Barnes. We described only a few of them in Section 5.1 (resp. 5.5). For other examples, we refer to the original paper [Bar5] and to Section 8.4, where we describe the families L_n^r and M_n^r in the more natural setting of Hermitian structures. Other generalizations of Barnes's families have been found by Scott ([Sco1], [Sco2]).

Unimodular lattices have been classified up to dimension $n = 25$ ([Bor1]; cf. [C-S], Chapter 16).

Kneser introduced the notion of a neighbour of a lattice in [Kn2] and proved the connectedness of the associated graph in the unimodular case. (More generally, connectedness is related to the partition of isometry classes into "spinor genera".) Kneser's proof relies on his difficult approximation theorem for algebraic groups, applied to the simply connected group Spin_n. (One needs a transition from Spin_n to O_n via SO_n.) This method can be generalized to lattices endowed with various algebraic structures, using other instances of Kneser's approximation theorem; see [Bac1], where she constructs an interesting lattice of dimension 32, norm 6, and determinant 2^{16} by applying Kneser's method to the Hurwitz ring $\mathfrak{M}_2$ of quaternions, which she would not have found (see [Bac-Ne1]) if she had only considered lattices which are integral *as* $\mathfrak{M}_2$-lattices.

Up to dimension 24, unimodular lattices are classified by their root systems, whence their appearance as Sections 5.6 and 5.7 of this chapter. This is not a general result.

In the next even case, namely dimension $n = 32$, only very partial results are known about the numerous possibilities (over 80 million, cf. [Se2], Chapter V, Section 2). The case of root systems with a maximal rank has been dealt with by Kervaire in [Ker]; there are 132 lattices for only 119 root systems. In the opposite direction, the partial classification of lattices with an empty root system done by Koch and Venkov ([K-V]) suggested that many such lattices (which are all extreme after [Ven3]) exist. King has proved in [Kg] that there are more than 10 million of such lattices. Most of the lattices have a root system which is neither empty nor of maximal rank.

It seems hopeless to classify all odd unimodular lattices beyond dimension $n = 25$, but those with an empty root system (and thus of norm 3) are known up to dimensions 28: none for $n \leq 22$ and also (Borcherds) for $n = 25$, one in each of the dimensions 23 and 24 (as we know) and 26 (Borcherds), then three in dimension 27, and 38 in dimension 28 ([Bc-V2]), among which $\bigwedge^2 \mathbb{E}_8$ and a lattice related to the symplectic group $\mathrm{Sp}_6(\mathbb{F}_3)$ described in [Bc-V1]. The number of such lattices then increases quickly with the dimension – see King's lower bounds in [Kg].

Beyond dimension 24, these lattices have not been tested for perfection.

6 Low-Dimensional Perfect Lattices

Introduction

The complete classification of perfect lattices up to dimension 5 was obtained in [K-Z3] by Korkine and Zolotareff. Their method relied on a study of a combinatorial nature of the possible configurations of minimal vectors, and yielded somewhat sharper results. An alternative proof of this result, relying an an algorithm due to Voronoi, will be given in the next chapter; this algorithm was used to classify perfect lattices in dimension 6 (Barnes) and 7 (Jaquet, obtaining in this way a confirmation of Kaye Stacey's list). Both Stacey and Jaquet needed heavy computer calculations; Barnes was able to work out by hand the Voronoi algorithm, but he reached there the very limit of hand calculations. We shall not give the proofs for dimensions 6 and 7 in this book, contenting ourselves with a description of the main results in Section 6.5.

The proofs we gave in the French edition were directly inspired by the original work of Korkine and Zolotareff. We shall here depart from this point of view for dimension 5, and give a much shorter proof, relying on methods of Watson, shortening considerably two steps of our proof: first we no longer need the inequalities proved in Section 2.8; moreover, we simplify the fifth section of the French edition. Note that Stacey's work relies on ideas of Watson, and that these ideas were also used later by Baranovskii and Ryshkov to handle dimension 6 ([Br-R]; see Section 6.5 below).

The starting point of the classification is a characterization of the $\mathbb{A}_n$ lattices that we prove in Section 6.1. The (complicated, though not really difficult) proof is an easy exercise up to dimension $n = 4$, so that the results of Section 6.2 devoted to these dimensions are really elementary. The classification in dimension 5 is given in Section 6.4.

Beforehand, we describe in Section 6.3 the classification of dual-extreme lattices up to dimension 4 (nothing is known for $n \geq 5$), giving the proof only for dimensions $n \leq 3$.

Section 6.5 is a purely descriptive one; it contains in particular the list of all inclusions between perfect lattices of dimension $s \leq 7$, relying on Barnes's and Stacey and Jaquet results.

The classification of perfect lattices is not known beyond dimension 7, and we can only give in Section 6.6 very partial results about the classification of

8-dimensional perfect lattices, concerning mainly those lattices which possess a perfect hyperplane section with the same norm. We also give a short proof, due to A-M. Bergé, of the fact that critical 8-dimensional lattices are similar to $\mathbb{E}_8$, relying only on the knowledge of the analogous result for dimensions 6 and 7. In this way we can find all critical lattices in dimension up to 8 without making use of the works of Blichfeldt and Vetchinkin, but this relies on extensive computer calculations in dimension 7.

6.1 A Combinatorial Characterization of the $\mathbb{A}_n$ Lattices

Recall (Section 4.2) that the lattice $\mathbb{A}_n$ is the intersection of $\mathbb{Z}^{n+1} \subset \mathbb{R}^{n+1}$ and the hyperplane $E = \{(x_0, \dots, x_n) \in \mathbb{R}^{n+1} \mid x_0 + x_1 + \cdots + x_n = 0\}$. Its $\frac{n(n+1)}{2}$ pairs of minimal vectors are represented by the n vectors $e_i = \varepsilon_0 - \varepsilon_i$, $1 \le i \le n$, which constitute a "Korkine–Zolotareff" basis for $\mathbb{A}_n$, and the $\frac{n(n-1)}{2}$ vectors $e_i - e_j$, $1 \le i < j \le n$. The set $S(\mathbb{A}_n)$ has the following property:

Proposition 6.1.1. *Any system of n independent vectors in $S(\mathbb{A}_n)$ is a basis for $\mathbb{A}_n$.*

Proof. This is a combinatorial property of any basis $\mathcal{B} = (e_1, e_2, \dots, e_n)$ for any free $\mathbb{Z}$-module M of rank n and of the subset S of M of the $n(n+1)$ vectors $\pm e_i$, $1 \le i \le n$ and $\pm(e_i - e_j)$, $1 \le i < j \le n$: any subset $S' = \{e'_1, e'_2, \dots, e'_n\}$ of n independent elements of S constitutes a basis for M. The proof is easier if we embed M into a free module L of rank $n+1$ in such a way that M be defined by the equation $\sum_{i=0}^{n} x_i = 0$ in some basis for L. One can take $L = \mathbb{Z}^{n+1}$ endowed with its canonical basis and then identify M with its image by the linear map defined by $e_i \mapsto \varepsilon_0 - \varepsilon_i$.

We use an induction argument on the number $n - r$ of vectors $\pm e_i$ which lie in S'. Since the differences $e_i - e_j$ do not generate $\mathbb{R} \bigotimes M$, r is nonzero. After changing if need be the indexing of the e'_i and their signs, we may assume that $e'_i = e_i$ for $1 \le i \le r$. The other elements of S' are of the form $\varepsilon_i - \varepsilon_j$, and at least one of the indices i, j must be $\le r$, for the set $\{\varepsilon_i - \varepsilon_j,\, j > i > r\}$ has rank at most $n - r - 1$. After performing a new permutation, we may assume that $e'_{r+1} = \varepsilon_0 - \varepsilon_j$ for some index $j > r$, and we may even assume that $j = r + 1$. The induction hypothesis then shows that S' is a basis for M. □

Theorem 6.1.2. (Korkine and Zolotareff.) *Let L be a lattice generated by the set $S(\Lambda)$ of its minimal vectors. Suppose that $S(\Lambda)$ has rank n, that it contains at least $\frac{n(n+1)}{2}$ pairs of vectors, and that any system of n independent vectors in $S(\Lambda)$ generate Λ. Then L is similar to $\mathbb{A}_n$.*

This statement is an easy consequence of the following one relative to free $\mathbb{Z}$-modules, which does not involve the notion of a lattice:

Lemma 6.1.3. *Let M be a free $\mathbb{Z}$-module of rank n and let S be a family of elements of M which satisfy the following five properties:*

1. $0 \notin S$.
2. $x \in S \Longrightarrow -x \in S$.
3. *S is of rank n.*
4. $|S| \geq n(n+1)$.
5. *n arbitrary independent elements of S generate M.*

Then S possesses exactly $\frac{n(n+1)}{2}$ pairs $\pm x$, from which one can extract a basis $(e_1, \dots, e_n)$ for M such that

$$S = \{\pm e_i \mid 1 \leq i \leq n\} \cup \{\pm(e_i - e_j) \mid 1 \leq i < j \leq n\}.$$

Proof of 6.1.2. We show that 6.1.2 is a consequence of 6.1.3. Lemma 6.1.3 shows that we can extract from $S(\Lambda)$ a basis $(e_1, e_2, \dots, e_n)$ for Λ such that the differences $e_i - e_j$, $i < j$ also belong to S. Hence $N(e_i - e_j) = N(e_k) = N(\Lambda)$ for all i, j, k with $j \neq i$, which implies that the scalar products $e_i \cdot e_j$ are equal to $\frac{1}{2}N(\Lambda)$ for $j \neq i$. Multiplying out all scalar products by $\frac{2}{N(\Lambda)}$, we transform Λ into a lattice Λ' which possesses a basis $\mathcal{B}' = (e'_1, \dots, e'_n)$ with $e'_i \cdot e'_i = 2$ and $e'_i \cdot e'_j = 1$ for $j \neq i$. Then $\mathrm{Gram}(\mathcal{B}')$ is equal to the Gram matrix of a Korkine–Zolotareff basis for $\mathbb{A}_n$, and Λ' is isometric to $\mathbb{A}_n$. □

Before entering into the proof of Lemma 6.1.3, we introduce the notion of a characteristic determinant:

Definition 6.1.4. *Let M be a free module of rank n endowed with a basis $\mathcal{B} = (e_1, \dots, e_n)$, let $S \supset \mathcal{B}$ be a finite subset of M, and let $s = |S|$. A* characteristic determinant *for the pair $(\mathcal{B}, S)$ is a determinant of order $r \leq n$ extracted from the $n \times s$ matrix of the components in $\mathcal{B}$ of the elements of S.* [For typographical reasons, we consider that these components are row-vectors rather than column-vectors.]

Lemma 6.1.5. *With the notation of Definition 6.1.4, let K be the maximum of the absolute values of determinants $\det_{\mathcal{B}}(x_1, \dots, x_n)$ for $x_1, \dots, x_n \in S$. Then the inequality $|d| \leq K$ holds for any characteristic determinant d of $(\mathcal{B}, S)$. In particular, the absolute values of the components in $\mathcal{B}$ of any element of S are bounded from above by K.*

Proof. A characteristic determinant d is the determinant of the matrix D of $r \leq n$ elements $x_1, \dots, x_r$ of S relatively to a system of r vectors $f_1, \dots, f_r$ of $\mathcal{B}$. After performing a suitable permutation of the e_i, we may assume that $f_i = e_i$. We then have

$$\pm d = \det{}_{(e_1, \dots, e_r)}(x_1, \dots, x_r) = \det{}_{(e_1, \dots, e_n)}(x_1, \dots, x_r, e_{r+1}, \dots, e_n),$$

and the absolute value of the last determinant is bounded above by K. □

Proof of 6.1.3. We use induction on n, Lemma 6.1.3 being obvious for $n = 1$, and we first look for convenience at the cases when $n = 2$ or $n = 3$ (a procedure which moreover renders quite elementary the results of Section 6.2 below). We denote by $(e_1, e_2, \ldots, e_n)$ a basis for M consisting of elements of S, and begin with a lemma that we shall use several times in this chapter:

Lemma 6.1.6. *Under the hypotheses of Lemma 6.1.3, we have:*

1. *If the components of two elements of S on two elements of $\mathcal{B}$ are nonzero, they are then equal or opposite.*
2. *There does not exist in S any 3-tuple $(e_i - e_j, e_i - e_k, e_j + e_k)$.*

Proof of 6.1.6. In the first case, after having changed if necessary the signs of e_1 and e_2, we would have the characteristic determinant

$$\det_{e_i,e_j}(e_i + e_j, e_i - e_j) = \begin{vmatrix} 1 & 1 \\ 1 & -1 \end{vmatrix} = -2\,;$$

similarly, we would have in the second case a third-order determinant

$$\det_{e_i,e_j,e_k}(e_i - e_j, e_i - e_k, ej + e_k) = \begin{vmatrix} 1 & 1 & 0 \\ -1 & 0 & 1 \\ 0 & -1 & 1 \end{vmatrix} = 2\,. \quad \square$$

If $n = 2$, a half-system of elements of S is $\{e_1, e_2, e_1 \pm e_2\}$. By Lemma 6.1.6, $e_1 + e_2$ and $e_1 - e_2$ cannot both belong to S. Negating e_2 if necessary, we obtain the half-system $\{e_1, e_2, e_1 - e_2\}$ that we need.

If $n = 3$, again by Lemma 6.1.6 (1), there can be in S at most one pair $\pm x$ with x of the form $\pm e_1 \pm e_2 \pm e_3$. By permutation and change of signs of e_1, e_2, e_3, we my assume that a half-system of elements of S consists of the five elements $e_1, e_2, e_3, e_4 = e_1 - e_2, e_5 = e_1 - e_3$ and a sixth one of the form $e_2 \pm e_3$ or $e_1 \pm e_2 \pm e_3$. Applying Lemma 6.1.5, we are left with the two possibilities $e_6 = e_2 - e_3$ or $e_6 = e_1 - e_2 - e_3$. In the first one we are done, and the second case reduces to the first case by taking $e_2' = e_1 - e_2$ instead of e_2.

We now turn to the proof of Lemma 6.1.3 for $n \geq 4$. Let M' be the submodule of M generated by $e_2, e_3, \ldots, e_n$, and let S' be the set of nonzero elements $x \in M'$ such that x or $\pm e_1 + x$ belongs to S. Chose a half-system $\overline{S}$ in S in such a way that the component on e_1 of any element of $\overline{S}$ be 0 or $+1$. One then obtains a map of $\overline{S} \smallsetminus \{e_1\}$ onto a half-system $\overline{S}'$ of S' by suppressing the component on e_1 of the elements of $\overline{S}$. The inverse image of an element of $\overline{S}'$ possesses one or two elements, this last possibility corresponding to *twin systems* $\{x, e_1 + x\}$. To be able to use induction from S' to S, we must prove the inequality $|\overline{S}'| \geq \frac{(n-1)n}{2}$. To this end, we now prove the following lemma:

Lemma 6.1.7. *The number of twin systems $\{x, e_1 + x\}$ in $\overline{S}$ is at most $n - 1$.*

Proof of 6.1.7. Lemma 6.1.6 (1) shows that two elements of S belonging to two twin systems may not have opposite nonzero components on one of the e_i, $i \geq 1$. Replacing e_i by $-e_i$ when need be, we may assume that all their components are non-negative on the e_i. Suppose that there be at least n twin systems, and let $x \in S'$ such that $(e_1, e_1 + x)$ is a twin system and that x has as many as possible components x_i equal to 1. After permuting the e_i if need be, we may assume that $x_2 = x_3 = 1$.

There do not exist elements $y, z \in S'$ belonging to distinct twin systems for which the components on e_2 and e_3 are respectively $(0,1)$ and $(1,0)$, for we would have

$$\det{}_{(e_1,e_2,e_3)}(x, e_1 + y, e_1 + z) = \begin{vmatrix} 0 & 1 & 1 \\ 1 & 0 & 1 \\ 1 & 1 & 0 \end{vmatrix} = 2\,.$$

We may thus assume that the system of components $(1,0)$ does not occur. Then replacing e_2 by $e_2 + e_3$ reduces the number of components equal to 1 for x, and does not increase it for the members of the other twin systems. If the component of x on e_4 is nonzero, we can in the same way consider the new basis obtained by replacing e_2 by $e_2 + e_4$, until each member of a twin system will possess a single nonzero component on the new basis. This contradicts the existence of at least n twin systems, since S' contains only $n - 1$ elements of this type. □

End of the proof of 6.1.3. The induction hypothesis first shows that S' contains exactly $\frac{(n-1)n}{2}$ pairs $\pm x$. Since one obtains S by adjoining to S' the elements $\pm e_1$ and at most $n - 1$ pairs $\pm(e_1 + x)$ with $x \in S'$, we see that S contains exactly $\dfrac{n(n+1)}{2}$ pairs $\pm x$, namely those of S', the pair $\pm e_1$, and $n - 1$ pairs $\pm(e_1 + x)$ with $x \in S'$.

There remains to prove that S contains a basis for M of the type we want. The induction hypothesis implies that up to a permutation of the elements of S', we may assume that S' contains a half-system of the form

$$\overline{S}' = \{e_i \mid 2 \leq i \leq n\} \cup \{e_i - e_j \mid 2 \leq i < j \leq n\}\,;$$

this implies that S contains the half-system $\overline{S}$ which one obtains from $\overline{S}'$ by adjoining e_1 and of $n - 1$ elements of the form $e_1 \pm e_i$ or $e_1 \pm (e_i - e_j)$.

Let r be the number of elements of $\overline{S}$ of the first kind. We shall show that we can restrict ourselves by a suitable permutation of the elements of S' to the case where $r = n - 1$.

We remark that the following transformations stabilize S' and thus allow a partition of S' into two half-systems, one of which is of the desired form:

- The simultaneous change of signs of $e_2, \ldots, e_n$.
- The permutations of $e_2, \ldots, e_n$.
- The transformations $p_{i,j}$ ($i, j \geq 2$ distinct) which map e_j onto $e_i - e_j$ and fix e_k for $k \neq j$.

Since $p_{i,j}$ transforms $e_1 \pm (e_i - e_j)$ into $e_1 \pm e_j$, we may assume that $r \geq 1$. By a permutation of $e_2, \ldots, e_n$, we restrict ourselves to the case where the r elements of the first kind are $e_1 \pm e_2, \ldots, e_1 \pm e_{r+1}$ for a suitable choice of the signs. If there exists in S an element $e_1 \pm (e_i - e_j)$ with $i, j > r+1$ applying $p_{i,j}$ allows us to increase r to $r+1$.

From now on, we suppose that such an element does not exist. Negating $e_2, \ldots, e_n$ if necessary, we may assume that there is a minus sign in front of e_2. Lemma 6.1.6 then shows that these r elements are $e_1 - e_2, \ldots, e_1 - e_{r+1}$.

If $r < n-1$, let $x = e_1 \pm (e_i - e_j)$ be another element of $\overline{S}$. We may assume that $i \leq r+1$. Lemma 6.1.6 shows that there must be a minus sign in front of e_i, hence that $x = e_1 - (e_i - e_j)$. Since there is a plus sign in front of e_j, we have $j > r+1$.

Since the transformation $p_{i,j}$ preserve the elements $e_1 - e_k$ for $2 \leq k \leq r+1$ and transforms $e_1 - (e_i - e_j)$ into $e_1 - e_j$, it can be used to increase again r to $r+1$. Iterating the process finally gives r the value $n-1$. □

6.2 Perfect Lattices up to Dimension 4

The definition of perfect lattices (Definition 3.2.2) immediately shows that a perfect lattice possesses at least $\frac{n(n+1)}{2}$ pairs of minimal vectors. We shall prove in this section that the converse holds up to dimension 4.

Theorem 6.2.1. *1. A lattice Λ of dimension $n \leq 4$ with $s(\Lambda) \geq \frac{n(n+1)}{2}$ is extreme, and similar to one of the lattices $\mathbb{Z}$, $\mathbb{A}_2$, $\mathbb{A}_3$, $\mathbb{D}_4$ or $\mathbb{A}_4$.*

2. We have $\gamma_1^1 = 1$, $\gamma_2^2 = \frac{4}{3}$, $\gamma_3^3 = 2$, and $\gamma_4^4 = 4$, and the corresponding critical lattices are similar to $\mathbb{Z}$, $\mathbb{A}_2$, $\mathbb{A}_3$, and $\mathbb{D}_4$ respectively.

Proof. We have proved in Section 4.6 that the lattices listed in assertion (1) are extreme. We shall now prove the converse, disregarding the trivial case of dimension 1.

By the Hermite inequality, namely $\gamma_n \leq (\frac{4}{3})^{(n-1)/2}$, we have $\gamma_n^{n/2} < 2$ for $n \leq 3$. Consequently, by Corollary 2.6.10, if $n = 2$ or if $n = 3$, any sublattice of Λ containing n independent minimal vectors is equal to the whole lattice Λ.

If $n = 2$, $S(\Lambda)$ contains effectively two independent vectors (because $s(\mathbb{Z}) = 1$). One can thus apply Theorem 6.1.2, which shows first that Λ is similar to $\mathbb{A}_2$, and then that $\gamma_2^2 = \frac{4}{3}$.

Let $n = 3$. The result above implies that the upper bound $s(L) \leq 3$ holds for any 2-dimensional lattice L, and thus that Λ contains three independent minimal vectors. One can again apply Theorem 6.1.2, which shows first that Λ is similar to $\mathbb{A}_3$, and then that $\gamma_3^3 = 2$.

Finally, let $n = 4$. By the Mordell inequality (Theorem 2.3.1), we have $\gamma_4^2 \leq 2$. Moreover, since a 3-dimensional lattice possesses at most six pairs of minimal vectors, Λ possesses four independent minimal vectors. Theorem 6.1.2 implies that Λ is similar to $\mathbb{A}_4$ except if $\gamma(\Lambda)^2 = 2$ and if Λ possesses four minimal vectors e_1, e_2, e_3, e_4 which generate a sublattice Λ' of index 2 in Λ. When this condition is satisfied, there exist four integers a_1, a_2, a_3, a_4 not all even such that $e = \frac{a_1e_1+a_2e_2+a_3e_3+a_4e_4}{2}$ belongs to Λ. By adding to e a suitable $\mathbb{Z}$-linear combination of the e_i, we may assume that the a_i are equal to 0 or 1.

To simplify the notation, we rescale Λ to norm 1. We then have $\Delta(\Lambda) = \gamma_n(\Lambda)^{-n/2} = \frac{1}{2}$, whence $\Delta(e_1, e_2, e_3, e_4) = 2\,\Delta(\Lambda) = 1$. The Hadamard inequality (Theorem 2.1.1) is then an equality for the system $\{e_1, e_2, e_3, e_4\}$. This shows that these vectors are pairwise orthogonal, hence that the lattice they generate is isometric to $\mathbb{Z}^4$. Since $N(e) = \frac{a_1^2+a_2^2+a_3^2+a_4^2}{4}$, the lower bound $N(e) \geq 1$ implies that $a_1 = a_2 = a_3 = a_4 = 1$. This shows that Λ must be the centred cubic lattice $\mathbb{D}_4^*$, which is similar to $\mathbb{D}_4$. This completes the proof of ssertion (1), and assertion (2) follows from the comparison of the Hermite invariants of $\mathbb{A}_4$ and $\mathbb{D}_4$: by results proved in Chapter 4, we have $\gamma_4^4(\mathbb{A}_4) = \frac{16}{5} < \gamma_4^4(\mathbb{D}_4) = 4$. □

Corollary 6.2.2. *For $n = 1$, 2, and 4, the squares of the γ'_n constants are $\gamma_1'^{\,1} = 1$, $\gamma_2'^{\,2} = \frac{4}{3}$, and $\gamma_4'^{\,2} = 2$, and the corresponding dual-critical lattices are similar to $\mathbb{Z}$, $\mathbb{A}_2$, and $\mathbb{D}_4$ respectively.*

Proof. By Proposition 2.8.2 (3), we have $\gamma'_n \leq \gamma_n$, equality holds if and only if there exists a critical lattice whose dual is also critical, and when equality holds, dual-critical lattices are precisely the lattices as above. When $n = 1$, 2, and 4, there is exactly one similarity class of critical lattices, which are moreover similar to their dual, whence the result. □

The following corollary proves a result announced in Section 2.6:

Corollary 6.2.3. *Representative vectors for the successive minima of a lattice of dimension $n \leq 4$ generate the lattice, except perhaps for lattices similar to $\mathbb{D}_4$ for which both indices 1 and 2 may occur, depending on the choice of the representatives for the successive minima.*

Proof. We know by Corollary 2.6.10 that the index, for a given lattice Λ, is bounded from above by $\lfloor \gamma_n(\Lambda)^{n/2} \rfloor$, hence by one if one excludes the $\mathbb{D}_4$-case. The particular case of $\mathbb{D}_4$ is obvious on its interpretation as a centred cubic lattice. □

Notice that Mordell's inequality applied with the known values of γ_4 (or of γ_3, this amounts to the same thing) yields the inequality

$$\gamma_n \leq 2^{(n-1)/6} \qquad (\forall n \geq 3)\,,$$

whence the upper bound

$$\gamma_5^{5/2} \le 2^{5/3} = 3,174\dots, \tag{6.2.4}$$

which is however not sharp enough to bound by 2 the index of a sublattice attached to the successive minima, a result to be proved later in Section 6.4.

Remark 6.2.5. The proof of Theorem 6.2.1 shows that the perfection rank r and the half-kissing number s of a lattice Λ of dimension $n \le 4$ are equal, except if $\Lambda \sim \mathbb{D}_4$, where $s = r + 2$. The two perfection relations which arise in this case are easily produced by three eutaxy relations defined using projections onto three systems of pairwise orthogonal lines: write $S(\mathbb{D}_4) = S_1 \cup S_2 \cup S_3$ with $S_1 = \{\pm\varepsilon_1 \pm\varepsilon_2, \pm\varepsilon_3 \pm\varepsilon_4\}$, $S_2 = \{\pm\varepsilon_1 \pm\varepsilon_3, \pm\varepsilon_2 \pm\varepsilon_4\}$, and $S_3 = \{\pm\varepsilon_1 \pm \varepsilon_4, \pm\varepsilon_2 \pm \varepsilon_3\}$; we then have

$$\mathrm{Id} = \sum_{x \in S_1/\{\pm\}} p_x = \sum_{y \in S_2/\{\pm\}} p_y = \sum_{z \in S_3/\{\pm\}} p_z\,.$$

6.3 Dual-Extreme Lattices up to Dimension 4

We describe in this section dual-extreme lattices of dimension $n \le 4$, but we shall not give the proof for dimension 4.

The 1-dimensional classification is trivial (all lattices are proportional to $\mathbb{Z}$). The 2-dimensional classification is Exercise 3.8.3, where more is proved. We give here a short proof.

Theorem 6.3.1. *A 2-dimensional lattice is dual-extreme if and only if it is similar to* $\mathbb{A}_2$.

Proof. The minimal vectors in such a lattice generate obviously critical 1-dimensional lattices having the same norm. By Corollary 3.8.10, they must be extreme, hence similar to $\mathbb{A}_2$ by Theorem 6.2.1. □

Before turning to dimension 3, we prove two lemmas of a general interest.

Lemma 6.3.2. *Two minimal vectors in a lattice Λ which are congruent modulo $2\,\Lambda$ are equal or opposite.*

Proof. Let $m = N(\Lambda)$. For $x, y \in \Lambda$ with $y \equiv x \mod 2\Lambda$, if both $y + x$ and $y - x$ are nonzero, their norm must be at least $4m$, and this contradicts the equality

$$N(x+y) + N(x-y) = 2\big(N(x) + N(y)\big)\,. \quad \square$$

Lemma 6.3.3. *Let Λ be an n-dimensional lattice, and let $x \in S(\Lambda)$ and $y \in S(\Lambda^*)$. Then, for all $n \le 4$, $x \cdot y$ is equal to* 0 *or* ± 1.

[The proof below, taking into account our knowledge of critical lattices up to dimension 8, actually shows that Lemma 6.3.3 holds for all $n \le 8$, except if $\Lambda \sim \mathbb{E}_8$ and if x, y are proportional.]

Proof. The scalar product $x \cdot y$ is an integer which satisfies the inequalities

$$|x \cdot y| \leq \|\Lambda\| \cdot \|\Lambda^*\| = \gamma'(\Lambda) \leq \gamma(\Lambda)\,.$$

For $n \leq 4$ (and even for $n \leq 8$ and $\Lambda \not\sim \mathbb{E}_8$), we have $\gamma(\Lambda) < 2$, whence the result. □

Recall that given E of dimension n, dual-extreme lattices satisfy the following properties: S and S^* generate E (and in particular $s \geq n$ and $s^* \geq n$), and $s + s^* \geq n + 1$, where we have written S^* for $S(\Lambda^*)$ and s^* for $s(\Lambda^*)$ (Proposition 3.8.2 and Corollary 3.8.7). We shall also restrict ourselves to the case where $s \geq s^*$ (this is possible, since Λ and Λ^* play symmetric rôles).

Theorem 6.3.4. *A 3-dimensional lattice is dual-extreme if and only if it is similar to* $\mathbb{A}_3$ *or to* $\mathbb{A}_3^*$.

Proof. We consider a dual-extreme lattice Λ in the 3-dimensional Euclidean space E. We have $s \geq s^*$, $s + s^* \geq 7$ (hence $s \geq 4$), and $E = \langle S \rangle = \langle S^* \rangle$.

If Λ contains a hexagonal lattice with the same norm, the argument we used in the proof of Theorem 6.3.1 works again: Theorem 6.2.1 implies that Λ must be extreme, whence similar to $\mathbb{A}_3$.

From now on, we suppose that Λ does not contain any minimal hexagonal section. Consider three independent minimal vectors e_1, e_2, e_3 of Λ. Then, because of the upper bound $\gamma_3^{3/2} < 2$, (e_1, e_2, e_3) is a basis for Λ, and a fourth minimal vector e_4 is of the form $\pm e_i \pm e_j$ or $\pm e_1 \pm e_2 \pm e_3$. The first possibility must be ruled out, for e_i, e_j would then generate a hexagonal lattice, and Lemma 6.1.6 (or Lemma 6.3.2) shows that there is at most one pair $\pm x$ of the second type. We thus have $s = 4$, and we may suppose after replacing some e_i by $-e_i$ that

$$S = \{\pm e_1, \pm e_2, \pm e_3, \pm e_4\} \ \text{ with } \ e_1 + e_2 + e_3 + e_4 = 0\,.$$

We must show that a lattice with $s^* \leq s = 4$ cannot be dual-extreme. To simplify some calculations, we rescale Λ to norm 1.

From the equality

$$\sum_{i=1}^{4} e_i = 0\,, \tag{1}$$

we deduce the following two relations in which (i, j, k, ℓ) may be any permutation of $(1, 2, 3, 4)$:

$$e_i \cdot e_j + e_j \cdot e_k + e_k \cdot e_i = -1 \tag{2}$$

and

$$e_i \cdot e_\ell + e_j \cdot e_\ell + e_k \cdot e_\ell = -1\,. \tag{3}$$

(The calculation of the norm of $e_i + e_j + e_k$ yields immediately the first one, and the second one follows by calculating the scalar product of the two sides in the first one.) Comparing (1) and (2), we obtain the further relation

$$e_i \cdot e_j = e_k \cdot e_\ell \,. \tag{4}$$

[This equality proves the existence of a reflection preserving Λ, see Exercise 6.3.7; this explains why s^* must be even in Lemma 6.3.5 below.]

Let $t_i = e_i \cdot e_4$ for $i = 1, 2, 3$; attach to any $x \in S(\Lambda^*)$ the element

$$(x \cdot e_1, x \cdot e_2, x \cdot e_3, x \cdot e_4) \in \mathbb{Z}^4 \,.$$

Lemma 6.3.5. *Let Λ be a lattice such that $S(\Lambda)$ and $S(\Lambda^*)$ span E, which satisfies the inequalities $s \geq 4$ and $s \geq s^* \geq 3$, and which does not contain any hexagonal lattice with the same norm. Then $s = s^* = 4$, and up to a permutation of the e_i, representatives for $S(\Lambda^*)/\{\pm\}$ are the four vectors f_1, f_1', f_2, f_2', whose scalar products with e_1, e_2, e_3, e_4 are $(1, -1, 0, 0)$, $(0, 0, 1, -1)$, $(1, 0, -1, 0)$, $(0, 1, 0, -1)$. Denoting by (e_1^*, e_2^*, e_3^*) the dual basis of (e_1, e_2, e_3), we have $f_1 = e_1^* - e_2^*$, $f_1' = e_3^*$, $f_2 = e_1^* - e_3^*$, and $f_2' = e_2^*$.*

Proof of 6.3.5. For any $f \in S(\Lambda^*)$, we have $f \cdot e_i = 0$ or ± 1 (Lemma 6.3.3) and $\sum_{i=1}^4 f \cdot e_i = 0$. The sequence of scalar products $f \cdot e_i \mod 2$ is thus either a permutation of $(1, 1, 0, 0)$, or the sequence $(1, 1, 1, 1)$. But two vectors in Λ^* attached to an identical sequence modulo 2 are congruent modulo $2\Lambda^*$. The lower bound $s^* \geq 3$ implies that the sequence $(1, -1, 0, 0)$ corresponds to some $f_1 \in S(\Lambda^*)$ (after a permutation of the e_i and a change of sign of f_1 if need be). Relation (3) shows that the vector f_1' corresponding to the sequence $(0, 0, 1, -1)$ is minimal in Λ^*, since the relative lattices orthogonal to f_1 and f_1' in Λ are isometric. For f corresponding to $(1, 1, 1, 1)$, we have $f \equiv f_1 + f_1' \mod 2\Lambda^*$, a relation which shows that f, f_1, f_1' do not generate Λ^*. Consequently, there are two other vectors f_2 and f_2' in $S(\Lambda^*)$, that we may suppose to correspond to sequences $(1, 0, -1, 0)$ and $(0, 1, 0, -1)$. We thus have $s^* \geq 4$, hence $s^* = 4$ and $S(\Lambda^*) = \{\pm f_1, \pm f_1', \pm f_2, \pm f_2'\}$. The last assertion is then immediate. □

Lemma 6.3.6. *Under the hypotheses of Lemma 6.3.5, the parameters t_i can be expressed as functions of $t = -t_2 \geq 0$ by the formulae $t_1 = -1 + 2t$ and $t_2 = t_3 = -t$.*

Proof of 6.3.6. Condition (1) reads $t_1 + t_2 + t_3 = -1$. We have $t_i \geq -\frac{1}{2}$ by the lower bound $N(e_j + e_k) \geq 1$, hence $t_i \leq 0$, whence $t \geq 0$. Since e_2^* and e_3^* have the same norm, their orthogonal lattices $\langle e_1, e_3 \rangle$ and $\langle e_1, e_2 \rangle$ in Λ have equal determinants, which shows that $1 - (e_1 \cdot e_3)^2 = 1 - (e_1 \cdot e_2)^2$, hence that $t_2 = t_3$. We thus have $t_2 = t_3 = -t$ and $t_1 = -1 - t_2 - t_3 = -1 + 2t$. □

End of the proof of 6.3.4. We complete the proof of Theorem 6.3.4 by calculating an extremum. Using (1), we obtain $e_1 \cdot e_2 = e_1 \cdot e_3 = -t$ and

$e_2 \cdot e_3 = -1 + 2t$. This gives us a Gram matrix for Λ and shows moreover, since e_3^* must be minimal, that $\gamma'(\Lambda)$ is equal to the function

$$\phi(t) = \frac{\begin{vmatrix} 1 & e_1 \cdot e_2 \\ e_1 \cdot e_2 & 1 \end{vmatrix}}{\det(\Lambda)} = \frac{1-t^2}{4t(1-t)^2} = \frac{1+t}{4(t-t^2)}\,.$$

Since e_1^* is not minimal, we have

$$\begin{vmatrix} 1 & e_2 \cdot e_3 \\ e_2 \cdot e_3 & 1 \end{vmatrix} - \begin{vmatrix} 1 & e_1 \cdot e_2 \\ e_1 \cdot e_2 & 1 \end{vmatrix} = (1-(-1+2t)^2) - (1-t^2) = (1-t)(3t-1) > 0,$$

hence $t > \frac{1}{3}$. Since $e_1 - e_2$ is not minimal in Λ, we have $N(e_1 - e_2) > 1$, i.e. $t < \frac{1}{2}$. On the interval $(\frac{1}{3}, \frac{1}{2})$, the numerator of ϕ' is $t^2 + 2t - 1$, whose unique root in $(\frac{1}{3}, \frac{1}{2})$ corresponds to a minimum of ϕ. Hence $\gamma'(\Lambda)$ has no local maximum with $s = 4 \geq s^* \geq 3$. □

In the previous section, we have calculated the constants γ_1', γ_2', and γ_4'. We can now solve the case $n = 3$ which was left open:

Corollary 6.3.7. *We have* ${\gamma_3'}^2 = \frac{3}{2}$*, and the* 3*-dimensional dual-critical lattices are those which are similar to* $\mathbb{A}_3$ *or to* $\mathbb{A}_3^*$*.*

Proof. These two lattices are the only dual-extreme lattices in dimension 3, and we know by Section 4.2 that $\gamma'(\mathbb{A}_n)^2 = \gamma'(\mathbb{A}_n^*)^2 = \frac{2n}{n+1}$. □

The constant γ_3' can be used to determine all constants $\gamma_{n,k}$ and $\gamma_{n,k}'$ of Definition 2.8.3 for $n \leq 4$. Indeed, we are left with the unique case $n = 4$, $k = 2$.

Corollary 6.3.8. *We have* $\gamma_{4,2} = \gamma_{4,2}' = \frac{3}{2}$*, and these values are attained exactly on lattices which are similar to* $\mathbb{D}_4$*.*

Proof. The results of Section 2.8 imply the equality $\gamma_{4,2} = \gamma_{4,2}'$ (Corollary 2.8.8), the upper bound $\gamma_{4,2}' \leq {\gamma_{3,1}'}^2 (= {\gamma_3'}^2)$ (Theorem 2.8.7 (2)), i.e. $\gamma_{4,2}' \leq \frac{3}{2}$ by Corollary 2.3.7, and the upper bound $\gamma_{4,1} \leq \gamma_{2,1}(\gamma_{4,2})^{1/2}$ (apply Theorem 2.8.6 with $n = 4$, $k = 1$, and $h = 2$), which then implies the lower bound $\gamma_{4,2} \geq \gamma_2^{-2}\gamma_4^2 = \frac{3}{4}\left(\frac{1}{2}\right)^{-1} = \frac{3}{2}$, from which we deduce the equalities $\gamma_{4,2} = \gamma_{4,2}' = \frac{3}{2}$.

We must now classify the lattices for which the invariants $\gamma_{4,2}$ and $\gamma_{4,2}'$ attain the value $\frac{3}{2}$. To this end, we must look at some details in some proofs in Section 2.8, in order to be able to characterize some cases of equalities which occur there. It is clear that equality in Proposition 3.8.6 can hold only for a lattice Λ with $\gamma_4(\Lambda) = \gamma_4$. By Theorem 6.2.1, such a lattice is similar to $\mathbb{D}_4$.

Moreover, the proof of Theorem 3.8.7 (1) shows that the inequality $\gamma_{4,2}(\Lambda) \leq \gamma_{4,2}'(\Lambda)$ holds for all lattices Λ on which these invariants attain their maximum values. Taking into account Proposition 3.8.5, we see that we have $\gamma'_{4,2}(\Lambda) = \gamma_{4,2}(\Lambda) = \frac{3}{2}$ on such a lattice, and the result we want for the invariant $\gamma'_{4,2}$ is a consequence of the corresponding one for $\gamma_{4,2}$ that we have just proved. □

The length of the proof of Theorem 6.3.4 gives an idea of the complexity of the analogous result for $n = 4$. We shall thus simply state the result, and we refer the reader to Section 5 of [B-M1] for a proof, which it will be highly desirable to simplify, and to Section 9.3, in which some extra indications can be found.

Theorem 6.3.9. *A 4-dimensional lattice is dual-extreme if and only if it is similar to* $\mathbb{D}_4 \sim \mathbb{D}_4^*$, $\mathbb{A}_4$ *or* $\mathbb{A}_4^*$. □

We shall prove later in Chapter 12 that for every dimension, there are up to similarity only finitely many dual-eutactic lattices whose minimal vectors span E. This could give an alternative way of proving Theorems 6.3.1, 6.3.4, and 6.3.9.

Exercise 3.8.3 yields the following result: *the dual-eutactic lattices of dimension* $n \leq 2$ *are those which are similar to* $\mathbb{Z}$, $\mathbb{A}_2$ *or* $\mathbb{Z}^2$. Each of them is similar to its dual.

For dimension 3, we can prove:

Theorem 6.3.10. *A dual-eutactic 3-dimensional lattice with* $s \geq 3$ *is similar to* $\mathbb{A}_3$, $\mathbb{A}_3^*$, $\mathbb{Z}^3$ *or to the lattice* Λ_0 *with* $s = s^* = 4$ *of Exercise 6.3.1, which is defined over* $\mathbb{Q}(\sqrt{2})$ *and is similar to its dual.*

Sketch of proof. This is obvious for $s = 6$. The case where $s = 5$ (resp. $s = 4$) is dealt with in Exercise 6.1.1 (resp. 6.3.1). Taking into account Exercise 6.3.2, we then recover Theorem 6.3.4. We are left with the case $s = 3$, for which we can use Bergé's classification (Theorem 9.2.2, Proposition 9.2.3 and Exercise 9.2.2) and the fact that S must generate E, which is proved using Exercise 3.8.4. □

The classification of dual-eutactic 4-dimensional lattices (even under the restriction that S should generate E) is not known.

6.4 Perfect Lattices in Dimension 5

In this section, we classify 5-dimensional perfect lattices by making use of combinatorial methods which go back to Korkine and Zolotareff. However, instead of following closely their method as in the French edition, we use a different strategy, relying on methods introduced by Watson. The proofs differ considerably at two points (see the notes).

As in Section 6.2, we obtain more than a mere classification: we characterize perfect lattices by properties of the configurations of their minimal vectors; see Theorems 6.4.1 and 6.4.2 below. We shall return to this classification for all dimensions $n \le 5$ in the next chapter, where we shall give a completely different proof relying on the Voronoi algorithm.

We still denote by E an n-dimensional Euclidean vector space, with $n = 5$ unless otherwise stated, for instance when we prove Watson's results in full generality.

Recall (Theorem 3.5.2) that a perfect n-dimensional lattice contains n independent minimal vectors outside any hyperplane section. The converse is true for 5-dimensional lattices having a $\mathbb{D}_4$-section with the same norm, indeed under a weaker hypothesis:

Theorem 6.4.1. *A 5-dimensional lattice which possesses a hyperplane section of the same norm similar to $\mathbb{D}_4$ and five minimal vectors outside this section is extreme and similar to $\mathbb{D}_5$.*

For lattices without such a $\mathbb{D}_4$-section, the lower bound $s \ge 15$ instead of $s \ge s(\mathbb{D}_4) + 5 = 17$ suffices:

Theorem 6.4.2. *A 5-dimensional lattice without any $\mathbb{D}_4$-section of the same norm and with $s \ge 15$ is extreme and similar to $\mathbb{A}_5$ or to $\mathbb{A}_5^3$.*

Corollary 6.4.3. *A 5-dimensional lattice is perfect if and only if it is similar to $\mathbb{D}_5$, $\mathbb{A}_5^3$, or $\mathbb{A}_5$.* □

Taking into account the values of the Hermite constants of these lattices $(\gamma_5(\mathbb{D}_5)^5 = 8 > \gamma_5(\mathbb{A}_5^3)^5 = \frac{512}{81} = 6,32\cdots > \gamma_5(\mathbb{A}_5)^5 = \frac{16}{3} = 5,33\ldots$; see Sections 4.2, 4.3, and 5.2), we obtain:

Corollary 6.4.4. *We have $\gamma_5^5 = 8$ and the critical 5-dimensional lattices are those which are similar to $\mathbb{D}_5$.* □

We now consider a lattice Λ of an arbitrary dimension $n \ge 1$ together with a (not necessarily unique) sublattice Λ' generated by representatives $e_1, \ldots, e_n$ for its successive minima (Definition 2.6.7). Watson's theory provides a method to study the quotient Λ/Λ'. Using deformation arguments of the kind of those we used to prove Minkowski's Theorem 2.6.8, we easily see that all possible finite Abelian groups occurring as a quotient Λ/Λ' also occur for a well–rounded Λ. We can thus restrict ourselves for the proofs to the case of equal successive minima; this device was indeed used (Corollary 2.6.10) to prove the general bound

$$[\Lambda : \Lambda'] \le \lfloor \gamma_n^{n/2} \rfloor .$$

Denote by $\operatorname{sgn}(x)$ the sign of the real number x, or 0 if $x = 0$.

Lemma 6.4.5. *Let $e_1, \ldots, e_n \in E$ be n independent vectors, let $a_1, \ldots, a_n$ be integers, let d be a positive integer, and let*

$$e = \frac{a_1 e_1 + \cdots + a_n e_n}{d}.$$

We then have the following formula:

$$\left(\left(\sum_{i=1}^{n} |a_i|\right) - 2d\right) N(e) = \sum |a_i| \big(N(e - \operatorname{sgn}(a_i) e_i) - N(e_i)\big).$$

Proof. Replacing e_i by $-e_i$ when a_i is negative, we may assume that $a_i \geq 0$ for all i. We have $N(e - e_i) - N(e_i) = N(e) - 2e \cdot e_i$. Multiplying both sides by a_i and adding yields the equality we want. □

Theorem 6.4.6. (Watson.) *With the notation of the preceding lemma, suppose that the e_i have equal norms. Let Λ' be the lattice that they generate, and let $\Lambda = \langle \Lambda', e \rangle$. If $N(\Lambda) = N(\Lambda')$, we have*

$$\sum_{i-1}^{n} |a_i| \geq 2d,$$

and equality holds if and only if $e - e_i$ is minimal for all i with $a_i \neq 0$. Moreover, when equality holds, if some a_j is equal to $\frac{d}{2}$, then all vectors e' deduced from e by replacing e_k by $-e_k$ on any set of indices k with $a_k = \frac{d}{2}$ are minimal, and so are all the $e' - e_i$ with $a_i \neq 0, \pm\frac{d}{2}$.
[The last assertion applies to e; however, e is not in general minimal.]

Proof. The inequality in the theorem and the characterization of the cases when equality holds follow immediately from Lemma 6.4.5. Since equality in Lemma 6.4.5 is preserved when one replace a_k by $-a_k$ for any number of indices k with $a_k = \pm\frac{d}{2}$, the last assertion is true for the vectors $e' - e_i$. One sees that it is also true for the vectors of the form e' by writing $e' = (e' \pm e_k) \mp e_k$ for some k with $a_k = \frac{d}{2}$ and a suitable choice of signs. □

Suppose that Λ/Λ' is cyclic of order d. After having chosen an appropriate indexing of the e_i, we can write, $\Lambda = \langle \Lambda', e \rangle$ with

$$e = \frac{a_1 e_1 + \cdots + a_m e_m}{d}$$

for some $m \leq n$, with $1 \leq a_i \leq \frac{d}{2}$ for $1 \leq i \leq m$.

If $n = 4$, the only possibility is $d = 2$, $m = 4$, and $a_1 = a_2 = a_3 = a_4 = 1$. The last assertion in Theorem 6.4.6 then shows the existence of 24 minimal vectors ($\pm e_i$ and $\dfrac{\pm e_1 \pm e_2 \pm e_3 \pm e_4}{2}$), and we recognize the $\mathbb{D}_4$-lattice, obtaining a new proof of the index 2 case in Theorem 6.2.1.

Before turning to the proofs of Theorems 6.4.1 and 6.4.2, we give a couple of definitions.

Definition 6.4.7. *We say that a lattice Λ is* of index at most d *if any system of representatives for the successive minima generates a sublattice Λ' of Λ of index $[\Lambda : \Lambda'] \leq d$, and that Λ is* of index d *if equality may hold.*

A restatement of Theorem 6.1.2 is that lattices of index 1 with $s \geq \frac{n(n+1)}{2}$ are similar to $\mathbb{A}_n$. A first specific application of Theorem 6.4.6 to dimension 5 is Proposition 6.4.8 below; this is a mere lemma for statements 6.4.1 to 6.4.4, since its validity will be obvious afterwards thanks to the upper bound $\gamma_5^{5/2} < 2$ which will result from Corollary 6.4.4.

Proposition 6.4.8. *A 5-dimensional lattice is of index 1 or 2.*

Proof. We keep the notation Λ, Λ', a_i, m, n. We have $\gamma_5^{5/2} \leq 2^{5/3} < 4$ by (6.2.4), hence $[\Lambda : \Lambda'] \leq 3$. We must show that index 3 does not occur. As we previously noticed, it suffices to consider well–rounded lattices. By Theorem 6.4.6, we can take $a_i = 1$ for $1 \leq i \leq m$, whence the lower bounds $n \geq m \geq 6$. □

Definition 6.4.9. *Let Λ be an n-dimensional, well–rounded lattice of index 2. The* (2-)length *of Λ is the smallest possible integer m which satisfies the following property: there exist minimal vectors $e_1, \dots, e_n$ of Λ such that*

$$\Lambda = \left\langle e_1, \dots, e_n, \frac{e_1 + \cdots + e_m}{2} \right\rangle.$$

By the results of Section 6.2, we have $4 \leq m \leq n$, and $\Lambda \cap \langle e_1, e_2, e_3, e_4 \rangle$ is similar to $\mathbb{D}_4$ if $m = 4$.

Proposition 6.4.10. *Suppose that $\Lambda = \langle \Lambda', e \rangle$, where $\Lambda' = \langle e_1, \dots, e_n \rangle$ and $e = \frac{e_1 + \cdots + e_m}{2}$, has index 2 and length m, and let $x = \sum_{j=1}^n a_j\, e_j \in S(\Lambda')$. Then:*

1. *The coefficients a_i belong to $\{0, \pm 1\}$.*
2. *Any minimal vector in Λ' of the form $\sum_{j=1}^m a_j e_j$ is one of the vectors $\pm e_i$.*

Proof. (1) Interpreting the a_i as characteristic determinants, we can attach to each nonzero a_i a lattice $\Lambda'' \subset \Lambda'$ of index $|a_i|$. We then have $[\Lambda : \Lambda''] = 2|a_i|$, whence $|a_i| = 1$.

(2) By permutation and changes of signs of the e_i with $i \leq m$, we may suppose that $x = e_1 + \cdots + e_k$ for some $k \leq m$. Writing $e = \frac{x + e_{k+1} + \cdots + e_m}{2}$, we see that the 2-length of Λ is at most $m - k + 1$. We thus have $m - k + 1 \geq m$, hence $k \leq 1$. □

We shall now prove successively Theorems 6.4.1 and 6.4.2. We consider a 5-dimensional lattice Λ of index 2 and 2-length m, with successively $m = 4$ and $m = 5$. We write $\Lambda = \langle \Lambda', e \rangle$ with $e = \frac{e_1 + \cdots + e_m}{2}$, and we rescale Λ to norm 1.

Proof of 6.4.1. We take $m = 4$, and we suppose that $S(\Lambda)$ contains at least five pairs of minimal vectors (including $\pm e_5$) other that the 12 pairs $\pm e_i$, $1 \le i \le 4$, and $\frac{\pm e_1 \pm e_2 \pm e_3 \pm e_4}{2}$. We must show that Λ is similar to $\mathbb{D}_5$. Let $L' = \langle e_1, e_2, e_3, e_4 \rangle \sim \mathbb{Z}^4$ and $L = \langle L', e \rangle \sim \mathbb{D}_4$.

We first investigate $S(\Lambda')$. Let $x \in S(\Lambda')$, $x \ne \pm e_i$. We have $\pm x = e_5 + y$ for some $y \in L'$. By permutation and changes of signs of the e_i, we may assume that $x = e_5 + y$ and that y is one of the vectors e_1, $e_1 + e_2$, $e_1 + e_2 + e_3$, or $e_1 + e_2 + e_3 + e_4$.

We must first exclude the last two cases, since we would then have $e = \frac{x + e_4 - e_5}{2}$ or $e = \frac{x - e_5}{2}$, which shows up an index 2 in dimension 3 or 2.

Next, if $y = e_1 + e_2$, we can write

$$e = \frac{(e_1 + e_2 + e_5) + e_3 + e_4 - e_5}{2},$$

which shows that e_3 and e_4 are orthogonal to e_5 and that $e_5 \cdot (e_1 + e_2 + e_5) = 0$, i.e. $e_1 \cdot e_5 + e_2 \cdot e_5 = -1$, whence $e_1 \cdot e_5 = e_2 \cdot e_5 = -\frac{1}{2}$ (because of the upper bounds $|e_i \cdot e_5| \le \frac{1}{2}$). Since each of the scalar products $e_i \cdot e_j$ may take at most one value, Λ is unique up to similarity, hence perfect; moreover, since $\sqrt{2}\,\Lambda$ is generated by norm 2 vectors, this is a root lattice which must obviously be isometric to $\mathbb{D}_5$ (and whose identification is anyway easy on the Gram matrix B_1 below).

Hence, if Λ is not similar to $\mathbb{D}_5$, the minimal vectors of $S(\Lambda') \smallsetminus L'$ must be of the form $\pm e_5$ or $\pm e_i \pm e_5$. Now $e_i + e_5$ and $e_i - e_5$ are not both minimal by Lemma 6.1.6 (1). If $e_i + e_5$ and $e_j + e_5$ were both minimal, we would have $e_i \cdot e_5 = e_j \cdot e_5 = -\frac{1}{2}$, whence $N(e_i + e_j + e_5) = 1$, and Λ would be again similar to $\mathbb{D}_5$ by what we just proved. Thus there can be besides $\pm e_5$ only one pair of minimal vectors in $S(\Lambda') \smallsetminus L'$.

We have thus proved that either $\Lambda \sim \mathbb{D}_5$ or Λ contains at most two pairs of minimal vectors in Λ' outside its $\mathbb{D}_4$-section. But when this situation occurs, there must then be at least three pairs of minimal vectors in $(e + \Lambda') \smallsetminus L$. This case now immediately reduces to the preceding one by considering the lattice $L'_1 = \langle e, e_2, e_3, e_4 \rangle$ and writing $L = \langle L'_1, e_1 \rangle$. □

The proof above shows that there are actually four pairs of minimal vectors outside L in each of the classes Λ' and $e + \Lambda'$, with possible representatives e_5, $e_1 + e_5$, $e_2 + e_5$, $e_1 + e_2 + e_5$, $e + e_5$, $e - e_3 + e_5$, $e - e_4 + e_5$, and $e - e_3 - e_4 + e_5$, and yields for Λ' and Λ the following Gram matrices (after having replaced e_5 by $-e_5$):

$$A_1 = \begin{pmatrix} 2&0&0&0&1 \\ 0&2&0&0&1 \\ 0&0&2&0&0 \\ 0&0&0&2&0 \\ 1&1&0&0&2 \end{pmatrix} \qquad \text{and} \qquad B_1 = \begin{pmatrix} 2&1&1&1&1 \\ 1&2&0&0&1 \\ 1&0&2&0&0 \\ 1&0&0&2&0 \\ 1&1&0&0&2 \end{pmatrix}.$$

Proof of 6.4.2. We now take $m = 5$, i.e. $e = \frac{e_1 + e_2 + e_3 + e_4 + e_5}{2}$, and we suppose that $S(\Lambda)$ contains at least 15 pairs of minimal vectors. By Proposition 6.4.10, we have $S(\Lambda') = \{\pm e_1, \pm e_2, \pm e_3, \pm e_4, \pm e_5\}$, and there are thus at least 10 pairs of minimal vectors in $e + \Lambda'$. Now all these vectors are of the form $x = \frac{1}{2}\sum_{j=1}^{5} a_j\, e_j$ for some odd integers a_j. Writing

$$e_1 = \frac{2x - a_2\, e_2 - a_3\, e_3 - a_4\, e_4 - a_5\, e_5}{-a_1}\,,$$

we see that $|a_1| \le 1$ (and similarly $|a_i| \le 1$ for all i), since Λ is of index 2.

We are thus reduced to prove the following assertion:

There exists up to similarity a unique lattice Λ possessing at least 10 minimal vectors among the 16 vectors $\dfrac{e_1 \pm e_2 \pm e_3 \pm e_4 \pm e_5}{2}$.

The identification with $\mathbb{A}_5^3$ will be then evident, since we shall know that there are exactly three similarity classes of perfect 5-dimensional lattices, and that our new lattice is not a root lattice.

After having replaced some e_i by $-e_i$, we may assume that e itself is minimal, and that there are at least nine minimal vectors among the 15 vectors $e - e_i$, $e - e_j - e_k$.

Lemma 6.4.11. *Let s' be the number of pairs of vectors of Λ of the form $e - e_j - e_k$, $j < k$. We have $s' \le 6$, and if $s' = 6$, up to a permutation of the e_i, they are the vectors $\pm(e - e_j - e_k)$, $j = 1, 2, 3$, $k = 4, 5$.*

Proof. For $1 \le j \le 5$, let $p(j)$ be the number of indices k such that $e - e_j - e_k$ is minimal (we have $s' = \frac{1}{2}\sum_j p(j)$), and let $p = \max_j p(j)$; we suppose that $p = p(5)$. If $p = 4$, Lemma 6.1.6 (2) bounds s' from above by 4. If $p = 2$ (resp. 1), we obviously have $s' \le \frac{1}{2}\binom{5}{2} = 5$ (resp. $s' \le 2$). If $p = 3$, we may suppose that $e - e_5 - e_2, e - e_5 - e_3, e - e_5 - e_4$ are minimal, and Lemma 6.1.6 (2) implies that the other minimal vectors of the form $e - e_j - e_k$ lie among the $e - e_4 - e_k$. □

Lemma 6.4.12. *There exist in Λ exactly three minimal vectors of the form $e - e_i$.*

Proof. Let s'' be the number of such minimal vectors. We have $s'' \ge 15 - 6 - s' \ge 3$, and the equality $N(e) = \sum_{i=1}^{5} N(e - e_i) - N(e_i)$ in Lemma 6.4.5 reads $\sum_{N(e-e_i)>1} N(e - e_i) = 6 - s''$, and implies the upper bound $s'' \le 4$.

If $s'' = 4$, we may assume that $e - e_1$, $e - e_2$, $e - e_3$, $e - e_4$ are minimal but that $e - e_5$ is not, and the preceding equality reduces to $N(e - e_5) = 2$, i.e. $e \cdot e_5 = 0$. Then, for $1 \le i \le 4$, we have $N(e - e_i - e_5) = 2 + 2e \cdot e_5 > 1$, which implies $s' \le 4$, whence $s(\Lambda) \le 6 + 4 + 4 < 15$. □

Lemma 6.4.13. *The lattice Λ possesses 15 pairs of minimal vectors which, up to a permutation and changes of signs of the e_i, can be represented by the 15 vectors e, e_i $(i \le 5)$, $e - e_i$ $(i \le 3)$ and $e - e_j - e_k$ $(j = 1, 2, 3,\ k = 4, 5)$.*

Proof. The e_i are of course minimal, and we have seen in Lemmas 6.4.12 and 6.4.13 that we may assume that $e, e - e_1, e - e_2, e - e_3$ also are. Then

$e - e_4$ and $e - e_5$ are not minimal, and we thus have (with the notation of Lemma 6.4.1) $s' \geq 6$, hence $s' = 6$ and $s = 15$, and also $p = 3$. Moreover, p attains this values on two indices j, which must now be shown to be $j = 4$ and $j = 5$. Otherwise, we would have, say, $p(1) = 3$, and the three vectors $e - e_1 - e_i$, $e - e_1 - e_j$, $e - e_1 - e_k$ would be minimal. Setting $e' = e - e_1$, we see that e', $e' + e_1$, $e' - e_i$, $e' - e_j$, $e' - e_k$ would be minimal, in contradiction with Lemma 6.4.12. □

End of the proof of 6.4.2. The lattice above, if it exists, is certainly perfect: otherwise, by a deformation preserving the configuration of minimal vectors, it would be possible to obtain a new lattice with the same index properties. Moreover, it is unique. We thus now know that there exists besides $\mathbb{D}_5$ and $\mathbb{A}_5$ at most one similarity class of 5-dimensional perfect lattices. Since there exists at least one (namely that of $\mathbb{A}_5^3$), the lattice above is similar to $\mathbb{A}_5^3$. □

One can easily make completely explicit the proof above by calculating all scalar products $e_j \cdot e_k$, which are well defined because Λ is perfect and that we have given ourselves the norms $e_i \cdot e_i$ (we have chosen for Λ the scale in which $e_i \cdot e_i = 1$ for all i). Let $x_i = e \cdot e_i$. We have $x_1 = x_2 = x_3 = \frac{1}{2}$ (because e, e_i and $e - e_i$ are minimal for $i = 1, 2, 3$), and the six conditions $N(e - e_j - e_k) = 1$ for $j = 1, 2, 3$ and $k = 4, 5$ imply the relations $e_j \cdot e_k = x_k - \frac{1}{2}$. We have $x_4 + x_5 = 1$ by the relation $N(e) = 1$, whence first $e_i \cdot e_4 + e_i \cdot e_5 = x_4 + x_5 - 1 = -\frac{1}{2}$ for $i = 1, 2, 3$, then $e_1 \cdot e_2 + e_1 \cdot e_3 = e_2 \cdot e_1 + e_2 \cdot e_3 = e_3 \cdot e_1 + e_3 \cdot e_2 = \frac{1}{2}$ by calculating $e \cdot e_i$, and moreover $e_1 \cdot e_2 = e_1 \cdot e_3 = e_2 \cdot e_3 = \frac{1}{4}$. Calculating now $e \cdot e_4$ and $e \cdot e_5$, we find $e_4 \cdot e_5 = \frac{1}{2} - x_4 = \frac{1}{2} - x_5$, whence finally $e_4 \cdot e_5 = x_4 = x_5 = \frac{1}{4}$ and $e_j \cdot e_k = -\frac{1}{4}$ ($j = 1, 2, 3$, $k = 4, 5$).

Changing e_4 into $-e_4$ and e_5 into $-e_5$, setting $e' = \frac{e_1 + e_2 + e_3 - e_4 - e_5}{2}$ and multiplying all vectors by 2, we obtain the following Gram matrices for $\mathcal{B}' = (e_1, e_2, e_3, e_4, e_5)$ and $\mathcal{B} = (e', e_2, e_3, e_4, e_5)$:

$$A_2 = \begin{pmatrix} 4 & 1 & 1 & 1 & 1 \\ 1 & 4 & 1 & 1 & 1 \\ 1 & 1 & 4 & 1 & 1 \\ 1 & 1 & 1 & 4 & 1 \\ 1 & 1 & 1 & 1 & 4 \end{pmatrix} \quad \text{and} \quad B_2 = \begin{pmatrix} 4 & 2 & 2 & -1 & -1 \\ 2 & 4 & 1 & 1 & 1 \\ 2 & 1 & 4 & 1 & 1 \\ -1 & 1 & 1 & 4 & 1 \\ -1 & 1 & 1 & 1 & 4 \end{pmatrix}.$$

The system of eigenvalues of the first one is $(8^1, 3^4)$, whence $\det(A_2) = 2^3 \cdot 3^4$ and $\det(B_2) = \frac{1}{4}\det(A_2) = 2 \cdot 3^4 = 162 = \det(\mathbb{A}_5^3)$.

Remark 6.4.14. *Theorems 6.1.2, 6.4.1, 6.4.2 and their proofs show that there exists exactly one 5-dimensional perfect lattice for each of the following three types, defined by the index d and the 2-length m if $d = 2$: $d = 1$ ($\mathbb{A}_5$), $d = 2$, $m = 4$ ($\mathbb{D}_5$), and $d = 2$, $m = 5$ ($\mathbb{A}_5^3$).*

Corollary 6.4.15. *Let Λ be a 5-dimensional lattice possessing a hyperplane section (of the same norm) similar to $\mathbb{A}_4$ and at least five pairs of minimal vectors outside this section. Then one of the following properties holds:*

1. *Λ has no $\mathbb{D}_4$-section, and is similar to $\mathbb{A}_5$.*

2. Λ has at least two $\mathbb{D}_4$-sections, and is similar to $\mathbb{D}_5$.
3. Λ has has exactly one $\mathbb{D}_4$-section and one $\mathbb{A}_4$-section, and $s(\Lambda) = 16$.

Proof. We have $s(\Lambda) \geq s(\mathbb{A}_4) + 5 = 15$. Denote by n_D and n_A the respective numbers of $\mathbb{D}_4$- and $\mathbb{A}_4$-sections of Λ.

If $n_D = 0$, Λ is similar to $\mathbb{A}_5$ or to $\mathbb{A}_5^3$, and the second possibility must be excluded, since $\mathbb{A}_5^3$ does not possess perfect norm 4 sections in dimensions 4 and 3. Hence Λ is similar to $\mathbb{A}_5$ (and we have $n_A = s(\mathbb{A}_5^*) = 6$).

If $n_D \geq 2$, without any other hypothesis, we have $s(\Lambda) \geq 2s(\mathbb{D}_4) - s(\mathbb{A}_3) = 24 - 6 = 18$, hence $\Lambda \sim \mathbb{D}_5$ by Theorem 6.4.1 (and moreover $n_D = 5$ and $n_A = 16$).

If $n_D = 1$, we have $s(\Lambda) \geq 12 + 10 - 6 = 16$, and indeed $s(\Lambda) = 16$, since $s(\Lambda) > 16$ occurs only when $\Lambda \sim \mathbb{D}_5$ by Theorem 6.4.1. This equality also shows that $n_A = 1$. [The case $n_D = 1$ has a nice interpretation in terms of the Voronoi graph of the next chapter: Λ must be on an edge connecting $\mathbb{A}_5$ to $\mathbb{D}_5$; this theory also shows that the number of $\mathbb{A}_4$-sections in a 5-dimensional lattice cannot be exactly 2, and is thus equal to 0, 1, 6 or 16, since $n_A \geq 3$ implies $s \geq 15$.] □

The classification of 5-dimensional dual-extreme lattices is not known. The only known examples are the perfect lattices and their duals as in dimensions $n \leq 4$, thus exactly six lattices up to similarity, and a pair (Λ, Λ^*) of irrational lattices discovered by A.-M. Bergé ([Ber4]), with

$$s = s^* = 10 \text{ and } \gamma'(\Lambda)^2 = \frac{2 + \sqrt{13}}{3} = 1.868\ldots$$

Conjecture 6.4.16. *We have ${\gamma'_5}^2 = 2$, and the 5-dimensional dual-critical lattices are $\mathbb{D}_5$, $\mathbb{D}_5^*$, $\mathbb{A}_5^3$ and $\mathbb{A}_5^2 \sim {\mathbb{A}_5^3}^*$.*

It should be noticed that this conjecture implies the strict inequality $\gamma'_9 < 2$ (use Corollary 2.8.10 together with an inspection of the cases when equality may hold); since the equality $\gamma'_8 = 2$ is known (see next section), we would obtain in this way an example where the strict inequality $\gamma'_{n+1} < \gamma'_n$ holds.

Applying then Theorem 2.8.7, one would obtain the upper bound $\gamma_9 \leq {\gamma'_9}^{2/9} \gamma_8^{8/9} \leq 2^{10/9} = 2,1602\ldots$, however slightly weaker than that of Rogers ([Rog]), namely $\gamma_9 \leq 2,1412\ldots$; notice that it is conjectured that $\gamma_9 = 2$, the only critical lattice being Λ_9.

The largest known value in dimension 9 of the invariant ${\gamma'}^2$ is $\frac{16}{5}$, attained on the pair of lattices $(\mathbb{A}_9^2, \mathbb{A}_9^5)$ and also on another pair discovered by Baril ([Bari]). If it were the value of ${\gamma'_5}^2$, we would have the improvement $\gamma_9 \leq 2,1072\ldots$ on Roger's bound.

We end this section with applications to dimensions 6 and 7.

Proposition 6.4.17. *1. We have $\gamma_6^6 \leq 2^{9/2}$ and $\gamma_7^7 \leq 2^{63/10}$.*

2. For $n = 6$ (resp. $n = 7$), the index in a given lattice of its sublattice generated by representatives for its successive minima is bounded from above by 4 (resp. by 8).

3. Unimodular lattices of dimension $n \leq 7$ are isometric to $\mathbb{Z}^n$.

[The bounds in (2) are optimal, as shown by the embeddings $\mathbb{A}_1^{\perp 6} \subset \mathbb{D}_6$ and $\mathbb{A}_1^{\perp 7} \subset \mathbb{E}_7$.]

Proof. By Corollary 2.3.2, we have for all $n \geq 5$ the upper bound $\gamma_n \leq 2^{3(n-1)/20}$. For $n = 6, 7$, this upper bound reduces to (1), and yields the bounds $\lfloor \gamma_6^{6/2} \rfloor \leq 4$ and $\lfloor \gamma_7^{7/2} \rfloor \leq 8$, whence (2) by Corollary 2.6.10

We now prove (3) by induction, starting with the trivial case of dimension 1. For Λ unimodular with $1 < n \leq 7$, we have $N(\Lambda) \leq \gamma_n < 2$, hence $N(\Lambda) = 1$, and any inclusion $\mathbb{Z} \subset \Lambda$ splits. We can thus write $\Lambda \simeq \mathbb{Z} \perp \Lambda'$ for some unimodular lattice Λ' of dimension $n - 1$, and we have $\Lambda' \simeq \mathbb{Z}^{n-1}$ by the induction hypothesis. □

6.5 Perfect Lattices in Dimensions 6 and 7

We have determined in Sections 6.2 and 6.4 all perfect lattices up to dimension 5, and we shall give an alternative proof in the next chapter, Sections 7.6 and 7.7. The only handy computational proof in print for larger dimensions is Barnes's 6-dimensional classification, from 1957, which relies on methods of the next chapter, and is too long to be reproduced in this book. The classification is also known for dimension 7, but all known proofs (that of Stacey ([Sta1], [Sta2]), relying on Watson's ideas, in particular on [Wat3], and which does not provide a complete identification of perfect forms, as well as its confirmation by Jaquet ([Ja2], [Ja5], completing [Ja3])) need heavy computational tools. Baranovskii and Ryshkov ([Br-R]) have written a very short note, in which they assert to have worked out the case of dimension 6; they indeed list the way perfect lattices share out among the possible index types. However, this note is too short to be considered as a published proof! I cannot imagine that one could avoid the use of a computer to deal with dimension 7. As for larger dimensions, they seem to be out of reach of presently available computers.

We consider, as previously, lattices in an n-dimensional Euclidean space E. We moreover make use of Conway and Sloane's notation in [C-S5], Section 6: the isometry classes of n-dimensional primitive, perfect lattices are denoted by P_n^i, where the upper index i runs from 1 to the number $i_{\max}$ of perfect lattices (up to similarity) of that dimension, in such a way that the Hermite invariant be a decreasing function of i. As these lattices are perfect, they are proportional to some primitive integral lattice, unique up to isometry. Its norm is a first invariant of the similarity class of the lattice.

We give in Table 6.5.1 for each $n \leq 7$ the number of perfect lattices, the number of extreme lattices, and the list of the possible norms.

Table 6.5.1. Statistics on the 48 perfect lattices of dimension $n \leq 7$.

Dimension	**1**	**2**	**3**	**4**	**5**	**6**	**7**
Perfect	1	1	1	2	3	7	33
Extreme	1	1	1	2	3	6	30
Norms	1	2	2	2	2, 4	2, 4	2, 3, 4, 6, 8, 10

Up to dimension 6, these lattices belong to infinite families which appear somewhere in this book: root lattices $\mathbb{A}_n$, $\mathbb{D}_n$, $\mathbb{E}_6$ or related lattices ($\mathbb{A}_5^3$, $\mathbb{E}_6^*$, $P_6 \simeq \mathbb{A}_6^{(2)}$; see Chapters 4, 5), or Barnes's lattices L_n^r and M_n^r to be defined later in Chapter 8 (Definitions 8.4.1, 8.4.4, 8.4.6) in connection with the Eisenstein integers, for which Conway and Sloane ([C-S5], Section 5) have found the unified notation $\mathbb{A}_{n,r'}$, whose definition we recall below:

Definition 6.5.2. *Let $r' < \frac{n}{2}$ be an integer, and let r be the integer defined by $n - r' = 2r$ or $n - r' = 2r - 1$. Let $\mathbb{A}_{n,r'} = L_n^r$ if $n - r'$ is even and $\mathbb{A}_{n,r'} = M_n^r$ otherwise.*

Hence $\mathbb{A}_{n,r'}$ is a member of the family L or of the family M according to whether $n-r'$ is even or odd. The unique non-extreme 6-dimensional perfect lattice is $\mathbb{A}_{6,2} \simeq L_6^2$, and we have $\mathbb{E}_6^* \sim \mathbb{A}_{6,0} \simeq L_6^3$.

Only nine out of the 33 perfect 7-dimensional lattices belong to one of the families above. The correspondence between the notation P_n^i and the previous families is displayed in Table 6.5.3 below.

Table 6.5.3. Correspondences between various notation.

P_1^1	P_2^1	P_3^1	P_4^1	P_4^2	P_5^1	P_5^2	P_5^3	P_6^1	P_6^2	P_6^3	P_6^4
$\mathbb{Z}$	$\mathbb{A}_2$	$\mathbb{A}_3$	$\mathbb{D}_4$	$\mathbb{A}_4$	$\mathbb{D}_5$	$\mathbb{A}_5^3$	$\mathbb{A}_5$	$\mathbb{E}_6$	$\mathbb{A}_{6,0} \sim \mathbb{E}_6^*$	$\mathbb{D}_6$	$\mathbb{A}_{6,2}$
P_6^5	P_6^6	P_6^7	P_7^1	P_7^2	P_7^3	P_7^4	P_7^5	P_7^{26}	P_7^{27}	P_7^{32}	P_7^{33}
$\mathbb{A}_6^{(2)}$	$\mathbb{A}_{6,1}$	$\mathbb{A}_6$	$\mathbb{E}_7$	$\mathbb{A}_7^4 \sim \mathbb{E}_7^*$	$\mathbb{A}_{7,1}$	$\mathbb{D}_7$	$\mathbb{A}_7^{(2)}$	$\mathbb{A}_{7,3}$	$\mathbb{A}_{7,0}$	$\mathbb{A}_{7,2}$	$\mathbb{A}_7$

We have collected in Tables 6.5.5 and 6.5.6 the main properties of the lattices P_n^i and of their duals. We refer to [Bt-M'] and to [C-S5] for definitions in terms of Gram matrices, and various data (roots, automorphism groups, ...) which are not displayed in the tables below.

We now state some results which are easy consequences of these data; lattices are defined up to similarity or up to isometry, depending on the statement.

Theorem 6.5.4.

1. *Every perfect lattice of dimension $n \le 7$ possesses a basis of minimal vectors.*
2. *All perfect lattices of dimension $n \le 7$ are even, except for $P_1^1 \simeq \mathbb{Z}$, of norm 1, and $P_7^2 \sim \mathbb{E}_7^*$, of norm 3.*
3. *For $1 \le n \le 7$, $\mathbb{A}_n$ is the least dense of the perfect lattices.*
4. *Perfect lattices of dimension $n \le 6$ are characterized by their Hermite invariant; the same holds in dimension 7 except for the two systems (P_7^4, P_7^5) and $(P_7^{26}, P_7^{27}, P_7^{28})$.*
5. *A perfect 7-dimensional lattice Λ is characterized by its Hermite invariant together with either its spectrum, or the isomorphism class of Λ^*/Λ (its Smith invariant).*
6. *A strict inclusion $\Lambda' \subset \Lambda$ between perfect lattices with same norm and same dimension $n \le 7$ is possible if and only if*
$$(\Lambda, \Lambda') \sim (P_7^1, P_7^{33}) \sim (\mathbb{E}_7, \mathbb{A}_7)\,.$$
7. *We have $\gamma_6 = 2 \cdot 3^{-1/6}$, $\gamma_7 = 2^{6/7}$, and the critical lattices in dimension 6 (resp. 7) are the lattices which are similar to $\mathbb{E}_6$ (resp. to $\mathbb{E}_7$).* □

Here are a few comments about this theorem.

(1) The fact that perfect lattices of dimension $n \le 7$ are generated by their minimal vectors was proved *a priori* by Rankin for $n \le 6$ and by Watson for $n \le 7$. In [Cs2], Csóka also proved *a priori* the slightly stronger result stated in (1), namely that these lattices possess a basis of minimal vectors. This can be verified *a posteriori* on the classification of perfect lattices, e.g. by applying an LLL reduction on the Gram matrices listed in [C-S5].

(2) The fact that $\sqrt{2}\,\mathbb{E}_7^*$ is odd implies that $\mathbb{E}_7^*$ does not possess any non-trivial perfect minimal section with the same norm; it is the only lattice having this property in the range $3 \le n \le 7$. Recall that its 28 directions of minimal vectors constitute an equiangular family of lines.

(3) That $\mathbb{A}_n$ might be for all n the less dense of all n-dimensional perfect lattices was conjectured by Coxeter in [Cox2]. We do not see any reason for this conjecture to be true, except experimentation in low dimensions.

(4) Up to dimension 5, characterizations of perfect lattices which do not involve explicitly the perfection property are known (see Sections 6.2 and 6.4). Dimension 6 can be dealt with by the same kind of arguments (see [Br-R]), but the case of dimension 7 looks much too complicated to be handled by the combinatorial methods of Korkine and Zolotareff, and Watson.

(5) The lattices $P_7^4 \sim \mathbb{D}_4$ and $P_7^5 \sim \mathbb{A}_7^{(2)} \sim P_7$ are easily distinguished by the smallest norm (2 or 4) which renders them integral, or by their s-invariant

(42 or 36), which only partially makes use of the information contained in the structure of the spectrum. The case of P_7^{26-28} is less easy: all of them are integral and primitive when they are rescaled to norm 4, and have determinant 648 and $s = 30$, but they can be distinguished by the annihilator of Λ^*/Λ, equal respectively to 18, 24 and 216.

(6) The existence of the inclusion $\mathbb{A}_7 \subset \mathbb{E}_7$ is a consequence of Coxeter's identification $\mathbb{E}_7 \simeq \mathbb{A}_7^2$. Conversely, we have $[\Lambda : \Lambda'] = \gamma(\Lambda)^{n/2}\,\gamma(\Lambda')^{-n/2}$, and it is easily verified that the inequality $\gamma(P_n^1)^{n/2}\,\left(\gamma(P_n^{i_{max}})\right)^{-n/2} \leq 2$ holds for all $n \leq 7$, and that it is strict unless $n = 7$.

Tables 6.5.5 and 6.5.6 below list the usual invariants of perfect lattices in dimension $n \leq 7$. To each primitive integral lattice, we attach the lattice $\Lambda' = \sqrt{m}\,\Lambda^*$ where m is the smallest possible integer such that Λ' is integral. The columns of the tables contain successively the integer N such that the similarity class of P_n^i contains a primitive (integral) lattice of norm N, thus defining the P_n^i up to isometry, the invariant $s = s(\Lambda)$, the determinant of Λ, then its Smith invariant (elementary divisors of (Λ^*, Λ)), whose first term is the annihilator of Λ^*/Λ; the notation $6 \cdot 3^3$ means that Λ^*/Λ is of type $(6, 3, 3, 3)$. In the fifth column, we list all possible structures for a quotient Λ/L, where L is generated by n independent minimal vectors in Λ; the notation is as above. The sixth column contains one of the letters C (for critical), E (for extreme but not critical), S for perfect and semi-eutactic but not extreme and P for perfect but not semi-eutactic. The seventh and eighth columns contain the norm $N(\Lambda')$ and the invariant $s(\Lambda') = s(\Lambda^*)$, and the last two columns the invariants $\gamma(\Lambda)$ and $\gamma(\Lambda')$ rounded down to three decimal places.

We have not displayed the value of the determinant of Λ', which is easily calculated using the formula

$$\det(\Lambda') = m^n \det(\Lambda)^{-1}.$$

As for the exact values of γ and γ', they can be obtained by the formulae

$$\gamma_n(\Lambda) = N \det(\Lambda)^{-1/n} \qquad \text{and} \qquad \gamma'_n(\Lambda) = \sqrt{NN'/m}\,.$$

Most of the data can be found in Conway and Sloane's paper [C-S5]; they have been verified and completed using programs due to Batut.

Tables 6.5.7 and 6.5.8 contain the "radiography" of perfect lattices up to dimension 7, i.e. the list (up to isometry) of perfect lattices with the same norm of the form $\Lambda \cap F$ where F is a subspace of E of dimension n' with $1 < n' < n$. Taking into account 6.5.4 (6), we then know all the strict inclusions $\Lambda' \subset \Lambda$ between perfect lattices with the same norm (just add the inclusion $P_7^{33} \subset P_7^1$). These data are extracted from the article [Bt-M]; the way they have been found is explained in Sections 13.1 and 13.2.

We end this section with a few results about eutaxy and dual-eutaxy (or dual-extremality; this amounts to the same thing since our lattices are perfect, hence dual-perfect).

By results of Chapters 4 and 5, the duals of the P_n^i are eutactic in the following cases: (a) $n \le 5$; (b) $n = 6$, $i \ne 4, 6$; (c) $n = 7$, $i = 1, 2, 4, 33$. Moreover, we know that these lattices are not weakly eutactic if $s < n$. The only undecided cases are thus those of the duals of P_7^{15}, P_7^{20}, and P_7^{27}, for which we have $s = 7, 9, 8$ respectively. We have verified that none of them is weakly eutactic. (By the way, they all have perfection rank s, and their automorphism groups act transitively on their sets of minimal vectors.)

The lattices P_n^i whose dual is eutactic happen to be themselves eutactic, hence dual-eutactic, whereas those for which $S({P_n^i}^*)$ does not generate E are not (Exercise 3.8.4). We are thus once more faced with the three lattices above. Baril has shown that P_7^{27} is dual-eutactic, but that P_7^{15} and P_7^{20} are not; P_7^{27} is the first found example of a dual-extreme lattice Λ such that one of the two lattices Λ, Λ^* is not eutactic.

Table 6.5.5. The invariants of the 15 perfect lattices of dimension ≤ 6.

	N	s	det	Smith	index	E	N'	s'	$\gamma \ge$	$\gamma' \ge$
P_1^1	1	1	1	1	1	C	1	1	1	1
P_2^1	2	3	3	3	1	C	2	3	1.154	1.154
P_3^1	2	6	4	4	1	C	3	4	1.259	1.224
P_4^1	2	12	4	2^2	$1,2$	C	2	12	1.414	1.414
P_4^2	2	10	5	5	1	E	4	5	1.337	1.264
P_5^1	2	20	4	4	$1,2$	C	4	5	1.515	1.414
P_5^2	4	15	162	$6 \cdot 3^3$	$1,2$	E	3	10	1.445	1.414
P_5^3	2	15	6	6	1	E	5	6	1.397	1.290
P_6^1	2	36	3	3	$1,2,3$	C	4	27	1.665	1.632
P_6^2	4	27	243	3^5	$1,2,3$	E	2	36	1.601	1.632
P_6^3	2	30	4	2^2	$1,2,2^2$	E	2	6	1.587	1.414
P_6^4	4	22	324	$18 \cdot 6 \cdot 3$	$1,2,3$	S	9	2	1.526	1.414
P_6^5	4	21	343	7^3	$1,2$	E	4	21	1.511	1.511
P_6^6	4	21	351	$39 \cdot 3^2$	$1,2$	E	18	2	1.506	1.358
P_6^7	2	21	7	7	1	E	6	7	1.446	1.309

6.6 Some Indications About 8-Dimensional Perfect Lattices

We first show how to characterize 8-dimensional critical lattices by making use of:

Table 6.5.6. The invariants of the 33 perfect lattices of dimension 7.

	N	s	det	Smith	index	E	N'	s'	$\gamma \geq$	$\gamma' \geq$
P_7^1	2	63	2	2	$1,2,3,4,2^2,2^3$	C	3	28	1.811	1.732
P_7^2	3	28	64	2^6	$1,2,3$	E	2	63	1.656	1.732
P_7^3	4	36	486	$18 \cdot 3^3$	$1,2,3,4$	E	9	1	1.652	1.414
P_7^4	2	42	4	4	$1,2,2^2$	E	4	7	1.640	1.414
P_7^5	4	36	512	$32 \cdot 4^2$	$1,2,3$	E	16	1	1.640	1.414
P_7^6	6	32	9216	$48 \cdot 12 \cdot 4 \cdot 2^2$	$1,2,3,4$	E	16	1	1.628	1.414
P_7^7	4	34	540	1803	$1,2,3,4$	E	100	3	1.628	1.490
P_7^8	4	32	576	$48 \cdot 6 \cdot 2$	$1,2,3,4$	E	27	4	1.613	1.5
P_7^9	10	28	356720	$7280 \cdot 7^2$	$1,2,3,4$	E	1568	2	1.609	1.467
P_7^{10}	4	32	588	$84 \cdot 7$	$1,2,3,4,2^2$	E	48	4	1.608	1.511
P_7^{11}	6	30	10080	$2520 \cdot 2^2$	$1,2,3$	E	963	4	1.607	1.514
P_7^{12}	6	28	10240	$40 \cdot 4^4$	$1,2,3$	E	15	6	1.604	1.5
P_7^{13}	8	28	76880	$2480 \cdot 31$	$1,2,3$	E	725	5	1.603	1.529
P_7^{14}	6	29	10336	$2584 \cdot 2^2$	$1,2,3$	E	912	1	1.602	1.455
P_7^{15}	8	28	77618	$394 \cdot 197$	$1,2,3$	E	115	7	1.601	1.528
P_7^{16}	6	30	10368	$72 \cdot 18 \cdot 2^3$	$1,2,3$	E	27	4	1.601	1.5
P_7^{17}	6	28	10528	$2632 \cdot 2^2$	$1,2,3,4$	E	988	2	1.597	1.500
P_7^{18}	8	28	78880	$19720 \cdot 2^2$	$1,2,3$	P	5675	2	1.597	1.517
P_7^{19}	6	28	10584	$168 \cdot 21 \cdot 3$	$1,2,3$	E	64	6	1.596	1.511
P_7^{20}	6	28	10658	$146 \cdot 73$	$1,2,3$	E	59	9	1.595	1.557
P_7^{21}	6	28	10752	$336 \cdot 16 \cdot 2$	$1,2,3$	E	124	4	1.593	1.488
P_7^{22}	4	30	630	$210 \cdot 3$	$1,2,3$	E	108	2	1.592	1.434
P_7^{23}	6	28	10780	$1540 \cdot 7$	$1,2,3$	E	560	1	1.592	1.477
P_7^{24}	6	28	10808	10808	$1,2,3$	E	3920	1	1.591	1.475
P_7^{25}	6	28	11008	$688 \cdot 4^2$	$1,2,3$	E	248	1	1.587	1.470
P_7^{26}	4	30	648	$18 \cdot 6^2$	$1,2,3$	S	9	3	1.586	1.414
P_7^{27}	4	30	648	$24 \cdot 3^3$	$1,2,3$	E	13	8	1.586	1.471
P_7^{28}	4	30	648	$216 \cdot 3$	$1,2,3$	E	112	3	1.586	1.440
P_7^{29}	4	29	684	$228 \cdot 3$	$1,2,3$	P	108	1	1.574	1.376
P_7^{30}	4	28	686	$98 \cdot 7$	$1,2,3$	E	49	1	1.573	1.414
P_7^{31}	4	28	720	$240 \cdot 3$	$1,2,3$	E	117	2	1.562	1.396
P_7^{32}	4	28	756	$84 \cdot 3^2$	$1,2$	E	39	2	1.551	1.362
P_7^{33}	2	28	8	8	1	E	7	8	1.485	1.322

Table 6.5.7. Perfect sections of perfect lattices ($n \leq 6$).
(The symbol $\times$ indicates that the dimension must be disregarded.)

P_3^1	P_2^1	$\times$	$\times$	$\times$
P_4^1	P_2^1	P_3^1	$\times$	$\times$
P_4^2	P_2^1	P_3^1	$\times$	$\times$
P_5^1	P_2^1	P_3^1	P_4^1, P_4^2	$\times$
P_5^2	P_2^1			$\times$
P_5^3	P_2^1	P_3^1	P_4^2	$\times$
P_6^1	P_2^1	P_3^1	P_4^1, P_4^2	P_5^1, P_5^3
P_6^2	P_2^1			P_5^2
P_6^3	P_2^1	P_3^1	P_4^1, P_4^2	P_5^1, P_5^3
P_6^4	P_2^1	P_3^1		P_5^2
P_6^5	P_2^1	P_3^1		
P_6^6	P_2^1	P_3^1		P_5^2
P_6^7	P_2^1	P_3^1	P_4^2	P_5^3

1. the characterization of 7-dimensional critical lattices as it results, e.g. from the Jaquet–Stacey classification,

and *one* of the following two results:

2 . the fact (first proved by Mordell) that a unimodular 8-dimensional lattice of norm $N \geq 2$ is isometric to $\mathbb{E}_8$,

or

2′. the characterization (due to Barnes) of 6-dimensional critical lattices, completing in this way all that concerns $\gamma_6, \gamma_7, \gamma_8$ without Blichfeldt's and Vetchinkin results (but with a heavy computer program for dimension 7).

Theorem 6.6.1. *We have $\gamma_8 = 2$, and the corresponding critical lattices are those which are similar to $\mathbb{E}_8$.*

Proof. The calculation of γ_8 is an immediate consequence of Mordell's inequality: we have on the one hand $\gamma_7 = 2^{6/7}$ (Theorem 6.5.4), hence $\gamma_8 \leq \gamma_7^{7/6}$ (Corollary 2,3.2), whence $\gamma_8 \leq 2$, and on the other hand, $\gamma_8 \geq \gamma(\mathbb{E}_8) = 2$ since $\mathbb{E}_8$ is unimodular of norm 2.

We now consider a lattice Λ of norm 2 with $\gamma(\Lambda) = 2$ and consequently determinant 1. Since Mordell's inequality is an equality, the orthogonal in Λ of a minimal vector in Λ^* is a 7-dimensional critical lattice, and Λ^* is also critical. Hence the orthogonal in Λ^* of a minimal vector in Λ is also a 7-dimensional critical lattice (Theorem 2.3.1). We also have $\det(\Lambda^*) = \det(\Lambda)^{-1} = 1$ and hence $N(\Lambda^*) = 2$ since Λ^* is critical.

Table 6.5.8. Perfect sections of perfect lattices ($n = 7$).

P_7^1	P_2^1	P_3^1	P_4^1, P_4^2	P_5^1, P_5^3	P_6^1, P_6^3, P_6^7
P_7^2					
P_7^3	P_2^1	P_3^1		P_5^2	P_6^2, P_6^4, P_6^6
P_7^4	P_2^1	P_3^1	P_4^1, P_4^2	P_5^1, P_5^3	P_6^3, P_6^7
P_7^5	P_2^1	P_3^1	P_4^1, P_4^2	P_5^1	
P_7^6	P_2^1	P_3^1			
P_7^7	P_2^1	P_3^1	P_4^1, P_4^2	P_5^2	P_6^4, P_6^6
P_7^8	P_2^1	P_3^1	P_4^1, P_4^2	P_5^2	P_6^4
P_7^9	P_2^1	P_3^1			
P_7^{10}	P_2^1	P_3^1	P_4^1, P_4^2		P_6^5
P_7^{11}	P_2^1	P_3^1	P_4^1, P_4^2		
P_7^{12}	P_2^1				
P_7^{13}	P_2^1	P_3^1			
P_7^{14}	P_2^1	P_3^1			
P_7^{15}	P_2^1	P_3^1			
P_7^{16}	P_2^1	P_3^1	P_4^1		
P_7^{17}	P_2^1	P_3^1			
P_7^{18}	P_2^1	P_3^1			
P_7^{19}	P_2^1	P_3^1	P_4^2		
P_7^{20}	P_2^1	P_3^1			
P_7^{21}	P_2^1	P_3^1	P_4^2		
P_7^{22}	P_2^1	P_3^1		P_5^2	P_6^4, P_6^6
P_7^{23}	P_2^1	P_3^1			
P_7^{24}	P_2^1	P_3^1			
P_7^{25}	P_2^1	P_3^1			
P_7^{26}	P_2^1	P_3^1	P_4^2	P_5^2	P_6^4
P_7^{27}	P_2^1	P_3^1		P_5^2	P_6^4
P_7^{28}	P_2^1	P_3^1	P_4^1, P_4^2	P_5^2	P_6^6
P_7^{29}	P_2^1	P_3^1	P_4^2	P_5^2	P_6^4, P_6^6
P_7^{30}	P_2^1	P_3^1			P_6^5
P_7^{31}	P_2^1	P_3^1	P_4^2	P_5^2	P_6^6
P_7^{32}	P_2^1	P_3^1	P_4^2	P_5^2	P_6^6
P_7^{33}	P_2^1	P_3^1	P_4^2	P_5^3	P_6^7

Let e and e' be minimal vectors in Λ and Λ^* respectively, let $M = \Lambda \cap (\mathbb{R}e')^\perp$ and $M' = \Lambda \cap (\mathbb{R}e)^\perp$, and suppose that e and e' are neither orthogonal nor proportional; such a pair (e, e') exists, for the perfect lattice Λ contains eight independent minimal vectors outside any hyperplane section. By (1), since the lattices M and M' are critical, they are similar to $\mathbb{E}_7$. Since $\det(M) = \det(\Lambda)\,N(e) = 2$, M is indeed isometric to $\mathbb{E}_7$. In particular, M is an integral lattice.

Lemma 6.6.2. *The lattice Λ is generated by M and e.*

Proof of 6.6.2. Since e is not orthogonal to e', it generates together with M an 8-dimensional lattice. Since e is not orthogonal to M (because e and e' are not proportional), the Hadamard inequality yields the strict inequality $\det(\Lambda') < \det(M)\,N(e) = 4$, hence the upper bound $[\Lambda : \Lambda']^2 = \frac{\det(\Lambda')}{\det(\Lambda)} < 4$. We thus have $[\Lambda : \Lambda'] = 1$. □

Lemma 6.6.3. *Λ is an integral lattice.*

Proof of 6.6.3. After Lemma 6.6.2, it suffices to prove that the scalar products $e \cdot x$ are integral for all $x \in M$. Let $x \in M$. Write $x = y + \lambda e$ with $y \in (\mathbb{R}e)^\perp$. We notice that y belongs to the orthogonal projection of Λ onto $(\mathbb{R}e)^\perp$, i.e. to M'^* (Proposition 1.3.4). Since $\det(M') = 2$, $\sqrt{2}\,M'^*$ is an integral lattice. We have

$$N(x) = N(y) + \lambda^2 N(e) = N(y) + 2\lambda^2 \quad \text{and} \quad e \cdot x = \lambda e \cdot e = 2\lambda\,,$$

which shows that $(e \cdot x)^2 = 4\lambda^2 = 2N(x) - 2N(y)$ is an integer. Now the scalar products of the vectors in Λ are rational numbers since Λ is a perfect lattice whose norm is rational (Proposition 3.2.11). Hence $e \cdot x$ is itself an integer. □

Proof of 6.6.1 (end). We now know that Λ is an integral lattice of norm 2 and determinant 1. This completes the proof of Theorem 6.6.1 under the hypotheses (1) and (2).

Otherwise, consider the plane lattice with basis (e, e'). The scalar product $e \cdot e'$ is a nonzero integer, and we have the strict inequality $|e \cdot e'| < \|e\| \cdot \|e'\|$, i.e. $|e \cdot e'| < 2$. Hence $e \cdot e' = \pm 1$, and the plane lattice $\mathbb{Z}e + \mathbb{Z}e'$ is isometric to $\mathbb{A}_2$. Since e and e' belong to $\Lambda = \Lambda^*$, its orthogonal in Λ is a 6-dimensional lattice $N \subset M$, with determinant $\det(\Lambda)\,\det(\mathbb{A}_2) = 3$ and norm ≥ 2. It thus follows from Barnes's classification of 6-dimensional perfect lattices that N is critical, hence isometric to $\mathbb{E}_6$ since $\det(N) = 3$. It is thus possible to choose a Coxeter basis $(e_2, e_3, \dots, e_8)$ for M such that $(e_3, \dots, e_8)$ is also a Coxeter basis for N. To these bases are attached the Dynkin diagrams of the respective root systems $\boldsymbol{E}_7$ and $\boldsymbol{E}_6$. We have $e \cdot e_i = 0$ for $i \geq 3$, $e \cdot e_2 \neq 0$ since Λ must be irreducible, and $|e \cdot e_2| < 2$ since e and e_2 are not proportional (otherwise, e would belong to M), hence $e \cdot e_2 = \pm 1$. Set $e_1 = -e$ if $e \cdot e_1 = +1$

and $e_1 = e$ if $e \cdot e_1 = -1$. We have $e_1 \cdot e_i = 0$ for $i \geq 3$ and $e_1 \cdot e_2 = 1$. This shows that the basis $(e_1, e_2, e_3, \ldots, e_8)$ for Λ defines the Dynkin diagram of the $\boldsymbol{E}_8$ system, hence that Λ is isometric to $\mathbb{E}_8$. □

Corollary 6.6.4. *We have $\gamma'_8 = 2$, and the corresponding dual-critical lattices are those which are similar to $\mathbb{E}_8$.* □

The remainder of this section is devoted to an overview of what is known about 8-dimensional perfect lattices.

The classification of perfect 8-dimensional lattices containing a perfect minimal section of dimension 7 is known:

Theorem 6.6.5. (Laïhem.) *Up to similarity, there are* 1175 8*-dimensional perfect lattices possessing a perfect hyperplane section with the same norm.* □

Laïhem ([Lah]) solved completely the case of a section not similar to a root lattice, using a variant of the Voronoi algorithm to be described in Chapter 13. He left aside those which had a section $\mathbb{A}_7$, $\mathbb{D}_7$ or $\mathbb{E}_7$, which have been dealt with later by Baril ([Bari]), thanks to an improvement of the algorithms, which allowed him to shorten the time needed to handle lattices which are contiguous to $\mathbb{E}_8$ in the sense of Chapter 13.

The lattices lying above the root lattice $\mathbb{E}_7$ (resp. $\mathbb{D}_7$, resp. $\mathbb{A}_7$) are $\mathbb{E}_8$ and $\mathbb{A}_8^2$ (resp. $\mathbb{E}_8$ and $\mathbb{D}_8$, resp. $\mathbb{E}_8$, $\mathbb{A}_8^2$, $\mathbb{D}_8$ and $\mathbb{A}_8$); see Section 5.1 for the definition of the Barnes's lattice $\mathbb{A}_8^2$.

Jaquet ([Ja7]) has verified that among the 1175 Laïhem lattices, only 383 are extreme, 21 being moreover semi-eutactic without being eutactic.

Jaquet ([Ja4]) has also determined the lattices contiguous to $\mathbb{D}_8$ in the sense of next chapter. His list contains 48 lattices, among which one finds the 3 irreducible root lattices, 44 Laïhem lattices, and a new lattice whose perfect minimal sections are similar to $\mathbb{A}_n$ for all $n \leq 6$.

Another family of lattices has been classified by Baril: those which possess a basis $(e_1, \ldots, e_8)$ such that $(e_1, \ldots, e_6)$ and (e_7, e_8) generate perfect lattices (the last one of course similar to $\mathbb{A}_2$). Also using methods of Chapter 13, he obtained the following result:

Theorem 6.6.6. (Baril.) *There are up to similarity exactly* 53 *perfect* 8*-dimensional lattices which are a direct sum of two perfect lattices with the same norm and respective dimensions* 6 *and* 2*, and which do not contain any perfect* 7*-dimensional section (with the same norm).*
[Baril's list contains Jaquet's lattice.] □

Then Napias ([Nap2]) has made an exploration using the Voronoi algorithm of Chapter 7. She started with the $1175 + 53 = 1228$ lattices of Theorems 6.6.5 and 6.6.6, picked out in this list those with $s = 36$ (for which the algorithm of contiguity is simpler), found all their Voronoi neighbours, added the new ones to the starting list, performed again the same calculations with the new lattices having $s = 36$, and so on, until the algorithm stopped. She finally found altogether **10770** perfect 8-dimensional lattices, a number which looks amazingly large when compared with the corresponding numbers in lower dimensions.

More recently, Napias's search was extended by Batut, who considered lattices with $s = 37, 38, 39$. He found in this way 146 new perfect lattices. In June 2000, **10916** perfect 8-dimensional lattices were known.

This list cannot be expected to be a complete one, for a few lattices might well be connected only with, say, $\mathbb{E}_8$, as $\mathbb{E}_7^*$ is only connected with $\mathbb{E}_7$; note that more than one half of the 10916 lattices are connected with $\mathbb{E}_8$. It seems plainly impossible to classify 8-dimensional perfect lattices.

The list of known perfect 8-dimensional lattices suggests that Coxeter's conjecture (according to which $\mathbb{A}_n$ should be the less dense among perfect lattices) holds in dimension 8. It also suggests the following conjecture, announced as a theorem by Csóka in [Cs2]:

Conjecture 6.6.7. *Every perfect 8-dimensional lattice possesses a basis of minimal vectors.*

[Such a result no longer holds from dimension 9 onwards, as shown for n odd (resp. even) by the lattices $\mathbb{A}_n^2$ (resp. $\mathbb{D}_n^+$). Notice, however, that because of the lower bound $s \geq \frac{n(n+1)}{2}$, bases of minimal vectors might well exist up to high dimensions for perfect lattices *which are generated by their minimal vectors*.]

The evolution of the characteristics of perfect lattices in dimensions growing from 2 to 8 (as far as we can have a good guess of what happens in dimension 8), suggests the following asymptotic conjecture:

Conjecture 6.6.8. *For n tending to infinity, almost all perfect lattices Λ have invariants $s(\Lambda) = \frac{n(n+1)}{2}$, $s(\Lambda^*) = 1$, and* $\mathrm{Aut}(\Lambda) = \{\pm \mathrm{Id}\}$.

One can reasonably conjecture more, for instance that for any given integer k, almost all n-dimensional perfect lattices have no perfect section with the same norm in codimension $n' \leq k$.

A look at the lists also suggests that almost no perfect lattice is eutactic, a somewhat natural fact: as far as we have both $s(\Lambda) = \frac{n(n+1)}{2}$ and $\mathrm{Aut}(\Lambda) = \{\pm \mathrm{Id}\}$, the eutaxy coefficients, which are unique, and which can be expected to have a more or less random distribution except for a unique relation (their sum is equal to $\frac{1}{n}$) should often take negative values. (Among Laïhem lattices, less that one third are eutactic.) The eutaxy condition thus appears to be a more severe restriction that could have been expected from the results in dimensions $n \leq 7$.

The set of norms of the 10916 known perfect 8-dimensional lattices (rescaled so as to be primitive) is the set of even integers in the interval

$[2, 32]$, except 28.

[*A priori* estimates of the largest norm of a primitive, integral perfect lattice have been given by Watson in [Wat6]; see also Rankin ([Ran2]) for a related problem; however, Watson's estimates are far beyond the expected maximal value of the norm.]

A very plausible conjecture is that all perfect 8-dimensional lattices are even; in particular, their norms should be even.

As for the invariant s, it takes all values in the interval $[36, 52]$, and moreover the four values 54, 58, 71, 120. Disregarding the case of $\mathbb{E}_8$, we conjecture that Barnes's lattice $\mathbb{A}_8^2$ is the only *perfect* lattice with $s \geq 71$. We also conjecture that $\mathbb{A}_8^2$ is the densest perfect 8-dimensional lattice which is not similar to E_8 (with $\gamma(\mathbb{A}_8^2) = 2^{5/4}3^{-1/4} = 1.807...$).

[**Warning:** Watson has proved in [Wat5] that the maximum value of s on *non-perfect* lattices is $s = 75 > s(\mathbb{A}_8^2) = 71$.]

6.7 Exercises for Chapter 6

6.1.1. Let Λ be a 3-dimensional lattice possessing 3 independent minimal vectors e_1, e_2, e_3.

1. Show that (e_1, e_2, e_3) is a basis for Λ.
2. Show that the components on (e_1, e_2, e_3) of minimal vectors of Λ are equal to 0 or ± 1.
3. Show that Λ contains at most one pair $\pm x$ of minimal vectors of the form $\pm e_1 \pm e_2 \pm e_3$.
4. Show that if $s(\Lambda) \geq 5$, it is possible to choose e_1, e_2, e_3 in such a way that $S(\Lambda) = \{\pm e_1, \pm e_2, \pm e_3, \pm(e_1 - e_2), \pm(e_2 - e_3)\}$.
5. Use this to give a direct proof of Theorem 6.2.1 for $n = 3$.

6.1.2. Prove that the number of sets of n independent vectors contained in a half-system of minimal vectors in $\mathbb{A}_n$ is equal to $(n+1)^{n-1}$. [To prove this result (which was conjectured by Batut), apply Exercise 3.2.2.]

6.3.1. With the notation of Lemmas 6.3.5 and 6.3.6, consider for $t \in [\frac{1}{3}, \frac{1}{2}]$ the lattice Λ_t of Lemma 6.3.6. Let $t_0 = \sqrt{2} - 1$ and let Λ_0 be the lattice corresponding to t_0.

1. Show that for t increasing from $\frac{1}{3}$ to $\frac{1}{2}$, the parameter t^* such that $(1+t)(1+t^*) = 2$ decreases from $\frac{1}{2}$ to $\frac{1}{3}$.
2. Show that the function $\gamma_3'(\Lambda_t)$ decreases on $[\frac{1}{3}, t_0]$ and increases on $[t_0, \frac{1}{2}]$.
3. Show that the linear map
$(e_1(t), e_2(t), e_3(t)) \mapsto (e_2^*(t^*), (e_1^* - e_2^*)(t^*), -e_3(t^*))$ is a similarity of Λ_t onto $\Lambda_{t^*}^*$.
[Represent L_t by the Gram matrix $A(t) = \begin{pmatrix} 1 & -t & -t \\ -t & 1 & -1+2t \\ -t & -1+2t & 1 \end{pmatrix}$ and verify that
$$\frac{4t(1-t)}{1+t} A(t^*)^{-1} = \begin{pmatrix} 2-2t & 1-t & 1-t \\ 1-t & 1 & t \\ 1-t & t & 1 \end{pmatrix}.]$$
4. Show that for $t = \frac{1}{3}$ (resp. $t = \frac{1}{2}$), Λ_t is similar to $\mathbb{A}_3^*$ (resp. to $\mathbb{A}_3$), and that Λ_0 is similar to its dual lattice.

6.3.2. With the notation above, denote moreover by p_i, q_j, q'_j ($i = 1,2,3,4$; $j = 1,2$) the orthogonal projections onto e_i, f_j, f'_j, so that Λ is dual-eutactic if and only if there exists a relation of the form $\sum_{i=1}^4 p_i = N(\Lambda^*)[\lambda\,(q_1+q'_1)+\mu\,(q_2+q'_2)]$ with $\lambda,\ \mu > 0$.

1. Show that the subgroup $G = \{1, (12)(34), (13)(24), (14)(23)\}$ of S_4 stabilizes $S(\Lambda)$.

2. Show that $G' = \{\mathrm{Id}\} \times G$ is a subgroup of $\mathrm{Aut}(\Lambda)$ which acts transitively on $S(\Lambda)$ and has two orbits on $S(\Lambda^*)$, namely $\{\pm f_1, \pm f'_1\}$ and $\{\pm f_2, \pm f'_2\}$.

3. Prove that $(\sum_{i=1}^4 p_i)(e_4)$ is of the form $t_1e_1 + t_2e_2 + t_3e_3 + e_4$, and verify the two equalities $(q_1 + q'_1)(e_4) = -\frac{1}{N(\Lambda^*)}e_3^*$ and $(q_2 + q'_2)(e_4) = -\frac{1}{N(\Lambda^*)}e_2^*$.

4. Calculate the scalar products with e_4 of the right–hand sides of the equalities above; use this to deduce the three formulae $2(t_1 + t_2t_3) = 0$, $2(t_3 + t_1t_2) = -\lambda$, and $2(t_2 + t_1t_3) = -\mu$.

5. Show that Λ_t is dual-eutactic if and only if $t = t_0$.

6.3.3. We keep the notation of the preceding two exercises.

1. Let $a_1, a_2, a_3, a_4 \in \mathbb{R}$ and let $u = a_1p_1 + a_2p_2 + a_3p_3 + a_4p_4$. Show that $u(e_i^*) = a_ie_i - a_4e_4$.

2. Show that p_1, p_2, p_3, p_4 are independent in $\mathrm{End}^s(E)$ and that the vector space they span does not contain q'_1 nor q'_2.

3. Show by considering the image of e_1 that a relation of the form
$$N(\Lambda^*)(\alpha q'_1 + \beta q'_2) = a_1p_1 + a_2p_2 + a_3p_3 + a_4p_4$$
implies $a_1 = (2t-1)\,d$ and $b = c = -\frac{2t-1}{t}\,d$.

4. Calculate the scalar product of e_1 with the image of e_2 using both sides of the equality above, and show that such an equality cannot hold unless $t = t_0$.

5. Show that Λ_t is dual-perfect for all $t \neq t_0$, but that Λ_0 is not; verify more precisely that the set of projections onto $S(\Lambda) \cup S(\Lambda^*)$ has perfection rank 5 if $t = t_0$.

6.3.4. Show that the group $\mathrm{Aut}(\Lambda_t)$ is the semi-direct product $\{\pm\,\mathrm{Id}\} \times H$, where H is the 2-Sylow subgroup of S_4 generated by the group G of Exercise 6.3.2, Question 1 and the transposition $(2,3)$.

6.3.5. (After Conway and Sloane, [C-S8].) Let $n \geq 3$ be an integer. In $\mathbb{R}^n$ endowed with its canonical basis $(\varepsilon_1, \varepsilon_2, \dots, \varepsilon_n)$, one considers the lattice $\mathbb{D}_n$ generated by its minimal vectors $\pm\varepsilon_i \pm \varepsilon_j$, $1 \leq i < j \leq n$. For $\lambda > 0$, let Λ_λ be the image $\mathbb{D}_n$ by the orthogonal affinity $(\varepsilon_1, \varepsilon_2, \dots, \varepsilon_{n-1}, \varepsilon_n) \mapsto (\varepsilon_1, \varepsilon_2, \dots, \varepsilon_{n-1}, \sqrt{\lambda}\,\varepsilon_n)$. For $1 \leq i \leq n-1$, let $e_i = \varepsilon_i + \lambda\varepsilon_n$ and $e'_i = -\varepsilon_i + \lambda\varepsilon_n$.

1. Show that the smallest vectors in Λ_λ having a nonzero projection onto $\mathbb{R}^{n-1}$ are the $2(n-1)$ pairs $\pm e_i, \pm e'_i$.

2. Deduce from this that $S(\Lambda_\lambda) = \{\pm e_i, \pm e'_i\}$ for $\frac{1}{3} < \lambda < 1$ and $N(\Lambda_\lambda) = \lambda + 1$ for $\frac{1}{3} \leq \lambda \leq 1$.

3. Show that $s(\Lambda_{1/3}) = 2n - 1$.

4. What is the spectrum of Λ_λ in the interval $(\frac{1}{3}, 1)$?

5. Suppose that $n = 3$ and that $\frac{1}{2} \leq \lambda \leq 1$. Show that
$$\mathcal{B} = (\varepsilon_1 + \varepsilon_2, \varepsilon_1 - \varepsilon_2 . \varepsilon_1 + \lambda\varepsilon_3)$$
is a basis for Λ_λ whose dual basis is
$$\mathcal{B}^* = \left(\tfrac{\varepsilon_1+\varepsilon_2-\sqrt{\lambda^{-1}}\varepsilon_3}{2}, \tfrac{\varepsilon_1-\varepsilon_2-\sqrt{\lambda^{-1}}\varepsilon_3}{2}, \sqrt{\lambda^{-1}}\varepsilon_3\right);$$

use this to deduce that on $\left(\frac{1}{2}, 1\right)$, we have $S(\Lambda_\lambda^*) = \left\{\frac{\pm\varepsilon_1 \pm \varepsilon_2 \pm \sqrt{\lambda^{-1}}\varepsilon_3}{2}\right\}$.

6. Show that Λ_λ is similar to the lattice with parameter $t = \frac{\lambda}{\lambda+1}$ defined in the proof of Theorem 6.3.4, in particular in Lemma 6.3.6.
[N.B. The fact that the vectors $e_1, e_1', -e_2, -e_2'$ add to zero suffices to prove that the family Λ_λ coincides with the family we used in the proof of Theorem 6.3.4.]

6.3.6. Let $M \subset \mathbb{R}^2$ be a lattice with at least two minimal vectors of norm 1. We study in this exercise norm 1 lattices $\Lambda \subset \mathbb{R}^3$ whose determinant is a local minimum among those which contain M as a hyperplane section. (This family contains the family of laminated lattices above M.) We denote by e_1, e_2 two minimal vectors in M whose angle is some $\theta \in \left[\frac{\pi}{3}, \frac{\pi}{2}\right]$.

1. Show that Λ contains three pairs of minimal vectors outside M.

2. Show that for a suitable choice of $e_3 \in S(\Lambda)$, we have
$S(\Lambda) = \{\pm e_1, \pm e_2, \pm e_3, \pm(e_1 - e_3), \pm(e_2 - e_3)\}$. [Use Exercise 6.1.1.]

3. Show that the parameter θ defines up to isomorphism a unique norm 1 lattice with $s(\Lambda) \geq 5$; verify that its determinant is equal to $\frac{1}{2}(1 + \cos\theta - 2\cos^2\theta)$.

4. Show that $\det(\Lambda)$ is minimal for $\theta = \frac{\pi}{3}$ and $\theta = \frac{\pi}{2}$, the two cases in which Λ is critical, and possesses on the interval $\left[\frac{\pi}{3}, \frac{\pi}{2}\right]$ a unique maximum, equal to $\frac{9}{16}$, and attained for $\theta = \cos^{-1}\frac{1}{4}$.

5. Show that the lattice corresponding to $\theta = \cos^{-1}\frac{1}{4}$, once rescaled to norm 4, is integral of determinant 36, that it can be defined by the Gram matrix $\begin{pmatrix} 4 & 2 & 1 \\ 2 & 4 & 2 \\ 1 & 2 & 4 \end{pmatrix}$, and that it is similar to the lattice Λ_λ of Exercise 6.3.5 for $n = 3$ and $\lambda = \frac{1}{3}$; prove that it is moreover eutactic.

6.3.7. Let Λ be 3-dimensional with $S(\Lambda) = \{\pm e_1, \pm e_2, \pm e_3, \pm e_4\}$, where $e_1 + e_2 + e_3 + e_4 = 0$.

1. Show that $\sigma : (e_1, e_2, e_3) \mapsto (-e_2, -e_1, -e_4,)$ is the reflection with respect to the hyperplane $\mathbb{R}(e_1 + e_2)^\perp = \langle e_1 - e_2, e_1 + e_3, e_2 + e_3\rangle$ and verify that Λ is invariant under σ.

2. Show that $s(\Lambda^*)$ is an even integer.

6.4.1. The aim of this exercise is to make Corollary 6.4.15 explicit. We consider a basis $(e_1, \dots, e_5)$ for E with $N(e_i) = 2$ and $e_j \cdot e_k = 0$ for $j < k \leq 4$, and the lattice Λ generated by $\Lambda' = \langle e_1, \dots, e_5\rangle$ and $e = \frac{e_1+e_2+e_3+e_4}{2}$. We suppose that Λ possesses exactly four pairs of minimal vectors outside $\Lambda = \langle e, e_2, e_3, e_4\rangle \simeq \mathbb{D}_4$, including $\pm e_5$. Let $e' = \frac{e_1-e_2+e_3+e_4}{2}$ and $e'' = \frac{e_1-e_2-e_3-e_4}{2}$.

1. Show that, up to an automorphism of L, we may assume that these pairs are represented by e_5, $e_5 + x$, $e_5 + y$, and $e_5 + z$ where x, y, z belong to distinct systems of pairwise orthogonal lines as defined in Remark 6.2.5.

2. Show that we can take $x = e_5$, $y = \pm e$, and $z = \pm e'$ or $\pm e''$.

3. Show that if $e_5 + e_1$ is minimal, then the vectors of the form $v = e_5 - \frac{e_1 \pm e_2 \pm e_3 \pm e_4}{2}$ are not. [Evaluate $e_5 \cdot v$ and observe that more minimal vectors would exist.]

4. Show that one must have $x = e_5$, $y = e$, and $z = e'$.

5. Show that Gram matrices of (e, e_2, e_3, e_4, e_5) are those which are of the form

$$A(t) = \begin{pmatrix} 2 & 1 & 1 & 1 & -1 \\ 1 & 2 & 0 & 0 & 0 \\ 1 & 0 & 2 & 0 & -t \\ 1 & 0 & 0 & 2 & -1+t \\ -1 & 1 & -t & -1+t & 2 \end{pmatrix}$$

with $0 < t < 1$.

6. Show that $L' = \langle e, e_2, e_3+e_4, e_5 \rangle$ is isometric to $\mathbb{A}_4$, that $S(\Lambda) = S(L) \cup S(L')$, and that $\det(\Lambda) = 4(1+t-t^2)$.

7. Show that Λ is eutactic if (and indeed only if) $t = \frac{1}{2}$.

8. Show that the lattices Λ corresponding to t and t' are isometric if and only if $t' = t$ or $t' = 1 - t$.

[As noticed by Baril and Bergé, Question 7 is a counter-example to a conjecture proposed by Coxeter in [Cox2], according to which any eutactic lattice with $s \geq \frac{n(n+1)}{2}$ should be extreme, a result which is true in dimension $n \leq 4$. A counter-example in dimension 32 is given in [C-S5].]

6.4.2. Show that a dual-extreme 5-dimensional lattice which possesses a minimal $\mathbb{D}_4$- (resp. $\mathbb{A}_4$)-section is similar to $\mathbb{D}_5$ (resp. to $\mathbb{D}_5$ or to $\mathbb{A}_5$). [Use Corollary 5.4.15.]

6.5.1. Let $\Lambda \subset \mathbb{R}^6$ be a lattice of norm 2 whose section Λ' by $\mathbb{R}^5$ is the lattice $\mathbb{D}_5$, generated by its minimal vectors $\pm\varepsilon_i \pm \varepsilon_j$, $1 \leq i < j \leq 5$. Recall the upper bound $\gamma_6 \leq 2^{3/4}$ (Proposition 6.4.17).

1. Let $e \in S(\Lambda)$ be of the form $\sum_{i=1}^{5} \lambda_i \varepsilon_i + \mu \varepsilon_6$ with $\mu > 0$. Show that the double inequality $2^{-1/4} \leq \mu \leq 2^{1/2}$ holds.

2. Use this to deduce that $\Lambda = \langle \Lambda', e \rangle$, and that μ does not depend on the choice of e.

3. Show that if $e = x + \mu \varepsilon_6$ and $e' = x' + \mu \varepsilon_6$ both belong to $S(\Lambda) \smallsetminus \Lambda'$, then $N(x' - x) = 2$ or $N(x' - x) = 4$.

4. We suppose in this question that $S(\Lambda)$ contains at least six vectors $\pm e_i$, $1 \leq i \leq 6$ outside $S(\Lambda')$ with a positive projection on $\mathbb{R}\varepsilon_6$. Show that there exists a pair (i, j) of indices such that $N(e_i - e_j) = 4$, and that if $e_i - e_j$ is of the form $2\varepsilon_k$, Λ is then isometric to $\mathbb{D}_6$.

5. We suppose in what follows that none of the vectors $e_i - e_j$ is of the form $2\varepsilon_k$. Show that up to a permutation of $\varepsilon_1, \dots, \varepsilon_5$, we may assume that $e_2 - e_1 = \varepsilon_1 + \varepsilon_2 + \varepsilon_3 + \varepsilon_4$.

6. Show that this is impossible unless e_1 is of the form

$$e_1 = -\tfrac{1}{2}(\varepsilon_1 + \varepsilon_2 + \varepsilon_3 + \varepsilon_4) + \lambda_5 \varepsilon_5 + \mu \varepsilon_6 .$$

[Consider the vectors $e_1 \pm (\varepsilon_i - \varepsilon_j)$, $1 \leq i, j \leq 4$.]

7. Construct a 1-parameter non-perfect family of 6-dimensional lattices possessing a hyperplane section with the same norm which is similar to $\mathbb{D}_5$ and moreover eight pairs of minimal vectors outside this section.

8. Show that a lattice having a hyperplane minimal section similar to $\mathbb{D}_5$ and possessing six *independent* minimal vectors outside this section is similar to $\mathbb{E}_6$ or to $\mathbb{D}_6$. [First show that under the hypotheses of Question 5, there exists up to a permutation of $\varepsilon_1, \dots, \varepsilon_5$ a minimal vector e_3 of one of the forms $e_1 + \varepsilon_1 + \varepsilon_5$ or $e_1 + \varepsilon_1 + \varepsilon_2 + \varepsilon_3 + \varepsilon_5$.]

[After Watson ([Wat5]), the lattices in Question 7 are characterized by the equality $s = 28$, and the inequality $s \geq 29$ holds only for lattices similar to D_6 or $\mathbb{E}_6$. These results also follow from the description of the 6-dimensional Voronoi graph to be given in the next chapter.]

6.5.2. (Watson.) Let $(e_1, \dots, e_n)$ be a basis for E all vectors of which have the same norm N, let Λ' be the lattice that they generate, and let $\Lambda = \langle \Lambda', e \rangle$ for some vector $e = \frac{e_1+\cdots+e_p+2e_{p+1}+\cdots+2e_q}{4}$. Suppose that $N(\Lambda) = N$.

1. Show that $p \geq 4$, and that moreover $q \geq 3$ whenever $p = 4$.
[Consider $e' = \frac{e_1+\cdots+e_p}{2}$.]

2. Prove that we have $n \geq 7$, and $p \in \{4, 5, 6\}$ if $n = 7$. [Use Theorem 6.4.6.]

3. Let $n = 6$ and let $\Lambda \supset \Lambda'$ be such that Λ' is generated by six independent minimal vectors in Λ. Show that $[\Lambda : \Lambda'] = 4$ occurs if and only if $(\Lambda, \Lambda') \sim (\mathbb{D}_6, \mathbb{A}_1^{\perp 6})$.

6.8 Notes on Chapter 6

When Korkine and Zolotareff began their research on the Hermite constant, its precise value was known only up to dimension 3. They worked in Saint Petersburg, a university where number theory was an active field of research, thanks in particular to Tchebychev (Zolotareff was in theory a student of Korkine who himself was a student of Tchebychev). They first solved in [K-Z1] (1872) the case of dimension 4 (without determining the quadratic forms on which the value of γ_4 is attained), by proving an inequality now known as the *Mordell inequality* (see the notes on Chapter 2) which allowed them to deduce the exact value of γ_4 from the previously known value (Gauss) of γ_3. Such a method does not yield a sharp upper bound for γ_5.

The notion of an extreme form appears at the very beginning of their second article ([K-Z2]), where they produce (without giving proofs) a list of extreme forms containing among others all forms which correspond to the densest known lattices up to dimension 9. But most of this paper is devoted to the Hermite reduction, which we have called the HKZ reduction in Chapter 2 because of [K-Z2]. They probably believed that it would have been possible to calculate all Hermite constants up to dimension 8 by making use of some of their reduction theorems (see Blichfeldt's comment in [Bl2]; the inequalities he refers to are those of the "outer" coefficients; see also, Section 2.9, and the notes on Chapter 2):

> ... This same relation, as well as the precise minima for 6, 7, and 8 variables to be established in this paper, would immediately follow from (2.2) and from a certain inequality that we would reasonably expect to be true, namely $A_{i+4} \geq \frac{1}{2} A_i$. But such an inequality is not always true as seen from the form ...

The paper ends with a proof of the *strict* inequality $\gamma_5^{5/2} < 3$, which they use to show that the minimal vectors of a well-rounded lattice generate a sublattice of index 1 or 2.

In the French edition, we somewhat simplified the proof of this difficult inequality by making use of the Bergé–Martinet invariant γ' (introduced in

[B-M1], dated 1989). Here, we have obtained the desired bound on the index above by making use of a remarkably simple argument of Watson

Sections 6.1 and 6.2 of this chapter come directly from the third paper [K-Z3] (the same is true for Section 4 of the French edition). We have somewhat shortened their proof (in Section 6.1) devoted to a characterization of $\mathbb{A}_n$ by the configuration of its minimal vectors), and replaced in Section 6.4 (thanks to the ideas of Watson) their lengthy proof (Coxeter said "prolix") by a much shorter one to obtain the 5-dimensional classification of perfect lattices.

Korkine and Zolotareff introduce the notion of a perfect form (without giving it a name, see the notes on Chapter 3), and they then no longer use the more restrictive notion of an extreme form.

Our Section 6.2 follows closely Korkine and Zolotareff's proof, and only makes use (as they themselves did) of the inequality $s \geq \frac{n(n+1)}{2}$ which is an immediate consequence of the perfection property. Similarly, we only use in Section 6.4 a weak form of the perfection property.

Section 6.3, devoted to the classification of 3-dimensional dual-extreme lattices follows [B-M1], to which we refer for the case of dimension 4.

In Section 6.4, we have deviated from the proof of [K-Z3] (which we largely followed in the French edition) to built a shorter one based on ideas of Watson. Proposition 6.4.17, which describes up to dimension 7 the best possible upper bounds for the index in a lattice Λ of a sublattice Λ' generated by $S(\Lambda)$, deserves a few comments.

First, we deduce from Theorem 6.6.1 that for $n = 8$, the inequality $[\Lambda : \Lambda'] \leq 16$ holds with equality only for $\Lambda \sim \mathbb{E}_8$ (and actually if and only if $(\Lambda, \Lambda') \sim (\mathbb{E}_8, \mathbb{A}_1^{\perp 8})$). Moreover, we would like to point out the following very interesting precisions found by Watson in [Wat4]: for $n = 7$ (resp. for $n = 8$), if Λ is not similar to $\mathbb{E}_7$ (resp. to $\mathbb{E}_8$), we have $[\Lambda : \Lambda'] \leq 4$ (resp. $[\Lambda : \Lambda'] \leq 9$). [When $n = 8$, index 9 occurs only for $(\Lambda, \Lambda') \sim (\mathbb{E}_8, \mathbb{A}_2^{\perp 4})$, with a non-cyclic quotient Λ/Λ', this quotient cannot be cyclic in case the index is 8, and index 7 never occurs ([Mar6]; Zahareva's paper [Zah] is not complete for large indices). For index up to 4, see Exercise 6.5.2.]

The fact that $\mathbb{A}_n$ is the only perfect lattice with index 1 suggests to consider the analogous problem for larger index. As a matter of fact, up to dimension 7, there are only six lattices with index 2, namely P_4^1, P_5^1, P_5^2, P_6^5, P_6^6, and P_7^{32} (see Section 6.6). No such example exists among the nearly 11 000 known 8-dimensional perfect lattices, and there are only two 8-dimensional lattices with index 3 ([Bt-M']). This suggests the following question:

Does there exist for all k an integer n_k such that any perfect lattice of dimension $n \geq n_k$ and index $\imath \leq k$ is similar to $\mathbb{A}_n$?

Except for Bergé's proof of Theorem 6.6.1, Sections 6.5 and 6.6 only contain statements of results obtained with the methods of the next chapter, with a possible alternative proof for dimension 6 succinctly described by Baranovskii and Ryshkov in [Br-R]. Cyclotomic constructions for 6-dimensional extreme lattices have been given by Craig in [Cra2].

A natural question is to obtain in higher dimension characterizations of some of the perfect lattices by properties of the configuration of their minimal vectors, analogous to those we proved for dimensions $n \leq 5$ in Sections 6.2 and 6.4.

In this direction, a detailed study of the kissing number has been done by Watson ([Wat5], 1971). In particular, he determined $s_{\max}$ up to $n = 9$ and also the maxima $s'_{\max}$ of s for non-perfect lattices up to $n = 8$. His results, together with Bannai–Odlyzko–Sloane's result for $n = 24$ (see [C-S], Chapters 13 and 14) are displayed in the table below.

dim	2	3	4	5	6	7	8	9	24
$s_{\max}$	3	6	12	20	36	63	120	136	98280
$s'_{\max}$	2	5	9	16	28	46	75	≥ 128	?

Watson actually proves much more than what we have accounted for above. He shows in particular that the inequality $s \geq 25$ for $n = 6$ (resp. $s \geq 42$ for $n = 7$) is possible only for lattices with a minimal section similar to $\mathbb{D}_5$ (resp. to $\mathbb{E}_6$), and determines the perfect 6-dimensional lattices with $s \geq 22$.

The theory of Section 9.1 implies that for every non-perfect lattice Λ, there exists a lattice Λ' with perfection corank 1 and an element u of $\mathrm{GL}(E)$ such that we have $u(\Lambda) \subset \Lambda'$ and $u(S(\Lambda)) \subset S(\Lambda')$. One shows that the invariants r, s and $s - r$ (r is the perfection rank) increase from Λ to Λ', and that corank-1 lattices are those which lie on a contiguity path in the sense of Chapter 7.

Thanks to Barnes and Jaquet, one knows all these paths for $n = 6$ and $n = 7$. We recover in this way for $n \leq 7$ the values of $s'_{\max}$ displayed in the previous table. Probably it would be possible to obtain further results thanks to our knowledge of the data for dimension 7, which were not available to Watson. In this respect, it would be very interesting to reconsider Watson's work [Wat4].

Finally, we quote the following Gram matrix

$$\begin{pmatrix} 4 & 2 & 2 & 2 & 2 & -2 & -2 & -2 \\ 2 & 4 & 2 & 2 & 2 & -2 & -2 & 0 \\ 2 & 2 & 4 & 0 & 2 & -2 & -2 & 0 \\ 2 & 2 & 0 & 4 & 0 & -2 & -2 & -2 \\ 2 & 2 & 2 & 0 & 4 & -2 & -2 & 0 \\ -2 & -2 & -2 & -2 & -2 & 4 & 2 & 1 \\ -2 & -2 & -2 & -2 & -2 & 2 & 4 & 1 \\ -2 & 0 & 0 & -2 & 0 & 1 & 1 & 4 \end{pmatrix}$$

which defines a non-perfect lattice of dimension 8 with $s = 75$. I found this lattice, which clearly contains an $\mathbb{E}_7$-section, inside the laminated lattice $\Lambda_{12}^{\mathrm{mid}}$.

7 The Voronoi Algorithm

Introduction

The aim of this chapter is to describe an algorithm due to Voronoi which attaches to a perfect quadratic form a finite number of *contiguous* (or *neighbouring*) perfect forms and allows in this way to find in a finite number of steps all perfect forms with given minimum and dimension. We shall moreover solve explicitly some particular cases and explain succinctly what can be done for non-perfect forms.

We also explain where possible the translation of the algorithm in terms of lattices. However, the setting of quadratic forms looks more suitable, because of the existence of a natural scalar product on the space of quadratic forms, and also for the algorithmic applications, where calculations are done by making use of a basis.

The complexity of the algorithm increases rapidly with the dimension, so it has been explicitly carried out only up to dimension 7.

It would be possible to read directly in Chapter 13 the theory developed in Sections 7.1–7.4, where "$\mathcal{T}$-algorithms" are considered for appropriate subspaces $\mathcal{T}$ of $\mathrm{Sym}_n(\mathbb{R})$. However, some special features of the general theory (e.g., Proposition 7.1.8 or the first assertion in Theorem 7.2.1) would be missed.

Another kind of extension appears in Chapter 9. Actually, the Voronoi algorithm produces a graph whose vertices are the perfect forms (up to proportionality) and edges are the contiguity paths. This graph can be viewed as a cell complex made of 0- and 1-dimensional cells. Chapter 9 deals with cells of arbitrary dimensions.

7.1 Voronoi Domains

We denote by E a Euclidean vector space of dimension $n > 0$, and we set $N = \frac{n(n+1)}{2}$. We denote by $\mathcal{L}$ the set of lattices in E and by $\mathcal{Q}$ the set of positive definite quadratic forms on $\mathbb{R}^n$; in this case, E stands for the space $\mathbb{R}^n$. We often identify the space Sym_n of symmetric real $n \times n$ matrices with the set of quadratic forms on $\mathbb{R}^n$, and apply to it the language of forms: in particular, we shall speak of positive (definite) matrices, and the minimum $m(A)$

of a matrix A denotes the minimum on $\mathbb{Z}^n$ of the form ${}^tXAX$ whenever A is positive definite. Similarly, we shall speak of positive (definite) symmetric endomorphisms, and often use the same notation Id for the identity in $\mathrm{End}^s(E)$ and for the unit matrix I_n of order n.

Definition 7.1.1. *The* Voronoi space of $\mathcal{L}$ *(resp. of* $\mathcal{Q}$*) is the vector space* $\mathrm{End}^s(E)$ *of symmetric endomorphisms of* E *(resp. the vector space of quadratic forms on* $\mathbb{R}^n$*), endowed with the* Voronoi scalar product $\langle u,v\rangle = \mathrm{Tr}(uv)$ *(resp.* $\langle Q,R\rangle = \mathrm{Tr}(AB)$*, where* A *and* B *denote the respective matrices of* Q *and* R*).*

If $A=(a_{i,j})$ and $B=(b_{i,j})$ are two symmetric matrices, we have

$$\langle A,B\rangle = \langle B,A\rangle = \mathrm{Tr}(AB) = \sum_{i,j} a_{i,j}b_{i,j} = \sum_i a_{i,i}b_{i,i} + 2\sum_{i<j} a_{i,j}b_{i,j}\,.$$

The Voronoi scalar product moreover satisfies the relations

$$\langle A,BC\rangle = \langle B,CA\rangle = \langle C,AB\rangle\,.$$

Definition 7.1.2. *A* cone *in a real vector space is a union of half-lines with origin* 0.

We thus non-necessarily consider hypersurfaces (we shall have to work with solid cones as well as with cones which lie in strict subspaces); in the case of hypersurfaces, the notion we use is sometimes named “half-cone”.

Each real quadratic form defines a point of the Voronoi space $\mathcal{Q}$, and the set of positive (resp. positive definite) forms is a closed cone (resp. an open cone, namely the inner part of the previous one). (On the contrary, there is no canonical procedure to attach to a lattice a point in the Voronoi space of lattices.)

Definition 7.1.3. *1. The (half-)cone of positive quadratic forms is the* Voronoi cone *in the Voronoi space of* $\mathcal{Q}$.

2. The circular (half-)cone with axis Id *(resp.* I_n*) and half-angle at the vertex equal to* $\alpha = \arccos\frac{1}{\sqrt{n}}$ *is the* projections cone *in the Voronoi space.*

In a finite–dimensional vector space V over an arbitrary field K, the rank 1 endomorphisms u are characterized by an element $\lambda\in K^*$ (the nonzero eigenvalue of u), a line D in V (the corresponding eigenline), and a complementary hyperplane H to D (the kernel of u), u being then obtained as the product of the homothetic transformation $x\to\lambda x$ and of the projection onto D parallel to H. Applied to E, this description shows that rank 1 symmetric positive endomorphisms are those which are of the form λp_x for some positive real number λ and some nonzero vector $x\in E$. These endomorphisms correspond to quadratic forms which are squares of nonzero linear forms. Moreover, if $x \in \mathbb{R}^n$ corresponds to the $n \times 1$ matrix X, then the projection onto $\mathbb{R}x$ corresponds to the $n \times n$ matrix $X{}^tX$ (see Proposition 3.1.13), but not to the number ${}^tXX$, which is the norm $N(x)$ of x, and we have for any symmetric matrix A the equalities

$$\langle A, X{}^tX\rangle = \mathrm{Tr}(AX{}^tX) = \mathrm{Tr}({}^tXAX) = {}^tXAX\,,$$

which show that the scalar product of a form Q with the rank-1 form with matrix $X{}^tX$ is nothing but the evaluation of Q at X. Note that if $Y = (y_1, \dots, y_n)$, the quadratic form with matrix $Y{}^tY$ is

$${}^tX(Y{}^tY)X = (y_1x_1 + y_2x_2 + \cdots + y_nx_n)^2.$$

The following proposition, which we only state for quadratic forms, explains the second part of Definition 7.1.3:

Proposition 7.1.4. *The Voronoi cone is the convex hull of the projections cone (the "solid projections cone"), and the projections cone is the set of positive quadratic forms of rank $r \leq 1$.*

Proof. We write in terms of matrices. First of all, the Voronoi cone is convex: indeed, if A and B are two positive matrices, the quadratic form ${}^tX(\lambda A + \mu B)X$ takes non-negative values for all λ, $\mu \geq 0$. The Voronoi cone thus contains the convex hull of the projections cone.

Let A be a nonzero symmetric positive matrix, with eigenvalues $\lambda_1, \dots, \lambda_n$. The angle β between A and Id is such that

$$\cos\beta = \frac{\langle A, \mathrm{Id}\,\rangle}{\|A\| \cdot \|\,\mathrm{Id}\,\|}\,.$$

We have the equalities

$$\langle A, \mathrm{Id}\,\rangle = \mathrm{Tr}(A) = \sum_i \lambda_i, \quad \|\,\mathrm{Id}\,\|^2 = \mathrm{Tr}(\mathrm{Id}) = n$$

and

$$\|A\|^2 = \mathrm{Tr}(A^2) = \sum_i \lambda_i^2,$$

which implies that $\cos\beta > 0$. This shows on the one hand that A lies in the side of the hyperplane $\mathbb{R}\,\mathrm{Id}^\perp$ which contains the projections, and on the other hand that the inequalities $\beta \leq \alpha$ and $\cos^2\beta \geq \cos^2\alpha$ are equivalent. Since

$$\cos^2\beta = \frac{1}{n}\frac{(\sum_i \lambda_i)^2}{\sum_i \lambda_i^2} = \frac{1}{n}\left(1 + 2\frac{\sum_{i<j}\lambda_i\lambda_j}{\sum_i \lambda_i^2}\right)\,.$$

we have $\cos^2\beta \geq \cos^2\alpha$, and equality holds exactly when at most one eigenvalue of A is nonzero, i.e. when A has rank 0 or 1.

[The calculations above moreover show that the solid projections cone is characterized in terms of matrices by the inequality $\|M\| \le \mathrm{Tr}(M)$; for a proof without trigonometry, see Section 13.3.] □

The formula $p_x(z) = \frac{x \cdot z}{x \cdot x}\, x$ shows that for nonzero $x, y \in E$, we have

$$p_x \circ p_y(z) = \frac{(x \cdot y)(y \cdot z)}{(x \cdot x)(y \cdot y)}\, x\,,$$

which proves that $p_x \circ p_y$ is a (non-symmetric) rank-1 endomorphism with kernel $\mathbb{R}y^\perp$, nonzero eigenvalue $\cos^2(x, y)$, and corresponding eigenvectors proportional to x, except when $x \cdot y = 0$, where $p_x \circ p_y = 0$. In particular, the angle $\widehat{p_x, p_y}$ in the Voronoi space is an acute angle, or a right angle when x and y are perpendicular. The images in the Voronoi space of the projections onto lines which have a given angle with a given line D thus belong to a cone of revolution, with axis $\mathbb{R}p_D$.

Definition 7.1.5. The Voronoi domain *of a lattice Λ (resp. of a positive definite quadratic form Q) is the convex hull in the Voronoi space of the set of half-lines with origin* 0 *which contain the projections onto the minimal vectors in Λ (resp. of Q).* Notation: *$\mathcal{D}_\Lambda$ (resp. $\mathcal{D}_Q$),*

$$\mathcal{D}_\Lambda = \Big\{ \sum_{x \in S(\Lambda)} \lambda_x p_x \mid \lambda_x \ge 0 \Big\} \quad \Big(\textit{resp. } \mathcal{D}_Q = \Big\{ \sum_{X \in S(Q)} \lambda_X X\, {}^tX \mid \lambda_X \ge 0 \Big\}\Big).$$

The Voronoi domain of a lattice or of a quadratic form is thus a closed cone with vertex 0. The interpretation of the perfection property in the Voronoi space is immediate:

Proposition 7.1.6. *For a lattice Λ (resp. a positive definite quadratic form Q), the following conditions are equivalent:*

1. *Λ is perfect (resp. Q is perfect).*
2. *The Voronoi domain of Λ (resp. of Q) is not contained in any hyperplane of the Voronoi space.*
3. *The interior of the Voronoi domain of Λ (resp. of Q) is not empty.* □

When the perfection property does not hold, the notion of the interior of the Voronoi space however exists, *relative to the subspace it generates* (whose dimension is clearly the perfection rank); this yields a general interpretation of the eutaxy property:

Proposition 7.1.7. *1. A lattice Λ (resp. a positive definite quadratic form Q with matrix A) is eutactic if and only if the identity endomorphism (resp. the matrix A^{-1}) belongs to the (relative) interior of $\mathcal{D}_\Lambda$ (resp. of $\mathcal{D}_Q$).*

2. *Any form which lies in the interior of the Voronoi domain of a positive definite form whose minimal vectors span* $\mathbb{R}^n$ *(and in particular of a perfect form) is itself positive definite.*

Proof. Assertion (1) is obvious for lattices, and the result for quadratic forms then follows from Theorem 3.2.5.

Let Q_0 be a form which satisfies the hypotheses of (2), and let $Q \in \mathrm{Int}(\mathcal{D}_{Q_0})$. Let $S(\mathbb{Q}_0) = \{\pm v_1, \dots, \pm v_s\}$. There exist strictly positive real numbers $\lambda_1, \dots, \lambda_s$ such that

$$Q(x) = \lambda_1 (v_1 \cdot x)^2 + \lambda_2 (v_2 \cdot x)^2 + \cdots + \lambda_s (v_s \cdot x)^2.$$

If $x \in \mathbb{R}^n$ is a zero of Q, it is orthogonal to all vectors v_i, whence $x = 0$. □

We thus see that the eutaxy condition, when expressed in terms of quadratic forms, involves the position of the reciprocal form of the form we consider; the position of the form itself does not seem to have any interpretation. As for the translation in terms of lattices, the lattice is involved only through the definition of its domain.

In the Voronoi space V of $\mathcal{L}$ or of $\mathcal{Q}$, let F be the affine hyperplane with equation $\langle u, \mathrm{Id} \rangle = 1$. This is the hyperplane which is orthogonal to Id and which contains the projections p_x onto the nonzero vectors in E; it intersects the axis of the projections cone at the point $u_0 = \frac{1}{n} \mathrm{Id}$, and its intersection with the cone itself is the sphere with centre u_0 and radius $\sqrt{\frac{n-1}{n}}$, since we have $\|\frac{1}{n} \mathrm{Id} - p_x\|^2 = \frac{n-1}{n}$ for all $x \in E$.

More generally, any cone in the Voronoi space whose axis is $\mathbb{R}$ Id is defined by its intersection with F, since it is the union of the set of lines through 0 which meet this intersection.

Let us apply this remark to the Voronoi domain $\mathcal{D}$ of a lattice Λ or of a positive definite quadratic form Q. The intersection $\mathcal{D} \cap F$ is the convex hull of the set of projections p_x for $x \in S(\Lambda)$ or $x \in S(\mathbb{Q})$. This is thus the polytope in F whose vertices (which belong to the finite set $\{p_x \mid x \in S(\Lambda)$ or $x \in S(Q)\}$) cannot be expressed as a barycentre of the others with non-negative coefficients. The domain $\mathcal{D}$ itself is the polyhedral cone whose edges are the half-lines through 0 which meet the vertices of the previous polytope. These vertices lie on a sphere, which is the intersection of F with the projections cone. But no point of a sphere is a barycentre with non-negative coefficients of some of its other points. Hence:

Proposition 7.1.8. *The Voronoi domain of a lattice or of a positive definite quadratic form is a polyhedral cone whose edges are* all *the half-lines through* 0 *which meet some* p_x *with* x *minimal. In particular, the minimal vectors of a lattice or of a form are well defined by the Voronoi domain of this lattice or of this form.* □

Recall that given a convex set C in a real vector space V, a *tac-plane* for C is an affine hyperplane H which meets C and which is such that C is contained in one of the two closed half-spaces defined by H. The intersections of C with its tac-planes are the *faces* of C. The *dimension of a face* is the dimension of the affine subspace it generates; it is a non-negative integer, and one often says *facet* instead of face when this codimension is 1. When C is strictly convex (like a ball in a Euclidean space), its faces reduce to a single point. When C is a finite intersection of closed half-spaces and spans V, which is the case of Voronoi domains of perfect forms, C is the intersection of half-spaces defined by facets. In the particular case of polyhedral cones with vertex 0 (e.g., the Voronoi domains), all the facets contain the origin, which is the unique 0-dimensional face, and the 1-dimensional faces are the edges.

We now consider the facets of the Voronoi domains of a perfect lattice (or of a perfect form); we denote by V the corresponding Voronoi space.

Definition 7.1.9. *Let $\mathcal{F}$ be a facet of the domain $\mathcal{D}$ of a perfect lattice Λ or of a perfect form Q. A* facet vector associated with *$\mathcal{F}$ is a nonzero vector R of V (thus R is an endomorphism or a quadratic form identified with its symmetric matrix), which is orthogonal to $\mathcal{F}$ and directed towards the interior of $\mathcal{D}$.*

Thanks to the perfection hypothesis, $\mathrm{Int}(\mathcal{D})$ is not empty and the orientation of the facet vectors is thus well defined. It is indeed easy to verify the following explicit characterization:

Proposition 7.1.10. *The facet vectors R associated with $\mathcal{F}$ are defined up to a positive homothetic transformation by the two conditions*

$$\langle R, Y\rangle = 0 \text{ for all edges } Y \text{ of } \mathcal{D} \text{ contained in } \mathcal{F}$$

and

$$\langle R, Y\rangle > 0 \text{ for all edges } Y \text{ of } \mathcal{D} \text{ not contained in } \mathcal{F}\,. \square$$

The following result, which does not extend without extra hypotheses to more general versions of Voronoi's theory, will be needed in Section 7.2.

Theorem 7.1.11. *Suppose that $n \geq 2$. Let $\mathcal{F}$ be a facet of the domain $\mathcal{D}$ of a perfect lattice or form, and let Λ' be a lattice in E. There exists a vector in Λ' which belongs to the open half-space defined by $\mathcal{F}$ which does not contain $\mathcal{D}$.*

Proof. We take the case where $\mathcal{D}$ is the domain of a form with matrix A. Let R be a vector orthogonal to $\mathcal{F}$ directed towards the exterior of $\mathcal{D}$. The form ${}^tXRX$ is not positive: otherwise, its isotropic cone (i.e., the set of column-matrices X such that ${}^tXRX = 0$) would coincide with the kernel of the quadratic form defined by R, which is contained in a hyperplane of E, and this would imply that the matrices $X{}^tX$ lie in a subspace of V of dimension $d \leq \frac{(n-1)n}{2} = N - n < N - 1$. This shows that there exists a column-matrix X satisfying the inequality ${}^tXRX < 0$. A topological argument shows then the existence of a column-matrix Y corresponding to a vector in Λ' and satisfying the same inequality: indeed, there exists $r > 0$ (for instance, the covering radius of Λ') such that every ball of E with radius r contains a point of Λ'. Centring these ball at points λX and making λ tend to infinity shows that for any $\beta > 0$, the cone with vertex 0, axis $\mathbb{R}$ Id, and half-angle β contains points of Λ'. For β sufficiently small, this cone is contained in the half-space defined by $\mathcal{F}$ which contains X. □

The following theorem, which will be completed in Section 7.4 (see Remark 7.4.2) makes more precise the relative positions of the Voronoi domains. We do not restrict ourselves to perfect forms or lattices, and the word "interior" has thus its *relative sense*.

Theorem 7.1.12. *1. The Voronoi domains are contained in the Voronoi cone.*

2. If a Voronoi domain contains a point which belongs to the interior of some other Voronoi domain, the first domain contains the second one.

3. A point which belongs to the interior of the Voronoi domain of a perfect form does not belong to the Voronoi domain of any other form.

Proof. We write in terms of forms and matrices.

(1) The forms which belong to the Voronoi domain of a form Q having s pairs of minimal vectors are sums $\sum_{i=1}^{s} \lambda_i Q_i$ with $\lambda_i \geq 0$, where the Q_i, which correspond to the projections onto the minimal vectors of Q, are squares of linear forms. They are thus positive.

(2) and (3) Let Q and Q' be two positive definite forms. and let $R \in \mathcal{D}_Q \cap \mathcal{D}_{Q'}$. Then R possesses a representation as a sum $R = \sum_{i=1}^{s} \lambda_i Q_i$ with $\lambda_i \geq 0$ and Q_i as in the proof of (1). Without loss of generality, we may assume that Q and Q' have the same minimum m. We have

$$\langle R, Q \rangle = \sum \lambda_i \langle Q, Q_i \rangle = m \sum \lambda_i$$

and

$$\langle R, Q' \rangle = \sum \lambda_i \langle Q', Q_i \rangle \geq m(Q') \sum \lambda_i = m \sum \lambda_i = \langle R, Q \rangle .$$

Exchanging the rôles of Q and Q', we obtain in the same way the opposite inequality $\langle R, Q' \rangle \geq \langle R, Q \rangle$, and hence the equality $\langle R, Q \rangle = \langle R, Q' \rangle$, which is equivalent to $\sum \lambda_i (\langle Q', Q_i \rangle - \langle Q, Q_i \rangle) = 0$.

If $R \in \mathrm{Int}(\mathcal{D}_Q)$, we have the strict inequalities $\lambda_i > 0$, and the inequalities $\langle Q', Q_i \rangle - \langle Q, Q_i \rangle \geq 0$ are indeed equalities. Any edge of $\mathcal{D}_Q$ is an edge of $\mathcal{D}_{Q'}$, and Proposition 7.1.8 then implies assertion (2).

If Q is moreover perfect, the Q_i span the Voronoi space, and we then have $Q' = Q$. □

7.2 Contiguity

The aim of this section is to attach to any perfect lattice (or form) a perfect lattice (or form), which is said to be contiguous to the previous one. We shall actually systematically consider forms, and just state the corresponding results for lattices. We also give a variant of this process which applies to non-perfect forms, yielding a perfect one after a finite number of steps.

We keep the notation of Section 7.1 and we moreover assume that $n \geq 2$.

Theorem 7.2.1. *Let Q be a perfect positive definite quadratic form on $\mathbb{R}^n$ with minimum m, let $\mathcal{F}$ be a facet of the Voronoi domain $\mathcal{D}$ of Q, and let R be a facet vector for $\mathcal{F}$. For all $\theta \in \mathbb{R}$, let $Q_\theta = Q + \theta R$.*

1. *There exists a unique $\rho > 0$ such that the form Q_θ is not perfect with minimum m for $0 < \theta < \rho$ and has a smaller minimum or is not positive definite for $\theta > \rho$.*
2. *For $0 < \theta < \rho$, the minimal vectors of Q_θ are the minimal vectors of Q whose images lie in $\mathcal{F}$.*
3. *For $\theta < 0$, Q_θ has a minimum smaller than m or is not positive definite.*
4. *The form Q_ρ is perfect with minimum m but is different from Q, and it is together with Q the unique perfect form with minimum m and such that its domain moreover contains the facet $\mathcal{F}$.*

Proof. Let $y \in \mathbb{R}^n$. According to whether $R(y)$ is positive, zero or negative, the affine function $\theta \mapsto Q_\theta(y)$ is strictly increasing, constant or strictly decreasing; this proves (3), as one sees by considering an $y \in S(Q)$ with image outside $\mathcal{F}$.

For strictly negative $R(y)$, $Q_\theta(y)$ tends to $-\infty$ with θ. By Theorem 7.1.11, such a y exists in $\mathbb{Z}^n$. This shows that when θ is sufficiently large, Q_θ takes strictly negative values on $\mathbb{Z}^n$, and in particular lower than m. Let ρ be the greatest lower bound of those θ which satisfy this property. For any $\theta \geq 0$, we have $Q_\theta(x) = Q(x)$ for all $x \in S(Q)$ with image in $\mathcal{F}$. Hence the form Q_θ cannot be negative definite. Consequently, for $\theta > \rho$, it can be indefinite, degenerate or positive definite with minimum $m_\theta < m$. For $\theta > 0$, we have on $S(Q)$:

(a) $Q_\theta(x) = Q(x)$ if the image of x lies in $\mathcal{F}$, or
(b) $Q_\theta(x) > Q(x)$ otherwise.

For sufficiently small θ, we have $S(Q_\theta) \subset S(Q)$, hence $m(Q_\theta) = m(Q)$ if θ is strictly positive and not too large. This proves the strict inequality $\rho > 0$. Let $\theta \in (0, \rho)$ and let $y \in S(Q_\theta)$. We have $m(Q_\theta) \geq m(Q)$ by the definition of ρ, hence $m(Q_\theta) = m(Q)$ since $R(x)$ is zero for x with image in $\mathcal{F}$. If $R(y)$ were strictly negative (resp. strictly positive), we would have $m(Q_{\theta'}) < m(Q)$ for every $\theta' > \theta$ (resp. for every $\theta' < \theta$). We thus have $R(y) = 0$, which proves (2) and completes the proof of (1).

Because of the inequality $m(Q_\theta) < m(Q)$ for all $\theta > \rho$, there exists $y \in S(Q_\rho)$ such that $R(y) < 0$. Since the image of y does not belong to $\mathcal{F}$

and since $\mathcal{F}$ generates a hyperplane in the Voronoi space, the form Q_ρ is perfect; since R is non-negative on $S(Q)$, Q_ρ is distinct from Q. Finally, let Q' be a perfect form whose domain $\mathcal{D}'$ contains $\mathcal{F}$ as a facet, and let $a \in \operatorname{Int}(\mathcal{F})$. The points sufficiently close to a which lie on the side of this hyperplane which contain $\mathcal{D}'$ indeed belong to $\mathcal{D}'$. Since they must lie in $\mathcal{D}_Q$ or in $\mathcal{D}_{Q_\rho}$, we have $Q' = Q$ or $Q' = Q_\rho$ by Theorem 7.1.12 (3). □

Definition 7.2.2. *The form Q_ρ defined in Theorem 7.2.1 is called the* contiguous form *(sometimes* neighbouring form*) to the form Q through the facet $\mathcal{F}$ (or* along the facet vector *R).*

Remark 7.2.3. *We know that Q is proportional to a rational form. Consequently, R can be chosen so as to be rational. If both Q and R are rational, then ρ is a rational number.*

We now turn to the behaviour of contiguity with respect to equivalence. Let Q with matrix A and Q' with matrix A' be two equivalent positive definite quadratic forms. This means that there exists a matrix $P \in \mathrm{GL}_n(\mathbb{Z})$ such that $A' = {}^tPAP$. The map $X \mapsto X' = P^{-1}X$ induces a bijection between $S(Q)$ and $S(Q')$, and the corresponding transformation on projection matrices transforms $X{}^tX$ into $X'{}^tX' = P^{-1}(X{}^tX)\,{}^tP^{-1}$. For any symmetric matrix B, we have

$$\langle {}^tPBP, X'{}^tX'\rangle = \operatorname{Tr}({}^tPB(X{}^tX){}^tP^{-1}) = \operatorname{Tr}({}^tP^{-1}({}^tPB)(X{}^tX)) = \langle B, X{}^tX\rangle\,.$$

We see in this way that if $\mathcal{F}$ is a facet of $\mathcal{D}_Q$ for which R is a facet vector, then $\mathcal{F}' = P^{-1}\mathcal{F}\,{}^tP^{-1}$ is a facet of $\mathcal{D}_{Q'}$ and the image R' of R by the transformation which maps Q onto Q' is a facet vector for it. This immediately implies:

Proposition 7.2.4. *With the notation above, if $Q + \rho R$ is contiguous to Q through the facet $\mathcal{F}$ of $\mathcal{D}_Q$, then $Q' + \rho R'$ is contiguous to Q' through the facet $\mathcal{F}'$ of $\mathcal{D}_{Q'}$.* □

Let us now consider a *non-perfect* positive definite quadratic form Q. Its Voronoi domain generates a subspace $\mathcal{F}$ of codimension ≥ 1 in V. By a *facet vector for Q*, we mean any nonzero vector F of the Voronoi space $V = \mathrm{Sym}_n(\mathbb{R})$ which is orthogonal to $\mathcal{F}$.

[This definition is not completely analogous to the one we gave for perfect forms.]

An adaptation to this situation of the proof of Theorem 7.2.1 easily shows:

Proposition 7.2.5. *Let Q be a non-perfect positive definite quadratic form with minimum m, let R be a facet vector for Q, and let $Q_\theta = Q + \theta R$. There exists a unique $\rho > 0$ such that Q_θ has minimum m for $0 \leq \theta \leq \rho$ and is indefinite, degenerate, or of minimum lower than m for $\theta > \rho$. The dimension of the Voronoi domain of Q_ρ is strictly larger than the dimension of that of Q.*

[For a more precise statement, see Theorem 9.1.9 (4)] □

It is worth noticing that the statement above remains valid if one replaces R by $-R$. Hence assertion (3) in Theorem 7.2.1 about negative values of θ does not generalize. Moreover, Proposition 7.2.5 does not yield any canonical choice for the perfect form whose existence is asserted.

We shall discuss shortly in Section 7.8 algorithms which allow the calculation of ρ. For the while, we now simply prove a lemma which can be used to produce an upper bound for ρ:

Lemma 7.2.6. *1. Let u be an endomorphism of rank r, with nonzero eigenvalues $\lambda_1, \dots, \lambda_r$. Then the polynomial $\det(\mathrm{Id} + Xu)$ has degree r and roots $-\frac{1}{\lambda_i}$.*
2. Let A be a nonsingular symmetric matrix and let B be a rank r symmetric matrix. Then the polynomial $\det(A + XB)$ is of degree r.

Proof. The second assertion follows from the first one since

$$A + XB = A\,(I_n + A^{-1}B)\,,$$

and the first one follows from the equality $(\mathrm{Id} + \theta u) = \theta(u - (-\theta)^{-1} I_n)$, which holds for any nonzero real θ. □

We end this section with a few remarks about the interpretation of contiguity in terms of lattices.

For $u \in \mathrm{End}^s(E)$ and $x \in E$, we have

$$N(u(x)) = u(x) \cdot u(x) = ({}^t u \circ u(x)) \cdot x = u^2(x) \cdot x\,.$$

The map $u \mapsto N(u(x))$ is thus a quadratic function of u. In order to obtain a linear map, it is natural to replace u (assumed to be positive definite) by its square root. The evaluation of the scalar products is then obtained via the formula $\sqrt{u}\,(x) \cdot \sqrt{u}\,(y) = u(x) \cdot y$.

Starting with a lattice Λ of E and a facet vector F for this lattice, we naturally define the contiguity property by considering the family of lattices $\Lambda_\theta = \sqrt{\mathrm{Id} + \theta F}\,(\Lambda)$. Let $\mathcal{B} = (e_1, e_2, \dots, e_n)$ be a basis for E and let $A = (a_{i,j})$ be its Gram matrix. That of the image of $\mathcal{B}$ by $\sqrt{\mathrm{Id} + \theta F}$ is the matrix A_θ whose entries are

$$(\mathrm{Id} + \theta F)(e_i) \cdot e_j = a_{i,j} + \theta F(e_i) \cdot e_j\,.$$

One can thus write $A_\theta = A + \theta R$, where R has entries

$$r_{i,j} = F(e_i) \cdot e_j = \sqrt{F}(e_i) \cdot \sqrt{F}(e_j)\,.$$

We can therefore attach to the facet vector F for Λ the facet matrix R for A. The contiguous matrix A_ρ to the matrix A along R is the image of the lattice Λ_ρ. These considerations justify the following definition:

Definition 7.2.7. *The* contiguous lattice to Λ along F *is the lattice*

$$\Lambda_\rho = \sqrt{\mathrm{Id} + \rho F}\,(\Lambda)\,,$$

where ρ denotes the smallest positive real number for which Λ_ρ is perfect and has the same norm as Λ.

The interpretation of the contiguity property in terms of lattices, as far as it produces a geometrical interpretation of it, is sometimes useful to understand how the minimal vectors transform when the parameter θ varies; Voronoi himself used this device to study some lattices contiguous to $\mathbb{D}_n$ (of course, he did not make use of the word "lattice"); see Section 7.6.

This is of particular interest when one considers two lattices Λ and Λ' with the same norm and the same sets of minimal vectors: they have the same Voronoi domain, but the notion of contiguity is not the same one; an example is provided by Exercise 7.5.1.

Notice, however, that the Voronoi domains of the corresponding quadratic forms *are not equal*, since their minimal vectors *must* be expressed on different bases.

7.3 Finiteness Results

The aim of this section, in which we keep the previous notation, is to provide a justification to Voronoi's sentence:

On démontrera *aisément* que le nombre des formes parfaites (4.) vérifiant l'inegalité $(\varphi, f) < P$ est limité,

(One will *easily* prove that perfect forms (4.) satisfying the inequality $(\varphi, f) < P$ are finite in number,)

whose meaning is easy to understand after Theorem 7.3.2 below, for which we shall give a proof due to A.-M. Bergé.

Recall that the Voronoi scalar product on $\mathrm{Sym}_n(\mathbb{R})$ is defined by $\langle M, M'\rangle = \mathrm{Tr}(MM')$ and the norm by $\|M\| = \sqrt{\langle M, M\rangle}$, and that these definitions are also used for the corresponding quadratic forms. Recall also that if M is symmetric positive (resp. positive definite), the quadratic forms obtained by diagonal extractions are also positive (resp. positive definite); in particular, the diagonal entries of M satisfy the inequalities $m_{i,i} \geq 0$ (resp. $m_{i,i} > 0$). We denote by $\|x\| = (\sum_{i=1}^n x_i^2)^{1/2}$ the usual norm on $\mathbb{R}^n$.

Lemma 7.3.1. *Let Q and Q_0 be two positive definite quadratic forms on $\mathbb{R}^n$ with respective matrices A and A_0, and let μ be the minimum of Q_0 on the unit sphere of $\mathbb{R}^n$. Then the eigenvalues $\lambda_1, \lambda_2, \ldots, \lambda_n$ of A satisfy the inequalities $0 < \lambda_i \leq \dfrac{1}{\mu}\langle Q, Q_0\rangle$.*

Proof. The eigenvalues are of course (strictly) positive. Let

$$\mathcal{B} = (\epsilon_1, \epsilon_2, \ldots, \epsilon_n)$$

be an orthonormal basis for $\mathbb{R}^n$ whose elements are eigenvectors of A, and let D (a diagonal matrix) and $M_0 = (m_{i,j})$ be the respective matrices of Q and of Q_0 in $\mathcal{B}$. We have

$$\langle Q, Q_0 \rangle = \mathrm{Tr}(DM_0) = \sum_{i=1}^{n} \lambda_i \, m_{i,i} = \sum_{i=1}^{n} \lambda_i \, Q_0(\epsilon_i) \geq \mu \sum_{i=1}^{n} \lambda_i \, .$$

This proves the required inequality, since the λ_i are positive. □

Theorem 7.3.2. *Let Q_0 be a positive definite quadratic form and let $K > 0$. Then the set of perfect forms Q with a given minimum m which satisfy the inequality $\langle Q, Q_0 \rangle \leq K$ is finite.*

Proof. Let Q be a form satisfying the hypotheses of Theorem 7.3.2. With the notation of Lemma 7.3.1, by making use of the Hermite inequality, we obtain the lower bound

$$\prod_{i=1}^{n} \lambda_i = \det(Q) \geq \frac{m^n}{\gamma_n^n} \, .$$

Lemma 7.3.1 then shows that there exists a bound $a > 0$ for the eigenvalues of Q. Let $x \in \mathbb{Z}^n$ be a minimal vector of Q and let $y_1, y_2, \ldots, y_n$ be its components in an orthonormal basis for $\mathbb{R}^n$ whose elements are eigenvectors of the matrix of Q. We have

$$m = Q(x) = \sum_{i=1}^{n} \lambda_i y_i^2 \geq a \sum_{i=1}^{n} y_i^2 = a\|x\|^2 \, ,$$

whence the inequality $\|x\| \leq (\frac{m}{a})^{1/2}$.

We have thus proved that $S(Q)$ is contained in a finite subset of $\mathbb{Z}^n$, namely in $\{x \in \mathbb{Z}^n \mid \|x\| \leq (\frac{m}{a})^{1/2}\}$, which in particular implies that there are only finitely many possible sets $S(Q)$. Since perfect forms with a given minimum are well defined by the sets of their minimal vectors, there are only finitely many of them. □

7.4 The Voronoi Graphs

We keep the notation of the first three sections. Recall that a *(non-oriented) graph* is a pair Γ of a set $v(\Gamma)$, whose elements are called the *vertices of the graph*, and of a family $e(\Gamma)$ of 2-element subsets of $v(\Gamma)$, whose elements are called the *edges* of Γ. We often write Γ for $v(\Gamma)$. Two vertices a and b such that $\{a, b\}$ is an edge are said to be *connected*, and Γ itself is said to be *connected* if for any pair $a, b \in v(\Gamma)$, there exists a finite sequence $c_0, c_1, \ldots, c_r$ of vertices such that $c_0 = a$, $c_r = b$, and c_{i-1}, c_i are connected for $1 \leq i \leq r$.

The contiguity relation being symmetric, the set of perfect forms with given minimum m (or up to proportionality) and given dimension n can be given a graph structure for which connected vertices correspond to contiguous forms. This graph is called the *(n-dimensional) Voronoi graph.*

In this section, we study the Voronoi graph, as well as its quotient by the equivalence relation for quadratic forms (which exists by Proposition 7.2.4), a finite graph which we also call *the Voronoi graph* (or, to be more precise, the *Voronoi graph of classes*); its vertices are indeed the classes (modulo $\mathrm{GL}_n(\mathbb{Z})$) of perfect quadratic forms up to proportionality, and can also be viewed as the similarity classes of perfect lattices.

Theorem 7.4.1. *Let Q be a perfect form and let Q' be a positive definite quadratic form. There exists a finite sequence $Q_0 = Q, Q_1, \ldots, Q_k$ of perfect forms such that Q' belongs to the domain of Q_k and that Q_{i+1} is contiguous to Q_i for $0 \leq i < k$.*

Proof. Let $Q_0 = Q$ and let $\mathcal{D}_0$ be its Voronoi domain. If $Q' \in \mathcal{D}_0$, we can choose $k = 0$. Otherwise (see Proposition 7.1.10) there exists a facet vector R for $\mathcal{D}_0$ such that $\langle R, Q' \rangle < 0$. Let $Q_1 = Q_0 + \rho R$ be the contiguous form to Q_0 along R and let $\mathcal{D}_1$ be its Voronoi domain. We have

$$\langle Q_1, Q' \rangle = \langle Q_0, Q' \rangle + \rho \langle R, Q' \rangle < \langle Q_0, Q' \rangle \,.$$

If $Q' \in \mathcal{D}_1$, we are done with $k = 1$. Otherwise, we construct similarly a form Q_2 with $\langle Q_2, Q' \rangle < \langle Q_1, Q' \rangle$ and so on, obtaining inductively a sequence $Q_0, Q_1, \ldots, Q_i$ of perfect forms with the same minimum, such that the sequence of real numbers $(\langle Q_i, Q' \rangle)$ is strictly decreasing. This first shows that these forms are pairwise distinct. Moreover, because of the inequalities $\langle Q_i, Q' \rangle < \langle Q, Q' \rangle$, Theorem 7.3.2 shows that the process stops after a finite number of steps. □

Remark 7.4.2. *By Theorem 7.1.12, assertions (1) and (3), we know that the Voronoi domains of two perfect forms are contained in the Voronoi cone and that their interiors are mutually disjoint. By Theorem 7.4.1, the union of the interiors of all Voronoi domains is equal to the open Voronoi cone. This property can be expressed by saying that the Voronoi domains constitute a* tessellation *of the* Voronoi cone *(open or closed). For an application to reduction theory, see the notes on this chapter.*

Theorem 7.4.3. *Let Q and Q' be two perfect forms. There exists a finite sequence $Q_0, Q_1, \ldots, Q_k$ of perfect forms such that $Q_0 = Q$, $Q_k = Q'$, and Q_{i+1} is contiguous to Q_i for $0 \leq i < k$.*

Proof. Choose arbitrarily a form $F \in \mathrm{Int}(\mathcal{D}_{Q'})$. Then F is positive definite by Proposition 7.1.7 (2). We can thus apply Theorem 7.1.1 to the pair (Q, F): there exists a finite sequence $Q_0 = Q, Q_1, \dots, Q_k$ of perfect forms such that F belongs to the domain of Q_k. Hence we have $\mathcal{D}_Q \cap \mathrm{Int}(\mathcal{D}_{Q'}) \neq \emptyset$. Since both Q_k and Q' are perfect, we have $Q_k = Q'$ by Theorem 7.1.12 (3). □

The results above can be restated in terms of graphs. We obtain:

Theorem 7.4.4. *The Voronoi graphs are connected; moreover, the Voronoi graph of classes is finite.* □

We are now going to describe the *Voronoi algorithm*, a deterministic one, which can be used (at least theoretically) to find all the similarity classes of perfect lattices in a given dimension. Its aim is to obtain a construction of the Voronoi graph of classes. (From an algorithmic point of view, the first one, which is infinite, is not interesting.)

1. Choose a perfect form Q_0 (e.g., a form corresponding to $\mathbb{A}_n$).
2. Determine one representative per orbit for $\mathrm{Aut}(Q_0)$ of facets of $\mathcal{D}_{Q_0}$.
3. Determine the contiguous form to each of the representatives above.
4. For any new–found contiguous form to Q_0, compare it for equivalence with previously found forms (Q_0 or a contiguous form to Q_0), and remove it in the case where an equivalence occurs.
5. Let $Q_1^1, \dots, Q_1^{k_1}$ be the forms which have not been removed (together with $Q_0^1 = Q_0$, this is a set of pairwise non-equivalent forms); then apply step 2–4 to each of the forms Q_1^j, producing a set $Q_2^1, \dots, Q_2^{k_2}$, then apply again steps 2–4 to the Q_2^j, ...
6. Stop when for an index ℓ, all forms which are contiguous to one of the Q_ℓ^j are equivalent to some $Q_i^{j'}$ with $0 \leq i \leq \ell$.

Since the Voronoi graph of classes is finite and connected, the algorithm actually terminates. However, no reasonable bound for the number ℓ (depending on the starting form, but experimentally a very small number relative to the number of perfect forms) is known.

To perform explicitly the Voronoi algorithm, one must be able to solve the following problems:

1. Given a perfect form, list the orbits of facets of its Voronoi domain under the action of its automorphism group.
2. Given a perfect form Q and a facet vector R for Q, find the real number $\rho > 0$ such that $Q + \rho R$ is the contiguous form to Q along R.
3. Decide whether two given forms are equivalent.

Question 3 is a fundamental one, of independent interest, but not a crucial one as far as solely the Voronoi algorithm is concerned, since we are not able to handle dimensions greater than 7. Question 2 will be discussed in Section 7.8. But the most important algorithmic problem arises from Question 1: the complexity increases quickly with the dimension n (because the dimension of the Voronoi space is quadratic in n) and also with the difference $s - \frac{n(n+1)}{2}$. It is because of the existence of $\mathbb{E}_8$ ($s = 120 >> \frac{n(n+1)}{2} = 36$) that it is radically impossible to run the Voronoi algorithm from dimension 8 onwards.

7.5 Lattices Contiguous to $\mathbb{A}_n$

Recall (Sections 4.1–4.3) that we denote by ε_i the canonical basis for $\mathbb{R}^{n+1}$ ($0 \le i \le n$) or of $\mathbb{R}^n$ ($1 \le i \le n$), and that $\mathbb{A}_n$ (resp. $\mathbb{D}_n$) is defined by the condition $\sum_i x_i\varepsilon_i = 0$ (resp. $\sum_i x_i\varepsilon_i \equiv 0 \mod 2$) on $\mathbb{Z}^{n+1}$ (resp. on $\mathbb{Z}^n$). Using the Korkine–Zolotareff basis $e_i = \varepsilon_0 - \varepsilon_i$ for $1 \le i \le n$ (resp. $e_1 = \varepsilon_1 + \varepsilon_2$ and $e_i = \varepsilon_1 - \varepsilon_i$ for $2 \le i \le n$), we associate with $\mathbb{A}_n$, $n \ge 2$, and $\mathbb{D}_n$, $n \ge 3$, the quadratic forms

$$A_n = 2\sum_i x_i^2 + 2\sum_{i<j} x_i x_j \quad \text{and} \quad D_n = A_n - 2x_1x_2\,.$$

We first consider a more general situation, namely the case of a lattice Λ for which $s = N$ (i.e., $s = \frac{n(n+1)}{2}$). To any edge of $\mathcal{D}_\Lambda$, there corresponds a well-defined facet, called the *opposite facet*, which is the only one which does not contain the edge we consider. We construct in this way a bijection between edges (or pairs of minimal vectors) and facets of $\mathcal{D}_\Lambda$, which is compatible with the action of $\mathrm{Aut}(\Lambda)$. To classify up to equivalence the lattices which are contiguous to Λ, it suffices to consider one minimal vector per orbit of $\mathrm{Aut}(\Lambda)$ (but it may happen that different orbits yield isometric contiguous lattices). When $\mathrm{Aut}(\Lambda)$ acts transitively on $S(\Lambda)$, things are even simpler: there is a unique isometry class of lattices contiguous to Λ. This is precisely what happens in the case where $\Lambda = \mathbb{A}_n$, where $\mathrm{Aut}(\mathbb{A}_n) \simeq \{\pm \mathrm{Id}\} \times S_{n+1}$ acts transitively on the set of the $2N$ vectors $(\varepsilon_i - \varepsilon_j)$.

Theorem 7.5.1. *The lattice $\mathbb{A}_n$ possesses a unique orbit of facets, and its contiguous lattices are isometric to $\mathbb{D}_n$ if $n \ge 3$ and to $\mathbb{A}_2$ if $n = 2$.*

Proof. A half-system of minimal vectors in $\mathbb{A}_n$ is provided by the vectors $\varepsilon_i - \varepsilon_j$, $0 \le i < j \le n$, and the rank-1 matrices $Q_{i,j}$ corresponding to the projections onto these vectors have entries $q_{k,\ell} = 0$ except for $q_{i,i} = q_{j,j} = 1$ and $q_{i,j} (= q_{j,i}) = -1$. The matrices $(y_{i,j})$ orthogonal to $Q_{i,j}$ are characterized by the relations $y_{i,i} + y_{j,j} - 2y_{i,j} = 0$. Writing down these conditions for all $(i,j) \ne (1,2)$, $i < j$, we see that the facet vectors relative to the facet opposite the edge $Q_{1,2}$ are proportional to the matrix $B = (b_{i,j})$ such that $b_{1,2} = b_{2,1} = -1$ and $b_{i,j} = 0$ otherwise, which defines the form $R(x) = -2x_1x_2$. Since $\langle B, Q_{1,2}\rangle = 4 > 0$, R is indeed a facet vector. We have $A_n + R = D_n$. For $n \ge 3$, D_n is perfect. Hence this is the contiguous form to A_n along R for $n \ge 3$. For $n = 2$, the matrix of $A_2 + 2R$ is $\left(\begin{smallmatrix} 2 & -1 \\ -1 & 2 \end{smallmatrix}\right)$, which defines the perfect form $x_1^2 - x_1x_2 + x_2^2 \sim A_2$. □

Corollary 7.5.2. *The perfect lattices of dimensions* 2 *and* 3 *are similar to* $\mathbb{A}_2$ *and* $\mathbb{A}_3$ *respectively.*

Proof. Taking into account the isometry $\mathbb{D}_3 \simeq \mathbb{A}_3$, this is an immediate consequence of Theorem 7.5.1. □

7.6 The Voronoi Algorithm in Dimension 4

We keep the notation of the preceding section. Our aim is to describe the Voronoi graph in dimension 4. However, for further use, we shall prove some results which apply for all lattices $\mathbb{D}_n$ (or forms D_n) with $n \geq 4$. The calculations we need will be made in the basis $(\varepsilon_1, \dots, \varepsilon_n)$ for $\mathbb{R}^n$, though it is not a basis for $\mathbb{D}_n$. This amounts to considering Euclidean structures on $\mathbb{R}^n$ which are not the usual ones, or to consider minimal vectors of forms on lattices which are not $\mathbb{Z}^n$.

Recall that the automorphism group of $\mathbb{Z}^n$ (the semi-direct product of S_n acting on the ε_i and of the group of the 2^n negations of these vectors) embeds in $\mathrm{Aut}(\mathbb{D}_n)$, and is equal to the whole group $\mathrm{Aut}(\mathbb{D}_n)$ for $n > 4$, but has index 3 for $n = 4$ (Section 4.3).

We represent the pairs of minimal vectors in $\mathbb{D}_n$ by the $N = n(n-1)$ vectors $\varepsilon_i \pm \varepsilon_j$, $1 \leq i < j \leq n$, with which we associate the projections $p_{i,j}^{\pm}$ and their matrices $Q_{i,j}^{\pm}$, whose elements $q_{k,\ell}$ are zero except $q_{i,i} = q_{j,j} = 1$ and $q_{i,j} = q_{j,i} = \pm 1$, and then the quadratic forms $Q'^{\pm}_{i,j} = x_i^2 + x_j^2 \pm 2x_i x_j$. The following lemma results from an easy calculation:

Lemma 7.6.1. *Let i, j, k be three distinct indices. The orthogonal complement to $Q_{i,j}^{+}$ (resp. to $\langle Q_{i,j}^{+}, Q_{i,j}^{-}\rangle$, resp. to $\langle Q_{i,j}^{\pm}, Q_{i,k}^{\pm}, Q_{j,k}^{\pm}\rangle$) in the Voronoi space is defined by the equations $x_{i,i} + x_{j,j} + 2x_{i,j} = 0$ (resp. $x_{i,i} + x_{j,j} = x_{i,j} = 0$, resp. $x_{i,i} = x_{j,j} = x_{k,k} = x_{i,j} = x_{i,k} = x_{j,k} = 0$).* □

We know from Section 4.3 that $\mathrm{Aut}(\mathbb{D}_n)$ acts transitively on $S(\mathbb{D}_n)$. The following lemma describes the orbits of $\mathrm{Aut}(\mathbb{D}_n)$ acting on pairs of minimal vectors.

Lemma 7.6.2. *The pairs (e, e') of minimal vectors in $\mathbb{D}_n$ having a given scalar product constitute a single orbit under the action of* $\mathrm{Aut}(\mathbb{D}_n)$ *except when $e \cdot e' = 0$ and $n \geq 5$, where there are two orbits, namely that of $(\varepsilon_1 + \varepsilon_2, \varepsilon_1 - \varepsilon_2)$ and that of $(\varepsilon_1 + \varepsilon_2, \varepsilon_3 + \varepsilon_4)$.*

Proof. We immediately restrict ourselves to the case where $e = \varepsilon_1 + \varepsilon_2$. If $e \cdot e' = \pm 2$, i.e. $e' = \pm e$, there is nothing to prove; if $e \cdot e' = 1$ or $e \cdot e' = -1$, we easily restrict ourselves by permutation and change of signs to the case where $e' = \pm(\varepsilon_1 + \varepsilon_3)$.

Suppose now that $e \cdot e' = 0$. Then e' is one of the vectors $\pm(\varepsilon_1 - \varepsilon_2)$ or $\pm\varepsilon_i \pm \varepsilon_j$, *lei* $< j \leq n$. Taking into account the group G of permutations and sign changes on the e_i, $i \geq 3$, we only need to consider the two cases $e'_1 = \varepsilon_1 - \varepsilon_2$ and $e'_2 = \varepsilon_3 + \varepsilon_4$. The second possibility does not exist for $n \leq 3$, and the orbits of e'_1 and e'_2 are equivalent under the special automorphism of $\mathbb{D}_4$, as one sees by considering the element $\sigma \in \mathrm{Aut}(\mathbb{D}_4)$ such that

$$\sigma(\varepsilon_1) = \frac{1}{2}(\varepsilon_1 + \varepsilon_2 + \varepsilon_3 + \varepsilon_4) \quad \text{and} \quad \sigma(\varepsilon_i) = -\frac{1}{2}(\varepsilon_1 + \varepsilon_i - \varepsilon_j - \varepsilon_k)$$

for $i = 2, 3, 4$, which maps $\varepsilon_1 + \varepsilon_2$ onto $\varepsilon_3 + \varepsilon_4$. For $n \geq 5$, since the stabilizer of e reduces to G, the orbits of (e, e'_1) and (e, e'_2) are clearly distinct. □

Lemma 7.6.3. *Let e and $e' \neq \pm e$ be two minimal vectors of $\mathbb{D}_n$. The projections onto the vectors in $S(\mathbb{D}_n)$ other than e and e' generate the Voronoi space, except if (e, e') is some pair $((\varepsilon_i + \varepsilon_j), (\varepsilon_i - \varepsilon_j))$ or if $n = 4$ and $e \cdot e' = 0$, where they generate a hyperplane which is not a facet of $\mathbb{D}_n$.*

Proof. Consider first the exceptional case. By Lemma 7.6.2, we may assume that $e = \varepsilon_1 + \varepsilon_2$ and $e' = \varepsilon_1 - \varepsilon_2$. Lemma 7.6.1 then shows that the subspace of the Voronoi space which is orthogonal to the plane defined by the complementary set of $\{e, e'\}$ is the set of forms proportional to $R = x_1 x_2$. This subspace itself is thus a hyperplane, but since the scalar products $\langle R, Q'^{+}_{1,2} \rangle$ and $\langle R, Q'^{-}_{1,2} \rangle$ have opposite values, it is not a facet of the Voronoi domain of $\mathbb{D}_n$. In the other cases, we may suppose that one of the following two conditions holds: either $(e, e') = (\varepsilon_1 + \varepsilon_2, \varepsilon_1 + \varepsilon_3)$, or $n \geq 5$ and $(e, e') = (\varepsilon_1 + \varepsilon_2, \varepsilon_3 + \varepsilon_4)$. Lemma 7.6.1 then shows what we need. □

We now restrict ourselves to dimension 4.

Theorem 7.6.4. *The Voronoi domain of $\mathbb{D}_4$ possesses two orbits of facets which all contain nine edges. The sets of three edges not in a facet are conjugate to the set of projections onto three minimal vectors e, e', e'' which generate either an $\mathbb{A}_2$-lattice, with contiguous lattice isometric to $\mathbb{D}_4$, or an $\mathbb{A}_3$-lattice, with contiguous lattice isometric to $\mathbb{A}_4$.*

Proof. A given facet must contain at least nine edges, because its perfection rank must be at least $\frac{n(n+1)}{2} - 1 = 9$, and at most nine by Lemma 7.6.3, which asserts that one must remove at least three pairs of minimal vectors, no two of which being orthogonal. Thus, up to conjugacy, we may assume that the set of removed pairs of minimal vectors contains $e = \varepsilon_1 + \varepsilon_2$, $e' = \varepsilon_1 + \varepsilon_3$, and a third vector $e'' \in \{\varepsilon_2 - \varepsilon_3,\ \varepsilon_2 + \varepsilon_3 \text{ or } \varepsilon_1 + \varepsilon_4\}$. If $e'' = \varepsilon_2 - \varepsilon_3$, then $\{\pm e, \pm e', \pm e''\}$ is the set of minimal vectors in an $\mathbb{A}_2$-lattice. Otherwise, the element $\sigma' \in \mathrm{Aut}(\mathbb{D}_4)$ such that $\sigma'(\varepsilon_1) = \sigma(\varepsilon_1)$ and $\sigma'(\varepsilon_i) = -\sigma(\varepsilon_i)$ preserves $\varepsilon_1 + \varepsilon_2$ and $\varepsilon_1 + \varepsilon_3$ (see the proof of Lemma 7.6.2), and maps $\varepsilon_2 + \varepsilon_3$ onto $\varepsilon_1 - \varepsilon_4$. This proves that the vectors $\varepsilon_2 + \varepsilon_3$ and $\varepsilon_1 + \varepsilon_4$ define the same orbit of facets, and that $\mathbb{D}_4$ possesses at most two isometry classes of contiguous lattices.

We first consider the facet corresponding to $e'' = \varepsilon_2 - \varepsilon_3$, and perform the calculations in the basis (ε_i), hence with the form $Q(x) = \sum_i x_i^2$, but considered as a form on $\mathbb{D}_4$ instead of $\mathbb{Z}^4$. Lemma 7.6.1 shows that a possible choice for a facet vector is

$$R(x) = x_1^2 + x_2^2 + x_3^2 - x_4^2 + 2x_1x_2 + 2x_1x_3 - 2x_2x_3\,.$$

Since the term x_4^2 in $Q + \theta R$ has coefficient 0 for $\theta = 1$, we have $0 < \rho < 1$. Since we expect the contiguous lattice to be a root lattice, it is natural to try $\rho = \frac{1}{2}$. It is clear that $Q' = Q_\rho$ takes integral even values on $\mathbb{D}_4$, which implies that its minimum is 2 as was that of Q, and that the three vectors $\varepsilon_1 - \varepsilon_2 - \varepsilon_3 \pm \varepsilon_4$ and $2\varepsilon_4$ are new norm 2 vectors. We have thus found the contiguous lattice, and we know that this is a root lattice with $s = 12$, hence isometric to $\mathbb{D}_4$.

Since we know by Theorem 7.5.1 that $\mathbb{A}_4$ is contiguous to $\mathbb{D}_4$, the lattices which are contiguous to $\mathbb{D}_4$ through a facet of the second orbit must be isometric to $\mathbb{A}_4$. □

Corollary 7.6.5. *A 4-dimensional perfect lattice is similar to $\mathbb{A}_4$ or to $\mathbb{D}_4$.* □

7.7 The Facets of $\mathbb{D}_n$ and the 5-Dimensional Perfect Lattices

We keep the notation of the previous section; unless otherwise stated, we assume that $n \geq 5$.

Lemma 7.7.1. *Let $\mathcal{F}$ be a facet of $\mathbb{D}_n$. For any pair (i,j) with $1 \leq i < j \leq n$, there exists $\alpha_{i,j} \in \{\pm 1\}$ such that $\mathcal{F}$ contains the projection $p'_{i,j}$ onto the minimal vector $\varepsilon_i + \alpha_{i,j}\,\varepsilon_j$, and the $\frac{n(n-1)}{2}$ projections obtained in this way are independent vectors in the Voronoi space.*

Proof. The existence of the $(\alpha_{i,j})$ follows immediately from Lemma 7.6.2.

Let $\sum_{i<j} \lambda_{i,j}\, p'_{i,j} = 0$ be a linear dependence relation between the $p'_{i,j}$. Evaluating the sum above on ε_n, we obtain the relation

$$\frac{1}{2} \sum_{i<n} \lambda_{i,n}\, \alpha_{i,n}\, (\varepsilon_i + \alpha_{i,n}\, \varepsilon_n) = 0\,.$$

Writing that the components of the left–hand side on the ε_i are zero, we see that the $\lambda_{i,n}$ are all zero. We conclude the proof by induction, starting with the trivial case $n = 1$. □

A symmetric matrix $X = (x_{i,j})$ is orthogonal to the projections $p'_{i,j}$ of Lemma 7.7.1 if and only if the following relations hold:

$$\forall\,(i,\,j) \text{ with } 1 \le i < j \le n\,,\; x_{i,j} = -\frac{1}{2}\,\alpha_{i,j}\,(x_{i,i} + x_{j,j})\,. \qquad (*)$$

To obtain a basis for a facet of $\mathbb{D}_n$ containing $\mathcal{F}$, one must add to the set of the projections $p'_{i,j}$ the set of the $n-1$ projections onto the vectors $\varepsilon_i - \alpha_{i,j}\,\varepsilon_j$ for a suitable set of $n-1$ pairs (i,j) of indices, which amounts to saying that $\mathcal{F}$ contains the $n-1$ corresponding *twin pairs* $(p^+_{i,j}, p^-_{i,j})$.

Lemma 7.7.2. *Let $\mathcal{F}$ be a facet of $\mathbb{D}_n$ and let $\mathcal{T}$ be the set of indices which correspond to a twin pair of $\mathcal{F}$. Then one of the pairwise exclusive following conditions holds:*

1. *$\mathcal{T} = \{1, 2, ..., n\}$.*
2. *There exists an index k such that $\mathcal{T}$ is the complementary set of $\{k\}$.*

In case (2), $\mathcal{F}^\perp$ is generated by the symmetric matrix $B = (b_{i,j})$ such that $b_{k,k} = 2$, $b_{i,k} = -\alpha_{i,k}$ for $i \ne k$, and $b_{i,j} = 0$ for $i, j \ne k$, and conversely the subspace orthogonal to the symmetric endomorphism defined by the matrix B above is a facet of $\mathbb{D}_n$ for which B is a facet vector and which contains all the $p^\pm_{i,j}$, except for the $n-1$ of them which correspond to the vectors $\varepsilon_i - \alpha_{i,k}\,\varepsilon_k$.

Proof. The space $\mathcal{F}^\perp$ is defined by the orthogonality relations $(*)$ together with the extra relations

$$x_{i,i} + x_{j,j} = 0 \text{ for every twin pair } (i,j), i < j\,. \qquad (**)$$

If some index k does not appear in the twin pairs $p^\pm_{i,j}$, it is then possible to choose arbitrarily the term $x_{k,k}$ without causing any modification in the set of solutions of the linear system defined by the relations $(*)$ and $(**)$. Since $\mathcal{F}$ has codimension 1 in the Voronoi space, such an index must be unique.

Suppose now that such an index exists. Since $\mathcal{F}$ has codimension 1, the linear system defined by the relations $(**)$ involving the unknowns $x_{i,i}$, $1 \le i < n$ possesses the only solution 0. This implies that $\mathcal{F}^\perp$ is defined by the relations $(*)$ together with the $n-1$ equations $x_{n,n} + 2\alpha_{i,n}\,x_{i,n} = 0$, and is thus generated by the endomorphism u with matrix B. Conversely, one easily verifies that u is orthogonal to the $(n-1)(n-2)$ projections $p^+_{i,j}$ and $p^-_{i,j}$ for $1 \le i < j < n$ and to the $n-1$ projections associated with the vectors $\varepsilon_i + \alpha_{i,j}\,\varepsilon_j$, and has positive scalar products with the $n-1$ projections associated with the vectors $\varepsilon_i - \alpha_{i,j}\,\varepsilon_j$. □

Definition 7.7.3. *For all $n \ge 4$, the facets of the Voronoi domain of $\mathbb{D}_n$ which satisfy condition (1) (resp. (2)) in Lemma 7.7.2 are called* ordinary facets *(resp.* special facets*).*

We now describe the ordinary facets. One must take into account both the conditions of a combinatorial nature which ensure that the system $(**)$ has rank $n-1$, and the sign conditions which prevent some hyperplanes generated by edges of $\mathbb{D}_n$ from being facets of $\mathbb{D}_n$, a problem we met previously in Lemma 7.6.2.

Lemma 7.7.4. *Let $\mathcal{F}$ be an ordinary facet of $\mathbb{D}_n$. Then $\mathcal{F}$ contains exactly $\frac{(n-1)(n+2)}{2}$ edges, and there exists an index k such that these edges are on the one hand the $2(n-1)$ projections $p^+_{i,k}$ and $p^-_{i,k}$, and on the other hand $\frac{(n-1)(n-2)}{2}$ projections $p^+_{i,j}$ or $p^-_{i,j}$ which must satisfy the only condition that there be exactly one such projection for each pair (i,j) with $i<j$ and $i,j\neq k$. Conversely, any family of edges satisfying the conditions above is the set of edges of some facet of $\mathbb{D}_n$.*

Proof. Let $B=(b_{i,j})$ be the matrix of a facet vector u for $\mathcal{F}$. There exists an index k for which $b_{k,k}$ is nonzero (otherwise, B would be the zero matrix). Since $\mathcal{F}$ is an ordinary facet, there exists an index ℓ such that $p^+_{k,\ell}$ and $p^-_{k,\ell}$ both belong to $\mathcal{F}$, which implies the equality $b_{k,k}+b_{\ell,\ell}=0$. Exchanging k and ℓ if need be, we may assume that $b_{k,k}$ is negative. For any index $i\neq k$, the two scalar products $\langle u,p^+_{k,i}\rangle=b_{k,k}+b_{i,i}+2b_{i,k}$ and $\langle u,p^-_{k,i}\rangle=b_{k,k}+b_{i,i}-2b_{i,k}$ are non-negative. This shows that the inequality $b_{i,i}>0$ holds for all $i\neq k$. Hence for $i,\ j\neq k$, $p^+_{i,j}$ and $p^-_{i,j}$ cannot both belong to $\mathcal{F}$, since $b_{i,i}+b_{j,j}$ would be zero otherwise. In view of its dimension, $\mathcal{F}$ must contain all projections $p^\pm_{i,k}$. This proves that we have found all edges of $\mathcal{F}$.

Conversely, let us give ourselves a list of edges as in the lemma, and denote by B the symmetric matrix defined by $b_{k,k}=-1$, $b_{i,i}=1$, $b_{i,k}=0$ for $i\neq k$, and $b_{i,j}=-\alpha_{i,j}$ for $i,j\neq k$, with $\alpha_{i,j}$ arbitrarily chosen in $\{\pm 1\}$. One immediately verifies that the symmetric endomorphism u with matrix B is orthogonal to all edges $p^+_{i,k}$, $p^-_{i,k}$, and also to $p'_{i,j}=p_{\varepsilon_i+\alpha_{i,j}\varepsilon_j}$, and has scalar products $\langle B,p''_{i,j}\rangle=2(1+\alpha^2_{i,j})>0$ with the projections $p''_{i,j}=p_{\varepsilon_i-\alpha_{i,j}\varepsilon_j}$. □

The following theorem describes all facets of $\mathbb{D}_n$, and moreover gives some indications about the orbits of facets under the operation of $\mathrm{Aut}(\mathbb{D}_n)$.

Theorem 7.7.5. *Let $n\geq 5$ be an integer.*

1. *The Voronoi domain of $\mathbb{D}_n$ contains $n\,2^{n-1}$ special facets and $n\,2^{(n-1)(n-2)/2}$ ordinary facets.*
2. *The special facets contain $(n-1)^2$ edges. They constitute a single orbit under the operation of $\mathrm{Aut}(\mathbb{D}_n)$, namely that of the facet having the quadratic form $2x_1(x_1-x_2-\cdots-x_n)$ as a facet vector in the basis $(\varepsilon_1,\dots,\varepsilon_n)$; this facet contains the facets of $\mathbb{D}_{n-1}$ embedded into $\mathbb{D}_n$ by the last $n-1$ coordinates.*
3. *The ordinary facets contain $\frac{(n-1)(n+2)}{2}$ edges. They are the images under $\mathrm{Aut}(\mathbb{D}_n)$ of the facets containing the projections associated with $\varepsilon_1+\varepsilon_j$ $(j\geq 2)$, $\varepsilon_1-\varepsilon_j$ $(j\geq 2)$, $\varepsilon_2+\varepsilon_j$ $(j\geq 3)$, and one vector $\varepsilon_i\pm\varepsilon_j$ in each pair (i,j) with $3\leq i<j\leq n$; two such facets belong to the same orbit if and only if one of them is the image of the other one by an element of S_{n-2} acting on $\{\varepsilon_3,\dots,\varepsilon_n\}$, or by the compositum of such an element with a transposition $(2,k)$ followed by the negations the ε_i for which the projection onto $\varepsilon_i-\varepsilon_k$ belongs to the facet.*

Proof. The result for special facets is a consequence of Lemma 7.7.2: performing if necessary some transposition $(1,k), k > 1$ followed by suitable sign changes of $\varepsilon_2, \dots, \varepsilon_n$, we may assume that the matrix B is such that $b_{1,i} = -1$ for $i > 1$ (we have $b_{1,1} = 2$ and $b_{i,j} = 0$ for $i,\ j > 1$).

In the case of an ordinary facet $\mathcal{F}$, one makes use of Lemma 7.7.4. There are n choices for the index k and $\binom{n-1}{2}$ choices for the $\alpha_{i,j} \in \{\pm 1\}$, whence the formula giving the number of ordinary facets. Finally, we can clearly restrict ourselves (as in the case of special facets) to the case where $k = 1$, and then to the case where $\alpha_{2,j} = +1$ by suitably negating some vectors among $\varepsilon_3, \dots, \varepsilon_n$. The other assertions of Theorem 7.7.5 are easy. □

A complete study of the Voronoi neighbours of $\mathbb{D}_n$ would need the classification up to automorphism of the ordinary facets of the type we considered in Theorem 7.7.5 (3). We just describe partial results, looking first at the orbits under S_{n-2} acting on $\{\varepsilon_2, \dots, \varepsilon_n\}$.

The number t of $\alpha_{i,j}$ with $j > i > 1$ which are equal to $+1$ is an invariant, which takes all values between 0 and $\frac{(n-2)(n-3)}{2}$, and it suffices to assume the inequality $2t \le \frac{(n-2)(n-3)}{2}$, the other facets being obtained by changing the signs of all $\alpha_{i,j}$. Let f_t be the number of corresponding facets. Clearly $f_0 = f_1 = 1$ (when $t = 1$, choose $\alpha_{i,j} = -1$ for $3 \le i < j \le n$ except $\alpha_{3,4} = +1$). When $t = 2$, one finds the orbit defined by $\alpha_{3,4} = \alpha_{3,5} = +1$ for all $n \ge 5$, and moreover the orbit defined by $\alpha_{3,4} = \alpha_{5,6} = +1$ for all $n \ge 6$. Hence $f_2 = 1$ if $n = 5$ and $f_2 = 2$ if $n \ge 6$.

The case where $t = 3$ is dealt with in the same way; one finds the following possibilities:

$$\begin{aligned}
&\alpha_{3,4} = \alpha_{3,5} = \alpha_{3,6} = +1 \ (\forall n \ge 6);\\
&\alpha_{3,4} = \alpha_{3,5} = \alpha_{4,5} = +1 \ (\forall n \ge 5);\\
&\alpha_{3,4} = \alpha_{3,5} = \alpha_{4,6} = +1 \ (\forall n \ge 6);\\
&\alpha_{3,4} = \alpha_{3,5} = \alpha_{6,7} = +1 \ (\forall n \ge 7);\\
&\alpha_{3,4} = \alpha_{5,6} = \alpha_{7,8} = +1 \ (\forall n \ge 8).
\end{aligned}$$

We thus have $t_3 = 1$ if $n = 5$, $t_3 = 3$ if $n = 6$, $t_3 = 4$ if $n = 7$, and $t_3 = 5$ if $n \ge 8$, a result which shows in particular that for $n = 4, 5, 6$, the number of S_{n-2}-orbits of ordinary facets of $\mathbb{D}_n$ is 1, 4, and 11 respectively.

Let us represent ordinary facets by triangular tables made of elements $\alpha_{i,j} \in \{\pm\}$ such that $i < j$, $3 \le i \le n-1$, and $4 \le j \le n$, corresponding to projections onto $\varepsilon_i + \alpha_{i,j}\,\varepsilon_j$. For $n = 5$, there are four possible tables, namely

$$T_0 = \begin{pmatrix} - & - \\ & - \end{pmatrix} \quad T_1 = \begin{pmatrix} - & - \\ & + \end{pmatrix} \quad T_2 = \begin{pmatrix} - & + \\ & + \end{pmatrix} \quad T_3 = \begin{pmatrix} + & + \\ & + \end{pmatrix};$$

for $n = 6$, we have the following 11 tables:

$$U_0 = \begin{pmatrix} - & - & - \\ & - & - \\ & & - \end{pmatrix} \quad U_1 = \begin{pmatrix} - & - & - \\ & - & - \\ & & + \end{pmatrix} \quad U_2 = \begin{pmatrix} - & - & - \\ & - & + \\ & & + \end{pmatrix} \quad U_2' = \begin{pmatrix} + & - & - \\ & - & - \\ & & + \end{pmatrix}$$

$$U_6 = \begin{pmatrix} + & + & + \\ & + & + \\ & & + \end{pmatrix} \quad U_5 = \begin{pmatrix} + & + & + \\ & + & + \\ & & - \end{pmatrix} \quad U_4 = \begin{pmatrix} + & + & + \\ & + & - \\ & & - \end{pmatrix} \quad U_4' = \begin{pmatrix} - & - & + \\ & - & + \\ & & - \end{pmatrix}$$

$$U_3 = \begin{pmatrix} + & + & + \\ & - & - \\ & & - \end{pmatrix} \quad U_3' = \begin{pmatrix} + & + & - \\ & + & - \\ & & - \end{pmatrix} \quad U_3'' = \begin{pmatrix} + & + & - \\ & - & - \\ & & - \end{pmatrix}.$$

Next, we must determine which pairs of orbits can be gathered by making use of a transposition $(2,k)$ followed by the necessary sign changes. We immediately see that the $\alpha_{i,k}$ are preserved, whereas the $\alpha_{i,j}$ $(i \neq j)$ are multiplied by $\alpha_{i,k}\,\alpha_{j,k}$ (we set $\alpha_{i,j} = \alpha_{j,i}$ for $j < i$). Tables T_0, T_5, U_0, U_6 are obviously preserved for every k, T_1 is transformed into T_2, which shows that the S_{n-2}-orbits of T_1 and of T_2 collapse into a single one under the operation of $\mathrm{Aut}(\mathbb{D}_n)$. With some care, one finds all the ways of regrouping pairs of orbits for $n = 6$, proving finally the following statement:

Proposition 7.7.6. *For $n = 5$ (resp. $n = 6$), there are 3 (resp. 7) S_{n-2}-orbits of ordinary facets, defined by T_0, $T_1 \equiv T_2$, and T_3 (resp. U_0, $U_1 \equiv U_3$, $U_2 \equiv U_4'$, $U_2' \equiv U_4$, $U_3' \equiv U_5$, U_3'', and U_6).* □

We now determine some contiguous lattices to $\mathbb{D}_n$, obtaining a complete list in dimension 5. We define facets by a facet vector, which is a quadratic form R that we represent in the basis $(\varepsilon_1, \dots, \varepsilon_n)$ by a matrix B of one of the following two types, according to whether we consider a special or an ordinary facet:

$$B = \begin{pmatrix} 2 & -1 & -1 & \dots & -1 \\ -1 & 0 & 0 & \dots & 0 \\ -1 & 0 & 0 & \dots & 0 \\ \vdots & \vdots & \vdots & \ddots & \vdots \\ -1 & 0 & 0 & \dots & 0 \end{pmatrix} \quad \text{or} \quad B = \begin{pmatrix} -1 & 0 & 0 & \dots & 0 \\ 0 & 1 & \pm 1 & \dots & \pm 1 \\ 0 & \pm 1 & 1 & \dots & \pm 1 \\ \vdots & \vdots & \vdots & \ddots & \vdots \\ 0 & \pm 1 & \pm 1 & \dots & 1 \end{pmatrix}.$$

It is indeed clear that every facet of $\mathbb{D}_n$ is the image by some automorphism of a facet defined by a matrix of one of the two types above.

The shape of the matrix above for a special facet shows that special facets are those which are compatible with the natural embedding $\mathbb{D}_{n-1} \hookrightarrow \mathbb{D}_n$.

Lemma 7.7.7. *Let R be a facet vector for $\mathbb{D}_n$ $(n \geq 4)$, defined in the basis $(\varepsilon_1, \dots, \varepsilon_n)$ by a matrix of one of the types above. Then the form Q_θ with matrix $A_\theta = I_n + \theta B$ takes integral even values on $\mathbb{D}_n$ if and only if $\theta \in \frac{1}{2}\mathbb{Z}$.*

Proof. It amounts to the same thing as saying that θR takes integral even values on $\mathbb{D}_n$. We prove the necessity of the condition "$2\theta \in \mathbb{Z}$" by evaluating θR on $\varepsilon_1 - \varepsilon_2$ in the special case, and on $\varepsilon_2 \pm \varepsilon_3$ in the ordinary case (the result is 4θ, and 0 or 4θ respectively).

It now suffices to show that the values of R on $\mathbb{D}_n$ are divisible by 4. This is obvious in the special case, where

$$R(x_1, x_2, \dots, x_n) = 2x_1(x_1 - x_2 - \dots x_n) \equiv 2x_1(x_1 + x_2 + \dots + x_n) \mod 4 .$$

In the ordinary case, we have

$$R(x) = -x_1^2 + (x_2^2 + \dots + x_n^2) + 2 \sum_{2 \le i < j \le n} a_{i,j}\, x_i x_j ,$$

hence

$$\begin{aligned} R(x) &\equiv -x_1^2 + (x_2^2 + \dots + x_n^2) + 2 \sum_{2 \le i < j \le n} x_i x_j \mod 4 \\ &\equiv -x_1^2 - (x_2^2 + \dots + x_n^2) + A_{n-1}(x_2, \dots, x_n) \mod 4 . \end{aligned}$$

(The notation A_n was introduced at the beginning of Section 7.5.) But

$$A_{n-1}(x_2, \dots, x_n) = (x_2 + \dots + x_n)^2 + x_2^2 + \dots + x_n^2 .$$

Hence, setting

$$a = x_1 \text{ and } b = x_2 + \dots + x_n ,$$

we obtain

$$R(x) \equiv (b^2 - a^2) \mod 4 ,$$

and Lemma 7.7.7 results from the fact that the integers a and b have the same parity. □

The following theorem, in the statement of which B and R are as in Lemma 7.7.7, gives a partial description of the lattices which are contiguous to $\mathbb{D}_n$ through a special facet. Recall that Q_θ is the quadratic form with matrix $A_\theta = I_n + \theta B$ (that we consider on $\mathbb{D}_n$, not on $\mathbb{Z}^n$). Thus

$$Q_\theta(x_1, x_2, \dots x_n) = x_1^2 + x_2^2 + \dots x_n^2 + 2\theta x_1(x_1 - x_2 - \dots - x_n) .$$

Theorem 7.7.8. *Let $n \ge 4$ be an integer, let $\mathcal{F}$ be a special facet of $\mathbb{D}_n$, let L_n be the corresponding contiguous lattice, and let $\rho > 0$ be such that L_n be isometric to $\mathbb{D}_n$ endowed with the Euclidean structure defined by Q_ρ.*

1. *We have $\rho < \frac{1}{\sqrt{n-1}}$ and* $\det(L_n) = 1 + 2\rho - (n-1)\rho^2$.
2. *For $n \ge 9$, L_n is not a root lattice.*
3. *For $n = 4, 5, 6, 7, 8$, we have $\rho = \frac{1}{2}$,* $\det(L_n) = 9 - n$, *and L_n is isometric to $\mathbb{A}_4$, $\mathbb{D}_5$, $\mathbb{E}_6$, $\mathbb{E}_7$, $\mathbb{E}_8$ respectively.*

Proof. The matrix of Q_θ is

$$A_\theta = \begin{pmatrix} 2\theta+1 & -\theta & -\theta & \dots & -\theta \\ -\theta & 1 & 0 & \dots & 0 \\ -\theta & 0 & 1 & \dots & 0 \\ \vdots & \vdots & \vdots & \ddots & 0 \\ -\theta & 0 & 0 & \dots & 1 \end{pmatrix} .$$

Developing $\det(A_\theta)$ with respect to the first row, or noticing that this determinant has degree 2 (the rank of B) with respect to θ, we see that $\det(A_\theta) = 1 + 2\theta - (n-1)\theta^2$, a polynomial whose unique positive root is $\theta_0 = \frac{\sqrt{n}+1}{n-1} = \frac{1}{\sqrt{n}-1}$; this proves (1).

Lemma 7.7.7 shows that L_n can be a root lattice only if ρ is half-integral, which implies by (1) first that ρ must be equal to $\frac{1}{2}$, and then that we must have $n < 9$, whence (2).

Take now $\theta = \frac{1}{2}$, and assume that $4 \le n \le 8$. It is clear that Q_θ is then positive definite and takes even values on $\mathbb{D}_n$. Moreover, $\varepsilon_1 - \varepsilon_2 - \varepsilon_3 - \varepsilon_4$ is a norm 2 vector for Q_θ, which is not a minimal vector in $\mathbb{D}_n$ for the usual Euclidean structure. We thus have $\rho = \frac{1}{2}$, and the corresponding lattice is generated by the vectors $\varepsilon_1 + \varepsilon_j$, $j \ge 2$ and $\varepsilon_1 - \varepsilon_2 - \varepsilon_3 - \varepsilon_4$, which all have norm 2 for Q_θ. Hence this is a root lattice, which is moreover irreducible since it is perfect. Its determinant is

$$\det(L_n) = \det(\mathbb{D}_n)\,(1 + 2\rho - (n-1)\rho^2) = 9 - n\,;$$

it is thus one of the lattices listed in (3). □

We now determine some more contiguous lattices; the results will then be used to recover the 5-dimensional classification of perfect lattices.

Theorem 7.7.9. *For all $n \ge 4$, the lattice which is contiguous to $\mathbb{D}_n$ through an ordinary facet whose entries $\alpha_{i,j}$ for $j > i \ge 2$ are all equal to $+1$ (resp. except for one of them) is isometric to $\mathbb{A}_n$ (resp. to $\mathbb{D}_n$).*

Proof. We keep the notation of Lemma 7.7.7.

In the first case, we have

$$B = \begin{pmatrix} -1 & 0 & 0 & \dots & 0 \\ 0 & 1 & 1 & \dots & \\ 0 & 1 & 1 & \dots & 1 \\ \vdots & \vdots & \vdots & \ddots & \vdots \\ 0 & 1 & 1 & \dots & 1 \end{pmatrix}.$$

This is clearly a rank 2 matrix, whose nonzero eigenvalues are -1 and $n-1$, which proves that $\det(A_\theta) = (1-\theta)(1+(n-1)\theta)$. By an argument previously used in the proof of Theorem 7.7.8, we see that for $\theta = \frac{1}{2}$, the lattice $\mathbb{D}_n$ endowed with the Euclidean structure defined by Q_θ is integral, even, perfect, and generated by vectors of norm 2 (notice that $2\,\varepsilon_1$ has norm 2 for Q_θ), hence is a root lattice. We thus have $\rho = \frac{1}{2}$. Moreover, this lattice has determinant

$$\det(\mathbb{D}_n)\ \det(A_\theta) = 4\,\frac{1}{2}\left(1 + \frac{n-1}{2}\right) = n+1\,,$$

hence is isometric to $\mathbb{A}_n$.

In the second case, we may assume that the unique entry which is equal to -1 is actually $\alpha_{n-1,n}$. We now have

$$B = \begin{pmatrix} -1 & 0 & \dots & 0 & 0 & 0 \\ 0 & 1 & \dots & 1 & 1 & 1 \\ \vdots & \vdots & \ddots & \vdots & \vdots & \vdots \\ 0 & 1 & \dots & 1 & 1 & 1 \\ 0 & 1 & \dots & 1 & 1 & -1 \\ 0 & 1 & \dots & 1 & -1 & 1 \end{pmatrix}.$$

This time, B has rank 4: let B_1 be the matrix obtained from B by deleting its first row and its first column; we verify that the determinant made with the last three rows and columns of B_1 is equal to -4. The nonzero eigenvalues of B are -1 (obvious from the first row) and the three nonzero eigenvalues of B_1, whose characteristic polynomial is

$$\chi(X) = X^{n-4}(X^3 - T_1X^2 + T_2X - T_3)\,,$$

where T_k denotes the sum of the diagonal determinants of order k extracted from B_1. We immediately see that $T_1 = n-1$ and $T_2 = 0$, and that the last $n-3$ diagonal determinants containing terms from the last two rows and from the last two columns of B_1 are nonzero, which shows that $T_3 = -4(n-3)$. We thus have

$$\begin{aligned}\chi(X) &= X^{n-4}(X^3 - (n-1)X^2 + 4(n-3)) \\ &= X^{n-4}(X-2)(X^2 - (n-3)X - 2(n-3))\,,\end{aligned}$$

and consequently

$$\det(A_\theta) = (1-\theta)(1+2\theta)(1+(n-3)\theta - 2(n-3)\theta^2)\,.$$

The positive roots of this polynomial are 1 and $\frac{\sqrt{n-3}+\sqrt{n+5}}{4\sqrt{n-3}} > \frac{1}{2}$, and we see as in the first case that for $\theta = \frac{1}{2}$, the vector $2\varepsilon_1$ has norm 2 for the Euclidean structure defined by Q_θ. This implies first that $\rho = \frac{1}{2}$, and then that the contiguous lattice is an irreducible root lattice. Since its determinant is equal to 4, it is isometric to $\mathbb{D}_n$. □

From Theorems 7.7.8 and 7.7.9, we obtain the list of all lattices contiguous to $\mathbb{D}_5$ through three out of the four existing orbits. The following result, though we could dispense with it as far as we merely wish to dispose of dimension 5, yields an interesting example, because of its analogy with Theorem 7.7.8:

Proposition 7.7.10. *For $n = 4,5,6,7,8$, the lattice contiguous to $\mathbb{D}_n$ through the facet defined by $\alpha_{i,j} = +1$ for $2 \le i < j \le n$ except $\alpha_{n-2,n} = \alpha_{n-1,n} = -1$ is isometric to $\mathbb{A}_4, \mathbb{D}_5, \mathbb{E}_6, \mathbb{E}_7, \mathbb{E}_8$ respectively.*

Proof. (*Sketch.*) A calculation of the kind that we did to prove Theorem 7.7.9 shows that the characteristic polynomial of B is

$$\chi(X) = (X+1)X^{n-4}(X^3 - (n-1)X^2 + 8(n-4))\,;$$

we deduce from this the equality

$$\det(A_\theta) = (1-\theta)(1+(n-1)\theta - 8(n-4)\theta^3)\,.$$

Hence, for $\theta = \frac{1}{2}$, the form Q_θ is positive definite, which implies, as in the proof of 7.7.9 that we have $\rho = \frac{1}{2}$ and then that the contiguous lattice has again determinant $9-n$. We can now complete the proof as that of 7.7.8. □

The missing facet of $\mathbb{D}_5$ is defined by $\alpha_{i,j} = -1$ for all $i, j \geq 2$. The eigenvalues of B are then -1 and $\frac{1}{n-3}$, of order 1, and 2, of order $n-2$, which gives the determinant of A_θ on $\mathbb{D}_n$ the value

$$\det(I_n + \theta B)(\mathbb{D}_n) = 4\,(1-\theta)(1-(n-3)\theta)\,(1+2\theta)^{n-2}\,,$$

a polynomial whose smallest positive root is $\frac{1}{n-3} < \frac{1}{2}$ for any $n \geq 5$. For $n \geq 5$, the contiguous lattice is no longer a root lattice. However, for $\theta = \frac{1}{4}$, the form Q_θ is positive definite, and then makes $\mathbb{D}_n$ a lattice which (once rescaled to an integral one) has norm 4 and determinant

$$2^n\,4\left(1-\frac{1}{4}\right)\left(\frac{3}{2}\right)^{n-2}\frac{7-n}{4} = 3^{n-1}(7-n)\,.$$

For $n=5$ and $n=6$, this determinant has values $2\cdot 3^4$ and 3^5 respectively. We recognize the determinants of $\sqrt{2}\,\mathbb{A}_5^3$ and of $\sqrt{3}\,\mathbb{E}_6^*$ respectively.

Proposition 7.7.11. *For $n = 5$ (resp. 6), the contiguous lattice to $\mathbb{D}_n$ through the ordinary facet defined by $\alpha_{i,j} = -1$ for all $i, j \geq 5$ is similar to $\mathbb{A}_5^3$ (resp. to E_6^*).*

Proof. (*Sketch.*) First, for $n=6$, the explicit calculation of the inverse of a Gram matrix of the lattice corresponding to $\theta = \frac{1}{4}$ allows us to identify $\mathbb{E}_6^*$; we then identify $\mathbb{A}_5^3$ by noticing that this is a hyperplane section of $\mathbb{E}_6^*$ possessing 15 pairs of minimal vectors. □

Corollary 7.7.12. *A perfect 5-dimensional lattice is similar to $\mathbb{A}_5$, $\mathbb{D}_5$ or $\mathbb{A}_5^3$.*

Proof. We perform the Voronoi algorithm, starting from $\mathbb{A}_5$. By Theorem 7.5.1, the first step uniquely yields $\mathbb{D}_5$. By Proposition 7.7.6, $\mathbb{D}_5$ possesses 4 orbits of facets. By 7.7.8, 7.7.9, and 7.7.11, the lattices which are contiguous to $\mathbb{D}_5$ are isometric to $\mathbb{A}_5$, $\mathbb{D}_5$ (twice), and $\mathbb{A}_5^3$. The only new lattice produced by the second step of the algorithm is thus $\mathbb{A}_5^3$. By Proposition 5.2.3 (1), we know that $\mathbb{A}_5^3$ possesses a single orbit of $15 = \frac{n(n+1)}{2}$ pairs of minimal vectors. Hence all its contiguous lattices are isometric to $\mathbb{D}_5$. □

[The use of Proposition 7.7.11, whose proof has just been sketched, could have been avoided by proving directly (see Exercise 7.7.1) that $\mathbb{D}_5$ is contiguous to $\mathbb{A}_5^3$, and noticing that there is at most one orbit of facets of $\mathbb{D}_5$ whose contiguous lattices are not root lattices.]

7.8 Determination of the Contiguous Form

In this section, we denote by Q a positive definite quadratic form on $\mathbb{R}^n$, by A its matrix (in the canonical basis $(\varepsilon_1, \dots, \varepsilon_n)$ for $\mathbb{R}^n$), by R a facet vector for Q, by $\mathcal{F}$ the orthogonal facet, and by B the matrix of R. For all $\theta \in \mathbb{R}$, we set $Q_\theta = Q + \theta R$ and $A_\theta = A + \theta B$, and we denote by m the minimum $m(Q)$ of Q (on $\mathbb{Z}^n$). We want to calculate the real number $\rho > 0$ such that Q_ρ is the contiguous form to Q along R.

The polynomial $\det(Q_\theta) = \det(A_\theta)$ has degree $r \leq n$. We can say more:

Proposition 7.8.1. *The degree r of $\det(Q_\theta)$ is equal to the rank of R. Its roots are all real, and there is at least one root of each sign. (Hence we have $r \geq 2$).*

Proof. The assertion on the degree is a consequence of Lemma 7.2.6. The fact that $\det(Q_\theta)$ has at least one positive root results from Theorem 7.1.11 when Q is perfect, and from its adaptation to the non-perfect case otherwise; see Proposition 7.2.5. In this last case, since $-R$ is also a facet vector, there exists a negative root. When Q is perfect, the existence of such a root follows from the fact that $Q(x)+(-\theta)R(x)$ is negative for sufficiently large θ whenever x is a minimal vector of Q whose image lies outside $\mathcal{F}$. □

Let θ_0 be the smallest positive root of $\det(Q_\theta)$. Suppose first that we know some $\theta' \in (\rho, \theta_0)$. From the upper (resp. lower) bound of θ', we know that $Q_{\theta'}$ is positive definite (resp. is of minimum $m(Q_{\theta'}) < m$). We can determine the finite set T_m of all $x \neq 0$ in $\mathbb{Z}^n$ such that $m(Q_{\theta'}(x)) < m$. Set $T_m = \{\pm x_1 \dots, \pm x_p\}$. For $1 \leq i \leq p$, let $m_i = Q_{\theta'}(x_i)$; we have $m_i = Q_{\theta'}(x_i) + \theta' R(x_i)$. Let θ_i be such that $Q_{\theta'}(x_i) + \theta_i R(x_i) = m$. We then have $m_i < m$, hence $R(x_i) < 0$, thus $Q(x_i) > m$ (because the inequality $R(x) \geq 0$ holds for all $x \in S(Q)$), hence finally $0 < \theta_i < \theta'$ for all i. Set $\phi = \min \theta_i$.

Let x be any nonzero vector in $\mathbb{Z}^n$. If $x \notin T_m$, we have $Q_{\theta'}(x) \geq m$ for all i, hence *a fortiori* $Q_\phi(x) \geq m$. If $x \in T_m$, let i such that $x = x_i$. We have $Q_\phi(x) \geq Q_{\theta_i}(x) = m$, which shows that $m(Q_\phi) = m$, hence that $\phi \leq \rho$. But if i is such that $\phi = \theta_i$, we have $Q_\theta(x_i) < Q_\phi(x_i) = m$ for all $\theta > \phi$, whence $\rho = \phi$.

Note that to determine the set T_m may need a great amount of time, especially if the minimum of $Q_{\theta'}$ is much smaller than m. A variant of this method consists in stopping the search of the small vectors of $Q_{\theta'}$ as soon as an $x \in \mathbb{Z}^n$ with $Q_{\theta'}(x) < m$ shows up, replacing θ' by θ'' such that $Q_{\theta''}(x) = m$, and starting again the calculation with θ'' instead of θ'. Since this procedure reduces the numbers of elements of T_m (at least, x is now missing), the algorithm stops after a finite number of steps.

We are finally reduced to the following problem: *to find a real number t' in the interval (ρ, θ_0).* (Of course, ρ is still not known.)

An obvious way of ensuring the inequality $\theta' < \theta_0$ is to calculate the smallest positive root of the polynomial $\det A+\theta B$, which amounts by Lemma 7.2.6

to calculating the negative eigenvalue of B which has the greatest absolute value. We now explain how to avoid such a calculation.

If we know elements $\theta_1 \in [0, \rho)$ and $\theta_2 \geq \theta_0$, we can find some $\theta' \in (\rho, \theta_0)$ by dichotomy: let $\theta'' = \frac{\theta_1+\theta_2}{2}$; give θ'' the value θ_2 if $\theta'' \geq \theta_0$, θ_1 if $\theta'' < \theta_0$, and compare again the new value of θ'' with θ_0, and so on. The comparison tests can be done by making use of an algorithm which, given a symmetric real matrix, first decides whether it is positive definite (which will permit us to compare θ' with θ_0), and lists in this case its vectors of norm least than m, which allows us to compare θ' with ρ; indeed, we have $\theta' > \rho$ if $m(Q_{\theta'}) < m$ and $\theta' \leq \rho$ if $m(Q_{\theta'}) = m$; moreover, the equality $\theta' = \rho$ is satisfied in this case if and only if $S(Q_{\theta'})$ strictly contains $S(Q)$. We then know when we must stop the algorithm.

Up to now, we have given an effective method to find some $\theta' \in [\rho, \theta_0)$ provided we know real numbers θ_1, θ_2 such that $0 \leq \theta_1 \leq \rho < \theta_0 \leq \theta_2$. To find in an effective way such a pair (θ_1, θ_2), we start with an arbitrary strictly positive θ. In case θ belongs to $[\rho, \theta_0)$, set $\theta' = \theta$. Otherwise, set $\theta_1 = \theta$ if $\theta < \rho$ and $\theta_2 = \theta$ if $\theta \geq \theta_0$. In the first case, we consider the sequence $\theta, 2\theta, 4\theta, \ldots$ and we stop as soon as we meet a number greater than θ_0; in the second case, we consider the sequence $\theta, \frac{\theta}{2}, \frac{\theta}{4}, \ldots$, and we stop as soon as we meet a number smaller than or equal to ρ.

To perform the method we explained above, we have to use an algorithm which, for a given positive definite form, lists all vectors of norm lower than some given bound. From an algorithmic point of view, this is a hard problem. To compare $\theta > 0$ with ρ, once we know the inequality $\theta < \theta_0$, it is useful to first perform some quick reduction algorithm, e.g. an LLL-algorithm. If it finds a vector with norm $m' < m$, we have located θ. Otherwise, we can either make a systematic search of vectors with norm $m' < m$, a calculation which is shorter once we know a reduced basis, or try directly some greater value of θ.

When the dimension is large, it is important to minimize the number of calculations of minimal vectors.

When the dimension is small, experiment shows that one can obtain fairly quickly the value of ρ by trial and error. Actually, we need to consider only rational (and even integral) forms Q, and we can then choose an integral facet vector R, for the pair (R, ρ) is defined only up to a transformation $(x, y) \mapsto (\lambda\, x, \frac{1}{\lambda}\, y)$. Since R and the contiguous form $Q + \rho R$ are rational, so is ρ. If a reasonable upper bound M for ρ is known, one easily finds ρ by testing a few rational numbers with a small denominator in the interval $(0, M)$.

7.9 Perfect Forms in Dimensions 6 and 7

For these dimensions, we just give a description of the results, due to Barnes ([Bar4]), and Stacey and Jaquet ([Sta1], [Sta2], [Ja2], [Ja5]) respectively. We also recall the results of Sections 7.5, 7.6, 7.7 concerning lower dimensions.

These lattices were "beautified" by Conway and Sloane ([C-S5]) and their automorphism groups and root systems determined. Conway and Sloane also reconstructed one perfect lattice found by Stacey that was omitted both from her published papers and even from her dissertation.

We make use of the notation of [C-S5] that we previously used in Section 6.5; perfect lattices up to similarity or perfect forms up to equivalence and proportionality are denoted by P_n^i, where $n \leq 7$ is as usual the dimension of E and the index i varies from 1 to some value $i_{\max}$ depending on n. (For the Voronoi algorithm to be meaningful, we must exclude the case $n = 1$.)

We now fix n. For every index i, we choose a representative for P_n^i and attach to it the set of facets modulo automorphism of the Voronoi domain of P_n^i. To each orbit, we then attach the equivalence class of the contiguous form through some facet in this orbit; it is a P_n^j for some index j. Finally, we associate with each pair (i, j) of indices the integer $n(i, j) \geq 0$ which is the number of orbits of facets through which the class of the contiguous form is the similarity class of P_n^j.

The results are given in the form of square tables in which the integer $n(i, j)$ is at the intersection of the row i and the column j. (However, for typographical reasons, we did dot write the zeros.) These tables are symmetric: this is a consequence of the symmetry of the contiguity relation when $n(i, j) = 0$. The general case, which is proved in Jaquet's thesis, is a theorem on graphs endowed with a group action; I do not know whether it is a general result in graph theory.

The sum $\sum_j n(i, j)$ of the numbers which lie in the row i is the number of orbits of facets of P_n^i.

If one takes into account solely the fact that two lattices are or are not contiguous, we can see the results on the (non-oriented) graph whose vertices are the perfect n-dimensional lattices and edges the pairs of contiguous perfect lattices. To display all informations of the tables on a graph, we should consider oriented graphs and moreover assign an integer (namely, $n(i, j)$) to each oriented edge.

The tables for dimensions $4, 5, 6, 7$ are at the end of this section (for dimensions 2 and 3, there is only one form and one orbit of facets). We now make some comments on the tables.

If the number of minimal vectors in P_n^i is the smallest possible, i.e. if $s(P_n^i) = \dfrac{n(n+1)}{2}$, and if $\mathrm{Aut}(P_n^i)$ acts transitively on $S(P_n^i)$, there exists exactly one orbit of facets (one can associate with each facet the unique direction of minimal vectors which it does not contain). The i-th row of the table then contains only one nonzero entry, which is equal to 1. This is the case for $\mathbb{A}_n$ (hence for $P_2^1, P_3^1, P_4^2, P_5^3, P_6^7, P_7^{33}$), for $\mathbb{A}_n^{(n+1)/2}$, $n \geq 5$ odd (hence for P_5^3 and $P_7^4 \sim \mathbb{E}_7^*$), and for $P_6 \sim \mathbb{A}_6^{(2)} \sim P_6^5$. These are the only perfect forms in dimensions $2 \leq n \leq 7$ which have a unique orbit of facets.

Table 7.9.1. Contiguity for dimensions 4 and 5.

	P_4^1	P_4^2
P_4^1	1	1
P_4^2	1	

	P_5^1	P_5^2	P_5^3
P_5^1	2	1	1
P_5^2	1		
P_5^3	1		

Table 7.9.2. Contiguity for dimension 6.

	P_6^1	P_6^2	P_6^3	P_6^4	P_6^5	P_6^6	P_6^7
P_6^1	3	1	3	2	1	1	
P_6^2	1		1	1			
P_6^3	3	1	1	1		1	1
P_6^4	2	1	1			1	
P_6^5	1						
P_6^6	1		1	1			
P_6^7			1				

In the other direction, the number of orbits of facets may be very large, either because s is much larger than its lower bound $\dfrac{n(n+1)}{2}$, or because there are many orbits of minimal vectors, For $n = 7$, one finds 157 orbits in the case of $P_7^1 \sim \mathbb{E}_7$ (for which $s = 63$), but $P_7^4 \sim \mathbb{D}_7$ (for which $s = 42$, the largest possible value of s except $\mathbb{E}_7$) has only 17 orbits of facets (the eleventh value), immediately followed by P_7^{24}, which has 16 orbits, but $s = 28$ (the smallest possible value).

From the point of view of complexity, the difficult step is to establish a list of facets containing exactly one facet per orbit. The naive algorithm, which consists in considering successively all systems of $\dfrac{n(n+1)}{2} - 1$ directions of minimal vectors, keeping only the systems which generate a hyperplane in the Voronoi space, and getting rid both of the repetitions and of the hyperplanes which do not contain facets, is inefficient when n or s is large. In the case of $\mathbb{E}_7$, one would have to examine $\binom{63}{28}$ #$6 \cdot 10^{17}$ combinations, a number beyond the possibilities of the existing computers, even taking into account the numerous automorphisms of $\mathbb{E}_7$. To deal with $\mathbb{E}_7$, whose Voronoi domain possesses more than 70 milliion facets, Jaquet had to use a new algorithm

Table 7.9.3. Contiguity for dimension 7.

	1	2	3	4	5	6	7	8	9	10	11	12	13	14	15	16	17	18	19	20	21	22	23	24	25	26	27	28	29	30	31	32	33
1	11	1	6	4	10	6	18	10	1	8	9	1	3	7	1	3	6	4	1	2	2	7	2	5	5	3	3	5	6	2	4	1	
2	1																																
3	6		3	1		3	3	2	1								1					1				1			1				
4	4		1	1				1	1			1							1			1				1	1		1		1	1	1
5	10				6	1	7	2		2	4		1	3		1	1	2	1	1	1		1	2									
6	6		3		1	2	2	1	1					1		1	1																
7	18		3		7	2	23	2		4	5			5	1	2	2	1		1	1	3	2	3	3		1	4			1		
8	10		2	1	2	1	2	1	1	1							1		1			1		1				1	1				
9	1		1	1		1		1				1																					
10	8				2		4	1		2	2							1					1	1				1		1			
11	9				4		5			2	4		1	1	1		1	1					1	1									
12	1			1					1																								
13	3				1						1						1																
14	7				3	1	5				1			1	1	1		1			1	1			1								
15	1						1				1			1																			
16	3				1	1	2							1				1				1			1		1						
17	6		1		1	1	2	1			1		1																				
18	4				2		1			1	1			1		1								1	1								
19	1			1	1			1																									
20	2				1		1																										
21	2				1		1							1																	1		
22	7		1	1			3	1						1		1						1			1		1		1		1		
23	2				1		2			1	1																						
24	5				2		3	1		1	1							1										1		1			
25	5						3							1		1		1				1					1	1					
26	3		1	1																									1				
27	3			1			1									1						1			1			1				1	
28	5						4	1		1														1	1		1	1	1				
29	6		1	1				1														1				1		1			1	1	
30	2									1														1									
31	4			1			1														1	1							1		1		
32	1			1																							1		1				
33				1																													

(the "cascade algorithm"), using projections on subspaces of the Voronoi space of dimensions between s and $\frac{n(n+1)}{2}$. We refer to Jaquet's thesis [Ja2] for a description of this algorithm.

The following diagrams are the Voronoi graphs for dimensions 2, 3, 4, 5, 6 of isometry classes of lattices (or equivalence classes of quadratic forms with a given minimum). They follow from the results we proved in Sections 7.5, 7.6, 7.7, and from Tables 7.9.1 and 7.9.2. Boxed lattices are those which are self-contiguous. To obtain a description of the contiguity graph which distinguishes also non-isomorphic Voronoi paths, one must add edges according to the previous tables. For instance, if $n = 2, 3, 4$ (resp. 5), one must add one edge (resp. two edges) connecting P_n^1 to itself.

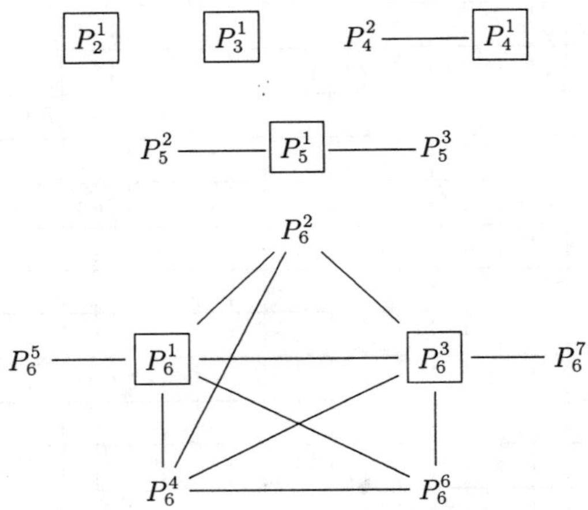

Figure 7.9.4. Voronoi graphs (dimensions 2 to 6).

7.10 Exercises for Chapter 7

7.1.1. The notation being that of Proposition 7.1.4, show that the projections cone is the set of extremal lines in the Voronoi cone.

7.2.1. Let Q and Q' be two positive definite quadratic forms with the same minimum m. For $t \in [0,1]$, set $Q_t(x) = t\,Q(x) + (1-t)\,Q'(x)$ (thus $Q = Q_1$ and $Q' = Q_0$), and let m_t be its minimum.

1. Show that the inequality $m_t \geq m$ holds for all $t \in [0,1]$.

2. Let $t \in (0,1)$. Show that $m_t = m$ if and only if $S(Q) \cap S(Q') \neq \emptyset$.

3. Assuming that $S(Q) \cap S(Q')$ is not empty, show that the perfection rank of Q_t is equal to that of $S(Q) \cap S(Q')$.

4. Suppose now that Q and Q' are perfect. Show that $t \mapsto Q_t$ defines a Voronoi path

connecting Q and Q' if and only if $S(Q) \cap S(Q')$ has perfection rank $\frac{n(n+1)}{2} - 1 = \frac{(n-1)(n+2)}{2}$, and that, for any $t \in (0,1)$, Q and Q' are then the two possible perfect forms which are attached to Q_t by Proposition 7.2.5.

7.5.1. (H. Napias.) Let $n \geq 9$ be an integer, so that $S(\mathbb{A}_n^2) = S(\mathbb{A}_n)$ (Section 5.2). Denote by e_i the vector $\varepsilon_0 - \varepsilon_i$ of $\mathbb{A}_n$, and let $e = \sum_{i=1}^{n} e_i$. Consider the basis $(e_1, \dots, e_{n-1}, e'_n)$ for $\mathbb{A}_n^2$ where $e'_n = \frac{1}{2}\,e$. Recall that the lattices $\mathbb{D}_n^t$ have been defined in Section 5.5.

1. Write down the matrices $X_{i,j}$ corresponding to the projections onto the minimal vectors $\varepsilon_i - \varepsilon_j$ of $\mathbb{A}_n^r$.

2. Suppose that n is odd. Let $B = b_{i,j}$ be the symmetric matrix whose entries are as follows: $b_{1,2} = -1$, $b_{k,\ell} = 0$ if $(k,\ell) \neq (1,2)$ and $k,\ell \neq n$, $b_{1,n} = b_{2,n} = b_{n,n} = -\frac{1}{2}$, $b_{k,n} = 0$ for $3 \leq n \leq n-1$. Show that $A + B$ defines a contiguous lattice to $\mathbb{A}_n^2$ which is isometric to the lattice $\mathbb{D}_{n,n-1}$.
[Consider the basis $\varepsilon_1+\varepsilon_2, \varepsilon_1-\varepsilon_2, \dots, \varepsilon_1-\varepsilon_{n-1}, -\frac{1}{2}\sum_{i=1}^{n-1}\varepsilon_i + \frac{n+1}{2}\varepsilon_1 + \frac{1}{2}(\varepsilon_2 \pm \varepsilon_n)$, where the sign is that of $(-1)^{(n-1)/2}$.]

3. Show that for even n, all lattices contiguous to $\mathbb{A}_n^2$ are isometric to $\mathbb{D}_{n,n-2}$.
[Replace the last vector in Question 2 by
$-\frac{1}{2}\sum_{i=1}^{n-2}\varepsilon_i + \frac{n+1}{2}\varepsilon_1 + \frac{1}{2}(\varepsilon_2 + \varepsilon_{n-1} \pm \varepsilon_n)$, where the sign is that of $(-1)^{n/2}$.]

7.6.1. Consider in $\mathbb{D}_4$ the following two sets of minimal vectors:
$T_1 = \{\pm(\varepsilon_1+\varepsilon_2), \pm(\varepsilon_1+\varepsilon_3), \pm(\varepsilon_2-\varepsilon_3)\}$, and
$T_2 = \{\pm(\varepsilon_1+\varepsilon_2), \pm(\varepsilon_1+\varepsilon_3), \pm(\varepsilon_2+\varepsilon_3)\}$.
Let $G_i \subset \mathrm{Aut}(\mathbb{D}_4)$ be the stabilizer of T_i, and let $H_i = G_i \cap \mathrm{Aut}(\mathbb{Z}^4)$.

1. Show that H_1 and H_2 are cyclic groups of order 24 which are both isomorphic to the direct product $S_3 \times C_2 \times C_2$.

2. Show that we have $G_1 = H_1$ and $[G_2 : H_2] = 3$.

3. Show that the Voronoi domain of $\mathbb{D}_4$ possesses 64 facets, among which 48 (resp. 16) have contiguous lattices isometric to $\mathbb{D}_4$ (resp. to $\mathbb{A}_4$).

7.7.1. Let the lattice $\mathbb{A}_5^3$ be represented (in its integral primitive form) by the following Gram matrix A of a basis $(e_1, \dots, e_5)$ of Proposition 5.2.3 (5):

$$A = \begin{pmatrix} 4 & 1 & -2 & -2 & 1 \\ 1 & 4 & 1 & -2 & -2 \\ -2 & 1 & 4 & 1 & -2 \\ -2 & -2 & 1 & 4 & 1 \\ 1 & -2 & -2 & 1 & 4 \end{pmatrix}.$$

1. Show that the cycle $(1,2,3,4,5)$ acts on the set of minimal vectors of $\mathbb{A}_5^3$ and that $e_1, e_1+e_3, e_1-e_2+e_3$ is a system of representatives for the orbits of pairs of minimal vectors.

2. Show that the matrix

$$R = \begin{pmatrix} 0 & -1 & 0 & 0 & -1 \\ -1 & 0 & 1 & 0 & 0 \\ 0 & 1 & 0 & -1 & 0 \\ 0 & 0 & -1 & 0 & 1 \\ 1 & 0 & 0 & -1 & 0 \end{pmatrix}$$

is a facet vector for A.

3. Show that for $\theta = 1$, the matrix $B = \frac{1}{2}(A+\theta R)$ defines a lattice Λ which is even

of determinant 4, hence isometric to $\mathbb{D}_5$.

4. Write down an explicit identification with $\mathbb{D}_5$.

[Consider the minimal vectors e_1, $-(e_1+e_3+e_4)$, e_4, e_2, $-(e_2+e_5)$ to construct an appropriate Dynkin diagram.]

7.7.2. (Voronoi.) With the notation of Theorem 7.7.8, let m be the integer such that $\sqrt{n}-1 \le m < \sqrt{n}$, and let $p = n - m^2$. (We thus have $0 < p \le 2m+1$.)

1. Show that the value of ρ which corresponds to the contiguous lattice to $\mathbb{D}_n$ through a special facet (along the facet vectors defined after Proposition 7.7.6) is given by the following rules:

$$p \ge 3 \text{ odd}: \qquad \rho = \frac{m^2+n-3}{2m(n-m-1)}\,.$$

$$p \text{ even}: \qquad \rho = \frac{(m-1)^2+n-3}{2(m-1)(n-m)}\,.$$

2. Use this to deduce that γ_{12} is strictly larger than 2.

[This was the first found example of a lattice with $\gamma > 2$.]

7.11 Notes on Chapter 7

Like his predecessors Korkine and Zolotareff, Voronoi studied in Saint Petersburg. After completing his thesis, he obtained a position in Warsaw, which belonged in those days to the Russian empire, as did his native country, the Ukraine. In 1907, he sent from Warsaw to Crelle's journal two long papers (287 pages altogether; the second one appeared in two parts ([Vo2] and [Vo3]). He says that these papers are the final result of 12 years of research. Twelve years earlier came his thesis, devoted to the famous "Voronoi algorithm", which is still a powerful tool to compute fundamental units in a cubic number field. (This algorithm was recently extended by Buchmann to all number fields with unit rank at most 2.)

The second paper is devoted to the so-called "Voronoi cells", briefly alluded to in Chapter 1 (Definition 1.8.3). The first paper ([Vo1]) is the most important one for the subject of this book.

First, it is in this paper that the formal definition of a perfect form appears for the first time, and that basic facts about them (finiteness theorem, characterization of the extremality property) are proved. We refer to the notes on Chapter 3 for a discussion of these topics. But above all, this paper develops the "second Voronoi algorithm", and most of the material of this chapter comes directly from [Vo1]: the definition of Voronoi domains and contiguous forms, the connectedness of the contiguity graph, the contiguity for $\mathbb{A}_n$ and for $\mathbb{D}_n$ through the "special facets", and the classification of perfect forms up to dimension 5.

Besides these fundamental results, we have proved (still following Voronoi) that contiguous forms exist for all facets and that all directions of minimal vectors are edges of the Voronoi domains (i.e., all are needed to generate their convex hull). These results are less important: we shall see in Chapter 13 that these properties are no longer true in some extensions of the Voronoi algorithm.

Voronoi also considered for general n non-special facets of $\mathbb{D}_n$ to construct a family of forms with $\gamma > 2$ for all $n \geq 12$, see Exercise 7.7.2. Actually, this happens from dimension 10 onwards, as shown by the lattices Λ_{10} and Λ_{11} (or K_{11}), but, as far as I know, these lattices were discovered much later (by Chaundy and by Barnes). Note that Voronoi really makes use (without any comment) of the notion of a lattice to handle the forms D_n, but comes back to the point of view of quadratic forms to state his results.

The structure of the Voronoi graph yields interesting precisions about Theorem 3.5.5, which asserts that the maximum of the invariant s is attained on a perfect lattice. Actually, given two contiguous perfect lattices Λ_0, Λ_1 with s-invariants s_0, s_1, the lattices in the open edge (Λ_0, Λ_1) have constant perfection rank $r = \frac{n(n+1)}{2} - 1 = \frac{(n+2)(n-1)}{2}$ and constant invariant $s < \min(s_0, s_1)$. It will be proved in Chapter 9 that the greatest value of the kissing number for non-perfect lattices is attained on some edge of the Voronoi graph.

The results of Sections 6.2 and 6.4 can be recovered by an inspection of the Voronoi graphs studied in Sections 7.5 to 7.7. Sharp results have been obtained by Watson in [Wat5], using solely some knowledge of dimensions $n \leq 6$, but applying nevertheless to dimensions 7 to 9; see the table in the notes on Chapter 6.

Voronoi had clearly begun to work on the 6-dimensional classification of perfect lattices, but had not enough time to complete his work. He did not write in [Vor1] any details of what he was doing for $n = 6$, except for the following:

> It is only from 6 variables onward that I have met positive definite quadratic forms which satisfy property (I) [i.e., which are perfect] and are not extreme forms.

One had to wait half a century for Barnes's 1957 paper [Bar4] before progress was made on the Voronoi algorithm. In this difficult paper, Barnes classifies *by hand* the 6-dimensional perfect lattices, obtaining in particular a confirmation of the statement above of Voronoi. This is probably the limit of what can be done without using a computer.

Using the Voronoi algorithm, Jaquet ([Ja2], [Ja5]) confirmed in 1990 Stacey's list of perfect 7-dimensional lattices that she had obtained in 1975, using combinatorial methods of Watson; Jaquet indeed computed the Voronoi graph for all $n \leq 7$. His calculation needed a considerable amount of computer time, despite the numerous improvements he did in the construction of the Voronoi domains. This is the most lengthy part of the Voronoi algorithm. Its complexity is measured by $\binom{s}{n}$. Most of the running time for dimension 7 is taken to determine the facets of the domain of $\mathbb{E}_7$. The existence of $\mathbb{E}_8$, for which $s = 120$, makes hopeless any attempt to construct the Voronoi graph in dimension 8.

Voronoi's results on the forms contiguous to $\mathbb{D}_n$ have been extended to dimension 6 by Barnes as part of his classification theorem, then by Jaquet and Sigrist to dimension 7 ([Ja-S1], 1989, now a part of [Ja5]), and finally by Jaquet to dimensions 8 and 9 ([Ja4], 1992). For the first time, the densest known lattice does not show up as a neighbour of $\mathbb{D}_9$. It might well be that $\mathbb{D}_n$ has no neighbour with $\gamma > 2$ below dimension 12, where Voronoi found an example.

Voronoi also explains in [Vo1], Definition 27, p. 143, how one can derive a new reduction theory from his algorithm. Consider a set $\mathcal{E}$ containing exactly one perfect form in each equivalence class modulo $\mathbb{R}^\times \mathrm{GL}_n(\mathbb{Z})$. Then any positive definite quadratic form is proportional to a form which belongs to the domain of some form of $\mathcal{E}$. One could call such a form a *Voronoi reduced form*. This notion can moreover be refined by considering the quotients of the Voronoi domains by their automorphism groups. Voronoi has written up the details for dimensions 2 (he then recovers Lagrange's reduction) and 3; a recent reference is Conway and Sloane's paper [C-S8]. As far as I know, higher dimensions have not been considered. It would be interesting to study dimension 4, where two classes of perfect forms exist.

8 Hermitian Lattices

Introduction

In this chapter, we construct lattices (often classical, but sometimes new, such as the K'_n series) inside semi-simple algebras with involutions, using the twisted reduced trace to produce Euclidean structures. The lattices are then modules over suitable subrings of these algebras.

We recall some general results in Section 8.1, and then study in Sections 8.2 to 8.5 various lattices constructed over the rings of Hurwitz quaternions or Eisenstein integers. This yields constructions of the laminated lattices Λ_n up to dimension $n = 16$, then of the K_n series and of an interesting K'_n series up to dimension 12; we obtain in particular the Barnes–Wall lattice $\mathrm{BW}_{16} = \Lambda_{16}$ and the Coxeter–Todd lattice K_{12}.

We make use in Section 8.6 of quaternion algebras over real quadratic fields, and apply the results we obtain to construct lattices in dimensions divisible by 8. As a consequence, we obtain the lattices Λ_n, K_n, and K'_n for all $n \leq 24$, and in particular the celebrated Leech lattice Λ_{24} and the Barnes–Wall lattice BW_{32}. Finally, Section 8.8 is devoted to lattices in dimensions 24 to 48.

8.1 Complex and Quaternionic Structures

We first explain in this section the general setting of Hermitian structures on a lattice, then look more precisely at the case where only commutative fields or skew-fields of rank 4 on their centre are involved, and finally consider the special situation afforded by imaginary quadratic fields or quaternion algebras with centre $\mathbb{Q}$.

Unless otherwise stated, modules are left modules, though considerations related to duality may sometimes force us to work with right modules. Recall that given a ring R, right R-modules are simply left R^{op}-modules where R^{op} is R with multiplication $(x, y) \mapsto yx$. More precise structures often occur. Given two rings R_1, R_2, we have the notion of an (R_1, R_2)-bimodule: an Abelian group endowed with structures of left R_1-modules and of right R_2-modules, which are assumed to satisfy the compatibility law $(\lambda_1 x)\mu_1 = \lambda_1(x\mu_1)$;

(R, R)-bimodules will be called *two-sided R-modules*. The left, right, two-sided submodules in R are its left, right, two-sided ideals respectively. The product IM of an ideal and a left module will be considered only if I is a right ideal for the ring on which the module structure of M is defined. Then, if I is a (R_1, R_2)-ideal and M is a *left* R_2-module, $I\,M$ is a left R_1-module. In particular, given a left R-module and a two-sided ideal I, $I\,M$ is defined, and is again a left R-module.

We now consider the following data:

- A semi-simple $\mathbb{Q}$-algebra K. We denote by $\overline{K}$ the $\mathbb{R}$-algebra $\mathbb{R}\bigotimes_{\mathbb{Q}}K$. We have $\mathrm{rk}_{\mathbb{R}}\,\overline{K} = \mathrm{rk}_{\mathbb{Q}}\,K$.
- An involution $x \mapsto \bar{x}$ on K, that we extend canonically to $\overline{K}$, keeping the same notation.
- A linear form $f : K \to \mathbb{Q}$ such that $T_f : (x, y) \mapsto f(x\bar{y})$ is a non-degenerate, positive definite, symmetric bilinear form on the $\mathbb{Q}$-vector space K. We still denote by f its extension to $\overline{K}$ (with values in $\mathbb{R}$).
- A free (left) $\overline{K}$-module E endowed with a bilinear form T related to T_f in the following way: E can be written as a direct sum $I_1e_1\bigoplus\dots\bigoplus I_ke_k$ where $I_1, \dots, I_k$ are (left) ideals in $\overline{K}$ and $e_1, \dots, e_k$ are elements of E; we assume that T is the orthogonal sum of the restrictions of T_f to the ideals I_ℓ.
- An *order* $\mathfrak{O}$ in K (subring of K which is a free sub-$\mathbb{Z}$-module of K with $\mathrm{rk}_{\mathbb{Z}}\,\mathfrak{O} = \mathrm{rk}_Q\,K$), invariant under the involution of K.
- A sub-$\mathfrak{O}$-module Λ of E which is free of rank $\dim_{\mathbb{Q}} E$ over $\mathbb{Z}$. Often, Λ will be a (left) fractional ideal $\mathfrak{a}$ of K relative to $\mathfrak{O}$.

Definition 8.1.1. *A* $\mathfrak{O}$-Hermitian lattice *is a Euclidean lattice constructed as the lattice* Λ *above.*

[The word "Hermitian" refers to a pair (R, M) of a ring R with involution and an R-module M, together with a map $h : M \times M \to R$ which is linear with respect to the first factor and such that $h(y, x) = \overline{h(x, y)}$. Here, the form h is derived from the obvious form $(x, y) \mapsto x\bar{y}$ on K.]

Comments.

• An involution is an element of order 1 or 2 in the linear group $\mathrm{GL}_{\mathbb{Q}}(K)$ such that $\overline{xy} = \bar{y}\bar{x}$.

• We refer to Appendix 1 for detailed explanations on the notions above. The hypothesis that K is semi-simple is equivalent to the fact that every K-module is isomorphic to a direct sum of ideals of K. Since we are in characteristic 0, this property extends to $\overline{K}$ and is moreover equivalent to the existence of a linear form f such that $f(xy)$ (or $f(x\bar{y})$) is non-degenerate.

• In the applications we shall consider, K will be of one of the three following types of algebras with involution:

1. A totally real number field, with involution the identity of K.

2. A *C.M. field*, i.e. a totally imaginary quadratic extension of a totally real number field K_0, with involution the conjugation over K_0.
3. A totally definite quaternion skew-field H with centre a totally real number field K_0, with involution the canonical quaternionic involution.

In case (2) (resp. (3)), K/K_0 has a basis $(1,\theta)$ (resp. $(1,i,j,k)$) with $\theta^2 = a$ (resp. $i^2 = a$, $j^2 = b$, $ij = -ji = k$) where a, b are totally negative elements of K_0. Given $q = x + y\theta \in K$ (resp. $q = x + yi + zj + tk \in H$), the involution is defined by $\bar{q} = x - y\theta$ (resp. $\bar{q} = x - yi - zj - tk$). The reduced trace of q is defined by the formulae

$$\mathrm{Trd}_{H/K_0}(q) = q + \bar{q} = 2x \quad \text{and} \quad \mathrm{Trd}_{H/Q} = \mathrm{Tr}_{K_0/\mathbb{Q}} \circ \mathrm{Trd}_{H/K_0} .$$

(In the quaternionic case, the reduced trace is half the trace; in the commutative case, we shall sometimes say " reduced trace" for "trace".) The *norm* (resp. the *reduced norm*) *of* q over K_0 is $\mathrm{N}_{K/K_0}(q) = q\bar{q} = x^2 - ay^2$ (resp. $\mathrm{Nrd}_{H/K_0}(q) = q\bar{q} = x^2 - ay^2 - bz^2 + abt^2$); they are positive definite quadratic forms over K_0.

• The form f will be a *twist* of the (reduced) trace form, i.e. a form $\mathrm{Tr}(\alpha x\bar{y})$ (or $\mathrm{Trd}(\alpha x\bar{y})$) for some totally positive $\alpha \in K_0$.

• We shall use explicitly ideals $I_\ell \subset \overline{K}$ and elements $e_\ell \in E$ only when E is free over $\overline{K}$, taking $I_\ell = \overline{K}\, e_\ell$.

• Most of the time, $\mathfrak{O}$ will be a *maximal order* (i.e., maximal with respect to inclusion among orders), often denoted by $\mathfrak{M}$. When K is commutative, such a maximal order is unique, equal to the integral closure of $\mathbb{Z}$ in K. In general, we can just say that there are only finitely many conjugacy classes of such orders. Note that to replace an order by one of its conjugates does not affect the isometry class of the Hermitian lattice.

Various examples of lattices constructed using Hermitian structures occurred in the preceding chapters. In particular, the Craig lattices (Section 5.4) correspond to the choice $K = \mathbb{Q}(\zeta_p)$ and $\mathfrak{O} = \mathbb{Z}_K$ for some odd prime p, taking for $\mathfrak{a}$ a power of the prime ideal of K lying above p. Several examples also occurred in Section 4.8 (see below).

A well-known theorem of Frobenius asserts that a skew-field $\mathbb{K}$ of finite rank over $\mathbb{R}$ is isomorphic to the field $\mathbb{R}$ of real numbers, the field $\mathbb{C}$ of complex numbers or the skew-field $\mathbb{H}$ of Hamilton quaternions. With the notation above, replacing K_0 by $\mathbb{R}$, we define $\mathbb{C}$ and $\mathbb{H}$ by taking $a = -1$ (resp. $a = b = -1$). When endowed with the (reduced) trace form, these three fields acquire a Euclidean structure for which left and right multiplications $x \mapsto q\,x$ and $x \mapsto x\,q$ are similarities of ratio $\mathrm{N}_{K/\mathbb{Q}}(q)$ (or $\mathrm{Nrd}_{K/\mathbb{Q}}(q)$).

From now on, K is of one of the three types of algebras over a totally real number field K_0 that we considered above, of rank 1, 2 or 4 over K_0. The R-algebra $\overline{K}$ is then a product of $[K_0 : \mathbb{Q}]$ copies of $\mathbb{R}$, $\mathbb{C}$ or $\mathbb{H}$. Because of the existence of the similarities quoted above, the case where $K_0 = \mathbb{Q}$ deserves special interest. It plays an interesting rôle, even if K_0 is larger than $\mathbb{Q}$, provided that K may be obtained as the compositum of K_0 with such a field of rank 1, 2, or 4.

For the rest of this section, we suppose that $K_0 = \mathbb{Q}$. *Hence* K *is either* $\mathbb{Q}$ *or an imaginary quadratic field or a totally definite quaternion algebra.*

An imaginary quadratic field is characterized by a negative integer, its *discriminant*. Similarly, it can be proved that totally definite quaternion algebras are characterized by their discriminant, of the form $-d^2$, where d can be (and also must be) any product of an *odd* number of primes; we denote by H_d the corresponding algebra, indeed a skew-field. Up to four exceptions, a maximal finite subgroup G of $K^\times$ is reduced to $\{\pm 1\}$ if K is commutative, and cyclic of order 2, 4, or 6 in the quaternion case. The four exceptions are the imaginary quadratic fields of discriminant -3 (the ring of Eisenstein integers) or -4 (the ring of Gaussian integers), and the quaternion fields H_2 and H_3, with groups the double cover $\widetilde{A}_4$ of A_4 and the quaternion group of order 12; in all four cases, (left) ideals are principal, maximal orders are conjugate to one of them, and any maximal order $\mathfrak{M}$ has a basis made of elements of the group, which is the group of all *units* (i.e., invertible elements) of $\mathfrak{M}$. (The quaternion group of order $4m$ is defined by two generators σ, τ and relations $\sigma^{2m} = \tau^2$, $\tau^4 = 1$, $\tau\sigma\tau^{-1} = \sigma^{-1}$.)

[Lattices over such orders are G-lattices in the sense of Chapter 11. A further example is the non-maximal order with basis $(1, i, j, k)$, $i^2 = j^2 = -1$, whose unit group is the quaternion group of order 8.]

A quadratic field K is defined by $K = \mathbb{Q}(\theta)$ and $\theta^2 = a \in \mathbb{Z}$, a square free. A basis for its ring of integers is $(1, \theta)$ if $a \equiv 2, 3 \mod 4$, and $(1, \omega)$ with $\omega = \frac{-1+\theta}{2}$ (hence $\omega^2 + \omega + \frac{1-a}{2} = 0$) if $a \equiv 1 \mod 4$. We write i instead of θ if $d = -4$. For $d = -3$ and $d = -4$, the respective groups of units are

$$\{\pm 1, \pm\omega, \pm\omega^2\} \quad \text{and} \quad \{\pm 1, \pm i\}\,.$$

The skew-field H_2 (resp. H_3) is defined by the relations $i^2 = j^2 = -1$, hence $k^2 = -1$ (resp. $i^2 = -1, j^2 = -3$, hence $k^2 = -3$). Let $\omega = \frac{-1+i+j+k}{2}$ (resp. $\omega = \frac{-1+j}{2}$ and $\omega' = i\omega = \frac{-i+k}{2}$); we have $\omega^2 + \omega + 1 = 0$ and ${\omega'}^2 = -1$, and $(1, i, j, \omega)$ (resp. $(1, i, \omega, \omega')$) is a basis over $\mathbb{Z}$ for a maximal order $\mathfrak{M}_2 \subset H_2$, called the *Hurwitz order* (resp. $\mathfrak{M}_3 \subset H_3$). The respective groups of units are

$$\left\{\pm 1, \pm i, \pm j, \pm k, \frac{\pm 1 \pm i \pm j \pm k}{2}\right\} \quad \text{and} \quad \{\pm 1, \pm\omega, \pm\omega^2, \pm i, \pm i\omega \pm i\omega^2\}\,.$$

The lattices defined by a maximal order in the four exceptional $\mathbb{Q}$-algebras described above are isometric to $\mathbb{A}_2$, $\mathbb{A}_1 \perp \mathbb{A}_1$, $\mathbb{D}_4$, and $\mathbb{A}_2 \perp \mathbb{A}_2$ respectively. (With the quaternion group of order 8, we obtain a lattice isometric to $\mathbb{A}_1^{\perp 4}$.)

We still consider a (skew-)field K which is $\mathbb{Q}$ itself, imaginary quadratic, or quaternion with centre $\mathbb{Q}$, its completion $\mathbb{K} = \mathbb{R} \otimes K$, which we identify with

$\mathbb{R}$, $\mathbb{C}$, or $\mathbb{H}$, and a (left) vector space E over $\mathbb{K}$ endowed with a positive definite Hermitian form h, and with the corresponding Euclidean scalar product

$$x \cdot y = h(x,y) + \overline{h(x,y)} = \mathrm{Trd}_{\mathbb{K}/\mathbb{R}}(h(x,y))\,.$$

Let $\mathfrak{O}$ be an order in K, still assumed to be invariant under the involution of K, and let Λ be an $\mathfrak{O}$-*lattice in* E, i.e. a lattice stable under left multiplications $x \mapsto \lambda\, x$ by elements $\lambda \in \mathfrak{O}$. As a $\mathbb{Z}$-lattice, Λ possesses a dual lattice $\Lambda^* = \{x \in E \mid x \cdot \Lambda \subset \mathbb{Z}\}$, and we say that Λ is integral if the scalar product takes integral values on Λ, a condition which is equivalent to the inclusion $\Lambda \subset \Lambda^*$. We now define corresponding notions with respect to $\mathfrak{O}$:

Definition 8.1.2. *The* $\mathfrak{O}$-dual *of* Λ *is*

$$\Lambda^\sharp = \{x \in E \mid \forall\, y \in \Lambda,\ h(x,y) \in \mathfrak{O}\}\,,$$

and we say that Λ *is* $\mathfrak{O}$-integral *if* $h(x,y) \in \mathfrak{O}$ *for all* x, $y \in \Lambda$.

Note that Λ^* and $\Lambda^\sharp$ are sub-$\mathfrak{O}$-modules of E, that Λ is $\mathfrak{O}$-integral if and only if $\Lambda \subset \Lambda^\sharp$, and that we always have the inclusion $\Lambda^\sharp \subset \Lambda^*$. This last inclusion can be made more explicit. Recall that the *codifferent of* $\mathfrak{O}$ is the two-sided ideal

$$\mathcal{C}_{\mathfrak{O}} = \{x \in K \mid \mathrm{Trd}(x\mathfrak{O}) \in \mathbb{Z}\}\,.$$

Its inverse, if it exists, is the *different of* $\mathfrak{O}$, denoted by $\mathcal{D}_{\mathfrak{O}}$. (This means that $\mathcal{D}$ is such that $\mathcal{C}\,\mathcal{D} = \mathfrak{O}$.) It always exists if $\mathfrak{O}$ is a maximal order, or if K is an imaginary quadratic field.

Proposition 8.1.3. *If the different* $\mathcal{D}$ *of* $\mathfrak{O}$ *exists, then* $\Lambda^* = \mathcal{D}_{\mathfrak{O}}^{-1}\Lambda^\sharp$.

Proof. This results from the equivalences

$$\begin{aligned} x \in \Lambda^* &\Longleftrightarrow \forall\, y \in \Lambda,\ \mathrm{Trd}(h(x,y)) \in \mathbb{Z} \\ &\Longleftrightarrow \forall\, y \in \Lambda,\ h(x,y) \in \mathcal{C} \\ &\Longleftrightarrow \forall\, y \in \Lambda,\ \mathcal{D}\, h(x,y) \subset \mathfrak{O} \\ &\Longleftrightarrow \forall\, y \in \Lambda,\ h(\mathcal{D}\, x,y) \subset \mathfrak{O} \\ &\Longleftrightarrow x \in \mathcal{D}^{-1}\Lambda^\sharp. \quad \square \end{aligned}$$

The following definition is an algebraic generalization of a notion which is quite usual over $\mathbb{Z}$, to which we shall return in Section 11.5. (The notion of a modular lattice is due to Quebbemann, who, however, considered a more restrictive definition. The generalization below occurred in [Bac1].)

Definition 8.1.4. *Let* $\mathfrak{a}$ *be a two-sided ideal of* $\mathfrak{O}$. *We say that an* $\mathfrak{O}$*-lattice* Λ *is* $\mathfrak{a}$-modular *if* $\Lambda = \mathfrak{a}\,\Lambda^\sharp$, *and that it is* modular *if it is* $\mathfrak{a}$*-modular for some* $\mathfrak{a}$. *If* $\mathfrak{a}$ *is a principal ideal* (λ) *(in the precise sense that* $\mathfrak{O}\,\lambda = \lambda\,\mathfrak{O} = \mathfrak{a}$*), we say* λ-modular, *and* unimodular *if* $(\lambda) = 1$. *For* $\mathbb{Z}$*-lattices, we choose the positive generator of* $\mathfrak{a}$. [To avoid ambiguities, it is important to give a precise definition for the ring $\mathfrak{O}$ we consider; however, the fact that $\mathfrak{a}$ be a two-sided ideal is not essential.]

Proposition 8.1.5. *If the different $\mathcal{D}$ of $\mathfrak{O}$ exists, an $\mathfrak{a}$-modular lattice is $|d_{\mathfrak{O}}|\operatorname{Nrd}(\mathfrak{a})$-modular when viewed as a $\mathbb{Z}$-lattice. In particular, $\mathfrak{O}$-unimodular lattices are $d_{\mathfrak{O}}$-modular $\mathbb{Z}$-lattices.* [Here, $d_{\mathfrak{O}}$ is the (reduced) discriminant; the reduced norm of an ideal is easily defined locally, using the equality $\operatorname{Nrd}((\alpha)) = (\operatorname{Nrd}(\alpha))$.]

Proof. We have $\Lambda = \mathfrak{a}\,\Lambda^\sharp = \mathfrak{a}\,\mathcal{D}\,\Lambda^*$, whence the result by taking the reduced norm and using the relation $d_{\mathfrak{O}} = \operatorname{Nrd}(\mathcal{D})$; see Appendix 1 for the last assertion. □

[An $\mathfrak{O}$-lattice may produce a $d_{\mathfrak{O}}$-modular $\mathbb{Z}$-lattice even if it is not $\mathfrak{O}$-integral. A remarkable example is Bachoc's lattice over $\mathfrak{M}_2$ odd dimension 32 over $\mathbb{Z}$ which is 2-modular, but not integral over $\mathfrak{M}_2$ ([Bac1], § 5; see also [Bac-Ne1]); other examples are provided by 3-modular lattices in dimension 24 over the ring of Eisenstein integers, discovered by Nebe ([Ne3]).]

Let $\mathrm{U}(E)$ be the unitary group of E with respect to h. The orthogonal group $\mathrm{O}(E)$ contains $\mathrm{U}(E)$. Its dimension is $[\mathbb{K}:\mathbb{R}]\times\dim \mathrm{U}(E)$. Hence unitary similarities, i.e. elements of the group $\mathbb{K}^\times\,\mathrm{U}(E)$, are in particular similarities in the usual sense, i.e. elements of $\mathbb{R}^\times\,\mathrm{O}(E)$; their ratio are the reduced norms of the unitary ratios. Note that if Λ is λ-modular, $x \mapsto \lambda x$ is a (unitary) similarity which maps $\Lambda^\sharp$ onto Λ.

8.2 Hurwitz Lattices: Enlargements of $\mathbb{D}_n$

In this section and in the next one, H is the field H_2 of usual quaternions over $\mathbb{Q}$, $\mathfrak{O} \subset H$ is the order with basis $1, i, j, k$ over $\mathbb{Z}$ and $\mathfrak{M}$ is the (maximal) Hurwitz order $\mathfrak{M}_2$, generated over $\mathbb{Z}$ by $\mathfrak{O}$ and $\omega' = \omega + 1 = \frac{1+i+j+k}{2}$. We denote by $\mathfrak{a}$ the two-sided ideal $\mathfrak{M}(1+i) = (1+i)\mathfrak{M}$. The skew-field $\mathbb{Q}\mathfrak{M} = H$ embeds canonically into $\mathbb{H}$, which we endow here with the scalar product $\frac{1}{2}\operatorname{Trd}(x\overline{y})$. We are going to construct new lattices by generalizing certain constructions of Section 4.8, where $\mathbb{E}_8$ and some lattices $\mathbb{D}_n$ were given complex or quaternionic structures. Let $m > 0$ be an integer.
[We insert provisionally the factor $\frac{1}{2}$ in front of Trd to recover the normalization to norm 2 used in Chapter 4 to define $\mathbb{D}_n$; we replace ω by the sixth-root of unity ω' (which has nothing to do with that defined inside H_3 in Section 8.1) to introduce some symmetries in the calculations.]

First, set $n = 2m$ and identify $\mathbb{R}^2$ with $\mathbb{C}$ by $(x, y) \mapsto x + iy$. This identifies $\mathbb{Z}^n \subset \mathbb{R}^n$ with $\mathbb{Z}[i]^m \subset \mathbb{C}^m$, and gives $\mathbb{D}_n$ a complex structure via its identification with $\{(z_1, \dots, z_n) \mid z_1 + \cdots + z_n \equiv 0 \mod (1+i)\}$.

Set now $n = 4m$. The identification of $\mathbb{R}^4$ with $\mathbb{H}$ via the map $(x, y, z, t) \mapsto x + yi + zj + tk$ identifies $\mathbb{Z}^n \subset \mathbb{R}^n$ with $\mathfrak{O}^m \subset \mathbb{H}^m$, and equips as above $\mathbb{Z}^n$ and $\mathbb{D}_n$ with $\mathfrak{O}$-structures. For $n > 4$, $\mathbb{D}_n$ is not an $\mathfrak{M}$-module, and our first task will be to study the lattice $\mathfrak{M}\,\mathbb{D}_n \subset \mathbb{H}^m$, which is known to be isometric to $\mathbb{E}_8$ in case $m = 2$.

For $1 \leq k \leq m$, we denote by I_k the interval $[4k-3, 4k]$ (the *block of index* k; here, k is an index, not a quaternion) and identify $\mathbb{Z}^n$ with the orthogonal sum of the m copies of $\mathfrak{O}$ whose bases are the vectors ε_p, $4k-3 \leq p \leq 4k$ of the canonical basis $(\varepsilon_1, \ldots, \varepsilon_n)$ for $E = \mathbb{R}^n$. The order $\mathfrak{O}$ acts by $i\,\varepsilon_{4k-3} = \varepsilon_{4k-2}$ and $j\,\varepsilon_{4k-3} = \varepsilon_{4k-1}$. For all indices k, we have $\pm\omega'\varepsilon_k = \frac{1}{2}(\sum_\ell \pm\varepsilon_\ell)$ where the sum is over the four indices of the block containing k, and contains 0, 2 or 4 minus signs; when $k = 1, 5, \ldots, 4m-3$, there are solely + signs. We now set

$$e_k = \frac{1}{2}\sum_{p\in I_k} \varepsilon_p \quad \text{and} \quad e_{k,\ell} = e_k + e_\ell \quad \text{for} \quad \ell \neq k\,.$$

Clearly $(e_1, \ldots, e_m)$ is an $\mathfrak{M}$-basis for $\mathfrak{M}\,\mathbb{Z}^n = \mathfrak{M}^m$.

The set of vectors of the form $\pm\varepsilon_p \pm \varepsilon_q$ with p and q in the same block is invariant by $\mathfrak{M}$. On the other hand, for $p \in I_k$ and $q \in I_\ell$ with $\ell \neq k$, $\omega'(\pm\varepsilon_p \pm \varepsilon_q)$ is of the form $\frac{1}{2}(\sum_{p\in I_k\cup I_\ell} \pm\varepsilon_p)$ with $0, 2, 4, 6$ or 8 minus signs, and we have $\omega'(\varepsilon_{4k-3} + \varepsilon_{4\ell-3}) = e_{k,\ell}$.

Theorem 8.2.1. *1. The lattice $\mathfrak{M}\,\mathbb{D}_n$ is the sublattice of $\mathfrak{M}^m$ generated by $\mathbb{D}_n$ together with the vectors $e_{k,\ell}$; the quotient $\mathfrak{M}\,\mathbb{D}_n/\mathbb{D}_n$ is the elementary Abelian group of order 2^{m-1} with basis over $\mathbb{F}_2$ the images of the vectors $e_{1,k}$, $k = 2, \ldots, m$.*

2. This lattice has norm 2, determinant 4^{2-m}, and possesses

$$s = 12m(4m-3) = 3n(n-3)$$

pairs of minimal vectors; it is half-integral, non-integral for $m \geq 3$, equal to $\mathbb{D}_4$ for $m = 1$ and to $\mathbb{E}_8$ for $m = 2$.

3. For $m \geq 3$, its dual lattice has norm 2, and its minimal vectors are the $12\,m$ pairs of minimal vectors in the orthogonal sum of the m sublattices $\mathbb{D}_4$ associated with each of the m blocks.

4. Let W' be the subgroup of index 2 in $\mathrm{Aut}(\mathbb{D}_4)$ of those elements which change an even number of signs in front of $\varepsilon_1, \varepsilon_2, \varepsilon_3, \varepsilon_4$. (We have $[W' : W(\mathbb{D}_4)] = 3$). Then, for $m \geq 3$, $\mathrm{Aut}(\mathfrak{M}\,\mathbb{D}_n)$ is the subgroup of the semi-direct product $\mathrm{Aut}(\mathbb{D}_4)^m \rtimes S_m$ such that none or all of its m components in $\mathrm{Aut}(\mathbb{D}_4)$ belong to W'. This is a group of unitary automorphisms, which acts transitively on the orbits of minimal vectors in the dual lattice, and splits $S(\mathfrak{M}\,\mathbb{D}_n)$ into two orbits, namely (1) the union of the sets $S(\mathbb{D}_4)$ associated with each block (which contains $12\,m$ pairs of minimal vectors), and (2) its complementary set (which contains $48m(m-1)$ pairs).

5. The lattice $\mathfrak{M}\,\mathbb{D}_n$ is extreme, strongly eutactic, and dual-extreme, and we have

$$\gamma(\mathfrak{M}\,\mathbb{D}_n)^n = 2^{6m-4} \quad \textit{and} \quad \gamma'(\mathfrak{M}\,\mathbb{D}_n)^2 = 4$$

for all $m \geq 2$ (i.e., $n \geq 8$).

Proof. (1) follows immediately from the definition of the vectors $e_{k,\ell}$, and shows that $\mathfrak{M}\,\mathbb{D}_n$ is the union of $\mathbb{D}_n$ and of the sets $e+\mathbb{D}_n$ where e runs through the set of sums of distinct vectors $e_{k,\ell}$. The norm on a set $e+\mathbb{D}_n$ is bounded from below by $\frac{t}{4}$ where t denotes the number of nonzero components of e on the ε_p. Since t is divisible by 8, we have $N(x)\ge 2$ on $e+\mathbb{D}_n$, and equality holds if and only if e is one of the $e_{k,\ell}$. The vectors of norm 2 in $e+\mathbb{D}_n$ are thus the vectors $\frac{1}{2}\sum\pm\varepsilon_p$, $p\in I_k\cup I_\ell$ with an even number of minus signs. Each set $S(e_{k,\ell}+\mathbb{D}_n)$ contributes 64 to s, which adds to $s(\mathbb{D}_n)=n(n-1)$, whence

$$s(\mathfrak{M}\,\mathbb{D}_n)=4m(4m-1)+64\binom{m}{2}=4m(4m-1)+32m(m-1)$$
$$=4m(12m-9)=3n(n-3)\,.$$

Since $\mathbb{D}_n$ has index 2^{m-1} in $\mathfrak{M}\,\mathbb{D}_n$, we have $\det(\mathfrak{M}\,\mathbb{D}_n)=4\,(2^{-(m-1)})^2=4^{2-m}$. Since the scalar products $\varepsilon_p\cdot e_{k,\ell}$ are half-integral, but are not integral for $p\notin I_k\cup I_\ell$, we have proved all assertions in (2).

To prove (3), we restrict ourselves to dimensions $n\ge 12$ (i.e., $m\ge 3$). The lattice is a sublattice of $\mathbb{D}_n^*=\mathbb{Z}^n\cup(e+\mathbb{Z}^n)$ where $e=\frac{1}{2}\sum_1^n\varepsilon_p$. Its norm is at least 2: indeed, we have $N(x)\ge\frac{n}{4}\ge 3$ for all $x\in e+\mathbb{Z}^n$, whence the inclusion $S((\mathfrak{M}\,\mathbb{D}_n)^*)\subset\mathbb{Z}^n$, and the vectors $\varepsilon_p\in\mathbb{Z}^n$ do not belong to $(\mathfrak{M}\,\mathbb{D}_n)^*$ (for $\varepsilon_k\cdot e_k=\frac{1}{2}$). Now the scalar products of $\pm\varepsilon_p\pm\varepsilon_q$ with the e_k are all integral if and only if p and q belong to the same block. This shows that $N((\mathfrak{M}\,\mathbb{D}_n)^*)=2$, and moreover that $S((\mathfrak{M}\,\mathbb{D}_n)^*)$ is the union of m pairwise orthogonal configurations isometric to $S(\mathbb{D}_4)$, whence (3). [We have $(\mathfrak{M}\,\mathbb{D}_n)^*=L\cup(e+L)$, where L is the sublattice of $\mathbb{Z}^n$ defined by the $m-1$ congruences $\sum_{p\in I_1\cup I_\ell}a_p\equiv 0 \mod 2$ for $2\ell\le m$.]

Assertion (3) shows that $\mathrm{Aut}(\mathfrak{M}\,\mathbb{D}_n)$ is a subgroup of the semi-direct product $\mathrm{Aut}(\mathbb{D}_4)^m\rtimes S_m$, where S_m acts as a permutation group on the set of blocks (Theorem 1.4.6). We are now left with the following problem: which elements of $\mathrm{Aut}(\mathbb{D}_4)\times\cdots\times\mathrm{Aut}(\mathbb{D}_4)$ preserve the parity of the number of minus signs in the sums $\sum_{i\in I_k\cup I\ell}\pm\varepsilon_i$? One easily sees that these elements are the sets $(u_1,\dots,u_m)$ such that all u_i belong to W' or to $\mathrm{Aut}(\mathbb{D}_4)\smallsetminus W'$. This proves (4), whose transitivity assertions are easy.

Since $S(\mathfrak{M}\,\mathbb{D}_n)$ contains $S(\mathbb{D}_n)$, $\mathfrak{M}\,\mathbb{D}_n$ is extreme and strongly eutactic because of the transitivity assertions of (4). Since its dual is moreover (strongly) eutactic (the set of its minimal vectors is a union of pairwise orthogonal eutactic configurations), it is dual-extreme. Finally, the calculation of the invariants γ and γ' results from (2) and (3). □

We have defined the lattice above as the smallest submodule of $\mathbb{H}^m$ containing $\mathbb{D}_n$. It can also be constructed as a sublattice of $\mathfrak{M}^m$, by a method similar to the one we use to construct $\mathbb{D}_n$ from $\mathbb{Z}^n$ in Section 4.3, which has the advantage of giving it directly the norm 4, which renders it integral when $\mathbb{H}$ is endowed with the reduced trace.

Proposition 8.2.2. *The sublattice*

$$\{(\lambda_1, \dots, \lambda_m) \in \mathfrak{M}^m \mid \lambda_1 + \dots + \lambda_m \equiv 0 \mod \mathfrak{a}\}$$

of $\mathfrak{M}^m$ is equal $\mathfrak{M}\,\mathbb{D}_n$, and its dual is

$$\{(\mu_1, \dots, \mu_m) \in \mathfrak{M}^m \mid \mu_1 \equiv \dots \equiv \mu_m \mod \mathfrak{a}\}\,.$$

For $m > 2$, the two orbits of minimal vectors are (1) the set of sequences of $m-2$ zeros and two quaternion units, and (2) the set of sequences of $m-1$ zeros and one quaternion of reduced norm 2.

Proof. Let L_1 and L_2 be the two lattices defined in Proposition 8.2.2. The first one contains all vectors $e_{k,\ell}$. Since $\mathrm{Nrd}(1+i) = 2$, it has index 4 in $\mathfrak{M}^n$. We thus have $\det(L_1) = (\frac{1}{4})^m 4^2 = 4^{2-m}$, and this proves that $L_1 = \mathfrak{M}\,\mathbb{D}_n$. Similarly, we have $[\mathfrak{M}^m : L_2] = 4^{m-1}$, hence

$$\det(L_2) = \left(\frac{1}{4}\right)^m 4^{2(m-1)} = 4^{m-2} = (\det(L_1))^{-1}.$$

Now for $(\lambda_1, \dots, \lambda_m) \in L_1$ and $(\mu_1, \dots, \mu_m) \in L_2$, we have $\sum_1^m \lambda_i \overline{\mu}_i \equiv \mu_1 \sum \lambda_i \equiv 0 \mod \mathfrak{a}$. Hence $\sum_1^m \lambda_i \overline{\mu}_i$ belongs to $\mathfrak{O}$, and consequently $\frac{1}{2}\,\mathrm{Tr}(\sum_1^m \lambda_i \overline{\mu}_i)$ belongs to $\mathbb{Z}$. This proves the inclusion $L_2 \subset L_1^*$, and indeed the equality $L_1^* = L_2$ since both sides have equal determinants. □

We now drop the quaternionic structures to construct lattices of all dimensions, using sections of the dual of $\mathfrak{M}\,\mathbb{D}_n$. Theorem 8.2.1 shows that $(\mathfrak{M}\,\mathbb{D}_n)^*$ possesses sections isometric to each of the lattices $\mathbb{A}_1$, $\mathbb{A}_2$, $\mathbb{A}_3$, and $\mathbb{D}_4$, acted on transitively by the automorphism group of $\mathfrak{M}\,\mathbb{D}_n$. Thus the isometry class of the orthogonal complement $\mathfrak{M}\,\mathbb{D}_n$ to one of the lattices above inside $(\mathfrak{M}\,\mathbb{D}_n)^*$ does not depend on the particular choice of the lattice; this is indeed obvious in the $\mathbb{D}_4$ case, where we obtain a lattice canonically isometric to $\mathfrak{M}\,\mathbb{D}_{n-4}$.

In order to obtain canonically defined lattices, we however fix a choice for A_1, A_2, A_3, choosing successively the lattices $\langle f_0\rangle$, $\langle f_0, f_1\rangle$ and $\langle f_0, f_1, f_2\rangle$ where $f_i = \varepsilon_{n-i} + \varepsilon_{n-i-1}$. Moreover, in order to handle integral lattices, we shall replace $\mathfrak{M}\,\mathbb{D}_n$ by its image by the similarity $x \mapsto x\,(1+i)$ (right multiplication preserves the left module structure). On the dual lattice, this transformation becomes left multiplication by $\frac{1-i}{2}$, which transforms $1, i, j, k$ into $\frac{1-i}{2}, \frac{1+i}{2}, \frac{j+k}{2}, \frac{-j+k}{2}$ respectively, and thus f_0 into $g_0 = \varepsilon_n$, f_1 into $g_1 = \frac{1}{2}(\varepsilon_{n-3} + \varepsilon_{n-2} + \varepsilon_{n-1} + \varepsilon_n)$ and f_2 into $g_2 = \varepsilon_{n-3}$.

[Note that, since left and right multiplications by $(1+i)$ transform $\mathfrak{M}$ into the same two-sided ideal $\mathfrak{a}$, we obtain the same lattices by using any of these transformations.]

Definition 8.2.3. *Let $n = 4h + \ell$, $\ell \in \{0, 1, 2, 3\}$ be a positive integer. We denote by J_n the lattice defined in the following way, where we set $n = 4h+4$ in the definitions of g_0, g_1, g_2:*

1. *For* $\ell = 0$, $J_n = (1+i)\,\mathfrak{M}\,\mathbb{D}_n = \mathfrak{M}\,\mathbb{D}_n\,(1+i)$.
2. *For* $\ell = 3$ *(resp.* $\ell = 2$*, resp.* $\ell = 1$*),* J_n *is the orthogonal in* J_{4h-4} *of* $\mathbb{Z}\,g_0$ *(resp. of* $\mathbb{Z}\,g_0 + \mathbb{Z}\,g_1$*, resp. of* $\mathbb{Z}\,g_0 + \mathbb{Z}\,g_1 + \mathbb{Z}\,g_2$*).*

[For $n \equiv 0 \mod 4$, an equivalent formulation of Definition 8.2.3 consists in endowing $\mathfrak{M}$ with the form $\mathrm{Tr}(\lambda\overline{\mu})$ and setting

$$J_n = \{\lambda \in \mathfrak{M}^h \mid \lambda_1 + \lambda_2 + \cdots + \lambda_h \equiv 0 \mod \mathfrak{a}\}. \qquad \mathbf{(8.2.3')}$$

Note that since $\mathfrak{M}\,(1+i) = (1+i)\,\mathfrak{M}$ is the same two-sided ideal $\mathfrak{a}$, J_n is a two-sided module. Note also that it is integral over $\mathfrak{M}$.]

Theorem 8.2.4. *For all* $n \geq 1$, J_n *is an even (integral) lattice of norm* 4*, which is a hyperplane section of* J_{n+1}*. It has the following invariants:*

$n = 4h$	$\det(J_n) = 2^{2h+4}$	$s = 12h(4h-3)$
$n = 4h+1$	$\det(J_n) = 2^{2h+5}$	$s = 4h(12h-7)$
$n = 4h+2$	$\det(J_n) = 3 \cdot 2^{(2h+4)}$	$s = 12h(4h-1)$
$n = 4h+3$	$\det(J_n) = 2^{2h+6}$	$s = 3(16h^2+4h+1)$

It is perfect for all n*, extreme for* $n \equiv 0$ *or* $1 \mod 4$*, and dual-extreme when* n *is divisible by* 4*.*

Proof. The definition of $\mathfrak{M}\,\mathbb{D}_n$ shows that for $n \equiv 0 \mod 4$, J_n is even of norm 4 and determinant $4^{2-h} \cdot 2^{4h} = 2^{2h+4}$, and that its dual lattice has norm 1. Hence all lattices J_n are even of norm at least 4; that equality holds for all n will be a consequence of the determination of the minimal vectors. For $\ell \neq 0$, J_n is the orthogonal section of a lattice isometric to $\frac{1}{\sqrt{2}}\,\mathbb{A}_\ell$, whose respective determinants for $\ell = 1,2,3$ are 1, $\frac{3}{4}$, and $\frac{1}{2}$. This proves the formulae for the determinants.

For the remainder of the proof, we return to the notation $\mathfrak{M}\,\mathbb{D}_n$ and use results of Theorem 8.2.1. For $n = 4h+3, 4h+2, 4h+1$ respectively, each of the $\frac{h(h-1)}{2}$ elements of norm 2 in $e_{k,\ell} + \mathbb{D}_n$ with $1 \leq k < \ell \leq h$ gives s a contribution equal to 64; that of $\mathbb{D}_n$ is equal to $1+(4h+2)(4h+1)$, $(4h+1)4h$, $4h(4h-1)$, and that of each of the h sets $e_{k,h+1} + \mathbb{D}_{4h+4}$ is equal to $32, 16, 8$. This gives for s the values we want. Moreover, the description of the minimal vectors shows that for all n, $S(J_n)$ contains n independent vectors outside J_{n-1}. The lattices J_n are thus relatively perfect, hence perfect for all $n \geq 8$ since J_8 is similar to $\mathbb{E}_8$. (For $n < 8$, we find up to scale the usual sections of $\mathbb{E}_8$ or of $\mathbb{D}_4$, i.e. the series $\mathbb{E}_7$, $\mathbb{E}_6$, $\mathbb{D}_5$, and $\mathbb{A}_3$, $\mathbb{A}_2$, $\mathbb{A}_1$.)

We have seen that J_n contains for $n \equiv 0 \mod 4$ a minimal section similar to $\mathbb{D}_n$; this shows that J_n is an extreme lattice. An argument of the same kind applies to $n = 4h+1$: by making use of an automorphism of order 3 of $\mathbb{D}_4$, we can map the vectors $\varepsilon_{4h+4} + \varepsilon_{4h+3}$, $\varepsilon_{4h+3} + \varepsilon_{4h+2}$, $\varepsilon_{4h+2} + \varepsilon_{4h+1}$

onto $\varepsilon_{4h+4} + \varepsilon_{4h+3}$, $\varepsilon_{4h+3} + \varepsilon_{4h+2}$, $\varepsilon_{4h+2} - \varepsilon_{4h+4}$ respectively, which shows that J_{4h+1} contains a lattice similar to $\mathbb{D}_{4h+1}$.

Finally, the fact that J_n is dual-extreme for $n \equiv 0 \bmod 4$ is part of assertion (5) of Theorem 8.2.1. □

For $n \leq 12$, the lattices J_n are laminated lattices as defined by Conway and Sloane (see Section 3.7). For $n \leq 8$, they are the root lattices $\{0\}$, $\mathbb{A}_1$, $\mathbb{A}_2$, $\mathbb{A}_3$, $\mathbb{D}_4$, $\mathbb{D}_5$, $\mathbb{E}_6$, $\mathbb{E}_7$, $\mathbb{E}_8$ rescaled to norm 4, whose determinants are $1, 4, 12, 32, 64, 128, 192, 256, 256$. For $n = 9, 10, 11, 12$, their determinants are 512, 768, 1024, 1024. Recall that these lattices are unique up to dimension $n = 24$, except for $n = 11, 12, 13$, where they are, however, characterized by their kissing number and denoted by one of the upper indices min, mid, max. Here are the values of s:

$n = 11$	$s(\Lambda_{11}^{\min}) = 216$		$s(\Lambda_{11}^{\max}) = 219.$
$n = 12$	$s(\Lambda_{12}^{\min}) = 312$	$s(\Lambda_{12}^{\mathrm{mid}}) = 316$	$s(\Lambda_{12}^{\max}) = 324.$
$n = 13$	$s(\Lambda_{13}^{\min}) = 444$	$s(\Lambda_{13}^{\mathrm{mid}}) = 445$	$s(\Lambda_{13}^{\max}) = 453.$

Proposition 8.2.5. *We have $J_n \simeq \Lambda_n$ for $n \leq 10$, $J_{11} \simeq \Lambda_{11}^{\max}$, $J_{12} \simeq \Lambda_{12}^{\max}$, and for $n > 12$, $\det(J_n) > \det(\Lambda_n)$.*

Proof. The case of dimensions $n \leq 8$ has been considered in the proof of Theorem 8.2.4. The calculation of the determinants for $n = 9, 10, 11, 12$ shows inductively that $J_9, J_{10}, J_{11}, J_{12}$ are laminated lattices, and we then identify J_{11} and J_{12} by their kissing numbers. Finally, the determinants of the laminated lattices decrease from dimension 12 onwards, whereas those of the J_n increase, and moreover strictly from J_{12} to J_{13}. □

The method we used in Section 5.5 to construct the lattices $\mathbb{D}_{n,t}$ by enlarging $\mathbb{D}_n$ can be adapted to obtain enlargements of J_n. Consider for $h \geq 0$ and $n \geq 4h$ the lattices $\langle \mathfrak{M}\,\mathbb{D}_{4h}, \mathbb{D}_n \rangle$. Rescaling them to norm 4, we obtain extreme integral lattices $J_{n,4h}$ with invariants $\det(J_{n,4h}) = 2^{n+4-2h}$ and $s(J_{n,4h}) = n(n-1) + 32h(h-1)$. For $n = 4h$ and $n = 4h+1$, they are isometric to J_n. For $h = 2$ and $n = 9$, we find a lattice similar to $\mathbb{D}_{9,8} \sim \Lambda_9$.

For $h = 3$ (and thus $n \geq 12$), we have $\det(J_{n,12}) = 2^{n-2}$. The method we used in Section 4.8 (with which we obtained $\mathbb{E}_7$ from $\mathbb{Z}^n$) applies here to enlarge this lattice for $n \geq 14$ by replacing the components 1 by pairs $(1,1)$ in the binary code ("repetition code"). We find for $n \geq 14$ the lattice

$$L = \left\langle J_{n,12}, \frac{1}{2}(\varepsilon_1 + \varepsilon_2 + \varepsilon_5 + \varepsilon_6 + \varepsilon_9 + \varepsilon_{10} + \varepsilon_{13} + \varepsilon_{14}) \right\rangle .$$

Rescaling it to norm 4, we obtain a lattice $\widehat{J}_{n,12}$ with invariants given by the following proposition, whose proof is left as an exercise:

Proposition 8.2.6. *For $n \geq 14$, $\widehat{J}_{n,12}$ is an integral extreme lattice of norm 4, determinant 2^{n-4}, and s-invariant $n(n-1)+448$. For $n = 14$, we have $s = 630$, and $S(\widehat{J}^*_{14,12}) \simeq S(\mathbb{Z}^{14})$.* □

The set of minimal vectors in $(\widehat{J}_{14,12})^*$ appears in a natural way as the union of seven pairwise orthogonal copies of $S(\mathbb{Z}^2)$, acted on by the semi-direct product $S_7 \rtimes \mathrm{Aut}(\mathbb{Z}^2)$. This group permutes transitively the minimal vectors, but it can be shown that the sublattices isometric to $\mathbb{Z}^2$ share out among two orbits.

Consider in $(\widehat{J}_{14,12})^*$ the images of the four vectors $x_1 = \varepsilon_{14} + \varepsilon_{13}$, $x'_1 = \varepsilon_{14} - \varepsilon_{13}$, $x_2 = \varepsilon_{12} + \varepsilon_{11}$, and $x_3 = \varepsilon_{14} + \varepsilon_{12} + \varepsilon_{10} \in L^*$, and set $R_1 = \langle x_1 \rangle$, $R_2 = \langle x_1, x'_1 \rangle$, $R'_2 = \langle x_1, x_2 \rangle$, $R_3 = \langle x_1, x'_1, x_2 \rangle$, and $R'_3 = \langle x_1, x_2, x_3 \rangle$. We again leave to the reader the proof of the following (easy, but somewhat tedious) proposition, which yields an explicit construction of some laminated lattices in dimensions 11 to 13:

Proposition 8.2.7. *The orthogonal complements to R_1, R_2, R'_2, R_3 and R'_3 inside $\widehat{J}_{14,12}$ are isometric to $\Lambda_{13}^{\max}$, $\Lambda_{12}^{\max}$, $\Lambda_{12}^{\mathrm{mid}}$, $\Lambda_{11}^{\max}$, and $\Lambda_{11}^{\min}$ respectively.* □

A further step consists in considering the lattice generated by J_{16} and $\widehat{J}_{14,12}$ embedded in $\mathbb{R}J_{16}$. This construction, analogous to that of $\mathbb{E}_8$ from $\mathbb{E}_7$ and $\mathbb{D}_8$ (the extended Hamming code is generated by suitable choices of the generating codes for $\mathbb{E}_7$ and $\mathbb{D}_8$; here, we should use codes over $\mathbb{F}_4$) yields a norm 4 lattice with determinant 1024. Scaled to norm 2, it can be viewed as the lattice obtained by adjoining to $\mathbb{D}_{16}$ the four vectors $\frac{1}{2}(\sum_\ell \varepsilon_\ell)$, with $\ell \in [1,8], [5,12], [9,16]$ or $\{1,2,5,6,9,10,13,14\}$. This lattice is not a sub-$\mathfrak{M}$-module of $\mathfrak{M}^4$, and the smallest submodule which contains it is obtained by adjoining two more vectors, corresponding to $\ell \in \{3,4,7,8,11,12,15,16\}$ and $\ell \in \{1,3,5,7,9,11,13,15\}$. When scaled to norm 4, the lattice we obtain in this way has determinant $2^8 = 256$, and is easily proved to be isometric to the Barnes–Wall lattice that we shall define in a more synthetic way in Section 8.3.

8.3 Hurwitz Lattices: Around Dimension 16

We keep in this section the notation of the previous two sections: orders $\mathfrak{O}$ and $\mathfrak{M}$ (the Hurwitz order), skew-fields H and $\mathbb{H}$ (still endowed with the scalar product $\frac{1}{2}\mathrm{Trd}(x\overline{y})$), … ; m and $n = 4m$ are positive integers, and $\mathfrak{a}$ is the two-sided ideal in both $\mathfrak{M}$ and $\mathfrak{O}$, generated *over* $\mathfrak{M}$ (on either sides) by $1+i$. We have $\mathfrak{a}^2 = (2)$ in $\mathfrak{M}$ but *not* in $\mathfrak{O}$ ($\mathfrak{a}$ is not invertible in $\mathfrak{O}$).

Definition 8.3.1. *Let $m \geq 2$ be even. We denote by J'_{4m} the sublattice of $\mathfrak{M}^{4m}$ whose elements $(\lambda_1, \dots, \lambda_m)$ satisfy the following system of congruences:*

$$\lambda_1 \equiv \lambda_2 \equiv \cdots \equiv \lambda_m \mod \mathfrak{a} \tag{C1}$$

$$\lambda_1 + \lambda_2 + \cdots + \lambda_m \equiv 0 \mod \mathfrak{a}^2. \tag{C2}$$

When $m = 4$ (and thus $n = 16$), this lattice is named the Barnes–Wall (16-dimensional) lattice, *and denoted by* BW_{16}.

[Using Definition 8.2.3′ for J_n, we obtain for J_n' the following equivalent formulation of Definition 8.3.1:

$$J_n' = J_n \cup J_n + v\,(1+i)^{-1}, \tag{\textbf{8.3.1}$'$}$$

where v denotes the vector $(1,\dots,1) \in \mathfrak{M}^m$. We recognize here an adaptation to the Hurwitz order of Kneser's neighbouring algorithm that we formally defined over $\mathbb{Z}$ in Section 5.7, see in particular Definition 5.7.5, and that we previously used in Chapter 4 to construct $\mathbb{D}_n^+$ as an enlargement of $\mathbb{D}_n$. Note that J_n' is again a two-sided module, and that multiplication on the left by $(1+i)^{-1}$ yields the same lattice. Note also that J_8' is similar to $\mathbb{E}_8$.]

From now on, we restrict ourselves to dimensions $n = 4m \geq 16$.

Theorem 8.3.2. *Let $m \geq 4$ be an integer.*

1. *The lattice J_n' is an even, norm 4 lattice, of determinant 2^{2m} and s-invariant $12m(4m-3)$ if $m \geq 6$, and $2160 = 2^4 \cdot 3^3 \cdot 5$ if $m = 4$; it is unimodular over $\mathfrak{M}$, hence 2-modular over $\mathbb{Q}$.*
2. *Right and left multiplications by $(1+i)^{-1}$ are similarities of ratio $\frac{1}{\sqrt{2}}$ of J_n' onto its dual, which is of norm 2.*
3. *The lattice J_n' is extreme, strongly eutactic, and dual-extreme, and $\gamma(J_n')^2 = \gamma'(J_n')^2 = 8$.*

Proof. Since m is even, the congruence $(\mathrm{C}'2): \lambda_1 + \lambda_2 + \cdots + \lambda_m \equiv 0 \mod \mathfrak{a}$ is a consequence of (C1). The lattice J_n' is thus in fact defined by m congruences, which are independent thanks to the assumption $m \geq 4$; hence it has an index 4^m in $\mathfrak{M}^m$, which proves that $\det(J_n') = 2^{2m}$.

Let $\lambda = (\lambda_1, \dots, \lambda_m) \in J_n'$. Congruence (C1) shows that either all or none of the λ_ℓ belong to $\mathfrak{a}$. In the first case, λ is of the form $\lambda'(1+i)$, and λ satisfies (C1) and (C2) if and only if λ' satisfies (C′2), which is equivalent by Proposition 8.2.2 to $\lambda \in J_n$. The norm of such an element is thus at least 4, and equality holds exactly on $S(J_n)$. In the second case, we have $N(\lambda) \geq m \geq 4$ and equality holds if and only if all λ_ℓ are units (= roots of unity) in $\mathfrak{M}$. The m-tuples which satisfy (C1) are those which are of the form $u(1, u_2, \dots, u_m)$ where u is one of the 24 units in $\mathfrak{M}$ and $u_2, \dots, u_m$ one of the eight units $\pm 1, \pm i, \pm j, \pm k \in \mathfrak{M}$, those which are congruent to 1 modulo $\mathfrak{a}$. If one chooses arbitrarily $u_3, \dots u_m$, u_2 is then determined up to its sign by (C2). There are thus $24 \cdot 8^{m-2}$ pairs $\pm\lambda$ of norm m defined by units. This shows first that $N(J_n') = 4$, and that $s(J_n') = s(J_n) = 12m(4m-3)$ if $m \geq 6$ and

$$s(J'_{16}) = s(J_{16}) + 24 \cdot 8^2 = 624 + 1536 = 2160\,.$$

Definition 8.3.1′ shows that J'_n is integral, thus in particular even as a $\mathbb{Z}$-lattice; since its determinant is that of $\mathfrak{M}^m$, it is unimodular as a lattice over $\mathfrak{M}$. This completes the proof of (1).

We have seen (Proposition 8.2.2) that the lattices L_1 and L_2 defined respectively by the congruences (C1) and (C2) are dual to one another. Adjoining congruence (C2) transforms L_1^* into

$$J_n'^{\,*} = \left\{\frac{1}{2}\,(2\lambda_1 + \lambda_0, \dots, 2\lambda_m + \lambda_0) \mid \lambda_\ell \in \mathfrak{M},\ \lambda_0 \in \mathfrak{a},\ \sum_1^m \lambda_\ell \in \mathfrak{a}\right\}$$

$$= \left\{(\lambda'_1, \dots, \lambda'_m) \in \mathbb{H}^m \mid \lambda'_\ell - \lambda'_1 \in \mathfrak{M},\ \sum_1^m \lambda'_\ell \in \mathfrak{a}\right\}.$$

The equality $(1+i)J_n'^{\,*} = J_n'^{\,*}(1+i) = J'_n$ is now clear, whence (2).

Since $N(J'_n) = 4$ and $N(J_n'^{\,*}) = 2$, we have $\gamma'(J'_n) = 8$, and also $\gamma(J'_n) = \gamma'(J'_n)$, because J'_n and $J_n'^{\,*}$ are similar. Finally, J'_n is extreme because $S(J'_n) \supset S(J_n)$, dual-extreme because $J_n'^{\,*}$, which is similar to J'_n, is obviously eutactic, and strongly eutactic for $n > 16$ because $S(J'_n) = S(J_n)$. The fact that J_{16} is strongly eutactic will result from the last assertion of Proposition 8.3.3 below. □

We now list a few properties of the automorphism group of J'_n.

Proposition 8.3.3. *1. The group* $\mathrm{Aut}(J_n)$ *stabilizes the lattice* J'_n*, and the corresponding injective homomorphism of* $\mathrm{Aut}(J_n)$ *into* $\mathrm{Aut}(J'_n)$ *is an isomorphism for any* $n \geq 24$.

2. For $n \geq 24$, $S(J'_n)$ *possesses two orbits under the action of* $\mathrm{Aut}(J'_n)$.

3. The group of unitary automorphisms of the lattice $J'_{16} = \mathrm{BW}_{16}$ *acts transitively on the set of its minimal vectors.*

Proof. The group $\mathrm{Aut}(J_n)$ contains a subgroup of index 2, which (in the notation of Proposition 8.2.2) acts as a permutation group on the set of λ_ℓ and moreover acts on each λ_ℓ by changing an even number of signs in front of $1, i, j, k$. This implies immediately the stability of J'_n under $\mathrm{Aut}(J_n)$. For $m > 4$ (i.e. $n \geq 24$), these two groups coincide (because $S(J'_n) = S(J_n)$), and we therefore find two orbits as in the case of J_{16}, whence (1) and (2).

For $m = 4$ (i.e. $n = 16$ and $J'_n = \Lambda_{16}$), the automorphism group may *a priori* be larger, and is indeed larger. We remark that there exist three orbits o_1, o_2, o_3 under the action of the unitary automorphism group of J_{16}, namely those of $(1,1,1,1)$, $(1+i, 1+i, 0, 0)$, $(2,0,0,0)$ respectively. To prove (3), it suffices to construct unitary automorphisms u_1 and u_2 for the Hermitian structure of J'_{16} mapping respectively $(1,1,1,1)$ and $(1+i, 1+i, 0, 0)$ onto $(2,0,0,0)$. This is checked using right multiplications by the following unitary matrices, which are easily seen to preserve congruences (C1) and (C2):

$$\frac{1}{2}\begin{pmatrix} +1 & +1 & +1 & +1 \\ +1 & +1 & -1 & -1 \\ +1 & -1 & +1 & -1 \\ +1 & -1 & -1 & +1 \end{pmatrix} \quad \text{and} \quad \frac{1}{2}\begin{pmatrix} 1-i & 1-i & 0 & 0 \\ 1-i & i-1 & 0 & 0 \\ 0 & 0 & 1-i & 1-i \\ 0 & 0 & 1-i & i-1 \end{pmatrix}. \ \square$$

The subgroup of elements of the form $(0,0,0,\lambda(1+i))$, $\lambda \in \mathfrak{M}$ is a 4-dimensional lattice isometric to $\mathbb{D}_4$, which is contained in BW_{16}^*. Its orthogonal is equal to $J_{12} \simeq \Lambda_{12}^{\max}$. Taking successively the orthogonal complements to $\langle 1+i \rangle$, $(1+i)\langle 1, \omega \rangle$, $(1+i)\langle 1, \omega, i \rangle$, we obtain norm 4 lattices containing J_{12}, with determinants 512, 768, 1024 and dimensions 15, 14, 13 respectively. Hence these lattices are laminated lattices. It is easy to see that $(1+i)\langle 1,i \rangle^\perp = \langle 1,i \rangle^\perp$ can be identified with the lattice $\widehat{J}_{14,12}$ that we met in Section 8.2; this proves that the lattice we obtained in dimension 13 is isometric to $\Lambda_{13}^{\max}$, a fact which can also be proved via the determination of its minimal vectors, the problem we are now going to consider.

Let $q = a + bi + cj + dk \in \mathbb{H}$. For $\lambda = x + yi + zj + tk \in \mathbb{H}$, we have $\frac{1}{2}\operatorname{Trd}(\lambda\bar{q}) = ax + by + cz + dt$. If $\lambda = (\lambda_1, \lambda_2, \lambda_3, \lambda_4) \in \mathbb{H}$, let us write $\lambda_\ell = x_\ell + y_\ell i + z_\ell j + t_\ell k$. For $q = 1 + i$, we observe that $\langle (0,0,0,q) \rangle^\perp$ is defined by the equation $x_4 + y_4 = 0$. This equation is satisfied by 12 units $(\pm j, \pm k, \pm\frac{1}{2}(1 - i \pm j \pm k))$ and six elements of norm 2 $(\pm j \pm k, \pm(1-i))$ in $\mathfrak{M}^\times$. The determination of the norm 4 elements in $\langle q \rangle^\perp$ is done as for the lattice Λ_{16} (u_ℓ stands for a unit of $\mathfrak{M}$):

- (u_1, u_2, u_3, u_4): 12 choices for u_4, 8 choices for u_3 and for u_2, and 2 choices for u_1, hence $2 \cdot 768$ elements;
- $(u_1, u_2, 0, 0)\,(1+i)$: $24 \cdot 8$ choices for (u_1, u_2), with 3 permutations, hence $3 \cdot 24 \cdot 8 = 2 \cdot 288$ elements;
- $(u_1, 0, 0, u_4)(1+i)$: $6 \cdot 8$ choices for (u_4, u_1), with 3 permutations, hence $2 \cdot 72$ elements;
- $2(u_1, 0, 0, 0)$: 24 choices for u_1, with 3 permutations, hence $2 \cdot 36$ elements;
- $2(0, 0, 0, u_4)$: 12 choices for u_4, hence $2 \cdot 6$ elements.

Altogether, we find $s = 768 + 288 + 72 + 36 + 6 = 1170 = 2 \cdot 3^2 \cdot 5 \cdot 13$, the known value for $s(\Lambda_{15})$.

We use the same kind of calculation with $\langle (1+i), (1+i)\,\omega \rangle$. Taking $q = (1+i)\,\omega$, we find for $\langle q \rangle^\perp$ the equation $y_4 + t_4 = 0$. The relations $x_4 + y_4 = y_4 + t_4 = 0$ are satisfied by six units $(\pm j, \pm\frac{1}{2}(1 - i \pm j + k))$ and no norm 2 element in $\mathfrak{M}^\times$. The five classes which we put in evidence above give the contributions 384, 288, 0, 36, 3 to s, hence 711 altogether, and this is the known value for $s(\Lambda_{14})$.

[A calculation of the same kind with $\langle 1, i \rangle^\perp$ (resp. with $\langle 1, i, (1+i)\,\omega \rangle^\perp$) gives $s = 256 + 288 + 48 + 36 + 2 = 630$ (resp. $s = 128 + 288 + 0 + 36 + 1 = 453$), which are precisely the values of $s(\widehat{J}_{14,12})$ (resp. of $s(\Lambda_{13}^{\max})$).]

Since the group of (quaternionic) unitary automorphisms of BW_{16}^* acts transitively on its set of minimal vectors, for all $e \in S(\mathrm{BW}_{16}^*)$, the lattices $\mathfrak{M}\,e$ are isometric; the same result holds for their orthogonal in BW_{16} itself, so that all are isometric to $\Lambda_{12}^{\max}$, indeed as $\mathfrak{M}$-lattices. This shows that to find laminated lattices in dimensions 13 and 12 different from those which were described in Section 8.2, we must consider the orthogonal complements to relative lattices of type $\mathbb{D}_4$ and $\mathbb{A}_3 \subset \mathbb{D}_4$ contained in BW_{16}^*, and such that $\mathbb{D}_4$ is not endowed with the quaternionic structure defined by BW_{16}^*. This we now do.

First consider the vectors $e_1 = (0,0,0,1+i)$, $e_1' = (0,0,-\omega,\omega)$, $e_2 = (0,0,0,1-i)$, $e_3 = (0,0,1,-i)$, $e_4 = (0,0,-1+j,0)$ in BW_{16}^*. We see immediately that the lattices $L_3 = \langle e_2, e_3, e_4 \rangle$, $L_4 = \langle e_1, L_3 \rangle$, and $L_4' = \langle e_1', L_3 \rangle$ are isometric to $\mathbb{A}_3$, $\mathbb{D}_4$, and $\mathbb{D}_4$ respectively.

The lattice $L_3^\perp$ is defined by the equations $x_3 = z_3 = x_4 = y_4$ We find for the various types of minimal vectors the following contributions to s: (u_1,u_2,u_3,u_4): $2 \cdot \frac{1}{2}(4 \cdot 4 \cdot 8 \cdot 2) = 256$; $(u_1,u_2,0,0)\,(1+i)$: $\frac{1}{2}\,24 \cdot 8 = 96$; $(u_1,0,u_3,0)\,(1+i)$ (4 permutations): $4 \cdot \frac{1}{2}\,4 \cdot 8 = 64$; $(0,0,u_3,u_4)\,(1+i)$: 0; $2\,(u_1,0,0,0)$ (2 permutations): $2 \cdot \frac{1}{2}\,24 = 24$; $2\,(0,0,u_3,0)$ (2 permutations): $2 \cdot \frac{1}{2}\,4 = 4$, which altogether gives $s = 256+96+64+0+24+4 = 444$. Hence $s(L_3^\perp) = s(\Lambda_{13}^{\min})$.

Next, considering the intersection with $\langle e_1 \rangle^\perp$, we obtain the equations $x_3 = z_3 = x_4 = y_4 = 0$, and a calculation as above gives us $s = 128 + 96 + 64 + 0 + 24 + 4 = 316$, whence $s(L_4^\perp) = s(\Lambda_{12}^{\mathrm{mid}})$.

Replacing finally e_1 by e_1', we find in the same way $s = 160 + 96 + 32 + 0 + 24 + 0 = 312 = s(\Lambda_{12}^{\min})$.

Now L_4 and L_4' are contained in the lattice defined inside BW_{16}^* by the eight real equations $\lambda_1 = \lambda_2 = 0$, a lattice isometric to $\mathbb{E}_8$. The orthogonal complements inside BW_{16} to this lattice and to its sections isometric successively to $\mathbb{E}_7$, $\mathbb{E}_6$, $\mathbb{D}_5$, $\mathbb{D}_4$, the last one being L_4 or L_4', are laminated lattices. Consequently, we have

$$L_4^\perp \simeq \Lambda_{12}^{\mathrm{mid}}, \quad L_4'^\perp \simeq \Lambda_{12}^{\min}, \quad \text{and} \quad L_3^\perp \simeq \Lambda_{13}^{\min}.$$

The (unique) quaternionic structures which exist on L_4 and on L_4' (as lattices isometric to $\mathbb{D}_4$) are not compatible with that of BW_{16}. Hence they induce $\mathfrak{M}$-module structures neither on $L_4^\perp$ nor on $L_4'^\perp$; this is indeed clear in the first case, since $s(L_4^\perp)$ is not divisible by 12.

Curiously, $\Lambda_{12}^{\min}$ can, however, be endowed with a quaternionic structure. Norm 4 $\mathfrak{M}$-lattices with determinant 1024 and with s invariant 312 have indeed been constructed (by F. Sigrist and by M. Laïhem, independently), for which it may be verified that the dual lattice has norm 1. The following lemma will allow us to identify these lattices:

Lemma 8.3.4. *An $\mathfrak{M}$-lattice of norm* 4 *and determinant* 1024 *whose dual lattice is of norm* 1 *is isometric to $\Lambda_{12}^{\min}$ or to $\Lambda_{12}^{\max}$.*

Proof. Let Λ be such a lattice. The existence of a quaternionic structure on its dual shows that Λ contains a section isometric to $\mathbb{D}_4^*$. The orthogonal in Λ to this section is a lattice with norm ≥ 4 and determinant $1024/4 = 256$, hence similar to $\mathbb{E}_8$ since it meets the Hermite constant in that dimension. This shows that Λ is a laminated lattice of dimension 12. Since $2s(\Lambda_{12}^{\mathrm{mid}})$ is not divisible by 24, Λ is isometric to $\Lambda_{12}^{\mathrm{min}}$ or to $\Lambda_{12}^{\mathrm{max}}$. □

We give now the unpublished data of Sigrist, which allow us to verify the assertions above. In the basis $(1, i, j, \mu = \frac{1+i+j+k}{2})$ for $\mathfrak{M}$, multiplications by μ and by i (which generate $\mathfrak{M}^\times$) correspond to the matrices

$$\begin{pmatrix} -1 & -1 & -1 & -1 \\ 1 & 1 & 0 & 1 \\ 1 & 1 & 1 & 0 \\ 1 & 0 & 1 & 1 \end{pmatrix} \quad \text{and} \quad \begin{pmatrix} -1 & -2 & 0 & 0 \\ 1 & 1 & 0 & 0 \\ 0 & 1 & 0 & -1 \\ 1 & 1 & 1 & 0 \end{pmatrix}.$$

The following four matrices D, U, V, W (one of which is symmetric and the other three antisymmetric) constitute a basis for the centralizer of $\mathfrak{M}$:

$$\begin{pmatrix} 2 & 1 & 1 & 1 \\ 1 & 2 & 0 & 0 \\ 1 & 0 & 2 & 0 \\ 1 & 0 & 0 & 2 \end{pmatrix}, \begin{pmatrix} 0 & -1 & 1 & 0 \\ 1 & 0 & 1 & 1 \\ -1 & -1 & 0 & 0 \\ 0 & -1 & 0 & 0 \end{pmatrix}, \begin{pmatrix} 0 & 0 & 0 & 1 \\ 0 & 0 & -1 & 1 \\ 0 & 1 & 0 & 0 \\ -1 & -1 & 0 & 0 \end{pmatrix}, \begin{pmatrix} 0 & -1 & 0 & 0 \\ 1 & 0 & 1 & 0 \\ 0 & -1 & 0 & 1 \\ 0 & 0 & -1 & 0 \end{pmatrix}.$$

The 12×12 matrix found by Sigrist is

$$A = \begin{pmatrix} 2D & D+U & V \\ D-U & 2D & D \\ -V & D & 2D \end{pmatrix}.$$

A calculation using a computer shows that the hypotheses of Lemma 8.3.4 are satisfied, proving indeed slightly more:

Proposition 8.3.5. *The symmetric matrix A is positive definite, and the lattice defined up to isometry by A is integral even of minimum* 4, *and determinant* 1024. *Moreover, the minimal vectors in Λ^* generate a lattice isometric to D_4^*; in particular, $\Lambda_{12}^{\mathrm{min}}$ is not dual-extreme.* □

Thanks to this proposition, we can identify Λ without appealing to Blichfeldt's theorem, by simply exhibiting a section similar to $\mathbb{E}_8$.

We remark that $\Lambda_{12}^{\mathrm{min}}$ and $\Lambda_{12}^{\mathrm{max}}$ have the same dual Hermite invariant, but that the latter one is not dual-extreme. Hence ${\gamma'_{12}}^2$ is certainly strictly larger than 4. The lattice $CT_{12} \simeq K_{12}$ to be constructed in Section 8.5 is an explicit example of a lattice with such a larger γ' invariant ($\frac{16}{3}$ instead of 4). (Another difference between $\Lambda_{12}^{\mathrm{min}}$ and $\Lambda_{12}^{\mathrm{max}}$: one must rescale $(\Lambda_{12}^{\mathrm{min}})^*$ to minimum 8 to make it integral, whereas minimum 4 suffices for $(\Lambda_{12}^{\mathrm{max}})^*$.)

Using the Voronoi algorithm for the group $G = \mathfrak{M}^\times$ of order 24, Sigrist has classified all G-perfect lattices in the sense of Chapter 13 up to dimension 12. His results imply in particular that there is a single $\mathfrak{M}$-structure on each of the lattices $\mathbb{D}_4$, $\mathbb{E}_8$, $\Lambda_{12}^{\mathrm{min}}$, and $\Lambda_{12}^{\mathrm{max}}$. Hence one of the following properties holds: either there are at least two possible $\mathfrak{M}$-structures on Λ_{16}, or $\Lambda_{12}^{\mathrm{min}}$ is a dead-end for the $\mathfrak{M}$-laminations, as $\Lambda_{13}^{\mathrm{mid}}$ is over $\mathbb{Z}$.

Probably the laminated lattices up to dimension 24 with $n \equiv 0 \mod 4$ and $2s \equiv 0 \mod 24$ possess a unique $\mathfrak{M}$-structure. If it is the case, the $\mathfrak{M}$-laminations are completely described up to dimension 24 by the inclusions

$$\{0\} \subset \Lambda_8 \subset \Lambda_{12}^{\mathrm{max}} \subset \Lambda_{16} \subset \Lambda_{20} \subset \Lambda_{24} \qquad \text{and} \qquad \Lambda_8 \subset \Lambda_{12}^{\mathrm{min}}.$$

8.4 Eisenstein Lattices: A Construction of Barnes

In this section, A denotes the subring $\mathbb{Z}[\omega]$ of $K = \mathbb{Q}(\omega) \subset \mathbb{C}$ where $\omega = \frac{-1+i\sqrt{3}}{2}$. Let $m > 0$ be an integer. Set $\pi = 1 - \omega$, and let $\mathfrak{a}$ be the ideal in A (of norm 3) generated by π; we have $\mathfrak{a}^2 = (3)$.

Let $\lambda = (\lambda_1, \lambda_2, \dots, \lambda_m) \in A^m$. We consider as in Section 8.3 the following systems of congruences between elements $\lambda_1, \dots, \lambda_m$ of A:

$$\lambda_1 \equiv \lambda_2 \equiv \cdots \equiv \lambda_m \mod \mathfrak{a}, \tag{C1}$$

$$\lambda_1 + \lambda_2 + \cdots + \lambda_m \equiv 0 \mod \mathfrak{a}, \tag{C'2}$$

$$\lambda_1 + \lambda_2 + \cdots + \lambda_m \equiv 0 \mod \mathfrak{a}^2. \tag{C2}$$

We endow $\mathbb{C}$ *with the scalar product* $T : (\lambda, \mu) \mapsto \mathrm{Tr}(\lambda\overline{\mu})$, leaving aside the coefficient $\frac{1}{2}$ of Sections 8.2 and 8.3. The ring A is thus isometric to the lattice $\mathbb{A}_2$, its units have Euclidean norm 2, and the Euclidean norms of elements of $\mathfrak{a}$ are divisible by 6. On K^m, the Hermitian scalar product is $h(x,y) = \sum_{k=1}^{n} x_k \bar{y}_k$.

In this section, we essentially consider the congruence (C′2).

Definition 8.4.1. *1. We denote by* L_{2m} *the lattice in* $\mathbb{C}^m$ *whose elements are the m-tuples* $\lambda = (\lambda_1, \dots, \lambda_m)$ *which satisfy the congruence* (C′2).
2. We denote by L_{2m-1} *the hyperplane section of* L_{2m} *defined by the condition* $\lambda_m \in \mathbb{R}$.

The notation L_n above is a provisional one: after Definition 8.4.4, the family L_n will appear as part of the two-parameter family L_n^r, indeed the lattices L_{2r}^r or L_{2r+1}^r. However (see Theorems 8.4.5 and 8.5.9 below), the lattices L_n possess special interesting properties.

Clearly L_1 and L_2 are similar to $\mathbb{A}_1$ and $\mathbb{A}_2$ respectively. To unify the forthcoming statements, we only consider from now on dimensions $n \geq 3$.

Theorem 8.4.2. *1. The lattices* L_n $(n \geq 3)$ *are even lattices of norm* 4, *integral over* A *when* n *is even, and possess the following invariants:*

$$n = 2h, \qquad \det(L_n) = 3^{h+2}, \qquad s(L_n) = \frac{9h(h-1)}{2},$$

$$n = 2h+1, \quad \det(L_n) = 2 \cdot 3^{h+2}, \quad s(L_n) = \frac{3h(3h-1)}{2}.$$

2. For n *even,* L_n^* *is the image by a homothetic transformation of ratio* $\frac{1}{3}$ *of the lattice defined by the congruence* (C1). *We have* $s(L_4^*) = 9$, $s(L_6^*) = 36$, *and* $s(L_n^*) = \frac{3n}{2}$ *for* $n \geq 8$.
3. L_5 *(resp.* L_6*) is isometric to* $\mathbb{A}_5^3$ *(resp. to* $\mathbb{E}_6^*$*).*
4. For n *even, the group of unitary automorphism of* L_n *acts transitively on* $S(L_n)$ *and on* $S(L_n^*)$.

5. *L_n is extreme for all $n \geq 5$. For all $n \geq 6$ even, L_n and L_n^* are strongly eutactic and dual-extreme, and $\gamma'_n(L_n)^2 = \frac{8}{3}$.*

Proof. The lattices L_n are clearly A-integral for n even, hence even, and this last property holds when n is odd because L_n is then a cross-section of L_{n+1}. Since nonzero vectors in L_n possess at least two nonzero components, we have $N(L_n) \geq 4$, and equality holds exactly on the $\frac{m(m-1)}{2}$ permutations of $\pm(\omega^i, -\omega^j, 0, \dots, 0)$ when n is even. This gives us immediately the value of s when n is even, and shows that

$$s(L_{2m-1}) = \frac{9(m-1)(m-2)}{2} + 3(m-1) = \frac{3(m-1)(3m-4)}{2}.$$

Replacing m by $h+1$ gives us the value of s for n odd.

Since $[A^m : L_{2m}] = 3$, we have $\det(L_{2m}) = 3^m \cdot 3^2 = 3^{m+2}$. Now for $\lambda \in A$, the condition "$\lambda \in \mathbb{Z}$" is equivalent to the fact that λ be orthogonal to the primitive element $\frac{\omega+2}{3}$ of the dual of A. We thus have

$$\det(L_{2m-1}) = \frac{2}{3}\det(L_{2m}) = 2 \cdot 3^{m+1} = 2 \cdot 3^{h+2},$$

whence (1).

The lattice L' defined by congruence (C1) has determinant $3^m \cdot 3^{2(m-1)} = 3^{3m-2}$. Hence

$$\det\left(\frac{1}{3}L'\right) = 3^{3m-2} \cdot 3^{-4m} = 3^{-m-2} = \det(L_{2m})^{-1}.$$

But if λ satisfies (C1) and μ satisfies (C$'$2), we have $\sum \lambda_i \overline{\mu}_i \equiv \lambda_1 \sum \overline{\mu}_i \equiv 0 \bmod \mathfrak{a}$. Since $\mathrm{Tr}(\mathfrak{a}) \subset 3\mathbb{Z}$, we have the inclusion $L_{2m}^* \supset L'_{2m}$, which is indeed an equality since both lattices have the same determinant. The remaining assertions of (2) are easily proved by inspection of congruence (C1) if $n \geq 8$ or if $n = 4$ (see also Remark 8.4.3 below), and the case where $n = 6$ follows from assertion (3), which we are now going prove.

The existence of a similarity $L_6 \sim \mathbb{E}_6^*$ has been proved in Section 4.8, and the section of $\mathbb{E}_6^*$ by the orthogonal of a minimal vector in $\mathbb{E}_6$ has been identified with $\mathbb{A}_5^3$ in Proposition 5.2.4; see also Chapter 4, proof of Theorem 4.7.2. This proves (3), and allows us to restrict ourselves to dimensions $n \geq 7$ for proving the last two assertions of Theorem 8.4.2.

The group $G_m = \{\pm \mathrm{Id}\} \times (C_3^m \rtimes S_m)$ is a unitary subgroup of $\mathrm{Aut}(\mathbb{A}_2^m)$ which preserves L_{2m} and clearly acts transitively on both $S(L_{2m})$ and $S(L_{2m}^*)$, hence (4).

We now prove the perfection of L_n ($n \geq 5$) by induction on n. The result is true for $n = 5$ thanks to the isomorphism $L_5 \simeq \mathbb{A}_5^3$. When we embed L_{n-1} inside L_n, we see immediately that $S(L_n) \smallsetminus S(L_{n-1})$ contains n independent vectors. This means that L_n is perfect relatively to L_{n-1} for all $n \geq 2$. If we assume that L_{n-1} is perfect, then L_n is also perfect by Proposition 3.5.3.

This proves that L_n is perfect, and moreover shows using (4) that L_n is strongly eutactic for all even $n \geq 6$. A better proof consists in observing that the group denoted by G_m in the proof of (4) acts irreducibly on the *real* vector space $\mathbb{C}^m$. This implies (Theorem 3.6.6) that both L_n and L_n^* are strongly eutactic for all even $n \geq 4$, and it is readily verified that the equality $\gamma'(L_n) = \frac{8}{3}$ holds for even n from $n = 6$ onwards.

We are left with the proof of eutaxy for odd $n \geq 7$, and we shall give a proof which works for all odd $n \geq 5$. Let $n = 2h + 1$. Embed L_n into $\mathbb{C}^h \times \mathbb{R}$ and denote by p_1 and p_2 the orthogonal projections onto $\mathbb{C}^h$ and $\mathbb{R}$. The group $\mathrm{Aut}(L_{2h})$ acts on L_{2h+1}, and the minimal vectors of L_{2m+1} share out among two orbits o_1 and o_2 under this action, namely those of the elements $(1, -1, 0, \dots, 0, 0)$ and $(1, 0, \dots, 0, 1)$, with respectively $\dfrac{9h(h-1)}{2}$ and $3h$ elements. There exists a eutaxy relation of the form $\mathrm{Id} = \alpha S_1 + \beta S_2$ where

$$S_1 = \sum_{x \in o_1/\{\pm \mathrm{Id}\}} p_x \quad \text{and} \quad S_2 = \sum_{y \in o_2/\{\pm \mathrm{Id}\}} p_y\,,$$

and L_n is eutactic if and only if the well-defined real numbers α and β are strictly positive. Applying p_2 to both sides of the eutaxy relation yields $p_2 = \beta p_2 \circ S_2$. Taking the trace on both sides of each of these relations, we obtain the two equations

$$2h + 1 = \frac{9h(h-1)}{2}\,\alpha + 3h\,\beta \quad \text{and} \quad 1 = \mathrm{Tr}(p_2 \circ S_2)\,\beta\,.$$

Let $e = (0, \dots, 0, 1)$ and let $y \in o_2$. Since $p_2 = p_e$ maps $\mathbb{C}^m \times \mathbb{R}$ onto $\mathbb{R}e$, the trace of $p_2 \circ p_y$ is the coefficient of e in $(p_2 \circ p_y)(e)$. We have $e \cdot e = 2$, $y \cdot y = 4$, $e \cdot y = -2$, hence $(p_2 \circ p_y)(e) = \dfrac{(e \cdot y)^2}{(y \cdot y)\,(e \cdot e)}\, e = \dfrac{1}{2}\, e$, whence $\mathrm{Tr}(p_2 \circ S_2) = \dfrac{3h}{2}$. We can now calculate β, then α, obtaining

$$\alpha = \frac{2(2h-1)}{9h(h-1)} > 0 \quad \text{and} \quad \beta = \frac{2}{3h} > 0\,. \ \square$$

Remark 8.4.3. *The lattice L_4 is eutactic and similar to its dual, but is not perfect, since $s(L_4) = 9$, nor dual-extreme: one can verify that the sets of projections onto the minimal vectors of L_4 and L_4^* span the same 9-dimensional subspace of* $\mathrm{End}^s(E)$.

The lattices L_n^r were defined by Barnes who applied the congruence (C′2) to an orthogonal sum of lattices equal to $\mathbb{A}_2$ or to $\mathbb{A}_1 \subset \mathbb{A}_2$. Here is a precise definition:

Definition 8.4.4. *Let r and t be non-negative integers. Let $m = r + t$ and $n = 2r + t$. (We thus have $0 \leq r \leq \frac{n}{2}$.) We denote by L_n^r the sublattice of A^m defined for $n \geq 1$ by the conditions*

$$\lambda_1 + \lambda_2 + \dots + \lambda_m \equiv 0 \mod \mathfrak{a} \quad \text{and} \quad \lambda_m, \lambda_{m-1}, \dots, \lambda_{m-t} = \lambda_r \in \mathbb{R}\,.$$

This family of lattices is a generalization of the family L_n of Definition 8.4.1. Indeed, we have $L_n = L_n^{n/2}$ if n is even and $L_n = L_n^{(n-1)/2}$ if n is odd. As previously, we only consider dimensions $n \geq 3$.

Theorem 8.4.5. *1. The lattices L_n^r are integral lattices of norm 4 and determinant $2^{n-2r} \cdot 3^{r+2}$.*
2. We have $s(L_n^r) = \frac{n(n+1)}{2} + (r-1)n + \frac{r(r-7)}{2}$.
3. The lattices L_n^r are perfect if and only if $n \geq 5$ and $r \geq 2$.

Proof. (1) The lattices L_n^r are obviously integral of norm 4. The calculation of the determinant follows that of the proof of Theorem 8.4.2, noting that L_n^r is the orthogonal in $L_{2(r+t)}$ of t pairwise orthogonal vectors of norm $\frac{2}{3}$, which implies that

$$\det(L_n^r) = \left(\frac{2}{3}\right)^t \det(L_{2(r+t)}) = 2^t \cdot 3^{r+2}.$$

(2) Up to the sign, the minimal vectors in L_n^r are the vectors with exactly two nonzero components equal to cube roots of unity, which must moreover be equal to 1 when the index of the component lies between $r+1$ and $r+t$, which immediately yields the formula

$$s(L_n^r) = 3^2 \binom{r}{2} + 3\,r\,t + 1 \cdot \binom{t}{2},$$

which can then be easily transformed into the formula of assertion (2).

(3) By the calculation above of $s(L_n^r)$, we see that for $r = 0, 1, 2$, we have $s = \frac{n(n+1)}{2} - n$, $s = \frac{n(n+1)}{2} - 3$, and $s = \frac{n(n+1)}{2} + n - 5$ respectively, which shows that L_n^r cannot be perfect unless $r \geq 2$ and $n \geq 5$.

Conversely, when these conditions are satisfied, we prove the perfection by induction on $t = n - 2r$: the cases $t = 0$ and $t = 1$ are those of Theorem 8.4.2; the general case follows from the observation $L_n^r = L_{2r+t}^r$ is perfect relatively to $L_{2r+t-1}^r = L_{n-1}^r$. □

Barnes ([Bar5], I, p. 64) has proved that a lattice L_n^r is eutactic only if it is a lattice L_n. The case of dimension 6 is of a particular interest: by Theorem 8.4.5, L_6^2 is a perfect lattice of determinant $4 \cdot 3^4 = 324$ with $s = 22$ which is not extreme (it is only semi-eutactic). It was proved by Barnes that it is the unique perfect, non-extreme lattice in dimension $n \leq 6$; hence it coincides with the perfect non-extreme 6-dimensional lattice mentioned by Voronoi in the introduction of [Vo1]. It can be shown ([B-M1], Theorem 4.4) that the lattices L_n^r are not dual-extreme except $L_5^2 \simeq \mathbb{A}_5^3$ and the L_{2r}^r, $r \geq 3$.

The lattice $L_n = L_n^m$ ($n = 2m$ even) plays with respect to the ring A of Eisenstein integers the rôle that plays the lattice $\mathbb{D}_n$ with respect to the ring $\mathbb{Z}$ of rational integers. A natural generalization of A_n would be to consider an A-hyperplane section of $\mathbb{A}^{m+1}$, but the lattices we obtain in this way are not perfect (see Exercise 8.4.1). As a consequence, we shall consider, following Barnes, various hyperplane sections of lattices L_n^r.

Definition 8.4.6. *Let n, r, and t be non-negative integers, such that $n+1=2r+t$. We denote by M_n^r the hyperplane section of L_{n+1}^r defined by the equation $\sum \lambda_i = 0$.*

It amounts to the same thing first, to consider the hyperplane section $\sum \lambda_i = 0$ of A^m, then to restrict ourselves to the $\mathbb{Z}$-module of those elements whose last t components are real, since congruence (C′2) is satisfied on the hyperplane $\sum \lambda_i = 0$. For $r=0$, we recover a scaled copy of $\mathbb{A}_n$. In what follows, r will be assumed to be strictly positive.

Recall that Conway and Sloane have given a unified definition of both the series L_n^r and M_n^r ([C-S5]; see Definition 6.5.2): for $0 \le r' \le \frac{n}{2}$, let $n-r' = 2r$ or $2r-1$; then $\mathbb{A}_{n,r'} = L_n^r$ if $n-r'$ is even and $\mathbb{A}_{n,r'} = M_n^r$ if $n-r'$ is odd. In this respect, Theorem 8.4.7 below, due to Barnes, can be viewed as an extension of Theorem 8.4.5. We describe the main properties of the M_n^r, but just sketch the proof, for the details of which we refer to [Bar1] and to [Bar5], I, no. 4, pp. 61–64.

Theorem 8.4.7. *Let r and $n \ge 2r-1$ be two positive integers.*

1. *The lattice M_n^r is integral, with norm 4 and determinant*
$$(n+2r+1)\cdot 2^{n-2r}\cdot 3^r\,.$$
2. *We have $s(M_n^r) = \frac{n(n+1)}{2} + \frac{r(r-3)}{2}$.*
3. *The lattice M_n^r $(r>0)$ is perfect if and only if $r \ge 3$.*

Proof. The fact that M_n^r is a norm 4 integral lattice is evident. The calculation of its determinant is carried out by interpreting the hyperplane $\sum \lambda_i = 0$ as the orthogonal in $\mathbb{R}L_{n+1}^r$ of the vector of components $1,\dots,1$, whose norm is $3r+(n+1-r) = n+2r+1$; taking into account the scaling factor $\frac{1}{3}$ which shows up in the transformation of $\mathbb{A}_2$ into its dual, we obtain the value of $\det(M_n^r)$. The calculation of s does not involve any difficulty, and shows that M_n^r is not perfect if $r<3$. Finally, the proof of the perfection assertion for $r \ge 3$ is done using relative perfection, reducing it to the case of M_6^3, which is proved to be perfect by showing that it is perfect relatively to a section isometric to A_5^3, using Theorems 6.4.1 and 6.4.2. □

Remark 8.4.8. *Barnes has moreover proved that the lattices M_n^r are extreme if and only if the inequalities*

$$r \ge 3 \quad \text{and} \quad 2r-1 \le n \le 4r-2$$

are satisfied by the pair (n,r). (Examples: M_6^3, M_7^3, M_7^4, M_8^3, $M_8^4,\dots$)

8.5 Eisenstein Lattices: The Coxeter–Todd Lattice

We keep the notation of Section 8.4: $A = \mathbb{Z}[\omega]$ $(\omega^3 = 1)$ is the ring of Eisenstein integers, m a strictly positive integer, $\pi = 1-\omega$, $\mathfrak{a} = (\pi)$, and K^m is endowed with its canonical Hermitian scalar product. For $\lambda = (\lambda_1, \lambda_2, \ldots, \lambda_m) \in A^m$, we still consider the congruences

$$\lambda_1 \equiv \lambda_2 \equiv \cdots \equiv \lambda_m \mod \mathfrak{a}, \tag{C1}$$

$$\lambda_1 + \lambda_2 + \cdots + \lambda_m \equiv 0 \mod \mathfrak{a}, \tag{C'2}$$

$$\lambda_1 + \lambda_2 + \cdots + \lambda_m \equiv 0 \mod \mathfrak{a}^2. \tag{C2}$$

We are going to study the lattices defined by the congruences (C1) and (C2). Remark that congruence (C1) implies that $N(\lambda) \equiv m\lambda_1 \mod \mathfrak{a}$. Consequently, when m is divisible by 3, the similarity $\lambda \mapsto (1-\omega)^{-1}\lambda$ preserves the integrality of the lattice defined by congruence (C1).

Definition 8.5.1. *For $n = 2m \equiv 0 \mod 3$ at least 12, we denote by CT_n the image by $x \mapsto \pi^{-1}x$ of the lattice defined by the congruences (C1) and (C2). The lattice CT_{12} (first constructed in [C-T]) is called the* Coxeter–Todd lattice, *and is denoted by K_{12}.*

[One obtains an A-isometric lattice by adjunction to L_n^m over A of the vector $\frac{1}{1-\omega}(1, 1, \ldots, 1)$: $\mathrm{CT}_n \simeq L_n^m \cup (L_n^m + \pi^{-1}v) \cup (L_n^m - \pi^{-1}v)$, **(8.5.1′)**
where v denotes the vector $(1, \ldots, 1) \in A^m$, compare (8.3.1′). As in Section 8.3, the construction of CT_n can be viewed as an extension to $A = \mathbb{Z}[\omega]$ of Kneser's neighbouring procedure.]

Theorem 8.5.2. *1. The lattice CT_n is unimodular as an A-lattice; as a $\mathbb{Z}$-lattice, it is even, 3-modular of norm 4 and determinant 3^m; for $n \geq 18$ (resp. for $n = 12$), it possesses $\frac{9m(m-1)}{2}$ (resp. $2 \cdot 3^3 \cdot 7 = 378$) pairs of minimal vectors.*
2. Multiplication by $\pi^{-1} = (1-\omega)^{-1}$ is a similarity of ratio $\frac{1}{\sqrt{3}}$ of CT_n onto its dual (of norm $\frac{4}{3}$).
3. The lattices CT_n are extreme, strongly eutactic, and dual-extreme, and

$$\gamma(\mathrm{CT}_n)^2 = \gamma'(\mathrm{CT}_n)^2 = \frac{16}{3}.$$

Proof. We shall work with the reformulation (8.5.1′) of Definition 8.5.1. The Hermitian scalar product takes integral values on $L_n^m \subset A^m$. Since 3 divides m, the Hermitian scalar products $h(x,y)$ are also integral for $y = \frac{1}{\pi}v$ and $x = \frac{1}{\pi}v$ or $x \in L_n^m$. Hence CT_n is A-integral, and thus even as a $\mathbb{Z}$-lattice (for $\theta \in A$, we have $\theta \cdot \theta = 2\,\mathrm{N}_{K/\mathbb{Q}}(x) \equiv 0 \mod 2$). Since its relative determinant is that of A^m, CT_n is A-unimodular, hence 3-modular as a $\mathbb{Z}$-lattice, of determinant $\det(A)^m = 3^m$. By Theorem 8.4.2, L_n^m possesses $\frac{9m(m-1)}{2}$ pairs of minimal vectors, with norm 4. The vectors in $\frac{1}{\pi} + L_n^m$ are the vectors of the form $\frac{1}{\pi}(\lambda_1, \ldots, \lambda_m)$ with $\lambda_1 \equiv \cdots \equiv \lambda_m \equiv 1 \mod \mathfrak{a}$ and $\sum \lambda_k \equiv 0 \mod \mathfrak{a}^2 = (3)$. This shows that the shortest vectors in $\frac{1}{\pi}L_n^m$ have norm $\frac{2m}{3} \geq 4$, and that there are 3^{m-1} such vectors. [These are the vectors $(\omega_1, \dots, \omega_m)$ with $\omega_i \in \{1, \omega, \omega^2\}$ and $\omega_1 + \cdots + \omega_m = 0$. One can choose arbitrarily $\omega_1, \dots, \omega_{m-1}$, and there is then a unique choice for ω_m.] If $n \geq 18$, we have $S(\mathrm{CT}_n) = S(L_n^m)$ whence $s(\mathrm{CT}_n) = \frac{9m(m-1)}{2}$. If $n = 12$, we find $s(\mathrm{CT}_{12}) = 135 + 243 = 378$. This completes the proof of (1), and also proves (2), which is a direct consequence of the fact that CT_n is A-unimodular.

Assertion (2) immediately implies that $\gamma(\mathrm{CT}_n)$ and $\gamma'(\mathrm{CT}_n)$ are equal and that $\gamma'(\mathrm{CT}_n)^2 = 4 \cdot \frac{4}{3} = \frac{16}{3}$. Since $S(\mathrm{CT}_n)$ contains $S(L_n^m)$ for all $n \geq 12$, which is an extreme configuration by Theorem 8.4.2, CT_n is extreme. It is also dual-extreme since $\mathrm{CT}_n^* \sim \mathrm{CT}_n$, and strongly eutactic for $n \geq 18$ since we then have $S(\mathrm{CT}_n) = S(L_n^m)$. There remains only to prove that CT_{12} is strongly eutactic. This is a consequence of Theorem 8.5.3 below (whose proof also shows that the group of unitary automorphisms of CT_{12} acts irreducibly on $\mathbb{R}^{12}$). □

We display for later use the two important symmetric sets of vectors of CT_n that we considered in the proof above, in the formulation (8.5.1′) of Definition 8.5.1; they contain respectively $\frac{9m(m-1)}{2}$ and 3^m pairs of vectors; The ω_i are cube roots of unity:

- Type I: $\pm(\omega_1, -\omega_2, 0, \dots, 0)$ and permutations;
- Type II: $\pm\frac{1}{1-\omega}(\omega_1, \omega_2, \dots, \omega_m)$ with $\sum \omega^{i_k} \equiv 0 \mod \mathfrak{a}^2 \iff \prod \omega_i = 1$.

The minimal vectors in CT_n are the vectors of type I if $n \geq 18$ and those of type I or II if $n = 12$.

Theorem 8.5.3. *The group of unitary automorphisms of* CT_n *acts transitively on its set of minimal vectors.*

Proof. Consider the group $G_m = \{\pm \mathrm{Id}\} \times (C_3^m \rtimes S_m)$ that we introduced in the proof of assertion (4), where C_3^m represents the set of multiplications $\lambda_k \mapsto \omega_k \lambda_k$ of components by cube roots of unity. Let G'_m be the subgroup of index 3 in G_m whose elements satisfy the condition $\prod \omega_k = 1$. It is clear that G'_m stabilizes CT_n and acts transitively on each of the sets of vectors of type I or II. This proves the theorem in dimensions $n \geq 18$. To complete the proof in dimension 12, it suffices to find a unitary transformation mapping some vector of type I onto a vector of type II. This is achieved by the unitary transformation defined by the matrix M' below:

$$M' = \begin{pmatrix} M & 0 \\ 0 & -M \end{pmatrix} \quad \text{with} \quad M = \frac{1}{\pi} \begin{pmatrix} 1 & 1 & 1 \\ 1 & \omega & \omega^2 \\ 1 & \omega^2 & \omega \end{pmatrix},$$

which maps $(1, 0, 0, -1, 0, 0)$ onto $\frac{1}{\pi}(1, 1, 1, 1, 1, 1)$. □

From now on, we consider CT_n only for $n = 12$, for which we use the notation K_{12}. We are going to embed in K_{12} the series of lattices $\{0\}$, $\mathbb{A}_1$, $\mathbb{A}_2$, $\mathbb{A}_3$, $\mathbb{D}_4$, $\mathbb{D}_5$, $\mathbb{E}_6$, rescaled to norm 4. (However, using Exercise 8.1.1, one sees that the resulting embedding of $\mathbb{D}_4$, and thus also of $\mathbb{E}_6 \supset \mathbb{D}_4$ cannot preserve the Hermitian structure of K_{12}.)

Lemma 8.5.4. *Consider in* K_{12} *the six vectors*

$$e_1 = (0,0,0,0,-1,1),\ e_2 = (0,0,0,-1,1,0),\ e_3 = (0,0,-1,1,0,0),$$
$$e_4 = \frac{1}{\pi}(\omega^2,\omega^2,1,1,\omega,\omega),\ e_5 = \frac{1}{\pi}(1,\omega,\omega^2,\omega^2,\omega^2,\omega^2),$$
$$\text{and}\ \ e_6 = (\omega,0,0,0,0,-1)\,.$$

For $0 \le m \le 6$, *let* M_i *be the sublattice of* K_{12} *generated by the vectors* e_i, $i \le m$ *(with* $M_0 = \{0\}$, *which is also given norm* 4*).*

1. *The lattices* $\widetilde{M_i} = \frac{1}{\sqrt{2}} M_i$ *are isometric to* $\mathbb{A}_0$, $\mathbb{A}_1$, $\mathbb{A}_2$, $\mathbb{A}_3$, $\mathbb{D}_4$, $\mathbb{D}_5$, $\mathbb{E}_6$ *respectively.*
2. *The respective determinants of the* M_i *are* 1, 4, 12, 32, 64, 128, *and* 192, *and those of their images* M_i' *in* K_{12}^* *by the map* $\lambda \mapsto \frac{1}{\pi}\lambda$ *are*
$$1,\ \frac{4}{3},\ \frac{4}{3},\ \frac{32}{27},\ \frac{64}{81},\ \frac{128}{243},\ \text{ and }\ \frac{64}{243}\,.$$
3. *The orthogonal of* M_6' *in* K_{12} *is isometric to* M_6.

Proof. The six vectors e_i have norm 4, and the scalar products $e_i{\cdot}e_j$ $(i<j)$ are equal to -2 if $(i,j) = (1,2),(2,3),(2,4),(4,5),(1,6)$, and to 0 otherwise. The identification of the M_i is now easy by making use of Dynkin diagrams for root lattices: e_2 is the central vertex if $m=3$ and the vertex with valence 3 if $m = 4,5,6$. This immediately implies assertions (2) and (3). We prove that $M_6'^{\perp} \simeq M_6$ by calculating explicitly a basis for $M_6'^{\perp}$, or by using the following short cut: we have $\gamma(M_6'^{\perp}) = \gamma(M_6) = \gamma_6$, hence $M_6'^{\perp} \simeq M_6 \sim \mathbb{E}_6$ by the Blichfeldt-Barnes theorem. □

Definition 8.5.5. *For* $0 \le m \le 12$, *let* K_m *be the lattice defined in the following way:*

1. *For* $6 \le m \le 12$, $K_m = (M_{12-m}')^{\perp}$.
2. *For* $0 \le m \le 6$, K_m *is the orthogonal complement to the image in* K_{12}^* *of* K_{12-m}.

Theorem 8.5.6. *1. For* $0 \le m \le 12$, *the sequence* K_m *is an increasing sequence of norm* 4 *sublattices of* K_{12} *whose determinants are the smallest possible.*

2. *For* $0 \le m \le 6$, *these lattices are similar to* $\mathbb{A}_0$, $\mathbb{A}_1$, $\mathbb{A}_2$, $\mathbb{A}_3$, $\mathbb{D}_4$, $\mathbb{D}_5$, $\mathbb{E}_6$.
3. *For* $6 \le m \le 12$, *the lattices* K_m *have the following invariants:*

Table 8.5.6′. The lattices K_n, $6 \le n \le 12$.

n	6	7	8	9	10	11	12
$\det(K_n)$	192	384	576	864	972	972	729
$s(K_n)$	36	46	66	90	138	216	378
$\mathrm{Smith}(K_n)$	3	$24 \cdot 2^4$	$12^2 \cdot 2^2$	$24 \cdot 6^2$	$18 \cdot 6 \cdot 3^2$	$12 \cdot 3^4$	3^6

4. *The lattices K_9, K_{10}, K_{11}, and K_{12} are perfect, but K_7 and K_8 have perfection default* 1.

Before proceeding to the proof, it is important to make a comment. The densest sections of K_{12} are not claimed to be unique up to isometry: there are indeed two orbits of norm 4 hexagonal lattices in K_{12}, one of which is the set of lattices $\langle x, \omega x\rangle$, $x \in S(K_{12})$, while the other one *is not* an A-lattice for the Hermitian structure induced by that of K_{12} (see Exercise 8.5.2). We shall prove below that the corresponding 10-dimensional lattices are not isometric, and that K_{10} is *the* choice for which the densest hyperplane section is *the* densest possible among all 9-dimensional sections of K_{12}.

Proof of 8.5.6. Assertion (2) is a direct consequence of Lemma 8.5.4, and implies assertion (1), directly for dimensions $m \le 6$, and then for the other dimensions because K_m is similar to the orthogonal of K_{12-m} in K_{12}^*. As for the data in assertion (3), only the determinants are easy consequences of that we previously proved, thanks to the formula $\det({M'_m}^{\perp}) = \det(K_n)\det(M'_m)^{-1}$. We have indeed asked the computer to tell us the Smith invariants as well as the perfection ranks of the K_m for $7 \le m \le 11$. Only the s invariants have been determined by hand; this we explain now for K_{11}, K_{10}, and K_9, referring to Exercises 8.5.3 and 8.5.4 for K_8 and K_7. Note that the description of minimal vectors in K_9 and K_{10} proves (using relative perfection) the perfection of K_{10} and of K_{11}, once that of K_9 is known.

For $i = 1, \dots, 6$, let $e'_i = \frac{1}{\pi}\, e_i \in K_{12}^*$.

Start with $e'_1 = \frac{1}{\pi}\,(0,0,0,0,-1,1)$. We have

$$(\lambda_1, \lambda_2, \lambda_3, \lambda_4, \lambda_5, \lambda_6) \cdot e'_1 = \frac{1}{3}\,\mathrm{Tr}((1-\omega)(\lambda_6 - \lambda_5))\,,$$

and we deduce from this calculation the list (up to the sign) of the vectors in $S(K_{12})$ which are orthogonal to e'_1. These are obtained by suitable permutations of the components of the vectors listed below:

Ia	54	pairs	$(.,.,.,.,0,0)$
Ib	12	pairs	$(.,.,.,.,0,\omega)$
Ic	12	pairs	$(.,.,.,.,\omega,0)$
Id	03	pairs	$(.,.,.,.,-\omega^{1+i},\omega^{1-i})$
IIa	81	pairs	$\frac{1}{\pi}(.,.,.,.,\omega^i,\omega^i)$
IIb	27	pairs	$\frac{1}{\pi}(.,.,.,.,\omega^2,\omega)$
IIc	27	pairs	$\frac{1}{\pi}(.,.,.,.,\omega,\omega^2)$

This shows that $s(K_{11}) = 216 = 2^3 \cdot 3^3$.

We determine in the same way the vectors of $S(K_{11}) \cap \langle e'_1, e'_2\rangle^\perp$, then those of $S(K_{10} \cap \langle e'_1, e'_2, e'_3\rangle^\perp$. In the first case, we find among type I (resp. type II) vectors 27 pairs of vectors of each of the forms $(.,.,.,0,0,0)$ and $(.,.,.,\omega,0,0)$, and 3 pairs of the form $(.,.,.,-\omega,\omega,0)$ (resp. 27 pairs of each of the forms $\frac{1}{\pi}(.,.,.,\omega^i,\omega^i,\omega^i)$, $\frac{1}{\pi}(.,.,.,\omega,\omega,\omega^2)$, and $\frac{1}{\pi}(.,.,.,\omega,\omega^2,\omega^2)$), whence $s(K_{10}) = 138 = 2 \cdot 3 \cdot 23$. In the second case, there are 9 pairs of the form $(.,.,0,0,0,0)$, 24 of the form $(.,.,\omega,0,0,0)$, and 6 of the form $(0,0,-\omega,\omega,0,0)$ (resp. 3 of the form $\frac{1}{\pi}(.,.,1,1,1,1)$ and 48 of the form $\frac{1}{\pi}(.,.,\omega^{\pm1},\omega^{\pm1},\omega^{\pm1},\omega^{\pm1})$), whence $s(K_9) = 90 = 2 \cdot 3^2 \cdot 5$.

Finally, we verify that the following 11 vectors are in $S(K_{11})$, but not in K_{10}, and are independent over $\mathbb{Q}$:
$(0,0,0,-1,0,\omega)$, $(0,0,0,-1,\omega,0)$, $(0,0,-1,1,0,0)$, $(0,0,-\omega,1,0,0)$, $(0,-1,0,1,0,0)$, $(0,-\omega,0,1,0,0)$, $(-1,0,0,1,0,0)$, $(-\omega,0,0,1,0,0)$, $(0,0,0,0,-1,\omega^2)$, $\frac{1}{\pi}(1,1,1,\omega,1,1)$, and $\frac{1}{\pi}(1,1,1,\omega^2,1,1)$.
This reduces the proof of the perfection of K_{11} to that of K_{10}. Similarly, we reduce the proof of the perfection of K_{10} to that of K_9 by considering the 10 vectors
$(0,0,-1,0,0,\omega)$, $(0,0,-1,0,\omega,0)$, $(0,0,-1,\omega,0,0)$, $(0,-1,1,0,0,0)$, $(0,-\omega,1,0,0,0)$, $(0,-1,\omega,0,0,0)$, $(-1,0,1,0,0,0)$, $(-\omega,0,1,0,0,0)$, $\frac{1}{\pi}(1,1,\omega,1,1,1)$, and $\frac{1}{\pi}(1,1,\omega^2,1,1,1)$.

We conclude the proof by a calculation of the perfection ranks of K_7, K_8, and K_9 on explicit Gram matrices. □

Remark 8.5.7. *Besides $\mathbb{D}_4$, K_{12} possesses a section similar to $\mathbb{A}_4$. The orthogonal of its image in K_{12}^* is a lattice K_8^a, of determinant $2^4 \cdot 3^2 \cdot 5 = 720$, with $s = 58$. Like K_7 and K_8, it has perfection default* 1.

We now define a new descending series K'_m of sections of K_{12}, which for even m are sections of K_{12} *over* A, and for odd m are sections of K_{m+1} with the largest possible Hermite invariant. To give a non-ambiguous definition, we use the $\mathbb{Z}$-basis $(v_1, \omega v_1, v_2, \dots, \omega v_6)$ for K_{12} defined in Exercise 8.5.1. Theorem 8.5.9 below will provide a canonical definition for the isometry classes of these lattices.

Definition 8.5.8. *For $12 \geq m \geq 0$, let K'_m be the sublattice of $K'_{12} = K_{12}$, a basis for which is provided by the first m vectors of the $\mathbb{Z}$-basis above for K_{12}.*

In the following theorem, the notation P^i_m is that of Section 5.5, introduced by Conway and Sloane in [C-S5].

Theorem 8.5.9. *1. The lattices K'_m are eutactic for all m, strongly eutactic for $m = 4, 5, 6, 8, 10, 12$, extreme for all $m \neq 3, 4$.*
2. The lattices K'_m, $0 \leq m \leq 12$ are well defined up to isometry as the relative lattices inside K_{12} having the largest possible Hermite invariant among those which constitute an increasing sequence whose even terms are A-sublattices of K_{12}.
3. For $m \leq 10$ and $r \leq \frac{m}{2}$, K_{12} contains relative lattices isometric to L^r_m.
4. For $4 \leq m \leq 8$, K'_m is isometric to L^r_m with $r = \lfloor \frac{m}{2} \rfloor$ (these are the lattices L_m of Definition 8.4.1).
5. We have $K'_{12} = K_{12}$, $K'_{11} \simeq K_{11}$, $K'_7 \simeq P^3_7$, $K'_6 \simeq P^2_6 \sim \mathbb{E}^*_6$, $K'_5 \simeq P^2_5 \simeq \mathbb{A}^3_5$, $K'_2 \sim \mathbb{A}_2$; K'_3 is isometric to the less dense lattice among 3-dimensional lattices with $s = 5$.
6. ${K'_{10}}^*$ is strongly eutactic and extreme, ${K'_9}^*$ is semi-eutactic, and both K'_{10} and K'_9 are dual-extreme.
7. The invariants of the lattices K'_m for $3 \leq m \leq 10$ are displayed in Table 8.5.9′ below:

Table 8.5.9′. The lattices K'_n, $3 \leq n \leq 10$.

Lattice	K'_3	K'_4	K'_5	K'_6	K'_7	K'_8	K'_9	K'_{10}
$\det(K'_n)$	36	81	162	243	486	729	972	972
$s(K'_n)$	5	9	15	27	36	54	81	135
$\mathrm{Smith}(K'_n)$	$12 \cdot 3$	$9 \cdot 3^2$	$6 \cdot 3^3$	3^5	$18 \cdot 3^3$	$9 \cdot 3^4$	$36 \cdot 3^3$	$6^2 \cdot 3^3$

Proof. Assertions (1), (4) (which imply (3) by the construction in Section 8.3 of the lattices L^r_n), (5), (6), and (7) are easily checked on a computer.

Only assertion (2) needs a more detailed discussion. For even m, it suffices to verify that the lattices K_m, $m = 2, 4, 6$, are characterized up to a unitary isometry as the sections of K_{12} on which the Hermite invariant attains its maximum *among A-lattices*: dimensions 8 and 10 are then dealt with thanks to the transformation $m \mapsto 12 - m$. The case of dimension 2 is trivial, and the proofs for dimensions 4 and 6 are not too difficult to carry out, using successive inclusions. However, one can prove a result stronger than (2), namely:

8.5.9 (2′) *The A-sublattices of K_{12} for which the Hermite invariant is maximum are A-isometric to the K_{2m}, $0 \leq 2m \leq 12$.*

Actually, it suffices once more to consider the case of dimensions 4 and 6, where we can use the following result, based on a reduction argument, for which we refer to [Mar1], Theorem 8.7.1:

An A-lattice which is integral over $\mathbb{Z}$ of dimension 4 (resp. 6) and whose determinant is bounded above by 81 (resp. by 243) is isometric to $\sqrt{2}\,\mathbb{D}_4$ or to $K_4 \simeq L_4^2$ (resp. to $\sqrt{2}\,\mathbb{E}_6$ or to $K_6 \simeq L_6^3 \sim \mathbb{E}_6^*$).

For an odd dimension, say, $2m+1$, we just determine the orbits of ${K'}_{2m+2}^{*}$, and keep, among the sections which are orthogonal to representatives for the orbits, those which moreover contain a lattice K'_{2m}. This completes the proof of Theorem 8.5.9 for even dimensions. □

In the following tables, we display some invariants of lattices dual to some relative lattices $L \subset K_{12}$ that we considered above: the smallest $m > 0$ such that $\sqrt{m}\,L^*$ is integral, the determinant of this rescaled lattice, its norm, the numbers $s = s(L^*)$ and $r = \mathrm{rk}_{\mathbb{Z}}\, S(L^*)$, and finally the perfection rank ρ of L^*.

Table 8.5.10 is devoted to the K_n, $7 \le n \le 11$ and to K_8^a, and Table 8.5.11 the K'_n, $3 \le n \le 10$.

We observe that among lattices which are dual to some K_n or K'_n with $n \le 11$, only $K_4^* \sim \mathbb{D}_4$, $K_6^* \simeq E_6^*$, ${K'_6}^{*} \sim \mathbb{E}_6$, and ${K'_{10}}^{*}$ have enough minimal vectors to have a chance to be perfect, and that they are then extreme, hence also dual-extreme.

Another remark concerns the eutaxy property for the lattices K_n^* and ${K'_n}^{*}$ (for $n > 6$). A look at Tables 8.5.10 and 8.5.11 shows that the s invariant attains the critical value n only for the lattices K_{11}^*, K_{10}^*, ${K'_{10}}^{*}$, ${K'_9}^{*}$, and ${K'_8}^{*}$. Using a program due to Batut, and leaving aside the lattices ${K'_{10}}^{*}$ and ${K'_8}^{*}$, which are known to be eutactic, we have verified that no eutaxy relation exists for K_{10}^*, and a that a single one exists for K_{11}^* and ${K'_9}^{*}$, of the form

$$\mathrm{Id} = p_0 + \frac{1}{4}\sum_{i=1}^{40} p_i \qquad \text{and} \qquad \mathrm{Id} = \sum_{i=1}^{12} p_i$$

respectively, where the p_i denote the orthogonal projections on the minimal vectors in a suitable indexing. These relations first show that K_{11}^* is eutactic, hence that K_{11} is dual-extreme, whereas ${K'_9}^{*}$ is only semi-eutactic. In both cases, a particular minimal direction shows up. In the case of K_{11}^*, it is the one for which the orthogonal hyperplane section is K'_{10}, the other 40 defining lattices isometric to K_{10}. In the case of ${K'_9}^{*}$, it is the one for which the orthogonal hyperplane section is $K'_8 \simeq L_8^4$, the other 12 defining lattices isometric to $P_8 \simeq \mathbb{A}_8^{(2)}$ (see Sections 8.5.3 and 8.5.4). In both cases, the existence of exactly two orbits of minimal vectors corresponds to the existence of exactly two orbits of minimal hexagonal lattices in $K_{12} \sim K_{12}^*$ and in K_{10}^*.

Table 8.5.10. The lattices K_n^*.

dual of	K_7	K_8	K_9	K_{10}	K_{11}	K_8^a
m	24	12	24	16	12	30
det	$2^{14} \cdot 3^6$	$2^{10} \cdot 3^6$	$2^{22} \cdot 3^6$	$2^8 \cdot 3^{15}$	$2^{20} \cdot 3^6$	$2^4 \cdot 3^6 \cdot 5^7$
norm	12	8	16	16	12	16
s	1	6	3	27	41	2
r	1	4	3	8	11	2
ρ	1	6	3	27	41	2

Table 8.5.11. The lattices $K_n'^*$.

dual of	K_3'	K_4'	K_5'	K_6'	K_7'	K_8'	K_9'	K_{10}'
m	12	9	6	3	24	9	36	6
det	$2^4 \cdot 3$	3^4	$2^4 \cdot 3$	3	$2 \cdot 3^5$	3^{10}	$2^{16} \cdot 3^{13}$	$2^8 \cdot 3^5$
norm	4	4	3	2	9	6	27	6
s	2	9	10	36	1	12	13	120
r	2	4	5	6	1	8	9	10
ρ	2	9	10	21	1	12	13	55

Finally, K_{10} (but not K_{10}') contains a perfect 9-dimensional lattice K_9'', say, with determinant 972 (like K_9'), but with $s = 82$ instead 81, discovered by Barnes ([Bar5]).

8.6 A General Construction of Hermitian Lattices

In this section, we construct new lattices from lattices which are modules over an order in a C.M. field or a quaternion division algebra over a totally real number field K. In particular, we shall explain generalizations, depending on a prime p, of the constructions performed in Sections 8.2 and 8.3 ($p = 2$) and 8.4 and 8.5 ($p = 3$), which may also apply to cases where p is unramified.

The notation is as follows: K is a totally real number field (we drop the subscript 0 we used in Section 8.1), and H is a K-algebra which is either a

C.M. quadratic extension of K, or a totally definite quaternion algebra with centre K. We moreover choose a maximal order $\mathfrak{M}$ in H.

Let $m \geq 0$ be an integer, let α be a totally positive element of K such that α^{-1} belongs to the codifferent $\mathcal{C}_K = \mathcal{D}_K^{-1}$ of K (more precisely, of $\mathfrak{M}$; in practice, α generates $\mathcal{D}_K$), and let $I_1, \dots, I_m$ be integral left ideals in $\mathfrak{M}$. The $\mathfrak{M}$-module $\bigoplus_j I_j$, endowed with the Hermitian form

$$T(\lambda, \mu) = \sum_{i=1}^{m} \mathrm{Trd}_{H/\mathbb{Q}}(\alpha^{-1}\lambda_i\,\overline{\mu}_i) = \sum_{i=1}^{m} \mathrm{Tr}_{K/\mathbb{Q}}\left(\alpha^{-1}\,\mathrm{Trd}_{H/K}(\lambda_i\,\overline{\mu}_i)\right),$$

becomes a lattice Λ in $E = \mathbb{R} \otimes H$.

Theorem 8.6.1. *With the notation above, let J be a left ideal in $\mathfrak{M}$, prime to all ideals I_j (i.e. we have $J + I_j = \mathfrak{M}$ for all j). Let us attach to Λ and J the two lattices*

$$\Lambda_{J,+} = \{\lambda \in \Lambda \mid \lambda_1 \equiv \cdots \equiv \lambda_m \mod J\}$$

and

$$\Lambda_{J,-} = \{\lambda \in \Lambda \mid \lambda_1 + \cdots + \lambda_m \equiv 0 \mod J\}.$$

Then:

1. *The lattices Λ, $\Lambda_{J,+}$ and $\Lambda_{J,-}$ are (integral) even.*
2. *Their determinants are*

$$\det(\Lambda_{J,+}) = \det(\Lambda) \cdot \mathrm{Nrd}_{H/\mathbb{Q}}(J)^{(m-1)[H:K]}$$

and

$$\det(\Lambda_{J,-}) = \det(\Lambda) \cdot \mathrm{Nrd}_{H/\mathbb{Q}}(J)^{[H:K]}.$$

3. *We have $\Lambda_{J,+}^* \subset \mathrm{Trd}(\alpha J^{-1})\Lambda_{\overline{J},-}$.*

(In conformity with the conventions of Section 8.1, we make the conjugate $\overline{J}$ of J a left $\mathfrak{M}$-module, using the external multiplication $\lambda \cdot x = x\,\overline{\lambda}$.)

Proof. Since α^{-1} belongs to the codifferent of H over $\mathbb{Q}$, assertion (1) is true for the lattice $\mathfrak{M}^m$, hence also for the three sublattices we consider.

Since J and I_j are coprime for all j, the index of a sublattice of Λ defined by a congruence modulo J is equal to the norm of J, which is the square of the reduced norm in the quaternionic case; this multiplies the determinant by the square of the index. Since the $m-1$ congruences which define $\Lambda_{J,+}$ are clearly independent, we obtain immediately the formulae in (2).

Finally, for $\lambda \in \Lambda_{J,+}$ and $\mu \in \Lambda_{\overline{J},-}$, we have

$$\sum_{j=1}^{m} \overline{\mu}_j\,\lambda_j = \sum_{j=1}^{m} \overline{\mu}_j\,(\lambda_j - \lambda_1) + \Big(\sum_{j=1}^{m} \overline{\mu}_j\Big)\,\lambda_1\,,$$

a sum of two terms the first of which lies in J and the second one in $\overline{J}$, two conjugate ideals which both have the same reduced trace $\mathrm{Trd}(J)$ over $\mathbb{Q}$. □

The precise determination of the duality relation which exists between $\Lambda_{J,+}$ and $\Lambda_{J,-}$ depends on calculations involving the reduced trace in $H/\mathbb{Q}$. These calculations, very simple when $K = \mathbb{Q}$, have been performed on the examples which were considered in Sections 8.2–8.5.

It is worth noticing that the hypothesis that the ideals I_j should be integral was used only to prove assertion (1) of the theorem. The proofs of assertions (2) and (3) are of a purely local nature, and they are valid provided that no common maximal ideal is involved in the composition series for J and for some I_j. Thanks to this remark, we can forget integrality conditions when calculating duals.

A natural generalization of the preceding two constructions consists in combining them by considering intersections $\Lambda_{J,+} \cap \Lambda_{J',-}$. We are going to push further the calculations when only one prime number is involved in J and J'. To this end, we first introduce some additional notation.

We consider a prime number p which does not ramify in K. We denote by $p\mathbb{Z}_K = \mathfrak{p}_1 \dots \mathfrak{p}_r$ its decomposition in $\mathbb{Z}_K$, and we moreover assume in the C.M. case that none of the $\mathfrak{p}_i$ is inert in H/K. Thus, for all i, one of the following properties (which do not depend on the choice of $\mathfrak{M}$) holds:

1. (*ramified case*) $\mathfrak{p}_i = \mathfrak{P}_i^2$ in $\mathfrak{M}$ for a unique two-sided ideal $\mathfrak{P}_i$ of $\mathfrak{M}$; we then set $\mathfrak{P}'_i = \mathfrak{P}_i^2$.
2. (*split case*) $\mathfrak{p}_i$ splits in H/K if $[H : K] = 2$, or the completion at $\mathfrak{p}_i$ of H is isomorphic to the 2×2 matrix algebra if $[H : K] = 4$. In this last case, we consider two distinct left ideals $\mathfrak{P}_i$ and $\mathfrak{P}'_i$ of $\mathfrak{M}$. (They will play the rôle that the two prime ideals lying above $\mathfrak{p}_i$ play when $[H : K] = 2$.)

Theorem 8.6.2. *With the notation above, let $\mathfrak{P} = \cap \mathfrak{P}_i$ and $\mathfrak{P}' = \cap \mathfrak{P}'_i$, and assume that m is divisible by p whenever at least one of the ideals $\mathfrak{p}_i$ ramifies in H/K. Consider the system of congruences*

$$\lambda_1 \equiv \cdots \equiv \lambda_m \mod \mathfrak{P} \tag{C1}$$

and

$$\lambda_1 + \cdots + \lambda_m \equiv 0 \mod \mathfrak{P}' . \tag{C2}$$

defined on Λ (the lattice we defined just before stating Theorem 8.6.1). Let Λ_1 (resp. Λ_2) be the submodule of Λ consisting of those elements of Λ which satisfy the congruences (C1) *for all i and* (C2) *for the indices i such that $\mathfrak{p}_i$ is unramified in H/K (resp. and* (C2) *for all i), endowed with the lattice structure induced by the bilinear form $\frac{1}{p}T$. (In other words, we apply the homothetic transformation of ratio $\frac{1}{\sqrt{p}}$ to the image of the submodules previously defined by the congruences alone.) Then:*

1. *The lattices Λ, Λ_1 and Λ_2 are even lattices of dimension $n = m\,[H : \mathbb{Q}]$.*

2. We have

$$\det(\Lambda_1) = \det(\Lambda) \prod_{\mathfrak{p}_i \text{ ramified}} \mathrm{N}_{K/\mathbb{Q}}(\mathfrak{p}_i)^{-1}$$

and

$$\det(\Lambda_2) = \det(\Lambda)\,.$$

[The form T is defined just before Theorem 8.6.1.]

Proof. Assertions (1) and (2) are of a local nature. and are easy locally at primes $q \neq p$. We may therefore prove them after having replaced $\mathbb{Z}$ by its localization $\mathbb{Z}_{(p)}$ at p. In particular, this allows us to take $I_i = \mathfrak{M}$ for all i.

(1) For all $\lambda = (\lambda_1, \ldots, \lambda_m) \in \Lambda$, we have $N(\lambda) = 2\sum_j \mathrm{Tr}_{K/\mathbb{Q}}(\alpha^{-1}\lambda_j\overline{\lambda}_j)$. Assertion (1) is evident for Λ; using the inclusion $\Lambda_2 \subset \Lambda_1$, we see that it is for Λ_2 a consequence of that for Λ_1. Since the $\mathfrak{p}_i$ are pairwise distinct (because p is unramified in K), the fact that Λ_1 should be integral and moreover even is equivalent to the fact that the congruence $\lambda_1\overline{\lambda}_1 + \cdots + \lambda_m\overline{\lambda}_m \equiv 0 \mod \mathfrak{p}_i$ should be satisfied on Λ_1 for all i.

When $\mathfrak{p}_i$ is ramified in H/K, since $\mathfrak{P}_i$ is invariant under conjugation, the congruences $\lambda_j \equiv \lambda_1 \mod \mathfrak{P}_i$ imply $\overline{\lambda}_j \equiv \overline{\lambda}_1 \mod \mathfrak{P}_i$, which shows that

$$\sum_j \lambda_j\overline{\lambda}_j \equiv m\,\lambda_1\overline{\lambda}_1 \equiv 0 \mod \mathfrak{P}_i\,,$$

and also that $\sum_j \lambda_j\,\overline{\lambda}_j \equiv 0 \mod \mathfrak{p}_i$, since $\mathfrak{p}_i = \mathfrak{P}_i \cap \mathbb{Z}_K$.

When $\mathfrak{p}_i$ splits and $[H:K]=2$, we have

$$\sum_j \lambda_j\,\overline{\lambda}_j = \sum_j (\lambda_j - \lambda_1)\,\overline{\lambda}_j + \lambda_1 \sum_j \overline{\lambda}_j\,,$$

and each sum is divisible by $\mathfrak{P}_i$: this is clear for the first one, and the result for the second one follows from the fact that $\mathfrak{P}_i'$ is conjugate to $\mathfrak{P}_i$. We thus again have $\sum_j \lambda_j\,\overline{\lambda}_j \equiv 0 \mod \mathfrak{p}_i$, since $\mathfrak{p}_i = \mathfrak{P}_i \cap \mathbb{Z}_K$.

We are left with the case where H is a quaternion skew-field in which $\mathfrak{p}$ is not ramified. After completion at $\mathfrak{p}$, for which we use the notation $\widehat{\ }$, H becomes an algebra $\widehat{H}$, isomorphic to the matrix algebra $\mathcal{M}_2(\widehat{K})$, and we may identify $\widehat{H}$ with $\mathcal{M}_2(\widehat{K})$ in such a way that

$$\widehat{\mathfrak{M}} = \mathcal{M}_2(\widehat{\mathbb{Z}}_K), \quad \widehat{\mathfrak{P}_i} = \begin{pmatrix} \widehat{\mathbb{Z}}_K & \widehat{\mathfrak{p}}_i \\ \widehat{\mathbb{Z}}_K & \widehat{\mathfrak{p}}_i \end{pmatrix}, \quad \text{and} \quad \widehat{\mathfrak{P}_i'} = \begin{pmatrix} \widehat{\mathfrak{p}}_i & \widehat{\mathbb{Z}}_K \\ \widehat{\mathfrak{p}}_i & \widehat{\mathbb{Z}}_K \end{pmatrix}.$$

We now identify λ_j with a matrix $\left(\begin{smallmatrix} x_j & z_j \\ y_j & t_j \end{smallmatrix}\right)$. Then congruences modulo $\mathfrak{P}_i$ or $\mathfrak{P}_i'$ become pairs of congruences modulo $\mathfrak{p}_i$, namely

$$z_j \equiv z_1 \mod \mathfrak{p}_i \ \text{ and } \ t_j \equiv t_1 \mod \mathfrak{p}_i$$

for $1 \leq j \leq m$, and

$$\sum_{j=1}^{m} x_j \equiv \sum_{j=1}^{m} y_j \equiv 0 \mod \mathfrak{p}_i\,.$$

Since the reduced norm of λ_j is equal for all j to the determinant of the corresponding matrix, we have

$$\sum_{j=1}^{m} \lambda_j \overline{\lambda}_j = \sum_{j=1}^{m} x_j t_j - y_j z_j \equiv \Big(\sum_{j=1}^{m} x_j\Big)\, t_1 - \Big(\sum_{j=1}^{m} y_j\Big)\, z_1 \equiv 0 \mod \mathfrak{p}_i\,,$$

which completes the proof of assertion (1).

(2) For all i, let q_i be the cardinal of the residue field of $\mathfrak{p}_i$. When $\mathfrak{p}_i$ is (resp. is not) ramified in H, the index in $\mathfrak{M}$ of the ideal $\mathfrak{P}_i$ (resp. of both $\mathfrak{P}_i$ and $\mathfrak{P}'_i$) is equal to q_i in the C.M. case and to q_i^2 in the quaternionic case, and the same result holds for the index of $\mathfrak{P}'_i$ in $\mathfrak{P}_i$ when $\mathfrak{p}_i$ is ramified. To introduce a congruence modulo $\mathfrak{P}_i$ or $\mathfrak{P}'_i$ amounts to multiplying the determinant by $q_i^{[H:K]}$ (notice that when $\mathfrak{p}_i$ is ramified, the congruence $\sum_j \lambda_j \equiv 0 \mod \mathfrak{P}_i$ is a consequence of congruence (C1) relative to the index i). In particular, the calculation of the determinant of Λ_1 reduces to that of Λ_2.

For each index i, there are m congruences, which define a lattice with determinant $\det(\Lambda)\, q_i^{m[H:K]}$. Consequently, the lattice Λ'_2 defined by all congruences (C1) and (C2) has determinant $\det(\Lambda)\,(\prod_i q_i)^{m[H:K]}$. But

$$\prod_i q_i = \prod_i \mathrm{N}_{K/\mathbb{Q}}(\mathfrak{p}_i) = \mathrm{N}_{K/\mathbb{Q}}\Big(\prod_i \mathfrak{p}_i\Big) = \mathrm{N}_{K/\mathbb{Q}}(p\,\mathbb{Z}_K) = p^{[K:\mathbb{Q}]}\,.$$

Hence $\det(\Lambda)$ is multiplied by $p^{\dim \Lambda}$, a value which is exactly balanced by the change of T into $\frac{1}{p}T$ (which amounts to replacing Λ'_2 by $\frac{1}{\sqrt{p}}\,\Lambda'_2$). □

Definition 8.6.3. *We denote by* $\Lambda\,[\mathfrak{P},\mathfrak{P}']$ *the lattice* Λ_2 *in Theorem 8.6.2, and we use the simplified notation* $\Lambda\,[p]$ *when there is no need to make precise* $\mathfrak{P}$ *and* $\mathfrak{P}'$.

That the ideals I_j be integral does not matter a great deal for the validity of (2). It indeed suffices to assume that they are of the form $a^{-1}J$ for some $a \in \mathfrak{M}$ with $\mathfrak{M}a$ and J prime to p. (Also, the assumption "α^{-1} belongs to $\mathcal{D}_K^{-1}$" is not an essential one.) These remarks allow us to state Definition 8.6.3 in an enlarged form. When p is unramified in H, this enlarged definition applies to the dual of the lattice Λ in Theorem 8.6.2, which gives sense to the following statement:

Proposition 8.6.4. *If* p *is unramified in* H, $(\Lambda\,[\mathfrak{P},\mathfrak{P}'])^* = (\Lambda^*)\,[\overline{\mathfrak{P}'},\overline{\mathfrak{P}}]$. *(The conjugate* $\overline{I}$ *of a left ideal* I *is again a left* $\mathfrak{M}$*-module for the external multiplication* $\lambda \cdot x = x\,\overline{\lambda}$.*)*

Proof. It suffices to check the equality locally. It is evident at primes $q \neq p$, since Λ and $\Lambda[p]$ have then the same localizations. To check the equality at p, we use the fact that Λ and Λ^* have now the same localizations ("Λ is p-unimodular"). If λ and μ satisfy the congruences $\lambda_j \equiv \lambda_1 \mod \mathfrak{P}$, $\sum_j \lambda_j \equiv 0 \mod \mathfrak{P}'$, $\mu_j \equiv \mu_1 \mod \overline{\mathfrak{P}'}$, and $\sum_j \mu_j \equiv 0 \mod \overline{\mathfrak{P}}$, we have $\sum_j \lambda_j \overline{\mu}_j \equiv \lambda_1 \sum_j \overline{\mu}_j \equiv 0 \mod \mathfrak{P}$ and $\sum_j \lambda_j \overline{\mu}_j \equiv (\sum_j \lambda_j) \overline{\mu}_1 \equiv 0 \mod \mathfrak{P}'$, which implies that this sum is divisible by p. The required equality then follows after rescaling by a factor $\frac{1}{\sqrt{p}}$ the modules attached to Λ and to Λ^* by congruences (C1) and (C2). □

The transformation $\Lambda \mapsto \Lambda[p]$, which preserves the determinant, may increase, reduce, or preserve the norm; when the norm is preserved, we cannot forecast whether $L[p]$ is or is not isomorphic Λ. Of course, we wish to get an as large as possible norm in order to improve on the Hermite invariant. Here is an example where we get a larger norm whenever L has norm 2.

Proposition 8.6.5. *With the hypotheses and notation of Theorem 8.6.2, if $m \geq 3$, then $\Lambda[2]$ has norm at least* 4.

Proof. Let $\mathfrak{P}$ and $\mathfrak{P}'$ be left ideals of $\mathfrak{M}$, and let $\Lambda[2] = \Lambda[\mathfrak{P}, \mathfrak{P}']$. For $\lambda = (\lambda_1, \dots, \lambda_m) \in \Lambda[2]$, let us identify $\lambda_j \in \mathfrak{M}$ with the element $\lambda \in \Lambda$ whose λ_j is the unique nonzero component. Then the norm of λ is $N(\lambda) = \sum_j \lambda_j \overline{\lambda}_j$; this is an even integer.

If no λ_j belongs to $\mathfrak{P}$, we have

$$N(\lambda) \geq \sum_j N(\lambda_j) \geq m \min N(\lambda_j) \geq 3 \min N(\lambda_j)$$

whence $N(\lambda) \geq 4$.

If one of the λ_j belongs to $\mathfrak{P}$, all λ_j are elements of $\mathfrak{P}$, and the products $\lambda_j \overline{\lambda}_j$ belong to $\mathfrak{P} \cap \mathbb{Z}_K = 2\mathbb{Z}_K$, whence the lower bound $N(\lambda) \geq 4$ as soon as at least two components among the λ_j are nonzero.

Finally, if λ is a permutation of $(\lambda_1, 0, \dots, 0)$, λ_1 belongs to $\mathfrak{P} \cap \mathfrak{P}' = 2\,\mathfrak{M}$. Writing $\lambda_1 = 2\,l_1'$, we obtain $N(\lambda_1) = 4\,N(\lambda_1') \geq 4$. □

We now apply the results above, taking $H = H_3$ (the quaternion algebra ramified at 3 and ∞) and $\mathfrak{M} = \mathfrak{M}_3$, see Section 8.1; recall that $\mathfrak{M}$ is unique up to conjugacy. We leave as an exercise the following proposition, which concerns $p = 3$:

Proposition 8.6.6. *The sublattice of $\mathfrak{M}^3$ defined by the congruences*

$$\lambda_1 \equiv \lambda_2 \equiv \lambda_3 \mod \mathfrak{P} \text{ (resp. } \lambda_1 + \lambda_2 + \lambda_3 \equiv 0 \mod \mathfrak{P})$$

is similar to $\mathbb{E}_6 \perp \mathbb{E}_6$ (resp. to $\mathbb{E}_6^* \perp \mathbb{E}_6^*$). □

Let $p = 2$ and $m = 3$. Restriction of scalars to the ring $A = \mathbb{Z}[j] \subset \mathfrak{M}$ of Eisenstein integers yields a 6-dimensional A-integral lattice with discriminant

3^6 over $\mathbb{Z}$. Hence this lattice is unimodular over A. A theorem of Feit ([Fe2]), which asserts the uniqueness of K_{12} as an A-lattice without norm 2 vectors, proves the following result, which can also be proved directly by interpreting the congruences above over A:

Theorem 8.6.7. *Let Λ be the lattice $\mathfrak{M}^3$ where $\mathfrak{M}$ is a maximal order in H_3. Then the lattice $\Lambda[2]$ is isometric (as an A-lattice) to the Coxeter–Todd lattice K_{12}.*
[Christine Bachoc has proved that this lattice is also unique as an $\mathfrak{M}$-lattice.] □

8.7 Quadratic Hermitian Structures

In this section, we study lattices which can be constructed by applying the methods of the first six sections to maximal orders in a quaternion algebra with centre a (totally real) quadratic field K. Such a field is characterized (within a given algebraic closure of $\mathbb{Q}$) by its discriminant $d = d_K > 0$.

We denote by H a totally definite quaternion algebra with centre K, by $\mathfrak{M}$ a maximal order in $\mathfrak{M}$, and by $\mathfrak{a}$ the product of the maximal ideals of K which ramify in H/K (i.e., in $\mathfrak{M}$, but $\mathfrak{a}$ does not depend upon the choice of $\mathfrak{M}$). The ideal $\mathfrak{a}^2$ is the (ideal) discriminant of the $\mathbb{Z}_K$-module $\mathfrak{M}$ with respect to the bilinear form $\mathrm{Trd}_{H/K}(xy)$, hence also with respect to $\mathrm{Trd}_{H/K}(x\overline{y})$.

Let α be a totally positive element of K. Extending the form

$$T(x,y) = \mathrm{Tr}_{K/\mathbb{Q}}\left(\alpha^{-1}\,\mathrm{Trd}\,H/K(x\overline{y})\right)$$

(which we still denote by T) to $E = \mathbb{R}\bigotimes H$, we turn E into a Euclidean space and $\mathfrak{M}$ into a lattice in E, with determinant

$$\det(\Lambda) = \left(d_{K/\mathbb{Q}}\,\mathrm{N}_{K/\mathbb{Q}}(\alpha)^{-1}\right)^4 \mathrm{N}_{K/\mathbb{Q}}(\mathfrak{a}^2)\,.$$

More generally, any fractional ideal of $\mathfrak{M}$ defines a lattice, whose determinant is $\mathrm{Nrd}(I)^4$ times that of $\mathfrak{M}$. [It suffices to consider the case where I is an integral ideal. Then the determinant is the square of the index, which is equal to the norm, which is itself the square of the reduced norm.]

The formula above (for which we need not suppose that K is quadratic) shows that there exist unimodular lattices as above if and only if H/K is unramified (a condition which, by Hasse's law of reciprocity, may hold if and only if K has even degree), and if moreover $\mathrm{Nrd}(I)\,\mathcal{D}_K$ is a principal ideal in the narrow sense. This last condition is easy to study when K is quadratic and $I = \mathfrak{M}$. We indeed have $\mathcal{D}_{K/\mathbb{Q}} = (\sqrt{d})$. The element $\sqrt{d}$ has signature $(+,-)$, and there exists a unit $u \in K$ with $u\sqrt{d} >> 0$ if and only if the fundamental unit ε of K has norm -1; $\varepsilon\sqrt{d}$ is then a totally positive generator of the different.

Since $\mathbb{E}_8$ is the unique 8-dimensional even unimodular lattice, we have proved:

Theorem 8.7.1. *Let K be a real quadratic field, let H be the quaternion skew-field with centre K which is unramified except at the two infinite places of K, let $\mathfrak{M}$ be a maximal order in H, and let I be a fractional left ideal of $\mathfrak{M}$ such that* $\mathrm{Nrd}(I)\mathcal{D}_K$ *possesses a totally positive generator α. Then the lattice* $(I, \mathrm{Tr}_{H/\mathbb{Q}}(\alpha^{-1}x\overline{y}))$ *is isometric to* $\mathbb{E}_8$. □

Theorem 8.7.2. *Let $\mathfrak{O}$ be a ring which is of one of the following four types:*

1. *an order in an imaginary quadratic field;*
2. *an order in a totally definite quaternion skew-field with centre $\mathbb{Q}$;*
3. *an order in a C.M. quartic field;*
4. *an order in a totally definite quaternion skew-field with centre a (real) quadratic field K, which is unramified at finite primes over its centre.*

Then $\mathbb{E}_8$ can be given a structure of Hermitian $\mathfrak{O}$-module.

Proof. Embedding $\mathfrak{O}$ in a maximal order in the algebra it generates, we restrict ourselves to the case where $\mathfrak{O}$ is itself maximal. We shall first prove the theorem under hypothesis (4), and then explain how the other cases reduce to this one.

Assuming hypothesis (4), we use the notation of Theorem 8.7.1, setting $\mathfrak{M} = \mathfrak{O}$. Since the reduced norm induces a surjective homomorphism of the set of left ideal classes onto Cl_K^+ (Eichler's theorem, see Appendix 1), there exists a left ideal ideal I in $\mathfrak{M}$ such that $\mathrm{Nrd}(I) \sim \mathcal{D}_K$, and we may apply Theorem 8.7.1 to it.

In case (3), the field of fractions L of $\mathfrak{O}$ contains a real quadratic field K. Let H be the quaternion field which is unramified at finite places of K. Then L splits H, hence embeds in H. We may thus embed $\mathfrak{O}$ into a maximal order in H, to which we may apply the result we proved in case (4).

The first case reduces in the same way to case (2), that we now consider.

Let $\mathfrak{M}_0$ be a maximal order in a totally definite quaternion field H_0 with centre $\mathbb{Q}$, and let S be the set of ramified primes in H_0. For all odd $p \in S$, let $n_p \in \mathbb{Z}$ be a quadratic non-residue modulo p. By Dirichlet's theorem on primes in an arithmetic progression, we may find a prime q such that $q \equiv 1 \mod 8$ and $q \equiv n_p \mod p$ for all odd $p \in S$. Set $K = \mathbb{Q}(\sqrt{q})$ and $H = K \bigotimes H_0$. Then H/K ramifies only at the two infinite primes of K, (because every $p \in S$ is inert in $K/\mathbb{Q}$). Since $\mathbb{Z}_K$ centralizes $\mathfrak{M}_0$, $\mathbb{Z}_K \mathfrak{M}_0$ is an order in H, which is contained in some maximal order $\mathfrak{M}$ in H, to which we may apply (4). □

Let us return to Theorem 8.7.1. Let u be a unit of $\mathfrak{M}$ of finite order (or of reduced norm $+1$ over K, this amounts to the same thing). For all $x \in H$, we have $\mathrm{Trd}_{H/\mathbb{Q}}((xu)(\overline{xu})) = \mathrm{Trd}(x\overline{x})$. Let U be the maximal finite subgroup of $\mathfrak{M}^*$ (such a subgroup is well defined by $\mathfrak{M}$, not by H). The elements of Λ with a given Euclidean norm share out among orbits under the action of U. To determine these orbits may be difficult, especially when U is small (U

may reduce to $\{\pm 1\}$). When $\mathfrak{M}$ contains a conjugate of the Hurwitz order, the order of U is divisible by 24, and indeed equal to 24 except if $K = \mathbb{Q}(\sqrt{5})$ (resp. if $K = \mathbb{Q}(\sqrt{2})$) where U is of order 120 (resp. 48), and this then does not depend on the choice of $\mathfrak{M}$, since both algebras have class number 1, which implies that their maximal orders are conjugate.

Let us determine the minimal vectors for the "usual" maximal orders of these two algebras. When $K = \mathbb{Q}(\sqrt{5})$, we have $d = 5$, $\mathcal{D}_K = (\sqrt{5})$, and $\varepsilon = \frac{1+\sqrt{5}}{2}$, so that we may choose $\alpha = \frac{5+\sqrt{5}}{2}$; when $K = \mathbb{Q}(\sqrt{2})$, we have $d = 8$, $\mathcal{D}_K = (2\sqrt{2})$, and $\varepsilon = 1 + \sqrt{2}$, so that we may choose $\alpha = 4 + 2\sqrt{2}$. The 240 minimal vectors in $\mathbb{E}_8$ share out among two orbits in the first case, and among five in the second case. An easy check (see Chapter 4, exercises on Section 4.8) then shows the following proposition:

Proposition 8.7.3. *Let K be a quadratic field.*

1. *If 2 is inert or ramified in K, the algebra of usual quaternions $K[i, j]$ ($i^2 = j^2 = -1$) over K is non-ramified at all finite primes of K.*
2. *If $K = \mathbb{Q}(\sqrt{5})$, the $\mathbb{Z}_K$-module $\mathfrak{M}$ with basis*

$$\left(1, i, \frac{i + \frac{1+\sqrt{5}}{2} j + \frac{1-\sqrt{5}}{2} k}{2}, \frac{1+i+j+k}{2}\right)$$

is a maximal order, and the minimal vectors in $\big(\mathfrak{M}, \mathrm{Trd}(\alpha^{-1} x\overline{y})\big) \simeq \mathbb{E}_8$, where $\alpha = \frac{5+\sqrt{5}}{2}$, are the $x\,u$, $u \in U$, $x \in \{1, \frac{1+\sqrt{5}}{2}\}$.

3. *If $K = \mathbb{Q}(\sqrt{2})$, the $\mathbb{Z}_K$-module $\mathfrak{M}$ with basis*

$$\left(1, \frac{1+i}{\sqrt{2}}, \frac{1+j}{\sqrt{2}}, \frac{1+i+j+k}{2}\right)$$

is maximal, and the minimal vectors in $\big(\mathfrak{M}, \mathrm{Trd}(\alpha^{-1} x\overline{y})\big) \simeq \mathbb{E}_8$, where $\alpha = 4+2\sqrt{2}$, are the $x\,u$, $u \in U$, $x \in \{1, 1+\sqrt{2}, \frac{1+\sqrt{2}+i}{\sqrt{2}}, \frac{1+\sqrt{2}+j}{\sqrt{2}}, \frac{1+\sqrt{2}+k}{\sqrt{2}}\}$. □

The notation being that of Theorem 8.7.1, we now construct lattices of dimension n divisible by 8, obtained as sub-$\mathfrak{M}$-modules of $\mathfrak{M}^m$ ($n = 8m$) defined by congruences modulo appropriate left ideals whose norms are powers of 2. Let $\mathfrak{p}$ be a prime ideal lying above 2 in K, and let $q = |\mathbb{Z}_K/\mathfrak{p}|$. Since H/K is unramified, we have $\mathfrak{M}/\mathfrak{p}\,\mathfrak{M} \simeq M_2(\mathbb{F}_q)$, and the maximal left ideals lying above $\mathfrak{p}$ in $\mathfrak{M}$, which actually lift maximal left ideals of $\mathcal{M}_2(\mathbb{F}_q)$, have norm q^2.
[Given any field F, a maximal left ideal of $\mathcal{M}_2(F)$ is isomorphic to the ideal $\left(\begin{smallmatrix} * & 0 \\ * & 0 \end{smallmatrix}\right)$.]

Theorem 8.7.4. *The notation being that of Theorem 8.7.1, suppose that 2 is inert (resp. splits) in K, and let $\mathfrak{P}$ be a (left) maximal ideal of $\mathfrak{M}$ lying above 2 (resp. the intersection of two maximal ideals lying above prime ideals of K*

above 2). Suppose moreover that I and $\mathfrak{P}$ are coprime. Then the sublattice $\Lambda' = \{(\lambda, \mu) \in I \times I \mid \lambda \equiv \mu \mod \mathfrak{P}\}$ of $I \times I$ is isometric to the Barnes–Wall lattice $\mathrm{BW}_{16} \simeq \Lambda_{16}$.

Proof. The left ideal $\mathfrak{P}$ and the right ideal $\overline{\mathfrak{P}}$ both have norm 16 (i.e., the quotient $\mathfrak{M}/\mathfrak{P}$ is of cardinality 16). Hence $\det(\Lambda') = 16^2 = 256$, and for any $x \in \mathfrak{P}$, $x\overline{x}$ belongs to the two-sided ideal $\mathfrak{P}\overline{\mathfrak{P}}$, which has norm 16^2, and thus coincides with $2\mathfrak{M}$. It results from this that the 8-dimensional lattice $\Lambda_0 = \{(\lambda, 0)\}$ is such that $\frac{1}{\sqrt{2}}\Lambda_0$ is integral, even, of determinant 1, hence isometric to $\mathbb{E}_8$, and that Λ has norm at least 4. The lattice Λ'^* identifies with $\frac{1}{2}\{(\lambda, \mu) \in I^{-1} \times I^{-1} \mid \lambda \equiv \mu \mod \overline{\mathfrak{P}}\}$. It has norm 2 and it contains $\frac{1}{2}\overline{\Lambda_0} \simeq \mathbb{E}_8$, which shows that it possesses an increasing sequence of sections of dimensions $0, 1, \ldots, 7, 8$ and determinants $1, 2, 3, 4, 4, 4, 3, 2, 1$, whose orthogonals in Λ' constitute a decreasing sequence of lattices of dimensions $16, 15, \ldots, 9, 8$ and determinants 256, 512, 768, 1024, 1024, 1024, 768, 512, 256. The last lattice is obviously similar to E_8, which proves that Λ' is a 16-dimensional laminated lattice, hence isometric to $\Lambda_{16} \simeq \mathrm{BW}_{16}$.
[When $I = \mathfrak{M}$ and $\mathfrak{P}$ is a principal ideal, i.e. of the form $\mathfrak{M}a$ (this is the case where $K = \mathbb{Q}(\sqrt{5})$, since the class number is then equal to 1), the map $x \mapsto x\,\frac{\overline{a}}{2}$ is a similarity of Λ' onto its dual.] □

As in Sections 8.2 to 8.4, we can construct families of lattices which generalize the lattice in Theorem 8.7.4, by considering congruences analogous to those denoted by (C1), (C2) and (C′2) in Section 8.4 (see also Theorem 8.6.1). Using an analogous to C1, we obtain lattices isometric to the lattices J'_n, defined in Section 8.3 for $n \equiv 0 \mod 4$, but endowed for $n \equiv 0 \mod 8$ with a structure over an order of dimension 8 instead of 4. The construction used in Theorem 8.2.4 then allows us to define lattices without any mod 8 restriction on the dimension. The proof of the following theorem, which follows closely that of Theorems 8.2.1 and 8.2.4, will be left to the reader; the notation L_n and L'_n is used only to state the theorem; the lattices L_n and L'_n might *a priori* depend for $n \not\equiv 0$ modulo 8 on the choice of the sequence of sublattices of $\mathbb{E}_8$ isometric to $\mathbb{E}_7$, $\mathbb{E}_6$, $\mathbb{D}_5$, $\mathbb{D}_4$, $\mathbb{A}_3$, $\mathbb{A}_2$, $\mathbb{A}_1$.

Theorem 8.7.5. *The notation and hypotheses are those of Theorem 8.7.4. Let $m \geq 0$ be an integer, and for $n = 8m$, let L_n (resp. L'_n) be the sublattice of $\mathfrak{M}^m$ defined by the congruence $\lambda_1 + \cdots + \lambda_m \equiv 0 \mod \mathfrak{P}$ (resp. by the $m-1$ congruences $\lambda_1 \equiv \cdots \equiv \lambda_m \mod \overline{\mathfrak{P}}$).*

1. *The right $\mathfrak{M}$-module $\frac{1}{2}L'_n$ is dual to L_n; L_8 and L'_8 are similar to $\mathbb{E}_8$.*
2. *The orthogonal in L_n of the submodule M of L'_n defined by the conditions $\lambda_1 = \cdots = \lambda_{m-1} = 0$ is canonically isometric to L_{n-8}.*
3. *For $0 \leq h \leq 8$, let L_{8m-8+h} be the orthogonal to a relative sublattice of M which is similar to the critical lattice in dimension $8-h$. Then the*

lattices L_n, $n \geq 0$ constitute an increasing sequence of perfect lattices of norm 4, similar to Λ_n for $n \leq 16$ ($\Lambda_n^{\max}$ for $n = 11, 12, 13$), and whose determinants are for $m \geq 1$ the sequence 256, 512, 768, 1024, 1024, 1024, 768, 512, 256 *when $h = 0, 1, \ldots, 8$.* □

We could have replaced $\mathfrak{M}^n$ by I^m as above; however, even in the particular case $I = \mathfrak{M}$, we cannot assert that the isometry classes of L_n and L'_n are independent of H and $\mathfrak{M}$ (and of the ideal which defines the congruence). Indeed, we have no analogue of Conway and Sloane's characterization of laminated lattices.

We now come to an analogue of constructions in Theorems 8.3.2 and 8.5.2:

Theorem 8.7.6. *The notation and hypotheses are those of Theorem 8.7.4. Let $U_n(\mathfrak{M}) = U_n$ be the lattice $\mathfrak{M}^m[\mathfrak{P}, \mathfrak{P}'] \simeq \mathfrak{M}^m[2]$ of Definition 8.6.3, i.e. the submodule of $\mathfrak{M}^m$ which satisfies the congruences*

$$\lambda_1 \equiv \lambda_2 \equiv \cdots \equiv \lambda_m \mod \mathfrak{P} \tag{C1}$$

and

$$\lambda_1 + \lambda_2 + \cdots + \lambda_m \equiv 0 \mod \mathfrak{P}' , \tag{C2}$$

endowed with the bilinear form $\frac{1}{2}\sum_{i=1}^m \mathrm{Trd}_{H/\mathbb{Q}}(\alpha^{-1}\lambda_i\overline{\mu}_i)$. Then:

1. *U_n is an even unimodular lattice of dimension $n = 8m$.*
2. *U_n is isometric to $\mathbb{E}_8$ for $n = 8$ and to $\mathbb{E}_8 \perp \mathbb{E}_8$ for $n = 16$.*
3. *For $n \geq 24$, U_n has norm 4.*
4. *U_{24} is isometric to the Leech lattice.*

Proof. Assertion (1) is a consequence of Theorem 8.6.2. It implies (2) if $n = 8$, and the case $n = 16$ is easily dealt with by making use of the map $(\lambda_1, \lambda_2) \mapsto (\frac{\lambda_1+\lambda_2}{2}, \frac{\lambda_1-\lambda_2}{2})$. If $n \geq 24$, the lower bound $N(U_n) \geq 4$ results from Proposition 8.6.5. Conway's characterization of the Leech lattice (an even unimodular lattice without norm 2 vectors) then implies assertion (4). [Variant: intersecting U_{24} with the lattices orthogonal to the laminated sublattices of the lattice defined by the equations $\lambda_1 = \lambda_2 = 0$, we construct U_{24} as a laminated lattice containing Λ_{16}; this moreover shows that the lattices L_n are perfect for $17 \leq n \leq 24$.] □

We are now going to study the minimal vectors in the lattices U_n. It will be seen that $s(U_n)$ does not depend on the choice of $\mathfrak{M}$, a result which is a consequence for $n = 24$, 32, 40 of a calculation with theta functions: indeed, the θ function of an even unimodular lattice is a modular form of weight $\frac{n}{2}$ for the whole modular group $\mathrm{SL}_2(\mathbb{Z})$ (see [Se2], Chapter 7, § 6), and these forms constitute a complex vector space of dimension $\lfloor \frac{n}{24} \rfloor$ (id., § 3, Corollary 1 to Theorem 4). Taking into account its value at 0 (we have $\Theta(0) = 1$), we see that the Θ function of these lattices is uniquely defined by the number of norm 2 vectors. However, the proof of assertion (2) that we give below under some restrictions also yields a useful description of $S(U_n)$ for these values of n. It applies in particular to the case where $K = \mathbb{Q}(\sqrt{5})$. Note also that the knowledge of $s(U_n)$ given by the theorem below for $n = 48, 56, 64$ determines uniquely the theta function of U_n.

Theorem 8.7.7. *1. We have* $s(U_n) = 15n(n-7)$ *for all* $n \geq 40$.

2. *For* $n = 24, 32, 40$, *the* n*-dimensional unimodular norm* 4 *lattices possess* $98280 = 2^3 \cdot 3^3 \cdot 5 \cdot 7 \cdot 13$, $73440 = 2^5 \cdot 3^3 \cdot 5 \cdot 17$, *and* $19800 = 2^3 \cdot 3^2 \cdot 5^2 \cdot 11$ *pairs of minimal vectors respectively.*

Proof. The vectors in U_n ($n = 8m$, $m \geq 3$) correspond to m-tuples $\lambda = (\lambda_1, \ldots, \lambda_m)$; their norms are equal to $\sum_i \lambda_i \overline{\lambda}_i$.

When λ_1 belongs to $\mathfrak{P}$, the nonzero λ_i are of norm $N \geq 2$, and the norm 4 vectors, up to a permutation of the indices, correspond to norm systems $(4, 0, \ldots, 0)$ or $(2, 2, 0, \ldots, 0)$. The number s_n of these vectors is of the form $am + b\frac{m(m-1)}{2}$, where a and b are the numbers of pairs of vectors of each type. It results from Theorems 8.7.1 and 8.7.4 that for $m \geq 3$, the lattice obtained by giving the value 0 to $m-1$ (resp. to $m-2$) elements λ_i is similar to $\mathbb{E}_8$ (resp. to Λ_{16}). We thus have

$$a = s(\mathbb{E}_8) = 120 \quad \text{and} \quad b = s(\Lambda_{16}) - 2s(\mathbb{E}_8) = 2160 - 240 = 120 \cdot 16\,,$$

whence $s_n = 120m\,(1 + 8(m-1)) = 15n(n-7)$.

When λ_1 does not belong to $\mathfrak{P}$, the λ_i are nonzero, and we then have $\sum_i \lambda_i \overline{\lambda}_i \geq m$. No minimal vectors arise in this way if $m \geq 5$. This proves (1), and also (2) when $n = 40$. In the case where the dimension is 24 (resp. 32), we must moreover consider systems of norms of the form $(1, 1, 2)$ and its three permutations (resp. $(1, 1, 1, 1)$).

The case of dimension 32 is the easier one. We study it in detail, assuming that 2 *is inert in* K and that *the elements of* $S(\mathbb{E}_8)$ *are equally distributed modulo the maximal ideals lying above* 2.

The 240 minimal vectors in $\mathbb{E}_8$ then share out among 16 vectors in each of the 15 nonzero classes. We verify that the congruence $\lambda_1 \equiv \lambda_0 \mod 2\mathfrak{M}$ has for given λ_0 the two solutions $\lambda_1 = \pm\lambda_0$. Let us choose arbitrarily λ_4 among the 120 pairs of minimal vectors in $\mathbb{E}_8$, then λ_3 and λ_2 among the elements of $\mathfrak{M}$ congruent to λ_4 modulo $\mathfrak{P}$ (16 choices for each of them). The congruences modulo $\mathfrak{P}$ and $\mathfrak{P}'$ are now equivalent to the single congruence $\lambda_1 \equiv \lambda_1 + \lambda_2 + \lambda_3 \mod 2\mathfrak{M}$, which has two solutions. The contribution to s of the 4-tuples with $\lambda_1 \notin \mathfrak{P}$ is therefore equal to $120 \cdot 16 \cdot 16 \cdot 2 = 61440$, to which we must add the $15 \cdot 32 \cdot 25 = 12000$ pairs with $\lambda_1 \in \mathfrak{P}$, which finally implies $s(U_{32}) = 73440$.

In the case of dimension 24, let us moreover assume besides the two hypotheses above, that $\mathfrak{P}$ *is a principal left ideal*, which ensures that among the $9 \cdot 120$ pairs of minimal norm 4 vectors in $\mathbb{E}_8$, exactly 120 lie in $\mathfrak{P}$.

Choose arbitrarily λ_3 among the $8 \cdot 120$ norm 2 vectors which do not lie in $\mathfrak{P}$; we then have as above 16 choices for λ_2 and two choices for λ_1.

Taking into account the three circular permutations $(\lambda_1, \lambda_2, \lambda_3)$, we find a contribution to $s(U_{24})$ equal to $8 \cdot 120 \cdot 16 \cdot 2 \cdot 3 = 92160$, which adds to the $15 \cdot 24 \cdot 17 = 6120$ pairs arising from systems in $\mathfrak{P}$, whence the equality $s(U_{24}) = 98320$, a number well-known to be half the kissing number of the Leech lattice. □

One could construct in the same way families $U_n(I)$ attached to any left ideal $\mathfrak{M}$ coprime with 2; we may thus construct even unimodular $\mathfrak{M}$-lattices for any of the orders $\mathfrak{M}$ considered in Theorem 8.7.2; in particular, we may endow the Leech lattice with $\mathfrak{M}$-structures; we shall not make use later of such a generalization. Even when $I = \mathfrak{M}$, we do not know beyond dimension 32 whether the isometry classes of lattices $U_n(\mathfrak{M})$ indeed depend on $\mathfrak{M}$ (the case of dimension 32 has been solved by Loyer and Solé in [Lo-So]: U_{32} is isometric to the Barnes–Wall lattice BW_{32}). It would be interesting to compare our lattices U_{40} with the various known 40-dimensional lattices.

The sublattice of $U_n(\mathfrak{M})$ defined by the equations $\lambda_1 = \dots \lambda_{m-1} = 0$ is isometric to $\Lambda_8 \sim \mathbb{E}_8$ and the orthogonal complements V_{n-p} $n \geq n-p \geq n-8$ to its sections by its sublattices Λ_p, $\leq p \leq 8$, which have dimensions n, $n-1, \dots, n-7$, $n-8$ and determinants 1, 4, 12, 32, 64, 128, 256, 256 can be identified with the lattices L_{n-p} of Theorem 8.7.5, and in particular with the laminated lattices $\Lambda_{24}, \Lambda_{23}, \dots, \Lambda 16 \simeq \mathrm{BW}_{16}$ when $n = 24$. We easily verify that $V_{n-8+h+1}$ is perfect relative to V_{n-8+h}, whence:

Proposition 8.7.8. *The lattices V_m, and in particular the lattices U_n are perfect.* □

Extending a technique we used before in lower dimensions, we now construct the lattices K_n and then K'_n in dimensions 12 to 24.

Let L be an n-dimensional lattice similar to its dual L^* by a similarity φ of ratio λ; if L is unimodular, we take $\varphi = \mathrm{Id}$. Let $\Lambda_0 = \{0\} \subset L_1 \subset \cdots \subset L_m$ be an increasing sequence of sublattices of L such that $L_i = L \cap \mathbb{R} L_i$ for $0 \leq i \leq m$. The sequence L'_{n-i} of the lattices orthogonal in L to the $\varphi(L_i)$ is a decreasing sequence of sublattices of L, with $L'_n = L$, and we have

$$\det(L'_j) = \det(L) \cdot \lambda^{n-j} \det(L_{n-j})$$

for $n \geq j \geq n-m$. If $n = 2m$ and if there exists an automorphism $u \in \mathrm{Aut}(L)$ such that $u(L_m) = L'_m$, we can extend the sequence L_i, $0 \leq i \leq n$ to a sequence L_i, $0 \leq i \leq n$ by setting $L_i = u^{-1}(L'_i)$ for $m \leq i \leq n$.

We previously used this construction taking $\Lambda = E_8$ and $\varphi = \mathrm{Id}$, with the sequence $\{0\} \subset \mathbb{A}_1 \subset \mathbb{A}_2 \subset \mathbb{A}_3 \subset \mathbb{D}_4$, and some of the constructions of sublattices of $\Lambda_{16} \simeq \mathrm{BW}_{16}$ performed in Section 8.3 can be interpreted in this way, with $m = 8$ and $L_m = \mathbb{E}_8$; when φ is right multiplication by $(1-i)^{-1}$ in the field of usual quaternions over $\mathbb{Q}$, we get the "principal series" $\Lambda_{11}^{\max} \subset \Lambda_{12}^{\max} \subset \Lambda_{13}^{\max}$; other identifications of L_8 with L'_8 yield constructions of other laminated lattices in dimensions 11 to 13 (with a map φ which does not respect the Hermitian structure). We also applied such a construction to K_{12} in Section 8.5, taking for φ multiplication by $\lambda = (1-\omega)^{-1}$ in the field of Eisenstein numbers, with the sequence

$$K_0 = \{0\} \subset K_1 \sim A_1 \subset \cdots \subset K_6 \sim \mathbb{E}_6 \,.$$

Let H_3 be the quaternion skew-field with centre $\mathbb{Q}$ which ramifies at 3 and at ∞ ($H = \mathbb{Q}[i,j]$, $i^2 = -1, j^2 = -3$), endowed with its usual maximal order $\mathfrak{M}_3$ with basis $1, i, \omega = \frac{-1+j}{2}, \omega' = i\omega$ over $\mathbb{Z}$, and let K be a real quadratic field in which 3 does not split. Then $H = K \otimes H_3$ is unramified over K except for the two infinite places of K. Choose a maximal order $\mathfrak{M}$ in H containing $\mathfrak{M}_3$ and two maximal left ideals $\mathfrak{P}$ and $\mathfrak{P}'$ of $\mathfrak{M}$ above 2, and construct the corresponding lattice $L = U_{24}(\mathfrak{M})$. Restricting to $\mathfrak{M}_3$ the congruences which define L identifies K_{12} endowed with its $\mathfrak{M}_3$-module structure of Section 8.6 with a sublattice of L, endowed with the $\mathfrak{M}_3$-module structure obtained by restriction from $\mathfrak{M}$ to $\mathfrak{M}_3$. The orthogonal complement to K_{12} in L is a $\mathfrak{M}_3$-lattice with determinant 3^6, hence isometric to K_{12}. The choice of such an isometry defines an extension of the sequence (K_j) to dimensions in the range 12 to 24, for which K_{24} is the Leech lattice. We cannot *a priori* assert that such a sequence is unique up to isometry, nor find an easy method to calculate $s(K_j)$; only the values of the determinants are evident, and also the fact that they are minimal among those sublattices of the Leech lattice which contain K_{12} or are contained in K_{12}: the second case was proved in Section 8.4, and it implies the first one.

Actually, uniqueness holds, thanks to results of Conway and Sloane, and Plesken and Pohst. For $18 \le j \le 24$, we have $K_j \simeq \Lambda_j$. Since the series K_i, $i \le 6$ coincides with the series L_i, it extends uniquely to all dimensions $i \le 24$ except in dimensions 11 to 13; this is proved in [C-S], Chapter 6. For $12 \le j \le 18$, uniqueness results from Plesken and Pohst's study of the "weakly laminated lattices for norm 4" (see Section 3.7), where it is proved that the sequence of weakly laminated lattices over K_{12} is unique up to isometry. The fact that the K_n lattices are for $n \le 12$ the densest sections of K_{12} ensures that they are the densest sections of Λ_{24} containing K_{12}. The perfection of the K_n is proved for $n > 12$ by checking that $S(K_{n+1}) \smallsetminus S(K_n)$ is of rank $n+1$. The following proposition summarizes the discussion above:

Proposition 8.7.9. *There exists up to isometry a unique increasing sequence K_j, $0 \le j \le 24$ of norm 4 lattices, which is the sequence K_j defined in Section 8.5 for $j \le 12$, and such that the orthogonal of K_j in Λ_{24} is isometric to K_{24-j}. Their determinants satisfy for all j the symmetry relation $\det(K_j) = \det(K_{24-j})$. We have $K_j \simeq \Lambda_j$ for $j \le 6$ and $j \ge 18$. These lattices are perfect to within the two exceptions of K_7 and K_8. They are the densest lattices inside Λ_{24} among those which contain or are contained in K_{12}.* □

Similarly, once an embedding of K_{12} and Λ_{24} endowed with their structures over the ring A of Eisenstein integers has been chosen (for instance by making use of the order $\mathfrak{M}_3$), we obtain by orthogonality a sequence K'_n extending for $n \geq 12$ the sequence we have constructed in Section 8.5. The lattice K_n coincides with Λ_n for $n \leq 2$ and $n \geq 22$, with K_n for $n = 11, 12, 13$, and carries an A-structure whenever n is even. By Plesken and Pohst's work, the K'_n are well defined up to isometry for all $n \leq 24$, except perhaps for $n = 16$, where there exists two lattices with the same determinant but with different kissing numbers and Smith invariant, say K'_{16} and K''_{16}. The lattice K'_9 possesses two non-isometric sections with the same determinant, namely K'_8 and a second section K''_8. We have indeed $K_8'^{\perp} \simeq K'_{16}$ by the very definition of the series K'_n, but setting $K''_{16} = K_8'^{\perp}$, we obtain a lattice which is not isometric to K'_{16}. This can be proved by observing that

$$\mathrm{Smith}(K'_{16}) = \mathrm{Smith}(K'_8) = 9 \cdot 3^4 \text{ and } \mathrm{Smith}(K''_{16}) = \mathrm{Smith}(K''_8) = 27 \cdot 9 \cdot 3,$$

and applying Theorem 1.9.10.

We end this section with two remarks.

First, by considering an embedding of $\Lambda_{12}^{\max}$ into Λ_{24} over the order $\mathfrak{M}$ of Hurwitz quaternions, we obtain the lattices Λ_m for m divisible by 4 (only $\Lambda_{12}^{\max}$ if $m = 12$; the $\mathfrak{M}$-structure on $\Lambda_{12}^{\min}$ described in Section 8.3 is not compatible with its embedding into Λ_{16}) endowed with $\mathfrak{M}$-structures, and all the Λ_m for $m \geq 14$ can be constructed as orthogonal lattices to Λ_{24-n}.

Finally, as far as I know, the eutaxy property has never been checked for the whole collections of Λ_n, K_n, K'_n. Of course, it would be astonishing to discover a non-extreme lattice among them (except for K'_3, K'_4, K_7 and K_8, which are not perfect).

8.8 Beyond Dimension 24

We consider an order $\mathfrak{O}$ in a totally real, C.M. or totally definite quaternion field K, endowed with its natural Hermitian form h, namely

$$h(x, y) = \mathrm{Trd}_{K/K_0}(x\bar{y})$$

where K_0 is the fixed field of the canonical involution of K; the form

$$(x.y) \mapsto x \cdot y = \mathrm{Trd}_{K/\mathbb{Q}}(x\bar{y}) = (\mathrm{Tr}_{K_0/\mathbb{Q}} \circ h)\,(x\bar{y})$$

is positive definite. The lattices we consider in this section are free left $\mathfrak{O}$-modules, which we view as lattices in $E = \mathbb{R} \otimes_{\mathbb{Z}} \mathfrak{O}^m$, endowed with the extension of the bilinear form defined above. Such lattices are *$\mathfrak{O}$-lattices*.

We moreover consider a left ideal $\mathfrak{P}$ of $\mathfrak{O}$. Its norm is $\mathrm{N}(\mathfrak{P}) = |\mathfrak{O}/\mathfrak{P}|$ (the square of its reduced norm in the quaternionic case).

Lemma 8.8.1. *Let Λ_0 be an $\mathfrak{O}$-lattice, of norm N_0 and rank m over $\mathfrak{O}$, such that $\Lambda'_0 = \mathfrak{P}\,\Lambda_0$ is of norm $N'_0 \geq 2N_0$. Let Λ be the sublattice of $\Lambda_0 \perp \Lambda_0$ whose elements are the pairs (x, y) such that $x \equiv y \bmod \mathfrak{P}$. Then Λ is a lattice of norm $N = \min(2N_0, N'_0)$ and determinant $\mathrm{N}(\mathfrak{P})^{2m} \det(\Lambda_0)^2$.*

Proof. We have $N((x,y)) = N(x) + N(y)$. If x and y are nonzero, this norm is bounded from below by $2N_0$, and the bound is attained exactly on some pairs of minimal vectors in Λ_0; if, for instance, y is zero, it is bounded by N_0', and this bound is attained exactly on the minimal vectors in Λ_0'. As for the determinant, it is given by the formula $\det(\Lambda) = \det(\Lambda_0)^2[\Lambda : \Lambda_0 \perp \Lambda_0]^2$, and the index above is equal to $[\mathfrak{O} : \mathfrak{P}]^m = \mathrm{N}(\mathfrak{P})^m$. □

The value of the invariant $s = s(\Lambda)$ depends on the relative values of N_0 and N_0'. Suppose that equality holds. Set $s_0 = S(\Lambda_0)$, $s_0' = S(\Lambda_0')$,

$$S_x = \{y \in S(L) \mid y \equiv x \mod \mathfrak{P}\}\,,$$

and $s_x = |S_x|$. We then have

$$s = 2s_0' + \sum_{x \in S(L)/\{\pm\,\mathrm{Id}\}} s_x\,.$$

The formula gets simpler if the group of Hermitian automorphisms of Λ_0 which stabilize $\mathfrak{P}$ acts transitively on $S(\Lambda_0)$, since s_x is then a constant. Moreover, when $\mathfrak{P}$ is principal, we have $s_0' = s_0$, as one sees using the formula $h(\alpha\, x, \alpha\, y) = \alpha\overline{\alpha}\, h(x,y)$ and taking into account the relations which exist between the norm (or the reduced norm) and the trace (or the reduced trace), see Appendix 1.

The calculations above immediately imply:

Proposition 8.8.2. *Suppose that $\mathfrak{P}$ is a principal ideal with reduced norm* 2. *Suppose moreover that the stabilizer of $\mathfrak{P}$ in $\mathrm{Aut}_h(\Lambda_0)$ acts transitively on $S(\Lambda_0)$. Let r be the value of s_x on $S(\Lambda_0)$. Then $s = (r+2)\, s_0$.* □

The determination of the dual lattice involves a modification of the preceding construction, consisting in replacing the congruence above by $y \equiv -x \mod \overline{\mathfrak{P}}$. It is worth noticing that in view of the formula $h(x, \lambda\, y) = h(\overline{\lambda}\, x, y)$, the operation of $\mathfrak{O}$ on the dual M^* (in the usual sense) of a lattice M is given by $\lambda \cdot x = \overline{\lambda}\, x$, which makes M^* a *right* $\mathfrak{O}$-module. The duality formula then reads

Proposition 8.8.3.

$$\Lambda^* = \overline{\mathfrak{P}}^{-1}\{(x,y) \in L_0^* \times L_0^* \mid y \equiv -x \mod \mathfrak{P}\}\,.$$

Proof. Indeed, for x, $z \in L_0^*$, x', $z' \in L_0$, $\alpha \in \mathfrak{P}$, and $\beta \in \overline{\mathfrak{P}}^{-1}$, we have

$$\begin{aligned}\overline{\beta}\, h\big((x, -x + \overline{\alpha}\, z), (x', x' + \alpha\, z')\big) &= \overline{\beta}\, h(-x, \alpha\, z') + h(\overline{\alpha}\, z, x') + h(\overline{\alpha}\, z, \alpha\, z') \\ &= \overline{(\alpha\beta)}\, \big(h(-x, z') + h(z, x') + h(z, \alpha\, z')\big)\,.\end{aligned}$$

Taking traces over $\mathbb{Q}$, we see that the left–hand side of the formula above is included in the right one, and we conclude by observing that the determinants of both sides are equal. [Note that when $\mathfrak{P}$ divides 2, the congruences $y \equiv -x \mod \mathfrak{P}$ and $y \equiv x \mod \mathfrak{P}$ are equivalent.] □

Recall (Definition 8.1.4) that a lattice is *q-modular* if it is integral, and if it is the image of its dual by a similarity of ratio $\sqrt{q}$; see also Section 11.5, in particular Definition 11.5.3; note that "1-modular" is equivalent to "unimodular". When $\mathfrak{P}$ is principal, a $\mathfrak{P}$*-modular* lattice, viewed as a mere $\mathbb{Z}$-lattice, is q-modular with $q = d_{\mathfrak{O}} \cdot \mathrm{Nrd}(\mathfrak{P})$; see Proposition 8.1.5 ($\mathfrak{P}$ need not be a two-sided ideal). The determinant of a q-modular lattice M is equal to $[M^* : M] = q^{n/2}$.

Proposition 8.8.4. *Suppose that $\mathfrak{P}$ is an ideal of reduced norm 2, and let L_0 be an $\mathfrak{O}$-lattice.*

1. *If L_0 is unimodular, then $N(\Lambda) = 2\,N(L_0)$ and $\Lambda = \overline{\mathfrak{P}}\,\Lambda^*$; if $\mathfrak{P}$ is principal, Λ is 2-modular.*
2. *If $L_0 = \overline{\mathfrak{P}}\,L_0^*$ (which implies that L_0 is 2-modular), $\overline{\mathfrak{P}}^{-1}\Lambda$ is a unimodular lattice with the same norm as L_0.*

Proof. By Proposition 8.8.3, we have $\Lambda = \overline{\mathfrak{P}}\,\Lambda^*$, which proves (1), and also shows that $\overline{\mathfrak{P}}^{-1}\Lambda$ is integral under the conditions of (2). Calculating the determinant then shows that is is unimodular. □

Suppose now that one of the hypotheses of Proposition 8.8.4 is satisfied, and denote by Λ' the lattice Λ if L_0 is unimodular, and the lattice $\overline{\mathfrak{P}}^{-1}\Lambda$ if L_0 is 2-modular.

Proposition 8.8.5. *Under the hypotheses of Proposition 8.8.4, if L_0 is even, Λ' is also even.*

Proof. This is immediate under hypothesis (1) in Proposition 8.8.4. Suppose now that hypothesis (2) holds, and consider a vector $(x, y) \in \Lambda$. Write $y = x + \alpha z$, $\alpha \in \mathfrak{P}$. We have

$$h(\alpha^{-1}(x,y), \alpha^{-1}(x,y)) = \frac{2}{\alpha\overline{\alpha}}\,h(x,x) + h(z,z) + h(x,z) + \overline{h(x,z)}\,,$$

which implies

$$N(\alpha^{-1}(x,y)) \equiv \mathrm{Trd}(h(x,z) + \overline{h(x,z)}) \mod 2\,,$$

and we conclude by noticing that $h(x,z)$ belongs to $\mathfrak{P}$. □

Here are two examples where $\mathfrak{O}$ is the Hurwitz order $\mathfrak{M}$ and $\mathfrak{P}$ is the two-sided ideal of reduced norm 2 (hence of norm 4), generated (on the left or on the right) by $(1+i)$ ($\mathfrak{O}$ could as well have been the ring of Gaussian integers and $\mathfrak{P}$ the ideal $(1+i)$).

Example 8.8.6. *Starting with* $\mathbb{D}_4$, *we get a lattice* Λ *of norm* 4 *and determinant* $4^2 \cdot 4^2 = 2^8$, *with* $s = (8+2) \cdot 12 = 120$. *(We have* $r = 8$, *the number of units of* $\mathfrak{M}$ *congruent to* 1 mod $\mathfrak{P}$.*) The lattice* Λ' *is obviously a scaled copy of* $\mathbb{E}_8$.

Example 8.8.7. *Starting with* $\mathbb{E}_8$ *(norm* 2, *determinant* 1, $s = 120$, $m = 2$*), we obtain a norm* 4, 16*-dimensional lattice of determinant* $1^2 \cdot 4^{2m} = 256$, *isometric to* $\mathrm{BW}_{16} = \Lambda_{16}$. *The calculation of* s *can be done in the following way: we must evaluate the number of pairs* $((x_1, x_2), (y_1, y_2))$ *such that* $(x_1, x_2) \equiv (y_1, y_2) \mod \mathfrak{P}$. *We have* $s(E_8) = 120$ *choices for* (x_1, x_2), *then* 8 *choices for* y_1, *and then* 2 *choices for* y_2 *(we must choose its sign). We obtain* $s = (8 \cdot 2 + 2) \cdot 120 = 18 \cdot 120 = 2160$, *as we know.*

Following Quebbemann, we now give a construction involving the Leech lattice Λ_{24}, which yields laminated lattices in dimensions 25 to 48.

Theorem 8.8.8. *The* 48*-dimensional lattice* Λ *obtained by applying Proposition 8.8.2 with* $L_0 = \Lambda_{24}$ *is the image of its dual under a similarity of ratio* $\sqrt{2}$, *and is transformed into a* 2*-modular laminated lattice by a homothetic transformation of ratio* $\frac{1}{\sqrt{2}}$.

Proof. Embed L_0 into Λ by the map $x \mapsto (1+i)\,x$. The orthogonal of L_0 in Λ is then isometric to L_0, and the sequence of lattices orthogonal to the sections Λ_{24-m} of L_0 is an increasing sequence of lattices of dimensions $n = 24 + m$. The orthogonality formula 1.3.5 shows that their determinants, once rescaled to norm 4, are equal to

$$2^{-(24+m)} 2^{24} \det(\Lambda_{24-n}) = 2^{24-n} \det(\Lambda_{48-n}),$$

which are the values of the determinants of the laminated lattices in dimensions $n \geq 24$ that we gave in Theorem 3.7.7. □

Using orthogonality in $\Lambda_{48}^* \sim \Lambda_{48}$, we can also define series K_n and K'_n in dimensions 25 to 48. By what we know in dimensions $n \leq 24$, we have $K_n = \Lambda_n$ for $0 \leq n \leq 6$, $18 \leq n \leq 30$, and $42 \leq n \leq 48$, $K'_n = K_n = \Lambda_n$ for $0 \leq n \leq 2$, $22 \leq n \leq 26$, and $46 \leq n \leq 48$, and moreover $K'_n = K_n$ for $n = 11, 12, 13, 35, 36, 37$, as well as the symmetry formulae

$$\det(K_n) = 2^{-n} \det(K_{48-n}) \quad \text{and} \quad \det(K'_n) = 2^{-n} \det(K'_{48-n})$$

for all $n \geq 24$. The densest sections of the laminated lattices are Λ_n for $0 \leq n \leq 8$ (because they are the densest possible among all lattices) hence also for $16 \leq n \leq 32$ and $40 \leq n \leq 48$. It is probable (though not proved) that they also are for $n = 9$, 10, 14, 15, hence also for $n = 33$, 34, 38, 39, and that the K_n are for $n = 11$, 12, 13, hence also for $n = 35$, 36, 37.

The construction of Λ_{48} can also be done over the orders $\mathfrak{O}$ of dimension 8 we considered in Section 8.7. This endows Λ_{48}, Λ_{40}, and Λ_{32} with

$\mathfrak{O}$-structures, from which we deduce the existence of $\mathfrak{M}_3$-structures on K'_n for $n \equiv 0 \mod 4$ (recall that $\mathfrak{M}_3$, unique up to isometry, denotes a maximal order in the quaternion algebra ramified at 3 and ∞). The existence of $\mathfrak{M}_2$- or $\mathfrak{M}_3$-structures implies that Λ_n and K'_n possess $\mathfrak{O}$-structures for all even n, where $\mathfrak{O}$ can be the ring of Eisenstein or of Gaussian integers.

Starting with a lattice L_m of dimension m over $\mathbb{Z}$ which satisfies one of the hypotheses of Proposition 8.8.4, we construct a sequence of lattices $L_{2^k m}$, of discriminants alternatively 1 and $2^{k-1}m$, and whose norm is doubled every two steps. When the norms of $L_{2^k m}$ and $L_{2^{k+1} m}$ are equal, the first lattice embeds in the second one; otherwise, we must first perform a homothetic transformation of ratio $\sqrt{2}$ to obtain such an embedding.

Starting from $\mathbb{D}_4$, we obtain a sequence D_4, $\mathbb{E}_8$, Λ_{16}, $U_{32}, \dots$, which is the sequence of the Barnes–Wall lattices BW_{2^k}, that we construct, e.g. over the Hurwitz order $\mathfrak{M}_2$, or over an 8-dimensional order from $n = 8$ onwards, but the important point is that we start with a lattice endowed with a structure over the *Gaussian ring* $\mathbb{Z}[i]$, which is equivalent to the existence of an automorphism σ such that $\sigma^2 = -\mathrm{Id}$. This is essentially the way the Barnes–Wall lattices were constructed. We dispose of a nice description of their minimal vectors via the *Reed–Muller* binary codes (see [McW-S] or [vL]). This description was used by Barnes and Wall to prove that they are extreme lattices, and recently by Venkov ([Ven3]; see Appendix 2) to prove the much more restrictive property of *strong perfection*. (The eutaxy property results from the irreducibility of their automorphism groups, which are extensions of $O_8^+(\mathbb{F}_2)$; see Broué and Enguehard's paper [B-E].)

Starting with Λ_{24}, we obtain lattices of dimensions 24, 48, 96 (unimodular of minimum 8),... By lack of a precise description of the set of minimal vectors, perfection properties are not known beyond dimension 48.

8.9 Exercises for Chapter 8

8.1.1. The aim of this exercise is to prove that there exists on $\mathbb{D}_4$ a unique structure over the ring $\mathbb{Z}[\omega]$ ($\omega^2 + \omega + 1 = 0$) of Eisenstein integers.

1. Show that the equalities $x \cdot \omega x = x \cdot \omega^2 x = -1$ hold for all $x \in S(\mathbb{D}_4)$.

2. Let o_1 and o_2 be two distinct orbits of the group $G = \{\pm 1, \pm\omega, \pm\omega^2\}$, and let $x \in o_1$. Show that there exists a unique $y \in o_2$ such that $x \cdot y = 0$ and $x \cdot \omega y = +1$.

3. Conclude that two $\mathbb{Z}[\omega]$-structures on $\mathbb{D}_4$ are isomorphic by a unitary isometry.

4. Show that the orbits of minimal vectors in $\mathbb{D}_4$ are o_1, o_2 together with the orbits o_3 of $x - \omega y$ and o_4 of $x + \omega^2 y$.

5. Construct such a structure by making use of the Hurwitz order.

6. Prove similarly that there exists on $\mathbb{D}_4$ a unique structure over the ring $\mathbb{Z}[i]$ of Gaussian integers.

8.1.2. This exercise is devoted to the study of the maximal orders in H_2 and H_3.

1. Prove that the Hurwitz order is the only maximal order in $\mathbb{H}_2$ containing i and j ($i^2 = j^2 = -1$).

We now consider the algebra H_3, defined over $\mathbb{Q}$ by $i^2 = -1$ and $j^2 = -3$.

2. Show that the order $\mathfrak{O}$ generated over $\mathbb{Z}$ by i and j is contained in exactly two maximal orders. (One of them contains $\frac{-1+j}{2}$ and the other one $\frac{-1+k}{2}$.)

3. Verify directly that these orders are conjugate.

[There is only one class (of left or right ideals) in both H_2 and H_3, which ensures that all maximal orders are conjugate.]

8.2.1. Show that the spectrum of J_{4m} is as follows:

For $48m(m-1)$ pairs $\pm x$:

$(3(16m^2 - 60m + 77), 0)$, $(128(m-2), 1)$, $(8(2m+3), 2)$;

For $12m$ pairs $\pm x$:

$(3(16m^2 - 28m + 13), 0)$, $(0, 1)$, $(8(6m-5), 2)$.

8.2.2. With the notation of Theorem 8.2.1, one considers for $1 \le i \le n$ the following vectors v_i in $\mathfrak{M}\,\mathbb{D}_n$: $v_i = \varepsilon_1 - \varepsilon_i$ for $i \not\equiv 1 \mod 4$, $v_1 = \varepsilon_1 + \varepsilon_2$, and $v_i = \frac{1}{2}(\varepsilon_{i-4} - \varepsilon_{i-3} + \varepsilon_{i-2} - \varepsilon_{i-1} + \varepsilon_i - \varepsilon_{i+1} + \varepsilon_{i+2} - \varepsilon_{i+3})$ for $i \ge 5$, $i \equiv 1 \mod 4$.

1. Show that for all $q \le n$, the image of $(v_1, v_2, \dots, v_q)$ under right multiplication by $1+i$ is a basis for J_q.

2. Show that for $n = 12$ and $3 \le p \le 12$, the matrices which consist of the first p rows and p columns of the Gram matrix of this basis for J_{12} are Gram matrices for Λ_p.

8.2.3. Let $n \ge 9$, let m_n be the smallest integer such that $\sqrt{m_n}\, J_n^*$ is an integer (m_n is the first term of the Smith invariant of J_n), let N_n be the norm of this lattice, and let s_n be the number of its pairs of minimal vectors.

1. Show that for $n \equiv 1, 2, 3 \mod 4$, the configuration of minimal vectors of J_n^* is similar to that of $\mathbb{A}_1^*$, $\mathbb{A}_2^*$, $\mathbb{A}_3^*$ respectively.

2. Show that the three invariants defined above are as follows:

$n \equiv 1 \mod 4$	$m_n = 8$	$N_n = 4$	$s_n = 1$
$n \equiv 2 \mod 4$	$m_n = 12$	$N_n = 8$	$s_n = 3$
$n \equiv 3 \mod 4$	$m_n = 4$	$N_n = 3$	$s_n = 4$

8.3.1. We consider the lattice J_n for $n = 4m \ge 12$, and a vector $v = (\lambda_1, \dots, \lambda_m) \in J_n \smallsetminus \mathfrak{a}\, J_n$. (We have $\lambda_1 + \dots + \lambda_m \equiv 0 \mod \mathfrak{a}$, $\mathfrak{a} = (1+i)$, and $\lambda_\ell \notin \mathfrak{a}$ for at least one index ℓ.)

1. Show that the $\mathfrak{M}$-lattice $L_v = \langle J_n, \frac{1}{1+i} v\rangle \subset H_2^m$ has determinant 4^m, and that its norm is $N \le 4$, with equality if and only if at least four components of v do not belong to $\mathfrak{a}$.

2. Show that if $N = 4$, L is $\mathfrak{M}$-isometric to the lattice for which the first four components of v are equal to 1 and the other are zero.

3. Show that for $n = 16$, L_v is isometric to J_{16}', as asserted in (8.3.1′).

4. We now assume that $n \ge 20$ and we consider the lattice

$$J_{v,v'} = \langle J_n, \tfrac{1}{1+i} v, \tfrac{1}{1+i} v'\rangle \subset H_2^m$$

with $v, v' \in J_{20}$. Show that if this lattice is of norm 4, then it is $\mathfrak{M}$-isometric to the lattice constructed with $v = (1,1,1,1,0)$ and $v' = (0,1,\omega,\omega^2,1)$.

5. Show that $L_{v,v'}$ is then isometric to the laminated lattice Λ_{20}.

8.4.1. Show that the section of A^m by the (complex) hyperplane $\sum_i z_i = 0$ is an $n = 2m-2$-dimensional lattice with $s = \frac{3n(n+2)}{8}$.

8.4.2. (The lattice L_4^2.)

1. Show that $L_4 = L_4^2$ is isomorphic to the 4-dimensional lattice of Exercise 8.4.1.
2. Show that L_4 is 9-modular and has Smith invariant $(9, 3^2, 1)$.
3. Show that L_4 is isometric to $\mathbb{A}_2 \bigotimes \mathbb{A}_2$.
4. Show that $|\operatorname{Aut}_{\mathrm{U}}(L_4)| = 36$ and that $|\operatorname{Aut}(L_4)| = 144$. (See Exercise 8.4.3.)

8.4.3. (About the automorphism group of L_n, $n \geq 3$.) Let $n = 2h \neq 6$ or $n = 2h+1 \neq 5$. The notation G_h is that the proof of assertion (4) in Theorem 8.4.2.

1. Show that the group $G_h = \{\pm \mathrm{Id}\} \times C_3^h \rtimes S_h$, of order $2 \cdot 3^h \cdot h!$ is the full group $\operatorname{Aut}_{\mathrm{U}}(L_{2h})$ of unitary automorphisms of L_{2h}.
[Consider the action of G_h on $S(L_{2h}^*)$.]

2. Show that $\operatorname{Aut}(L_{2h})$ is the semi-direct product $\operatorname{Aut}_{\mathrm{U}}(L_{2h}) \rtimes C_2^h$ where the k-th factor of the cyclic group acts by conjugation on the k-th component of A^m. Hence for $h \neq 3$, $|\operatorname{Aut}(L_{2h})| = 2^{h+1}\, 3^h\, h!$.

3. Show that $\operatorname{Aut}(L_{2h+1})$ stabilizes L_{2h}, and that restriction to L_{2h} induces an isomorphism $\operatorname{Aut}(L_{2h+1}) \simeq \operatorname{Aut}_{\mathrm{U}}(L_{2h}) \rtimes C_2^h$ for all $h \neq 2$.

4. Show that L_n is strongly eutactic if and only if $n = 5$ or n is even, and that for $n = 3$ or $n \geq 7$ odd, there are two orbits of minimal vectors under the action of $\operatorname{Aut}(L_n)$.

5. (Refers to Section 8.5.) Show that G'_h (see the proof of Theorem 8.5.3) is the group of all unitary automorphisms of $\mathbb{C}^h$ which stabilizes CT_{2h}, thus that $\operatorname{Aut}_{\mathrm{U}}(\mathrm{CT}_{2h}) = G'_h$ for any $n = 2h \geq 18$.

8.4.4. Write $n = 2m$ and consider integers k with $2 \leq k \leq m$. Let S'_{2k} be the set of vectors of norm $2k$ which are sums of k minimal vectors of L_n and set $s'_{2k} = \frac{1}{2}\,|S'_{2k}|$.

1. Show that
$$s'_k = 3^k \binom{m}{k} \frac{\eta_k}{2} \text{ with } \eta_k = |\{(\pm 1, \dots, \pm 1) \in \{-\}^k \mid \textstyle\sum \pm k \equiv 0 \mod 3\}|.$$

2. Prove the equalities $s_4 = s'_4 = 9\,\frac{m(m-1)}{2}$, $s_6 = 3m + s'_6 = 3m\,\frac{3m^2-9m+8}{2}$, and $s_8 = s'_8 = 3^5\,\frac{m(m-1)(m-2)(m-3)}{24}$.

3. Show that the respective numbers of orbits of vectors of norm $4, 6, 8$ are $1, 2, 1$ for $m \geq 4$, and $1, 1, 0$ for $m = 2, 3$. [Representatives for the two orbits above are $(\omega - \omega^2, 0, 0, 0, \dots, 0)$ and $(1, 1, 1, 0, \dots, 0)$; to prove that they are distinct, one can calculate the numbers of minimal vectors orthogonal to each of the vectors above.]

4. Show that a congruence modulo 2 between vectors of norm $N \leq 8$ occurs exactly on pairs $\pm x$ and on some systems of three pairs of vectors of norm 8 with pairwise orthogonal directions.

5. Show that there exist representatives for norm $N \leq 8$ for L_n modulo 2 if and only if $n \leq 8$.

8.4.5. (The annihilator of L_n.)

1. Show that from $n = 5$ onwards, the sequence of the annihilator a_n of L_n is periodic with period 6, equal to $6, 3, 18, 9, 18, 9, \dots$

2. Show that the sequence of norms of $\sqrt{a_n}\Lambda_n^*$ is also periodic with period 6, equal

to $3, 2, 9, 6, 9, 6, \ldots$.

3. Show that $s(L_n^*) = 1$ for $n \geq 7$ odd.

8.5.1. (A basis for K_{12}.)

1. Show that if $n > 12$, the minimal vectors in CT_n generate a lattice of index 3 in CT_n.

2. Show that the following six vectors
$v_1 = (0,0,0,0,-1,1)$, $v_2 = (0,0,0,-1,1,0)$, $v_3 = \frac{1}{\pi}(1,1,1,1,1,1)$, $v_4 = (0,-1,0,1,0,0)$, $v_5 = (0,-\omega,1,0,0,0)$, and $v_6 = (0,0,-1,1,0,0)$ constitute a basis for K_{12} over $\mathbb{Z}[\omega]$.

8.5.2. The notation is that of Exercise 8.5.1.

1. Show that the number (up to the sign) of type I (resp. type II) vectors whose scalar product with v_1 is equal to $0, \pm 1, \pm 2$ is equal to $54, 52, 28$ (resp. $81, 108, 54$). Conclude that all minimal vectors in K_{12} have the spectrum $(135, 0)$, $(160, 1)$, $(82, 2)$.

2. Let v and v' be two minimal vectors in K_{12} which generate a rank-2 module over $A = \mathbb{Z}[\omega]$. Show that the lattice generated by $v, \omega v, v'$ (resp. by $v, \omega v, v', \omega v'$) has determinant $d \geq 36$ (resp. $d \geq 81$), and possesses five (resp. nine) pairs of minimal vectors when its determinant is a minimum; show that the corresponding lattice is then isometric to L_4^3 (resp. L_3^1).

3. Use Question 2 to deduce that the A-structure of $\mathbb{D}_4$ (see Exercise 8.1.1) and that of K_{12} (see Definition 8.5.1) are not compatible with the inclusion $\mathbb{D}_4 \hookrightarrow K_{12}$.

4. Show that the smallest possible determinant of a rank-3 sublattice of K_{12} over A (resp. of a hyperplane section of such a lattice) is 243 (resp. 162). [Such a lattice is isometric to $\mathbb{E}_6^* \simeq L_6^3$ (resp. to $A_5^3 \simeq L_5^2$); see Section 4.7, and [Mar1].]

5. Show that K'_{10} is characterized up to isometry among the A-sublattices of K_{12} by its determinant and its dimension, and that all the K'_m ($m \leq 12$) are well defined up to isometry by the condition that K'_m is a hyperplane section of K'_{m+1}, and is an A-lattice if m is even, or has the smallest possible determinant if m is odd. (Recall that $K'_m = K_m$ for $m = 11, 12$.)

8.5.3. Verify that the vectors in K_8 lie among those in K_9 which are orthogonal to $(\omega, \omega, \omega^2, \omega^2, 1, 1)$, and show that those which are in $S(K_8)$ are described as follows among the 90 pairs of vectors (29 of type I and 51 of type II) in $S(K_9)$:

Ia. $(\omega^i, \omega^j, 0, 0, 0, 0)$, $(i,j) = (0,0), (1,1), (2,2), (0,2), (2,0)$ (5 pairs out of 9).

Ib. Permutations of the coordinates $1, 2$ and $3, 4, 5, 6$ of $(-\omega^i, 0, 0, 0, 0, \omega)$, $i = 0, 2$ (16 pairs out of 24).

Ic. Permutations of the coordinates $3, 4, 5, 6$ of $(0, 0, -\omega, \omega, 0, 0)$ (6 pairs out of 6).

IIa. $\frac{1}{1-\omega}(\omega^i, \omega^j, 1, 1, 1, 1)$, $(i,j) = (0,0), (1,2), (2,1)$ (3 pairs out of 3).

IIb. 36 pairs among the 48 pairs $\frac{1}{1-\omega}(\omega^i, \omega^j, \omega^{\pm 1}, \omega^{\pm 1}, \omega^{\pm 1}, \omega^{\pm 1})$ depending on three choices for (i,j):

- $\frac{1}{1-\omega}(\omega, \omega, \omega, \omega, \omega, \omega)$ and $\frac{1}{1-\omega}(1, 1, \omega^{-1}, \omega^{-1}, \omega^{-1}, \omega^{-1})$ (2 pairs);
- Permutations of the coordinates $3, 4$ and $5, 6$ of $\frac{1}{1-\omega}(\omega^i, \omega^j, \omega, \omega^{-1}, \omega, \omega^{-1})$; $(i,j) = (0,0), (1,2), (2,1)$ (12 pairs);
- Permutations of the blocks $\{3,4\}$ and $\{5,6\}$ of $\frac{1}{1-\omega}(\omega^i, \omega^j, \omega, \omega, \omega^{-1}, \omega^{-1})$, $(i,j) = (0,0), (1,2), (2,1)$ (6 pairs);
- Permutations of the coordinates $5, 6$ of $\frac{1}{1-\omega}(\omega^i, \omega^j, \omega, \omega, \omega, \omega^{-1})$,

$(i,j) = (0,1), (1,0)$ (4 pairs);

- Permutations of the coordinates $3,4$ of $\frac{1}{1-\omega}(\omega^i, \omega^j, \omega^{-1}, \omega, \omega, \omega)$, $(i,j) = (0,1), (1,0)$ (4 pairs);
- Permutations of the coordinates $5,6$ of $\frac{1}{1-\omega}(\omega^i, \omega^j, \omega^{-1}, \omega^{-1}, \omega^{-1}, \omega)$, $(i,j) = (0,2), (2,0)$ (4 pairs);
- Permutations of the coordinates $3,4$ $\frac{1}{1-\omega}(\omega^i, \omega^j, \omega, \omega^{-1}, \omega^{-1}, \omega^{-1})$, $(i,j) = (0,2), (2,0)$ (4 pairs).

(We thus have $s(K_8) = 66$, and $S(K_8)$ contains 27 pairs of type I and 39 pairs of type II vectors.)

8.5.4. The notation is that of Exercise 8.5.3. Show that the minimal vectors in K_7 are to be chosen in the following way among the 66 pairs of vectors (27 of type I and 39 of type II) of $S(K_8)$: 2 (resp. 11, resp. 6) pairs among those of type Ia (resp. Ib, resp. Ic), hence 19 pairs of type I, 1 pair among the 3 pairs of type IIa, and finally 26 pairs among the 36 pairs of type IIb, hence 27 pairs of type II.

8.5.5. Show that the hyperplane sections of norm 4 of K_8^a whose determinant is minimum are isometric to K_7.

8.5.6. The notation is that of Exercise 4.8.1, where we gave a Hermitian construction for $\mathbb{E}_6$ on the field $\mathbb{Q}(\zeta)$ of ninth roots of unity, endowed with the form $\frac{1}{9}\,\mathrm{Tr}(x\overline{y})$.

1. Identify $L = \{(x,y) \in \mathfrak{p}^2 \times \mathfrak{p}^2 \mid x \equiv y \mod \mathfrak{p}^3\}$ with a lattice in $\mathbb{E}_6 \times \mathbb{E}_6$ containing $\mathbb{A}_2^{\perp 6}$ to index 3.

2. Show that $L' = \{(x,y) \in \mathfrak{p}^2 \times \mathfrak{p}^2 \mid x \equiv y \mod \mathfrak{p}^4\}$ has norm 4 and determinant 3^6.

3. Identify L' (as an Eisenstein lattice) with K_{12}. [Use the identification of $\sqrt{3}\,\mathbb{E}_6^*$ in Exercise 4.8.1.]

8.5.7. (K_{12} modulo 2.)

1. Show that K_{12} contains 378, 2016, and 10206 vectors of norm 4, 6, and 8 respectively, and that there is a single orbit in each case. [Use Exercise 8.4.4 and the proof of Theorem 8.5.2.]

2. Show that the vectors of norm 8 in K_{12} which are congruent modulo 2 to $x = (1,1,-1,-1,0,0)$ are, up to the sign, the six pairwise orthogonal vectors x, $(1,-1,1,-1,0,0)$, $(1,-1,-1,1,0,0)$, and $\frac{1}{1-\omega}\,(\omega^2, \omega^2, \omega^2, \omega^2, -2\omega^k, -2\omega^\ell)$ with $(k,\ell) = (0,1)$, $(1,0)$, $(2,2)$.

3. (Quebbemann.) Prove that representatives for $K_{12} \mod 2K_{12}$ can be chosen to be one vector out of each pair $\pm x$ of vectors of norm 4 or 6 and one vector out of each system as above of six pairs of vectors of norm 8. (For a more conceptual proof, see [Mar8].)

8.5.8. Let C be a ternary code of length m, dimension d and weight $wt(C) = w$, and let Λ_C be the lattice of dimension $n = 2m$ generated by L_n and the vectors $\frac{1}{1-\omega}\,u$, $u \in C$.

1. Show that $\det(L_C) = 3^{m-2(d-1)}$.

2. Show that $N(L_C) = \min(4, \frac{2}{3}\,w)$.

3. Suppose that $w \geq 6$, and let r be the number of words of weight 6 in C. Show that $s(L_C) = 3^5\,r + 9\,\frac{m(m-1)}{2}$.

4. Let $C = \begin{pmatrix} 1 & 1 & 1 & 1 & 1 & 1 & 0 & 0 & 0 & 0 \\ 0 & 0 & 1 & 1 & -1 & -1 & 1 & 1 & 0 & 0 \\ 0 & 1 & 0 & 1 & 0 & 1 & 1 & -1 & 1 & 0 \\ 0 & 0 & 0 & 1 & 1 & 1 & 0 & 1 & -1 & 1 \end{pmatrix}$, and let $\mathcal{C}$ be the ternary code with generator matrix C. Show that the lattices $\Lambda_{\mathcal{C}'}$ attached to the codes defined by 1, 2, 3 or 4 lines of C, of dimensions 12, 16, 18, 20 have determinants 3^6, 3^6, 3^5, and 3^4 and invariants $s = 378$, $s = 1224$, $s = 3240$, and $s = 7695$ respectively. [The first one is of course K_{12}, the other three are K'_{16}, K'_{18}, and K'_{20} to be defined in Section 8.6.]

8.6.1. The notation is that of Theorem 8.6.7.

1. Show that there exist three maximal left ideals $\mathfrak{P}, \mathfrak{P}', \mathfrak{P}''$ lying above 2 in $\mathfrak{M}$, and that a possible choice for them is $\mathfrak{P} = \mathfrak{M}\,(1+i)$, $\mathfrak{P}' = \mathfrak{M}\,(1+\omega i)$, $\mathfrak{P}'' = \mathfrak{M}\,(1+\omega^2 i)$.

2. Let U be the group (of order 12) of units of $\mathfrak{M}$. Show that each of the ideals $\mathfrak{P}'$ and $\mathfrak{P}''$ possesses a generator congruent to 1 modulo $\mathfrak{P}$, unique up to its sign. In the following questions, we denote by w such a generator, and by u (resp. v) an arbitrary element of U (resp. of $\{\pm 1, \pm i\}$).

3. Show that the 378 pairs of minimal vectors in K_{12} correspond to the following permutations of the sets $(\lambda_1, \lambda_2, \lambda_3)$ of elements of $\mathfrak{M}^3$:

- $2\,u\,(1,0,0)$ (18 pairs);
- $u\ (1+i, v\,(1+i), 0)$ (72 pairs);
- $u\,(\pm 1, v, w)$, $w \in \mathfrak{P}'$ (144 pairs);
- $u\,(\pm 1, v, w)$, $w \in \mathfrak{P}''$ (144 pairs).

8.7.1. Let K be a real quadratic field, of discriminant d, and let H be the "usual" quaternion algebra over K, such that $i^2 = j^2 = -1$. Denote by $x \mapsto x'$ the non-trivial automorphism of K.

1. Show that if $d \equiv 4 \mod 8$, the units of K have norm $+1$.

2. Show that if $d \equiv 1 \mod 8$, H is ramified at the two primes lying above 2 in K, and that $(1, i, j, \frac{-1+i+j+k}{2})$ is a $\mathbb{Z}_K$-basis for a maximal order in H.

3. From now on, suppose that $d \equiv 0$ or $5 \mod 8$. Show that H is unramified over K.

4. Suppose that $d \equiv 5 \mod 8$. Let $\tau = \frac{1+\sqrt{d}}{2}$. Show that the $\mathbb{Z}_K$-module with basis $(1, i, \frac{i+\tau j+\tau' k}{2}, \frac{-1+i+j+k}{2})$ is a maximal order in H.

5. Suppose finally that $d \equiv 0 \mod 8$. Write $(2) = \mathfrak{p}^2$ in K, and let $\pi \in \mathbb{Z}_K$ such that $v_{\mathfrak{p}}(\pi) = 1$. Show that the $\mathbb{Z}_K$-module generated by the elements of H of the form $\frac{\alpha}{\pi}(1+u)$ with $u \in \{i, j, k\}$ and $\alpha \in \mathbb{Z}_K$ such that $v_{\mathfrak{q}}(\alpha) \geq v_{\mathfrak{q}}(\pi)$ for all $\mathfrak{q} \neq \mathfrak{p}$ is a maximal order in H.

8.7.2. Let I be a (left or right) ideal in a ring R.

1. Show that $G = \{u \in R^\times \mid u - 1 \in I\}$ is a subgroup of $R^\times$.

2. Let R be the maximal order defined in Proposition 8.7.3 (2) and let I be the left ideal generated by $1+i$. Show that G is the group generated by the eight units $\pm 1, \pm i, \pm j, \pm k$ and their products by $\varepsilon = \frac{1}{2}(-1+i+j-k)$.

3. Show that the minimal vectors in the lattice $L' \simeq \mathrm{BW}_{16}$ in Theorem 8.7.4. are on the one hand, 240 pairs of vectors of the form $(\lambda, 0)$ or $(0, \lambda)$, and on the other hand $120 \cdot 16 = 1920$ pairs of vectors of the form $x\,(y, 1)$ with $x \in S(\mathbb{E}_8)$ and $y \in G$.

8.7.3. The notation is that of Proposition 8.7.3; in particular, $\varepsilon = \frac{1+\sqrt{5}}{2}$ and $\mathbb{Z}_K = \mathbb{Z}[\varepsilon]$. Recall that H has class number 1. For every nonzero left ideal $\mathfrak{a}$ in $\mathbb{Z}_K$, let $N_{\mathfrak{a}}$ be the number of ideals of $\mathfrak{M}$ with norm $\mathfrak{a}$.

1. Show that we have $N_{\mathfrak{ba}} = N_{\mathfrak{b}}\, N_{\mathfrak{a}}$ if $\mathfrak{b}$ is two-sided and $(\mathfrak{a}, \mathfrak{b}) = 1$.

2. Show that $N_{\mathfrak{p}} = \mathrm{N}_{K/\mathbb{Q}}(\mathfrak{p}) + 1$ if $\mathfrak{p}$ is a maximal ideal. [Enumerate the maximal ideals of $\mathcal{M}_2(\mathbb{Z}_K/\mathfrak{p})$.]

3. Let $\lambda = \frac{u+v\sqrt{5}}{2} \in \mathbb{Z}_K$. Show that $\mathrm{Tr}_{K/\mathbb{Q}}(\alpha^{-1}\lambda) = \frac{u-v}{2}$.

4. Let $\lambda \in \mathbb{Z}_K \smallsetminus \{0\}$ have trace m over $\mathbb{Q}$. Prove the following equivalences: $\lambda \in \mathrm{Nrd}(\mathfrak{M}) \Longleftrightarrow \lambda >> 0 \Longleftrightarrow \frac{m-\sqrt{5m^2-4}}{2} \le v \le \frac{m+\sqrt{5m^2-4}}{2}$.

5. Express the number of pairs of norm $2m$ vectors in $\mathbb{E}_8$ as a sum $60 \sum_v N_{\frac{u+v\sqrt{5}}{2}}$ where $u = v + 2m$; compare this expression with the value $(120\,\sigma_3(m))$ obtained from the theta function of $\mathbb{E}_8$; make the calculation explicit for $2m = 4$.

8.8.1. Show that we obtain the lattice U_{32} when applying the construction of Proposition 8.8.2 to BW_{16}, and recover the calculation of $s(U_{32})$ in Theorem 8.7.7.

8.8.2. Let $k = \mathbb{Q}(\sqrt{m})$, $m \in \mathbb{Z}, m \equiv 2 \mod 4$, m square free, be a real quadratic field. Write $(2) = \mathfrak{p}^2$, and assume that $\mathfrak{p}$ possesses a totally positive generator α. Let H be the quaternion algebra with centre k whose ramified places are the two real places of k, and let $\mathfrak{M}$ be a maximal order in H.

1. Show that H is the "usual" quaternion algebra.

2. Show that $(\mathfrak{M}, \frac{1}{2}\,\mathrm{Trd}(\alpha^{-1} x\overline{y}))$ is an even lattice of norm 2 and determinant 1, hence isometric to $\mathbb{E}_8$. (Compare Exercise 4.8.3.)

3. Show that $\Lambda = \{(x, y) \in \mathfrak{M}^2 \mid x \equiv y \mod \mathfrak{p}\,\mathfrak{M}\}$ is isometric to Λ_{16}.

8.10 Notes on Chapter 8

In his fundamental paper "Extreme forms" ([Cox2], 1951), Coxeter introduces two new ideas: the use of lattices in the theory of extremality (see the notes on Chapter 3), and the use of congruences (on the ring of Eisenstein integers) to construct new extreme lattices. These ideas were then generalized in the paper [C-T], published in 1953 with Todd, in which they construct the lattice which bears their names, now usually called K_{12}, and then by Barnes ([Bar5]). We have developed the methods of Coxeter, Todd and Barnes in Sections 8.2 to 8.5, first on the rings of Hurwitz quaternions (Sections 8.2 and 8.3), then on the ring of Eisenstein integers in Sections 8.4 and 8.5, which is really the first ring that occurred in this setting.

We construct the series K_n for $n \le 12$ and also introduce in these dimensions the companion series K'_n. It should be noticed that the conjecturally densest lattice in dimension 11, namely K_{11}, was not discovered by Coxeter and Todd, but by Barnes ([Bar5, II], p. 221) by means of a "lamination" above L_{10}^4; Barnes also gave its alternative interpretation as a section of K_{12}.

Sections 8.2 and 8.3 consist in transposing over the Hurwitz order the methods above. The lattices that we denoted by J_n play the rôle of Barnes's L_n, the Barnes–Wall lattice Λ_{16} that of the Coxeter–Todd lattice K_{12}, and the laminated lattices $\Lambda_n^?$, $n \le 16$ except for $\Lambda_{13}^{\mathrm{mid}}$, that of the K_n, K'_n, $n \le 12$.

However, Λ_{16} (together with its hyperplane section Λ_{15}) was discovered by Barnes and Wall ([B-W], 1959; see also [Ha]) as part of a series BW_{2^m} of alternatively 2- and 1-modular lattices, beginning with $\mathbb{D}_4$ and $\mathbb{E}_8$, see Section 8.8. (The method used in Section 8.8 to double the dimension of a lattice can be viewed of a special case of the method described in [Sig2].)

We have made use of the quaternionic construction of Λ_{16} to construct all laminated lattices up to dimension 16 except the dead-end $\Lambda_{13}^{\mathrm{mid}}$, which is not contained in Λ_{16}, as one can see by listing the norm 4 sections of Λ_{16} which are similar to $\mathbb{A}_3$. Plesken and Pohst have proved in [Pl-P2] that $\Lambda_{13}^{\mathrm{mid}}$ embeds in Λ_{17}.

The complete description of the laminated lattices up to dimension 25 is due to Conway and Sloane, whose original paper in the 1982 issue of *Annals of Math.* is Chapter 6 of [C-S]; the principal series was known to Leech (cf. [C-S], Chapter 6, Section 1).

We describe here a method which yields an embedding of $\Lambda_{13}^{\mathrm{mid}}$ inside the Leech lattice Λ_{24}, and which can indeed be used to construct all sequences

$$M_{13} = \Lambda_{13}^{\mathrm{mid}} \subset M_{14} \subset \cdots \subset \Lambda_{24},$$

each term of which has the smallest possible determinant among the sublattices of Λ_{24} containing the preceding one.

Since we have the inclusion $\Lambda_{12}^{\mathrm{mid}} \subset \Lambda_{13}^{\mathrm{mid}}$, and since $\Lambda_{12}^{\mathrm{mid}}$ is isometric to its orthogonal in Λ_{24}, we are reduced to studying the "*antilaminations*" of $\Lambda_{12}^{\mathrm{mid}}$ (decreasing sequences of sections, each of which is of minimal determinant). We find three sections of determinant 1024, namely $\Lambda_{11}^{\mathrm{min}}$, $\Lambda_{11}^{\mathrm{max}}$ and the one which we are interested in, which we denote by $\Lambda_{11}^{\mathrm{mid}}$ (although we have $s(\Lambda_{11}^{\mathrm{mid}}) = 211 < s(\Lambda_{11}^{\mathrm{min}}) = 216$). This is a perfect lattice which again possesses three sections of minimal determinant, namely 1024, which are perfect, with $s = 138$, $s = 135$, and $s = 130$ respectively. (Note that $\Lambda_{11}^{\mathrm{mid}}$ *is not* a laminated lattice, though its norm and determinant are those of the corresponding laminated lattices.)

The first two lattices both possess a perfect section in dimension 9 of determinant 768, with $s = 99$. These sections contain the non-perfect 8-dimensional lattice with $s = 75$ quoted in the notes on Chapter 6, and we finally reach $\Lambda_7 \sim \mathbb{E}_7$, whose orthogonal is Λ_{17}.

The third one is a remarkable lattice discovered by Souvignier, and denoted by Q_{10} in [C-S4]; see Section 14.5. This is an extreme norm 4 lattice similar to its dual, with $s = 130$; its γ' invariant is the largest known in dimension 10 (together with those of $\mathbb{D}_{10}^{+}$, K'_{10}, and $K_{10}'^{*}$). Two 8-dimensional Laïhem lattices occur as norm 4 sections of Q_{10}, namely Lh_6 (with $s = 54$) and Lh_8 (with $s = 52$), both of determinant 768.

By an exploration of the numerous bifurcations which show up at successive sections, we recover all of Plesken–Pohst's weakly laminated lattices above $\Lambda_{13}^{\mathrm{mid}}$, but this method cannot prove that we have indeed found all of them.

The existence of a Hurwitz structure on a lattice having the same determinant and the same kissing number as $\Lambda_{12}^{\mathrm{min}}$ was found independently by Sigrist in a search (using a generalized Voronoi algorithm) for lattices with an automorphism group containing $\widetilde{A}_4$ ([Sig1]) and by Laïhem (private communication). The uniqueness of Λ_{16} *as a unimodular lattice over the Hurwitz order* was first proved by Quebbemann (private communication), using a mass formula; it is also proved in Bachoc's paper [Bac1], using an adaptation of Kneser's neighbouring algorithm. Note that the uniqueness of Λ_{16} as *a unimodular lattice over* $\mathfrak{M}$ does not imply that no other structure may exist: we cannot exclude the existence of non-$\mathfrak{M}$-integral models for Λ_{16}.

Various uniqueness results (about $\Lambda_{12}^{\max}$ and $\Lambda_{12}^{\min}$ over the Hurwitz order, $\mathbb{E}_6^*$, L_4^2 over the ring of Eisenstein integers, etc.) can be found in [Mar1].

We find in papers by Tits ([Ti]) and by Conway and Sloane ([C-S], Chapter 8, Section 2.2) a construction of the Leech lattice over the usual quaternion algebra with centre $\mathbb{Q}(\sqrt{5})$ which could be interpreted in terms of a pair maximal ideals as we did in Sections 8.6 and 8.7. From the strict point of view of lattices, the choice of this particular skew-field is not important, except that its large unit groups may yield a simpler description of the set of minimal vectors (as in the case of $\mathbb{E}_8$, see Exercise 8.7.3). However, this choice is important for group theoretical reasons: the group of the corresponding Hermitian automorphisms is a double cover of Hall–Janko's simple group J_2 ([Ti]; [ATLAS], p. 42.)

The possibility of endowing K_{12} with a $\mathfrak{M}_3$-structure was proved in a non-explicit way by Gross in [Gro], 1989.

The lattices K_n, $n \leq 24$ were discovered by Leech, (see [C-S], Chapter 6, Section 1). Up to dimension 24, the densest known lattices are the Λ_n, except curiously in the three dimensions (11, 12, 13) where they are not unique, but this supremacy of the K_n breaks down in dimensions $n = 10$ and $n = 14$, for which K'_n has the same density as K_n.

The principal invariants of the lattices Λ_n, K_n, and K'_n have been calculated by Plesken and Pohst in [Pl-P2], together with those of all norm 4 weakly laminated lattices; see also [Bt-M'].

I discovered the possibility of constructing easily an even 2-modular lattice of norm 8 in dimension 48 through an e-mail of Quebbemann to Bachoc, motivated by the constructions of [Bac1].

The densest known 32-dimensional lattices are 2-modular of norm 6, hence of determinant 2^{16} and Hermite constant $3\sqrt{2} = 4,242\ldots$ Four such lattices are known, denoted by Qb_{32}, Qb'_{32}, Bac_{32} and Ne_{32}. They have distinct automorphism groups, see [Pl-S] for the first three. They were found by Quebbemann ([Q4], [Q3]), Bachoc ([Bac1]), and Nebe ([Ne5]); Qb_{32} is constructed over $\mathbb{Z}[\zeta_8]$ (see [C-S], Chapter 8, Section 4), and Bac_{32} over the Hurwitz order. (Nebe–Sloane's WEB page ([Ne-Sl]) on lattices refers to a lattice discovered by Elkies, with the same minimum and determinant, but apparently not modular; it has however the same Hermite invariant.)

Using the techniques of Proposition 8.8.2, we can attach to the four lattices above four 64-dimensional unimodular lattices. One such lattice was found by Quebbemann (see [C-S], Chapter 8, Section 3); we obtain in this way probably (at least) three new lattices, but I do not see how to prove that they are distinct. Their cross-sections in dimensions $32+r$ and $32-r$, which have the same density, produce examples of dense lattices around dimension 32; see Section 14.4.

9 The Configurations of Minimal Vectors

Introduction

We say that two lattices Λ and Λ' with respective sets of minimal vectors S and S' are *minimal-equivalent* if there exists $u \in \mathrm{GL}(E)$ which maps Λ onto Λ' and $S(\Lambda)$ onto $S(\Lambda')$.

In this chapter, we study this equivalence relation as well as an analogue involving duality. We prove in Section 9.1 finiteness theorems for both these notions. Sections 9.2 and 9.3 are then devoted to classification results in dimensions $n \leq 4$, in connection with some results of Chapter 6.

The notion of a minimal class is then used in Section 9.4 to prove the finiteness of the class number (up to similarity) of eutactic lattices (Avner Ash's theorem), in a more general form, which only requires weak eutaxy. The method of proof will give us a classification theorem in dimension $n \leq 4$ which we describe in Section 9.5, where we also state and verify for small dimensions Avner Ash's "mass formula with signs".

We finally introduce the notion of a *perfect pair* used to prove a theorem of A.-M. Bergé, which says in particular that there exist only finitely similarity classes of dual-extreme lattices in each dimension.

9.1 Minimal-Equivalent Lattices

The group $\mathrm{GL}(E)$ acts on the set of lattices and also on the set of pairs (Λ, Λ^*) consisting of a lattice and its dual. (The dual lattice is transformed according to the rule $\Lambda^* \mapsto {}^t u^{-1}(\Lambda^*)$.)

Definition 9.1.1. *We say that two lattices Λ and Λ' are* minimal-equivalent *if there exists $u \in \mathrm{GL}(E)$ which maps Λ onto Λ' and $S(\Lambda)$ onto $S(\Lambda')$. Similarly, we say that two pairs (Λ, Λ^*) and (Λ', Λ'^*) are* minimal-equivalent *if there exists $u \in \mathrm{GL}(E)$ which maps Λ onto Λ' and such that moreover we have the two equalities $u(S(\Lambda)) = S(\Lambda')$ and ${}^t u^{-1}(S(\Lambda^*)) = S(\Lambda'^*)$. We denote by $\Lambda \prec \Lambda'$ the relation obtained by replacing the equality $u(S(\Lambda)) = S(\Lambda')$ by the inclusion $u(S(\Lambda)) \subset S(\Lambda')$, and we define in the same way the relation $(\Lambda, \Lambda^*) \prec (\Lambda', \Lambda'^*)$.*

The classes of lattices Λ (resp. of pairs (Λ, Λ^*)) for the equivalence relation above are called minimal classes *(resp.* dual-minimal classes*).*

In both cases, the relations we defined above are clearly equivalence or quasi-ordering relations, which are compatible with similarity, which justifies the definition below:

Definition 9.1.2. *We denote by $\prec$ the ordering relation induced on the set of equivalence classes of lattices (resp. of pairs (Λ, Λ^*) of lattices) by the relations "$\prec$" above.*

To say that two lattices Λ and Λ' are minimal-equivalent amounts to saying that there exist bases for Λ and Λ' in which the components of the vectors in $S(\Lambda)$ and in $S(\Lambda')$ define the same subspaces of $\mathbb{Z}^n$. A similar remark applies to pairs (Λ, Λ^*). We obtain in this way analogues to Definitions 9.1.1 and 9.1.2 in the setting of positive definite quadratic forms on $\mathbb{R}^n$ (or of positive definite real symmetric matrices); these definitions are compatible with equivalence under $\mathrm{GL}_n(\mathbb{Z})$. The notation Q^* stands for the *inverse of the form* Q (i.e., if A is the matrix of Q, that of Q^* is A^{-1}; one could have replaced the inverse form by the *adjoint form*, whose matrix is the adjoint matrix of A.

Definition 9.1.1′. *We say that two quadratic forms Q and Q' are* minimal-equivalent *(resp. that the pairs (Q, Q^*) and (Q', Q'^*) are* minimal-equivalent*) if there exists a form Q'_1 equivalent to Q' such that $S(Q'_1) = S(Q)$ (resp. $S(Q'_1) = S(Q')$ and $S(Q_1'^*) = S(Q^*)$).*

Definition 9.1.2′. *Same definition as above, replacing everywhere "$=$" by "$\prec$".*

It should be noticed that the relation $\Lambda \prec \Lambda'$ amounts to the existence of an inclusion $S \subset S'$ between the set of components *in suitable bases* of minimal vectors in Λ and in Λ', but that, however, given the set $S' \subset \mathbb{Z}^n$ of minimal vectors of a quadratic form, an arbitrary subset S of S' is not necessarily the set of minimal vectors of a quadratic form. A 4-dimensional counter-example is provided by $S' = S(\mathbb{D}_4)$, taking for S any subset of S' with 11 or 10 elements, see Theorem 6.2.1.

Given a lattice L, since for all $u \in \mathrm{GL}(E)$ sufficiently close to the identity, vectors in $S(u(L))$ are images under u of vectors in $S(L)$, we have:

Proposition 9.1.3. *Let Λ be a lattice in E. There exists a neighbourhood $\mathcal{V}$ of Λ on which every lattice Λ' satisfies the two inequalities $\Lambda' \prec \Lambda$ and $(\Lambda', \Lambda'^*) \prec (\Lambda, \Lambda^*)$.* □

Lemma 9.1.4. *Let Λ be a lattice, let $\mathcal{B} = (e_1, \dots, e_n)$ be a basis for Λ, and let $K_{\mathcal{B}}$ the real number defined by*

$$N(e_1) \cdots N(e_n) = K_{\mathcal{B}}^2 \det(\Lambda).$$

Let moreover $\mathcal{B}^* = (e_1^*, \dots, e_n^*)$ be the dual basis to $\mathcal{B}$. Then:

1. *For $i = 1, \dots, n$, we have $N(e_i)\,N(e_i^*) \le K_{\mathcal{B}}^2$.*
2. *The absolute values of the components in $\mathcal{B}$ of the minimal vectors of Λ and of those in $\mathcal{B}^*$ of the minimal vectors in Λ^* are bounded from above by $K_{\mathcal{B}}$.*

Proof. For each index $i \in [1, n]$, let F_i be the hyperplane in E orthogonal to e_i^* and let $\Lambda_i \subset F_i$ be the lattice $\oplus_{j \ne i} \mathbb{Z}\, e_j$. We have $\Lambda_i = \Lambda \cap F_i$, hence $\det(\Lambda_i) = \det(\Lambda)\, N(e_i^*)$ (Corollary 1.3.5). By the Hadamard inequality, we have

$$\det(\Lambda_i) \le \prod_{j \ne i} N(e_j)\,,$$

which implies

$$N(e_i)\,N(e_i^*) \le \prod_i N(e_i)\ \det(\Lambda)^{-1} \le K_{\mathcal{B}}^2$$

by definition of $K_{\mathcal{B}}$. This proves (1). We then deduce (2) from (1) by observing that for $x = \sum x_i\, e_i \in S(\Lambda)$ and $y = \sum y_i\, e_i^* \in S(\Lambda^*)$, we have

$$x_i^2 = (x \cdot e_i^*)^2 \le N(e_i)\,N(e_i^*) \quad \text{and} \quad y_i^2 = (y \cdot e_i)^2 \le N(e_i^*)\,N(e_i)\,. \ \square$$

As a consequence, we obtain easily the finiteness statement below:

Theorem 9.1.5. (A.-M. Bergé.) The numbers of minimal classes of lattices and of pairs of dual lattices in a given dimension are finite.

Proof. We use the Hermite inequality (Theorem 2.2.1): every lattice has a basis $(e_1, \dots, e_n)$ for which the constant $K_{\mathcal{B}}$ in Lemma 9.1.4 is bounded from above by $(\frac{4}{3})^{n(n-1)/2}$. This shows that we can find for any lattice Λ a basis $\mathcal{B}$ such that the components in $\mathcal{B}$ of the minimal vectors in Λ and the components in $\mathcal{B}^*$ of the minimal vectors in Λ^* are bounded from above by a constant depending solely on the dimension. □

Since perfect lattices are determined up to similarity by the components of their minimal vectors in a basis, Theorem 9.1.5 implies Voronoi's finiteness theorem (Theorem 3.5.4). Actually, the proof we gave in Chapter 3 is essentially an adaptation of the proof above to *perfect classes*, i.e. classes which consist in a similarity class of perfect lattices; see Theorem 9.1.9 (2) below.

A minimal class $\mathcal{C}$ is clearly the disjoint union of classes of pairs (Λ, Λ^*) for Λ running through $\mathcal{C}$. To determine the corresponding partition of $\mathcal{C}$ is a difficult problem. Indeed, the minimal classification of lattices (resp. of pairs of lattices) is known only in dimensions ≤ 5 (resp. ≤ 3), see Sections 9.2 and 9.3 below.

The following statement will be made more precise in the case of usual minimal classes: indeed, it will result from Theorem 9.1.11 that the set $\widetilde{\mathcal{C}}$ defined below is in fact the *topological closure* of $\mathcal{C}$.

Proposition 9.1.6. *Let $\mathcal{C}$ be a minimal class of lattices or of pairs of lattices, and let $\widetilde{\mathcal{C}}$ be the union of the classes which contain $\mathcal{C}$. Then $\widetilde{\mathcal{C}}$ is a closed subset in the space of lattices, and $\mathcal{C}$ is an open subset in $\widetilde{\mathcal{C}}$. Moreover, if the minimal vectors in the lattices in $\mathcal{C}$ span E, $\widetilde{\mathcal{C}}$ is compact modulo proportionality.*

Proof. • Let $\mathcal{C}'$ be a class containing $\mathcal{C}$. Let us show that its closure $\overline{\mathcal{C}'}$ (in the space of lattices in E) is included in $\widetilde{\mathcal{C}}$. Let $L_0 \in \overline{\mathcal{C}'}$, and let $\mathcal{C}_0$ be its class. The lattices L' sufficiently close to L satisfy the condition $L' \prec L$; applied to a lattice $L' \in \mathcal{C}'$ we obtain the inclusion $\mathcal{C}' \prec \mathcal{C}_0$, hence in particular $\mathcal{C} \prec \mathcal{C}_0$, which shows that L_0 belongs to the union $\widetilde{\mathcal{C}}$ of the classes containing $\mathcal{C}$. Applying this result to the classes contained in $\widetilde{\mathcal{C}}$, we prove that $\widetilde{\mathcal{C}}$ is closed.
• To prove that $\mathcal{C}$ is open in $\widetilde{\mathcal{C}}$, consider a lattice $L \in \mathcal{C}$ and a neighbourhood $\mathcal{V}(L)$ of L on which $L' \prec L$. Its intersection with $\widetilde{\mathcal{C}}$ is a neighbourhood of L in $\widetilde{\mathcal{C}}$ on which we have $L' \prec L$ and $L \prec L'$. Hence $\mathcal{V}(L) \cap \widetilde{\mathcal{C}}$ is contained in the class $\mathcal{C}$ of L.
• Identifying the set of lattices modulo proportionality with the set of lattices with a given norm, we may apply Mahler's compactness lemma in the form of Corollary 2.4.3. This immediately implies that the closed set $\widetilde{\mathcal{C}}$ is indeed compact. □

In the discussion that follows, we restrict ourselves to ordinary classes, for lack of corresponding results in the case of pairs (Λ, Λ^*).

Let $\mathcal{C}$ a minimal class of lattices. It amounts to the same thing to consider either similarity classes of lattices in $\mathcal{C}$, or isometry classes of lattices in $\mathcal{C}$ with a given minimum m. By the dictionary lattices – quadratic forms (Section 1.7), we construct a bijection between $\mathcal{C}$ and a union of equivalence classes (i.e., of orbits under $\mathrm{GL}_n(\mathbb{Z})$) of quadratic forms with minimum m whose set of minimal vectors is a conveniently given subset S of $\mathbb{Z}^n$ (see Definition 9.1.1′). Since quadratic forms correspond to points in $\mathbb{R}^{n(n+1)/2}$, we have the notions of a *convex set* and of a *convex* (or *concave*) *function* on a convex set, which will give sense to the statements in Theorems 9.1.10 and 9.1.11.

The following proposition allows us to attach an important invariant to a minimal class. Recall (Definitions 3.2.2 and 3.2.8) that the perfection rank of a lattice is the dimension of the span in $\mathrm{End}^s(E)$ of the projections onto its minimal vectors, and (Definition 3.2.9 and Proposition 3.1.13) that the perfection rank of a positive definite quadratic form Q is the dimension of the span in $\mathrm{Sym}_n(\mathbb{R})$ of the matrices $X\,{}^tX$, $\pm X \in S(Q)$.

Proposition 9.1.7. *The perfection rank is constant on minimal classes.*

Proof. Let Λ and $\Lambda' = u(\Lambda)$, $u \in \mathrm{GL}(E)$ be two lattices in a class $\mathcal{C}$. Choose a basis $\mathcal{B}$ for Λ. Then the Gram matrices of $\mathcal{B}$ and of $u(\mathcal{B})$ have the same set S of minimal vectors, hence the same perfection rank. □

Definition 9.1.8. *The* perfection rank of a minimal class *of lattices (or of quadratic forms) is the perfection rank of one particular lattice (or form) in the class.*

Theorem 9.1.9. *1. The perfection rank is a strictly increasing function on the set of minimal classes.*
2. A minimal class of lattices (resp. of quadratic forms having a given minimum) is a maximal element in the set of minimal classes if and only if it is the set of lattices which are similar to a given perfect lattice (resp. the set of quadratic forms which are equivalent modulo $\mathrm{GL}_n(\mathbb{Z})$ *to a given perfect form).*
3. For every minimal class $\mathcal{C}$*, there exists a perfect class* $\mathcal{C}'$ *such that* $\mathcal{C} \prec \mathcal{C}'$*.*
4. Given minimal classes $\mathcal{C}$ *and* $\mathcal{C}'$ *such that* $\mathcal{C} \prec \mathcal{C}'$ *with respective perfection ranks* t *and* $t' > t$*, there exists an increasing sequence*

$$\mathcal{C}_0 = \mathcal{C} \prec \mathcal{C}_1 \prec \cdots \prec \mathcal{C}_r = \mathcal{C}'$$

of classes such that $\mathcal{C}_i$ *has perfection rank* $t + i$*.*

Proof. We write it in terms of quadratic forms, restricting ourselves to forms with a given minimum m, and identify forms with their corresponding symmetric matrix.

• Let $\mathcal{C}$ and $\mathcal{C}' \succ \mathcal{C}$ be two minimal classes. Replacing the forms in one of these classes by their images under a conveniently chosen element of $\mathrm{GL}_n(\mathbb{Z})$, we may represent these classes by sets of forms Q with minimum m and such that $S(Q) = S$ and $S(Q) = S'$, where S and $S' \supset S$ are two finite subsets of $\mathbb{Z}^n$. The inequality $\mathcal{C} \prec \mathcal{C}'$ is then strict if and only if the inclusion $S \subset S'$ is strict.

It is clear that if the perfection rank of $\mathcal{C}'$ is strictly larger than that of $\mathcal{C}$, then S' contains strictly S.

Let us show that conversely, if the perfections ranks are equal, then $S = S'$. The matrices of forms in $\mathcal{C}$ are those positive definite matrices A such that $\langle A, X^t X\rangle \geq m$ for all nonzero $X \in \mathbb{Z}^n$ and that equality holds if and only if $X \in S$. This last equality defines an affine subvariety of Sym_n, whose dimension is the perfection corank of $\mathcal{C}$. Replacing S by S', we obtain an affine subvariety of the preceding one, which is non-empty (because S' corresponds to the class $\mathcal{C}'$), and which therefore has the same dimension (because $\mathcal{C}$ and $\mathcal{C}'$ have equal perfection ranks). Hence these two varieties are identical, which shows that every element in S' belongs to S.

• By definition of perfection, the perfection rank attains its maximum (equal to $\frac{n(n+1)}{2}$) exactly on perfect forms. Hence classes containing a perfect form are maximal, and since a perfect form is well defined by its minimum and the set of its minimal vectors, a class which contains a perfect form Q, is in fact the collection of all forms which are equivalent modulo $\mathrm{GL}_n(\mathbb{Z})$ to a form proportional to Q.

Conversely, we prove that a class whose elements are non-perfect forms is not a maximal one by making use of the variant of the Voronoi algorithm described in Proposition 7.2.5: if the perfection rank of a form Q (i.e., the dimension of its Voronoi domain) is not maximal, there exists a form R such that $S(R)$ strictly contains $S(Q)$, so that the class of R is strictly larger (for the $\prec$ relation) than that of Q.

• We now prove (3) by induction on the perfection corank. The result is obvious when the corank is zero, since the class is then perfect. Let $\mathcal{C}$ be a class whose perfection rank is not maximal, and let $Q \in \mathcal{C}$. The preceding process allows us to construct a class $\mathcal{C}' \succ \mathcal{C}$ whose perfection corank is strictly smaller than that of Q.

• To prove (4), we interpret $\mathcal{C}$ and $\mathcal{C}'$ as classes of quadratic forms with minimum m and sets of minimal vectors S and $S' \supset S$. Let $Q \in \mathcal{C}$ and $Q' \in \mathcal{C}'$, and let $\mathcal{D}$ and $\mathcal{D}'$ be the corresponding Voronoi domains in Sym_n (for Q and Q', or for $\mathcal{C}$ and $\mathcal{C}'$, this amounts to the same thing). Let $\mathcal{H}$ be a (hyperplane) facet of $\mathcal{D}'$ containing $\mathcal{D}$. Such a facet exists, since

$$X \mapsto \langle Q' - Q, X{}^tX \rangle = Q'(X) - Q(X)$$

is zero on S and takes negative values on $S' \smallsetminus S$. Let F be a facet vector for $\mathcal{H}$ in $\mathcal{D}'$, i.e. a nonzero quadratic form in $\mathcal{D}'$ such that the scalar products $\langle F, X{}^tX \rangle = F(X)$ are zero for $X{}^tX \in \mathcal{H}$ and are positive for $X \in S'$, $X{}^tX \notin \mathcal{H}$. Consider now the forms $Q_\theta = Q' + \theta F$ with positive θ. On S', the Voronoi scalar product $\langle Q_\theta, X{}^tX \rangle = Q_\theta(X)$ is strictly larger than m if $X{}^tX \notin \mathcal{H}$, and is zero otherwise (and in particular, it is zero on S). For sufficiently small θ, since the minimal vectors of Q_θ come from those in S', we have $Q_\epsilon(X) > m$ for all nonzero X outside S'. The perfection rank of Q_θ is thus equal to $t' - 1$, and an induction argument allows us to obtain any value of the perfection rank in the range $[t, t']$. □

Theorem 9.1.10. *Let $m > 0$, let S be a finite subset of $\mathbb{Z}^n$, and let $\mathcal{Q}$ be the set of positive definite quadratic forms Q such that* $\min Q = m$ *and* $S(Q) = S$. *Then $\mathcal{Q}$ is a convex set, and the determinant is a strictly logarithmically concave function on $\mathcal{Q}$.*

Proof. Let Q and R be two distinct forms in $\mathcal{Q}$ and let $t \in [0,1]$. For any nonzero $x \in \mathbb{Z}^n$, we have $t\,Q(x) + (1-t)\,R(x) \geq tm + (1-t)m = m$, with equality if $x \in S$ and strict inequality otherwise, which shows that $\mathcal{Q}$ contains the segment $[Q, R]$, hence that $\mathcal{Q}$ is a convex set.

[More generally, we have $[Q, R) \subset \mathcal{Q}$ under the hypotheses $S(Q) = S$ and $S(R) \supset S$.]

We now study the determinant, using the method known as "double diagonalization". Let A and B be the matrices of Q and R respectively. Since A is positive, it is of the form ${}^tUU$ (we can choose $U \in \mathrm{Sym}_n$, for instance the square root of A), and we have $B = {}^tUCU$ for $C = {}^tU^{-1}BU^{-1}$, whence the identity

$$\det(t\,A + (1-t)\,B) = \det(A)\,\det(t\,I_n + (1-t)\,C)$$

on the whole segment $[Q, R]$. We then transform C into a diagonal matrix D using an orthogonal base change matrix V. We have

$$\det(t\,I_n + (1-t)\,C) = \det(t\,I_n + (1-t)\,VDV^{-1}) = \det(t\,I_n + (1-t)\,D)\,,$$

hence denoting by $\lambda_1, \lambda_2, \ldots, \lambda_n$ the eigenvalues of D and of C,

$$\det(t\,A + (1-t)\,B) = \det(A)\,\prod_{i=1}^{n}(t + (1-t)\lambda_i)\,.$$

The second derivative of the logarithm of the product above is

$$\sum_{i=1}^{n} -\frac{(1-\lambda_i)^2}{[(1-\lambda_i)t + \lambda_i]^2}\,.$$

This function of t takes only non-positive values, and vanishes only if $\lambda_1 = \lambda_2 = \cdots = \lambda_n = 1$, a condition which is equivalent to $C = I_n$, i.e. $B = A$, a case that we have excluded when assuming $Q \neq R$. □

The method we used to prove Theorem 9.1.10 allows us to get more precise information about the union $\widetilde{\mathcal{C}}$ of classes containing a given class $\mathcal{C}$; we prove in particular that $\widetilde{\mathcal{C}}$ is the closure of $\mathcal{C}$:

Theorem 9.1.11. *Let $m > 0$, let S be a finite subset of $\mathbb{Z}^n$ stable under the symmetry $x \mapsto -x$, and let $\mathcal{Q}_m(S)$ be the set of positive definite quadratic forms Q with $\min Q = m$ and $S(Q) = S$. Assume that $\mathcal{Q}_m(S)$ is not empty. Then its closure is equal to*

$$\widetilde{\mathcal{Q}}_m(S) = \bigcup_{T \supset S} \mathcal{Q}_m(T)\,.$$

This is a convex set, the extremal points of which are the perfect forms whose sets of minimal vectors contain S, and it is the convex hull of its extremal points. In particular, the closure $\overline{\mathcal{C}}$ of a minimal class $\mathcal{C}$ of lattices is the union $\tilde{\mathcal{C}}$ of the minimal classes $\mathcal{C}' \succ \mathcal{C}$.

Proof. • That $\overline{\mathcal{Q}_m(S)}$ be contained in $\widetilde{\mathcal{Q}}_m(S)$ is just a variant of the first assertion in Proposition 9.1.6: if (Q_p) is a sequence of forms in $\mathcal{Q}_m(S)$ which converges to a form Q, we have $Q(x) \geq m$ for all nonzero $x \subset \mathbb{Z}^n$, with equality if $x \in S$.
• To prove that $\widetilde{\mathcal{Q}}_m(S)$ is convex, we consider two forms Q and Q' in this set, with sets of minimal vectors T and T'. For $0 \leq t \leq 1$, let

$$Q_t = t\,Q + (1-t)\,Q'\,.$$

For any nonzero $x \in \mathbb{Z}^n$, we have

$$Q_t(x) = t\,Q(x) + (1-t)\,Q'(x) \geq tm + (1-t)m = m\,,$$

and equality holds if and only if both the equalities $t\,Q(x) = tm$ and $(1-t)\,Q'(x) = (1-t)m$ hold, which implies $Q(x) = Q'(x) = m$ for all $t \in (0,1)$, hence $S(Q_t) = T \cap T' \supset S$.

• Let $Q' \in \widetilde{\mathcal{Q}}_m(S)$. For any $Q \in \mathcal{Q}_m(S)$ and $0 < t \leq 1$, Q_t belongs to $\mathcal{Q}_m(S)$, which shows that Q' is an element in the closure of $\mathcal{Q}_m(S)$.

• If a form Q_t as above is perfect, since $S(Q_t)$ is maximal among the subsets in $\mathbb{Z}^n$ which are possible sets of minimal vectors, we have $T = T'$, which proves that T and T' are also maximal and thus that Q and Q' are perfect. Since a perfect form is defined by its set of minimal vectors, we then have $Q_t = Q = Q'$. Hence perfect forms are extremal points of $\widetilde{\mathcal{Q}}_m(S)$.

• On the contrary, a non-perfect form Q is not an extremal point; we prove this by making use of Proposition 7.2.5: for any nonzero form R orthogonal in the Voronoi space to the span of the domain of Q, there exist strictly positive real numbers ρ and ρ' such that $Q_\rho = Q + \rho R$ and $Q_{\rho'} = Q - \rho' R$ have the same minimum than Q and have Voronoi domains of dimensions strictly larger than that of the Voronoi domain of Q (i.e., the perfection ranks of $Q + \rho R$ and of $Q - \rho R'$ are strictly larger than that of Q) Since Q lies in the interior of the segment $[Q_\rho, Q_{\rho'}]$, it is not an extremal point for $\widetilde{\mathcal{Q}}_m(S)$.

• To complete the proof of the theorem, it suffices to prove that every form $Q \in \widetilde{\mathcal{Q}}_m(S)$ can be written as the barycentre with non-negative coefficients of perfect forms in $\widetilde{\mathcal{Q}}_m(S)$. We use induction on the perfection corank of Q, observing that the result is evident if this corank is zero, since Q is then perfect. If Q is not perfect, we write it as a barycentre with non-negative coefficients of two forms Q_ρ and $Q_{\rho'}$ as above, which can themselves be written (since they have smaller perfection coranks) as barycentres with non-negative coefficients of perfect forms. □

Remark 9.1.12. *When S spans $\mathbb{R}^n$, there exists only finitely many forms in $\widetilde{\mathcal{Q}}_m(S)$ which are equivalent to a given form $Q \in \widetilde{\mathcal{Q}}_m(S)$. The number of finite subsets of $\mathbb{Z}^n$ containing S and which are sets of minimal vectors for some form is finite (Lemma 9.1.4), and $\widetilde{\mathcal{Q}}_m(S)$ contains in particular only finitely many perfect forms. The set $\widetilde{\mathcal{Q}}_m(S)$ is thus a convex, compact polytope which is the convex hull of its vertices.*

On the contrary, when it does not span $\mathbb{R}^n$, S is preserved by infinitely many elements of $\mathrm{GL}_n(\mathbb{Z})$. Then $\mathcal{Q}_m(S)$ contains infinitely many forms which are equivalent modulo $\mathrm{GL}_n(\mathbb{Z})$ to a given form Q in $\mathcal{Q}_m(S)$, and the number of maximal finite subsets T in $\mathbb{Z}^n$ containing S which are sets of minimal vectors is infinite. In this case, $\widetilde{\mathcal{Q}}_m(S)$ possesses infinitely many extremal points, and it is not bounded. In particular, it is not a compact set.

Besides the kissing number and the perfection rank, we can attach various invariants to any minimal class $\mathcal{C}$. Here are a few examples:

• For $\Lambda \in \mathcal{C}$, the structure of the quotient Λ/Λ_0, and in particular the index $[\Lambda : \Lambda_0]$, where Λ_0 denotes the sublattice of Λ generated its minimal vectors. [Indeed, this index is a locally constant function.]
• The set of possible structures of Λ/Λ', where $\Lambda \in \mathcal{C}$ and Λ' contains n independent minimal vectors in Λ.
• The set of $(n-1)$-dimensional classes containing a lattice $\Lambda_1 \subset \Lambda$ such that $\Lambda_1 \cap S(\Lambda)$ has rank $n-1$.

The main problem in the classification of dual-minimal classes is that of the decomposition of a given minimal class $\mathcal{C}$ as a union of (finitely many) dual-minimal classes $\mathcal{C}_1, \dots, \mathcal{C}_\ell$. Explicit splittings are completely known only up to $n = 3$. There is obviously a single class when $\mathcal{C}$ is *perfect*, i.e. is the similarity class of some perfect lattice Λ. We shall see below other examples in which a minimal class $\mathcal{C}$ constitutes a single dual-minimal class, related to the existence of similarities which map every lattice in $\mathcal{C}$ onto its dual. Here is another kind of example:

Proposition 9.1.13. *Let L be a critical lattice of dimension $n-1$ and let $\mathcal{C}$ be a minimal class such that all lattices $\Lambda \in \mathcal{C}$ possess a hyperplane section Λ_0 which is similar to L and has the same norm as Λ. Then $S(\Lambda^*)$ is the set of primitive vectors in Λ^* which are orthogonal to an L_0 as above, and $\mathcal{C}$ constitutes a single dual-minimal class.*

Proof. The first assertion is clear, since critical minimal sections of Λ have the smallest possible determinant, and thus generate the hyperplanes which are orthogonal to vectors in $S(\Lambda^*)$. The second assertion then follows from the equality $u(x) \cdot {}^t u^{-1}(y) = x \cdot y$. □

9.2 Classes of Dimension $n \leq 3$.

We prove or state some classification results for minimal and dual-minimal classes in dimension $n \leq 3$. For the applications to perfect (resp. dual-extreme lattices), one only needs to consider configurations S which span E (resp. pairs (S, S^*) such that both S and S^* span E and which satisfies the inequalities $s \geq s^*$ and $s + s^* \geq \frac{n(n+1)}{2} + 1$). For this reason, we do not take much account of classes which are not well rounded.

For $n \leq 4$, any well–rounded lattice is generated by four arbitrary independent minimal vectors except if $\Lambda \sim \mathbb{D}_4$. Disregarding this special case, we shall describe a configuration by a $n \times s$ matrix with entries $0, \pm 1$, from which it is possible to extract the identity matrix I_n any time we restrict to well–rounded lattices.

There is nothing to say for $n = 1$. The classification of 2-dimensional lattices follows immediately from Proposition 9.1.14 together with the following statement:

Proposition 9.2.1. *Any 2-dimensional lattice is similar to its dual by a direct similarity of angle $\frac{\pi}{2}$.*

Proof. Let Λ be a 2-dimensional lattice. Replacing Λ by a proportional lattice, we may assume that $\det(\Lambda) = 1$. We now choose a basis (e_1, e_2) for Λ, then an orthonormal basis $(\varepsilon_1, \varepsilon_2)$ for E in such a way that $e_1 = a\,\varepsilon_1$ and $e_2 = b\,\varepsilon_1 + \frac{1}{a}\,\varepsilon_2$ for conveniently chosen real numbers a and b. We have $e_1^* = \frac{1}{a}\,\varepsilon_1 - b\,\varepsilon_2$ and $e_2^* = a\,\varepsilon_2$, and we immediately see that the rotation $\varepsilon_1 \mapsto \varepsilon_2, \varepsilon_2 \mapsto -\varepsilon_1$ maps e_1 onto e_2^* and e_2 onto $-e_1^*$. □

As a consequence, we see that minimal (resp. dual-minimal) classes only depend on the invariant s, and that the corresponding set S (resp. the pair (S, S^*) is defined by $1 \le k \le 3$ columns in the left matrix (resp. in both matrices) below:

$$S: \begin{pmatrix} 1 & 0 & 1 \\ 0 & 1 & 1 \end{pmatrix} \qquad S^*: \begin{pmatrix} 1 & 0 & 1 \\ 0 & 1 & -1 \end{pmatrix}.$$

The minimal classification in dimension 3 is not difficult, though the number of minimal vectors no longer suffices. We easily deduce from the results of Section 6.2 that the (well–rounded) minimal classes are defined by the first k columns $(3 \le k \le 6)$ of the left matrix below for configurations containing that of a hexagonal lattice, and by the first k columns $(1 \le k \le 4)$ of the right matrix below otherwise.

$$\begin{pmatrix} 1 & 0 & 1 & 0 & 1 & 0 \\ 0 & 1 & -1 & 0 & 0 & -1 \\ 0 & 0 & 0 & 1 & -1 & -1 \end{pmatrix} \qquad \begin{pmatrix} 1 & 0 & 0 & 1 \\ 0 & 1 & 0 & 1 \\ 0 & 0 & 1 & 1 \end{pmatrix}.$$

In particular, well–rounded classes share out among five types, denoted by a_3, a_4, b_4, a_5, a_6 (the index is s, also equal to the perfection rank for any $n \le 3$), where a_4 is the only class without a hexagonal lattice.

The dual classification is much more difficult, in particular when non-well–rounded minimal classes are involved. For the sake of completeness, we state in full generality Bergé's results to be found in [Ber1] (where we have not however written pairs (S, S^*) with $s < s^*$). All that concerns well–rounded lattices will be proved below.

Theorem 9.2.2. *Up to the exchange of S and S^*, the 3-dimensional dual-minimal classes can be represented with respect to a suitable pair $(\mathcal{B} = (e_1, e_2, e_3), \mathcal{B}^*)$ of bases by exactly one of the following systems of components:*

(s, s^*)	S	S^*
$(1,1)$	$\{e_1\}$	$\{e_1^*\}$ *or* $\{e_3^*\}$
$(2,1)$	$\{e_1, e_2\}$	$\{e_1^*\}$ *or* $\{e_3^*\}$
$(2,2)$	$\{e_1, e_2\}$	$\{e_1^*, e_2^*\}$ *or* $\{e_1^*, e_3^*\}$

(3,1)	$\{e_1, e_2, e_3\}$ *or* $\{e_1, e_2, e_1 + e_2\}$	$\{e_3^*\}$
(3,2)	$\{e_1, e_2, e_3\}$	$\{e_2^*, e_3^*\}$
(3,3)	$\{e_1, e_2, e_3\}$	$\{e_1^*, e_2^*, e_3^*\}$
(4,1)	$\{e_1, e_2, e_1 + e_2, e_3\}$	$\{e_3^*\}$
(4,2)	$\{e_1, e_2, e_3, e_1 + e_2 + e_3\}$	$\{e_1^*, e_2^* - e_3^*\}$
(4,4)	$\{e_1, e_2, e_3, e_1 + e_2 + e_3\}$	$\{e_1^*, e_2^* - e_3^*, e_2^*, e_1^* - e_3^*\}$
(5,2)	$\{e_1, e_2, e_3, e_1 + e_2, e_1 + e_3\}$	$\{e_2^*, e_3^*\}$
(6,4)	$\{e_1, e_2, e_3, e_1 + e_2, e_1 + e_3, e_2 - e_3\}$	$\{e_1^*, e_2^*, e_3^*, -e_1^* + e_2^* + e_3^*\}$

The three families of lattices scaled to norm 1 defined by the conditions $s = 5$, $s = s^* = 4$, and $s = s^* = 3$ are connected and depend on a single parameter. They can be represented respectively by the Gram matrices below (see Exercises 6.1.1 and 6.3.1 for the first two and Proposition 9.2.3 below for the third one):

$$\begin{pmatrix} 1 & -\frac{1}{2} & -\frac{1}{2} \\ -\frac{1}{2} & 1 & t \\ -\frac{1}{2} & t & 1 \end{pmatrix}, \quad \begin{pmatrix} 1 & -1+2t & -t \\ -1+2t & 1 & -t \\ -t & -t & 1 \end{pmatrix}, \quad \begin{pmatrix} 1 & t & t \\ t & 1 & t \\ t & t & 1 \end{pmatrix},$$

where the parameter t satisfies the respective inequalities

$$0 < t < \frac{1}{2}, \quad \frac{1}{3} < t < \frac{1}{2}, \quad -\frac{1}{3} < t < \frac{1}{2}.$$

The end-points of the intervals above correspond to lattices similar to $\mathbb{A}_3$ or $\mathbb{A}_3^*$; in all cases the Hermite invariant possesses a unique minimum, namely K_3' (Definition 8.5.8), obtained for $t = \frac{1}{2}$, $\mathbb{A}_3^*$, and $\mathbb{Z}^3$; see Example 9.5.1 (6) below; for the γ' invariant, see Exercise 9.6.2.

Proof of Theorem 9.2.2 for well–rounded lattices. We first observe that $s = 6$ corresponds to the similarity class of the perfect lattice $\mathbb{A}_3$, whose dual lattice $\mathbb{A}_3^*$ has $s^* = 4$. Next, the classes b_4 and a_5 contain hexagonal minimal sections, so that they constitute a whole dual-minimal class, with respectively $s^* = 1$ and $s^* = 2$. We are thus left with the two classes a_3 and a_4.

Proposition 9.2.3. *Let Λ be well rounded with $s = 3$, and let*

$$S(\Lambda) = \{\pm e_1, \pm e_2, \pm e_3\}.$$

Then $S(\Lambda^*)$ is contained in $\{\pm e_1^*, \pm e_2^*, \pm e_3^*\}$.

Proof. We may restrict ourselves to lattices Λ of norm 1. Set $a = e_1 \cdot e_2$, $b = e_1 \cdot e_3$ and $c = e_2 \cdot e_3$. The Gram matrix A of the basis $\mathcal{B} = (e_1, e_2, e_3)$ and its adjoint B are

$$A = \begin{pmatrix} 1 & a & b \\ a & 1 & c \\ b & c & 1 \end{pmatrix} \quad \text{and} \quad B = \begin{pmatrix} 1-c^2 & bc-a & ac-b \\ bc-a & 1-b^2 & ab-c \\ ac-b & ab-c & 1-a^2 \end{pmatrix}.$$

The minimal vectors in Λ have components $0, \pm 1$ on $\mathcal{B}$. The hypothesis $s = 3$ is equivalent to the fact that the vectors $e_i \pm e_j$ and $e_1 \pm e_2 \pm e_3$ are of norm $N > 1$, itself equivalent to the set of inequalities

$$(1) \quad |a| < \frac{1}{2}, |b| < \frac{1}{2}, |c| < \frac{1}{2} \quad \text{and} \quad (2) \quad \pm a \pm b \pm c > -1,$$

with an even number of minus signs in (2). Let Λ' be the lattice proportional to Λ^* having a basis $\mathcal{B}' = (e'_1, e'_2, e'_3)$ proportional to $\mathcal{B}^*$, with Gram matrix B. We have

$$\begin{aligned} N(e'_1 + e'_2) - N(e'_3) &= 1 - c^2 + 1 - b^2 + 2(bc - a) - (1 - a^2) \\ &= 1 + a^2 - 2a - [c^2 + b^2 - 2bc] \\ &= (1 - a + b - c)\,(1 - a - b + c) \\ &= \frac{1}{4}\,[N(e_1 - e_2 + e_3) - 1]\,[N(e_1 - e_2 - e_3) - 1] > 0, \end{aligned}$$

and the five equalities obtained from the previous one by permutation and change of signs of the e_i. Thus the vectors $e_i^* \pm e_j^*$ are never minimal if $s = 3$. Similarly, we have

$$\begin{aligned} N(e'_1 + e'_2 + e'_3) - N(e'_1) &= 1 - a^2 + 1 - b^2 + 2\,(bc - a + ac - b + ab - c) \\ &= -(a - b)^2 + 2\,(1 - c)\,(1 - a - b) \\ &= -|a - b|^2 + 2\,|a - b|\,(1 - a - b) + 2\,(1 - c - |a - b|)\,(1 - a - b) \\ &= |a - b|\,(2\,(1 - a - b) - |a - b|) + 2\,(1 - c - |a - b|)\,(1 - a - b)\,. \end{aligned}$$

Here, $2(1 - a - b) - |a - b| = 2 - 3a - b$ or $2 - a - 3b$ is strictly greater than $2 - 1/2 - 3/2 = 0$, and $(1 - c - |a - b|)\,(1 - a - b)$ is also strictly positive by (1) and (2). Thus no vector among $\pm e_1^* \pm e_2^* \pm e_3^*$ can be minimal. □

This completes the proof of Theorem 9.2.2 for $\mathcal{C} = a_3$. If $\mathcal{C} = a_4$, some inequalities in the proof above may become equalities. However, we still have the strong inequalities $N(\pm e_1^* \pm e_2^* \pm e_3^*) > N(e_i^*)$ because the strict inequalities $|a|, |b|, |c| < \frac{1}{2}$ still hold, but for $i = 1, 2, 3$ and $j, k \neq i$, we now have $N(e_j^* - e_k^*) = N(e_i^*)$ because of the equality $a + b + c = -1$, so that S is a union $S_2 \cup S_4 \cup S_6$ of three dual-minimal classes indexed by s^*. To complete the proof of Theorem 9.2.2 for well–rounded lattices, we just need to exhibit examples of each possibility. By permutation and sign changes, we may assume that a, b, c have the same sign and are ordered in the form $a \geq b \geq c$ in both the cases $\mathcal{C} = a_3$ and $\mathcal{C} = a_4$, and we obtain the various values of s^* by taking successively $a > b > c$, $a > b = c$ and $a = b = c$. □

Note that we have proved for well–rounded lattices of dimension $n \leq 3$ the implication $s = n \Longrightarrow s^* \leq n$. This is no longer true for larger dimensions; see Exercise 9.5.1 where we give an example with $s = n = 4$ and $s^* = 8$.

9.3 Classification in Dimension 4

We assume in this section that E has dimension 4, and we first consider the classification of lattices from the point of view of minimal equivalence. We restrict ourselves to the case of lattices Λ possessing four independent minimal vectors, with $s(\Lambda) \leq 9$: we are reduced to dimension 3 if $S(\Lambda)$ does not span E, and we know by Theorem 6.2.1 that a lattice with $s \geq 10$ is similar to $\mathbb{A}_4$ or to $\mathbb{D}_4$.

We denote by $a_i, b_i, \ldots$ the classes with $i \geq 4$ minimal vectors, where the increasing alphabetic order corresponds to the increasing number of minimal hexagonal sections. Since we leave aside the similarity class of $\mathbb{D}_4$, four arbitrary minimal vectors in $S(\Lambda)$ generate Λ (Corollary 6.2.3), and this implies that the determinants extracted from the matrices of components of minimal vectors are equal to 0 or ± 1 (Lemma 6.1.5). We denote by $\pm e_i$ $(1 \leq i \leq s)$ the vectors in $S = S(\Lambda)$ and we express the e_i on the basis (e_1, e_2, e_3, e_4).

There exists obviously a single configuration with $s = 4$, denoted by a_4, and three configurations with $s = 5$, which can be represented in the following way:

$$a_5 \,:\, e_5 = e_1 + e_2 + e_3 + e_4; \quad b_5 \,:\, e_5 = e_1 + e_2 + e_3; \quad c_5 \,:\, e_5 = e_1 + e_2;$$

the inclusions are $a_4 \prec a_5$, $a_4 \prec b_5$, and $a_4 \prec c_5$.

The only enlargements of a_5 are obtained by adjoining to it a vector e_6 of the form $e_i + e_j$ or $e_i + e_j + e_k$ (the + signs are necessary); in both cases, a hexagonal minimal lattice shows up. This proves that the only configuration with $s \geq 6$ without any minimal hexagonal lattice is equivalent to $a_6 = \{b_5, e_6\}$, where $e_6 = e_1 + e_2 + e_4$. Completing c_5 by a vector e_6 equal to $e_1 + e_3 + e_4$, $e_1 + e_3$, or $e_3 + e_4$, we obtain three more configurations with $s = 6$, denoted by b_6, c_6, d_6. These four configurations are characterized by their number h of minimal hexagonal lattices ($h = 0$, 1, or 2), together with the fact that if $h = 2$, either the intersection of the two planes is a line containing a minimal vector, or it reduces to $\{0\}$. The list for $s = 6$ is as follows:

$$\begin{array}{llll}
a_6 & (h=0) & : e_5 = e_1 + e_2 + e_3, & e_6 = e_1 + e_2 + e_4; \\
b_6 & (h=1) & : e_5 = e_1 + e_2, & e_6 = e_1 + e_3 + e_4; \\
c_6 & (h=2) & : e_5 = e_1 + e_2, & e_6 = e_1 + e_3; \\
d_6 & (h=2) & : e_5 = e_1 + e_2, & e_6 = e_3 + e_4.
\end{array}$$

To describe completely all relations between configurations with $s \leq 6$, there remains to list the possible extensions of a_5, b_5, c_5 to one of the a_6, b_6, c_6, d_6. In the case of a_5, we must consider up to equivalence the adjunction of $e_1 + e_2$ or $e_1 + e_2 + e_3$. Replacing e_1 by $e_1 + e_2$ or $e_1 + e_2 + e_3$, we obtain a configuration equivalent to b_6. In the case of b_5, one may adjoin $e_1 + e_2$, $e_1 + e_4$, $e_1 + e_2 + e_4$, or $e_1 + e_2 + e_3 + e_4$, which are equivalent to c_6, b_6, a_6, and b_6 respectively. Finally, in the case of c_5, we must consider the possible adjunctions of $e_1 + e_3$, $e_3 + e_4$, $e_1 + e_2 + e_3$, $e_1 + e_3 + e_4$, and $e_1 + e_2 + e_3 + e_4$, equivalent to c_6, d_6, c_6, b_6, b_6 respectively. The complete list of inclusions is thus

$$a_5 \prec b_6, \quad b_5 \prec a_6, b_6, c_6, \quad c_5 \prec b_6, c_6, d_6.$$

To classify the configurations with seven vectors, we first show that all contain c_6, by looking at all possible extensions of a_6, a_6, b_6, c_6, d_6. Only the case of a_6 deserves a proof. Starting with $e_5 = e_1+e_2+e_3$ and $e_6 = e_1+e_2+e_4$, we first observe that it is impossible to adjoin $e_1 + e_3 + e_4$, for this would imply the existence of a characteristic determinant equal to ± 2; we then deal with $e_3 + e_4$: replacing e_1 by $e_1' = e_1 + e_2$ and e_3 by $e_3' = e_3 + e_4$ yields the system $e_1' + e_3' - e_4$, $e_1' + e_4$, $e_3' - e_4$.

In what follows, to classify the configurations with $s \geq 7$, we assume that they contain the vectors $e_5 = e_1 + e_2$ and $e_6 = e_1 + e_3$.

Adjoining a vector with no component on e_4 yields the configuration $S(\mathbb{A}_3) \cup \{\pm e_4\}$, with $h = 4$. The case of vectors with exactly two nonzero components reduces to e_1+e_4 or to e_2+e_4, and the other possibilities amount to the previous one if $h > 2$, and to adjoining $-e_2+e_3+e_4$ otherwise. Finally, we find the following four possible types:

$$\begin{array}{lllll}
a_7 & (h = 2) & : e_5 = e_1 + e_2, & e_6 = e_1 + e_3, & e_7 = -e_2 + e_3 + e_4; \\
b_7 & (h = 3) & : e_5 = e_1 + e_2, & e_6 = e_1 + e_3, & e_7 = e_1 + e_4; \\
c_7 & (h = 3) & : e_5 = e_1 + e_2, & e_6 = e_1 + e_3, & e_7 = e_2 + e_4; \\
d_7 & (h = 4) & : e_5 = e_1 + e_2, & e_6 = e_1 + e_3, & e_7 = e_2 - e_3,
\end{array}$$

and the inclusions between configurations with six and seven minimal vectors are

$$a_6 \prec a_7, b_7, \quad b_6 \prec a_7, c_7, \quad c_6 \prec a_7, b_7, c_7, d_7, \quad d_6 \prec c_7.$$

Up to a permutation of the minimal vectors in $S(\mathbb{A}_3) \subset d_7$ and up to the sign in front of e_4, the only extension of d_7 is $b_8 = d_7 \cup \{\pm(e_1 + e_4)\}$. We find the same thing in the case of b_7 by checking that all possible adjunctions are of the form $e_i - e_j$ or $e_1 + e_i + e_j$,with $1 < i < j$, and the existence of a configuration $S(\mathbb{A}_3)$ becomes evident.

The adjunction to c_7 of vectors with three or four nonzero components reduces by base change to the case of only two nonzero components, and we then have the choice between b_8 or a configuration equivalent to $a_8 = c_7 \cup \{\pm(e_3 + e_4)\}$.

Extending a_7 by vectors with two nonzero components can *a priori* be done by choosing $e_1 \pm e_4$ or $e_2 \pm e_3$, and we then obtain a configuration containing b_7 or d_7, hence equivalent to b_8, or by $e_2 - e_4$ or $e_3 + e_4$, and we see by considering one of the bases e_1, e_2, e_3, e_2-e_4 or e_1, e_2, e_3, e_3+e_4 that we obtain a_8. There remains the possibility of extensions equivalent to adjoining $e_1 + e_3 + e_4$, for which we again get a_8: making use of the transformation

$e_3 \mapsto -(e_1+e_3+e_4)$, $e_4 \mapsto -e_4$, we find the system $e_1, e_2, e_3, e_4, e_1+e_2, e_1+e_3, e_3+e_4, e_1+e_2+e_3+e_4$, which reduces to a_8 by $e_2 \mapsto e_1+e_2$, $e_4 \mapsto e_3+e_4$. Finally, the configurations with eight minimal vectors are of one of the following two types:

$$\begin{array}{llll} a_8 & (h=4): c_6 \text{ and } & e_7 = e_2+e_4, & e_8 = e_3+e_4; \\ b_8 & (h=5): c_6 \text{ and } & e_7 = e_1+e_4, & e_8 = e_2-e_3. \end{array}$$

and the inclusions between configurations with 7 and 8 minimal vectors are

$$a_7,\ c_7 \prec a_8, \quad a_7,\ b_7,\ c_7,\ d_7 \prec b_8.$$

Passing from eight to nine vectors is easier, as there are only a few possible extensions. Up to equivalence, we find a unique extension of b_8 and two extensions of a_8, obtained by adjunction of a vector e_9:

$$a_9\,(h=6): \ a_8 \text{ and } e_9 = e_1+e_2+e_3+e_4; \quad b_9\,(h=7): \ b_8 \text{ and } e_9 = e_2-e_4,$$

with the following inclusions which complete the description of 4-dimensional minimal classes:

$$a_8 \prec a_9,\ b_9,\ b_8 \prec b_9,\ a_9 \prec a_{10} = S(\mathbb{D}_4),\ b_9 \prec b_{10} = S(\mathbb{A}_4),\ a_{10} = S(\mathbb{D}_4).$$

Lattices which realize each of the configurations $a_4, a_5, b_5, \dots, a_9, b_{10}, a_{10}$ are constructed up to similarity using a (symmetric) Gram matrix whose diagonal entries are equal to a same positive (arbitrary) real number m, and whose other entries are affine functions of $10-s$ parameters, which we calculate by writing down that e_i must have norm 1 for $5 \le i \le s$. We give below the Gram matrices for the families a_8 and a_9, taking $m=2$ for convenience; the conditions that must be satisfied by the family a_8 are $e_1 \cdot e_2 = e_1 \cdot e_3 = e_2 \cdot e_3 = e_2 \cdot e_4 = -1$, and we take $u = e_1 \cdot e_4$ and $v = e_2 \cdot e_3$ as parameters; we obtain a_9 by imposing the extra condition $(e_1+e_2)\cdot(e_3+e_4) = -1$, equivalent to $u+v=1$:

$$A_8(u,v) = \begin{pmatrix} 2 & -1 & -1 & u \\ -1 & 2 & v & -1 \\ -1 & v & 2 & -1 \\ u & -1 & -1 & 2 \end{pmatrix} \qquad A_9(t) = \begin{pmatrix} 2 & -1 & -1 & t \\ -1 & 2 & 1-t & -1 \\ -1 & 1-t & 2 & -1 \\ t & -1 & -1 & 2 \end{pmatrix}$$

where $A_9(t) = A_8(t, 1-t)$. It remains to determine the range of the parameters u, v, t, which we do by writing that the vectors $\alpha_1 e_1 + \alpha_2 e_2 + \alpha_3 e_3 + \alpha_4 e_4$ have norm $N > 2$ for all $\alpha_1, \alpha_2, \alpha_3, \alpha_4 \in \{0, \pm 1\}$ except if they belong to $S(A_8)$ or $S(A_9)$. Consideration of the vectors e_1+e_4, e_2+e_3 and $e_1+e_2+e_3+e_4$ yields the conditions $|u|, |v| < 1$, $u+v > 1$, i.e. $0 < u, v < 1$ and $u+v > 1$; we check that they are sufficient and that they imply the result we want for the family $A_9(t)$ (for which Proposition 9.3.1 below gives an alternative proof).

The following proposition shows a remarkable property of the family a_9:

Proposition 9.3.1. *Let $t \in [0,1]$. Then $A_9(t)$ is the Gram matrix of a lattice L_t with minimum 2 and determinant $(2+t)^2(1-t)^2$, which is a Hermitian* *lattice over the ring $\mathfrak{O}$ of Eisenstein integers; L_t and L_{1-t} are isometric by an orthogonal symmetry which fixes a plane, and L_t is similar to L_t^* by the product of two planar rotations with the same angle $\frac{\pi}{2}$ and orthogonal planes. We have $s(L_t) = 9$ for $0 < t < 1$, whereas L_0 and L_1 are isometric to $\mathbb{D}_4$. The lattice $L_{1/2}$ is similar to Barnes's lattice $L_4^2 \simeq \mathbb{A}_1 \bigotimes \mathbb{A}_2$ (Section 8.4).*

Proof. It is easily verified that

$$\det(A_9(t)) = (2-t)^2(1+t)^2$$

and

$$(2-t)(1+t)A_9(t)^{-1} = \begin{pmatrix} 2 & 1 & 1 & 1-t \\ 1 & 2 & t & 1 \\ 1 & t & 2 & 1 \\ 1-t & 1 & 1 & 2 \end{pmatrix}.$$

We recognize $\mathbb{D}_4$ when giving t the value 0, and this proves that $A_9(t)$ is positive definite on $[0,1]$, since its determinant does not vanish on this interval. Let us interpret $A_9(t)$ as a matrix of scalar products $e_i \cdot e_j, 1 \le i,j \le 4$. The map σ which sends (e_1,e_2,e_3,e_4) onto $(e_2,-e_1-e_2,e_4,-e_3-e_4)$ is an isometry with minimal polynomial X^2+X+1, hence defines a $\mathfrak{O}$-structure on L_t. The nine vectors $e_1,e_2,-e_1-e_2,e_3,e_4,-e_3-e_4,e_1+e_3,e_2+e_4,-e_1-e_2-e_3-e_4$ (regrouped in three hexagonal lattices) have norm 2. The map $(e_1,e_2,e_3,e_4) \mapsto (e_1^*,e_2^*,e_3^*,e_4^*)$ transforms L_t into a lattice isometric to L_{1-t}^*, and the double exchange $e_1 \leftrightarrow e_2$ and $e_3 \leftrightarrow e_4$ transforms L_t into L_{1-t}. Hence the similarity $\tau : (e_1,e_2,e_3,e_4) \mapsto (e_2^*,-e_1^*,-e_4^*,e_3^*)$ transforms L_t into L_t^*.
[Let $\lambda = (2-t)(1+t)$. The matrix P of τ with respect to the bases $\mathcal{B}$ and $\mathcal{B}^*$ represents a similarity because of the relation ${}^tP = \lambda^2 A P^{-1} A$, and τ is represented in $\mathcal{B}$ by the matrix $A^{-1}P$, whose square is $\lambda^{-2} I_4$.]

It is clear that $2A_9(t)$ defines an integral primitive lattice only if $t = \frac{1}{2}$, which suffices to prove that $\sqrt{2}\,L_{\frac{1}{2}}$ is similar to the lattice L_4^2, which is also a norm 4 integral primitive lattice in the class a_9.

Let us prove directly that the nine norm 2 vectors above are indeed minimal on the whole interval $[0,1]$. If it were not true, there would exist a value of t in $(0,1)$ for which a tenth minimal vector would show up. The lattice L_t would them be one of the root lattices $\mathbb{A}_4$ or $\mathbb{D}_4$ (Theorem 6.2.1), which is impossible, since t is not integral. □

Lattices which belong to the class a_9 possess many automorphisms. Curiously, all lattices in a configuration containing a_6 possess at least one non-trivial automorphism:

Proposition 9.3.2. *Let Λ be lattice whose configuration of minimal vectors contains that of the class a_6, let D be the line $\mathbb{R}(e_1+e_2)$, and let $H = D^\perp$. Then $H = \langle e_1-e_2, e_1+e_3, e_1+e_4\rangle$, and the orthogonal reflection σ with respect to H (which maps (e_1,e_2,e_3,e_4) onto $(-e_2,-e_1,e_5,e_6)$) is an automorphism of Λ.*

Proof. In the scale for which $N(\Lambda) = 2$, the conditions $N(e_5) = N(e_6) = 2$ amount to $e_1 \cdot e_2 + e_1 \cdot e_i + e_2 \cdot e_i = -2$ for $i = 3, 4$. This shows that the three vectors $e_1 + e_i$ span H, which then implies that σ stabilizes $S(\Lambda)$, hence also Λ.

To extend σ to classes α which strictly contain a_6, we must make explicit all base changes used to prove the relation $a_6 \prec \alpha$. We shall just consider the example of $\alpha = b_7$; the other cases are dealt with in a similar way. Enlarging a_6 by $e_1 + e_2$, we obtain b_7 in the notation above by the base change $(e_1, e_2, e_3, e_4) \mapsto (e_1 + e_2, -e_1, e_3, e_4)$, whence $\sigma((e_1, e_2, e_3, e_4)) = (-e_1, e_1 + e_2, e_1 + e_3, e_1 + e_4)$. □

We know only very incomplete results about the "dual" classification (the classification of minimal configurations of pairs (Λ, Λ^*)). The perfect classes a_{10} and b_{10} obviously constitute a single class in the dual sense, with $s^* = 12$ and $s^* = 6$ respectively; see Propositions 4.2.3 and 4.3.4.

This result generalizes to the other configurations containing a minimal hyperplane cross-section of type $\mathbb{A}_3$ (namely, d_7, b_8, b_9). If there exist two minimal hyperplane sections Λ_1, Λ_2 of type $\mathbb{A}_3$, we indeed have

$$\frac{1}{2}|S(\Lambda_1 \cup \Lambda_2)| = s(\Lambda_1) + s(\Lambda_2) - s(\Lambda_1 \cap \Lambda_2) = 12 - s(\Lambda_1 \cap \Lambda_2) \geq 9\,.$$

Hence $s^* \geq 2$ implies $s \geq 9$, from which we deduce that $s^* \geq 3$ implies $s \geq 10$, which happens only in the case of A_4 and $\mathbb{D}_4$ (Theorem 6.2.1). This shows that each of the classes d_7, b_8, b_9 constitute a unique minimal class in the dual sense, with $S^* = \{\pm e_4^*\}$, $S^* = \{\pm e_4^*\}$, and $S^* = \{\pm e_3^*, \pm e_4^*\}$ respectively.

Similarly, Proposition 9.3.1 shows that the class a_9 again constitutes a single minimal class in the dual sense, for which $s^* = 9$.

The minimal classification in dimension 4 is the basic tool for the classification of dual-extreme lattices obtained in [B-M1] and stated as Theorem 6.3.9. By Proposition 3.8.2 and Corollary 3.8.7, we can limit ourselves to consider lattices whose minimal classes satisfy the following conditions:

1. S spans E,
2. S^* spans E,
3. $s \geq s^*$,
4. $s + s^* \geq 11$, hence $s \geq 6$,

and the proof consists in showing that among these lattices, only $\mathbb{A}_4$ and $\mathbb{D}_4$ are dual-extreme. This is obtained by somewhat intricate calculations of extrema.

It would be interesting to investigate in detail the minimal classification of pairs (Λ, Λ^*), especially when $s = s^*$ to get characterizations of lattices which are similar to their duals, but also to simplify the proofs in [B-M1]. We do not know ever the possible values for (s, s^*). Various constructions of lattices which are similar to their duals show the existence of minimal classes in the dual sense with $4 \le s = s^* < 9$, but again we do not know the possible values for s in this case.

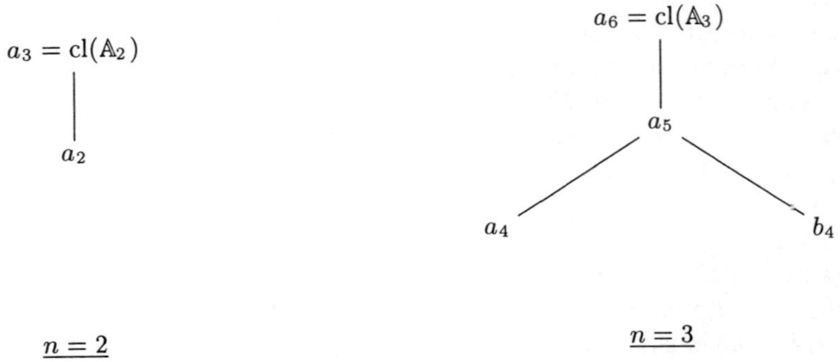

Figure 9.3.3 (a). Inclusion graphs for minimal classes in dimensions 2 and 3.

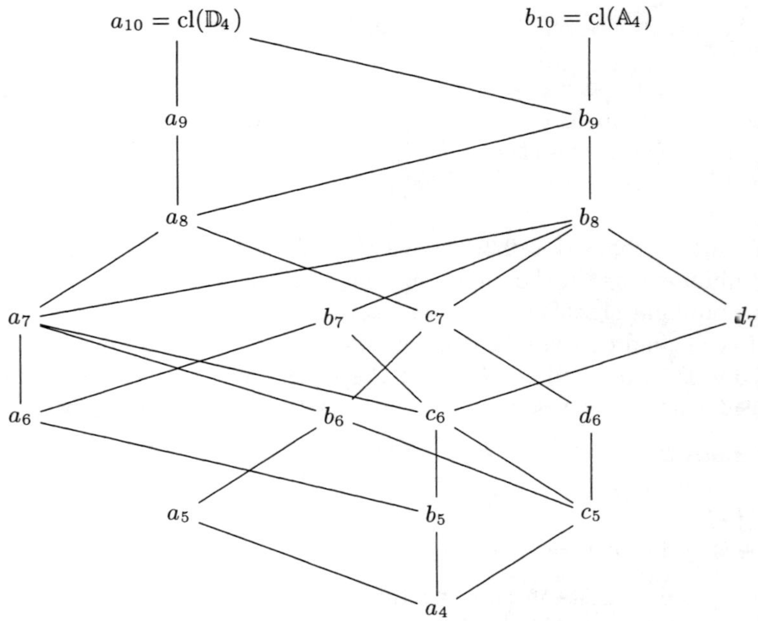

Figure 9.3.3 (b). Inclusion graphs for minimal classes in dimension 4.

After the French edition of this book was written, the minimal classification for dimension 5 has been obtained by Batut. We shall state Batut's main results in Sections 9.4 and 9.7 below. The proof, which we cannot reproduce here (there are 136 well–rounded classes!), needs the help of a computer. The strategy is not at all the one we used in dimensions $n \leq 4$. Instead of using induction on the perfection rank, he starts with the knowledge of the Voronoi graph that we determined in Chapter 7, and then considers the possibility of removing minimal vectors from a generic lattice in a Voronoi path. The class $\mathbb{D}_5 \longleftrightarrow \mathbb{D}_5$ with $s = 16$ is easy to handle, using the classification results established in Chapter 6: we find one class for each value of s between 13 and 16. Things are, however, much more complicated, with classes lying below one of the three other paths, for which $s = 14$. He proves in particular that all classes can be obtained by deleting minimal vectors from lattices belonging to the path $\mathbb{D}_5 \longleftrightarrow \mathbb{D}_5$ with $s = 14$, i.e. any class lying below $\mathbb{D}_5 \longleftrightarrow \mathbb{A}_5$ or $\mathbb{D}_5 \longleftrightarrow \mathbb{A}_5^3$ also lies below $\mathbb{D}_5 \longleftrightarrow \mathbb{D}_5$.

9.4 Weakly Eutactic Lattices in a Minimal Class

In this section, E again denotes an arbitrary n-dimensional Euclidean space. We first prove a characterization of weak eutaxy which will prove useful for the study Hermite's invariant:

Theorem 9.4.1. *In a minimal class $\mathcal{C}$, weakly eutactic lattices are those for which the Hermite invariant attains a minimum. If $\mathcal{C}$ is not perfect, the Hermite invariant has no maximum in $\mathcal{C}$.*

Proof. The second assertion is an obvious consequence of the convexity Theorem 9.1.10: if $\mathcal{C}$ does not reduce to a single similarity class, since this is an open subset in its closure, for every lattice in $\mathcal{C}$, there exist lattices which are arbitrarily close to it and have the same norm, but a strictly smaller determinant. We now prove the first assertion. Let Λ be a norm 1 lattice in some class $\mathcal{C}$, and let $u \in \mathrm{GL}(E)$. For u sufficiently close to the identity, the minimal vectors in $\Lambda' = u(\Lambda)$ come from those of Λ (more precisely, we have $S' \subset u(S)$); then a necessary and sufficient condition for $\Lambda' = u(\Lambda)$ to belong to $\mathcal{C}$ and to have norm 1 is that the equality $N(u(x)) = N(x)$ should hold for all $x \in S$; in view of the formula

$$\forall x \in E,\, N(u(x)) - N(x) = u(x) \cdot u(x) - x \cdot x = ({}^t uu - \mathrm{Id})(x) \cdot x\,,$$

this is equivalent to

$$\forall x \in S,\, \varphi_x({}^t uu - \mathrm{Id}) = 0$$

(see Definition 3.2.1 for the notation φ_x). In other respects, if $\Lambda' = u(\Lambda)$ and Λ are two lattices with the same norm, we then have, setting ${}^t uu - \mathrm{Id} = v$,

$$\left(\frac{\gamma(u(\Lambda))}{\gamma(\Lambda)}\right)^{-n} = \frac{\det(u(\Lambda))}{\det(\Lambda)} = (\det(u))^2 = \det(\mathrm{Id} + v)\,,$$

which amounts, using the characteristic polynomial

$$X^n + \sum_{1\leq i\leq n}(-1)^i T_i(v) X^{n-i}$$

of v, to

$$\left(\frac{\gamma(u(\Lambda))}{\gamma(\Lambda)}\right)^{-n} = 1 + T_1(v) + T_2(v) + \cdots + T_n(v)\,.$$

Developing this relation to the second order, we obtain

$$\left(\frac{\gamma(u(\Lambda))}{\gamma(\Lambda)}\right)^{-n} = 1 + T_1(v) + T_2(v) + o(v^2)\,, \qquad (*)$$

where we have set $T_1(v) = \mathrm{Tr}(v)$ and $2T_2(v) = (T_1(v))^2 - T_1(v^2)$. The term $T_1(v^2)$, which is the sum of the squares of the eigenvalues of the symmetric endomorphism v, is strictly positive unless v is zero.

• Suppose first that Λ is weakly eutactic, and let $\Lambda' = u(\Lambda)$ be a norm 1 lattice belonging to the class $\mathcal{C}$ of Λ, but not isometric to Λ. The symmetric endomorphism $v = {}^t uu - \mathrm{Id}$ is then nonzero. By the formulae above, if v is sufficiently close to 0 (and hence Λ' sufficiently close to Λ), we have $\varphi_x(v) = 0$ for all $x \in S$, which implies, since the linear form trace is a linear combination of the φ_x, $\mathrm{Tr}(v) = 0$. The expansion $(*)$ then becomes

$$\left(\frac{\gamma(u(\Lambda))}{\gamma(\Lambda)}\right)^{-n} = 1 + T_2(v) + o(v^2), \quad \text{with} \quad T_2(v) < 0\,,$$

from which we deduce that for sufficiently small v, we have $\gamma(u(\Lambda)) > \gamma(\Lambda)$.

• Suppose on the contrary that Λ is not weakly eutactic. Then there exists a symmetric endomorphism v such that $\varphi_x(v) = 0$ for all $x \in S$ and $T_1(v) \neq 0$. Changing v into $-v$ if need be, we may assume that $T_1(v) > 0$; let $\varepsilon > 0$ and let $u = (\mathrm{Id} + \varepsilon\, v)^{-1/2}$. Then, for sufficiently small ε, $u(\Lambda)$ is a norm 1 lattice which belongs to $\mathcal{C}$, and by development (*), its Hermite invariant is smaller than that of Λ. Hence $\gamma(\Lambda)$ is not a local minimum. □

Corollary 9.4.2. *On a minimal class consisting of lattices whose minimal vectors do not span E (a "non-well-rounded class"), the Hermite invariant has no maxima nor minima.*

Proof. We know that local maxima of the Hermite invariant are attained only on perfect lattices, and that perfect lattices are well rounded (Theorems 3.4.6 and 3.5.2 (3)), and Theorem 9.4.1 above together with Theorem 3.6.10 (2) shows that local minima may not exist. □

The theorem below proves among other results a finiteness theorem for weakly eutactic lattices, which contains finiteness results both for perfect lattices (Voronoi's Theorem 3.5.4) and for eutactic lattices (Avner Ash, [Ash1]).

Theorem 9.4.3. *1. A minimal class contains at most one similarity class of weakly eutactic lattices.*

2. *Up to similarity, there are only finitely many weakly eutactic lattices in a given dimension.*

Proof. Since there are only finitely many minimal classes by Theorem 9.1.5, (2) is a consequence of (1).

Let Λ and Λ' be two non-similar lattices belonging to a same class $\mathcal{C}$. Replacing Λ' by a scaled copy if need be, we may assume that Λ and Λ' have the same minimum m. By Definition 9.1.1, there exists a non-orthogonal linear transformation u of E such that $\Lambda' = u(\Lambda)$ and $S(\Lambda') = u(S(\Lambda))$. Let $\mathcal{B}$ be a basis for Λ and let $\mathcal{B}' = u(\mathcal{B})$ the corresponding basis for Λ'. Let A and A' be the Gram matrices of $\mathcal{B}$ and $\mathcal{B}'$. They are distinct (since they define non-isometric lattices) and quadratic forms attached to them both have minimum m, attained on the same subset S of $\mathbb{Z}^n$. This contradicts Theorem 9.1.10 above, according to which the logarithm of the determinant is a strictly concave function on any minimal class. □

Unlike perfect lattices, eutactic lattices are not necessarily proportional to integral lattices, and we must content ourselves with the following weaker statement.

Theorem 9.4.4. *Every eutactic lattice is proportional an algebraic lattice, i.e. a lattice such that all scalar products $x \cdot y$, $x,\ y \in \Lambda$ are algebraic numbers.*

Proof. Let $\mathcal{C}$ be a minimal class, and let $\mathcal{E}$ be the set of lattices in $\mathcal{C}$ whose minimum is a given rational number m. The class $\mathcal{C}$ is defined by s column-vectors $X_1, X_2, \dots, X_s$, namely the columns of the components of a half-system $e_1, e_2, \dots, e_s$ of minimal vectors in lattices $\Lambda \in \mathcal{C}$ in suitable bases $\mathcal{B} = (f_1, \dots, f_n)$ for these lattices. For $1 \le i \le s$, denote by $x_{i,j}$ the components in $\mathcal{B}$ of the vectors X_i, and let $E_i = X_i\,{}^tX_i$.

The lattices in $\mathcal{E}$ are defined up to isometry by the $\dfrac{n(n+1)}{2}$ scalar products $u_{k,\ell} = f_k \cdot f_\ell$, submitted to the s conditions $e_i \cdot e_i = m$ for $1 \le i \le s$, which are equivalent to the linear system

$$\sum_{k,\ell} x_{k,i} x_{\ell,i}\, u_{k,\ell} = m\,, \quad i = 1, 2, \dots, s$$

in the unknowns $u_{k,\ell}$. The rank of this system is the perfection rank r of the family $(e_1, e_2, \dots, e_s)$.

Solving this system, we express the unknowns as functions of $q = \dfrac{n(n+1)}{2} - r$ of them, say, $u_1, \dots, u_q$, which we consider as indeterminates. Set $(u_1, u_2, \dots, u_q) = U$. The lattices in $\mathcal{E}$ possess Gram matrices of the form $A = (f_i \cdot f_j) \in \mathcal{M}_n(\mathbb{Q}[U])$. For a suitable choice of m, we obtain a matrix belonging to $\mathcal{M}_n(\mathbb{Z}[U])$. [Note that the existence of $\mathcal{E}$ is not *a priori* evident: it depends on the possibility of specializing U so as to get a matrix which defines a positive definite quadratic form with minimum m, whose only minimal vectors are the $\pm X_i$, which amounts to writing inequalities on the u_i; to make them explicit is a problem of reduction of quadratic forms, which may be difficult to handle.]

Let us return to weak eutaxy. Let B be the adjoint matrix to A; it belongs to $\mathcal{M}_n(\mathbb{Z}[U])$, and so do the matrices E_i, which more precisely lie in $\mathcal{M}_n(\mathbb{Z})$. We are searching for real numbers $\lambda_1, \lambda_2, \dots, \lambda_s$ which satisfy the condition

$$\sum_{i=1}^{s} \lambda_i E_i = B\,,$$

which is equivalent to a system of $\frac{n(n+1)}{2}$ equations in the s unknowns $\lambda_1, \lambda_2, \dots, \lambda_s$. The well-known conditions of possibility for linear systems show that the existence of the λ_i is equivalent to the fact that $(u_1, u_2 \dots, u_q)$ runs through the set of real zeros of a family of polynomials in $\mathbb{Z}[U]$. By the finiteness Theorem 9.4.3, the algebraic variety defined in this way possesses only finitely many real points. Thus it has dimension 0 over $\mathbb{C}$, and this implies that the u_i are algebraic numbers, thanks to the following lemma (to which we give, in view of further application, a slightly more general form than what we need here):

Lemma 9.4.5. *Let N be a positive integer, and let X be the variety of zeros of a finite set of rational polynomials in N indeterminates. Then $X(\mathbb{R})$ possesses only finitely many isolated real points, and these points are defined over the algebraic closure $\overline{\mathbb{Q}}$ of Q in $\mathbb{R}$.*

Proof. We argue by induction on $\dim X$, the lemma being evident in case X has dimension 0. The implicit functions theorem shows that the set of isolated points in the real locus $Y(\mathbb{R})$ of an irreducible real algebraic variety of strictly positive dimension is a subset of the singular locus of Y, whose dimension is strictly smaller than that of Y. Applied to the irreducible components of positive dimension of X, this remark allows us to restrict ourselves to a smaller dimension. □

Note that if a lattice Λ is algebraic, the field K generated over $\mathbb{Q}$ by the values of the pairwise scalar products of its vectors is a number field, since it is generated by the entries of the Gram matrix of one of its bases. Replacing Λ by a scaled copy $\sqrt{\lambda}\,\Lambda$, we replace its norm by $\lambda\, N(\Lambda)$. Choosing λ conveniently in K, we obtain a lattice generating the same field K, whose minimum is an arbitrary rational number m. We can choose $m = 1$ without modifying the field K. Given Λ, there is thus a smallest possible number field K such that the scalar products take values in K on an appropriate scaled copy of Λ. This field K is called the *field of definition of Λ*.

Let $\mathcal{C}$ be a well–rounded class (i.e., the minimal vectors in all lattices in $\mathcal{C}$ span E). Proposition 9.1.6 shows that its closure is compact *modulo homothetic transformations*; hence the Hermite invariant possesses a lower bound on $\mathcal{C}$, which is attained on at least one class $\mathcal{C}' \prec \overline{\mathcal{C}}$, and on $\mathcal{C}$ itself if and only if it contains a weakly eutactic lattice.

We shall see in next section that from dimension 4 onwards, there exist classes which do not contain any weakly eutactic lattice, or which contain weakly eutactic, non-eutactic lattices, or which contain eutactic lattices that are not proportional to an integral lattice.

Here is an application of the search for the minima of the Hermite invariant on a class:

Proposition 9.4.6. *Let $\mathcal{C}$ be a $(n-1)$-dimensional class and let*

$$\gamma_{\mathcal{C}} = \inf_{\Lambda \in \mathcal{C}} \gamma(\Lambda)\,.$$

Let $\mathcal{C}'$ be an n-dimensional class whose lattices possess a hyperplane section belonging to $\mathcal{C}$. Let $\Lambda \in \mathcal{C}$, let $\mathcal{B}$ be a basis for Λ, let $\Lambda' \in \mathcal{C}'$ have Λ as a hyperplane section, and let $e' \in S(\Lambda') \smallsetminus \Lambda$. Then the index in Λ' of the lattice generated by Λ and e' is bounded from above by $\dfrac{\gamma_n^{n/2}}{{\gamma_{\mathcal{C}}}^{(n-1)/2}}$.

Proof. We may limit ourselves to norm 1 lattices. Then among the lattices M in E having Λ as a section, the one with the smallest Hermite invariant is $\Lambda \perp \mathbb{Z}$, whose invariant is $\gamma_{\mathcal{C}}$. Proposition 9.4.6 is now clear from the lower bound $\gamma(M) \le \gamma_n$. □

By making use of the classification of weakly eutactic lattices in dimensions 3 and 4 given in Chapter 14 and sketched in the next section, and taking into account the fact that $\gamma_{\mathcal{C}}$ has the same value on the classes a_6 and b_7, we obtain the following lower bounds for $\gamma_{\mathcal{C}}$, which we display in the two tables below, relating to dimensions 3 and 4; the (too long) list in dimension 5 can be extracted from [Bt].

9.5 The Classification of Eutactic Lattices

We still denote by E an n-dimensional Euclidean space. The minimal classes we consider are always assumed to be well rounded, a harmless restriction (see Corollary 3.2.6 and Theorem 3.6.10). Moreover, it should be understood that lattices and classes are defined up to similarity.

In this section, we explain without entering into numerical details the basic principles which can be used to get the effective classification of weakly eutactic lattices in a given dimension. Such a classification relies on the knowledge of minimal classes. Since minimal classes have not been classified beyond dimension 5, (weakly) eutactic lattices are known only for $n \le 5$; the results for $n = 2, 3, 4$ can be found in Chapter 14 (for dimension 5, we just give a summary at the end of this section).

Table 9.4.7. Lower bounds for $\gamma_{\mathcal{C}}$ (dimension 3).

$\mathcal{C}$	a_3	a_4	b_4	a_5	a_6
$\gamma_{\mathcal{C}}$	1	1.190	1.100	1.211	1.259

Table 9.4.8. Lower bounds for $\gamma_{\mathcal{C}}$ (dimension 4).

$\mathcal{C}$	a_4	a_5	b_5	c_5	a_6	b_6	c_6	d_6	a_7
$\gamma_{\mathcal{C}}$	1	1.196	1.139	1.074	1.240	1.224	1.154	1.154	1.274
$\mathcal{C}$	b_7	c_7	d_7	a_8	b_8	a_9	b_9	a_{10}	b_{10}
$\gamma_{\mathcal{C}}$	1.240	1.240	1.189	1.301	1.277	1.333	1.316	1.414	1.337

We also state Ash's mass formula with signs for eutactic lattices, giving it an explicit form for $n = 2, 3, 4$; we also consider Bavard's formula, which applies to the set of all weakly eutactic lattices.

The uniqueness up to similarity of weakly eutactic lattices in a given class (Theorem 9.4.3) allows us to describe without specific calculations for certain classes $\mathcal{C}$ the weakly eutactic lattice in $\mathcal{C}$, provided it belongs to a known family. Here are a few examples:

Example 9.5.1. *(1) A perfect class contains a unique weakly eutactic lattice. For $n \leq 7$, there are 48 such classes, among which 44 contain a eutactic lattice. (There are four exceptions, one in dimension 6 and three in dimension 7; see Section 6.5.)*

(2) The class defined by n minimal vectors which constitute a basis for its lattices contains the eutactic lattice $\mathbb{Z}^n$; this is Theorem 3.6.10 (3).

(3) The class defined by $n+1$ minimal vectors e_i, $1 \leq i \leq n+1$ which generate the lattice, with relation $e_{n+1} = e_1 + e_2 + \cdots + e_m$ $(2 \leq m \leq n)$ contains the eutactic lattice $\sqrt{\frac{m+1}{m}}\mathbb{A}_m^* \perp \mathbb{Z}^{n-m}$. This results from Theorem 4.7.2 and Proposition 4.2.3.

(4) The class defined by $n \geq 5$ independent minimal vectors whose half sum together with $n-1$ of them constitute a basis, contains the eutactic lattice $\mathbb{D}_n^*$ (centred cubic lattice); see Proposition 4.3.4.

(5) The two 2-dimensional classes contain the eutactic lattices $\mathbb{Z}^2$ and $\mathbb{A}_2$.

(6) The five 3-dimensional classes contain the eutactic lattices $\mathbb{Z}^3$, $\mathbb{A}_2 \perp \mathbb{A}_1$, $\mathbb{A}_3^*$, K_3' and $\mathbb{A}_3$ (see Section 8.5 for the definition of K_3', called the axial centred-cuboidal lattice in [C-S9], and the fragile lattice in the reference to K.L. Fields given there).

(7) Among the eighteen 4-dimensional classes which are listed in Section 9.3, ten contain a eutactic lattice that we met before: $\mathbb{Z}^4 \in a_4$, $\mathbb{A}_4^* \in a_5$, $\mathbb{A}_3^* \perp \frac{\sqrt{3}}{2}\mathbb{Z} \in b_5$, $\mathbb{A}_2 \perp \mathbb{A}_1 \perp \mathbb{A}_1 \in c_5$, $K_3' \perp 2\mathbb{Z} \in c_6$, $\mathbb{A}_2 \perp \mathbb{A}_2 \in d_6$,

$\mathbb{A}_3 \perp \mathbb{A}_1 \in d_7$, $L_4^2 \in a_9$, $\mathbb{A}_4 \in b_{10}$ and $\mathbb{D}_4 \in a_{10}$. (For Barnes's L_4^2, see Remark 8.4.3 and Definition 8.4.4.)

We now give on two examples in dimension 4 a few details about the methods which can be used to find the weakly eutactic lattice (if any) in a minimal class defined up to proportionality by a family of Gram matrices. They rely on the proof of Lemma 3.4.4 and on Theorem 3.6.10 respectively. They have been chosen so as to provide an example of a non-rational lattice, and an example of a semi-eutactic lattice belonging to the closure of the class we consider. Whatever the example, we made use of a package (*PARI*) which computes the adjoint of a matrix depending on a few parameters.

In the remainder of this section, we denote by $\mathcal{C}$ a minimal class and by r its perfection rank. Let $N = \dfrac{n(n+1)}{2}$; this is the maximal possible value for the perfection rank of a lattice. We rescale all lattices to a fixed minimum m, so that the maxima of the Hermite invariant correspond to the minima of the determinant.

When $N - r$ is small, it is rather easy to calculate with the help of a package for formal calculus the zeros on $\mathcal{C}$ (or on its closure) of the partial derivatives determinant. The example below dedicated to the a_8 class could have been easily solved by this procedure, which we actually used to verify our calculations.

It results from Section 9.3 that $\mathcal{C}$ is characterized by the existence of a basis (e_1, e_2, e_3, e_4) of minimal vectors such that the four extra vectors $e_5 = e_1 + e_2$, $e_6 = e_1 + e_3$, $e_7 = e_2 + e_4$, and $e_8 = e_3 + e_4$ are also minimal. Taking $m = 2$, we can construct a Gram matrix A depending on two parameters t and u, namely with $a_{i,i} = 2$, $a_{1,4} = a_{4,1} = u$, $a_{2,3} = a_{3,2} = v$, and $a_{i,j} = -1$ otherwise.

Since no combination of e_1, e_4 nor of e_2, e_3 occurs in the list of minimal vectors, the entries $b_{1,4}$ and $b_{2,3}$ in the adjoint matrix B of A must be zero, which yields the system of two equations in two unknowns

$$(2 - v)(2 - 2u - vu) = 0 \quad \text{and} \quad (2 - u)(2 - 2v - vu) = 0\,.$$

Taking into account the upper bounds $|u| < 1$ and $|v| < 1$, which simply mean that $e_1 + e_4$ and $e_2 + e_3$ have norm strictly larger than 2, we successively find the conditions $v = \dfrac{2(1-u)}{u}$, then $\dfrac{2}{u}(u^2 + 2u - 2) = 0$, whence $u = -1 \pm \sqrt{3}$, then $u = -1 + \sqrt{3}$, and we finally have $v = u = -1 + \sqrt{3}$. Using for instance a decomposition into squares, we control that A indeed belongs to a_8 when $u = v = -1 + \sqrt{3}$, that the eutaxy coefficients (which are unique, since $r = s$) exist in $\mathbb{Q}(\sqrt{3})$, and by numerical approximation, that they are strictly positive.

To compute the adjoint of a matrix depending on many parameters, which is the case where $N - r$ is large, is not very efficient. When r is small, we get significant simplifications by making use of the orthogonality properties in

Theorem 3.6.10. As an illustration, we now handle the class a_6, characterized by the existence of a basis (e_1, e_2, e_3, e_4) of minimal vectors such that the two extra vectors $e_5 = e_1 + e_2 + e_3$ and $e_6 = e_1 + e_2 + e_4$ are also minimal.

We observe that the only minimal vectors lying outside the hyperplane spanned by e_1, e_2, e_3 are $\pm e_4$ and $\pm(e_1+e_2+e_4)$. In the scale where $m = 2$, we can write down a Gram matrix, depending on the six parameters $x = e_1 \cdot e_2$, $y = e_1 \cdot e_3$, $z = e_1 \cdot e_4$, $u = e_2 \cdot e_3$, $v = e_2 \cdot e_4$, and $w = e_3 \cdot e_4$, submitted to the two relations $u = -2 - x - y$ and $v = -2 - x - z$.

Let $(e_1^*, e_2^*, e_3^*, e_4^*)$ be the dual basis to $\mathcal{B}$. It results from Theorem 3.6.10 that there exists a relation of the form $e_4^* = ae_4 + b(e_1 + e_2 + e_4)$. Taking scalar products with e_1, e_2, e_3, we obtain the three equations

$$az + b(2 + x) = 0\,,\ av + b(2 + x) = 0\,,\ aw + b(y + u) = 0\,.$$

We have $|x| < 1$, hence $a \neq 0$, and $a(z-v) = 0$, hence $v = z$ and $x = -2(z+1)$, and also $z(y + u) - w(2 + x) = 0$. Exchanging the rôles played by e_3 and e_4, we obtain similarly the relations

$$u = y,\ x = -2 - 2y,\ \text{ and }\ y(z + v) = w(2 + x)\,,$$

which finally allow us to calculate all the unknowns as functions of one parameter t: $x = -2 - 2t$, $y = z = u = v = t$, $w = -t$. We display below the corresponding Gram matrix A together with $B = -2t(t + 2)A^{-1}$:

$$A = \begin{pmatrix} 2 & -2(t+1) & t & t \\ -2(t+1) & 2 & t & t \\ t & t & 2 & -t \\ t & t & -t & 2 \end{pmatrix} \qquad B = \begin{pmatrix} 1-t & 1 & -t & -t \\ 1 & 1-t & -t & -t \\ -t & -t & -2t & 0 \\ -t & -t & 0 & -2t \end{pmatrix}.$$

In a eutaxy relation $B = \sum_{i=1}^{6} \lambda_i E_i$, we must have $\lambda_5 + \lambda_6 = b_{1,2} = 1$, $\lambda_5 = b_{1,3} = -t$ and $\lambda_6 = b_{1,4} = -t$, whence $t = -\frac{1}{2}$. But for this value of t, the vector $e_1 + e_2$ has norm 2, which implies that the lattice we found does not belong to the class a_6.

A look at the classification in Section 9.3 shows that the lattice above lies in b_7. Hence this is the weakly eutactic lattice in b_7. Calculating the eutaxy coefficients (which are unique), we verify that it is semi-eutactic, but not eutactic.

To complete the list of weakly eutactic lattices in dimension 4, it remains to study the five classes b_6, a_7, c_7, b_8, b_9. We shall not do it, and we content ourselves to sum up the main known results, including those obtained by Batut for dimension 5, for which we can only refer to [Bt]. We only consider well–rounded classes.

Theorem 9.5.2. *1. For n = 1, 2, 3, 4, 5, there are* 1, 2, 5, 18, *and* 136 *well–rounded minimal classes.*

2. Up to dimension 3*, each class contains a (unique up to similarity) weakly eutactic lattice, and this lattice is eutactic and rational (i.e. proportional to an integral lattice).*

3. *In dimension* 4, *each class contains a (unique) weakly eutactic lattice, to within the exception of* a_6. *These lattices are eutactic, except for that in the class* b_7 *which is only semi-eutactic, and they are rational, except for those in in the classes* a_7 *and* a_8, *whose fields of definition are* $\mathbb{Q}(\sqrt{13})$ *and* $\mathbb{Q}(\sqrt{3})$ *respectively.*
4. *In dimension* 5, *there are* 127 *weakly eutactic lattices, among which* 118 *are eutactic, and five are semi-eutactic, but not eutactic.*
5. *Among the* 127 *weakly eutactic lattices,* 69 *are rational. The other* 58 *lattices are defined over totally real number fields, with degrees* 2, 3, 4, 5, *or* 9. □

The comparison in dimensions 1 to 5 between the numbers of eutactic lattices given in Theorem 9.5.2 and the number of perfect lattices (1, 1, 1, 2, 3 respectively) suggests that the number of eutactic 6-dimensional lattices must be enormous. The classification has not been done.

In [Ash2], Avner Ash has given a mass formula with signs for lattices in a given dimension n. We state it in the theorem below:

Theorem 9.5.3. (Ash.) *The set of similarity classes of n-dimensional eutactic lattices satisfies the formula*

$$\sum_L \frac{(-1)^{i(L)}}{|\operatorname{Aut}^+(L)|} = \chi(\mathrm{SL}_n(\mathbb{Z}))\,. \square$$

In this formula, L runs through the set of similarity classes of eutactic lattices (we consider *eutaxy*, not *weak eutaxy*); the index $i(L)$ is the codimension of the *Voronoi domain of* L, equal to $\frac{n(n+1)}{2} - r$ where r is the perfection rank of L; $\operatorname{Aut}^+(L)$ is the group of automorphisms of determinant $+1$ of L (group of positive isometries of E which preserve L); finally, the notation $\chi(G)$ stands for the *Euler characteristic of a group G*.
(For a definition of $\chi(G)$, see J-P. Serre, Cohomologie des groupes discrets, Ann. of Math. Studies **70** (1971), 77–169, = [Se4], vol. 2, no. 88, 593–685.)
[Ash's original formula contained a factor 2 in its left hand side for even n. This is an error, but it shows up only if $n = 2$: indeed, $\chi(\mathrm{SL}_n(\mathbb{Z}))$ is zero for all $n \geq 3$, see Serre, *no. 3.7, p. 156, Example (i)*. Dimension 2 is an exception: we have $\chi(\mathrm{SL}_2(\mathbb{Z})) = -\frac{1}{12}$ $(= \zeta_{\mathbb{Q}}(-1))$, see Serre, no. 3.7, p. 159, Example (iv).]

As for the index $i(\Lambda)$, it is interpreted by Ash as the index of a critical point of a certain Morse function on the space of real quadratic forms, i.e. the number of $+$ signs in the signature of a matrix of second partial derivatives.

To verify directly the formula, we need to compute the automorphism groups of eutactic lattices. The calculation is a standard one for rational lattices, and we can use a computer to do it. We have $|\operatorname{Aut}^+(\Lambda)| = \frac{1}{2}|\operatorname{Aut}(\Lambda)$ for all lattices when n is odd, and we can check on the classification that this property also holds for eutactic lattices in dimensions $n = 2$ and $n = 4$.

We describe now briefly a method to handle non-rational lattices. We first consider automorphism groups of quadratic forms on $\mathbb{Z}[u_1, \dots, u_q]$; specializing the indeterminates to appropriate rational numbers, we obtain a rational lattice, whose automorphism group contains the previous group. We can guess in this way the group we want by a machine computation, and we then in practice easily verify that we have got the good one. We then specialize the parameters to the algebraic values we need, choose a bound M such that vectors of norm up to M generate the lattice, and ensure by looking at the spectrum that no new automorphisms show up by specialization (or sometimes find explicitly the short list of new automorphisms).

In dimension 2, there are two eutactic lattices, namely $\mathbb{Z}^2$ and $\mathbb{A}_2$, whose automorphism groups are dihedral of orders 8 and 12 respectively, and the formula to be verified is simply

$$-\frac{1}{4} + \frac{1}{6} = -\frac{1}{12} .$$

We display below the analogous weighted formulae for dimensions 3 and 4, taken from [B-M5]:

$$-\frac{1}{24} + \frac{1}{24} + \frac{1}{12} - \frac{1}{8} + \frac{1}{24} = 0 ;$$

$$\frac{1}{192} - \frac{1}{120} - \frac{1}{48} - \frac{1}{48} + \frac{1}{12} + \frac{1}{16} + \frac{1}{144}$$
$$-\frac{1}{8} - \frac{1}{8} - \frac{1}{48} + \frac{1}{8} + \frac{1}{8} - \frac{1}{72} - \frac{1}{12} + \frac{1}{120} + \frac{1}{576} = 0 .$$

We refer to [Bt] for the 5-dimensional formula, which would occupy about half a page.

We have only a poor knowledge of dual-eutactic lattices. The classification is known only for dimensions $n \leq 3$; we propose it as Exercise 9.6.1 (for the case of lattices which are not well rounded) and 9.6.2 for the more interesting case of well–rounded lattices. Finiteness and algebraicity results are known, which are best understood in terms of actions of closed subgroups of $\mathrm{GL}_n(\mathbb{R})$. We shall return to this question in Section 10.5.

9.6 Perfect Pairs of Lattices

The aim of this section is to develop rigid notions for dual pairs that will play the rôle that perfection plays in the cases of a single lattice. Indeed dual-perfection (see Definition 8.8.3) is not really an analogue to perfection, since there exist families of dual-perfect lattices whose sets of similarity classes constitute manifolds of positive dimensions; an example is provided by the set of 2-dimensional lattices with $s = 2$, which are dual perfect except for the square lattices, and depend modulo similarity on an angle $\theta \in (\frac{\pi}{3}, \frac{\pi}{2}]$; see Exercise 3.8.3; a 3-dimensional example is provided by Exercise 6.3.3; the class a_9 is a 4-dimensional example.

Thanks to the notion of a *perfect pair*, we shall notably obtain analogues to Proposition 3.2.11 and Theorem 3.5.4. As a consequence, we obtain finiteness and algebraicity theorems, which solve questions posed in Chapter 3. Recall (Definition 3.8.1) that a lattice is *dual-extreme* if the invariant γ' (the geometric mean of the Hermite invariant of the lattice and its dual) attains a local maximum on it. The following theorem will be proved at the end of this section as a consequence of the new notions we introduce here. We shall also state more general results, to be proved later in Chapter 10.

Theorem 9.6.1. (A-M. Bergé.) *There are only finitely many similarity classes of dual-extreme lattices, and dual-extreme lattices are proportional to algebraic lattices.*

Actually, such a statement holds more generally for well–rounded dual-eutactic lattices (see Proposition 10.5.3). The important point is that we introduce restricted notions of perfection on pairs (Λ, Λ^*) which (contrary to the notion of dual-perfection) hold only on a set of isolated lattices (modulo similarity).

Let Λ be a lattice, let r be its perfection rank (indeed, the rank of the system of linear forms $\varphi_x, x \in S(\Lambda)$; see Definition 3.2.1), and let $x_0 \in S(\Lambda)$. Then the forms $\varphi_x - \varphi_{x_0}$ constitute a system whose rank must be equal to r or $r-1$, and since this rank cannot exceed $N = \dfrac{n(n+1)}{2}$, Λ is perfect if and only if this rank is equal to $N-1$. We shall see that analogous considerations apply to the system

$$\varphi_x,\ x \in S(\Lambda) \cup S(\Lambda^*)\,,$$

provided there exists an appropriate *dual-eutaxy relation*

$$\sum_{x \in S(\Lambda)} \rho_x\, \varphi_x = \sum_{y \in S(\Lambda^*)} \rho_y^*\, \varphi_y\,,$$

which is precisely the case for dual-extreme lattices.

Definition 9.6.2. *Set* $N = \frac{n(n+1)}{2}$. *We say that a pair* (Λ, Λ^*) *is a* perfect pair *if for all* $x_0 \in S(\Lambda)$ *and* $y_0 \in S(\Lambda^*)$, *the system of linear forms*

$$\varphi_x - \varphi_{x_0},\ \varphi_y - \varphi_{y_0},\ x \in S(\Lambda),\ y \in S(\Lambda^*)$$

is of rank $N-1$.

Proposition 9.6.3. *A lattice is dual-extreme if and only if it is dual-eutactic and* (Λ, Λ^*) *is a perfect pair.*

Proof. Recall that a lattice is dual-extreme if and only if it is dual-eutactic and dual-perfect; see Definition 3.8.3 and Theorem 3.8.5; notice that all these definitions, as well as Definition 9.6.2 above, are relative to the pair (Λ, Λ^*).

In a perfect pair, the system of forms (φ_x, φ_y), $x \in S(\Lambda), y \in S(\Lambda^*)$ has rank $\rho > N - 1$. Hence its rank is exactly N, since φ_{x_0} does not vanish on Id, whereas the differences $\varphi_x - \varphi_{x_0}, \varphi_y - \varphi_{y_0}$ do. This shows that a perfect pair is dual-perfect.

In the rest of the proof, we limit ourselves to dual-eutactic pairs (Λ, Λ^*). Denote by Φ and Φ_0 the following systems of linear forms on $(\mathrm{End}^s(E))$:

$$\Phi = \{\varphi_x, \varphi_y\},\ x \in S,\ y \in S^*,\ \ \Phi_0 = \{\varphi_x - \varphi_{x_0}, \varphi_y - \varphi_{y_0}\},\ x \in S,\ y \in S^*.$$

Since the identity Id is orthogonal to Φ_0 but not to Φ, we have in $\mathrm{End}^s(E)$ the inclusions $\Phi^\perp \oplus \mathbb{R}\mathrm{Id} \subset \Phi_0^\perp$. Making use of the dual-eutaxy property, we now prove an inclusion in the other direction. Let $\sum_x \rho_x \varphi_x = \sum_y \rho_y^* \varphi_y$ $(\rho_x > 0,\ \rho_y^* > 0)$ be a dual-eutaxy relation, and let $v \in \Phi_0^\perp$. We have $\varphi_x(v) = \varphi_{x_0}(v)$ for all $x \in S$ and $\varphi_y(v) = \varphi_{y_0}(v)$ for all $y \in S^*$. Using the translation $v \mapsto v + \lambda\,\mathrm{Id}$ (which stabilizes $\Phi_0^\perp$), we restrict ourselves to the case when $\varphi_{x_0}(v) = 0$. Then from the relation $(\sum_x \rho_x)\varphi_{x_0}(v) = (\sum_y \rho_y^*)\varphi_{y_0}(v)$, in which $\sum_y \rho_y^*$ is strictly positive, we deduce the equality $\varphi_{y_0}(v) = 0$, and finally, the φ_x and the φ_y all vanish on v. Consequently, v belongs to $\Phi^\perp$, which proves the inclusion $\Phi_0^\perp \subset \Phi^\perp \oplus \mathbb{R}\mathrm{Id}$, from which we immediately deduce that (Λ, Λ^*) is a perfect pair. □

Another characterization of perfect lattices is that the kissing number gets smaller under a sufficiently small deformation which is not a similarity. The analogous definition for pairs (Λ, Λ^*) yields a notion which is slightly less restrictive than that of a perfect pair, but which however suffices to prove algebraicity results.

Definition 9.6.4. *We say that a pair (Λ, Λ^*) is a* quasi-perfect pair *if the kissing number of at least one of the lattices Λ, Λ^* gets smaller under any sufficiently small deformation which is not a similarity.*

This amounts to saying that the pair (Λ, Λ^*) is isolated (modulo similarity) in its minimal class in the dual sense of Definition 9.1.1, or that on a sufficiently small neighbourhood of Λ, we have $(M, M^*) \prec (\Lambda, \Lambda^*)$, the relation being strict if M is not similar to Λ.

Theorem 9.6.5. *Up to similarity, there are only finitely many quasi-perfect lattices, and they possess algebraic representatives.*

Proof. For $1 \le i \le j \le n$, let P_{ij} be the polynomial in the $N = \frac{n(n+1)}{2}$ indeterminates $x_{p,q}$, $1 \le p \le q \le n$ which expresses the cofactor relative to (i, j) in a symmetric matrix $(x_{p,q})$ (an entry of the adjoint matrix).

By the finiteness Theorem 9.1.5, we may restrict ourselves to lattices belonging to a given class $\mathcal{C}$ (in the dual sense), represented by the subsets $\Sigma = \{\pm\xi^{(1)}, \cdots, \pm\xi^{(s)}\}$ and $\Sigma^* = \{\pm\eta^{(1)}, \cdots, \pm\eta^{(s^*)}\}$ of $\mathbb{Z}^n$. We moreover restrict ourselves to lattices with minimum $m = 1$ (but we may not fix the norm of its dual!). Every lattice Λ in $\mathcal{C}$ possesses a Gram matrix $A = (a_{ij})$ such that

$$\begin{cases} {}^t\xi^{(k)} A\, \xi^{(k)} = 1 & (k = 1, 2, \cdots, s)\,, \\ {}^t\eta^{(\ell)} A^{-1} \eta^{(\ell)} = {}^t\eta^{(1)} A^{-1} \eta^{(1)} & (\ell = 2, 3, \cdots, s^*)\,, \end{cases}$$

or explicitly,

$$\begin{cases} \displaystyle\sum_{ij} a_{ij}\, \xi_i^{(k)} \xi_j^{(k)} = 1 & (k = 1, 2, \cdots, s)\,, \\ \displaystyle\sum_{ij} P_{ij}(a_{pq})\, (\eta_i^{(\ell)} \eta_j^{(\ell)} - \eta_i^{(1)} \eta_j^{(1)}) = 0 & (\ell = 2, 3, \cdots, s^*)\,. \end{cases}$$

Thus the N-tuple $(a_{ij},\, 1 \le i \le j \le n)$ is a common zero to the $r = s + s^* - 1$ rational polynomials $f_1, \cdots, f_r$. In fact, the N-tuples corresponding to lattices in $\mathcal{C}$ constitute an *open* subset in the set $Z(\mathbb{R})$ of common zeros of the polynomials $f_1, \cdots, f_r$, which are in particular defined over $\mathbb{R}$. The perfect pairs in $\mathcal{C}$ correspond to isolated points in $Z(\mathbb{R})$. Theorem 9.6.5 is now a consequence of Lemma 9.4.5. □

In the rest of this section, we endow the real vector space $\mathrm{End}^s(E)$, of dimension $N = \frac{n(n+1)}{2}$, of the norm defined by $\|w\| = \max_i |\lambda_i|$, where the λ_i denote the (real) eigenvalues of w. Recall that a positive definite symmetric endomorphism w (i.e., w has strictly positive eigenvalues) possesses a unique positive definite square root, that we denote by $\sqrt{w}$ (see Definition 3.1.3 and Theorem 3.1.6).

To every automorphism $u \in \mathrm{GL}(E)$, we attach the symmetric endomorphisms

$$v = {}^t u u - \mathrm{Id} \quad \text{and} \quad v^* = u^{-1}\, {}^t u^{-1} - \mathrm{Id}\,.$$

We remark that the eigenvalues of v and v^* are larger than -1 (because ${}^t u u$ and its inverse are positive definite), and that conversely, under this condition, any $v \in \mathrm{End}^s(E)$ comes from an automorphism u (for instance, from $u = \sqrt{\mathrm{Id} + v}$); in this correspondence, u is a similarity if and only if v (or v^*, this amounts to the same thing) is a multiple of the identity. Moreover, if u is sufficiently close to the identity, two lattices Λ and $u(\Lambda)$ in E are similar if and only if u is itself a similarity.

Theorem 9.6.6. *Let (Λ, Λ^*) be a pair of dual lattices, with sets of minimal vectors S and S^*, and let $x_0 \in S$ and $y_0 \in S^*$. Then (Λ, Λ^*) is quasi-perfect if and only if the only elements v in $\mathrm{End}^s(E)$ such that*

$$\forall\, x \in S\,, \forall\, y \in S^*\,, \forall\, p \geq 1\,,\; (\varphi_x - \varphi_{x_0})(v) = 0 \;\; and \;\; (\varphi_y - \varphi_{y_0})(v^p) = 0$$

are scalar multiples of the identity transformation.

To prove Theorem 9.6.6, we need two lemmas.

Lemma 9.6.7. *Let $\mathcal{P} = (\Lambda, \Lambda^*)$, S, S^*, x_0, y_0 be as above, and let $\mathcal{P}' = (\Lambda', \Lambda'^*)$ be another pair of dual lattices.*

1. *We have $\mathcal{P} \prec \mathcal{P}'$ if and only if there exists u with $\Lambda' = u(\Lambda)$ such that the endomorphisms $v = {}^t uu - \mathrm{Id}$ and $v^* = u^{-1}\,{}^t u^{-1} - \mathrm{Id}$ satisfy the conditions*

$$\forall\, x \in S,\; \forall\, y \in S^* : \varphi_x(v) = m' - m \;\; and \;\; \varphi_y(v^*) = m'^* - m^*,$$

in which m, m^*, m', m'^* denote the respective minima of Λ, Λ^*, Λ', Λ'^*.

2. *If $\mathcal{P}'$ is equivalent to $\mathcal{P}$, then for u, v, v^* as above, we have*

$$\forall\, x \in S,\; \forall\, y \in S^* :\; (\varphi_x - \varphi_{x_0})(v) = 0 \;\; and \;\; (\varphi_y - \varphi_{y_0})(v^*) = 0\,.$$

If moreover $\|u - \mathrm{Id}\,\|$ is sufficiently small, the converse holds.

Proof. (1) It results from the definitions of v and φ_x that for all $x \in S$, we have

$$N(u(x)) = u(x) \cdot u(x) = x \cdot {}^t uu(x) = x \cdot (\mathrm{Id} + v)(x) = N(x) + \varphi_x(v) = m + \varphi_x(v)\,,$$

which implies

$$u(S) \subset S' \Longleftrightarrow \forall\, x \in S,\; N(u(x)) = m' \Longleftrightarrow \forall\, x \in S,\; \varphi_x(v) = m' - m\,,$$

and similarly

$${}^t u^{-1}(S^*) \subset S'^* \Longleftrightarrow \forall\, y \in S^*,\; \varphi_y(v^*) = m'^* - m^*.$$

Assertion (1) is now a consequence of the very definition of the ordering $\prec$.

(2) The first assertion in (2) is an immediate consequence of (1).

Conversely, suppose that $\Lambda' = u(\Lambda)$ for some u sufficiently close to the identity in order that the inclusions $S' \subset u(S)$ and $S'^* \subset {}^t u^{-1}(S^*)$ hold, and moreover such that the corresponding forms $\varphi_x(v)$ (resp. $\varphi_y(v^*)$) are constant on S (resp. on S^*). These last two hypotheses mean that $N(u(x))$, equal to $m + \varphi(x)$, is constant on S, and that the norm of $u(y)$ does not depend on y. The inclusion $S' \subset u(S)$ implies that there exists at least one vector in S, say, x_1, with $u(x_1) \in S'$. Since all vectors $u(x)$ have the same norm, they all belong to S'. We thus have $u(S) \subset S'$, and finally $u(S) = S'$. Similarly, we have ${}^t u^{-1}(S^*) = S'^*$. This proves that $\mathcal{P}'$ and $\mathcal{P}$ are equivalent. □

Lemma 9.6.8. *Let φ be a linear form on $\mathrm{End}^s(E)$. For $v \in \mathrm{End}(E)$, set $v^* = (\mathrm{Id} + v)^{-1} - \mathrm{Id}$. Then there exists $\varepsilon > 0$ such that for all $v \in \mathrm{End}^s(E)$ with $\|v\| < 1$, we have:*

(i) if $\varphi(v^p) = 0$ for all $p \geq 1$, then $\varphi(v^*) = 0$;

(ii) otherwise, for $0 < \|v\| < \varepsilon$, then $\varphi(v^*) \neq 0$.

Proof. For $\|v\| < 1$, we have the series expansions

$$v^* = -v + v^2 + \cdots + (-1)^p v^p + \cdots$$

and

$$\varphi(v^*) = -\varphi(v) + \varphi(v^2) + \cdots + (-1)^p \varphi(v^p) + \cdots$$

Assertion (1) is evident on the expansion above. We now prove (2). Let v^p be the smallest power of v on which φ does not vanish. We then have sufficiently

$$\varphi(v^*) = (-1)^p \varphi(v^p) + \|v\|^p \epsilon(\|v\|, v/\|v\|)\,,$$

where $\epsilon(t, w)$ tends to 0 uniformly on the compact set $\|w\| = 1$ as $t \to 0$. Hence for $\|v\| < \varepsilon$ (ε sufficiently small), we have $\varphi(v^*) \sim (-1)^p \varphi(v^p) \neq 0$. □

Proof of 9.6.6. It results from Lemmas 9.6.7 and 9.6.8 that for u sufficiently close to the identity transformation, the pairs $(u(\Lambda), (u(\Lambda))^*)$ and (Λ, Λ^*) are equivalent if and only if the endomorphism v that we attached to u satisfies the following condition:

$$\forall\, x \in S,\, \forall\, y \in S^*,\, \forall\, p \geq 1\,, (\varphi_x - \varphi_{x_0})(v) = 0 \quad \text{and} \quad (\varphi_y - \varphi_{y_0})(v^p) = 0\,.$$

Theorem 9.6.6 now results from the fact that for u sufficiently close to Id, the lattice $u(\Lambda)$ is similar to Λ if and only if v is proportional to the identity. □

Theorem 9.6.6 immediately implies:

Corollary 9.6.9. *A perfect pair is quasi-perfect.* □

Proof of 9.6.1. If Λ is a dual-extreme lattice, (Λ, Λ^*) is a perfect pair by Proposition 9.6.3, hence also a quasi-perfect pair by Corollary 9.6.9, and we conclude using Theorem 9.6.5. □

One-parameter families of dual-eutactic lattices exist from dimension $n = 3$ onwards; see Exercise 9.6.1. To state finiteness and algebraicity theorems, we must restrict ourselves to well–rounded lattices, a property which is automatic for eutactic lattices. The following theorem will be proved in Section 10.5, except for the algebraicity assertion, which can be proved using the methods we previously used to prove Theorem 9.6.5.

Theorem 9.6.10. *In a given dimension, there are only finitely many similarity classes of well–rounded dual-eutactic lattices. In a given well–rounded minimal class (in the dual-sense), dual-eutactic lattices are those on which the invariant γ' attains a minimum; up to similarity, they are unique in their class, and are proportional to algebraic lattices.* □

9.7 Complements

This short, purely descriptive section is intended to report on results, extracted most of the time from Batut's paper [Bt], which appeared after the French edition was written. We begin with some comments on the dictionary relating lattices and quadratic forms, considering on the left lattices modulo proportionality and on the right quadratic forms with a fixed given minimum m, say, $m = 1$, and restricting ourselves to *well–rounded lattices* and *forms*, i.e., we always assume that minimal vectors in lattices (resp. of forms) span E (resp. $\mathbb{R}^n$).

First consider the space $\mathcal{Q}$ of (well–rounded, positive definite) quadratic forms. Call a finite subset S in $\mathbb{Z}$ *admissible* if it is the set of minimal vectors of some form in $\mathcal{Q}$. [Note that S must be symmetric; however, in practice, we consider an ordered set $S' = (e_1, \dots, e_s)$ such that $S = S' \cup (-S')$ and $S \cap S' = \emptyset$.] Given an admissible set S, denote by $\mathcal{C}_S$ the minimal class whose elements are the forms $Q \in \mathcal{Q}$ such that $S(Q) = S$. Let r be the perfection rank of $\mathcal{C}$ and let $t = \frac{n(n+1)}{2} - r$ be its perfection corank. Then $\mathcal{C}$ is an open convex set in a t-dimensional affine subspace in $\mathbb{R}^{n(n+1)/2}$ (hence a *t-dimensional cell*), and we endow $\mathcal{Q}$ in this way with a *cell complex* structure; perfect lattices are the 0-dimensional cells, open Voronoi paths connecting two perfect forms are the 1-dimensional cells, etc.

It is possible to extract from this infinite cell complex C a finite subcomplex C_0 in such a way that every minimal class possesses at least one representative in C_0. If we require that C_0 should be minimal, it is then unique up to equivalence.

On the side of lattices, a Voronoi path Q_1—Q_2 as above becomes something we write as Λ_1—Λ_2, where Λ_1 and Λ_2 are perfect lattices. If Λ_1 and Λ_2 are not isometric, there is nothing more to say: we still have the union of one 1-cell and two 0-cells.

However, if Λ_1 and Λ_2 are isometric, they define the same 0-dimensional minimal class, and we are led to fold the class defined by the open Voronoi path, obtaining a half open interval; see Exercise 9.1.1 below.

The viewpoint of quadratic forms, which yields cell complexes rather than "orbifolds", is easier to understand. Moreover, as we know, it also has another advantage, namely that the cells carry natural structures of convex sets. For these reasons, and also for computational reasons, we shall put emphasis in the sequel on quadratic forms rather than on lattices, except when geometrical definitions give a better insight.

Let $\mathcal{C}$ be a minimal class, first of lattices. The definition of equivalence of lattices Λ and Λ' (namely, the existence of an element u in $\mathrm{GL}_n(\mathbb{Z})$ such that $u(\Lambda) = \Lambda$ and $u(S(\Lambda)) = S(\Lambda)$) applies to the case where $L' = L \in \mathcal{C}$. We define in this way the group

$$G(L) = \{u \in \mathrm{GL}(E) \mid u(L) = L \text{ and } u(S(L)) = S(L)\},$$

which is finite because $\mathcal{C}$ is assumed to be well rounded, which contains the group $\mathrm{Aut}(L)$, and which varies inside $\mathcal{C}$ according to the rule

$$G(v(L)) = v\,G(L)\,v^{-1}\,.$$

On the side of quadratic forms, things become simpler, *once we have chosen a finite set* $S \subset \mathbb{Z}^n$ such that $\mathcal{C} = \mathcal{C}_S$: indeed, the choice of S allows us to define canonically the *automorphism group of* $\mathcal{C}$ *(with respect to* S*)*, by

$$\mathrm{Aut}_S(\mathcal{C}) = \{u \in \mathrm{GL}_n(\mathbb{Z}) \mid u(S) = S\}\,.$$

Note that there may be a great discrepancy between the automorphism group of $\mathcal{C}$ and that of an individual form (or lattice) in $\mathcal{C}$. For instance, when $\mathcal{C}$ is the class of $\mathbb{Z}^n$, $\mathrm{Aut}(\mathcal{C}) = \mathrm{Aut}(\mathbb{Z}^n)$ has order $2^n\,n!$, whereas $\mathrm{Aut}(\Lambda)$ reduces to $\pm\,\mathrm{Id}$ generically on $\mathcal{C}$; various other groups (e.g., $S_n \times \{\pm\,\mathrm{Id}\}$) may occur.

We now fix a class $\mathcal{C}$ and a finite subset S in $\mathbb{Z}^n$ such that $\mathcal{C} = \mathcal{C}_S$, and attach to S the set of quadratic forms Q with $S(Q) = S$. Set

$$\mathcal{Q}_S = \{Q \mid S(Q) = S\}\,.$$

Proposition 9.7.1. (Batut.) *Let A be the matrix of a form $Q \in \mathcal{Q}_S$. Let M be the arithmetic mean of the transforms of A by* $\mathrm{Aut}(S)$, *i.e.*

$$M = \frac{1}{|\,\mathrm{Aut}(S)|}\sum_{P\in\mathrm{Aut}(S)} {}^t\!PAP\,.$$

1. *M belongs to $\mathcal{Q}_S$ and* $\mathrm{Aut}(M) = \mathrm{Aut}(S)$.
2. *We have* $\det(M) \geq \det(A)$, *and equality holds if and only if $M = A$.*
3. *The set of forms Q with $S(Q) = S$ and* $\mathrm{Aut}(Q) = \mathrm{Aut}(S)$ *is the intersection of $\mathcal{Q}_S$ with an affine subspace in $\mathbb{R}^n$; it is non-empty.*
4. *For the weakly eutactic form Q_e in $\mathcal{Q}_S$ (if any), we have* $\mathrm{Aut}(Q_e) = \mathrm{Aut}(S)$. □

This proposition shows that the calculation of $\mathrm{Aut}(Q_e)$, which is needed to test Ash's formula, can be performed using only rational quadratic forms, even if Q_e is not proportional to a rational form. We can even go further, by exhibiting an almost canonical positive definite matrix with automorphism group equal to $\mathrm{Aut}(S)$. The set S is well defined by s representatives for the vectors $\pm x$, $x \in S$, with which we associate an $n \times s$ matrix whose columns are the s vectors above (in any order), which we denote by S'. Following Bacher, set

$$T = S'\,{}^tS'\,.$$

This is a positive definite $n \times n$ matrix, which up to a permutation of rows and columns, solely depends on S.

Proposition 9.7.2. (Bergé.) *We have* $\mathrm{Aut}(S) = \mathrm{Aut}(T^{-1})$. □

Bavard has recently proved a mass formula with signs which applies to the set of *all* well–rounded minimal classes in a given dimension n. It looks like Ash's formula, with denominators equal to the automorphism groups of the classes, thus the same denominator for classes containing a eutactic lattice, as shown in Proposition 9.7.1 (4). We just state it here, referring to [Bav2] for a proof; the notation is that of Theorem 9.5.3.

Theorem 9.7.3. *The set of well–rounded minimal classes satisfies the formula*

$$\sum_{\mathcal{C}} \frac{(-1)^{i(\mathcal{C})}}{|\,\mathrm{Aut}^{+}(\mathcal{C})|} = \chi(\mathrm{SL}_n(\mathbb{Z}))\,. \ \square$$

To deduce Ash's formula from Bavard's or vice-versa amounts to proving that the sum in the formula above on classes containing no eutactic forms is zero. The formulae are the same if $n = 2$ or 3, and differ by two terms which obviously cancel if $n = 4$. Things are less clear if $n = 5$, where 18 classes are to be considered. Some connections between classes without any weakly eutactic form on the one hand, and with weakly eutactic, non-eutactic forms on the other hand, are described in [Bt], assertion (3) in Theorem 3.2. It might well be that a more detailed study of these connections could account for the relation between Ash's and Bavard's formulae.

As for the fields of definition, they have quite surprising degrees and discriminants, which we are unable to explain. It was observed in Theorem 9.5.2 that they are totally real if $n \le 5$. This is no longer true in higher dimensions: counter-examples have been found by Batut for $n = 6$ and $n = 7$ on Voronoi paths, including the simplest possible one, namely a cubic field with one real and one complex place; actually, such a field occurs as the field of definition of a 4-dimensional *dual*-eutactic lattice; see Exercise 9.5.1. (By the way, Batut has done a complete exploration of the Voronoi graph in dimensions 6 and 7; no strongly eutactic lattices showed up except for the two previously known on the paths $\mathbb{E}_6$ — $\mathbb{E}_6^*$ and $\mathbb{E}_7$ — $\mathbb{E}_7^*$, quoted in [Mar-V].)

Questions 9.7.4. *Can any number field with at least one real place be the field of definition of a eutactic lattice? Of a dual-eutactic lattice? Of a dual-extreme lattice?*

Batut has also proved ([Bt], Proposition 3.4) that the real conjugates to a weakly eutactic form are not positive definite. It would be interesting to find a geometrical interpretation of the signature. The question of an interpretation within Morse theory (for eutactic lattices) has been raised by Avner Ash.

9.8 Exercises for Chapter 9

9.1.1. Let $\mathcal{C}$ be a minimal class with perfection rank $\frac{n(n+1)}{2}-1$. Let Q be a quadratic form in $\mathcal{C}$, let R be a nonzero form orthogonal to a facet of the Voronoi domain of Q for the Voronoi scalar product, and let ρ_1 and ρ_2 be the positive real numbers such that $P_1 = Q - \rho_1 R$ and $P_2 = Q + \rho_2 R$ are perfect, with the same minimum as Q (see Proposition 7.2.5). Let $S = S(Q)$.

1. Show that the set of quadratic forms in $\mathcal{C}_S$ (notation of Section 9.7) having the same minimal vectors as Q is $\mathcal{F} = \{Q + \theta R \mid -\rho_1 < \theta < +\rho_2\}$.

2. Show that P_1 and P_2 are two contiguous perfect forms (in the sense of Voronoi), and that every class in $\mathcal{C}$ of corank 1 is the set of classes modulo $\mathrm{GL}_n(\mathbb{Z})$ of forms in a Voronoi path as above; use this to deduce that $\mathcal{C}$ modulo similarities is homeomorphic to an open interval (a 1-dimensional cell), and that $\overline{\mathcal{C}}$ is homeomorphic to a compact interval or to a bounded half open interval.

3. Show that $\mathcal{C}$ solely depends on the orbit under $\mathrm{Aut}(P_1)$ of the facet of $\mathcal{D}_{P_1}$ which is orthogonal to R.

4. Show that if $\mathcal{C}$ does not contain any weakly eutactic form, we then have $\gamma(P_1) \neq \gamma(P_2)$, say, $\gamma(P_1) < \gamma(P_2)$, and that under these conditions, P_1 is not extreme.

5. Use Question 3 and the contiguity results in Chapter 7 to show that the number of corank-1 classes in dimensions $2, 3, 4, 5$ is $1, 1, 2, 4$ respectively.
[To prove that two orbits of facets of $\mathbb{D}_5$ having $\mathbb{D}_5$ as a contiguous lattice define distinct classes, note that among the four equivalence classes of Voronoi paths of type $\mathbb{D}_5 — \Lambda$, there is a $\mathbb{D}_4$ section for one of them (which has $s = 16$) whereas $s = 14$ holds for the other three; compare Exercise 6.4.1.]

6. Verify that the unique weakly eutactic lattice contained in each of the four classes above (which exists by Question 3) is indeed eutactic.

9.1.2. Let $\mathcal{C}$ and $\mathcal{C}' \succ \mathcal{C}$ be to minimal classes.

1. Suppose that both $\mathcal{C}$ and $\mathcal{C}'$ contain weakly eutactic lattices. Prove the strict inequality $\gamma_{\mathcal{C}'} > \gamma_{\mathcal{C}}$.
[Consider a segment with end-points a quadratic form in each class.]

2. Show that in all cases, we have the inequality $\gamma_{\mathcal{C}'} \geq \gamma_{\mathcal{C}}$, with equality if and only if $\gamma_{\mathcal{C}}$ and $\gamma_{\mathcal{C}'}$ are attained on a class $\mathcal{C}'' \succ \mathcal{C}'$ (then $\gamma_{\mathcal{C}}$ is not attained on $\mathcal{C}$, and $\gamma_{\mathcal{C}'}$ is attained on $\mathcal{C}'$ only if $\mathcal{C}'' = \mathcal{C}'$).

9.1.3. Suppose that $n \geq 3$. Let $A(t)$ $(0 \leq t \leq 1)$ be the matrix whose entries $a_{i,j}(t)$ are equal to 2 on the diagonal and to 1 outside, except for $a_{1,2}(t) = a_{2,1}(t) = t$.
[These matrices describe the Voronoi path $\mathbb{A}_n — \mathbb{D}_n$ in the Korkine–Zolotareff basis $(e_1, \dots, e_n)$, $e_i = \varepsilon_0 - \varepsilon_i$ for $\mathbb{A}_n$ through the facet of $\mathbb{A}_n$ obtained by deleting the direction of $e_1 - e_2$, see Section 7.5.]
Let M_t be a lattice with Gram matrix $A(t)$.

1. Show that $d(t) = \det(A(t))$ is a polynomial of degree at most 2, which takes for $t = 0,\ 1,\ 2$ the values $4, n+1, 0$; prove that $d(t) = (2-t)[(n-1)t+2]$.

2. Show that $d(t)$ possesses a unique maximum, attained for $t = 1 - \frac{1}{n-1}$ and equal to $\frac{n^2}{n-1}$.

3. Show that the primitive integral lattice L_n proportional to M_t for $t = 1 - \frac{1}{n-1}$

has norm $2(n-1)$ and determinant $n^2(n-1)^{n-1}$.

4. Show that for $0 < t < 1$, we have $\mathrm{Aut}(M_t) \simeq \{\pm \mathrm{Id}\} \times \{\pm\tau\} \times S_{n-1}$, where τ denotes the transposition which exchanges ε_1 and ε_2, S_{n-1} acting canonically on $\{0, 3, 4, \dots, n\}$. Verify that the minimal vectors in M_t share out among two orbits with $2(n-1)$ and $\frac{(n-1)(n-2)}{2}$ pairs of vectors respectively.

5. Show that L_n is eutactic, with eutaxy coefficients on each orbit equal to $\frac{2}{n^2}$ and $\frac{2(n-2)}{n^2(n-1)}$ respectively.

6. Prove that M_t possesses exactly two sections isometric to $\mathbb{A}_{n-1}$, and that $s(M_t^*) = 2$.

9.1.4. Consider an integer $n \le 5$. Let $\mathcal{C}$ be a minimal class, with perfection rank r. Show that we have $s = r$, except in the following three cases:
(1) $\mathcal{C}$ is the class of $\mathbb{D}_4$; then $s = r + 2$;
(2) $\mathcal{C}$ is the class of $\mathbb{D}_5$; then $s = r + 5$;
(3) We have $n = 5$, $\mathcal{C}$ is not the class of $\mathbb{D}_5$, and the lattices in $\mathcal{C}$ possess a hyperplane section with the same norm which is similar to $\mathbb{D}_4$; then $s = r + 2$.
[Use Exercise 9.1.1, Question 5 together with the hint given after this question.]

9.2.1. This exercise is intended to constructing 3-dimensional lattices Λ such that $(S(\Lambda), S(\Lambda^*))$ (or $(S(\Lambda^*), S(\Lambda))$) may represent every dual-minimal class for which none of the lattices Λ, Λ^* is well rounded (see Theorem 9.2.2).

1. Show that the family $t\,\mathbb{Z} \perp \mathbb{Z} \perp t'\,\mathbb{Z}$ $(0 < t \le 1,\ t' > 1)$ contains lattices belonging to the two classes $\{e_1; e_3^*\}$ and $\{e_1, e_2; e_3^*\}$.

2. Show that the family $t\,\mathbb{A}_1 \perp \mathbb{A}_2$ contains lattices belonging to the two classes $\{e_1; e_1^*\}$ and $\{e_1, e_2, e_1 + e_2; e_3^*\}$.

3. Let M be the 2-parameter family of 2-dimensional lattices with Gram matrices $\left(\begin{smallmatrix} 1 & b \\ b & m \end{smallmatrix}\right)$, $b \ge 0$, where m is the second minimum of M. Show that the family $t\,\mathbb{Z} \perp M$ contains lattices belonging to the classes $\{e_1, e_2, ; e_1^*\}$ and $\{e_1, e_2, ; e_1^*, e_3^*\}$.

4. Prove that the class $\{e_1, e_2, ; e_1^*, e_2^*\}$ contains no reducible lattice. Show that the matrix $\begin{pmatrix} 1 & 0 & b \\ 0 & 1 & b \\ b & b & d \end{pmatrix}$ is a Gram matrix for a lattice in the class above whenever $d > 1$, $0 < b < \frac{1}{2}$, and $d - b^2 < 1$. [Use the fact that in dimension 3, given a basis (e_1, e_2, e_3) with $N(e_1) \le N(e_2) \le N(e_3)$, the minimal norm is equal to $N(e_1)$ provided that the inequality $N(x) \ge N(e_1)$ holds for all vectors x of the form $e_i \pm e_j$ and $e_1 \pm e_2 \pm e_3$.]

9.2.2. Show that the lattices which are similar to $\mathbb{Z}^n$ are the only dual-eutactic lattices in the 3-dimensional dual-minimal class with $s = s^* = 3$ described in Theorem 9.2.2. [Use Proposition 3.2.3 (2) and Exercise 3.8.6.]

9.3.1. Let C_5 be the binary code of length 5 generated on $\mathbb{F}_2$ by the row vectors of the matrix $\left(\begin{smallmatrix} 1 & 1 & 1 & 0 & 0 \\ 0 & 0 & 1 & 1 & 1 \end{smallmatrix}\right)$. We consider the lifting of C_5 to $\mathbb{Z}^5$.

1. Show that we obtain an integral lattice L_5 with norm 3 and determinant 64, generated by its norm 3 vectors, with $s = 8$, which does not contain any hexagonal lattice with the same norm.

2. Show that its densest hyperplane sections constitute an integral norm 3 lattice L_4 with $s = 6$ and determinant 32, having no hexagonal norm 3 section, and that the densest sections of L_4 are similar to $\mathbb{A}_3^*$.

9.3.2. Let C_7 be the extension of the code C_5 of the preceding exercise generated by the rows of the matrix $\begin{pmatrix} 1\,1\,1\,0\,0\,0\,0 \\ 0\,0\,1\,1\,1\,0\,0 \\ 1\,0\,0\,0\,1\,1\,0 \\ 0\,0\,1\,0\,0\,1\,1 \end{pmatrix}$.

1. Show that the lifting to $\mathbb{Z}^7$ of C_7 is an integral norm-3 lattice similar to $\mathbb{E}_7^*$, having no hexagonal minimal section.

2. Show that this lattice contains a hyperplane section L_6 with determinant 64 and $s = 16$, containing the lattice L_5 of Exercise 9.3.1, associated with a code C_6 extracted from C_7.

3. Show that L_6 is similar to $\mathbb{D}_6^+$.

9.4.1 Let Λ be a 5-dimensional lattice containing 5 independent minimal vectors $e_1, \dots, e_5$. Let $\mathcal{C}$ be the class of the relative lattice generated by e_1, e_2, e_3, e_4.

1. Show that if $\gamma_{\mathcal{C}} > 2^{1/4} = 1,189\dots$, then $(e_1, e_2, e_3, e_4, e_5)$ is a basis for Λ.

2. Deduce from the question above that $(e_1, e_2, e_3, e_4, e_5)$ is a basis for Λ except possibly if $\mathcal{C}$ is one of the classes a_4, b_5, c_5, c_6, d_6, or d_7.

9.4.2 Let Λ be a 5-dimensional lattice containing five independent minimal vectors $e_1, \dots, e_5$. Denote by Λ' the sublattice of Λ generated by $S(\Lambda)$. Recall that we have $[\Lambda : \Lambda'] \le 2$ (Proposition 6.4.8).

1. Show that if Λ possesses a minimal section of type $\mathbb{D}_4$, it also possesses a basis of minimal vectors.

2. We discard this case in Questions 2, 3 and 4. Show that we have $a_i \in \{0, \pm 1, \pm 2\}$ for every minimal vector e in Λ of the form $a_1e_1 + a_2e_2 + a_3e_3 + a_4e_4 + a_5e_5$, and that if one of the components a_i is equal to ± 2, the others are equal to ± 1; show moreover that e and $-e$ are then the only vectors which satisfy this condition, and that $(e_1, e_2, e_3, e_4, e_5)$ is a basis for Λ (see Lemma 6.1.5 and Proposition 6.4.8.)

3. Suppose now that $(e_1, e_2, e_3, e_4, e_5)$ is not a basis for Λ. Let $e = \frac{1}{2}(e_1 + e_2 + e_3 + e_4 + e_5)$. Show that (e_1, e_2, e_3, e_4, e) is then a basis for Λ.

4. Show that if Λ possesses at least two pairs of minimal vectors outside any hyperplane, it then possesses a basis of minimal vectors. [Hint: otherwise, Λ would contain the vector e of Question 3 and a minimal vector $e_6 = a_1e_1 + a_2e_2 + a_3e_3 + a_4e_4 + e_5$ with $a_i \in \{0, \pm 1\}$, and the vector $\frac{1}{2}(e_6 + \sum_{a_i=0} e_i)$ would then belong to Λ.]

5. Show that a lattice Λ containing five independent minimal vectors belongs to one of the following mutually exclusive four types:

(a) The minimal vectors in Λ are linear combinations with coefficients $0, \pm 1$ of minimal vectors which constitute a basis for Λ. In this case, five arbitrary independent minimal vectors constitute a basis for Λ.

(b) Λ possesses a basis of the form (e_1, e_2, e_3, e_4, e) where $e = \frac{1}{2}(e_1 + e_2 + e_3 + e_4 + e_5)$, the vectors e_i are minimal vectors, e is not minimal, and the minimal vectors other than $\pm e_5$ are contained in the hyperplane generated by e_1, e_2, e_3, e_4.

(c) Λ possesses a basis $\mathcal{B} = (e_1, e_2, e_3, e_4, e_5)$ such that e_1, e_2, e_3, e_4 generate a lattice similar to $\mathbb{D}_4$.

(d) Λ does not contain any minimal section similar to $\mathbb{D}_4$; it possesses a basis $\mathcal{B} = (e_1, e_2, e_3, e_4, e_5)$ of minimal vectors, $e = e_1 + e_2 + e_3 + e_4 + 2e_5$ is minimal, and the minimal vectors in Λ other than $\pm e$ have components 0 or ± 1 on $\mathcal{B}$.

6. Show that $s(\Lambda) \ge 6$ is a sufficient condition for Λ to possess a basis of mini-

mal vectors. [This last result is due to Batut ([Bt]), who found it as a byproduct of its classification of 5-dimensional minimal classes.]

9.5.1. The aim of this exercise is to study dual-eutaxy in the class a_8. With the notation of Section 9.5, choosing parameters $u = e_1 \cdot e_4$ and $v = e_2 \cdot e_3$, we obtain for $\Lambda \in a_8$ the Gram matrix

$$A = \begin{pmatrix} 2 & -1 & -1 & u \\ -1 & 2 & v & -1 \\ -1 & v & 2 & -1 \\ u & -1 & -1 & 2 \end{pmatrix}$$

1. Prove that $\Lambda \in a_8 \iff 0 < u, v < 1$ and $u + v > 1$.
[Consider the vectors $e_i \pm e_j$, $e_1 + e_2 + e_3$, $e_1 + e_2 + e_4$, and $e_1 + e_2 + e_3 + e_4$.]

2. Prove that the adjoint matrix to A is

$$B = \begin{pmatrix} 2(-v^2+v+2) & uv-2u-2v+4 & uv-2u-2v+4 & uv^2-4u-2v+4 \\ uv-2u-2v+4 & 2(-u^2+u+2) & vu^2-2u-4v+4 & uv-2u-2v+4 \\ uv-2u-2v+4 & vu^2-2u-4v+4 & 2(-u^2+u+2) & uv-2u-2v+4 \\ uv^2-4u-2v+4 & uv-2u-2v+4 & uv-2u-2v+4 & 2(-v^2+v+2) \end{pmatrix}.$$

3. Evaluate the quantities $N(z) - N(e_1^*)$ for $z = e_1^* - e_2^*$, $e_1^* - e_4^*$, $e_1^* - e_2^* - e_3^*$, and $e_1^* - e_2^* - e_3^* + e_4^*$.

4. Show that $S(\Lambda^*)$ is one of the sets $S_2 = \{\pm e_1^*, \pm e_4^*\}$, $S_2' = \{\pm e_2^*, \pm e_3^*\}$, or $S_4 = S_2 \cup S_2'$, according to whether $u - v$ is < 0, > 0, or 0, and that a_8 splits into two dual minimal classes.

5. Show that if $u \leq v$, then $\gamma'(\Lambda)^2 = \frac{4(u+1)}{(2-v)(uv+2u+2v)}$.

6. Let a_8' be the subset of a_8 on which $s(\Lambda^*) = 4$. Show that γ' has a unique minimum in a_8', attained on the real root $\theta = 0.835\ldots$ of the polynomial $p(x) = 2x^3 + 5x^2 + 4x - 8$; prove that a_8 contains a unique weakly dual-eutactic lattice, and that this lattice is dual-eutactic, but not dual-perfect. (It *is not* the eutactic lattice in the class, which is defined over $\mathbb{Q}(\sqrt{3})$.)

7. Prove that p defines the cubic subfield of the Hilbert class field of $\mathbb{Q}(\sqrt{-61})$, of discriminant -244.

9.6.1. The aim of this exercise is to classify dual-eutactic lattices Λ of dimension $n \leq 3$ such that $S(\Lambda)$ *does not* span E.

1. Show that such lattices do not exist if $n \leq 2$. [Use Proposition 9.2.1.]

2. Suppose from now on that $n = 3$. Show that Λ necessarily belongs to one of the classes $(\{e_1\}, \{e_1^*\})$ or $(\{e_1, e_2\}, \{e_1^*, e_2^*\})$ of Theorem 9.2.2. [Use Exercise 3.8.4.]

3. Show that in the class $(\{e_1\}, \{e_1^*\})$, dual-eutactic lattices are those which are similar to the lattices of the form $\Lambda_0 \perp \mathbb{Z}$ for which $N(\Lambda_0) > 1$ and $N(\Lambda_0^*) > 1$. [Use Exercise 2.8.5; up to similarity, these lattices constitute a two-parameter family; see Exercise 3.8.3.]

4. Suppose now that Λ belongs to the class $(\{e_1, e_2\}, \{e_1^*, e_2^*\})$ of Theorem 9.2.2. Show that e_3 is perpendicular to e_1 and e_2, and that e_3^* is proportional to e_3.

5. Deduce from this that up to similarity, these lattices are those of the one parameter family $\mathbb{Z}^2 \perp t\mathbb{Z}$ $(t > 1)$.

6. Show that one-parameter families of pairwise non-similar dual-eutactic lattices exist in all dimensions $n \geq 3$.

9.6.2. (Written in collaboration with Anne-Marie Bergé.) Show that the only 3-dimensional dual-eutactic lattices possessing three independent minimal vectors are $\mathbb{Z}^3$, $\mathbb{A}_3^*$, $\mathbb{A}_3$, and the *ccc* (or *mcc*) lattice defined in Proposition 11.6.1; verify that all these lattices are strongly dual-eutactic by proving that their automorphism group acts transitively on their sets of minimal vectors.
[Use Theorem 9.6.10, Exercises 9.6.1 and 9.2.2, and the hints given there; dual-eutaxy follows for the *ccc* lattice from the fact that it is σ-extreme for an isoduality σ, see Section 11.11.]

9.9 Notes on Chapter 9

The notion of a minimal class appears without a name in [B-M1], as a tool for classifying dual-extreme lattices; indeed, it had been considered earlier by the Russian school, and in particular by Štogrin (= Shtogrin), see [St], and various articles by Baranowskii, Ryškov, and Štogrin; a few of them are given in the reference list. An analysis of the structure of the proofs in [B-M1] led A.-M. Bergé to introduce in [Ber1] the corresponding dual notion of a dual-minimal class, for which she could extend some of the results previously known for ordinary classes, notably the finiteness of the set of dual-extreme lattices.

Section 9.1 stems essentially from [Ber1] and [B-M5]; Sections 9.2 and 9.3 are a synthesis of results from [Ber1] and [B-M1].

Most of the results proved in Sections 9.4 and 9.5 are extracted from [B-M5], an article which was inspired by Avner Ash's paper [Ash1]. In comparison with [Ash1], the following improvements are to be noted: the replacement of eutaxy by the less restrictive property of weak eutaxy, a characterization of weak eutaxy by a property of extremality on minimal classes (a connection which had not been noticed in Štogrin's paper [St]), and the classification of (weakly) eutactic lattices up to dimension 4. We also account briefly for the much more complicated 5-dimensional classification recently obtained by Batut ([Bt]).

The mass formula with signs for eutactic lattices appeared in [Ash2], and Bavard's modification for all minimal classes in [Bav2]. The Euler–Poincaré characteristic of a group G is defined by the formula

$$\chi(G) = \frac{1}{[G:H]} \sum_{i=0}^{+\infty} \operatorname{rg} H^i(H, Z)\,,$$

which makes sense when G contains a subgroup H of finite index and finite cohomological dimension (then the result does not depend on the choice of H). Formulae for the Euler–Poincaré characteristic of certain "arithmetical" groups, involving the value at negative integers of zeta functions of number fields, have been proved by Harder ("A Gauss–Bonnet formula for discrete arithmetically defined groups", Ann. Sc. E.N.S. 4 (1971), 409–455). In his theory, the vanishing of $\chi(\mathrm{SL}_n(\mathbb{Z}))$ for $n \geq 3$ comes from the vanishing of $\zeta_{\mathbb{Q}}$ at even negative integers. It would be interesting to exhibit families of lattices endowed with extra algebraic structures for which mass formulae with nonzero right–hand sides show up in larger dimensions.

Results of Section 9.6 originate from [Ber1], Sections 3 and 4. However, we called "quasi-perfection" what she called "perfection" for pairs (Λ, Λ^*), and named "perfect pair" the notion which naturally shows up in the study of local extrema from the point of view of Lie groups; see Chapter 10 and [B-M4]. All these notions have been recently generalized by Bavard ([Bav4]) in the setting of Riemannian geometry. Some of our somewhat artificial definitions look quite natural in this new theory.

Section 9.7 (and Exercise 9.5.1) did not exist in the French edition. They have been written to take into account some recent developments of the theory.

10 Extremal Properties of Families of Lattices

Introduction

In this chapter (except in the last section) we consider families $\mathcal{F}$ of lattices which are orbits of one of them under the action of a subgroup $\mathcal{G}$ of $\mathrm{GL}(E)$, which we assume to be closed, invariant under transposition, such that the connected component of its unit element has finite index, and which moreover satisfies the following condition: either $\mathcal{G}$ has determinant 1, or $\mathcal{G}$ contains all positive homothetic transformations. We study *$\mathcal{F}$-extreme* lattices, i.e. lattices in $\mathcal{F}$ on which the Hermite invariant attains a local maximum among lattices which belong to $\mathcal{F}$. Thanks to our hypotheses on $\mathcal{G}$, we can use its structure as a *Lie group*, and more precisely the existence of a tangent space at the origin and of an exponential map.

A particularly important example is that of the *G-lattices*, the set of lattices whose automorphism group contains a given finite subgroup G of the orthogonal group $\mathrm{O}(E)$ of E. This example contains as a special case the set of all lattices in E; another example whose importance showed up recently is that of lattices which are similar to their duals under a given similarity of E. We shall consider closely these two examples in Chapter 11.

However, we will omit other important families, such as lattices having a given section (with the same norm), or "patchwork lattices", which we will study in Chapter 12 (and in Chapter 13 from the point of view of the Voronoi algorithm.) The families that we study in this chapter are subvarieties of the space of lattices. In the case of lattices with a given section, this is a closed convex subset in an affine variety. To deal simultaneously with all these situations would seem to involve complicated properties of manifolds with boundaries, and we have not tried to develop such a generalization (see however assertion (3) in Lemma 10.3.4.)

To study $\mathcal{F}$-extreme lattices, we need to consider the more restrictive notion of *strictly $\mathcal{F}$-extreme lattices*: these are those lattices Λ_0 such that any lattice Λ in a sufficiently small neighbourhood Λ_0 inside $\mathcal{F}$ has a strictly smaller Hermite invariant than that of Λ_0 except if it is similar to Λ_0. In the usual case of all lattices, these two notions coincide. However, this property does not always hold, and it is the notion of a strictly extreme lattice which can be characterized in terms of appropriate notions of eutaxy and perfection.

After two technical sections in which we explain a few elementary properties of Lie groups and add a few complements to results on perfection established in Chapter 3, we prove in Section 10.3 a characterization of strictly extreme lattices in terms of perfection and eutaxy. Section 10.4 is devoted to a generalization of some of the results about minimal classes that we proved in Chapter 9, and we complete in Section 10.5 the study of dual-extreme lattices that we began in Section 2.8. Section 10.6 deals with a variant of the methods of Section 10.3 which applies to the Rankin invariants we introduced in Definition 2.8.3.

Sections 10.3 to 10.5 essentially come from the article [B-M4]; Section 10.6 just consists of statements of the main results of Coulangeon ([Cou3]).

10.1 Some Elementary Results on Lie Groups

An n-dimensional *(real) Lie group* consists of a *differential manifold (i.e. C^∞)* $\mathcal{G}$ endowed with a group structure for which the two maps $(x,y) \mapsto xy$ of $\mathcal{G} \times \mathcal{G}$ onto $\mathcal{G}$ and $x \mapsto x^{-1}$ of $\mathcal{G}$ onto $\mathcal{G}$ are also differentiable. A concrete example is provided by the group $\mathrm{GL}(E)$, of dimension n^2, and the Lie groups we consider in this chapter are indeed subgroups of some group $\mathrm{GL}_m(\mathbb{R})$.

A Lie group $\mathcal{G}$ is in particular a *topological group*, which implies the following properties: the closure of a subgroup is a subgroup, an open subgroup is a closed subgroup, inner automorphisms of $\mathcal{G}$ are continuous, the conjugate of an open (resp. closed) subgroup is an open (resp. closed) subgroup. As a consequence, the *connected component $\mathcal{G}^\circ$ of $\mathcal{G}$* (i.e., the connected component of the unit of $\mathcal{G}$) is an open, normal subgroup of $\mathcal{G}$.

One can show (see [Bou3], Chapter 3, Section 8, Theorem 2) that any closed subgroup of a Lie group possesses a canonical structure of Lie group. In particular, the connected component of a Lie group is a Lie group.

Let $\mathcal{G}$ be a Lie group. As does every differential manifold, $\mathcal{G}$ possesses a *tangent space* at every point, and in particular at the origin, which we simply call the tangent space to $\mathcal{G}$, and denote in this section by $\mathcal{T}_1$.

A special case of a great practical importance is that of the algebras with involution, for the proof of which we refer to [Bou3], Chapter 3 p. 145, Proposition 37:

Theorem 10.1.1. *Let $\mathcal{A}$ be a subalgebra of* $\mathrm{End}(E)$*, and let $\imath$ be an involution of $\mathcal{A}$, i.e. an isomorphism of $\mathcal{A}$ onto its inverse algebra such that $\imath^2 = \mathrm{Id}$. Let*

$$\mathcal{G} = \{u \in \mathcal{A} \mid u\,\imath(u) = 1_{\mathcal{A}}\}\,.$$

Then $\mathcal{G}$ is a Lie subgroup of $\mathrm{GL}(E)$*, and its tangent space is the subspace*

$$\{v \in \mathcal{A} \mid \imath(v) = -v\}$$

of A. □

The essential property we shall need is the existence of the *exponential map*, defined on $\mathcal{T}_1$, with values in $\mathcal{G}$, which is a local homeomorphism (continuous bijection between appropriate neighbourhoods of 0 in $\mathcal{T}_1$ and of 1 in $\mathcal{G}$), and whose image is the connected component of $\mathcal{G}$. This map is easy to define on subgroups of $\mathrm{GL}(E)$: it is simply the restriction to $\mathcal{T}_1 \subset \mathrm{End}(E)$ of the usual exponential map

$$\exp(u) = \sum_{p=0}^{\infty} \frac{u^p}{p!},$$

a series which converges everywhere, since the corresponding power series has an infinite radius of convergence (see Exercise 3.1.5).

As we shall see in the following sections, the properties of the exponential map will enable us to dispense with some of the explicit convexity calculations that we used in Chapter 3.

A calculation with eigenvalues easily shows the following proposition:

Proposition 10.1.2. *For all $u \in \mathrm{End}(E)$, we have*

$$\det(\exp(u)) = \exp(\mathrm{Tr}(u)). \quad \square$$

In view of the formula $u(x) \cdot u(y) = {}^t\!uu(x) \cdot y$, the map $u \mapsto {}^t\!uu$ of $\mathcal{G} \subset \mathrm{End}(E)$ into $\mathrm{End}^s(E)$ plays as in Chapter 3 an important rôle. This is a differentiable map (but not a homomorphism), and consequently, its image is a submanifold X of $\mathrm{GL}(E)$, endowed with a distinguished point, namely the unit of G. In particular, it possesses a tangent space at the origin.

Definition 10.1.3. *We denote by $\mathcal{T}$ the tangent space to X at the origin.*

The following statement results from elementary properties of tangent spaces:

Proposition 10.1.4. *The space $\mathcal{T}$ is the* symmetrized space *of the tangent space $\mathcal{T}_1$ of $\mathcal{G}$: we have*

$$\mathcal{T} = \{v + {}^t v \mid v \in \mathcal{T}_1\} = \{v \in \mathcal{T}_1 \mid {}^t v = v\}. \quad \square$$

The tangent space (at the origin) of a Lie group $\mathcal{G}$ is endowed with a structure of *Lie algebra*, a non-associative algebra denoted by $L_{\mathcal{G}}$, whose multiplicative law, the *Lie bracket* $[x, y]$, satisfies the following two properties:

1. $\forall x \in L_{\mathcal{G}},\ [x, x] = 0$;
2. $\forall x, y, z \in L_{\mathcal{G}},\ [[x, y], z] + [[y, z], x] + [[z, x], y] = 0$ (Jacobi's identity).

This structure can be used to handle extremality problems; see [Qi-Z] for an example. For the group $\mathcal{G} = \mathrm{GL}(E)$, $L_{\mathcal{G}}$ is canonically isomorphic to $\mathrm{End}(E)$ endowed with its bracket $[v, w] = vw - wv$. The following statement, which we leave as an exercise, interprets the Lie bracket in terms of the exponential map:

Proposition 10.1.5. *For all v, $w \in \operatorname{End}(E)$, we have*

$$\exp(v)\exp(w)\exp(-v)\exp(-w) = \operatorname{Id} + [v,w] + O(t^3)\,,$$

where $t = \max(\|v\|, \|w\|)$. □

10.2 Perfection and Eutaxy

In this section, we denote by $\mathcal{T}$ a vector subspace of $\operatorname{End}^s(E)$; moreover, *when* Id *belongs to* $\mathcal{T}$, we denote by $\mathcal{T}_0$ the kernel of the restriction to $\mathcal{T}$ of the trace map. This is a subspace of codimension 1 in $\mathcal{T}$. We assume that n (the dimension of E) is at least 2. The aim of this section is to make more precise some of the perfection and eutaxy properties relative to $\mathcal{T}$ and $\mathcal{T}_0$.

Let x be a nonzero vector in E. Recall that φ_x denotes the linear form $u \mapsto u(x)\cdot x$ on $\operatorname{End}^s(E)$ and p_x the orthogonal projection onto the line $\mathbb{R}x$, and that the notions of perfection and eutaxy relative to $\mathcal{T}$ are defined using the restrictions to $\mathcal{T}$ of the trace and the forms φ_x. More precisely (Definition 3.2.2), we say that a finite set $S \subset E$ is *$\mathcal{T}$-perfect* if the restrictions to $\mathcal{T}$ of the forms φ_x, $x \in S(\Lambda)$ span the dual $\mathcal{T}^*$ of $\mathcal{T}$, and that it is *$\mathcal{T}$-eutactic* if there exist strictly positive coefficients ρ_x, $x \in S(\Lambda)/\{\pm\operatorname{Id}\}$ such that $\operatorname{Tr}_{|\mathcal{T}} = \sum_x \rho_x\,\varphi_{x\,|\mathcal{T}}$; for a lattice Λ, these notions are applied to the set $S = S(\Lambda)$.

Using the duality defined by the Voronoi scalar product $\langle u,v\rangle = \operatorname{Tr}(uv)$ on $\operatorname{End}^s(E)$ (we have $\varphi_x(u) = (x\cdot x)\operatorname{Tr}(p_x \circ u)$, see Proposition 3.1.10), we immediately obtain:

Proposition 10.2.1. *For all nonzero $x \in E$, let ω_x be the orthogonal projection of p_x onto $\mathcal{T}$ (in $\operatorname{End}^s(E)$), and let Ω be the orthogonal projection onto $\mathcal{T}$ of* $\operatorname{Id} \in \operatorname{End}^s(E)$. *Then a necessary and sufficient condition for a lattice to be $\mathcal{T}$-perfect (resp. $\mathcal{T}$-eutactic), is that the ω_x, $x \in S(\Lambda)$ span $\mathcal{T}$ (resp. that there exists a linear combination $\Omega = \sum_{x\in S(\Lambda)/\{\pm \operatorname{Id}\}} \rho'_x\,\omega_x$ with strictly positive coefficients ρ'_x).* □

The following proposition will be proved by adapting to an arbitrary subspace $\mathcal{T}$ of $\operatorname{End}^s(E)$ some calculations of Section 3.4.

Proposition 10.2.2. *Let $\mathcal{T}$ be a vector subspace of E, and let S be a finite set of nonzero vectors in E. Then the following conditions are equivalent:*

1. *S is both $\mathcal{T}$-perfect and $\mathcal{T}$-eutactic,*
2. *$v = 0$ is the only solution in $\mathcal{T}$ to the system of linear inequalities*

$$\varphi_x(v) \geq 0 \text{ for all } x \in S \quad\text{and}\quad \operatorname{Tr}(v) \leq 0\,.$$

Proof. Suppose first that S satisfies (1), and let $v \in \mathcal{T}$ be such that

$$\varphi_x(v) \geq 0 \text{ for all } x \in S \text{ and } \operatorname{Tr}(v) \leq 0\,.$$

In the $\mathcal{T}$-eutaxy relation applied to v:

$$\operatorname{Tr}(v) = \sum_{x \in S} \rho_x\, \varphi_x(v)\,,\ \ \rho_x > 0 \text{ for all } x\,,$$

the left–hand side is negative and the right one is positive; hence they are both zero. Since all $\rho_x\, \varphi_x(v)$ are positive and all ρ_x are strictly positive, we must have $\varphi_x(v) = 0$ for all $x \in S$, which implies $v = 0$ since S is $\mathcal{T}$-perfect. Hence condition (2) is satisfied.

Conversely, suppose that S satisfies (2), and let us prove that S is then $\mathcal{T}$-perfect. Let $v \in \mathcal{T}$ such that $\varphi_x(v) = 0$ for all $x \in S$; since $-v$ also satisfies this hypothesis, we may assume (changing v into $-v$ if need be) that $\operatorname{Tr}(v) \leq 0$. By (2), v is then zero.

Finally, $\mathcal{T}$-eutaxy is proved by applying Stiemke's Theorem 3.3.1 to the set of forms $-\operatorname{Tr}$ and φ_x, $x \in S$ on $V = \mathcal{T}$. □

In the remainder of this section, we assume that $\mathcal{T}$ contains the identity endomorphism Id, which ensures that $\mathcal{T}_0$ has codimension 1 in $\mathcal{T}$.

Proposition 10.2.3. *Let $\mathcal{T}$ be a subspace of $\operatorname{End}^s(E)$ containing the identity, let $\mathcal{T}_0$ be the hyperplane in $\mathcal{T}$ consisting of the endomorphisms of trace zero, and let $S = \{x_1, \cdots, x_s\}$ a finite set of unit vectors of E. Denote by φ^i the restriction to $\mathcal{T}$ of the linear form φ_{x_i}, and let φ_0^i be the restriction of φ^i to $\mathcal{T}_0$. Then:*

1. *$\mathcal{T}$ is the orthogonal sum of $\mathcal{T}_0$ and of the line $\mathbb{R}\mathrm{Id}$.*
2. *$\mathcal{T}$-eutaxy and $\mathcal{T}_0$-eutaxy are equivalent.*
3. *A necessary and sufficient condition for S to be $\mathcal{T}$-perfect is that it should be $\mathcal{T}_0$-perfect and that the forms φ_0^i satisfy a relation of the shape*

$$\sum_{1 \leq i \leq s} \alpha_i\, \varphi_0^i = 0, \qquad \textit{with} \qquad \sum_i \alpha_i \neq 0\,.$$

4. *S is $\mathcal{T}$-perfect and $\mathcal{T}$-eutactic if and only if it is $\mathcal{T}_0$-perfect and $\mathcal{T}_0$-eutactic.*

Proof. Assertion (1) is evident.

(2) Suppose that S satisfies a $\mathcal{T}_0$-eutaxy relation $\sum_i \rho_i\, \varphi_0^i = 0$ with $\rho_i > 0$. Then it also satisfies the $\mathcal{T}$-eutaxy relation

$$\sum_i \frac{n\rho_i}{\sum \rho_i}\, \varphi^i = \operatorname{Tr}\,.$$

Indeed, let $v \in \mathcal{T}$ and let $v_0 = v - \frac{1}{n}\operatorname{Tr}(v)\,\mathrm{Id}$ be its orthogonal projection onto $\mathcal{T}_0$; we have $\sum_i \rho_i\, \varphi^i(v_0) = 0$, i.e.

$$\sum_i \rho_i \varphi^i(v) = \frac{\sum_i \rho_i}{n} \operatorname{Tr}(v)\, \varphi^i(\mathrm{Id}) = \frac{\sum_i \rho_i}{n} \operatorname{Tr}(v)\,;$$

the converse is evident.

(3) If S is $\mathcal{T}$-perfect, it is trivially $\mathcal{T}_0$-perfect; moreover, the restriction Tr to $\mathcal{T}$ of the trace form can be expressed on the set of the φ^i (which generates $\mathcal{T}^*$ by our hypothesis) in the form $\operatorname{Tr} = \sum_i \alpha_i \varphi^i$. Applied to Id, this relation implies that $n = \sum_i \alpha_i$, hence by restriction to $\mathcal{T}_0$, $0 = \sum_i \alpha_i \varphi_0^i$.

Conversely, suppose that there exists a relation $\sum_{1 \le i \le s} \alpha_i \varphi_0^i = 0$ with $\sum_i \alpha_i \neq 0$; let $v = v_0 + \frac{1}{n} \operatorname{Tr}(v) \mathrm{Id}$, $v_0 \in \mathcal{T}_0$, be an element of $\mathcal{T}$ such that $\varphi^i(v) = 0$ for all i. We then have $\varphi_0^i(v_0) + \frac{1}{n} \operatorname{Tr}(v) = 0$ for all i, which implies $\sum_i \alpha_i \varphi_0^i(v_0) + \frac{\sum_i \alpha_i}{n} \operatorname{Tr}(v) = 0$, where $\sum_i \alpha_i \varphi_0^i(v_0) = 0$ and $\sum_i \alpha_i \neq 0$. Hence $\operatorname{Tr}(v)$ is zero, and $v = v_0$ belongs to $\mathcal{T}_0$. If S is $\mathcal{T}_0$-perfect, the relation $\varphi^i(v) = 0$ for all i shows that $v = 0$, which proves that S is $\mathcal{T}$-perfect.

(4) is now an easy consequence of (2) and (3), since any $\mathcal{T}_0$-eutaxy relation $\sum_i \rho_i \varphi_0^i = 0$ with $\rho_i > 0$ is such that $\sum \rho_i \neq 0$. □

10.3 Extremality

In this section, we consider a family $\mathcal{F}$ of lattices which is an orbit of one of them under the operation of a closed subgroup $\mathcal{G}$ of $\mathrm{GL}(E)$, which is moreover assumed to satisfy the following conditions:

(a) the connected component $\mathcal{G}^\circ$ of $\mathcal{G}$ is of finite index in $\mathcal{G}$;
(b) $\mathcal{G}$ is stable under transposition in $\mathrm{GL}(E)$;
(c) either $\mathcal{G}$ has determinant 1,
or $\mathcal{G}$ contains all positive homothetic transformations.

We denote by $\mathcal{T}$ the symmetrized subspace of the tangent space (at the origin of $\mathcal{G}$; see Definition 10.1.3). We endow $\mathrm{End}^s(E)$ with an arbitrary norm, denoted by $\|.\|$.

Since the connected component $\mathcal{G}^\circ$ is an open subgroup of finite index in $\mathcal{G}$, we can without loss of generality replace $\mathcal{G}$ by $\mathcal{G}^\circ$ to study the local maxima of the Hermite invariant on $\mathcal{F}$. (We might have assumed $\mathcal{G}$ to be connected, but such an hypothesis would not have been suitable, since some important subgroups of the linear group are not connected.)

Definition 10.3.1. *We say that a lattice $\Lambda \in \mathcal{F}$ is* (strictly) $\mathcal{F}$-extreme *if the Hermite invariant of Λ is a (strict modulo similarity) local maximum among the lattices in $\mathcal{F}$, and that it is* $\mathcal{F}$-perfect *(resp.* $\mathcal{F}$-eutactic*) if it is $\mathcal{T}$-perfect (resp. $\mathcal{T}$-eutactic). We also say* (strictly) $\mathcal{G}$-extreme, $\mathcal{G}$-perfect, $\mathcal{G}$-eutactic.

Lemma 10.3.2. *The image of $\mathcal{G}^\circ$ (resp. of $\mathcal{G}$) by the determinant map is one of the subgroups $\{1\}, \mathbb{R}^{*+}$ (resp. $\{1\}$, $\{\pm 1\}$, $\mathbb{R}^{*+}$, $\mathbb{R}^*$) of $\mathbb{R}^*$.*

Proof. The image of $\mathcal{G}^\circ$ is a connected subgroup of $\mathbb{R}^*$, hence is equal to $\mathbb{R}^{*+}$ or is reduced to $\{1\}$. If $\det(\mathcal{G})$ does not contain $\mathbb{R}^{*+}$, it is a finite subgroup of $\mathbb{R}^*$, hence is contained in $\{\pm 1\}$. □

Remark 10.3.3. *To study extremality properties, we may replace $\mathcal{G}$ either by the larger group $(\mathbb{R}^{*+})\,\mathcal{G}$, or by the kernel of the determinant map, i.e., we may restrict ourselves to groups which either contain the positive homothetic transformations, or have determinant* 1.

The following lemma will allow us to estimate the variation of the norm of lattices in $\mathcal{F}$ using only calculations in $\mathcal{T}$.

Lemma 10.3.4. *1. For all $u \in \mathrm{GL}(E)$ and $x \in E$, we have*
$$N(u(x)) = N(x) + \varphi_x({}^t\!uu - \mathrm{Id}).$$
2. For all $v \in \mathrm{End}^s(E)$ and for all $x \in E$, we have $\varphi_x(\exp(v) - \mathrm{Id}) \geq \varphi_x(v)$, and equality holds if and only if $v(x) = 0$ (then both sides of the inequality are zero).
3. Let S be a finite set of nonzero vectors in E and let F be a closed cone in $\mathrm{End}^s(E)$ such that $\min_{x\in S} \varphi_x(v)$ is negative for all $v \neq 0$ in F. Then there exists $\alpha > 0$ such that, for all $v \in F$ with $0 < \|v\| < \alpha$, we have $\min_{x\in S} \varphi_x(\exp(v) - \mathrm{Id}) < 0$.
4. Let Λ be a lattice and let S be the set of its minimal vectors. For $u \in \mathrm{GL}(E)$ sufficiently close to the identity, we have
$$N(u(\Lambda)) = N(\Lambda) + \min_{x\in S} \varphi_x({}^t\!uu - \mathrm{Id}).$$

Proof. Assertion (1) is a consequence of the equalities (see Section 3.4)
$$u(x)\cdot u(x) - x\cdot x = {}^t\!uu(x)\cdot x - x\cdot x = ({}^t\!uu - \mathrm{Id})(x)\cdot x = \varphi_x({}^t\!uu - \mathrm{Id})\,.$$

We shall prove (2) and (3) using an argument of convexity. Let Σ be the unit sphere in $\mathrm{End}^s(E)$. We remark that for all $w \in \Sigma$ and all $x \in E$, $f_w : t \mapsto f_w(t) = \varphi_x(\exp(tw) - \mathrm{Id})$ is a convex function, and that $f_w(0) = 0$ and $f_w{}'(0) = \varphi_x(w)$. Indeed, denoting by λ_i the eigenvalues of w and by (ε_i) an orthonormal basis for E consisting of eigenvectors for w, and setting $x = \sum_i \xi_i\,\varepsilon_i$, we have $f_w(t) = \sum \xi_i^2(e^{t\lambda_i} - 1)$, whence $f_w{}'(t) = \sum \xi_i^2 \lambda_i\, e^{t\lambda_i}$ and $f_w{}''(t) = \sum \xi_i^2 \lambda_i^2\, e^{t\lambda_i} \geq 0$, with equality if and only if $w(x) = 0$.

(2) Let $x \in E$ and $v \neq 0$. Set $\|v\| = t$ and $w = \dfrac{v}{t} \in \Sigma$. Since f_w is a convex function, we have
$$\varphi_x(\exp v - \mathrm{Id}) = \varphi_x(\exp(tw) - \mathrm{Id}) \geq t\,\varphi_x(w) = \varphi_x(v)\,,$$
and equality may not hold unless $w(x) = v(x) = 0$.

(3) Let $w \in F \cap \Sigma$. By our hypotheses, there exists $x \in S$ such that $\varphi_x(w) < 0$. By the convexity of the corresponding function f_w, we see that there exists $t_w > 0$ such that $f_w(t)$ in negative for all $t \in (0, t_w)$. The same property holds for $M_w(t) = \min_x(\varphi_x(\exp(tw) - \mathrm{Id}))$, and more precisely, if

M_w is negative at some point t_0, it is then negative on the whole interval $(0, t_0)$.

Since the function $w' \mapsto M_{w'}(t_w)$ is continuous on $F \cap \Sigma$, there exists an open neighbourhood $V(w)$ of w in $F \cap \Sigma$ such that for all $w' \in V(w)$, $M_{w'}$ is negative at t_w, hence also on the whole interval $(0, t_w)$. From the covering $\bigcup_{w \in F \cap \Sigma} V(w)$ of the compact set $F \cap \Sigma$, we may extract a finite one, say $\bigcup_{1 \le i \le r} V(w_i)$. Set $\alpha = \min(t_{w_1} \cdots t_{w_r})$, let $v \in F$ such that $0 < \|v\| < \alpha$, and let $w = \frac{1}{\|v\|} v \in \Sigma$. There exists an index i such that w belongs to $V(w_i)$. This shows that $M_w(t)$ is strictly negative on the interval $(0, \alpha) \subset (0, t_{w_i})$.

(4) For u sufficiently close to the identity (modulo the orthogonal group), the minimal vectors in $u(\Lambda)$ come from minimal vectors in S, so that $N(u(\Lambda)) = \min_{x \in S} N(u(x))$, whence the result, using (1). □

The following lemma will be needed to decide whether an $\mathcal{F}$-extremal lattice is strictly $\mathcal{F}$-extremal:

Lemma 10.3.5. *Let Λ be a lattice, and let $u \in \mathrm{GL}(E)$ such that $u(\Lambda)$ is similar to Λ. Then if u is sufficiently close to the identity, it is a similarity.*

Proof. The similarity ratio λ_u of the two lattices satisfies the condition

$$\lambda_u^{2n} = \frac{\det(u(\Lambda))}{\det(\Lambda)} = (\det u)^2 \,,$$

hence tends to 1 when u tends to Id. Replacing u by $\lambda_u^{-1} u$, we may assume that the two lattices are isometric. Then there exists an isometry f with $(fu)(\Lambda) = \Lambda$. Hence fu belongs to the discrete subgroup $\mathrm{GL}(\Lambda)$ of $\mathrm{GL}(E)$, and ${}^t\!uu = {}^t(fu)(fu)$ belongs to the discrete set of the ${}^t\!vv$, $v \in \mathrm{GL}(\Lambda)$. For u sufficiently close to the identity, we therefore have ${}^t\!uu = \mathrm{Id}$, which precisely means that u is an isometry. □

The theorem below and its first corollary, which yield a characterization *à la Voronoi* of strictly extreme lattices, are the central results of this chapter.

Theorem 10.3.6. *Let $\mathcal{F}$, $\mathcal{G}$, and $\mathcal{T}$ be as above. Let Λ be a lattice belonging to the family $\mathcal{F}$, and let S be the set of its minimal vectors. Assume as usual either that $\mathcal{F}$ is invariant under positive homothetic transformation, or that the lattices in $\mathcal{F}$ have equal determinant. Then:*

1. *Λ is strictly extreme in $\mathcal{F}$ if and only if it is $\mathcal{T}$-perfect and $\mathcal{T}$-eutactic.*
2. *If Λ is extreme but is not strictly extreme, there exists in $\mathcal{F}$ a path with origin Λ, whose elements are pairwise non-similar extreme lattices, whose Hermite invariants are equal to that of Λ, and whose sets of minimal vectors, to within the only exception of Λ, all lie inside a strict subspace of E.*

Proof. To study the Hermite invariant near Λ, we may replace $\mathcal{F}$ by the normalized family $\mathcal{F}_0 = \{\Lambda' \in \mathcal{F} \mid \det(\Lambda') = \det(\Lambda)\}$, and thus replace $\mathcal{T}$ by $\mathcal{T}_0 = \{v \in \mathcal{T} \mid \mathrm{Tr}(v) = 0\}$. The Hermite invariant is then proportional to the norm of the lattice.

Suppose first that S is $\mathcal{T}$-perfect and $\mathcal{T}$-eutactic. By Proposition 10.2.2, we have $\min_{x\in S}\varphi_x(v) < 0$ for all $v \neq 0$ in $\mathcal{T}_0$ (because $\mathrm{Tr}(v) = 0$). By Lemma 10.3.4 (3) (applied to S and the cone $F = \mathcal{T}_0$), there exists $\alpha > 0$ such that

$$v \in \mathcal{T}_0 \text{ and } 0 < \|v\| < \alpha \Longrightarrow \min_{x\in S}\varphi_x(\exp(v/2) - \mathrm{Id}) < 0\,.$$

Similarly, by Lemma 10.3.4 (4), there exists $\beta > 0$ such that

$$\|v\| < \beta \Longrightarrow N\big(\exp(v/2)(\Lambda)\big) - N(\Lambda) = \min_{x\in S}\varphi_x\big(\exp(v/2) - \mathrm{Id}\big)\,.$$

Let $\varepsilon = \min(\alpha,\beta)$. For every lattice $\Lambda' \neq \Lambda$ belonging to the neighbourhood $\mathcal{U} = \{\exp(\frac{1}{2}v)(\Lambda),\ v \in \mathcal{T}_0,\ \|v\| < \varepsilon\}$ of Λ in $\mathcal{F}_0$, we have $N(\Lambda') - N(\Lambda) < 0$, i.e. $\gamma(\Lambda') < \gamma(\Lambda)$: in $\mathcal{U}$, $\gamma(\Lambda)$ is a strict local maximum. Therefore, Λ is strictly extreme.

Suppose, conversely, that the Hermite invariant attains a maximum at $\Lambda \in \mathcal{F}$ on some neighbourhood $\mathcal{U}$ of Λ in $\mathcal{F}$, which we assume to be sufficiently small, in order that the minimal vectors of the lattices belonging to it come from those of Λ, and let $v \in \mathcal{T}_0$ such that

$$\min_{x\in S}(\varphi_x(v)) \geq 0\,. \tag{10.3.a}$$

For $t > 0$, consider

$$u_t = \exp(\frac{t}{2}v) \in \mathcal{G} \cap \mathrm{GL}^+(\mathbb{R}) \quad \text{and} \quad \Lambda_t = u_t(\Lambda) \in \mathcal{F}_0\,. \tag{10.3.b}$$

Suppose that t has been chosen sufficiently small, so as to ensure that Λ_t belongs to $\mathcal{U}$, and that u_t satisfies the hypothesis of Lemma 10.3.5. Since $\mathrm{Tr}(v) = 0$, we have $\det u_t = 1$ (Proposition 10.1.2), hence $\det(\Lambda_t) = \det(\Lambda)$, and for t sufficiently small, condition (10.3.a) implies (Lemma 10.3.4 (4),(3))

$$\begin{aligned}\det(\Lambda)^{1/n}\,(\gamma(\Lambda_t) - \gamma(\Lambda)) = N(\Lambda_t) - N(\Lambda) &= \min_{x\in S}\,\varphi_x(\exp(tv) - \mathrm{Id})\\ &\geq t\min_{x\in S}\,\varphi_x(v) \geq 0\,.\end{aligned}$$

Since $\gamma(\Lambda)$ is a maximum on $\mathcal{U}$, the inequalities above are indeed equalities, which shows that $\gamma(\Lambda_t) = \gamma(\Lambda)$. Moreover, the minimal vectors in Λ_t are the vectors $u_t(x)$ for $x \in S$ such that $\varphi_x(\exp(tv) - \mathrm{Id})) = t\,\varphi_x(v) = 0$, i.e., by Lemma 10.3.4 (2), $v(x) = 0$, hence $u_t(x) = x$. Consequently, we have

$$S(\Lambda_t) = S \cap \mathrm{Ker}(v)\,. \tag{10.3.c}$$

If $\gamma(\Lambda)$ is a strict maximum in $\mathcal{U}$, the relation $\gamma(\Lambda_t) = \gamma(\Lambda)$ may not hold unless Λ_t is similar to Λ. Then by Lemma 10.3.5, u_t must be an isometry (recall that $\det(u_t) = 1$), which is equivalent to $v = 0$. Thus if γ attains a strict maximum at Λ, condition (10.3.a) implies that $v = 0$, which shows that Λ is $\mathcal{T}_0$-perfect and $\mathcal{T}_0$-eutactic. Taking into account Proposition 10.2.3, this completes the proof of assertion (1).

If $\gamma(\Lambda)$ is not a strict maximum, by the preceding proof, S is not both $\mathcal{T}_0$-perfect and $\mathcal{T}_0$-eutactic, so that there exists in $\mathcal{T}_0$ an element $v \neq 0$ which satisfies condition (10.3.a). The lattices Λ_t constructed using v are then pairwise non-similar, and satisfy the properties of (2). □

Corollary 10.3.7. *If the minimal vectors in every $\mathcal{F}$-extreme lattice span E, every $\mathcal{F}$-extreme lattice in $\mathcal{F}$ is strictly $\mathcal{F}$-extreme, hence both $\mathcal{F}$-perfect and $\mathcal{F}$-eutactic.*

Proof. Every neighbourhood of an extreme, non-strictly extreme lattice contains an extreme lattice whose minimal vectors do not span E.

Corollary 10.3.8. *If a lattice Λ is strictly extreme for a group $\mathcal{G}$, the number s of its pairs of minimal vectors satisfies the lower bound*

$$s \geq \dim(\mathcal{G}),$$

and even

$$s \geq \dim(\mathcal{G}) + 1$$

if all elements of $\mathcal{G}$ have determinant $\{\pm 1\}$.

Proof. We have $s \geq \dim(\mathcal{T}) = \dim(\mathcal{G})$ by the $\mathcal{T}$-perfection of $S(\Lambda)$; if moreover $\det(\mathcal{G}) \subset \{\pm 1\}$, $\mathcal{T}$ is contained in $\ker \mathrm{Tr}$, so that $\mathcal{T}$-eutaxy implies the existence of a non-trivial relation between the $\varphi_x, x \in S(\Lambda)$, whence the lower bound $s \geq \dim(\langle \varphi_x, x \in S(\Lambda) \rangle) + 1$. [Variant: if $\det(\mathcal{G}) \subset \{\pm 1\}$, Λ is $\mathcal{G}'$-extreme for the group $\mathcal{G}' = \mathbb{R}^* \mathcal{G}$, of dimension $\dim(\mathcal{G}) + 1$.] □

We end this section with a succinct description of an example of a group $\mathcal{G}$ for which there are extreme, but not strictly extreme lattices.

We suppose that E has dimension 3. Let D be a line in E, and let σ be a rotation of angle $\frac{\pi}{2}$ about the line D. Let $\mathcal{F}$ be the set of σ-*isodual* lattices, i.e. lattices Λ such that $\Lambda^* = \sigma(\Lambda)$. It is not difficult to verify that the family $\mathcal{F}$ is the set of those lattices Λ which are the orthogonal sum of a lattice $\Lambda_0 \subset D$ and a lattice $\Lambda_1 \subset P = D^\perp$, both having determinant 1 (thus $\Lambda_0 \simeq \mathbb{Z}$), and that $\mathcal{F}$ is the orbit of one of its lattices under the action of the group $\mathcal{G} = \{u \in \mathrm{GL}(E) \mid \sigma u = u \sigma^{-1}\}$ (see Exercise 10.3.1, Proposition 11.7.2, and Exercise 11.5.5).

We have

$$\gamma(\Lambda) = N(\Lambda) = \min\left(N(\Lambda_0), N(\Lambda_1)\right).$$

Choosing Λ_1 of norm ≥ 1 (see Exercise 10.3.2), we obtain a family of $\mathcal{G}$-extreme lattices with $\gamma = 1$, of dimension 2 modulo similarities.

10.4 Minimal Classes

In this section, we keep the notation E, $\mathcal{F}$, $\mathcal{G}$ and the hypotheses of Section 10.3. We are going to generalize some results on minimal classes that we proved in the first section of the preceding chapter.

Definition 10.4.1. *We say that two lattices Λ and Λ' in $\mathcal{F}$ are* minimal-equivalent for $\mathcal{F}$ *(or for $\mathcal{G}$) if there exists $u \in \mathcal{G}$ which maps Λ onto Λ' and $S(\Lambda)$ onto $S(\Lambda')$. We denote by $\Lambda \prec \Lambda'$ the relation obtained by replacing the equality $u(S) = S'$ by the inclusion $u(S) \subset S'$.*

These relations are respectively an equivalence and a quasi-ordering relation, and the last one induces an ordering relation on the set of classes, still denoted by $\prec$. We recover the definitions of Chapter 9 when $\mathcal{G}$ is the whole linear group $\mathrm{GL}(E)$; see next section for the case of dual-minimal classes.

The following theorem shows in particular that classes in the sense of Definition 10.4.1 contain at most one strictly $\mathcal{G}$-extreme lattice.

Theorem 10.4.2. *Let $\mathcal{C}$ be a $\mathcal{G}$-class and let $\Lambda \in \mathcal{C}$ be a $\mathcal{G}$-eutactic lattice.*

1. *The Hermite invariant attains on Λ its minimum in the union $\widetilde{\mathcal{C}}$ of classes $\mathcal{C}' \succ \mathcal{C}$.*
2. *If $S(\Lambda)$ spans E, or if Λ is $\mathcal{G}$-perfect, all $\mathcal{G}$-eutactic lattices in $\mathcal{C}$ are similar to Λ.*

[If the number of classes if finite (as in the examples quoted in the introduction), Theorem 10.4.2 implies the finiteness of the set of lattices which are strictly extreme for $\mathcal{G}$, and even of the set of well-rounded $\mathcal{G}$-eutactic lattices.]

Proof. We may assume for the proof that $\mathcal{G}$ has determinant 1. Let $\Lambda' = u(\Lambda) \in \mathcal{F}$, $u \in \mathcal{G}$, be a lattice such that $S' \supset u(S)$. We have $N(u(x)) = N(\Lambda')$ for all $x \in S$, i.e. (Lemma 10.3.4 (1)),

$$\forall x \in S\,,\ \varphi_x({}^t\!uu - \mathrm{Id}) = N(\Lambda') - N(\Lambda)\,.$$

Moreover, since $\mathcal{G}$ is connected, there exists $v \in \mathcal{G}$ (with of course $\mathrm{Tr}(v) = 0$) such that ${}^t\!uu = \exp(v)$. Hence we have

$$\varphi_x(\exp(v) - \mathrm{Id}) = N(\Lambda') - N(\Lambda) \ \text{ for all } \ x \in S\,. \qquad \mathbf{(10.4.a)}$$

For all $x \in S$, set

$$\psi_x(v) = \varphi_x(\exp(v) - \mathrm{Id}) - \varphi_x(v)\,.$$

By Lemma 10.3.4 (2), we have the lower bound $\psi_x(v) \geq 0$, with equality if and only if $v(x) = 0$, and also

$$\forall x \in S\,,\ N(\Lambda') - N(\Lambda) = \varphi_x(v) + \psi_x(v)\,. \qquad \mathbf{(10.4.b)}$$

Since S is $\mathcal{G}$-eutactic, there exist strictly positive coefficients ρ_x such that

$$\sum_{x\in S} \rho_x\, \varphi_x(v) = \mathrm{Tr}(v) = 0\,.$$

Using linear combinations of the relations (10.4.b), we obtain

$$\Big(\sum_{x\in S} \rho_x\Big)\Big(N(\Lambda') - N(\Lambda)\Big) = 0 + \sum_{x\in S} \rho_x\, \psi_x(v) \geq 0\,, \qquad \textbf{(10.4.c)}$$

hence $N(\Lambda') - N(\Lambda) \geq 0$, whence $\gamma(\Lambda') \geq \gamma(\Lambda)$, which proves (1).

To prove (2), we moreover assume that Λ' is $\mathcal{G}$-eutactic and belongs to the class $\mathcal{C}$ (i.e., we have $S(\Lambda') = u(S)$). Exchanging the rôles played by Λ' and Λ, we see that $N(\Lambda') - N(\Lambda) = 0$ (i.e., $\gamma(\Lambda') = \gamma(\Lambda)$), hence by (10.4.c), that $\psi_x(v) = 0$, i.e. that $v(x) = 0$ for all $x \in S$. We thus have $S \subset \operatorname{Ker} v$. This implies that $v = 0$: this is clear if S spans E, and if Λ' is $\mathcal{G}$-perfect, it results from the equalities $\varphi_x(v) = 0$ for all $x \in S$. Finally, we have ${}^t u u = \mathrm{Id}$, hence u is an isometry. □

Corollary 10.4.3. *A strictly $\mathcal{G}$-extreme lattice is isolated (modulo similarity) in its class $\mathcal{C}$; in particular, if it is a (strict) absolute maximum, this lattice is unique modulo similarity in the union $\widetilde{\mathcal{C}}$ of the $\mathcal{G}$-classes which contain $\mathcal{C}$.* [Other formulation: a lattice as above loses minimal vectors under any sufficiently small deformation which is not a similarity.]

Proof. Indeed, the Hermite invariant attains on it a relative (or absolute) maximum in $\mathcal{F}$, hence in $\mathcal{C}$, and also by Theorem 10.4.2 an absolute minimum in $\mathcal{C}$. □

10.5 Dual-Extreme Lattices

In this section, we denote by E' the Cartesian product $E \times E$, of dimension $m = 2n$, endowed with its natural Euclidean structure, which identifies it with the orthogonal sum of two copies of E, and we associate with any lattice $L \subset E$ the lattice $\Lambda = L \perp L^* \subset E'$. We denote by $\mathcal{F}$ the family of lattices of this form inside E'.

Recall (Section 3.8) that the invariant γ' of a lattice L is the geometrical mean of the Hermite invariants of L and of L^*, so that we say that L is *dual-extreme* if γ' attains a local maximum at L, *dual-perfect* if the set of orthogonal projections onto the minimal vectors in L and L^* spans $\mathrm{End}^s(E)$, and *dual-eutactic* if there exists a relation $\sum_{x\in S(L)} \rho_x p_x = \sum_{y\in S(L^*)} \rho'_y p_y$ with strictly positive coefficients.

We now give a new proof of Theorem 3.8.5, relying on the theory developed in the first four sections.

Lemma 10.5.1. *1. If Λ is $\mathcal{F}$-extreme, L and L^* have the same norm.*
2. If L and L^* have the same norm, then $\gamma'(L) = \gamma(\Lambda)$.

Proof. (1) Let $\Lambda = L \perp L^*$. We have $\det(\Lambda) = 1$, hence

$$\gamma(\Lambda) = N(\Lambda) = \min(N(L), N(L^*)) .$$

Let $\lambda = (\frac{N(L^*)}{N(L)})^{1/2}$, and let $\Lambda_\lambda = \lambda L \perp (\lambda L)^*$. For all $t > 0$, we have $(tL)^* = t^{-1}L^*$, and consequently, $\gamma(\Lambda_\lambda) = \max(\lambda, \lambda^{-1})\gamma(\Lambda)$. If the norms of L and of L^* were different, there would exist a path in the space of lattices along which the Hermite invariant would increase strictly from its value on Λ.

(2) Indeed, we have

$$\gamma(\Lambda) = N(L) = N(L^*) = \left(\gamma(L)\,\gamma(L^*)\right)^{1/2} = \gamma'(L) . \quad \square$$

The proposition below immediately follows from the preceding lemma:

Proposition 10.5.2. *A lattice L is dual-extreme if and only if it is proportional to a lattice L_1 such that $\Lambda = L_1 \perp L_1^*$ is $\mathcal{F}$-extreme.* $\square$

Using the formula $u(L)^* = {}^t u^{-1}(L^*)$ (Proposition 1.3.2 (2)), we see that the family $\mathcal{F}$ constitutes a unique orbit under the action of the closed subgroup

$$\mathcal{G} = \{(u, {}^t u^{-1}) \mid u \in \mathrm{GL}(E)\}$$

of $\mathrm{GL}(E')$, which is clearly invariant under transposition. In other respects, we know (Proposition 3.8.2) that if L is dual-extreme, $S(L)$ and $S(L^*)$ both span E. Proposition 10.5.2 and Corollary 10.3.7 show that dual-extreme lattices are indeed strictly dual extreme, and characterized as those lattices which are both $\mathcal{T}$-perfect and $\mathcal{T}$-eutactic, where $\mathcal{T}$ stands for the symmetrized space of the tangent space of $\mathcal{G}$.

We may thus derive from the results proved in Sections 10.1 and 10.2 a characterization dual-extreme lattices; it suffices to determine $\mathcal{T}$ and then to write down explicitly the conditions of $\mathcal{T}$-perfection and $\mathcal{T}$-eutaxy.

The group $\mathcal{G}$ is the set of elements u in the algebra

$$\mathcal{A} = \mathrm{End}(E) \times \mathrm{End}(E) \subset \mathrm{End}(E') ,$$

endowed with the involution $\imath : (v_1, v_2) \mapsto ({}^t v_2, {}^t v_1)$, which satisfy the condition $u\,\imath(u) = \mathrm{Id}$. Theorem 10.1.1 shows that the tangent space of $\mathcal{G}$ is the set of elements of the form $(v, -{}^t v)$ of $\mathcal{A}$. Our symmetrized space is therefore

$$\mathcal{T} = \{(v, -v) \mid v \in \mathrm{End}^s(E)\} .$$

Given elements $u \in \mathrm{End}^s(E')$ and $v \in \mathcal{T}$, we can represent u, v, and uv by matrices of the following form, corresponding to the decomposition of E' as the orthogonal sum of $E \times \{0\}$ and $\{0\} \times E$:

$$\begin{pmatrix} u_1 & u' \\ {}^t u' & u_2 \end{pmatrix}, \qquad \begin{pmatrix} v & 0 \\ 0 & -v \end{pmatrix}, \qquad \text{and} \qquad \begin{pmatrix} u_1 v & -u'v \\ {}^t u' v & -u_2 v \end{pmatrix}$$

(with $u' \in \operatorname{End}(E)$ and $u_1, u_2, v \in \operatorname{End}^s(E)$). As a consequence, $\mathcal{T}^\perp$ is the set of matrices such that $u_2 = +u_1$, and the components on $\mathcal{T}$ and $\mathcal{T}^\perp$ of p_x ($x \in E \times \{0\}$), p_y ($y(\in \{0\} \times E)$, and $\operatorname{Id} \in \operatorname{End}^s(E')$ have the following forms:

$$\begin{pmatrix} \frac{1}{2}p_x & 0 \\ 0 & -\frac{1}{2}p_x \end{pmatrix} + \begin{pmatrix} \frac{1}{2}p_x & 0 \\ 0 & \frac{1}{2}p_x \end{pmatrix}, \ \begin{pmatrix} -\frac{1}{2}p_y & 0 \\ 0 & \frac{1}{2}p_y \end{pmatrix} + \begin{pmatrix} \frac{1}{2}p_y & 0 \\ 0 & \frac{1}{2}p_y \end{pmatrix}, \ \text{and} \ \left(\begin{smallmatrix} 0 & 0 \\ 0 & 0 \end{smallmatrix}\right) + \left(\begin{smallmatrix} I_n & 0 \\ 0 & I_n \end{smallmatrix}\right).$$

This shows that for a lattice $\Lambda = L \perp L^*$ with $N(L) = N(L^*)$, $\mathcal{T}$-perfection (resp. $\mathcal{T}$-eutaxy) is expressed by the fact that the set of p_x, $x \in S(L) \cup S(L^*)$ spans $\operatorname{End}^s(E)$ (resp. that there exists a relation of the form

$$0 = \sum_{x \in S(L)} \rho_x p_x - \sum_{y \in S(L^*)} \rho'_y p_y$$

with strictly positive coefficients ρ_x, ρ'_y).
We have thus recovered Theorem 3.8.5.

Proposition 10.5.3. *The number of similarity classes of dual-eutactic lattices whose minimal vectors span E is finite.*

Proof. Indeed, if Λ is such a lattice, and if $S(\Lambda)$ spans E, then $S(\Lambda^*)$ also spans E (see Exercise 3.8.4). Taking into account Theorem 9.1.5, we may then apply Theorem 10.4.2.
[The hypothesis "S spans E" is essential; see Exercise 9.6.1.] □

10.6 The Rankin Invariants

We keep the notation of the previous sections. The *Rankin invariant of order k of an n-dimensional lattice Λ* was introduced in Definition 2.8.3: given an integer k between 1 and $n-1$, we set

$$\gamma_{n,k}(\Lambda) = \frac{\delta_k(\Lambda)}{\det(\Lambda)^{k/n}},$$

where $\delta_k(\Lambda)$ denotes the minimum of the determinant of k independent vectors in Λ. For $k = 1$ (resp. $k = n-1$), this is the Hermite invariant of Λ (resp. of Λ^*). We also set

$$\gamma_{n,k} = \sup_{\Lambda} \gamma_{n,k}(\Lambda)$$

(*the Rankin constant of order k in dimension n*). We have $\gamma_{n,k}(\Lambda^*) = \gamma_{n,n-k}(\Lambda)$, hence $\gamma_{n,n-k} = \gamma_{n,k}$.

[**Warning**: in contrast to our convention concerning Hermite's invariant and Hermite's constant, we incorporate an index (here, $_{n,k}$) in both cases.]

In this section, we consider the local maxima of the Rankin invariants. We state the essential results obtained by Coulangeon, and refer to the original paper [Cou3] for the proofs.

Definition 10.6.1. *We say that a lattice Λ is* k-extreme *if $\gamma_{n,k}$ attains a local maximum at Λ, and that it is* strictly k-extreme *if there exists a neighbourhood of Λ on which the invariant $\gamma_{n,k}$ of any lattice which is not similar to Λ is strictly lower than that of Λ. We say that Λ is* k-critical *if it concerns an absolute maximum.*

The determinant (it is indeed one of its possible definitions) can be calculated in the exterior algebra $\bigwedge E$ of E. Indeed, once $\bigwedge^k E$ is endowed with its Euclidean structure defined in Theorem 1.10.3, we have

$$\delta_k(x_1, \dots, x_k) = N(x_1 \wedge \cdots \wedge x_k)\,,$$

a formula which allows us, *provided the norm of $\bigwedge^k \Lambda$ is attained on split k-vectors*, to give an interpretation of the invariant $\gamma_{n,k}$ of Λ as the Hermite invariant of $\bigwedge^k \Lambda$. Unfortunately, we do not know of any general result which would ensure *a priori* that there do not exist any non-split system of k-vectors with a norm strictly smaller than $\delta_k(\Lambda)$. For this reason it is not possible to apply directly the results of Section 10.3 to the orbit in $\bigwedge^k E$ of one lattice $\bigwedge^k \Lambda$ under the group $\mathcal{G} = \{\overset{k}{\wedge} u \mid u \in \mathrm{GL}(E)\}$. This problem induces various difficulties, the details of which we cannot enter into. We shall content ourselves with the following result, whose proof is left to the reader, which makes more precise the interpretation of determinants as elements of the exterior algebra.

Proposition 10.6.2. *If two k-tuples $(x_1, \dots, x_k)$ and $(y_1, \dots, y_k)$ of vectors in E have determinant $\delta_k(\Lambda)$, then $x_1 \wedge \cdots \wedge x_k = \pm y_1 \wedge \cdots \wedge y_k$.* □

We now define the suitable notions of perfection and eutaxy for invariants of order k:

Definition 10.6.3. *Let Λ be a lattice.*

1. *We denote by $S_k(\Lambda)$ the set of subspaces of E which possess a basis of k vectors $x_1, \dots, x_k$ of Λ with determinant $\delta_k(\Lambda)$.*
2. *We say that Λ is* k-perfect *if the orthogonal projections p_X onto the elements X of $S_k(\Lambda)$ span* $\mathrm{End}^s(E)$.
3. *We say that Λ is* k-eutactic *if* Id_E *is a linear combination with strictly positive coefficients of the projections p_X, $X \in S_k(\Lambda)$.*

Definitions 10.6.1 and 10.6.3 reduce to those of Chapter 3 when $k = 1$, taking for S_1 the set of *directions* of vectors in S.

With these definitions, we can state a generalizations of Voronoi's theorem:

Theorem 10.6.4. (Coulangeon.) *Let Λ be a lattice in E, and let $k \in [1, n-1]$ be an integer. The following conditions are equivalent:*

1. *Λ is k-extreme.*
2. *Λ is strictly k-extreme.*
3. *Λ is k-perfect and k-eutactic.* □

To test whether a given lattice is k-perfect, we must first list its *k-minimal sections* (k-dimensional sublattices of determinant $\delta_k(\Lambda)$). This can be done using inequalities "à la Hermite", for instance those we established in Theorems 2.2.1 and 2.6.11, which explicitly provide constants K_r such that any r-dimensional lattice L possesses a basis $(e_1, \dots, e_r)$ which satisfies the inequality $\prod_{j=1}^r N(e_j) \le K_r \det(\Lambda)^r$. Applying this upper bound with $r = k$ after having ordered $e_1, \dots, e_k$ in increasing norms, we can find all k-dimensional lattices contained in Λ with a determinant smaller than any given norm, by giving successive upper bounds for the possible norms of e_1, then of e_2, ... , as in the proof of Theorem 9.1.5.

This not very efficient procedure is however a very simple one when δ_k is attained on k minimal vectors in Λ which generate a critical lattice, since it then suffices to make an exploration of the k-tuples in $S(\Lambda)$ to find all k-minimal sections.

Example 10.6.5. *The irreducible root lattices of dimension $n \ge 4$ are 2-perfect. (See Exercise 10.6.3; however, $\mathbb{A}_3$ is not to 2-perfect, since $\mathbb{A}_3^*$ is not perfect.)*

Example 10.6.6. "Watson lattices" W_n, $n = 1, \dots, 7$ *(integral lattices of minimum 3 having the largest possible kissing number; their kissing number is then the largest possible one among* all *lattices which do not contain any hexagonal lattice generated by minimal vectors), defined in [Wat7], are obtained as cross-sections of $\sqrt{2}\,\mathbb{E}_7^*$. Scalar products of pairwise non-proportional minimal vectors are equal to ± 1; their δ_2 invariant is equal to 8; we have $W_7 \sim \mathbb{E}_7^*$, $W_6 \sim \mathbb{D}_6^+$, $W_5 \sim \mathbb{A}_5^3$, $W_3 \sim \mathbb{A}_3^*$. These lattices are 2-perfect in the range $n \in [3, 7]$. However, for $n \ge 4$, their $\gamma_{n,2}$ invariants are slightly less good than those of the critical lattices:*

$$\sqrt{2} = 1.414\ldots < \gamma_{4,2}(\mathbb{D}_4) = \tfrac{3}{2}\,,$$
$$2^{7/5}3^{-2/5} = 1.700\ldots < \gamma_{5,2}(\mathbb{D}_5) = 3 \cdot 2^{-4/5} = 1.723\ldots\,,$$
$$2 < \gamma_{6,2}(\mathbb{E}_6) = 3^{2/3} = 2.080\ldots\,,$$
$$2^{9/7} = 2.438\ldots < \gamma_{7,2}(\mathbb{E}_7) = 3 \cdot 2^{-2/7} = 2.461\ldots$$

Example 10.6.7. *The invariants $\gamma_{2k,k}$ are the same for a lattice and its dual. Hence k-extreme lattices of dimension $n = 2k$ show up in pairs (Λ, Λ^*). For $n = 4$ and $k = 2$, five 2-extreme lattices are known, namely $\mathbb{A}_4$, $\mathbb{A}_4^*$, $\mathbb{D}_4 \sim \mathbb{D}_4^*$, W_4, W_4^*. Here are Gram matrices for W_4 and $\sqrt{8}\,W_4^*$:*

$$\begin{pmatrix} 3 & 1 & 1 & 1 \\ 1 & 3 & -1 & 1 \\ 1 & -1 & 3 & 1 \\ 1 & 1 & 1 & 3 \end{pmatrix} \quad \textit{and} \quad \begin{pmatrix} 4 & -2 & -2 & 0 \\ -2 & 5 & 3 & -2 \\ -2 & 3 & 5 & -2 \\ 0 & -2 & -2 & 4 \end{pmatrix}.$$

One can see that $\delta_2(\sqrt{8}\,W_4^*)$ is attained on two families of pairwise non-isometric lattices, which can be defined by the Gram matrices $\left(\begin{smallmatrix} 4 & 0 \\ 0 & 4 \end{smallmatrix}\right)$ and $\left(\begin{smallmatrix} 4 & 2 \\ 2 & 5 \end{smallmatrix}\right)$; lattices in the second family are not generated by minimal vectors.

We define as in Section 10.4 and in Chapter 9 the notion of a *minimal k-class*: we say that two lattice Λ and Λ' are *k-minimal-equivalent* if there exists $u \in \mathrm{GL}(E)$ such that $u(\Lambda) = \Lambda$ and $u(S_k(\Lambda)) = S_k(\Lambda)$. Using calculations similar to that of Section 9.1, we can one prove:

Theorem 10.6.8. *There are only finitely many minimal k-classes. In particular, modulo similarity, there are only finitely many k-perfect lattices.* □

In view of the finiteness theorem above, the classification of k-perfect or k-extreme lattices in a given dimension n is a natural problem. When $k = 1$ or $k = n-1$, this is the usual problem for the Hermite invariant. For $k \neq 1, n-1$, we are faced with real Rankin invariants. No classification result is known, and the only known non-trivial Rankin constant is $\gamma_{4,2}$ (see Corollary 6.3.8), a rather easy case, thanks to the fact that the Mordell inequality for $n = 3$ yields the exact value of γ_4. Certainly, the most important open classification questions in small dimensions are the following ones:

Problem 10.6.9. 1. Classify 2-extreme lattices in dimension 4.
2. Calculate the constant $\gamma_{5,2}$.
[The answer to the second question is probably $\gamma_{5,2}(\mathbb{D}_5) = 3 \cdot 4^{-2/5} = 1.723\ldots$]

Finally we quote the following problem, "dual" to that of the Rankin constants:

Problem 10.6.10. Characterize the local maxima of the invariant $\gamma'_{n,k}$ (Definition 2.8.3).

10.7 Exercises for Chapter 10

10.1.1. Prove the formula $\det(\exp(u)) = \exp(\mathrm{Tr}(u))$.
[First consider the case of trigonal matrices with complex entries.]

10.1.2. Let $\mathcal{G}$ be a closed finite subgroup $\mathrm{GL}(E)$, invariant under transposition, and let $u \in \mathcal{G}^\circ$.

1. Show that any positive symmetric element u of $\mathcal{G}^\circ$ is of the form $\exp(v)$ for a unique element v of $\mathcal{T}$. [Use Exercise 3.1.5.]

2. Show that the symmetric and orthogonal components of u belong to $\mathcal{G}^\circ$. [Apply Question 1 to ${}^t uu$ (or to $u{}^t u$) and consider $\frac{v}{2}$.]

10.1.3. Show that the Lie algebra $\mathrm{SL}(E)$ is the set $\mathfrak{sl}(E)$ of elements of $\mathrm{End}(E)$ with trace zero.

2. Show that the Lie algebra of $\mathrm{O}(E)$ is $\mathrm{End}^a(E)$ (Definition 3.1.1).

10.3.1. We suppose in this exercise that $\dim E = 3$. Let D be a line with orthogonal plane P, and let τ be the orthogonal symmetry whose D is the set of fixed points. Given a lattice Λ, set $\Lambda_0 = D \cap \Lambda$ and $\Lambda_1 = P \cap \Lambda$.

1. Show that if $\tau(\Lambda) = \Lambda$, Λ contains $\Lambda_0 \perp \Lambda_1$ to index 1 or 2, and that if the index is 2, $\Lambda = \langle \Lambda_0 \perp \Lambda_1, \frac{1}{2}(e_0 + e_1)\rangle$, where e_0 and e_1 are primitive vectors in Λ_0 and Λ_1 respectively. [Consider the maps $x \mapsto x \pm \tau x$.]

2. Let σ be a rotation of angle $\frac{\pi}{2}$ about D. Suppose that $\sigma(\Lambda) = \Lambda^*$, and that the index above is equal to 2. Calculate $\frac{1}{2}(e_0+e_1)\cdot\sigma(\frac{1}{2}(e_0+e_1))$; conclude that $\frac{e_0}{2} \in \Lambda^*$.

3. Show that if Λ is σ-isodual, it is then the orthogonal sum of Λ_0 and Λ_1.

10.3.2. Let $K \subset \mathbb{R}^2$ be the compact set defined in polar coordinates by the inequalities $\lambda \geq 1$ and $\frac{\pi}{3} \leq \theta \leq \arcsin\frac{1}{\lambda}$. Show that we obtain a parametrization by K of the set of 2-dimensional lattices Λ of determinant 1 and norm $N(\Lambda) \geq 1$ by choosing a basis (e_1, e_2) where e_1, e_2 are the successive minima of Λ, $N(e_1) = \lambda$, and $\widehat{e_1, e_2} = \theta$.

10.6.1. Given a k-dimensional subspace X of E, denote by φ_X the linear form $v \mapsto \mathrm{Tr}(p_X \circ v)$ on $\mathrm{End}^s(E)$. Let $(x_1, \dots, x_k)$ be a basis for X.

1. Show that $\varphi_X(v) = \frac{1}{\delta_k(X)}(x_1 \wedge \cdots \wedge x_k)\cdot\sum_{i=1}^k x_1 \wedge \cdots \wedge v(x_i) \wedge \cdots \wedge x_k$.

2. Show that a lattice is k-perfect (resp. k-eutactic) if and only if the φ_X, $X \in S_k$ span $\mathrm{End}^s(E)^*$ (resp. there exists a relation $\mathrm{Tr} = \sum_{X \in S_k} \rho'_X \varphi_X$ with strictly positive coefficients ρ'_X).

10.6.2. Let Λ be a lattice.

1. Show that if the representation of $\mathrm{Aut}(\Lambda)$ afforded by E is $\mathbb{R}$-irreducible, Λ is k-eutactic for all k.

2. Show that all irreducible root lattices as well as their duals are k-eutactic for all k.

10.6.3. Show that irreducible root lattices of dimension at least 4 are 2-extreme. [Use the description of minimal hexagonal sections given in Exercises 4.2.1, 4.3.4, 4.4.6, and 4.5.3.]

10.8 Notes on Chapter 10

The results of Sections 10.2, 10.3, 10.4, and 10.5 essentially come from [B-M4], the only noticeable change being the adjunction of a few new developments in Section 10.5. Section 10.6 summarizes the third part of Coulangeon's thesis, published as [Cou3].

The theory of this chapter was later generalized by Bavard, first in [Bav1], then recently in [Bav4], who developed it within the framework of Riemannian geometry. Due to the large amount of new notions which would have been needed, it has not been possible to account for this theory in this book.

11 Group Actions

Introduction

The aim of this chapter is to study the extremal property of lattices of one of the following two types:

1. *G-lattices*, which are invariant under a given finite subgroup G of the orthogonal group $\mathrm{O}(E)$.
2. σ-isodual lattices, which are transformed into their dual by a given element σ of finite order in $\mathrm{O}(E)$.

In both cases, the results of this chapter heavily rely on those of Chapter 10.

We first recall some results about group representations, then consider the case of G-lattices in the spirit of Chapter 10. In Section 11.4 we study the case where G is cyclic and defines an irreducible representation over $\mathbb{Q}$ ("cyclotomic lattices"), and some related situations.

Sections 11.5–11.8 are devcted to isodual lattices and to a related property that we call *normality* (Section 11.6), which is satisfied by all isodual lattices, and plays an important rôle for the classification problems considered in Section 11.8. Before this, we develop in Section 11.7 the notions of an isodual orthogonal (or symmetric) lattice and of a symplectic lattice, which are of great interest in various applications.

Finally, in Section 11.9, we consider rationality and finiteness questions, for both G-lattices and isodual lattices.

11.1 Rational and Integral Representations

Let G be a finite group and let K be a field. Traditionally, a *representation of G over K* consists of a finite–dimensional K-vector space V and a homomorphism $\rho : G \to \mathrm{GL}(V)$; choosing a basis for V, we can transform ρ into a homomorphism of G into $\mathrm{GL}_n(K)$ ($n = \dim V$). It amounts to the same thing as considering a $K[G]$-module structure on V, i.e. an action $(s, x) \mapsto sx$ of G on V, with ρ defined by $\rho_s(x) = sx$. We shall adopt the latter point of view.

We refer to [Se3] or [Se'3] for the main properties of representations. Recall that the representation defined by V is *irreducible* if $\{0\}$ and V are the only stable subspaces of V, that it is *absolutely irreducible* if it remains irreducible when one extends the scalars to an algebraic closure of K, and that it is *rational over a subfield* K_0 *of* K if $V = K\,V_0$ ($\simeq K \bigotimes V_0$) for some invariant K_0-subspace V_0 of V with $\dim_{K_0} V_0 = \dim_K V$, i.e. if ρ can be defined by matrices with entries in K_0. Recall also that two representations ρ and ρ' are *equivalent* if the $K[G]$-modules that they define are isomorphic; in terms of matrices, this amounts to the existence of a matrix $P \in \mathcal{M}_n(K)$ such that $\rho_s = P\rho'_s P^{-1}$ for all $s \in G$. (A $K[G]$-isomorphism φ is such that $\varphi \circ s = s \circ \varphi$.)

We say that a representation is *faithful* if the corresponding homomorphism $\rho : G \to \mathrm{GL}(V)$ is injective. Replacing G by $G/\ker\rho$, we restrict ourselves to this case, and can then identify G with a subgroup of $\mathrm{GL}(V)$. We shall very often make use of this possibility.

[However (see below), it is generally no longer possible to split a representation into a direct sum of irreducible ones if we restrict ourselves to faithful representations.]

The first fundamental result in the theory of representations is *Maschke's theorem*, which asserts that all representations over K are direct sums of irreducible subrepresentations if and only if the characteristic of K does not divide the order of G. (In other words, the condition "$\mathrm{Char}\,K \not|\, |G|$" is equivalent to the condition "$K[G]$ is semi-simple".) When these equivalent properties hold, the representation splits uniquely *up to isomorphism*. The proof of the complete reducibility (for which we may not restrict ourselves to faithful representations) over $\mathbb{C}$ can be done using an averaging argument, which is a model for some of our proofs.

[Consider an arbitrary positive definite Hermitian form h_0 on the vector space of the representation, and let

$$h(x,y) = \tfrac{1}{|G|}\textstyle\sum_{s\in G} h_0(sx,sy)\,.$$

Then h is invariant under G, and the orthogonal for h of a stable subspace is again a stable subspace; such an argument also works over $\mathbb{R}$ and over the skew-field $\mathbb{H}$ of Hamilton quaternions.]

For the applications to lattice theory, we are chiefly interested in rationality questions over $\mathbb{Q}$ or $\mathbb{R}$. Rationality over $\mathbb{R}$, which has some importance in relation to the eutaxy property (Theorem 3.6.6), will be used to compute dimensions of spaces of lattices. As for rationality over $\mathbb{Q}$, it has a fundamental importance in view of the following result, in the statement of which we do not assume that G is finite:

Theorem 11.1.1. *Let E be a Euclidean space, and let G be a subgroup of* $\mathrm{O}(E)$. *There exists a lattice invariant under G if and only if G is* finite *and if the representation it defines is* rational *over* $\mathbb{Q}$.

Proof. Suppose first that there exists a G-invariant lattice Λ. Since G is a subgroup of $\mathrm{Aut}(\Lambda)$, G is finite (Theorem 1.4.2). Then $V = \mathbb{Q}\Lambda$ is a $\mathbb{Q}$-subspace of E, invariant under G, of dimension n, and such that $E = \mathbb{R}V$.

Hence the representation of G afforded by Λ comes from a representation which is rational over $\mathbb{Q}$.

Conversely, assume that G is finite and rational over $\mathbb{Q}$, i.e. that E is of the form $\mathbb{R}V$ for some n-dimensional Q-vector space $V \subset E$ invariant under G. Let $\mathcal{B}$ be a basis for V. The $\mathbb{Z}$-module Λ generated by the finite set of vectors $s\,e$, $s \in G$, $e \in \mathcal{B}$, is a finitely generated sub-$\mathbb{Z}$-module of E, of rank n, hence a lattice in E, which is clearly invariant under G. □

Another type of argument, known as "averaging argument" allows us to deal with any real representation which is rational over $\mathbb{Q}$, and even to preserve integral representations: start with an integral representation, afforded by a rank-n, torsion-free $\mathbb{Z}[G]$-module M (see below); let $E' = \mathbb{R}\bigotimes M$; endow E' first with an arbitrary scalar product $x * y$, then with the invariant Euclidean structure defined by the scalar product

$$(x, y) \mapsto x \cdot y = \frac{1}{|G|} \sum_{s \in G} sx * sy\,.$$

For integral representations, this answers a question of Louis Michel ([Mic]).

We now definitely assume that G is a finite *subgroup of* $\mathrm{O}(E)$.

The notion of an *integral representation* is analogous to that of a rational one, except that we must replace the base field $\mathbb{Q}$ by its subring $\mathbb{Z}$. We can consider either a homomorphism of G in the automorphism group of a (finitely generated, torsion-free) $\mathbb{Z}$-module, that is, once a basis for M has been chosen, a homomorphism of G into $\mathrm{GL}_n(\mathbb{Z})$ (n being the rank of M), or a finitely generated, torsion-free $\mathbb{Z}[G]$-module. We shall consider the latter point of view, for which usual equivalence of integral representations corresponds to isomorphism between the corresponding $\mathbb{Z}[G]$-modules.

The classification results concerning integral representations are very sparse. We have a finiteness theorem for any fixed $\mathbb{Z}$-rank; in particular, there are only finitely many isomorphism classes of $\mathbb{Z}[G]$-modules M such that $\mathbb{Q}\bigotimes M$ defines a given representation of G over $\mathbb{Q}$. In return, there exist groups for which the ranks of the irreducible integral representations are not bounded, a phenomenon which never occurs for maximal orders or for representations over a field, where the degrees of the irreducible characters are bounded (in zero characteristic, we have $\sum_{\chi \text{ irred.}} \chi(1)^2 = |G|$).

11.2 G-Lattices

In this section, we still consider an n-dimensional Euclidean space E and a finite subgroup G of the orthogonal group $\mathrm{O}(E)$.

Definition 11.2.1. *A* G-lattice *is a lattice in E which is stable under G. We denote by $\mathcal{L}_G$ the set of G-lattices, endowed with the topology induced by that of the space $\mathcal{L}$ of all lattices in E (Section 1.1). We denote by $\mathcal{G}_G$ (or simply by $\mathcal{G}$) the centralizer of G in* $\mathrm{GL}(E)$*:*

$$\mathcal{G} = \{u \in \mathrm{GL}(E) \mid \forall s \in G,\, us = su\}\,,$$

and by $\mathrm{End}^s_G(E)$ *the set of all symmetric endomorphisms of* E *which commute with* G*. A* G-isometry *(resp. a* G-similarity*) is an isometry (resp. a similarity) which commutes with* G*; these groups are denoted by* $\mathrm{O}_G(E)$ *and* $\mathrm{Sim}_G(E)$ *respectively. We often drop the reference to* E *in this notation, and we also write* $\mathcal{G}_\sigma$, $\mathrm{End}^s_\sigma(E)$*, etc. when* G *is a cyclic group generated by an element* $\sigma \in G$*.*

It should be noticed that the notion of a G-lattice depends on the orthogonal representation defined by G, i.e. on the conjugacy class of G in $\mathrm{O}(E)$, not only on the isomorphism class of the group G.

Let Λ be a lattice, and let $\mathcal{B}$ be a basis for Λ, with which we associate its Gram matrix $A = \mathrm{Gram}(\mathcal{B})$ and the quadratic form $Q(X) = {}^t\!XAX$ on $\mathbb{R}^n$. For all $u \in \mathrm{GL}(E)$, let $P_u = \mathrm{Mat}(u, \mathcal{B})$. Then u belongs to $\mathrm{Aut}(\Lambda)$ if and only if ${}^tP_uAP_u = A$. Using the map $s \mapsto P_s$, we transform the notion of a G-lattice into that of a G-quadratic form:

Definition 11.2.1′. Let H be a subgroup of $\mathrm{GL}_n(\mathbb{Z})$. We say that the symmetric matrix A is H-invariant, or that the form $Q = {}^t\!XAX$ is an H-quadratic form, if we have ${}^tPAP = A$ for all matrices $P \in H$.

Let us return to lattices. The group $\mathcal{G}_G$ is the group of invertible elements of the centralizer C_G (subalgebra of $\mathrm{End}(E)$) of the real representation afforded by the action of G on E. Let χ be the character of this representation, and for all irreducible character φ contained in χ, let K_φ be the centralizer of the representation defined up to isomorphism by φ. This is a skew-field, isomorphic to $\mathbb{R}$, $\mathbb{C}$, or $\mathbb{H}$ (Appendix 1), and we have an isomorphism

$$\mathcal{G} \simeq \prod_\varphi \mathrm{GL}_{m_\varphi}(K_\varphi)\,.$$

In particular, $\mathcal{G}$ has 2^t connected components, where t denotes the number of characters φ with centralizer $\mathbb{R}$. From the existence of this decomposition, we deduce:

Proposition 11.2.2. *1. The group* $\mathcal{G}$ *is stable under transposition.*

2. Every element u *in* $\mathcal{G}$ *splits uniquely as a product*

$$u = \sigma f \ \textit{with}\ \sigma \in \mathrm{O}(E) \cap \mathcal{G} \ \textit{and}\ f \in \mathrm{End}^s_{\mathcal{G}}(E)\,,$$

and the map $u \mapsto (\sigma, f)$ *induces a surjective homeomorphism*

$$\mathcal{G} \to \mathrm{End}^{s++}(E) \times \mathrm{O}_G(E)\,. \quad \square$$

Remark 11.2.3. *Assertion (1) also results from the following calculation, valid for all* $u \in \mathcal{G}$ *and* $s \in G$*:*

$${}^t\!u\, s = {}^t({}^ts\, u) = {}^t(s^{-1}\, u) = {}^t(u\, s^{-1}) = {}^ts^{-1}\,{}^t\!u = s\, {}^t\!u\,,$$

and assertion (2) then follows from assertion (1) by a calculation which applies under the general hypotheses of Chapter 10.

It results from Proposition 11.2.2 that the connected components of $\mathcal{G}$ are in one-to-one correspondence with that of O_G, which implies:

Corollary 11.2.4. *The space $\mathcal{G}$ is connected modulo G-isometries. In particular, the orbits under $\mathcal{G}$ of a G-lattice are connected modulo G-isometries.* □

The following statement makes more precise the algebraic and topological properties of the operation of G on $\mathcal{L}_G$:

Proposition 11.2.5. *The orbits of $\mathcal{G}$ acting on $\mathcal{L}_G$ are both open and closed, and are sets of G-lattices which are integrally equivalent to one of them.*

Proof. Let Λ and Λ' be two G-lattices. We must first show that if Λ' is sufficiently close to Λ, it is of the form $u(\Lambda)$ for some $u \in \mathcal{G}$. Write $\Lambda' = u(\Lambda)$ for a suitable $u \in \mathrm{GL}(E)$. For all $s \in G$, we have $su(\Lambda) = s(\Lambda') = \Lambda' = u(\Lambda)$, whence $usu^{-1} \in \mathrm{GL}(\Lambda)$, so that the commutator $[u,s] = usu^{-1}s^{-1}$ also belongs to $\mathrm{GL}(\Lambda)$. But this commutator tends to 1 as u tends to the identity. Since $\mathrm{GL}(\Lambda)$ is discrete, there exists a neighbourhood $\mathcal{V}_s$ of Id_E on which $[u,s] = \mathrm{Id}$, i.e. $us = su$. Let $\mathcal{V} = \cap_{s\in G}\mathcal{V}_s$, and let $\mathcal{U} = \{u(\Lambda) \mid u \in \mathcal{V}\}$. This is a neighbourhood of Λ in $\mathcal{L}_G$ on which all lattices belong to the orbit of Λ. This proves that the orbits are open. Since they are open, they are closed.

Clearly, for all $\Lambda \in \mathcal{L}_G$ and $u \in \mathcal{G}$, u is an isomorphism of $\mathbb{Z}[G]$-modules of Λ onto $u(\Lambda)$. Conversely, let Λ and Λ' be two G-lattices which define equivalent integral representations, and let φ be an isomorphism of $\mathbb{Z}[G]$-modules of Λ onto Λ'. This means that φ is an isomorphism of $\mathbb{Z}$-modules, and that we have $\varphi \circ s = s \circ \varphi$ for all $s \in G$. The linear continuation φ' of φ to $V = \mathbb{Q}\Lambda$ is an isomorphism of $\mathbb{Q}$-vector spaces of V onto $V' = \mathbb{Q}\Lambda'$, which commutes with s. Since V is dense in E, there exists a unique continuous continuation ψ of φ' to an element of $\mathrm{GL}(E)$, and ψ still satisfies for all $s \in G$ the commutation relation $\psi \circ s = s \circ \psi$. Hence ψ belongs to $\mathcal{G}$. □

It is evident that a G-lattice is in particular an H-lattice for every subgroup H of G, and that $\mathcal{L}_H = \mathcal{L}$ when $H = \{\mathrm{Id}\}$ or $H = \{\pm\mathrm{Id}\}$. In the other direction, it may happen that G-lattices all are indeed G'-lattices for some subgroup G' of $\mathrm{O}(E)$ which contains strictly G. This phenomenon leads us to the following definition:

Definition 11.2.6. *The* Bravais group *of a subgroup G of $\mathrm{O}(E)$ is the intersection of the automorphism groups of all G-lattices. In other words, the Bravais group of G is*

$$G' = \bigcap_{\Lambda\in\mathcal{L}_G} \mathrm{Aut}(\Lambda)\,.$$

The Bravais group of a group G always contains the group $G \cup -G$ generated by G and $\{\pm\mathrm{Id}\}$. Here is a less trivial example, to which we shall return in Section 11.4.

We consider a cyclic subgroup G of order $q \geq 3$ of $\mathrm{O}(E)$, such that the representation afforded by E is rational and irreducible over $\mathbb{Q}$. Let σ be a generator of G and let e be a nonzero vector in E. The vectors $\sigma^i e$, $0 \leq i < \varphi(q)$ (φ stands for the Euler function) constitute a basis for E. Let $\tau \in \mathrm{GL}(E)$ be such that $\tau\sigma^i e = \sigma^{-i} e$ for $0 \leq i < \varphi(q)$. Since $x \cdot \sigma^i x = x \cdot \sigma^{-i} x$ for all $x \in E$, τ is an element of $\mathrm{O}(E)$. In view of the relations $\tau^2 = \mathrm{Id}$ and $\tau\sigma\tau^{-1} = \sigma^{-1}$, G and τ generate a dihedral group G', of order $2q$. Since every $x \in E$ is of the form $\alpha\, e$ for a suitable $\alpha \in \mathbb{Z}[G]$, it is clear that τ stabilizes all lattices invariant under σ, hence that the Bravais group of G contains G'. (This is indeed equal to G' or to $G' \cup -G'$.)

Let Λ be a G-lattice and let $\mathcal{B}$ be a basis for Λ. Using the action of G on the vectors in $\mathcal{B}$, we may identify G with a subgroup of $\mathrm{GL}_n(\mathbb{Z})$. We attach in this way a *G-quadratic form* to any G-lattice. This shows that the theory of G-lattices is equivalent to that of the positive definite quadratic forms whose automorphism group contains a given finite subgroup of $\mathrm{GL}_n(\mathbb{Z})$, well defined by G up to conjugacy.

11.3 G-extreme Lattices

We keep the notation of the preceding section. We want to study the local maxima of the Hermite invariant on the family $\mathcal{L}_G$ of G-lattices. The tangent space at the origin of the group $\mathcal{G}$ is clearly the centralizer $\mathrm{End}_G(E)$ of G in $\mathrm{End}(E)$, and its symmetrized space is the space $\mathrm{End}_G^s(E)$ of Definition 11.2.1. In view of the definitions of Chapter 10, the notions of $\mathcal{T}$-perfection, $\mathcal{T}$-eutaxy, etc. that we need must be relative to $\mathcal{T} = \mathrm{End}_G^s(E)$.

Definition 11.3.1. *1. We say that a G-lattice Λ is* (strictly) G-extreme *if the Hermite invariant attains a (strict) local maximum at Λ on the set $\mathcal{L}_G$ of G-lattices.*

2. We say that a G-lattice is G-perfect *(resp.* G-eutactic, *resp.* weakly G-eutactic*) if it is $\mathcal{T}$-perfect (resp. $\mathcal{T}$-eutactic, resp. weakly $\mathcal{T}$-eutactic) for $\mathcal{T} = \mathrm{End}_G^s(E)$.*

Recall that the properties above, relative to a given subspace $\mathcal{T}$ of $\mathrm{End}^s(E)$, can be defined in two equivalent ways, by considering either the restrictions to $\mathcal{T}$ of the linear forms φ_x, or dually, the *orthogonal projections ω_x onto $\mathcal{T}$* (for the Voronoi scalar product in $\mathrm{End}^s(E)$) of the orthogonal projections p_x (in E) onto the lines $\mathbb{R}x$; see Proposition 10.2.1.

Proposition 11.3.2. *1. For every nonzero $x \in E$,*

$$\omega_x = \frac{1}{|G|} \sum_{s \in G} s p_x s^{-1} = \frac{1}{|G|} \sum_{s \in G} p_{sx}\,.$$

2. The projections ω_x only depend on the orbit of x.

3. *A G-lattice Λ is G-perfect if and only if the* ω_x, $x \in S(\Lambda)$ *span* $\mathrm{End}^s_G(E)$.

Proof. Assertions (2) and (3) are easy consequences of assertion (1), and the equality of the two sums in assertion (1) results from the formula $\sigma p_x \sigma^{-1} = p_{\sigma x}$, which applies to every $\sigma \in \mathrm{O}(E)$.

For all $x \in E$, $s \in G$, and $u \in \mathrm{End}^s_G$, we have

$$\langle p_x, u\rangle = \mathrm{Tr}(px \circ u) = \mathrm{Tr}(sp_x s^{-1}\, sus^{-1}) = \mathrm{Tr}(p_{sx} u) = \langle p_{sx}, u\rangle\,.$$

This proves that the projection of p_x onto End^s_G is equal to the average value of the projections on the orbit of x. Since this average value belongs to End^s_G, this is indeed the projection of p_x onto End^s_G. □

For any subgroup H of G, we have the inclusion $\mathrm{End}^s_G(E) \subset \mathrm{End}^s_H(E)$. This proves that H-perfection implies G-perfection and that H-eutaxy implies G-eutaxy, etc. This remark applies in particular to the cases where $H = \{\mathrm{Id}\}$ or $H = \{\pm\,\mathrm{Id}\}$, where these notions are the usual ones. The converse holds true for the eutaxy property (but not for the perfection property, see Example 11.3.9 below), as shown by Proposition 3.6.5:

Proposition 11.3.3. *A lattice is (weakly) G-eutactic if and only if it is (weakly) eutactic.* □

The following proposition plays an important rôle in classifying extreme G-lattices. We must however point out that its third assertion is only a lemma for proving Theorem 11.3.5 below, which will imply that we can use indifferently either of the assertions (1) and (2) of the proposition.

Proposition 11.3.4. *Let Λ be a G-lattice which satisfies one of the following three properties:*

1. *Λ is G-eutactic.*
2. *Λ is G-perfect.*
3. *Λ is G-extreme.*

Then the minimal vectors in Λ span E.

Proof. Taking into account Proposition 11.3.3, we see that assertion (1) is a consequence of Theorem 3.6.10 (2).

To prove (2), we consider the span F of $S(\Lambda)$ in E; it is stable under G, and so is its orthogonal $F^\perp$. Since the orthogonal projection $p_{F^\perp}$ onto $F^\perp$ is an element of End^s_G and Λ is Λ is G-perfect, there exist coefficients ρ_x, $x \in S(\Lambda)$ such that

$$p_{F^\perp} = \sum_{x\in S(\Lambda)} \rho_x\, p_x\,.$$

Applying this identity to an element y of $F^\perp$, we obtain the new equality

$$y = p_{F^\perp}(y) = \sum_{x \in S(\Lambda)} \rho_x \frac{x \cdot y}{x \cdot x} x \,,$$

which shows that y belongs to F. Consequently, we have $F^\perp \subset F \cap F^\perp = \{0\}$, hence $F = E$.

To prove (3), we still consider the space F above. For $\lambda > 0$, let u_λ be the endomorphism which is the identity on F and the homothetic transformation of ratio λ on $F^\perp$. This is an element of $\mathcal{G}$ which preserves $S(\Lambda)$. We have $\det(u(\Lambda)) = \lambda^{\dim F^\perp}$, and for λ sufficiently close to 1, we have $N(u_\lambda(\Lambda)) = N(\Lambda)$, since the minimal vectors in $u_\lambda(\Lambda)$ come from those of Λ. Choosing moreover $\lambda < 1$, we see that the inequality $\gamma(u_\lambda(\Lambda)) \leq \gamma(\Lambda)$ does not hold, except if $\dim F^\perp = 0$, i.e. if $F = E$. □

We are now in a position to prove the central result of this section:

Theorem 11.3.5. *For a G-lattice Λ, the following conditions are equivalent:*

1. *Λ is G-extreme.*
2. *Λ is strictly G-extreme.*
3. *Λ is G-perfect and G-eutactic.*

Proof. (2) $\Longrightarrow$ (1) is trivial, and (2)$\Longleftrightarrow$(3) is a consequence of Theorem 10.3.6, since $\mathcal{G}$ is stable under transposition (Proposition 11.2.2) and homothetic transformations. Finally, (1) $\Longrightarrow$ (2) results from the fact that minimal vectors in a G-extreme lattice span E by Proposition 11.3.4. □

The search for perfect lattices by the methods of Chapter 6 makes use in a crucial way of an evaluation of the number of minimal vectors. To adapt these methods to the case of general G-lattices, we must calculate lower bounds for the number of G-orbits of minimal vectors in G-perfect lattices. To this end, we must calculate the dimension of $\mathrm{End}_G^s(E)$, in order to be able to use the following obvious proposition, which reduces to the lower bound $s \geq \frac{n(n+1)}{2}$ when $G = \{1\}$:

Proposition 11.3.6. *The number of orbits of pairs $\pm x$ of minimal vectors in a G-perfect lattice is at least* $\dim \mathrm{End}_G^s(E)$. □

Consider a decomposition of E as an orthogonal sum of stable subspaces:

$$E = E_1 \perp E_2 \perp \cdots \perp E_r \,.$$

An endomorphism u of E can then be written in a unique way as a sum

$$u = \sum_{1 \leq j,k \leq r} u_{j,k} \quad \text{with} \quad u_{j,k} \in \mathrm{Hom}(E_j, E_k) \,,$$

in which we have identified $u_{j,k} \in \mathrm{Hom}(E_j, E_k)$ with its continuation to the whole space $\mathrm{End}(E)$, giving it the value 0 on the components E_ℓ, $\ell \neq j$;

explicitly, denoting by p_i the orthogonal projection $E \to E_i$, we have $u_{j,k} = p_k \circ u \circ p_j$.

Lemma 11.3.7. *1.* $u \in \mathrm{End}^s(E) \Longleftrightarrow \forall\, j, \forall\, k, {}^t u_{j,k} = u_{k,j}$ *; in particular,* $\forall\, j, u_{j,j} \in \mathrm{End}^s(E_j)$.

2. u commutes with G if and only if the endomorphisms $u_{j,k}$ commute with G.

3. For all i and all $x \in E_i$, and for all $u \in \mathrm{End}^s(E)$, $\varphi_x(u) = \varphi_x(u_{i,i})$.

Proof. The first assertion is evident, and the second one is an immediate consequence of the uniqueness of the decomposition $u = \sum u_{j,k}$. To prove the third one, we observe that in the decomposition $\varphi_x(u) = \sum_{j,k} u_{j,k}(x) \cdot x$, at most one term among the n^2 terms of the sum may be nonzero, because $u_{j,k}(x)$ is zero for $j \neq i$ and E_k is orthogonal to E_i for $k \neq i$. □

For any absolutely irreducible character φ of G, denote by E_φ the *isotypic component* of E corresponding to the $\mathbb{R}$-irreducible character φ, $\varphi + \overline{\varphi}$, or 2φ, according to whether φ is real, complex, or quaternionic (see [Se'3], Propositions 38 and 39). (A representation is *isotypic* if it is a sum of isomorphic irreducible representations.) The canonical decomposition of the representation of G defined by E is then $E = \oplus_{(\varphi,\overline{\varphi})} E_\varphi$, The group G acts on the spaces $E\varphi$, defining in this way by restriction new representations of G, and thus faithful representations of quotients G_φ of G, that we identify with subgroups of $\mathrm{O}(E_\varphi)$.

Theorem 11.3.8. *Let χ be the character of G acting on E. For every $\mathbb{R}$-irreducible character φ of G, let a_φ be the number of times that χ contains φ, and led $K\varphi$ be the centralizer of the absolutely irreducible character (defined up to complex conjugacy) which is contained in φ.*

1. *E is the orthogonal sum of its subspaces E_φ.*
2. *We have* $\dim_{\mathbb{R}} \mathrm{End}^s_G(E) = \sum_{\varphi \text{ irred}} \dim_{\mathbb{R}} \mathrm{End}^s_{G_\varphi}(E_\varphi)$.
3. *We have*

$$\dim_{\mathbb{R}} \mathrm{End}^s_{G_\varphi}(E_\varphi) = \begin{cases} \dfrac{a_\varphi(a_\varphi+1)}{2} & \text{if } K_\varphi \simeq \mathbb{R}\ ; \\ a_\varphi^2 & \text{if } K_\varphi \simeq \mathbb{C}\ ; \\ a_\varphi(2a_\varphi - 1) & \text{if } K_\varphi \simeq \mathbb{H}\ . \end{cases}$$

Proof. (1) Let φ be an irreducible character of G. Since G acts orthogonally and E_φ is stable, the orthogonal space F to E_φ is stable. Since E_φ contains all the subspaces of E with character φ, the character of F does not contain φ. Hence F is the direct sum of the E_ψ with ψ irreducible and distinct from φ. This proves (1), using induction on $\dim E$.

(2) Let φ and ψ be two irreducible characters without common irreducible components. The space $\mathrm{Hom}_G(E_\varphi, E_\psi)$ reduces then to 0, and Lemma 11.3.7

shows that the centralizer of G in $\mathrm{End}(E)$ reduces to its diagonal terms. Consequently, we can identify $\mathrm{End}^s_G(E)$ with the direct product of the algebras $\mathrm{End}^s_{G_\varphi}(E_\varphi)$.

(3) Set $b_\varphi = \dim_{\mathbb{R}} K_\varphi$, and let $z \mapsto \overline{z}$ be the conjugation in K_φ (the identity if $K_\varphi = \mathbb{R}$). The centralizer of the representation can be identified with a matrix algebra $\mathcal{M}_m(K_\varphi)$, whose symmetric part is then the set of self-adjoint matrices, i.e. matrices M such that ${}^tM = \overline{M}$. These matrices are characterized by the conditions $m_{i,i} \in \mathbb{R}$ and $\overline{m_{i,j}} = m_{j,i}$ for $j < i$. Hence the dimension we want to calculate is equal to $m + b_\varphi \frac{m(m-1)}{2}$, i.e. to $\frac{m(m+1)}{2}$, m^2, or $m(2m-1)$, according to whether the representation is real, complex, or quaternionic. To complete the proof, we just need to show that $m = a_\varphi$. This is obvious when $a_\varphi = 1$; we can then check this equality in the general case by extending the base field to $\mathbb{C}$ and noticing that the centralizer of $m\varphi$ is the tensor product by $\mathcal{M}_m(\mathbb{C})$ of that of φ. □

From the point of view of extremality properties, the case of $\mathbb{R}$-irreducible representations is not an interesting one: the algebra End^s_G is 1-dimensional, and $\mathcal{G}$ is thus a group of similarities. Consequently, modulo G-similarities, G-lattices are isolated, hence G-extreme.

Example 11.3.9. *If E has dimension 2 and if G contains a rotation of order 4 (resp. 6), the G-lattices are the square (resp. hexagonal) lattices. They are of course eutactic (by Theorem 3.6.6), but only the hexagonal ones are perfect.*

We have seen in Chapter 1 (Theorem 1.4.5 of Eichler and Kneser) that a lattice possesses a unique decomposition as an orthogonal (direct) sum of irreducible (nonzero) lattices, its *irreducible components*. The group G permutes these components, and orthogonal sums on the various orbits of G constitute the *G-irreducible components* of the lattice.

We know by Theorem 3.5.2 that a reducible lattice is not perfect. For G-perfection, the situation is somewhat different:

Proposition 11.3.10. *Let Λ be a G-lattice, with G-irreducible components $\Lambda_1, \dots, \Lambda_r$. Then Λ is G-perfect (resp. G-extreme) if and only if the following three conditions are satisfied:*

1. *The lattices Λ_i have the same norm.*
2. *The lattices Λ_i are G-perfect (resp. G-extreme).*
3. *The restrictions to the subspaces E_i of the representation of G defined by E have no common irreducible components.*

Proof. Condition (1) means that the minimal vectors in Λ never lie in the orthogonal complement to one of the spaces Λ_i. It is thus a necessary condition, since the set of minimal vectors of a G-perfect lattice generates E.

From now on, we assume that condition (1) holds. Then Λ is eutactic if and only if all Λ_i are (Proposition 3.6.13). Hence it suffices to consider the perfection assertions. Moreover, for all i, the projection p_i onto E_i is an element of End^s_G, and so is $p_i \circ u$ for all $u \in \mathrm{End}^s_G$. If Λ is G-perfect, $p_i \circ u$ can be written as a linear combination of the p_x, $x \in S(\Lambda)$. Since $p_i \circ p_i = p_i$ and $p_i \circ p_x$ is the null map for $x \in S(\Lambda_j)$, $j \neq i$, we see that $p_i \circ u$ is a linear combination of the p_x, $x \in S(\Lambda_i)$, which precisely means that all elements in $\mathrm{End}^s_G(E_i)$ are linear combination of projections p_x, $x \in S(\Lambda_i)$. This proves that condition (2) is also a necessary one.

From now on, we assume that both conditions (1) and (2) hold. Then the set of projections p_x, $x \in S(\Lambda)$ is the union of the sets of projections onto the minimal vectors in the $S(\Lambda_i)$. These projections span the subspace $\bigoplus \mathrm{End}^s_G(\mathbb{R}\Lambda_i)$ of $\mathrm{End}^s_G(E)$. Since these two spaces are equal if and only if condition (3) is satisfied, we are done. □

11.4 Cyclotomic Lattices

We keep the notation E, n, G of the previous sections, but we assume, *unless otherwise stated*, that G is a *cyclic group of order* $q \geq 3$. We denote by σ a generator of G, by ζ (or sometimes ζ_q) a root of unity of order q in $\mathbb{C}$, by $K_q \subset \mathbb{C}$ the field $\mathbb{Q}(\zeta)$ and K'_q its maximal real subfield, and by ψ an injective homomorphism of G into $\mathbb{C}^*$. This is a faithful irreducible complex character of G. The other characters of this type are of the form $\psi \circ u$ with $u \in \mathrm{Aut}(G) \simeq (\mathbb{Z}/q\mathbb{Z})^\times$ or, which amounts to the same thing, $s \circ \psi$ with $s \in \mathrm{Gal}(K_q/\mathbb{Q})$. Indeed, the field K_q is a $\mathbb{Q}$-vector space of dimension $\varphi(q)$ (Gauss's theorem; φ is the *Euler function*, i.e. the cardinality of $(\mathbb{Z}/q\mathbb{Z})^\times$), which defines the unique (up to isomorphism) $\mathbb{Q}$-irreducible rational representation of G; its character ψ' is the sum of the conjugates of ψ.

As usual in the theory of cyclotomic fields, we assume that q is either odd, or divisible by 4: if $q \equiv 2 \mod 4$, replacing ζ by $-\zeta$ transforms q into $\frac{q}{2}$.

We moreover assume that *unless otherwise stated*, the character defined by E is a multiple $\chi = m\psi'$, $m > 0$ of the representation above. In this case, G-lattices, which we call *cyclotomic lattices*, are endowed with a structure of module over $\mathbb{Z}[\zeta]$, the ring of integers in K_q, via the formula $(\sum a_i \zeta^i)x = \sum a_i \sigma^i x$. This justifies the title of this section, and allows us to denote indifferently by $x \mapsto \sigma x$ or by $x \mapsto \zeta x$ the action of σ on E. Two integral representations are equivalent if and only if the corresponding $\mathbb{Z}[\zeta]$-modules are isomorphic. Since $\mathbb{Z}[\zeta]$ is Dedekind domain, integral representations are classified by the Steinitz class of the corresponding modules. [A torsion-free module M of rank $n > 0$ over a Dedekind domain R is well defined up to isomorphism by the class of the ideal $\mathfrak{a}$ such that $M \simeq R^{n-1} \oplus \mathfrak{a}$, named the *Steinitz class of* M.]

Note that $\varphi(q)$ (whence also n) is even.

Theorem 11.4.1. *Let Λ be a G-lattice in E and let V be the K_q-vector space $\mathbb{Q}\Lambda \simeq \mathbb{Q} \bigotimes_{\mathbb{Z}} \Lambda \simeq K_q \bigotimes_{\mathbb{Z}[\zeta]} \Lambda$; thus $E = \mathbb{R}V \simeq \mathbb{R} \bigotimes_{\mathbb{Q}} V$.*

1. *There exist a basis $(e_1, \dots, e_m)$ for V over K_q, totally positive elements $\alpha_1, \dots, \alpha_m$ in K'_q, and fractional ideals $\mathfrak{A}_1, \dots, \mathfrak{A}_m$ of K_q such that Λ is isometric to the orthogonal sum of the ideals $\mathfrak{A}_k$, each of which being endowed the symmetric $\mathbb{Q}$-bilinear form $(x, y) \mapsto \mathrm{Tr}_{K_q/\mathbb{Q}}(\alpha_k x\overline{y})$.*
2. *With this notation, we have*

$$\det(\Lambda) = d_{K_q} \prod_{k=1}^{m} \mathrm{N}_{K_q/\mathbb{Q}}(\mathfrak{A}_k)^2 \prod_{k=1}^{m} \mathrm{N}_{K'_q/\mathbb{Q}}(\alpha_k)^2 \,.$$

3. *The determinant of Λ is of the form df^2, where $d = q$ or $d = 1$ according to whether q is or is not a prime power, and f is a rational number, which is the product of the norm of an ideal of K_q and of the norm of a* principal *ideal of K'_q; in particular, f is a product of rational numbers of the form $p^{\pm a}$, with $a > 0$ and $p^a \equiv \pm 1 \mod q$.*

Proof. Applying Steinitz's theorem (see above) to the $\mathbb{Z}[\zeta]$-module Λ shows that Λ is the direct sum of the submodules $\mathfrak{A}_k e_k$. Thanks to this remark, we may assume that $m = 1$ for the rest of the proof. For all nonzero $e \in V$, we then have $V = K_q\, e$ and $\Lambda = \mathfrak{A}\, e$, where $\mathfrak{A} = \{\lambda \in K_q \mid \lambda e \in \Lambda\}$.

We are now going to prove that there exists a unique element $\alpha \in K'_q$ which satisfies the identity $\mathrm{Tr}_{K_q/\mathbb{Q}}(\alpha\lambda\overline{\mu}) = \lambda e \cdot \mu e$. Such an α is then necessarily totally positive, since the form $\mathrm{Tr}(\alpha\lambda\overline{\mu})$ is positive definite.

Let $\omega_i = \zeta^i + \overline{\zeta}^i$. The identity $\mathrm{Tr}(\alpha\lambda\overline{\mu}) = \lambda e \cdot \mu e$ needs a proof only for pairs $(\lambda, \mu) = (\zeta^i, \zeta^j)$, and in view of the invariance of both sides under conjugation, is equivalent to the following system of $\frac{n}{2}$ equations:

$$\mathrm{Tr}_{K_q/\mathbb{Q}}(\alpha\, \omega_i) = e \cdot \omega_i e \quad \text{for} \quad 0 \le i \le \frac{n}{2} - 1 \,.$$

Expressing α as a $\mathbb{Q}$-linear combination of the ω_i, we obtain the linear system

$$\sum_{0 \le j < n/2} (\omega_i \omega_j)\, x_j = e \cdot \omega_i e \,.$$

whose determinant, which is the discriminant of the basis (ω_i) for K'_q over $\mathbb{Q}$, is nonzero.

This proves (1), and (2) is a consequence of standard algebraic properties of discriminants, whereas (3) is a consequence of (2), taking into account the structure of discriminants and the decomposition laws of primes in cyclotomic fields. [For more details, see Bayer and Martinet's paper [Bay-M], Theorem 2.3] □

We now turn to G-perfect G-lattices.

Proposition 11.4.2. *The number of orbits of pairs $\pm x$ of minimal vectors in a G-perfect G-lattice Λ is at least $\frac{1}{2}\varphi(q)m^2$. When q is odd (resp even), we have $s(\Lambda) \ge \frac{1}{2}q\varphi(q)m^2$ (resp. $s(\Lambda) \ge \frac{1}{4}q\varphi(q)m^2$).*

Proof. The first assertion is a consequence of Proposition 11.3.6 and Theorem 11.3.8: χ contains $\frac{1}{2}\varphi(q)$ pairs of irreducible characters $(\psi^k, \overline{\psi}^k)$ of complex type, each with multiplicity m. To prove the second one, we remark that if $s \in G$ stabilizes a pair $\pm x$, $x \neq 0$, we have $s^2 x = x$, and that the restriction of ψ' to the line $\mathbb{R}x$ is a degree 1 character, hence a conjugate of ψ, whose kernel contains s^2. Hence we have $s^2 = \mathrm{Id}$, i.e. $s = \mathrm{Id}$ or $s = \sigma^{q/2}$, and this last possibility occurs if and only if q is even. □

These lower bounds suffice to prove easily a first classification result:

Theorem 11.4.3. *Suppose that* $\dim E = 4$. *If* G *is cyclic of order* 3 *(resp.* 5*), the* G*-lattices which satisfy the previous lower bound, and in particular those which are* G*-perfect, are similar to* $\mathbb{D}_4$ *(resp. to* $\mathbb{A}_4$*).*

Proof. Let Λ be such a lattice. We have $s \geq 12$ (resp. $s \geq 10$ and $s \equiv 0 \mod 5$), which implies by Theorem 6.2.1 that Λ must be similar to $\mathbb{D}_4$ (resp. to $\mathbb{A}_4$). □

Since $\mathbb{A}_4$ is the unique G-extreme lattice when G is of order 5, we immediately obtain:

Corollary 11.4.4. *On the set of* 4*-dimensional lattices which possess an automorphism of order* 5*, the Hermite invariant is bounded from above by*

$$2 \cdot 5^{1/4} = 1.337\ldots < \gamma_4 = 1.414\ldots. \quad \square$$

Before going further into the classification of G-perfect lattices, we first look at a few properties related to cyclotomy, which will prove useful for explicit calculations.

The minimal polynomial of the endomorphism $x \mapsto \sigma x = \zeta x$ is the cyclotomic polynomial ϕ_q, of degree $\varphi(q)$, whose roots are the conjugates ζ^a ($a \mod q$, $(a,q) = 1$) of ζ; its characteristic polynomial is ϕ_q^m. Hence the G-lattices that we consider in this section are exactly those which have been studied (from a different point of view) by Eva Bayer in [Bay1] and [Bay2].

Let $x, y \in E$. To calculate the pairwise scalar products of the vectors in their orbit, it suffices to calculate the $\varphi(q)$ scalar products $\sigma^k x \cdot y$, $0 \leq k < \varphi(q)$: this is achieved by making use of the formula

$$\sigma^i x \cdot \sigma^j y = \sigma^{i-j} x \cdot y\,,$$

first replacing in it $i - j$ by its representative $k \mod q \in [0, q-1]$, and then σ^k for $k \geq \varphi(q)$ by the remainder of the Euclidean division of X^k by $\phi_q(X)$.

When x and y belong to the same orbit, the calculation above can be pushed one step further: we have

$$\sigma^k x \cdot x = x \cdot \sigma^{-k} x = \sigma^{-k} x \cdot x = \frac{1}{2}(\sigma^k x + \sigma^{-k} x) \cdot x\,,$$

which allows us to perform the Euclidean division by the minimal polynomial ϕ_q^* of $\zeta+\zeta^{-1}$, of degree $\frac{\varphi(q)}{2}$.

We can express $\sigma^a+\sigma^{-a}$ ($a \le \frac{\varphi(q)}{2}$) in the form $P_a(\sigma+\sigma^{-1})$, where P_a is a polynomial in $\mathbb{Z}[X]$ of degree a, even if a is even, otherwise odd. For $a \ge 3$, the determination of P_a can be done inductively, by reduction to that of P_{a-2}, using the identity

$$P_{a+2}(X) = (X^2-2)P_a(X) - P_{a-2}(X)$$

together with the initial values $P_0 = 2$, $P_2 = X^2-2$, and $P_{-1} = P_1 = X$.

Let us set provisionally $\varphi(q) = 2d$, and let $\phi_q(X) = \sum_{i=0}^{2d} a_i X^{2d-i}$. This is a symmetric polynomial, i.e. it satisfies the identity $X^{2d}\phi_q(X^{-1}) = \phi_q(X)$, for its roots occur in pairs $(\lambda, \bar\lambda = \lambda^{-1})$. The equality $\phi_q(\zeta) = 0$ can then be written in the form

$$a_d + \sum_{i=1}^{d} a_{d-i}\,(\zeta^i + \zeta^{-i}) = 0\,,$$

which allows us to calculate fairly easily the polynomial ϕ_q^*.

Let Λ be a lattice in E, of character $m\psi'$, $m > 0$. Since Λ carries a structure of $\mathbb{Z}[\zeta]$-module, there exist m vectors $e_1, e_2, \dots, e_m$ in E and an ideal $\mathfrak{a}$ of $\mathbb{Z}[\zeta]$ such that $\Lambda = \mathbb{Z}[\zeta]e_1 \oplus \mathbb{Z}[\zeta]e_2 \oplus \cdots \oplus \mathbb{Z}[\zeta]e_{m-1} \oplus \mathfrak{a}e_m$. When Λ is free over $\mathbb{Z}[\zeta]$, one can take $\mathfrak{a} = (1)$ and choose $e_1, e_2, \dots, e_m$ in Λ. Then $(\zeta^i e_j)$, $0 \le i \le \varphi(q)-1$, $1 \le j \le m$ is a $\mathbb{Z}$-basis for Λ.

[When Λ is not free over $\mathbb{Z}[\zeta]$, we can construct $\mathbb{Z}$-bases for Λ by performing a reduction to Hermite normal form ("HNF reduction").]

The case where Λ is free of rank 1 (then $m = 1$) is specially interesting. We can find $e \in \Lambda$ such that $\Lambda = \mathbb{Z}[\zeta]\,e$, and the choice of e is unique up to a unit of $\mathbb{Z}[\zeta]$. For small q, it is not difficult to determine the unit group of $\mathbb{Z}[\zeta]$ (see [Was]), and to find all units in $\varepsilon \in \mathbb{Z}[\zeta]^\times$ such that $\varepsilon\, e$ is a minimal vector is often possible by constructing *a priori* a not too long list of units which must contain all those we need. The classification results for G-perfect lattices to be given below have been obtained using this technique. We shall see in Chapter 13 how to obtain such results by making use of an extension of the Voronoi algorithm.

Lemma 11.4.5. *Let $G \subset \mathrm{O}(E)$ be a cyclic group of order q such that the representation of G afforded by E is a multiple of a faithful $\mathbb{Q}$-irreducible representation, let Λ be a G-lattice, let $e_1, e_2, \dots, e_m$ be minimal vectors in Λ which are independent over $\mathbb{Q}(\zeta)$, and let $\Lambda' = \bigoplus \mathbb{Z}[G]\,e_i$. Then Λ' is a sublattice of Λ for which the index $[\Lambda : \Lambda']$ is bounded above by $\gamma_n^{n/2}$ and is a product of prime powers p^f with either $p \mid q$, or $p^f \equiv 1 \mod q$.*

Proof. For all i, the vectors $\zeta^j e_i$ generate a $\mathbb{Z}$-module of rank $\varphi(q)$. This implies that Λ' has rank $m\varphi(q) = n$ over $\mathbb{Z}$, hence that this is a full sublattice of Λ, and we know by Corollary 2.6.10 that its index is bounded above by $\gamma_n^{n/2}$.

Now the index of Λ' in Λ can be expressed by a formula of the form

$$[\Lambda : \Lambda'] = \mathrm{N}_{\mathbb{Q}(\zeta)/\mathbb{Q}}(\chi_{\mathbb{Z}[\zeta]}(\Lambda, \Lambda'))\,,$$

where χ is the relative invariant defined in [Se1], Section 3.1 (a generalization to general Dedekind domains of the invariant which occurs in Proposition 1.1.5). Let $\mathfrak{p}$ a prime divisor of $\chi_{\mathbb{Z}[\zeta]}(\Lambda, \Lambda')$ in $\mathbb{Z}[\zeta]$, lying above some prime $p \in \mathbb{Z}$, with residue degree f. We have $\mathrm{N}_{\mathbb{Q}(\zeta)/\mathbb{Q}}(\mathfrak{p}) = p^f$. If $\mathfrak{p}$ is ramified in $\mathbb{Q}(\zeta)$, then $p \mid q$. Otherwise, f is the smallest integer such that $p^f \equiv 1 \mod q$. □

In Table 11.4.5′ below, we list a few pairs (n, q) among which the values of the index $[\Lambda : \Lambda']$ are to be found; we have taken into account the exact calculation of the residue degree in the ramified cases, and the fact that for $n \le 8$ even, the equality $[\Lambda : \Lambda'] = \gamma_n^{n/2}$ holds only if $n = 4$, $\Lambda = \mathbb{D}_4$, $\Lambda' = \mathbb{Z}^4$, or if $n = 8$, $\Lambda = \mathbb{E}_8$, $\Lambda' = \mathbb{Z}^8$. For $n > 8$, we made use of the upper bounds for γ_n given in [C-S], p. 13 (in terms of the "centred density" $\delta = 2^{-n}\gamma_n^{n/2}$), obtained by the method developed by Rogers in [Rog]. (From dimension 12 onwards, the bounds are too large to be of any practical use. For instance, we have $\lfloor \gamma_{12}^6 \rfloor \le 268$ by [Rog], and certainly $\lceil \gamma_{12}^6 \rceil \ge \lceil \gamma(K_{12})^6 \rceil \ge 152$.)

Table 11.4.5′. A few *a priori* possible values for the index.

n	4	6	6	8	8	8	8	10
q	3, 5, 8	7	9	15	16	20	24	11
index	1	1	1, 3	1	1, 2, 4, 8	1, 5	1, 4, 9	1, 11, 23

The following theorem gives in particular the classification of 6-dimensional, G-perfect lattices for any cyclic, $\mathbb{Q}$-irreducible group G. We shall prove it only for groups of order 7, which is the only case for which the bound for the Hermite invariant improves on the general bound; see [B-M3], Theorem 4.6 for groups of order 9; G-perfect lattices can also be classified by means of the extended Voronoi algorithm (see Chapter 13).

Recall that for $q = 7$ (resp. $q = 9$), the lattices $\mathbb{A}_6$ and $\mathbb{A}_6^{(2)}$ (resp. $\mathbb{E}_6$ and $\mathbb{E}_6^*$) are G-lattices: this results from Proposition 5.3.6 (resp. from Exercise 4.8.1).

Theorem 11.4.6. *Let E be 6-dimensional, and let G be a subgroup of order 7 (resp. 9) of the orthogonal group $\mathrm{O}(E)$ of E. Then the following conditions relative to a G-lattice $\Lambda \subset E$ are equivalent:*

1. *Λ is G-perfect.*
2. *Λ is extreme.*
3. $s(\Lambda) \geq 21$ *(resp.* $s(\Lambda) \geq 27$*).*
4. *Λ is similar to* $\mathbb{A}_6$ *or to* $\mathbb{A}_6^{(2)} \simeq P_6$ *(resp. to* $\mathbb{E}_6$ *or to* $\mathbb{E}_6^*$*).*

Proof. This follows the scheme (4) ⇒ (2) ⇒ (1) ⇒ (3) ⇒ (4).

(4) ⇒ (2). We know that the lattices listed in (4) are extreme (Theorems 4.7.2 and 5.3.8).

(2) ⇒ (1). A G-lattice which is extreme is in particular G-extreme, hence G-perfect.

(1) ⇒ (3). This is Proposition 11.4.2.

(3) ⇒ (4). Recall that we only consider the case where $q = 7$. The strategy of proof consists in showing that up to G-similarity, there exist at most two lattices. (The same kind of proof would work in the case where $q = 9$.) We know that Λ has at least three orbits of minimal vectors. Let x, y, z be representatives for these orbits. By Table 11.4.5′, each of the vectors x, y, z constitutes a $\mathbb{Z}[\zeta]$-basis for Λ. There thus exist units η, $\eta' \in \mathbb{Z}[\zeta]$ such that $y = \eta x$ and $z = \eta' x$. For $i = 1, 2, 3$, let $\omega_i = \zeta^i + \zeta^{-i}$; $\{\omega_1, \omega_2\}$ is a fundamental system of units for $\mathbb{Z}[\zeta]$, and we have the relations $\omega_1\omega_2\omega_3 = 1$ and $\omega_1 + \omega_2 + \omega_3 = -1$. Changing, if necessary, representatives for the orbits of y and z, we may assume that η and η' are of the form $a\omega_1 + b\omega_2 + c\omega_3$ for some coprime integers a, b, c.

For $\eta = a\omega_1 + b\omega_2 + c\omega_3$, the matrix of the components of $y, \zeta y, \zeta^2 y, \zeta^3 y$ on the basis $\zeta^i x$ $(i = 1, -1, 2, -2, 3, -3)$ for Λ is

$$\begin{pmatrix} a & a & b & b & c & c \\ -a & b-a & 0 & c-a & b-a & c-a \\ a-b & c-b & -b & c-b & a-b & 0 \\ b-c & 0 & a-c & b-c & -c & a-c \end{pmatrix}.$$

The bound $\gamma_6^3 \leq 4$ implies that the absolute values of the determinants extracted from this matrix are bounded above by 4. Consideration of those of order 2 shows that $|ab|$, $|b(b-a)|$, $|b(c-a)|$, $(b-c)^2$, $|(a-b)(a-c)|$, and the analogous quantities obtained by circular permutations of a, b, c are bounded above by 4, which shows that a, b, c are equal to $0, \pm 1, \pm 2$, and must moreover have the same sign when the value ± 2 is attained. (Note that the circular permutations of $\omega_1, \omega_2, \omega_3$ are the elements of $\mathrm{Gal}(K'_q/\mathbb{Q})$.) Finally, we find that modulo circular permutations and admissible changes of signs, η and η' are to be chosen among the following six units:

$$\omega_1,\ \omega_1 + \omega_2 = \omega_1^{-1},\ \omega_1 + 2\omega_2 = -\omega_2\omega_1^{-1},\ \omega_1 - \omega_2 - \omega_3 = -\omega_1^2\omega_2^{-1},$$
$$2\omega_1 + \omega_2 + \omega_3 = \omega_1\omega_2^2,\ 2\omega_1 + 2\omega_2 + \omega_3 = -\omega_2^2.$$

Under the six permutations of x, y, z, the pair (η, η') takes the six values

$$(\eta, \eta'),\ (\eta'^{-1}, \eta\eta'^{-1}),\ (\eta'\eta^{-1}, \eta^{-1}),\ (\eta', \eta),\ (\eta\eta'^{-1}, \eta'^{-1}),\ (\eta^{-1}, \eta'\eta^{-1}).$$

Replacing η by its inverse exchanges the families (1) and (2) on the one hand, and (3) and (5) on the other hand, but does not stabilize the set of families (4) and (6), which must consequently be rejected.

We now calculate the scalar products in E as functions of the parameters $\alpha_i = x \cdot \sigma^i x$. Rescaling Λ to norm 1, we may assume that $\alpha_0 = 1$, and as functions of $\alpha = \alpha_1$ and $\beta = \alpha_2$, we find that $\alpha_6 = \alpha_1$, $a_5 = \alpha_2$, and $\alpha_3 = \alpha_4 = -\frac{1}{2} - \alpha - \beta$.

Writing that the vectors $\sigma x + \sigma^{-1} x$ and $x + \sigma^3 x + \sigma^{-3} x$ have norm $N \geq 1$, we obtain the inequalities

$$\beta \geq -\frac{1}{2} \quad \text{and} \quad \alpha + 2\beta \leq 0\,.$$

But we have

$$N(\omega_1 x + 2\omega_2 x) = 2 - 8\alpha - 14\beta = 2 - 8(\alpha + 2\beta) + 2\beta \geq 2 + 2\beta \geq 1\,,$$

with equality only if $\alpha + 2\beta = 0$ and $\beta = -\frac{1}{2}$, which implies for α the impossible value 1. Hence $\omega_1 x + 2\omega_2 x$ cannot be a minimal vector, which excludes families (3) and (5). Taking into account permutations of x, y, z and of $\omega_1, \omega_2, \omega_3$, we see that we can take $\eta = \omega_1$, and that η' must then belong to the set $\{\omega_2,\, \omega_3 = \omega_1^{-1}\omega_2^{-1},\, \omega_1^{-1},\, \omega_2^{-1},\, \omega_1\omega_2\}$, which gives for $\eta'\eta^{-1}$ one of the values $\omega_1^{-1}\omega_2,\, \omega_1^{-2}\omega_2^{-1},\, \omega_1^{-2},\, \omega_1^{-1}\omega_2^{-1}$, or ω_2. We verify that only the first and the last one are admissible, and we are left with at most two G-lattices. [Writing explicitly that the vectors in one of the sets $\{x, \omega_1 x, \omega_2 x\}$ or $\{x, \omega_1 x, \omega_1\omega_2 x\}$ are minimal (we have $\omega_1\omega_2 x = (\omega_2 + \omega_3)x\}$), we determine the possible values for α and β; the result is $(\alpha, \beta) = (-\frac{1}{2}, 0)$ or $(\alpha, \beta) = (-\frac{1}{2}, -\frac{1}{4})$. The first case corresponds to a lattice which is similar to $\mathbb{A}_6$, since it is integral when rescaled to norm 2.] □

Corollary 11.4.7. *For a 6-dimensional G-lattice Λ possessing an automorphism of order 7, we have* $\gamma(\Lambda) \leq \frac{4}{7^{1/2}} < 1.512 < \gamma_6 = \frac{2}{3^{1/6}} = 1.665\ldots$

Proof. Indeed, the maximum of γ on these G-lattices is attained on one of the two G-perfect G-lattices of Theorem 11.4.6, and we have

$$\gamma(\mathbb{A}_6) = \frac{2}{7^{1/6}} = 1.446\ldots < \gamma(\mathbb{A}_6^{(2)}) = \frac{4}{7^{1/2}} = 1.511\ldots \quad \square$$

We end this section with brief description of two samples of classification results based on lower bounds for the number of orbits of minimal vectors; we do not give the proofs, which could be made easier using the methods of Chapter 13. Only the first one is strictly cyclotomic.

The first one concerns cyclotomic lattices associated with groups of orders 15, 16, and 24 in dimension 8. Here, the lower bounds by 4 of the number of orbits given by Theorem 11.3.8 do not suffice, but it can be proved that G-perfect lattices always have at least five orbits of minimal vectors, and that this better lower bound actually suffices.

The lattice $\mathbb{E}_8$ does not possess any automorphism of order 16, but $\mathbb{D}_8$ does (this results from [ATLAS] or from [Bay1]). We deduce from this a non-trivial upper bound for the Hermite invariant of G-lattices with $|G| = 16$, namely

$$\gamma(\Lambda) \leq 2^{11/4} 241^{-1/4} < 1.708 < \gamma_8 = 2\,.$$

The second sample concerns groups G of prime order $n = q$. In this case, there are $2h$ integral rank-1 representations, where h denotes the number of ideal classes in $\mathbb{Z}[\zeta_q]$: half of them come from modules over the maximal order $\mathfrak{M} \simeq \mathbb{Z} \times \mathbb{Z}[\zeta_q]$, and the others are represented by projective $\mathbb{Z}[G]$-modules M, characterized up to isomorphism by the $\mathfrak{M}$-module $\mathfrak{M} \otimes_{\mathbb{Z}[G]} M$; if $h = 1$, M corresponds to the regular representation of G over $\mathbb{Z}$. Lattices of the first kind are similar to orthogonal sums of $\mathbb{Z} \perp L$ where L is cyclotomic.

Let us now consider those of the second kind. It can be proved that for $q = 3$ and $q = 5$, the lower bound by $\frac{q+1}{2}$ of Theorem 11.3.8 for the number of orbits of minimal vectors ensures G-perfection. For $q = 3$, the lattices with two orbits of minimal vectors are similar to $\mathbb{A}_3$ $(s = 2 \times 3)$ or to $\mathbb{A}_3^*$ $(s = 1 \times 3 + 1 \times 1)$. For $q = 5$, we find the three extreme lattices $\mathbb{D}_5$, $\mathbb{A}_5^2$, $\mathbb{A}_5$ with respectively $4, 3, 3$ orbits with five elements, and one "exotic", non-perfect lattice with $s = 11$, a kind of analogue of $\mathbb{A}_3^*$.

11.5 Isodual Lattices, Modular Lattices, and Normal Lattices

In this section and the next one, we still denote by E an n-dimensional Euclidean space, and consider lattices in E which are similar to their duals.

We met previously such examples among perfect lattices, e.g. $\mathbb{Z}$, $\mathbb{A}_2$, $\mathbb{D}_4$, $\mathbb{A}_6^2$ (the only examples in dimensions $n \leq 7$), $\mathbb{E}_8$, $\mathbb{D}_{2m}^+$ $(2m \geq 10)$, $\mathbb{A}_{p-1}^{((p+1)/4)}$ ($p \equiv 3 \mod 4$ a prime), K_{12}, Λ_{16}, Λ_{24}, and some higher–dimensional lattices.

The relation $(\lambda\Lambda)^* = \lambda^{-1}\Lambda^*$ shows that a lattice Λ which is similar to its dual is proportional to a lattice which is isometric to its dual: it suffices to replace Λ by $\lambda\Lambda$ in such a way that $\det(\lambda\Lambda) = 1$, which is achieved by taking $\Lambda = \det(\Lambda)^{1/2n}$. The lattices which are isometric to their duals have been called *isodual lattices* by Conway and Sloane (Appendix 2 to [Bu-Sa]; [C-S9]). We now state more precisely the appropriate definitions:

Definition 11.5.1. *Let $\sigma \in \mathrm{O}(E)$. We say that a lattice Λ is* σ-isodual *if $\sigma(\Lambda) = \Lambda^*$, and that it is* isodual *if it is σ-isodual for some $\sigma \in \mathrm{O}(E)$. The* isoduality group of Λ *is the group*

$$\mathrm{Aut}^{\#}(\Lambda) = \{\sigma \in \mathrm{O}(E) \mid \sigma(\Lambda) = \Lambda \text{ or } \sigma(\Lambda) = \Lambda^*\}\,.$$

The elements of $\mathrm{O}(E)$ which map Λ onto Λ^* are the isodualities of Λ

The group $\mathrm{Aut}^{\#}(\Lambda)$ contains $\mathrm{Aut}(\Lambda)$ to index 1 or 2, and the index is 2 if and only if Λ is isodual, but not unimodular. (Note that except if Λ is unimodular, $\mathrm{Aut}^{\#}(\Lambda)$ does not contain only isodualities.) Consideration of the group $\mathrm{Aut}^{\#}$ suggests the following more general definition:

Definition 11.5.2. *Let $G^{\#}$ be a (finite) subgroup of $\mathrm{O}(E)$ and let G be a subgroup of index 2 in $G^{\#}$. We say that a lattice Λ is* $(G^{\#}, G)$-isodual *(or simply* $G^{\#}$-isodual*) if Λ is a G-lattice, and if the elements of $G^{\#} \setminus G$ map Λ onto Λ^*.*

[Note that it is essential to mention the subgroup G in this definition, and that σ-isodual lattices in the sense of Definition 11.5.1 are exactly those which are isodual for the group generated by σ; of course, if $G^{\#}$ is cyclic, there is no ambiguity on G.]

It should be noted that for every isoduality σ and every odd integer m, the isometries $\pm\sigma^m$ are still isodualities. As a consequence, we see that an isodual lattice is always isodual for some transformation whose order is a power of 2, and that it possesses isodualities of determinant $+1$ whenever the dimension of E is odd.

By abuse of language, and in order to avoid introducing irrationalities in their definition, we shall often represent isodual lattices by lattices which are only similar to their duals: for instance, to make $\mathbb{A}_2$, $\mathbb{D}_4$, and $\mathbb{A}_6^{(2)}$ isodual in the narrow sense, we would need to introduce square roots of 3, 2, and 7 respectively, so we prefer to consider them in their standard form.

The following notion, due to Quebbemann, is an important source for constructing isodual lattices; see also Definition 8.1.4:

Definition 11.5.3. (Compare Definition 8.1.4.) *Let q be a positive integer. We say that a lattice Λ is* q-modular *if it is integral, and if there exists a similarity of ratio $\sqrt{q}$ (we also say of* norm q*, because of the formula $\sigma x \cdot \sigma y = q\,(x \cdot y)$) which maps Λ^* onto Λ. We say that Λ is* modular *if it is q-modular for some q. We then say that q is the* level of Λ.

1-modular lattices are unimodular lattices. All examples of isodual lattices given above are modular (because they are proportional to integral lattices; see Proposition 11.5.5 below). The lattices $\mathbb{D}_4$ and Λ_{16} are 2-modular; the lattices $\mathbb{A}_{p-1}^{((p+1)/4)}$ ($p \equiv 3 \mod 4$ a prime) are p-modular; K_{12} (and $\mathbb{A}_2$) are 3-modular; the lattices $\sqrt{2}\,\mathbb{D}_{4m+2}^{+}$ are 4-modular; an order in an imaginary quadratic field of discriminant $-d$ is d- or $\frac{d}{2}$-modular (see Exercise 11.5.3). Numerous examples of modular lattices can be found in [Bt-Q-S] and in [Sch-SP]. Examples of isodual, sometimes not modular lattices occurred in Chapter 9: the 2-dimensional lattices (Proposition 9.2.1), and the 4-dimensional lattices belonging to the a_9 class (Proposition 9.3.1).

When working on duality, we meet two natural normalizations for a given lattice Λ: one for which Λ and Λ^* have the same norm, and one for which they have the same determinant (necessarily equal to 1). These two normalizations coincide on isodual lattices.

Definition 11.5.4. *We say that a lattice Λ is* normal *if Λ and Λ^* have the same Hermite invariant.* Equivalent definition*: Λ is* normal *if and only if Λ is proportional to a lattice which has the same norm and the same determinant as its dual.*

Among the various notions considered in this section, we have the implications

$$\text{modular} \Longrightarrow \text{similar to its dual} \Longrightarrow \text{normal.}$$

In the following proposition, we state some consequences of the definitions above:

Proposition 11.5.5. *Let Λ be a lattice.*

1. *Λ is normal if and only if* $\det(\Lambda)^2 = \left(\dfrac{N(\Lambda)}{N(\Lambda^*)}\right)^n$.
2. *An integral lattice which is similar to its dual is modular.*
3. *For an odd dimensional q-modular lattice, q is a square.*
4. *The Smith invariant of a* primitive *q-modular lattice is invariant under the symmetry $(a_1, a_2, \ldots, a_{n-1}, a_n) \mapsto (q/a_n, q/a_{n-1}, \ldots, q/a_2, q/a_1)$.*

Proof. (1) Let Λ be a lattice, and let $\lambda > 0$ be such that $\det(\lambda\Lambda) = 1$. Then $\lambda^{2n}\det(\Lambda) = 1$ and $\lambda^2 N(\Lambda) = \lambda^{-2} N(\Lambda^*)$.

(2) If σ is a similarity which maps Λ^* onto Λ, we have $\det(\Lambda) = \det(\sigma)^2\det(\Lambda^*)$. Hence the ratio of similarity λ of σ is the n-th root of the determinant of Λ. But we also have the relation $\lambda^2 N(\Lambda)^* = N(\Lambda)$, which implies that λ^2 is rational. Since λ^n is an integer, λ^2 is itself an integer, and Λ is λ^2-modular.

(3) Since the lattices Λ and $\sqrt{q}\,\Lambda^*$ are isometric, we have $\det(\Lambda)^2 = q^n$.

(4) Recall (Definition 1.9.1) that the Smith invariant of a lattice Λ proportional to an integral lattice is the sequence of the invariant factors (or elementary divisors) of the pair (L^*, L), where L denotes the primitive integral lattice proportional to Λ. We know (Proposition 1.9.4) that the Smith invariant of the dual is obtained by replacing the sequence (a_i) of elementary divisors of the pair (L^*, L) by the sequence (a_1/a_{n-i+1}). Since L and L^* are similar, they share the same Smith invariant by Proposition 1.9.2. □

Remark 11.5.6. *When q is square-free, we recover from assertion (4) of the proposition above that n is even, say, $n = 2m$, and moreover we see that the Smith invariant contains m terms equal to q and m terms equal to 1. When q is not square-free, different possibilities may exist for the Smith invariant. For instance, in the following examples with $n = 10$ and $q = 4$, the Smith invariant of $\sqrt{2}\,\mathbb{D}_{10}^+$ is equal to $(4, 2^8, 1)$, whereas that of Souvignier's lattice Q_{10} is $(4^4, 2^2, 1^4)$.*

Remark 11.5.7. *Applying the normality criterion in the proposition above, we immediately see from Tables 6.5.5 and 6.5.6 that the isodual lattices $\mathbb{Z}$,*

$\mathbb{A}_2$, $\mathbb{D}_4$, and $\mathbb{A}_6^{(2)}$ are indeed the only normal perfect lattices in dimensions $n \le 7$.

11.6 Normal Lattices

In this section, we prove four classification results about normal lattices, which we shall use later to classify some families of isodual lattices. We begin with a result about dimension 3:

Proposition 11.6.1. *Let Λ be a normal 3-dimensional lattice such that both $S(\Lambda)$ and $S(\Lambda^*)$ span E. Then either $s(\Lambda) = s(\Lambda^*) = 3$, and Λ is similar to $\mathbb{Z}^3$, or $s(\Lambda) = s(\Lambda^*) = 4$, and Λ is similar to the lattice ccc with Gram matrix*

$$\begin{pmatrix} 1 & -3+2\sqrt{2} & 1-\sqrt{2} \\ -3+2\sqrt{2} & 1 & 1-\sqrt{2} \\ 1-\sqrt{2} & 1-\sqrt{2} & 1 \end{pmatrix}.$$

These lattices are isodual, and their Hermite invariants are 1 and $\dfrac{1}{2(5\sqrt{2}-7)^{1/3}} = 1.207\dots$ respectively.

[*ccc* (notation of [C-S9]) stands for *central centred-cuboidal* lattice; this lattice is also denoted by *mcc* (for mean-centred cuboidal) in [C-S"], p. xxvi.]

Proof. We have $(s, s^*) = (6,4)$, $(4,4)$ or $(3,3)$ by Theorem 9.2.2.

The first case is that of the lattice $\mathbb{A}_3$, for which we have

$$\det(\Lambda)^2 = 16 \neq \left(\frac{N(\Lambda)}{N(\Lambda^*)}\right)^3 = \left(\frac{8}{3}\right)^3.$$

To deal with the other two cases, we calculate the adjoint matrices of the Gram matrices (corresponding to $N(\Lambda) = 1$) which are displayed after Theorem 9.2.2.

In the second case, we find

$$N(\Lambda) = 1,\ \ N(\Lambda^*) = \frac{1+t}{4t(1-t)},\ \ \det(\Lambda) = 4t(1-t)^2\,,$$

and using assertion (1) of Proposition 11.5.5, we see that t must be a zero of the polynomial $t^4 + 2t^3 + 2t - 1 = (t^2+1)(t^2+2t-1)$, whose unique root in the interval $(\frac{1}{3}, \frac{1}{2})$ is $\sqrt{2}-1$.

Finally, in the third case, we find

$$N(\Lambda) = 1,\ \ N(\Lambda^*) = \frac{1+t}{(1-t)(1+2t)},\ \ \det(\Lambda) = (1+2t)(1-t)^2,$$

and Proposition 11.5.5 now shows that t must be a zero of the polynomial $t^3(t+2)$, whose unique root in the interval $(-\frac{1}{3}, \frac{1}{2})$ is $t = 0$.

The calculation of the Hermite invariants is trivial, and there just remains to prove that the normal lattice Λ with $s = s^* = 4$ is isodual. Now the minimal class (in the sense of duality) of the lattices such that $s = s^* = 4$ is stable under duality, and for the extreme values of the parameter $t \in [\frac{1}{3}, \frac{1}{2}]$, we find the lattices $\mathbb{A}_3$ and $\mathbb{A}_3^*$. This shows that the open interval $(\frac{1}{3}, \frac{1}{2})$ contains at least one value of t for which the corresponding lattice is isodual. Since the *ccc* is the unique normal lattice in the class, it is isodual. □

The second result is related to the cases where equality holds in Mordell's inequality (Section 2.3).

Theorem 11.6.2. *Let Λ be a lattice of dimension $n \geq 3$ having a critical hyperplane section with the same norm. Then Λ is normal if and only if it is critical and if "Mordell's equality" $\gamma_n = \gamma_{n-1}^{(n-1)/(n-2)}$ holds. If $n \leq 8$, the normal lattices having a critical hyperplane section with the same norm are $\mathbb{A}_2$, $\mathbb{D}_4$, and $\mathbb{E}_8$.*

Proof. The second part of the theorem is an immediate consequence of the first one, using the known values of the Hermite constant, and dealing directly with the special case $n = 2$. Let us now prove its first part.

Set $m = N(\Lambda)$, $m^* = N(\Lambda^*)$, $d = \det(\Lambda)$, and let M be a critical hyperplane section of norm m of Λ. We have

$$\det(M) = \frac{N(M)^{n-1}}{\gamma(M)^{n-1}} = \frac{m^{n-1}}{\gamma_{n-1}^{n-1}}.$$

The minimum of Λ^* is attained on the primitive vectors orthogonal to M. Hence

$$m^* = \frac{\det(M)}{\det(\Lambda)} = \frac{1}{d}\,\frac{m^{n-1}}{\gamma_{n-1}^{n-1}}.$$

By Proposition 11.5.5 (1), we also have $m^* = \dfrac{m}{d^{2/n}}$. Writing that the two values found above for m^* are equal, we obtain the equalities

$$\gamma_{n-1}^{n-1} = \frac{m^{n-2}}{d^{(n-2)/n}} = \gamma_n(\Lambda)^{n-2}.$$

The Mordell inequality $\gamma_n \leq \gamma_{n-1}^{(n-1)/(n-2)}$ (Theorem 2.3.1) then implies that Λ is critical, and that this inequality is indeed an equality. □

Our next result concerns uniquely even–dimensional lattices, in connection with half-dimensional sections. As it allows us to detect some of the minimal vectors in the dual lattice, it yields an easy calculation of the minimum of the dual lattice, the difficult step in the calculation of the invariant γ'_n.

Theorem 11.6.3. *Let Λ be a normal lattice of even dimension $n = 2m$, and let M be a critical m-dimensional cross-section of Λ, with the same norm* *as Λ. Then the orthogonal $M^\perp$ of M in Λ^* is critical and has the same norm as Λ^*. Moreover, if $\dim \Lambda = 4$ and if $\mathcal{B}$ is a basis for Λ made of minimal vectors, the first two of which generate M, then the last two vectors in $\mathcal{B}^*$ are minimal and generate $M^\perp$.*

Proof. The calculation of the determinant of a section (Corollary 11.3.5) and the normality criterion in Proposition 11.5.5 (1) imply the two equalities

$$\det(M) = \det(\Lambda)\det(M^\perp) \quad \text{and} \quad \frac{\det(M)}{\det(M^\perp)} = \frac{N(\Lambda)^m}{N(\Lambda^*)^m},$$

from which we deduce the further relations

$$\frac{\gamma(M^\perp)^m}{\gamma(M)^m} = \frac{\det(M)}{\det(M^\perp)} \cdot \frac{N(M^\perp)^m}{N(M)^m} = \frac{N(\Lambda)^m}{N(M)^m} \cdot \frac{N(M^\perp)^m}{N(\Lambda^*)^m} = \frac{N(M^\perp)^m}{N(\Lambda^*)^m} \geq 1\,.$$

Since the maximum γ_m of the m-dimensional Hermite invariant is attained on M, the inequality above is indeed an equality. This implies that $M^\perp$ is critical and has the same norm as Λ, which proves the first assertion of the theorem.

Suppose now that $n < 8$, and consider a basis $\mathcal{B} = (e_1, \dots, e_n)$ of minimal vectors in Λ. (Such a basis certainly exists if $n = 4$.) For all minimal vector $x' \in \Lambda^*$, the components in $\mathcal{B}^*$ of x' are the scalar products $x' \cdot e_i$; they are equal to 0 or ± 1 by Lemma 6.3.3.

We now apply the results above to a vector $x' \in M^\perp$ and to its components in the basis $(e^*_{m+1}, \dots, e^*_n)$ for $M^\perp$, taking $n = 4$. Then $M^\perp$ is similar to A_2, hence possesses three pairs of minimal vectors. But $x = e^*_3 + e^*_4$ and $y = e^*_3 - e^*_4$ cannot be simultaneously minimal, since

$$N(x) + N(y) = 2(N(e^*_3) + (Ne^*_4)) \geq 4N(\Lambda)\,.$$

Hence e^*_3 and e^*_4 (and also one of the two vectors $e^*_3 \pm e^*_4$) are minimal. □

Our last result concerns 4-dimensional normal lattices having at least two minimal hexagonal sections. It relies on the preceding theorem, but its proof needs some calculations which can reasonably be left to a computer. For this reason, we shall not give all the details of the proof. The notation d_6, a_9, and a_{10} are the one we used to describe 4-dimensional minimal classes in Section 9.3. The result will be applied later only to lattices with $s \geq 7$. However, we nevertheless consider the class d_6, which contains an interesting 2-parameter family of isodual lattices.

Theorem 11.6.4. *Normal 4-dimensional lattices having at least two hexagonal sections with the same norm constitute (modulo similarities) the closure of the 2-parameter sub-family of the class d_6, defined by the Gram matrices*

$$\begin{pmatrix} 2 & -1 & x & y \\ -1 & 2 & y & -x-y \\ x & y & 2 & -1 \\ y & -x-y & -1 & 2 \end{pmatrix}$$

with $|x| < 1$, $|y| < 1$, and $|x+y| < 1$, whose boundary is $\overline{a_9} = a_9 \cup a_{10}$.

Proof. Lattices which satisfy the hypotheses of the theorem either belong to one of the 12 classes c_6, d_6, a_7, b_7, c_7, d_7, a_8, b_8, a_9, b_9, a_{10}, a_{10}, which we described in Section 9.3, or their minimal vectors lie in a hyperplane. In the latter case, and also in that of the classes d_7, b_8, b_9, and a_{10}, the lattices we consider all have a critical hyperplane section, and Theorem 11.6.2 shows that they are not normal. We are thus left with the classes c_6, d_6, a_7, b_7, c_7, and a_8, which will be dealt with by the following method: we first write down Gram matrices which represent the classes above; they depend on a number of parameters equal to the perfection corank of the class, indeed $10 - s$ in the cases we consider. Applying Theorem 11.6.3, we obtain some relations between the parameters, which correspond to the fact that normal lattices lie at the intersection of classes with some algebraic varieties. In the cases we consider, these varieties happen to split; thanks to this property, we are able to carry out explicit calculations. For the five classes other than d_6 that we must study, we verify that these conditions force the existence of extra minimal vectors, that is to say, the lattices we want to consider belong to a bigger class (for the $\succ$ ordering). We shall not write explicitly the details, and content ourselves with the observation, that for the classes c_7 and a_8, the result is a consequence of the study to be completed below of the class d_6 (because we have $a_8 \succ c_7 \succ d_6$).

Let us now consider lattices belonging to d_6. Up to similarity, representatives for these lattices are described by the Gram matrices

$$D_6 = \begin{pmatrix} 2 & -1 & x & y \\ -1 & 2 & z & u \\ x & z & 2 & -1 \\ y & u & -1 & 2 \end{pmatrix},$$

where the parameters x, y, z, u are submitted to various *inequalities of reduction* that we need not make explicit, for conditions $|x|, |y|, |z|, |u| \leq 1$ will suffice. Theorem 11.6.3 shows that between the entries $d'_{i,j}$ of the adjoint matrix D'_6 of D_6, the following relations must hold:

$$(1)\ d'_{1,1} - d'_{2,2} = 0,\ (2)\ d'_{3,3} - d'_{4,4} = 0,\ (3)\ 2d'_{1,2} = \pm d'_{2,2},\ (4)\ 2d'_{3,4} = \pm d'_{4,4}$$

(with a suitable choice of signs).

Making explicit (1) and (2), we obtain the two equations

$$(y^2 - z^2) + (x^2 - u^2) + xy - zu = 0 \ \text{ and } \ (z^2 - y^2) + (x^2 - u^2) + xz - yu = 0,$$

whose sum and difference yield the two new equations

$$(x - u)\big(2(x + u) + y + z = 0\big) \ \text{ and } \ (y - z)\big(2(y + z) + x + u\big) = 0.$$

Exchanging e_1 and e_2 induces the double transposition $(x, z)(y, u)$, so that it suffices to consider the two possibilities

$$z = y,\ u = x \quad \text{and} \quad z = y,\ u = -x - y.$$

The second one is the result we want to prove. As for the first one, we verify that (3) becomes either $2x^2 + 5xy + 2y^2 = 0$, or $xy = -2$, according to the chosen sign. The second equality is plainly impossible, whereas the first one implies $x = y = z = u = 0$, which is just a particular member of the family in the statement of the theorem.

Writing that all nonzero vectors with components $0, 1, -1$ on the basis have a norm $N \geq 2$, we prove that the lattices belong to the class d_6 if and only if $|x| < 1$, $|y| < 1$, and $|x + y| < 1$. The circular permutation $(x, y, -x-y)$, which is induced by the circular permutation $(e_1, e_2, -e_1-e_2)$, permutes transitively these three inequalities. The elements lying on the boundary of d_6 are thus equivalent to elements with $|x| = 1$, and even with $x = -1$, as we see by negating e_1. Setting $x = -1$ and $y = t$, we recover the matrices $A_9(t)$ of the class a_9 displayed in Section 9.3. □

11.7 Extreme Symmetric and Symplectic Lattices

We keep the notation of the previous section. We denote by σ an orthogonal transformation, and we consider a σ-isodual lattice Λ, to which we will apply the results of Chapter 10. However, the existence of continuous families of pairwise non-similar lattices on which the Hermite invariant takes a constant value shows that we cannot characterize extreme-isodual lattices in full generality, using only appropriate notions of perfection and eutaxy. For this reason, we shall essentially restrict ourselves to the isodualities which occur in Definition 11.7.5 below.

Definition 11.7.1. *1. We denote by $\mathcal{F}_\sigma$ the set of σ-isodual lattices in E.*
2. We denote by $\mathcal{G}_\sigma$, or simply $\mathcal{G}$, the subgroup of $\mathrm{GL}(E)$ *defined by*

$$\mathcal{G}_\sigma = \{u \in \mathrm{GL}(E) \mid {}^t u \sigma u = \sigma\}.$$

3. We denote $\mathcal{T}_\sigma$, or simply $\mathcal{T}$, the subspace of $\mathrm{End}^s(E)$ *associated with $\mathcal{F}_\sigma$ (or with $\mathcal{G}_\sigma$) by Proposition 10.1.4.*
4. We denote by b_σ the bilinear form $(x, y) \mapsto x \cdot \sigma y$.

Proposition 11.7.2. *Let $\Lambda \in \mathcal{F}_\sigma$.*

1. *The connected component of any Λ in $\mathcal{F}_\sigma$ is contained in the orbit of Λ under the action of $\mathcal{G}_\sigma$.*
2. *The group $\mathcal{G}_\sigma$ is stable under transposition.*
3. *$\mathcal{G}_\sigma$ is the orthogonal group of the bilinear form b_σ.*
4. *The space $\mathcal{T}$ associated with $\mathcal{F}_\sigma$ is*

$$\mathcal{T} = \{v \in \mathrm{End}^s(E) \mid \sigma v = -v\sigma\} \subset \mathrm{Ker}\,\mathrm{Tr}\,.$$

Proof. (1) Let $\Lambda \in \mathcal{F}_\sigma$ and $u \in \mathrm{GL}(E)$. We then have

$$u(\Lambda) \in \mathcal{F}_\sigma \iff (u(\Lambda))^* = \sigma(u(\Lambda)) \iff {}^t u^{-1}(\Lambda^*) = \sigma(u(\Lambda)) \\ \iff {}^t u^{-1}(\sigma(\Lambda)) = \sigma(u(\Lambda)) \iff \sigma^{-1}\, {}^t u \sigma u \in \mathrm{GL}(\Lambda)\,,$$

from which we deduce the relation $\sigma^{-1}\,{}^t u \sigma u = \mathrm{Id}$, valid for any u sufficiently close to the identity transformation.

(2) Since the transformation σ is orthogonal, we have

$$u \in \mathcal{G}_\sigma \iff {}^t u^{-1} \in \mathcal{G}_\sigma \iff {}^t u \in \mathcal{G}_\sigma\,.$$

(3) This is a consequence of the equivalence, which holds for all $u \in \mathrm{GL}(E)$, $x \in E$, $y \in E$:

$$u(x) \cdot \sigma u(y) = x \cdot \sigma(y) \iff x \cdot {}^t u \sigma u(y) = x \cdot \sigma y\,.$$

(4) We make use of Theorem 10.1.1, applied to the involution $u \mapsto u^\iota = \sigma\, {}^t u\, \sigma^{-1}$. Indeed, we have $(u^\iota)^\iota = \sigma^2 u \sigma^{-2}$, and u commutes with σ^2, since σ-isodual lattices are G-lattices for the group G generated by σ^2. □

[To handle more generally $(G^\#, G)$-isodual lattices, we must consider the intersections of $\mathcal{F}_\sigma$, $\mathcal{G}_\sigma$ and $\mathcal{T}_\sigma$ for σ running through $G^\# \smallsetminus G$. Note that a $(G^\#, G)$-lattice is in particular a σ-isodual lattice for all $\sigma \in G^\# \smallsetminus G$.]

To deal with extremal questions, we must make use of the orthogonal projection onto $\mathcal{T} = \cap_{\sigma \in G^\# \smallsetminus G} \mathcal{T}_\sigma$ (in $\mathrm{End}^s(E)$, for the Voronoi scalar product) of the projections (in E) onto the minimal vectors in a lattice. We give below a method to calculate this projection in the Voronoi space:

Proposition 11.7.3. *Let $\varphi : G^\# \to \{\pm 1\}$ be the character equal to $+1$ on G and to -1 on $G^\# \smallsetminus G$. For any nonzero $x \in E$, let p_x be the projection onto the line $\mathbb{R}x$. Then the orthogonal projection onto $\mathcal{T}$ of p_x is equal to*

$$\omega'_x = \frac{1}{|G^\#|} \sum_{s \in G^\#} \varphi(s)\, s p_x s^{-1} = \frac{1}{|G^\#|} \sum_{s \in G^\#} \varphi(s)\, p_{sx}\,.$$

Proof. The projection ω'_x of p_x onto $\mathcal{T}$ is that of ω_x onto $\mathcal{T}$ in $\mathrm{End}^s_G(E)$, where ω_x, the projection of p_x onto $\mathrm{End}^s_G(E)$, is equal to

$$\omega_x = \frac{1}{|G|} \sum_{s \in G} \varphi(s)\, s p_x s^{-1} = \frac{1}{|G|} \sum_{s \in G} \varphi(s)\, p_{sx}$$

by Proposition 11.3.2. Now the orthogonal in $\mathrm{End}^s(E)$ of $\mathcal{T}$ is the centralizer $\mathcal{T}'$ of $G^\#$. Indeed, given $v \in \mathcal{T}$ and $w \in \mathrm{End}^s_G$, we have

$$\mathrm{Tr}(vw) = \mathrm{Tr}(\sigma v \sigma^{-1} \sigma w \sigma^{-1}) = -\,\mathrm{Tr}(v \sigma w \sigma^{-1})$$

for all $\sigma \in G^\# \setminus G$, which shows the inclusion $\mathcal{T}^\perp \subset \mathcal{T}'$. That equality holds is then proved by calculating dimensions: we have $\mathcal{T} \cap \mathcal{T}' = \{0\}$, and also $\mathcal{T} + \mathcal{T}' = \mathrm{End}_G^s$, because we have the decomposition

$$u = \frac{1}{2}(u + \sigma u \sigma^{-1}) + \frac{1}{2}(u - \sigma u \sigma^{-1})$$

for all $u \in \mathrm{End}_G^s$; hence the projection onto $\mathcal{T}$ in End_G^s is the map $q : u \mapsto \frac{1}{2}(u + \sigma u \sigma^{-1})$; finally, calculating $q \circ \omega_x$ yields the formula we wanted to prove. □

We now return to the case where $G^\#$ is a cyclic group, which we endow with a fixed generator σ. The following proposition, whose second assertion is a general result when one considers elements belonging to the connected component of the unit element of $\mathcal{G}_\sigma$ (see Exercise 10.1.2), allows us to make a precise description of $\mathcal{G}_\sigma$:

Proposition 11.7.4. *1. A necessary and sufficient condition for a positive symmetric endomorphism $v \in \mathrm{GL}(E)$ to belong to $\mathcal{G}_\sigma$ is that for all eigenvalue λ of v, σ exchanges the eigenspaces of u relative to λ and λ^{-1}.*
2. Let $u \in \mathrm{GL}(E)$, written as a product $u = fs$, $f \in \mathrm{O}(E)$, $s \in \mathrm{End}^{s++}(E)$. Then u belongs to $\mathcal{G}_\sigma$ if and only if both f and s belong to $\mathcal{G}_\sigma$.

Proof. (1) Let $v \in \mathrm{End}^{s++}(E)$. For $v \in \mathcal{G}_\sigma$ and $x \in E$, if $v(x) = \lambda x$, we have

$$v\sigma(x) = \sigma v^{-1}(x) = \sigma(\lambda^{-1}(x)) = \lambda^{-1}\sigma(x)\,.$$

Conversely, the same calculation shows that if σ exchanges the eigenspaces relative to λ and λ^{-1}, the equality $v\sigma v(x) = \sigma(x)$ holds on all eigenspaces of v, hence on the whole space E.

(2) The relation $u = fs$ implies that $s^2 = {}^t\!uu$ belongs to $\mathcal{G}_\sigma$. Since the eigenvalues of s^2 are the squares of that of s, and since this correspondence between eigenvalues preserves the eigenspaces, assertion (1) shows that s also belongs to $\mathcal{G}_\sigma$. Then so does $f = us^{-1}$. □

Definition 11.7.5. *We say that σ is of* symmetric *(or* orthogonal*) type if the form b_σ is symmetric, and that σ is of* symplectic *type if b_σ is alternating. When σ is of symmetric type (which amounts to saying that σ is a symmetric endomorphism), the signature of the form b_σ is called the* signature of σ.

Proposition 11.7.6. *1. The bilinear form b_σ has determinant ± 1, equal to the determinant of σ.*
2. A necessary and sufficient condition for σ to be of symmetric (resp. symplectic) type is that $\sigma^2 = +\mathrm{Id}$ (resp. $\sigma^2 = -\mathrm{Id}$).
3. When σ is of symmetric type with signature (p, q), p and q are the numbers of eigenvalues of σ which are equal to $+1$ and to -1 respectively, and b_σ is definite if and only if $\sigma = \pm\mathrm{Id}$.

Proof. (1) Thanks to the existence of the Euclidean structure, the determinant of a bilinear form b is defined absolutely (not only modulo squares) as $\det_{\mathcal{B}}(b(\varepsilon_i,\varepsilon_j))$ where $\mathcal{B}=(\varepsilon_1,\dots,\varepsilon_n)$ is any orthonormal basis for E. For any $\sigma\in\mathrm{GL}(E)$ and b_σ defined as in Definition 11.7.1, we have

$$\det_{\mathcal{B}} b_\sigma = \det_{\mathcal{B}^*}\sigma(\mathcal{B}) = \det_{\mathcal{B}}\sigma(\mathcal{B})\det_{\mathcal{B}^*}\mathcal{B} = \det(\sigma)\cdot\det(\mathrm{Gram}(\mathcal{B}))\,.$$

Since $\mathrm{Gram}(\mathcal{B})=I_n$, $\det(b_\sigma)=\det(\sigma)$.

(2) Set $\epsilon=+1$ (resp. $\epsilon=-1$). For all $x,\,y\in E$, we have

$$b_\sigma(y,x)=y\cdot\sigma x = x\cdot\sigma^{-1}y\,,$$

which implies

$$\forall\, x,\,y\in E,\; b_\sigma(y,x)=\epsilon\, b_\sigma(x,y)\Longleftrightarrow \forall x,y\in E,\; x\cdot(\sigma y-\epsilon\sigma^{-1}y)=0\,,$$

whence $\sigma=\epsilon\sigma^{-1}$, i.e. $\sigma^2=\epsilon\,\mathrm{Id}$.

(3) Since σ is both symmetric and orthogonal, its eigenvalues are equal to ±1. If λ is an eigenvalue of σ, we have $b_\sigma(x,\sigma x)=\lambda b_\sigma(x,x)$ on the corresponding eigenspace E_λ, and assertion (3) is now clear on the decomposition $E=E_{+1}\perp E_{-1}$ of E into eigenspaces for σ. □

The theorem we are now going to state yields a description of σ-isodual lattices of symmetric or symplectic type. However, the bases which show up have generally nothing to do with minimal vectors, so they are not very helpful in solving the extremal problems related to σ-isodual lattices. Moreover, unimodular lattices, which play a special rôle because of the equality $\mathrm{Aut}^{\#}(\Lambda)=\mathrm{Aut}(\Lambda)$, must be excluded in what follows.

We shall need a classification of indefinite and alternating integral symmetric bilinear forms of determinant ±1. The case of symmetric forms, a description of which we gave in Theorem 5.6.4, after Milnor and Serre, reads as follows: *an indefinite quadratic $\mathbb{Z}$-module of signature (p,q) is isometric to an orthogonal sum $p\mathbb{Z}^+ + q\mathbb{Z}^-$ if it is odd, and $p'U+q'\mathbb{E}_8^{\pm}$ when it is even.* (U denotes the hyperbolic plane over $\mathbb{Z}$.)

We have an analogous (and much easier) result in the alternating case (see [M-H], Chapter I, Corollary 3.5, or Exercise 11.7.1 below): *a symplectic $\mathbb{Z}$-module of determinant 1 is isomorphic to an orthogonal direct sum of rank-2 free modules equipped with the form $(x,y)\mapsto x_1y_2-x_2y_1$.*

It is clear that the form b_σ is integral on σ-isodual lattices, so that in the symmetric case, we may say that an isodual lattice is of *even* or of *odd type* according to whether the form b_σ is even or odd.

Making use of the results we just recalled, we are now able to prove:

Theorem 11.7.7. *Let $\sigma\in\mathrm{O}(E)$ of square $\pm\mathrm{Id}$. In the symmetric case (i.e., if $\sigma^2=+\mathrm{Id}$), let (p,q) be its signature, and assume that σ is not equal to $\pm\mathrm{Id}$. Then the family $\mathcal{F}_\sigma$ reduces to a single orbit under the action of $\mathcal{G}_\sigma$*

(represented by $\mathbb{Z}^n$ endowed with a suitable automorphism), except for lattices of symmetric type with $p \equiv q \mod 8$ (which may exist only for even $n \geq 10$), for which there exists a second orbit, represented by the lattice $\mathbb{D}_n^+$.

Proof. Since $\mathcal{G}_\sigma$ is the orthogonal group of b_σ, two lattices belong to the same orbit of $\mathcal{G}_\sigma$ if and only if the corresponding forms b_σ are isometric. Since b_σ is integral on σ-isodual lattices and has determinant ± 1 (Proposition 11.7.6), the classification of symmetric or alternating bilinear forms that we recalled above shows that there is at most one orbit of σ-isodual lattices under the action of $\mathcal{G}_\sigma$, except possibly if b_σ is symmetric and such that $p \equiv q \mod 8$, where two orbits may exist.

It suffices now to construct in each case at least one example of a σ-isodual lattice. To this end, we consider in $\mathbb{R}^n$ endowed with its canonical basis $(\varepsilon_1, \varepsilon_2, \dots, \varepsilon_n)$ the vectors

$$\varepsilon = \frac{\varepsilon_1 + \varepsilon_2 + \cdots + \varepsilon_n}{2} \quad \text{and} \quad \varepsilon' = \frac{-\varepsilon_1 + \varepsilon_2 + \cdots + \varepsilon_n}{2},$$

and also the lattices $\mathbb{Z}^n \subset \mathbb{R}^n$, $\mathbb{D}_n \subset \mathbb{Z}^n$, $\mathbb{D}_n^+ = \langle \mathbb{D}_n, \varepsilon \rangle$, and $\mathbb{D}_n^- = \langle \mathbb{D}_n, \varepsilon' \rangle$. (These last two lattices, which are indeed isometric, are defined only for even n; see Sections 4.1, 4.3 and 4.4.) The lattice $\mathbb{Z}^n$ is unimodular, and duality fixes or exchanges $\mathbb{D}_n^+$ and $\mathbb{D}_n^-$, according to whether n is or is not divisible by 4.

For $n = 2m$ even, we give $\mathbb{Z}^n$ and $\mathbb{D}_n^\pm$ the structure of an isodual lattice of symplectic type by setting $\sigma(\varepsilon_i) = \varepsilon_{i+1}$ if i is odd and $\sigma(\varepsilon_i) = -\varepsilon_{i-1}$ if i is even.

For any n and any $p, q \geq 0$ with $p + q = n$, we give $\mathbb{Z}^n$ a structure of isodual lattice of symmetric type and signature (p, q) by setting $\sigma(\varepsilon_i) = \varepsilon_i$ for $i \leq p$ and $\sigma(\varepsilon_i) = -\varepsilon_i$ for $i > p$.

Finally, let $n = 2m \geq 10$ and $p, q \geq 0$ with $p + q = n$ and $q - p \equiv 0 \mod 8$. Let $k \equiv m \mod 2$ and let $\sigma \in \mathrm{O}(E)$ be defined by $\sigma(\varepsilon_i) = +\varepsilon_i$ for $i \leq k$ and $\sigma(\varepsilon_i) = -\varepsilon_i$ for $i > k$. The form b_σ is even on $\mathbb{D}_n$, for

$$(\varepsilon_i \pm \varepsilon_j) \cdot \sigma(\varepsilon_i \pm \varepsilon_j) = \pm \varepsilon_i \cdot \varepsilon_i \pm \varepsilon_j \cdot \varepsilon_j \equiv 0 \mod 2,$$

and we have $b_\sigma(\varepsilon, \varepsilon) = \frac{k-(n-k)}{4} = \frac{k-m}{2}$. For $k = p$, we have $\frac{k-m}{2} = \frac{p-q}{4} \equiv 0 \mod 2$, which shows that $\mathbb{D}_n^+$ is a σ-isodual lattice of symmetric type. □

We now consider extremal problems on spaces of isodual lattices. Recall that a "Voronoi theory" cannot be expected to hold in full generality, because of the following remark:

Remark 11.7.8. *There exist isodual lattices for which local maxima are not strict; see Exercise 11.5.5.*

Definition 11.7.9. *Let Λ be a σ-isodual lattice.*

1. *We say that Λ is* (strictly) σ-extreme *if its Hermite invariant attains a (strict) local maximum at Λ on $\mathcal{F}_\sigma$.*
2. *We say that Λ is* σ-perfect *(resp.* σ-eutactic*) if it is $\mathcal{T}$-perfect (resp. $\mathcal{T}$-eutactic) in the sense of Definition 3.2.8, with $\mathcal{T} = \mathcal{T}_\sigma$ as in Definition 11.7.1.*

Theorem 11.7.10. *Let $\sigma \in \mathrm{O}(E)$ with $\sigma^2 \pm \mathrm{Id}$, $\sigma \neq \pm \mathrm{Id}$.*

1. *The minimal vectors in σ-extreme lattices generate E.*
2. *Every σ-extreme lattice is strictly σ-extreme.*
3. *A σ-isodual lattice is σ-extreme if and only if it is σ-perfect and σ-eutactic.*

Proof. We know (Theorem 10.3.6) that condition (1) implies condition (2) and that conditions (2) and (3) are equivalent. To prove (1), we make use of the following fact: if $v \in \mathcal{T}$ is such that $\varphi_x(v) \geq 0$ for all $x \in S = S(L)$, there exists an extreme lattice (of the form $(\exp(tv/2))(L)$ for sufficiently small $t > 0$) with set of minimal vectors $S \cap \operatorname{Ker} v$; see Formula (10.3.c). In all cases, we shall restrict ourselves to the case where the minimal vectors in the lattices L we consider are contained in a σ-stable subspace of codimension ≥ 2 in E.

We first consider the case of a symplectic lattice L. If S is contained in a hyperplane H of E, let $F = H \cap \sigma(H)$ be the maximal σ-stable subspace of H. In a suitable basis $\mathcal{B}$ for $P = F^\perp$, the matrix of the restriction of σ to P is $\left(\begin{smallmatrix} 0 & 1 \\ -1 & 0 \end{smallmatrix}\right)$. Consider the endomorphism $v \in \mathcal{T}$ which is zero on F and whose restriction to P has matrix $\left(\begin{smallmatrix} 0 & 1 \\ 1 & 0 \end{smallmatrix}\right)$ in $\mathcal{B}$. Then $x \mapsto \varphi_x(v)$ is either the null map, or has a constant sign on H, hence on S. Indeed, let $x = \lambda e + y$, $\lambda \in \mathbb{R}$, $y \in F$, be an element of H. We have $\varphi_x(v) = v(x)\cdot x = \lambda v(e)\cdot(\lambda e) = \lambda^2 v(e)\cdot e$ (because $v(e)\cdot y = e \cdot v(y) = 0$); hence, replacing v by $-v$ if need be, we may assume that $\varphi_x(v) \geq 0$ for all $x \in H$. We have constructed in this way a σ-extreme lattice whose set of minimal vectors is $\operatorname{Ker} v \cap S \subset F$. We may thus assume that S is contained in F. Let K be a hyperplane in F which is disjoint from S, let $G = K \cap \sigma(K)$ be its corresponding maximal stable subspace, and let Q be the orthogonal complement to G in F (a plane), so that $E = P \perp Q \perp G$. Consider $v \in \mathcal{T}$ which is zero on G and which exchanges the planes P and Q; in a suitable basis for $P \perp Q$, matrices for σ and v are

$$\mathcal{M}(\sigma) = \begin{pmatrix} 0 & 1 & 0 & 0 \\ -1 & 0 & 0 & 0 \\ 0 & 0 & 0 & 1 \\ 0 & 0 & -1 & 0 \end{pmatrix} \quad \text{and} \quad \mathcal{M}(v) = \begin{pmatrix} 0 & 0 & 1 & 0 \\ 0 & 0 & 0 & -1 \\ 1 & 0 & 0 & 0 \\ 0 & -1 & 0 & 0 \end{pmatrix}.$$

We have $\varphi_x(v) = 0$ for all $x \in F = Q \perp G$, for $v(x)$ belongs to $P = F^\perp$. Hence we can use the endomorphism v to construct a (σ-extreme) lattice whose set of minimal vectors is $S \cap \operatorname{Ker} v \subset S \cap G \subset S \cap K = \emptyset$, which is plainly impossible.

From now on, we assume that $\sigma^2 = \mathrm{Id}, \sigma \neq \pm\mathrm{Id}$, and we denote by E^+ and E^- the eigenspaces of σ for the eigenvalues $+1$ and -1. Let L be a σ-extreme lattice whose set S of minimal vectors is contained in a hyperplane H of E. There exists a σ-stable hyperplane F in H whose orthogonal plane $P = F^\perp$ contains a norm 1 eigenvector e of σ for the eigenvalue $+1$ and a norm 1 eigenvector e' for the eigenvalue -1:

- If H is not stable under σ, it suffices to take $F = H \cap \sigma(H)$, as in the symplectic case; indeed, if the plane $F^\perp$ were contained for instance in E^+, the line $H^\perp \subset F^\perp$ would also be stable, and consequently H itself would be σ-stable; this shows that the restriction of σ to $F^\perp$ is $\neq \pm\mathrm{Id}$.
- If H is stable, for instance if $H^\perp$ is contained in E^+, we consider the span P of the line $H^\perp$ and of a nonzero vector in E^- (our hypothesis implies that such a vector exists); the σ-stable subspace $F = P^\perp$ is then a stable subspace of codimension 2.

The endomorphism v which is zero on F and which exchanges e and e' belongs to $\mathcal{T}$, and we may, as in the symplectic case, first assume that $\varphi_x(v) \geq 0$ for all $x \in H$, and then construct a σ-extreme lattice L' whose set $S' = S \cap \operatorname{Ker} v$ of minimal vectors is contained in F.

Denote from now on by L a σ-extreme lattice whose set S of minimal vectors is contained a σ-stable subspace G of F of minimal dimension. Since G is stable under σ and nonzero, it contains at least a norm 1 eigenvector a for σ, say, $a \in E^+$. The endomorphism v which exchanges the vectors e' and a and which is zero on the orthogonal of the plane $\langle a, e' \rangle$ belongs to $\mathcal{T}$ (if a were in E^-, we should replace e' by e). Moreover, $S \subset G$ is contained in the hyperplane $\operatorname{Ker} v \perp \mathbb{R}a$, on which $\varphi_x(v)$ is zero or has a constant sign, and we can then construct a new σ-extreme lattice, whose set of minimal vectors $S \cap \operatorname{Ker} v$ is contained in the σ-stable subspace $G \cap \operatorname{Ker} v \subset_{\neq} G$, because a does not belong to $\operatorname{Ker} v$ (same proofs as in the symplectic case). This contradicts the minimal property of G. □

Taking into account Theorem 11.7.7, the description of the set of σ-isodual lattices for an endomorphism σ of symmetric or symplectic type reduces to the description of the group $\mathcal{G}_\sigma$. The following proposition describes the symmetrized space of the tangent $\mathcal{T}_\sigma$ at the origin of $\mathcal{G}_\sigma$, and allows us in particular to calculate its dimension, an important tool for the classification of σ-extreme lattices.

Proposition 11.7.11. *Let $\sigma \in \mathrm{O}(E)$ with $\sigma^2 = \pm\mathrm{Id}$. If σ is of symmetric type (i.e., if $\sigma^2 = \mathrm{Id}$), let E^+ (resp. E^-) be the eigenspace of σ for the eigenvalue $+1$ (resp. -1). Set $p = \dim E^+$ and $q = \dim E^-$ (thus (p, q) is the signature of b_σ).*

1. *If σ is of symmetric type, we have*

$$\mathcal{T}_\sigma = \{v \in \mathrm{End}^s(E) \mid v(E^+) \subset E^- \text{ and } v(E^-) \subset E^+\};$$

in particular, $\dim \mathcal{T}_\sigma = pq$.

2. *If σ is of symplectic type, let $n = 2m$, and let $\mathcal{B}$ be an orthonormal basis for E in which the matrix of b_σ consists of diagonal blocks of the form $\left(\begin{smallmatrix} 0 & 1 \\ -1 & 0 \end{smallmatrix}\right)$. Then the elements of $\mathcal{T}$ are those whose matrices in $\mathcal{B}$ are the symmetric matrices consisting of blocks of the form $\left(\begin{smallmatrix} a & b \\ b & -a \end{smallmatrix}\right)$; in particular,* $\dim \mathcal{T} = m^2 + m$.

[Note that if σ is of symmetric type with trivial signature, then $\dim \mathcal{T}_\sigma = 0$: in this case, the space of isodual lattices is totally disconnected, and the theory of σ-extreme lattices is trivial.]

Corollary 11.7.12. *A $\mathcal{T}$-perfect isodual lattice of symmetric type and signature (p,q) (resp. of symplectic type and dimension $2m$) possesses at least $pq + 1$ (resp. $m^2 + m + 1$) pairs of minimal vectors.*

Indeed, since $\mathcal{G}_\sigma$ has determinant ± 1, the lower bound of s for such lattices is $s \geq \dim \mathcal{G}_\sigma + 1$. □

Proof of 11.7.11. The assertions about dimensions result immediately from the determination of the set of $v \in \mathrm{End}^s(E)$ such that $\sigma v = -v\sigma$. In the symmetric case, we characterize the matrices by writing that this relation holds on the eigenspaces of σ. In the symplectic case, we perform explicit calculations with matrices. □

11.8 Isodual Lattices: Examples and Classification Results

We keep the notation of Section 11.7; the dimension n of E, when even, is denoted by $2m$.

The isodualities of symmetric or symplectic type which may exist on $\mathbb{Z}^n$ or on $\mathbb{D}_n^+$ have been described in the proof of Theorem 11.7.7. Constructions over maximal orders in C.M. fields or in totally definite quaternion skew-fields produce numerous examples of modular lattices, for which the similarity σ of ratio $\sqrt{q}$ which maps Λ^* onto Λ has square $-q\,\mathrm{Id}$, and thus defines after rescaling even symplectic σ-isodual lattices, among which some are extreme in the usual sense, hence in particular σ-extreme.

Example 11.8.1. **(1)** $\underline{q = 3}$. *The Coxeter–Todd lattice constructed over the quadratic field $\mathbb{Q}(\sqrt{-3})$ or over a maximal order in the quaternion field ramified at 3 and ∞ (Section 8.5) is modular of level 3 and extreme.*

(2) $\underline{q = p}$, $p \equiv -1 \bmod 4$. *The Craig lattice $\mathbb{A}_{p-1}^{((p+1)/4)}$ is modular of level p (Proposition 5.4.7) and conjecturally extreme (see Problem 5.4.10).*

(3) *$q = 2$. Constructions using $\mathbb{Z}(\sqrt{-2})$ or $\mathbb{Z}[i]$, and more specially larger orders ($\mathbb{Z}[\zeta_8]$, the Hurwitz order) allow the construction of lattices which are most of the time 2-modular of symplectic type, although this cannot be proved a priori. Examples are provided by $\mathbb{D}_4$, $\mathbb{E}_8$, the Barnes–Wall lattice Λ_{16}, the Leech lattice Λ_{24}, and various 32-dimensional lattices (Quebbemann, Bachoc, Nebe). All these lattices are perfect; a general proof of this fact follows from work by Bachoc and Venkov ([Ven3], [Bac-V]).*

[It is an interesting question to decide whether some lattices constructed over the order $\mathbb{Z}[i]$, say, can be given structures related to the scalar product over larger orders, e.g. $\mathbb{Z}[\zeta_8]$ or the Hurwitz order.]

The following proposition was pointed out to me by Coulangeon.

Proposition 11.8.2. *For all n and for all r with $0 < r < n$, the lattices $\bigwedge^r \Lambda$ and $\bigwedge^{n-r} \Lambda^*$ are similar. In particular, for even $n = 2r$, $\bigwedge^r \Lambda \subset \bigwedge^r E$ are isodual lattices of dimension $\binom{n}{r}$.* □

Applying this proposition to an integral lattice Λ, we obtain a modular lattice. If Λ is moreover modular, the lattices $\bigwedge^r \Lambda$ are modular without any hypothesis on r. For instance, $\bigwedge^2 \mathbb{E}_8$ is a 28-dimensional unimodular lattice, which can be shown to be perfect of norm 3.

The lattice $\bigwedge^2 \mathbb{D}_4$ is a 6-dimensional norm 3 lattice, isometric to the densest sections of $P_7^2 \sim \mathbb{E}_7^*$, and similar to $\mathbb{D}_6^+$; it has $s = 16$, the maximum possible value of s among 6-dimensional lattices having no hexagonal sections with the same norm (Watson, [Wat7]).

The lattice $\bigwedge^3 \mathbb{E}_6$ is a 20-dimensional, 3-modular even lattice of norm 4, extremal in the sense that its norm is the greatest possible one among lattices with these parameters. (See Table 2.4 in [Sch-SP].)

We now prove two classification results, beginning with dimension 3.

In the following two lemmas, we denote by τ the rotation of order 2 (i.e., of angle π) about some line D. Let $P = D^\perp$.

Lemma 11.8.3. *Let Λ be a τ-stable lattice. Then $\Lambda' = (\Lambda \cap D) \perp (\Lambda \cap P)$ is a sublattice of Λ, and there exist bases (e_1) for $\Lambda \cap D$ and (e_2, e_3) for $\Lambda \cap P$ such that (e_1, e_2, e_3) or $(e_1, \frac{e_1+e_2}{2}, e_3)$ is a basis for Λ.*

Proof. For $x \in E \smallsetminus D$ (resp. $x \in E \smallsetminus P$), $x + \tau x$ (resp. $x - \tau x$) is a nonzero vector in D (resp. of P), which shows that $\Lambda \cap D$ and $\Lambda \cap P$ are lattices. In view of the decomposition $x = \frac{x+\tau x}{2} + \frac{x-\tau x}{2}$, if $\Lambda' \neq \Lambda$, we can find a basis (e_2, e_3) for $\Lambda \cap P$ such that $\frac{e_1+e_2}{2} \in \Lambda$. Then $\frac{e_1+e_3}{2}$ does not belong to Λ, since otherwise, $\Lambda \cap P$ would contain $\frac{e_2 \pm e_3}{2}$. □

Lemma 11.8.4. *Let Λ be a lattice and let σ be an isoduality of Λ, which is a rotation of order a power of 2. Then σ has order 1, 2, or 4, and in this last* *case, Λ is the orthogonal sum of a lattice isometric to $\mathbb{Z}$ contained in the axis D of σ and of a lattice of determinant* 1 *contained in $P = D^{\perp}$.*

Proof. The order m of σ^2 is such that $\varphi(m) \leq 3$. Hence $m = 1$, 2, or 4. In the last two cases, Λ is stable under a rotation τ (σ or σ^2) of order 2, to which we may apply Lemma 11.8.3. With the notation of this lemma, let (e'_2, e'_3) be the dual basis to (e_2, e_3). Then $(\frac{e_1}{N(e_1)}, e'_2, e'_3)$ is a basis for Λ'^*. Since $\sigma(\Lambda) = \Lambda^*$, we have $\Lambda^* \cap D = \Lambda \cap D = \mathbb{Z}e_1$, whence $N(e_1) = 1$. This implies that we indeed have $\Lambda = \Lambda'$, because $e_1 \cdot \sigma(\frac{e_1+e_2}{2}) = e_1 \cdot \frac{e_1}{2} = \frac{1}{2}N(e_1)$ cannot be an integer unless $N(e_1)$ is an even integer.

Finally, if σ were of order 8, $\Lambda \cap P$ would be preserved by a rotation of order 4, hence isometric to the unimodular lattice $\mathbb{Z}^2$, which has no automorphism of order 8, a contradiction. □

Theorem 11.8.5. *A 3-dimensional σ-extreme σ-isodual lattice for a rotation σ of angle π is isometric to the ccc lattice of Proposition 11.6.1.*

Proof. Let Λ be such a lattice. Since it is of symmetric type, $S(\Lambda)$ spans E (Theorem 11.7.10). Since Λ is normal, it is isometric to one of the lattices $\mathbb{Z}^3$ or *ccc* (Proposition 11.6.1), and a calculation with matrices (see Exercise 11.8.1) immediately shows that the $\mathcal{T}_\sigma$-perfection rank of $\mathbb{Z}^3$ is $< 2 = \dim \mathcal{T}_\sigma$. □

Corollary 11.8.6. (Conway and Sloane.) *The lattice ccc is the densest 3-dimensional isodual lattice.*

Proof. Indeed, an isodual lattice Λ is σ-isodual for some rotation σ (if $\det(\sigma) = -1$, then $\det(-\sigma) = +1$), whose order is 1, 2, or 4. In the first case, it is unimodular, hence isometric to $\mathbb{Z}^3$, and we have $\gamma(\Lambda) = 1$. In the third case, we again have the inequality $\gamma(\Lambda) \leq 1$ (Lemma 11.8.4 and Exercise 11.5.1). The maximum of the Hermite invariant is thus attained in the second case, for which we may apply Theorem 11.8.5. □

The second result concerns dimension 4.

Theorem 11.8.7. *A 4-dimensional σ-isodual σ-extreme lattice of symplectic type is similar to $\mathbb{D}_4$.*

Proof. We know that such a lattice possesses at least seven pairs of minimal vectors (Corollary 11.7.12). By Theorem 11.6.4, this lattice belongs to the class a_9 or to its boundary, represented by $\mathbb{D}_4$. Since the Hermite invariant has no maximum on a non-perfect class (Theorem 9.4.1), an extreme σ-isodual lattice must be similar to $\mathbb{D}_4$. □

The remainder of this section is devoted to constructions of isodual lattices (sometimes without determining an explicit isoduality, see Exercise 11.8.2), obtaining some lattices having a fairly high density. The method consists in considering a path $t \mapsto \Lambda_t$, $t \in [\alpha, \beta] \subset \mathbb{R}$ in the space of lattices of minimum 1 connecting a lattice L (often a critical one) to another one which is isometric to L^*, such that the set of Λ_t is stable under duality and that the determinant is a strictly monotone function on $[\alpha, \beta]$. Then for all t, Λ_t^* is isometric to Λ_{t^*} for a unique value t^* of the parameter. Since the function $t \mapsto \det(\Lambda_t)$ takes inverse values at the end–points of the interval $[\alpha, \beta]$, there exists a unique value of t for which Λ_t is isodual.

In practice, the map $t \mapsto t^*$ will be a linear fractional function, and for a suitable parametrization, we shall have $tt^* = \alpha\beta$; we then obtain the isodual lattice for $t = \sqrt{\alpha\beta}$, the geometric mean of the end–points of the interval. It is this way the (not yet) *ccc* lattice was found in [B-M1], p. 34, proof of Theorem 5.1; see Lemma 6.3.5. The idea used by Conway and Sloane in [C-S9] is a different one: they connect a lattice to its dual by adjoining *glue vectors* to a one parameter family of the form $L_1 \perp \sqrt{t}\, L_2$.

Our first example concerns dimension 6. We consider a real parameter $t > 0$ and the lattice $L_t = \mathbb{D}_4 \perp \sqrt{t}\,\mathbb{A}_2 \subset \mathbb{R}^7$. We represent the nonzero classes of $\mathbb{D}_4^*/\mathbb{D}_4$ and of $\mathbb{A}_2/2\mathbb{A}_2$ by the vectors $u_1 = \varepsilon_1$, $u_2 = \frac{1}{2}(-\varepsilon_1 + \varepsilon_2 + \varepsilon_3 + \varepsilon_4)$, $u_3 = -u_1 - u_2$, and $v_1 = \varepsilon_5 - \varepsilon_6$, $v_2 = \varepsilon_6 - \varepsilon_7$, $v_3 = \varepsilon_7 - \varepsilon_5$, and we denote by Λ_t the lattice generated by L_t and the vectors $w_i(t) = u_i + \frac{1}{2}\sqrt{t}\, v_i$. We refer to Exercise 11.8.3 for the proof of the following proposition:

Proposition 11.8.8. *Set* $t^* = \frac{4/3}{t}$.

1. *The path* Λ_t, $\frac{2}{3} \le t \le 2$, *connects lattices isometric to* $\mathbb{E}_6^*$ *and to* $\mathbb{E}_6$.
2. *The lattices* Λ_t^* *and* Λ_{t^*} *are isometric.*
3. *For* $\frac{2}{3} < t < 2$, *we have*
   $N(\Lambda_t) = 1 + \frac{t}{2} = N(u_i + \sqrt{t}\, v_i)$, $s(\Lambda_t) = s(\Lambda_t^*) = 24$, *and* $\det(\Lambda_t) = \frac{3}{4}t^2$.
4. *The lattice* $\Lambda_{\sqrt{4/3}}$ *is isodual, with Hermite invariant* $1+\frac{1}{\sqrt{3}}=1.577\ldots$ □

We end this section by quoting a second example, concerning dimension 5. This is the isodual lattice Λ_1 of symmetric type, which is constructed in Exercise 11.8.4, of norm $\frac{7}{5}$ (hence Hermite invariant $\frac{7}{5} = 1.4$), proportional to an integral lattice of norm 7 and determinant 5^5, with $s(\Lambda_1) = 10$ and $\mathrm{Aut}(\Lambda_1) \simeq C_2 \times S_5$.

This lattice and the previous one first occurred in [C-S9].

11.9 Rationality and Finiteness Questions

We keep the notation of Sections 11.2 and 11.5. According to whether we consider G-lattices or isodual lattices, the group $\mathcal{G}$ corresponds to a group G or to a pair $(G^\#, G)$.

For G-perfect G-lattices, we can prove a result similar to that of Korkine and Zolotareff (Proposition 3.2.11 and Corollary 3.4.7), which reduces to this last result when G is trivial:

Theorem 11.9.1. *A G-perfect G-lattice is proportional to an integral lattice.*

Proof. We want to prove that the Gram matrix A of a basis $\mathcal{B}$ for a G-perfect G-lattice Λ of rational norm m has rational entries.

Since the representation of G afforded by Λ is rational over $\mathbb{Q}$, the $\mathbb{R}[G]$-module $V = \mathrm{End}_G^s(E)$ comes from a $\mathbb{Q}[G]$-module W. (This means that we have $V = \mathbb{R}W \simeq \mathbb{R}\bigotimes W$ and that the action of G on V extends that of G on W.) Attaching to elements of V and of W their matrices in the basis $\mathcal{B}$, we define a G-isomorphism of W onto a subspace W' of $\mathrm{Sym}_n(\mathbb{R})$ endowed with an action of G, which induces a G-isomorphism of V onto $V' = \mathbb{R}V \simeq \mathbb{R}\bigotimes W'$.

Let $B_1, B_2, \dots, B_m$ be matrices in $\mathrm{Sym}_n(\mathbb{Q})$ which form a $\mathbb{Q}$-basis for W'; they also form an $\mathbb{R}$-basis for V', on which A can be written as an $\mathbb{R}$-linear combination $A = \sum_{i=1}^m \lambda_i B_i$.

In the isomorphism above between W and W', the averages ω_x of the orthogonal projections onto the minimal vectors of Λ become matrices $Y = \dfrac{1}{|G|}\sum_{s\in G} s(X^tX)$ (Proposition 11.3.2). Denote by $Y_1, Y_2, \dots, Y_s$ the matrices which correspond to the column-matrices $X_1, X_2, \dots, X_s$ of the components on $\mathcal{B}$ of the minimal vectors in Λ. Since Λ is G-perfect, these matrices generate V' over $\mathbb{R}$, hence W' over $\mathbb{Q}$. Suppose that $(Y_1, Y_2, \dots, Y_m)$ is a $\mathbb{Q}$-basis for W'. Since Λ is a G-lattice, A is stable under the action of G. The equalities $\langle A, X_i{}^tX_i\rangle = m$ (which ensure that X_i represents for all i a vector in $S(\Lambda)$) imply the relations $\langle A, Y_i\rangle = m$ for $1 \le i \le s$. Restricting to the indices $i \le m$, we get a system which is equivalent to the system

$$\sum_{j=1}^{m}\sum_{i=1}^{m} \langle Y_i, B_j\rangle = m,\ \ 1 \le i \le m\,,$$

whose coefficients are rational. This is a Cramer system, since it has only one solution. Hence this unique solution $(\lambda_1, \lambda_2, \dots, \lambda_m)$ is rational, which shows that the matrix A itself has rational entries. □

In the case of $(G^\#, G)$-isodualities, perfect $(G^\#, G)$-lattices, even when they are *strictly* $(G^\#, G)$-extreme, can be irrational, as can be seen on the example of the 3-dimensional *ccc* lattice. However, it is likely that strictly extreme lattices are algebraic, exactly as the ordinary dual-extreme lattices are. Such a result holds true in the case of symmetric or symplectic isodual lattices.

We now turn to finiteness results, and we first consider the case of G-lattices, for which Voronoi's finiteness theorem (Theorem 3.5.4) generalizes, in a general form analogous to Theorem 9.1.5. This statement makes use of the notion of a minimal class relative to G-lattices. In accordance with the general definitions of Chapter 10, we set the following definition:

Definition 11.9.2. *We say that two G-lattices Λ and Λ' are G-*minimal-equivalent *if there exists an element u in the centralizer of G in $\mathrm{GL}(E)$ such that $u(\Lambda) = \Lambda'$ and $u(S(\Lambda)) = S(\Lambda')$.*

The fact that a Gram matrix for a G-perfect G-lattice with a given minimum is uniquely determined by the projections ω_x onto $\mathcal{T}_G$ for $x \in S(\Lambda)$ immediately implies the following generalization of the remark which follows Theorem 9.1.5:

Proposition 11.9.3. *The minimal G-class of a G-perfect G-lattice is the set of G-lattices which are G-similar to it in the sense of Definition 11.2.1.* □

Results of Sections 9.1 and 9.4 apply almost word to word to G-lattices. The convexity result of Theorem 9.1.10, can be stated as it stands there, and immediately implies that a minimal class in the sense of G-lattices contains (up to G-similarity) at most one G-eutactic lattice. (Note: "G-eutactic" is equivalent to "eutactic", but we consider here classification *modulo* $\mathrm{Sim}_G(E)$).

Theorem 11.9.4. *The number of minimal G-classes whose minimal vectors span E is finite.*

We immediately deduce from this the following two corollaries, which are analogues to Theorems 3.5.4 and 9.4.3 (2).

Corollary 11.9.5. (Jaquet, [Ja6].) *The number of G-similarity classes of G-perfect G-lattices is finite.* □

Corollary 11.9.6. *The number of G-similarity classes of (G-)eutactic G-lattices is finite.* □

Proof of 11.9.4. Let Λ be a G-lattice. Choose a basis $\mathcal{B} = (e_1, e_2, \dots, e_n)$ for Λ such that, with the notation of Lemma 9.1.4 and Theorem 9.1.5, we have the inequality $N(e_i)N(e_i^*) \le K_n^2$. Consider the matrices $P_s = (s(e_j) \cdot e_i^*)$, $s \in G$ (indeed, the matrices in $\mathcal{B}$ of the automorphisms $s \in G$). Since L is invariant under G, these matrices are integral; moreover, their entries are bounded (Lemma 9.1.4). But there are only finitely many integral rank-n representations of G. The proof follows from these two remarks together with the fact that if two bases $\mathcal{B}$ and $u(\mathcal{B})$ ($u \in \mathrm{GL}(E)$) for E define the same integral representation $s \mapsto P_s$ of G, the base change u belongs to the centralizer $\mathcal{G}$ of G: indeed, we have

$$s(u(e_j)) \cdot (u(e_i))^* = s(u(e_j)) \cdot {}^t u^{-1}(e_i^*) = (u^{-1} s u)(e_j) \cdot e_i^* \,,$$

and the condition on u reads $u^{-1}su = s$ for all $s \in G$. □

We now consider a pair $(G^{\#}, G)$ where $G^{\#}$ is a finite group and G a subgroup of index 2 in $G^{\#}$. Let φ be the character of $G^{\#}$ which is $+1$ on G and -1 on $G^{\#} \setminus G$. Let

$$\mathcal{G} = \{u \in \mathrm{GL}(E) \mid \forall s \in G^{\#},\ su = \varphi(s)\, us\}\,.$$

In conformity with the general definitions of Chapter 10, we set the following definition:

Definition 11.9.7. *Two $(G^{\#}, G)$-lattices Λ and Λ' are* minimal-equivalent *for the pair $(G^{\#}, G)$ if there exists $u \in \mathcal{G}$ such that*

$$u(\Lambda) = \Lambda' \text{ and } u(S(\Lambda)) = S(\Lambda')\,.$$

Theorem 11.9.8. *The $(G^{\#}, G)$-isodual lattices whose minimal vectors span E share out among finitely many $\mathcal{G}$-classes in the sense of Definition 11.9.7.*

Corollary 11.9.9. *Up to $\mathcal{G}$-similarity, there are only finitely many $\mathcal{G}$-eutactic lattices whose minimal vectors span E. In particular, the set of similarity classes of σ-extreme lattices of orthogonal or symplectic type is finite.*

Indeed, up to $\mathcal{G}$-similarity, each class contains at most one $(\mathcal{G})$-eutactic lattice whose minimal vectors span E (Theorem 10.4.2), and the minimal vectors in σ-isodual lattices for which $\sigma^2 = \pm\,\mathrm{Id}$ span E (Theorem 11.7.10), except perhaps when $\sigma = \pm\,\mathrm{Id}$, but finiteness is then a consequence of the fact that they are unimodular. □

Proof of 11.9.8. We first remark that the Hermite invariant of a lattice whose minimal vectors span E is at least 1: indeed, applying the Hadamard inequality to the lattice Λ' generated by n vectors of $S(\Lambda)$, we obtain the inequalities

$$\gamma(\Lambda)^n \geq \gamma(\Lambda')^n = N(\Lambda')^n \det(\Lambda')^{-1} \geq 1\,.$$

Here, we have $\det(\Lambda) = 1$, hence $N(\Lambda) \geq 1$. As in the proof of Theorem 11.9.4, let us choose for each lattice Λ a basis $\mathcal{B} = (e_1, e_2, \dots, e_n)$ which satisfies the conditions of the statements 9.1.4 and 9.1.5. In view of the lower bound $N(\Lambda) \geq 1$, there exists a constant K_n which bounds from above the components in $\mathcal{B}$ of the minimal vectors in Λ, and for all $\sigma \in G^{\#}$, $\sigma \notin G$, we also have

$$|b_\sigma(e_i, e_j)| \leq \|e_i\| \cdot \|\sigma(e_j)\| = \|e_i\| \cdot \|e_j\| \leq K_n\,.$$

Hence there are only finitely many possible choices for the set of components in $\mathcal{B}$ of the vectors in $S(\Lambda)$, and also for the set of matrices B_σ in $\mathcal{B}$ of the forms b_σ.

Let Λ_1 and Λ_2 be two $(G^\#, G)$-isodual lattices which have the same set of components and the same matrices B_σ with respect to suitable bases $\mathcal{B}_1$ and $\mathcal{B}_2$ made with minimal vectors. Let u be the element of $\mathrm{GL}(E)$ which maps $\mathcal{B}_1$ onto $\mathcal{B}_2$; we have $u(\Lambda_1) = u(\Lambda_2)$. The first condition shows that we also have $u(S(\Lambda_1)) = u(S(\Lambda_2))$, and the second one that u belongs to $\mathrm{O}(b_\sigma)$ for all $\sigma \in G^\# \smallsetminus G$. This proves that u belongs to $\mathcal{G}$, hence that Λ_1 and Λ_2 lie in the same $\mathcal{G}$-class. □

11.10 Exercises for Chapter 11

11.4.1. Show that for a multiple of a $\mathbb{Q}$-irreducible cyclotomic representation of G, the Bravais group is the dihedral group of order $2q$ (or $4q$ if q is odd), generated by $\pm\sigma$ (σ denoting a generator of G) and τ, where τ is defined in the following way: given an arbitrary G-lattice Λ and an arbitrary $\mathbb{Q}(\zeta)$-basis $\mathcal{B}$ for $\mathbb{Q}\Lambda$, τ acts on $\sigma^i x \in \mathcal{B}$ by $\tau\sigma^i x = \sigma^{-i} x$.

11.4.2. Let G be of order 5 and and let $\Lambda = \mathbb{Z}[G]\,x$ for some $x \in E$ of norm 2.

1. Show that the Gram matrix of $\mathbb{Z}[G]\,x$, as a function of $t = x \cdot \sigma x$, is equal to

$$M(t) = \begin{pmatrix} 2 & t & -1-t & -1-t \\ t & 2 & t & -1-t \\ -1-t & t & 2 & t \\ -1-t & -1-t & t & 2 \end{pmatrix}.$$

We denote by Λ_t a lattice with Gram matrix $M(t)$.

2. Show that x is minimal in Λ_t if and only if $-1 \le t \le 0$, that $\gamma(\Lambda_t)$ is maximum exactly at the edges of this interval, and that Λ_t is then isometric to $\mathbb{A}_4$. [Argue as in the proof of Proposition 9.3.1.]

3. Show that the map $\sigma \mapsto \sigma^2 \in \mathrm{Aut}(G)$ corresponds to the transformation $t \mapsto -1-t$ of the parameter.

4. Show that $\det(\Lambda_t) = 5(1-t-t^2)^2$; use this to deduce that Λ_t and $\Lambda_{t'}$ are isometric if and only if $t' = t$ or $t' = -1-t$.

5. Show that $\Lambda_{-1/2}$ is similar to $\mathbb{A}_4^*$.

6. Show that up to similarity, there exists a unique G-lattice similar to its dual. [For the calculation of the corresponding parameter t, see Exercise 11.8.2.]

7. Show that for $t \neq 0, -\frac{1}{2}, -1$, $\mathrm{Aut}(\Lambda_t) = \{\pm \mathrm{Id}\} \times H$, where H denotes the dihedral group D_5 generated by σ and $\tau : (e_1, e_2, e_3, e_4) \mapsto (e_1, e_3, e_2, e_4)$.

11.4.3. We say that a lattice Λ is *G-dual-extreme* if the invariant γ' attains a local maximum at Λ on the set $\mathcal{L}_G$ of G-lattices in E, and that it is *G-dual-perfect* if $S(\Lambda) \cup S(\Lambda^*)$ spans $\mathrm{End}_G^s(E)$.

1. Show that each of the sets of minimal vectors in a G-dual-extreme lattice and in its dual span E. [Argue as in Proposition 3.8.2.]

2. Deduce from this that a G-dual-extreme lattice is strictly G-dual-extreme.

3. State and prove for the property of being G-dual-extreme a characterization which reduces to that of Section 11.5 when $G = \{\mathrm{Id}\}$.

4. Show that a lattice is G-dual-extreme if and only if it is G-dual-perfect and (G-) dual-eutactic.

11.5.1. 1. Show that an isodual lattice can be uniquely written as an orthogonal sum of isodual lattices, which are irreducible *as isodual lattices*, and that these lattices are either irreducible isodual lattices, or orthogonal sums $L \perp L^*$ where L is irreducible, but not isodual.

2. Deduce from this that the 3-dimensional reducible isodual lattices are exactly the orthogonal sums $\mathbb{Z} \perp L$, where L is an arbitrary 2-dimensional lattice of determinant 1.

11.5.2. Let L/K be a quadratic extension, let R be a Dedekind domain with quotient field K, let B be an order of R in L, and let I be a fractional (nonzero) ideal in B. Let $C = \{\lambda \in L \mid \lambda I \subset I\}$ be the order associated to I.

1. Verify that C is indeed an order of R in L.

2. Show that I is locally free on C, hence invertible as an ideal of C.

11.5.3. Let Λ be a primitive 2-dimensional integral lattice, which is q-modular for some integer $q \geq 1$. Let u be a similarity of ratio $\sqrt{q}$ which maps Λ^* onto Λ. Let $\mathcal{B} = (e_1, e_2)$ be a basis for Λ, let $\mathcal{B}^* = (e_1^*, e_2^*)$ be its dual basis, and let a, b, c, d be such that $u(e_1^*) = ae_1 + be_2$ and $u(e_2^*) = ce_1 + de_2$.

1. Show that Λ has Smith invariant $(q, 1)$, and that $ad - bc = q$.

2. Let θ be a root of the characteristic polynomial of the matrix $\left(\begin{smallmatrix} a & b \\ c & d \end{smallmatrix}\right)$, acting on on E by $\theta\, x = u(x)$. Show that the order $C = \mathbb{Z}[\theta]$ maps Λ^* onto Λ and that Λ is a rank-1 module over C.

3. Let I be a fractional ideal of C such that Λ is isomorphic to I as a C-module. Show that I is invertible. [Use Exercise 11.5.2.]

4. Show that every 2-dimensional modular lattice is isometric to an invertible fractional ideal of an order in an imaginary quadratic field endowed with a bilinear symmetric form proportional to the form $\mathrm{Tr}(x\bar{y})$.

11.5.4. Show that a 3-dimensional lattice with $s \geq 5$ is not normal. [Use Theorem 9.2.2.]

11.5.5. Consider in a 3-dimensional Euclidean space E the following data: a line D, the plane $P = D^\perp$, a unit vector $e \in D$, the lattice $L_1 = \mathbb{Z}\, e$, and a lattice $L_2 \subset P$ of determinant 1. Let $L = L_1 \perp L_2$, and let σ be the rotation with angle $\frac{\pi}{2}$ about the line D; hence by Exercise 10.3.1, the set of lattices L is the set of all σ-isodual lattices in E.

1. Let $e_2 \in S(L_2)$ and let e_3 be minimal in $L_2 \smallsetminus \mathbb{Z}\, e_2$, chosen in such a way that the angle θ of e_2 and e_3 belongs to $[0, \frac{\pi}{2}]$. Prove that the condition $1 \leq N(e_2) \leq \frac{1}{\sin\theta}$ is necessary and sufficient to ensure that $N(L_2) \geq 1$.

2. Prove that the σ-similarity class of (σ, σ^2)-extreme lattices contains a 2-parameter family with constant minimum 1.

11.7.1. Let R be a principal ideal domain, K its quotient field, V a K-vector space of dimension n, endowed with a non-degenerate alternating form b (hence $n = 2m$), and M a free R-module of rank n, on which b takes values in R.

1. For any nonzero $x \in M$, let $I_x = \{b(x, y) \mid y \in M\}$. Show that I_x is a nonzero ideal in R.

2. Show that if I_x is maximal among ideals of this kind, x is indivisible in M.

3. Let e_1 be a vector as above, let $a \in R$ such that $I_{e_1} = (a)$, and let e_2 such that $b(e_1, e_2) = a$. Show that $I_{e_1} = I_{e_2}$.

4. Let M_1 be the submodule of M generated by e_1 and e_2. Show that M/M_1 is

torsion-free, that M is the orthogonal sum of M_1 and $M_1^\perp$.

5. Show that M has a basis $(e_1, e_2, \ldots, e_n)$ such that M is the orthogonal sum of its submodules $M_i = \langle e_i, e_{i+1}\rangle$ for $i = 1, 3, \ldots, n-1$.

6. Show that each module M_i has a basis in which the matrix of b is of the form $\left(\begin{smallmatrix} 0 & -d_i \\ d_i & 0 \end{smallmatrix}\right)$, and that the determinant of M is equal to $D = (d_1 \cdot d_2 \cdots d_n)^2$.

7. Deduce from this that if $\det(M) = 1$, there exists a basis $(e_1, e_2, \ldots, e_n)$ for M in which b is of the form $\sum_{i\,\text{odd}} x_i y_{i+1} - x_{i+1} y_i$.

11.7.2. Let Λ be a σ-isodual lattice.

1. Show that in dimension $n \le 8$, the form b_σ takes values $0, \pm 1$ on the pairs of minimal vectors in Λ, except for pairs of proportional minimal vectors in $\mathbb{E}_8$.

2. Show that for lattices $\mathbb{D}_n^+$, $n \ge 10$, the value 2 is attained.

11.7.3. Let $\sigma \in \mathrm{O}(E)$, and let $\mathcal{C}$ be a *σ-isodual minimal class*, i.e. a class whose all lattices are σ-isodual. (Examples: the minimal 2-dimensional classes, the a_9 class in dimension 4.) Show that a lattice in $\mathcal{C}$ is dual-eutactic if and only if it is σ-eutactic.

11.8.1. The aim of this exercise is to prove the assertion about $\mathbb{Z}^3$ which we used in the proof of Theorem 11.8.5.

1. Show that the automorphisms of order 2 of $\mathbb{Z}^3$ are conjugate to

$$-\mathrm{Id},\ (e_1, e_2, e_3) \mapsto (-e_1, -e_2, e_3),\ \text{or}\ (e_1, e_2, e_3) \mapsto (-e_2, e_1, e_3).$$

[One can use the identifications $\mathrm{Aut}^+(\mathbb{Z}^3) = \mathrm{Aut}^+(\mathbb{D}_3) = \mathrm{Aut}^+(\mathbb{A}_3) = S_4$.]

2. Show that the spaces $\mathcal{T}_\sigma$ attached to the last two transformations above are defined by the matrices $\begin{pmatrix} 0 & 0 & a \\ 0 & 0 & b \\ a & b & 0 \end{pmatrix}$ and $\begin{pmatrix} a & b & 0 \\ b & -a & 0 \\ 0 & 0 & 0 \end{pmatrix}$.

3. Show that the $\mathcal{T}_\sigma$-perfection ranks of $\mathbb{Z}^3$ are 0 and 1 respectively.

11.8.2. The notation is that of Exercise 11.4.2.

1. Verify that the adjoint matrix to $M(t)$ is

$$M'(t) = (1 - t - t^2)\begin{pmatrix} 2t{+}6 & -t{+}2 & t{+}3 & 3t{+}4 \\ -t{+}2 & -2t{+}4 & -3t{+}1 & t{+}3 \\ t{+}3 & -3t{+}1 & -2t{+}4 & -t{+}2 \\ 3t{+}4 & t{+}3 & -t{+}2 & 2t{+}6 \end{pmatrix}.$$

2. Show that the dual lattice to Λ_t corresponds to $t^* = -\frac{4+3t}{3+t}$ or to $t^* = -\frac{1-3t}{2-t}$, according to whether $t \le -\frac{1}{2}$ or $t \ge -\frac{1}{2}$.

3. Show that Λ_t is isodual exactly for the two values

$$t_0 = 2 - \sqrt{5} \quad \text{and} \quad t_1 = -3 + \sqrt{5}\,(= -1 - t_0)$$

of the parameter t.

4. Deduce from the previous questions that there exists (up to isometry) a unique 4-dimensional isodual lattice L having an automorphism of order 5, and that its Hermite invariant is $\gamma(L) = 2\frac{(2+\sqrt{5})^{1/2}}{5^{3/4}} = 1.231\ldots$

5. Show that for $t = t_0$, $\varphi : (e_1, e_2, e_3, e_4) \mapsto (e_2^*, -e_1^* + e_3^*, -e_2^* + e_4^*, -e_3^*)$ defines an isometry of Λ_t onto Λ_t^*, and that $\varphi^2 = -\mathrm{Id}$; deduce from this that Λ_t is of symplectic type, and that $\mathrm{Aut}^{\#}(\Lambda_t)$ is of order 40, isomorphic to $C_4 \times D_5$.

11.8.3. The notation is that of Proposition 11.8.8. Consider the matrices

$$M(t) = \begin{pmatrix} 2 & 0 & 1 & 1 & 1 & -1 \\ 0 & 2 & 1 & 1 & 1 & 0 \\ 1 & 1 & 2 & 1 & 1 & 0 \\ 1 & 1 & 1 & 2 & 1 & 0 \\ 1 & 1 & 1 & 1 & 1{+}t/2 & -1/2{-}t/4 \\ -1 & 0 & 0 & 0 & -1/2{-}t/4 & 1{+}t/2 \end{pmatrix} \quad \text{and} \quad P = \begin{pmatrix} 0 & 0 & 1 & 1 & 0 & 1 \\ 2 & 0 & 1 & 1 & 1 & -1 \\ 1 & 1 & 1 & 2 & 1 & 0 \\ 1 & 1 & 2 & 1 & 1 & 0 \\ 1 & 0 & 1 & 1 & 1 & 0 \\ 0 & 1 & 0 & 0 & 0 & 0 \end{pmatrix}.$$

1. Show that $\mathcal{B} = (\varepsilon_1 - \varepsilon_2, \varepsilon_1 + \varepsilon_2, \varepsilon_1 - \varepsilon_3, \varepsilon_1 - \varepsilon_4, w_1(t), w_2(t))$ is a basis for Λ_t with Gram matrix $M(t)$.

2. Show that ${}^tPM(t)^{-1}P = M(t^*)$; deduce from this that Λ_t is isometric to the dual of Λ_{t^*}.

3. Show that for $t = 2$ (resp. $t = \frac{2}{3}$), Λ_t is isometric to $\mathbb{E}_6$ (resp. E_6^*).

4. For $i = 1, 2, 3$, set $w_i' = u_i - \frac{\sqrt{t}}{2} v_i$. For $f \in \mathrm{Aut}(\mathbb{D}_4)$, let j be the index such that $f(u_i) \equiv u_j \mod \mathbb{D}_4^*$. Show that there exists a unique automorphism $\tilde{f}$ (resp. $\tilde{f}'$) of Λ_t whose restriction to $\mathbb{D}_4$ is equal to f, and such that $\tilde{f}(w_i) = w_j$ (resp. $\tilde{f}'(w_i') = w_j'$).

5. Suppose from now on that $\frac{2}{3} < t < 2$. Show that $S(\Lambda_t)$ is equal to the orbit of w_1 under the action of $\mathrm{Aut}(\Lambda_t)$ (hence we have $N(\Lambda_t) = 1 + \frac{t}{2}$), and that the components in $\mathcal{B}$ of the minimal vectors in Λ_t do not depend on t.

6. Show that $\mathrm{Aut}(\Lambda_t)$ is of order $2304 = 2^8 \cdot 3^2$ and is isomorphic to the direct product $C_2 \times \mathrm{Aut}(D_4)$.

7. Let $Q = M(\sqrt{4/3})^{-1}P$, and let $Q' = Q^2$. Show that

$$Q' = \begin{pmatrix} -1 & 1 & 0 & 0 & -1 & 1 \\ -1 & 0 & 0 & 0 & -1 & 1 \\ 1 & 0 & 1 & 0 & 1 & -1 \\ 1 & 0 & 0 & 1 & 1 & -1 \\ 0 & 0 & 0 & 0 & 0 & 1 \\ 0 & 0 & 0 & 0 & -1 & 1 \end{pmatrix} \quad \text{and} \quad Q'^3 = \begin{pmatrix} 1 & 0 & 0 & 0 & 1 & -2 \\ 0 & 1 & 0 & 0 & 1 & -1 \\ 0 & 0 & 1 & 0 & 0 & 1 \\ 0 & 0 & 0 & 1 & 0 & 1 \\ 0 & 0 & 0 & 0 & -1 & 0 \\ 0 & 0 & 0 & 0 & 0 & -1 \end{pmatrix}.$$

Verify that P is an isoduality of order 12 of $\Lambda_{\sqrt{4/3}}$, that P^3 is not of symplectic type, and that P^6 is not of orthogonal type.

8. Deduce from Question 5 that the perfection rank of Λ_t is constant on the interval $(\frac{2}{3}, 2)$. Taking for granted that this rank is equal to 20 (which can be proved by calculating it for *one* (rational) value of t), show that after rescaling L_t to norm 2 and choosing a suitable parameter, the family Λ_t can be identified with a segment connecting contiguous forms (in the sense of Voronoi) representing $\mathbb{E}_6^*$ and $\mathbb{E}_6$.

9. Show that Λ_1 is the unique eutactic lattice inside the interval $(\frac{2}{3}, 2)$.

11.8.4. Let $L_t = \mathbb{A}_4 \perp \sqrt{5t}\,\mathbb{Z}$, let $u = \frac{4\varepsilon_0 - \varepsilon_1 - \varepsilon_2 - \varepsilon_3 - \varepsilon_4}{5} \in \mathbb{A}_4$ (a generator for $\mathbb{A}_4/\mathbb{A}_4^*$), let $v = \sqrt{5t}\,\varepsilon_5$, and let $w = u + \frac{2}{5}v$. Denote by Λ_t the lattice generated by L_t and w.

1. Show that $(\varepsilon_0 - \varepsilon_1, \varepsilon_0 - \varepsilon_2, \varepsilon_0 - \varepsilon_3, w, v)$ is a basis for Λ_t.

2. Show that its Gram matrix is $\begin{pmatrix} 2 & 1 & 1 & 1 & 0 \\ 1 & 2 & 1 & 1 & 0 \\ 1 & 1 & 2 & 1 & 0 \\ 1 & 1 & 1 & \frac{4(t+1)}{5} & 2t \\ 0 & 0 & 0 & 2t & 5t \end{pmatrix}$.

3. Show that the path $[\frac{1}{4}, 4] : t \mapsto \Lambda_t$ connects lattices similar to $\mathbb{D}_4$ and to $\mathbb{D}_4^*$, and that the map $t \mapsto t^* = \frac{1}{t}$ transforms Λ_t into a lattice similar to Λ_t^*.

4. Show that Λ_1 is isodual, with invariants $N = \frac{7}{5}$, $s = 10$, $\det = 1$.

5. Show that for $t = \frac{2}{3}$, Λ_t is similar to $\mathbb{A}_5^3$.

6. Show that the values of $s(\Lambda_t)$ are as follows: $t = 4$: $s = 20$; $t = \frac{2}{3}$: $s = 15$; $\frac{2}{3} < t < 4$ or $t > 4$: $s = 10$; $\frac{1}{4} \le t < \frac{2}{3}$: $s = 5$; $t < \frac{1}{4}$: $s = 1$; prove that $s = s^* = 10$ for $\frac{2}{3} < t < \frac{3}{2}$.

7. Show that except for $t \in \{\frac{1}{4}, \frac{2}{3}, \frac{3}{2}, 4\}$, $\mathrm{Aut}(\Lambda_t)$ is isomorphic to $C_2 \times S_5$, and that there exists a single orbit of minimal vectors.

8. Verify that the perfection rank of Λ_t is constant and equal to 10 on the interval

$\frac{2}{3} < t < 4$. [Argue as in Exercise 11.8.3, Questions 5 and 7.]
[The family L_t is a path through $\mathbb{A}_5^3$ and $\mathbb{D}_4$ for the generalization of the Voronoi algorithm to the set of G-lattices, G cyclic of order 5; see Chapter 13.]

9. Show that the unique eutactic lattice in the interval $(\frac{2}{3}, 4)$ is $\Lambda_{3/2} \sim A_5^2$.

11.8.5. Let $H = \langle \sigma \rangle$ be a cyclic subgroup of odd order $q \geq 3$ of $\mathrm{O}(E)$, whose character is a multiple of the cyclotomic character. (We thus have $n = m\varphi(q)$ for some $m \in \mathbb{Z}$.) Let moreover $\tau \in \mathrm{O}(E)$ be an element of order 4 in the centralizer of σ. Let Λ be a τ-isodual H-lattice.

1. Show that $G^{\#} = \langle \tau \rangle$ is of order $4q$, and that Λ is a σ'-isodual lattice for $\sigma' = \tau\sigma$.
2. For every character χ of H of order q, let E_χ be the subspace of E with character $\chi + \overline{\chi}$. Denote by p_χ^+ (resp. p_χ^-) the multiplicity of the eigenvalue $+i$ (resp. $-i$) of τ acting on E_χ. Show that the set of $v \in \mathrm{End}^s(E_\chi)$ satisfying the three conditions (1) $v\sigma = \sigma v$, (2) $v\tau = -\tau v$, and (3) ${}^t\overline{v} = v$ is a subspace of E_χ of dimension $2p_\chi^+ p_\chi^-$.
3. Show that the σ'-isodual lattices which are σ'-extreme possess at least $1 + \sum_{\chi+\bar{\chi}} 2p_\chi^+ p_\chi^-$ pairs of minimal vectors.
4. Show that the maximum dimension of the variety of σ'-isodual lattices is equal to $\frac{\varphi(q)}{2} \lfloor \frac{m^2}{2} \rfloor$. (This dimension depends on the action of τ.)
5. Show that for $m = 2$ and $q = 3$, the σ'-isodual lattices are the lattices in the class d_6 that we described in Theorem 11.6.4.
6. Show that if $m = 1$, the σ'-isodual lattices are isolated.

11.11 Notes on Chapter 11

The notion of a G-lattice was introduced with the aim of improving lower bounds for regulators of Galois extensions of number fields obtained using the geometrical methods initiated by Remak; references can be obtained from the reference list in Bergé and Martinet's paper [B-M2], in which the method of Remak is extended to relative extensions and the general theory of G-lattices is outlined in Section 11.7. Beforehand, the possibility of improving on lower bounds for regulators by making use of a Galois action had been exploited by Schoof and Washington in the 1988 paper [S-W] devoted to cyclic extension of degree 5 of $\mathbb{Q}$, where they calculate the extremum which occurs at the beginning of Exercise 11.4.2.

Most of the results of Sections 11.1 to 11.4 are extracted from [B-M3], which, however, we made more precise in some places (e.g., structure of the determinants of cyclotomic lattices). The progress obtained from the extension to G-lattices of the Voronoi algorithm is studied in Chapter 13.

The notion of an isodual lattice is a more recent one. The fact that certain important lattices are similar ("equivalent") to their duals is most certainly in no way a recent observation. However, the importance of isodual lattices was not fully recognized among the specialists of lattice theory before the 1994 article [Bu-Sa] of Buser and Sarnak and its appendix by Conway and Sloane, where they begin to be considered independently of the applications to Abelian varieties, for the complex theory of which symplectic lattices play a major rôle (they describe "principally polarized Abelian varieties"). Maybe, methods from algebraic geometry will allow us in the future to give somewhat more natural reasons for the existence of some important symplectic lattices, by constructing directly the corresponding Abelian varieties. For recent expository papers on isodual lattices, see [Ber3] and [Mar3].

For the questions of extremality, we have followed [B-M4], leaving aside the extension of the problem of algebraicity that we encountered in Chapter 3 (Bergé's Theorem 3.8.11). Isodual lattices of orthogonal or symplectic type are indeed proportional to algebraic lattices. This result, together with more general ones which have been proved by Bavard in the setting of Riemannian geometry, can be found in his preprint [Bav5]; see also his Crelle paper [Bav1].

Quebbemann's original definition of modular lattices (cf. Definition 11.5.4) contains the extra hypothesis that modular lattices should be even. Our definition, which was that of the French edition, is also Conway and Sloane's in their preface to the third edition [C-S"] of [C-S].

The 3-dimensional example of Exercise 11.5.5 shows that the Voronoi-like Theorem 11.7.10 relating extremality to perfection and eutaxy does not extend to $(G^{\#}, G)$-extreme lattices without some extra hypothesis on the pair $(G^{\#}, G)$. A possibility might be to restrict ourselves to pairs $(G^{\#}, G)$ affording a $\mathbb{Q}$-irreducible representation of G. This hypothesis, which holds for σ-isodual lattices of symplectic or orthogonal type (assuming $\sigma \neq \pm\mathrm{Id}$), could perhaps also suffice to obtain finiteness and algebraicity results.

Our examples of dense isodual lattices in dimensions 5 and 6 are due to Conway and Sloane ([C-S9]). They will be again considered in Chapter 13, where they appear as special points on paths attached to an equivariant generalization of the Voronoi algorithm; this method is outlined in some of the exercises above.

It is natural to seek a theory of lattices over rings of algebraic integers (or even more generally over maximal orders in semi-simple algebras over $\mathbb{Q}$), and to try to extend to this setting the theory of cyclotomic lattices. A reasonable notion can be found in Eva Bayer's paper [Bay5]; see also [Bay3], and [Bay4] for specific questions related to modular lattices. However, it seems that an infinity of connected components show up for these spaces of "$\mathbb{Z}_K$-lattices", and that no finiteness theorem can be expected (as in Jaquet and Sigrist's example in [Ja-S2]). In this respect, cyclotomic lattices (together with generalizations involving finite quaternion groups) look somewhat exceptional.

12 Cross-Sections

Introduction

In the whole chapter, we denote by E an n-dimensional Euclidean vector space. By a *cross-section* (for short, a *section*) *of a lattice* Λ, we mean the intersection of Λ with some subspace F of E. In general, it is to be understood that this intersection is indeed a (full) lattice in F.

We have seen in Section 3.7 that there exists an easy test to decide whether a lattice having a perfect hyperplane section is itself perfect: it suffices to test whether its minimal vectors lying outside the hyperplane span E. We shall more generally define a notion of *perfection relative to a section*, which is not *a priori* assumed to be of codimension 1, nor to be perfect (but relative perfection will imply usual perfection only in case of a perfect section); we shall also define a notion of *relative eutaxy*, and obtain a characterization *à la Voronoi* of lattices which are in an obvious sense *relatively extreme*.

We first study in Sections 12.1 and 12.2 a kind of inverse problem, namely that of finding all perfect sections (with the same norm) of a given lattice. We then study in Section 12.3 how to pass from relative perfection to proper perfection, and give in Section 12.4 a characterization of relative extremality. Section 12.5 is devoted to a problem closely related to the previous ones, that of *patchwork lattices*, those which are direct sums of lattices isometric to given lattices, all having the same norm. When one chooses n times the lattice $\mathbb{Z}$, this problem is that of determining perfect lattices having a basis of minimal vectors. (Note, however, that up to dimension 7, all perfect lattices do have such a basis, a result proved *a priori* by Csóka and confirmed afterwards by the classification of perfect lattices. This makes clear that classifying perfect patchwork lattices may be a very difficult problem when the pieces to be glued together have small dimensions).

12.1 Embedding a Lattice in a Larger One

We consider an integral lattice Λ in E. We denote by N its norm, by D its determinant, and by s the number of its pairs of minimal vectors. Similarly, we denote by N^*, $D^* = D^{-1}$, and s^* the analogous invariants of its dual Λ^*, and by q the annihilator of Λ^*/Λ; q is the smallest of the integers a such that $\sqrt{a}\,\Lambda$ is integral. The rescaled lattice $\Lambda' = \sqrt{q}\,\Lambda^*$ is integral, and its corresponding invariants are $N' = qN^*$, $D' = q^n D^{-1}$, and $s' = s^*$.

We moreover consider a lattice M, also integral, of dimension $m \leq n$ and determinant d, and we ask ourselves the following:

Problem: Under what condition does the lattice M, once rescaled to the norm of Λ, embed in Λ?

A necessary condition for such an embedding to exist is that the norm of M should divide that of Λ. This remark applies to the lattice $P_7^2 \sim \mathbb{E}_7^*$, which is integral primitive of norm 3, and hence does not contain any perfect section with the same norm of dimension $m > 1$, since integral perfect lattices of dimensions $m \in [2,6]$ are even (see Table 6.5.5).

In what follows, we shall assume that the norm of M divides that of Λ, and we even restrict ourselves by rescaling M to the case where $N(M) = N$. We moreover exclude the trivial case where $N(M) = 1$.

Suppose that M is embedded as a relative lattice inside Λ. Let $F = \mathbb{R}M$ be its span in E, and let $\widetilde{M} = F \cap \Lambda$; this is a lattice in F, which contains M as a subgroup of finite index. We shall say that $\widetilde{M}$ is a *(cross-)section* of Λ.

The map which attaches to a section L of Λ its orthogonal $L^\perp$ in Λ^* induces a one-to-one correspondence between m-dimensional sections of Λ and $n-m$-dimensional sections of Λ^*. The determinant of $L^\perp$ is easily calculated thanks to the formula

$$\det(L) = \det(\Lambda)\det(L^\perp)$$

(Corollary 1.3.5). Since replacing Λ^* by Λ' multiplies the determinant of $L^\perp$ by q^{n-m}, the formula above can also be written in the following form, in which $M^\perp_{\Lambda'}$ denotes the orthogonal of M in Λ':

$$\frac{\det(M)}{[\widetilde{M} : M]^2} = q^{-(n-m)} \det(M^\perp_{\Lambda'})\,.$$

Since the product $[\widetilde{M} : M]^2 \cdot \det(M^\perp_{\Lambda'})$ is an integer, we have:

Theorem 12.1.1. *If an integral lattice M with the same norm as Λ embeds in Λ, then the determinant of Λ divides the product $q^{n-m}\det(M)$.* □

The data we shall use in the following examples can be found in Sections 5.2, and 8.5.

Example 12.1.2. *If $\Lambda = K_{12}$, we have $N = 4$, $D = 3^6$, and $q = 3$. This implies that an integral lattice M of norm 2 or 4 and dimension $m \geq 7$ does not embed in K_{12} if its determinant is not divisible by 3. In particular the scaled copies to norm 4 of the root lattices $\mathbb{E}_7$, $\mathbb{D}_7$, and $\mathbb{A}_7$, whose determinants are powers of 2, are not sublattices of the Coxeter–Todd lattice K_{12}. Note that* *the existence of an inclusion* $\mathbb{A}_7 \subset \mathbb{A}_8$ *shows that* $\sqrt{2}\,\mathbb{A}_8$ *is not a sublattice of* K_{12}*, a result which cannot be proved using the preceding argument, since* $\det(\mathbb{A}_8) = 9$.

Example 12.1.3. *Similarly, since* $\mathbb{A}_6^{(2)} = P_6^5$ *has determinant* 7^3*, and the determinants of the perfect lattices* M_m^i *in smaller dimensions are not divisible by* 7*, the lattice* P_6^5 *has no perfect section with the same norm in dimensions* 4 *and* 5*. However,* $\mathbb{A}_2$ *and* $\mathbb{A}_3$ *both embed in* P_6^5.

Example 12.1.4. *In the case of* $P_5^2 \sim \mathbb{A}_5^3$*, we have* $D = 2 \cdot 3^4$ *and* $q = 6$*, which shows that* P_5^2 *has no perfect section (with the same norm) in dimensions* 3 *and* 4*. In the case of* $P_6^2 \sim \mathbb{E}_6^*$*, we have* $D = 3^5$ *and* $q = 3$*, which shows that* P_6^2 *has no other perfect sections than* P_2^1 *and* P_5^2*, which are indeed sections. The same kind of argument shows that the perfect sections of* P_6^4 *and of* P_6^6 *are* P_2^1*,* P_3^1*, and* P_5^2.

The examples above together with Theorem 4.6.7 on inclusions among root lattices immediately prove:

Proposition 12.1.5. *In dimensions* $m \leq 6$*, the inclusions among perfect lattices scaled to the same norm are exactly those of Table 6.5.7.* □

Consideration of the spectra (Definition 1.4.3) of M and Λ may also be used to prove some impossibilities. Indeed, if the embedding $M \subset \Lambda$ is possible, then for every $x \in S(M)$, the number of vectors in $S(M)$ having a given scalar product with x is bounded above by the corresponding number of vectors in $S(\Lambda)$. This yields an ordering relation on the set of spectra, and the spectrum of Λ must be larger than or equal to that of M. (An obvious example is the lower bound $s \geq s(M)$.)

A useful special case concerns the value 0 of the scalar products: if all scalar products on $S(\Lambda)$ are nonzero, the same property must hold on $S(M)$.

Example 12.1.6. *Root lattices of dimension* $n \geq 3$ *all have pairs of orthogonal minimal vectors, whereas the lattices* P_5^3 *and* P_6^2 *do not. We recover in this way the results of Example 12.1.4.*

Example 12.1.7. *Similarly, scalar products are nonzero on* $S(P_7^{12})$*. Since* P_7^{12} *has norm* 6*, its maximal perfect sections of norm* 6 *must be similar to* $\mathbb{A}_2$ *(and* $\mathbb{A}_2$ *does embed in* P_7^{12}*).*

The methods we used until now do not suffice to establish up to dimension 7 the exact list of inclusions among perfect lattices with the same norm. Table 6.5.8 was obtained thanks to a general C-program for inclusions written by Batut, the basic principles of which we now briefly explain.

Consider as above two integral lattices Λ and M. First rescale M to the norm of Λ, then test the integrality properties of Theorem 12.1.1 and compare the spectra of M and Λ. If no contradiction shows up, we make use of a *back tracking* algorithm (analogous to the ones which are used to test isometry between two given lattices): we try to map a given basis $(e_1, \dots, e_m)$ for M onto a sequence $(f_1, \dots, f_m)$ of vectors in Λ having the same spectrum. In practice, the e_i (hence also the f_i) are minimal vectors, and we improve the efficiency of the algorithm by first performing an LLL reduction on the basis for M.

This program has been used in particular when M is one of the 47 perfect lattices of dimension $m \in [2, 7]$, using as Gram matrices reduced forms of those which are displayed in [C-S5]. It allows us to test in a reasonable amount of time the embedding of a P_m^i lattice in a given lattice Λ provided that Λ be of dimension, say, at most 12. One first lists once and for all the inclusions among perfect lattices (which certifies Table 6.5.8), and then makes use of the fact that any obstruction to an embedding $P_m^i \hookrightarrow \Lambda$ is also an obstruction to an embedding $P_{m'}^{i'} \hookrightarrow \Lambda$ any time the embedding $P_{m'}^{i'} \hookrightarrow P_m^i$ is impossible.

The method we just sketched also applies to the case where M and Λ have the same dimension. Note that by Theorem 6.5.4, an inclusion $P_m^i \hookrightarrow P_m^j$ occurs only if $m = 7$ and $(i, j) = (1, 33)$, and then corresponds to the inclusion $\mathbb{A}_7 \subset \mathbb{E}_7 \simeq \mathbb{A}_7^2$.

Table 12.1.10 below contains the list of all inclusions $P_m^i \hookrightarrow \Lambda$ where Λ is one of the "classical" lattices that we met along the previous chapters, with dimension at most 12 (or exceptionally 13).

Theorem 12.1.8. *All perfect lattices of dimension $m \leq 7$ which are integral when scaled to the minimum 4, and in particular all perfect lattices of dimension $m \leq 6$, embed in the Leech lattice Λ_{24}.*

Proof. We see in the table that all these lattices embed in K_{10} or in $\Lambda_{12}^{\min}$, except for P_7^{30}, which embeds in $\Lambda_{13}^{\mathrm{mid}}$ (and we remark that K_{10} contains all perfect lattices of dimension $m \leq 6$.) □

Remark 12.1.9. *Proposition 12.3.2 below will show that if Λ is perfect, and if a lattice M of dimension $m < n$ with s' pairs of minimal vectors is contained in Λ and has the same norm as Λ, we then have the lower bound $s - s' \geq \frac{(n-m)(n+m+1)}{2}$. In particular, if s has its minimal possible value, namely $\frac{n(n+1)}{2}$, we must have $s' \leq \frac{m(m+1)}{2}$. For instance, if M is itself perfect, of dimension 6, and if s is minimal, M must be one of the lattices P_6^5, P_6^6, or P_6^7.*

12.2 X-Rays of Lattices

As in the preceding section, we consider an integral lattice $\Lambda \subset E$, of dimension n and norm N, and we aim to list *all* perfect lattices of norm N and dimension $m \geq 2$ which are contained in Λ. We restrict ourselves to *well-rounded* lattices (those whose minimal vectors span E).

Table 12.1.10. Perfect sections of dimension $m \leq 7$ of some lattices of dimension ≥ 7.

Lattice	2	3	4	5	6	7
K_7	1	1	1,2	1,3	1,3	impossible dimension
K_8	1	1	1,2	1,3	1,3	
K_9	1	1	1,2	1,2,3	1,3,4,5,6,7	7,10,26,27,28,32
K_{10}	1	1	1,2	1,2,3	1,2,3,4,5,6,7	3,7,10,26,27,28,32
K_{11}	1	1	1,2	1,2,3	1,2,3,4,5,6,7	3,7,10,26,27,28,32
K_{12}	1	1	1,2	1,2,3	1,2,3,4,5,6,7	3,7,10,26,27,28,32
K'_8	1	1		2	2,4,6	3,27
K'_9	1	1	1,2	2,3	2,4,6	3,7,27,28
K'_{10}	1	1	1,2	2,3	2,4,6	3,7,27,28
${K'_{10}}^*$	1	1	1			6,16
Λ_8	1	1	1,2	1,3	1,3,7	1,4,33
Λ_9	1	1	1,2	1,3	1,3,7	1,4,33
Λ_{10}	1	1	1,2	1,3	1,3,7	1,4,5,33
$\Lambda_{11}^{\min}$	1	1	1,2	1,2,3	1,3,4,6,7	1,4,5,7,8,28,29,31,32,33
$\Lambda_{11}^{\max}$	1	1	1,2	1,3	1,3,7	1,4,5,33
$\Lambda_{12}^{\min}$	1	1	1,2	1,2,3	1,2,3,4,6,7	1,3,4,5,7,8,22,27,28,29,31,32,33
$\Lambda_{12}^{\mathrm{mid}}$	1	1	1,2	1,2,3	1,3,4,6,7	1,4,5,7,8,27,28,29,31,32,33
$\Lambda_{12}^{\max}$	1	1	1,2	1,3	1,3,7	1,4,5,33
$\Lambda_{13}^{\min}$	1	1	1,2	1,2,3	1,2,3,4,6,7	1,3,4,5,7,8,22,27,28,29,31,32,33
$\Lambda_{13}^{\mathrm{mid}}$	1	1	1,2	1,2,3	1,3,4,5,6,7	1,4,5,7,8,10,22,26,27,28,29,30,31,32,33
$\mathbb{A}_8^{(2)}$	1	1	1,2	2	2,4,6	3,7,28
$\mathbb{A}_9^{(2)}$	1	1	1,2	1,2,3	4,6	7,22,31
$\mathbb{A}_{10}^{(2)}$	1	1	1,2	2	4,5,6	7,8,10,22,30,31
$\mathbb{A}_{10}^{(3)}$	1					
$\mathbb{A}_{10}^{(4)}$	1	1				
$\mathbb{A}_{11}^{(2)}$	1	1	1,2	1,2,3	4	7
$\mathbb{A}_{12}^{(2)}$	1	1	1,2	2,3	4,5,6	7,8,10,22,27,28,29,30,31,32
$\mathbb{A}_{12}^{(3)}$	1	1	1,2			

[The columns of this table, after the name of a lattice Λ, contain for $m = 2, 3, 4, 5, 6$ the list of the indices i such that P_m^i embeds in Λ. Except for K_7 and K_8, all lattices that are displayed in the table are perfect.]

The embedding problem for perfect lattices of dimension $m \le 7$ is solved by the inclusion program referred to in Section 12.1. This no longer works if $\dim \Lambda \ge 9$, and we are faced with the problem of finding perfect lattices of small codimension in Λ, for which no classification result is available.

In theory, it is possible to list all lattices contained in Λ with a given dimension $m \le n$, whose determinants are bounded above by a given positive constant K. Indeed, let M be such a lattice. We know by the Hermite inequality that M has a basis $(e_1, \dots, e_m)$ which satisfies an inequality of the form $N(e_1)N(e_2)\dots N(e_m) \le C_m K$, where we can take $C_m = \left(\frac{4}{3}\right)^{m(m-1)/2}$, and even to $C_m = \gamma_m$ if $m \le 4$ (see Theorems 2.2.1 and 2.6.11). We may moreover assume that $N(e_1) \le N(e_2) \le \dots \le N(e_m)$, which bounds $N(e_1)$ by $(C_m K)^{1/m}$, then $N(e_2)$ by $\left(\frac{C_m K}{N(e_1)}\right)^{1/(m-1)}$, ... (a common method to prove that there are only finitely many integral lattices with a given determinant). Since we can list all vectors in Λ whose norm is smaller than a given constant K', we are left with finitely many possibilities: we consider the set of systems of m vectors satisfying the inequalities above, we calculate in each case the determinant of the lattice they generate, and we remove the lattices which have too large a determinant.

We can now *in theory* find all perfect lattices of dimension m and determinant $d \le K$ which can be embedded in Λ, by testing the perfection of all lattices in the list and keeping only one lattice per isometry class.

To work out the procedure above, it is necessary to find an upper bound K_m for the determinants of m-dimensional perfect lattices having a given norm N. For $m \le 7$, we know that we can take $K_m = (m+1)(\frac{N}{2})^m$ (the determinant of $\mathbb{A}_m$ scaled to norm N). That $\mathbb{A}_m$ is the less dense perfect lattice also beyond dimension 7 was conjectured by Coxeter. This conjecture, which might well be false for large n, is probably true in low dimensions, say, in the range 8–12. We shall return later to this question.

If we do not accept Coxeter's conjecture, an exhaustive search needs the consideration of fairly large determinants, since we are restricted to applying the Hadamard inequality, which gives $K_m = N^m$. In other words, rescaling lattices to norm 1, we must replace $\frac{m+1}{2^m}$ by 1, and we obtain exponential bounds which are too large to be of any practical use.

At any rate, the theoretical process to list relative lattices inside Λ we have described above is useless in practice. However, thanks to the fact that $n-m$ is relatively small, something can be done: indeed, the formula

$$\det(M) = q^{-(n-m)} \det(\Lambda) \det(M^{\perp}_{\Lambda'})$$

proved in Section 12.1 allows us to swap m and $n-m$.

The case of codimension 1 can be dealt with by the following algorithm, which describes the isometry classes of perfect sections of a given lattice Λ, whose determinant is bounded above by some given constant K:

• Find all vectors in Λ' of norm at most $Kq^{n-m}/\det(\Lambda)$.
• Determine the orbits of these vectors under the action of $\mathrm{Aut}(\Lambda')$ (or of $\mathrm{Aut}(\Lambda)$, this amounts to the same thing), and select one vector per orbit.
• Compute a basis for the orthogonal in Λ of each of these vectors.
• Keep one representative for these lattices in each isometry class.
• Test the perfection of these lattices.

Due to limitations of computational time and of storage capacity, the algorithm was applied only to lattices of dimension at most 12, and with a relatively small bound K; even the $\mathbb{A}_m$ bound K_m proves too large from dimension 11 onwards. Hence the results we obtained are certain only for dimensions 9 and 10, but even in these dimensions, they rely on the validity of the Coxeter conjecture.

In Table 12.2.1 below, we consider a few norm 4 lattices in dimensions 9 and 10. For each of them, we give the number of their norm 4 hyperplane sections up to a certain bound of their determinant, depending on the lattice, and then the number of those sections which are perfect. The sections are ordered according to increasing values of their determinant. We display in the last line the ranks of the perfect sections in the list; figures beyond 20 always correspond to an $\mathbb{A}_{n-1}$ section.

Table 12.2.1. Hyperplane sections of lattices in dimensions 9 and 10.

Lattice	Λ_9	K_9	K_9'	Λ_{10}	K_{10}	K_{10}'	$K_{10}'^{*}$
sect.	29	57	26	141	189	26	16
perf.	4	6	5	4	10	3	2
nb	1,3,6,28	3–6,12,21	1–4,6	1–4	1–5,7,12,14,15,28	1–3	1,2

The determinants of the perfect sections non-similar to $\mathbb{A}_{n-1}$ appear to be much smaller than that of $\mathbb{A}_{n-1}$ (scaled to the norm of Λ). This observation supports Coxeter's conjecture in dimensions 9 and 10.

The algorithm in codimension $k > 1$ is of the same kind, but listing k-dimensional lattices contained in Λ' raises important storage problems which drastically limit the dimension.

12.3 Lattices with a Fixed Cross-Section

In this section, we consider a subspace $\mathbb{E}_0$ of E, of dimension $m < n$, and a lattice $\Lambda_0 \subset E_0$ of norm N. We denote by $\mathcal{F}$ the set of lattices Λ in E such that $\Lambda \cap E_0 = \Lambda_0$ and $N(\Lambda_0) = N$.

Given a lattice Λ_1 with $\Lambda_1 \cap E_0 = \Lambda_0$, the equality $\Lambda \cap E_0 = \Lambda_0$ is equivalent to the fact that Λ be of the form $u(\Lambda_1)$ for some element u of the group

$$\mathcal{G}' = \{u \in \mathrm{GL}(E) \mid u_{|E_0} \in \mathrm{GL}(\Lambda_0)\}.$$

To study $\mathcal{F}$ locally, it is more suitable to consider the group

$$\mathcal{G} = \{u \in \mathrm{GL}(E) \mid u_{|E_0} = \mathrm{Id}_{E_0}\},$$

which is the connected component of the identity of the previous one. In a basis for E which is the union of bases for E_0 and $E_0^{\perp}$, the groups $\mathcal{G}'$ and $\mathcal{G}$ are represented by the sets of matrices of the form

$$\begin{pmatrix} M_0 & M' \\ M'' & M_1 \end{pmatrix} \quad \text{and} \quad \begin{pmatrix} I_m & M' \\ M'' & M_1 \end{pmatrix}$$

where the "small" matrices satisfy the conditions

$$M_0 \in \mathrm{GL}_m(\mathbb{R}),\ M_1 \in \mathrm{GL}_{n-m}(\mathbb{R}),\ M' \in \mathcal{M}_{m,n-m}(\mathbb{R}), M'' \in \mathcal{M}_{n-m,m}(\mathbb{R}).$$

Since Λ/Λ_0 is torsion-free, every basis $\mathcal{B}_0$ for Λ_0 extends to a basis $\mathcal{B}$ for Λ, which allows us to represent the lattices in the orbit of Λ under $\mathcal{G}$ by Gram matrices of the form

$$A = \begin{pmatrix} A_0 & B \\ {}^tB & A' \end{pmatrix}$$

with $A_0 = \mathrm{Gram}(\mathcal{B}_0)$, $A' \in \mathrm{Sym}_n(\mathbb{R})$, and $B \in \mathcal{M}_{n,n-m}(\mathbb{R})$; because Λ belongs to $\mathcal{F}$, the matrices A' and B must be chosen in such a way that A be positive definite, and have the same minimum as A_0.

Definition 12.3.1. *We say that a lattice Λ is* (strictly) extreme relative to Λ_0, *or* (strictly) $\mathcal{F}$-extreme, *if its Hermite invariant is a (strict) local maximum among those of lattices in $\mathcal{F}$ (which are not similar to Λ)*

The methods of Chapter 10 do not directly apply to this situation: on the one hand, the groups $\mathcal{G}$ and $\mathcal{G}'$ are not stable under transposition; on the other hand, the fact that an element $u \in \mathrm{GL}(E)$ belongs to $\mathcal{G}$ or to $\mathcal{G}'$ does not ensure that u stabilizes $\mathcal{F}$. Nevertheless, the relative extremality of a lattice $\Lambda \in \mathcal{F}$ can be tested using deformations of Λ by elements of $\mathcal{G}$ which are close to the identity and which preserve the norm of Λ; this last condition amounts to saying that u belongs to a suitable cone in $\mathcal{G}$. The norm of a vector after deformation can be calculated by means of the equalities

$$N(u(x)) = u(x) \cdot u(x) = {}^tuu(x) \cdot x = \varphi_x({}^tuu);$$

as usual, it is more suitable to replace $\mathcal{G}$ by its *symmetrized manifold*

$$X = \{{}^tuu \mid u \in \mathcal{G}\}.$$

Proposition 12.3.2. *The tangent space at the identity of X is the vector subspace*

$$\mathcal{T} = \{v \in \mathrm{End}^s(E) \mid v(E_0) \subset E_0^{\perp}\}$$

of $\mathrm{End}^s(E)$. *Its dimension is* $\frac{(n-m)(n+m+1)}{2}$.

Proof. The identity of X possesses a fundamental system of neighbourhoods made of elements $u \in \mathcal{G}$ represented by matrices of the form $\begin{pmatrix} I_m & \varepsilon M' \\ \varepsilon M'' & I_{n-m}+\varepsilon M_1 \end{pmatrix}$, where the matrices M', M'', M_1 have a fixed norm (for arbitrary chosen norms on the matrix spaces to which they belong), and ε is a sufficiently small positive real number. The product ${}^t u u$ corresponds to the matrix

$$I_n + \varepsilon \begin{pmatrix} 0 & M' + {}^tM'' \\ {}^tM' + M'' & M_1 + {}^tM_1 \end{pmatrix} + \varepsilon^2 \begin{pmatrix} {}^tM'' M'''' & {}^tM'' M_1 \\ {}^tM_1 M'' & {}^tM' M' + {}^tM_1 M_1 \end{pmatrix},$$

whose first–order term runs through the space $\mathcal{T}$ when $\varepsilon, M', M'', M_1$ vary.

The assertion on the dimension is then clear, since the matrices above depend on $m(n-m) + \frac{(n-m)(n-m+1)}{2}$ parameters. □

We are thus led to the following definition, which will prove to be an appropriate one to adapt the theory of Korkine & Zolotareff and Voronoi to our situation.

Definition 12.3.3. *We say that a lattice is* perfect *(resp.* eutactic*) relative to* Λ_0 *if it is* $\mathcal{T}$*-perfect (resp.* $\mathcal{T}$*-eutactic); we also say* $\mathcal{F}$-perfect *(resp.* $\mathcal{F}$-eutactic*).*

Lemma 12.3.4. *Let* p_0 *be the orthogonal projection onto* E_0.

1. *The orthogonal of* $\mathcal{T}$ *in* $\mathrm{End}^s(E)$ *(for the Voronoi scalar product) is* $\mathcal{T}^{\perp} = \{v \in \mathrm{End}^s(E) \mid v(E_0) \subset E_0$ *and* $v(E_0^{\perp}) = \{0\}\}$.
2. *The projection onto* $\mathcal{T}$ *of* $v \in \mathrm{End}^s(E)$ *is* $v - p_o \circ v$.
3. *A lattice* $\Lambda \in \mathcal{F}$ *is relatively perfect if and only if the set of projections onto the minimal vectors in* $\Lambda \smallsetminus E_0$ *is* $\mathcal{T}$*-perfect.*
4. *For every relative lattice* $\Lambda_0' \subset \Lambda_0$ *having the same norm as* Λ_0*, the lattices in* $\mathcal{F}$ *which are perfect (resp. extreme) relative to* Λ_0' *are perfect (resp. extreme) relative to* Λ_0*; in particular, a perfect (resp. extreme) lattice is perfect (resp. extreme) relative to all its sections of a lower dimension having the same norm.*

Proof. The formula for $\mathcal{T}^{\perp}$ is easily checked on the respective representations in matrix form of the two sets, namely

$$\begin{pmatrix} 0 & * \\ {}^t* & * \end{pmatrix} \quad \text{and} \quad \begin{pmatrix} * & 0 \\ 0 & 0 \end{pmatrix}.$$

The second assertion is equivalent to the fact that the projection onto $\mathcal{T}^\perp$ of $v \in \mathrm{End}^s(E)$ is equal to $p_o \circ v$. This equality is clear, for $p_0 \circ v$ belongs to $\mathcal{T}^\perp$ and coincides with v for $v \in \mathcal{T}^\perp$.

The third assertion is a consequence of the second one, since $p_x = p_0 \circ p_x$ on E_0.

The part of assertion (4) relative to extremality is evident, and that relative to perfection is a consequence of the obvious inclusion $\mathcal{T} \subset \mathcal{T}'$ between the spaces $\mathcal{T}$ and $\mathcal{T}'$ attached to E_0 and E_0'. □

The theorem below, the first assertion of which is just a lemma for the proof of Theorem 12.4.1, contains statements 3.5.3 and 3.7.2.

Theorem 12.3.5. *Let $\Lambda \in \mathcal{F}$. Set $S' = S(\Lambda) \smallsetminus S(\Lambda_0)$.*

1. *If Λ is relatively extreme, S' spans E.*
2. *If Λ is relatively perfect, S' spans E.*
3. *If E_0 is of codimension 1 in E and if S' spans E, Λ is relatively perfect.*
4. *If Λ_0 is perfect, Λ itself is perfect if and only if it is relatively perfect.*

Proof. (1) Let $(e_1, e_2, \ldots, e_n)$ be a basis for Λ which extends a basis for Λ_0 and let Λ_1 be the relative lattice in E generated by $e_1, e_2, \ldots, e_{n-1}$. Its norm is equal to that of Λ. By Lemma 12.3.4, Λ is extreme relative to Λ_1, and (1) follows from Proposition 3.7.2.

(2) Otherwise, the vectors in $S(\Lambda) \smallsetminus S(\Lambda_0)$ are contained in a hyperplane H of E. Let $e \in H^\perp$. For all $x \in S'$, we have $e \cdot x = 0$, hence $p_x \circ p_e = 0$. Since Λ is relatively perfect, Lemma 12.3.4 shows that every $v \in \mathcal{T}$ can be written as an $\mathbb{R}$-linear combination

$$v = \sum_{x \in S'} \lambda_x \left(p_x - p_0 \circ p_x\right),$$

which implies that $v \circ p_e$ is zero. Taking $v = p_e - p_0 \circ p_e$, we see that $p_e = p_0 \circ p_e$, i.e. that e belongs to E_0.

Let $e' \in E_0^\perp$. Consider a basis $\mathcal{B}_0 = (e_1, \ldots, e_m)$ for E_0 (resp. $\mathcal{B}_0' = (e_{m+1}, \ldots, e_n)$ of $E_0^\perp$), with $e_1 = e$ (resp. $e_{m+1} = e'$) and $e_i \cdot e = 0$ for $2 \le i \le m$ (resp. $e_j \cdot e' = 0$ for $m + 1 \le j \le n$). The endomorphism v of E which exchanges e and e' and maps the other vectors of $\mathcal{B}_0 \cup \mathcal{B}_C'$ onto 0 belongs to $\mathcal{T}$. Hence we have $v \circ p_e = 0$, which contradicts the equality $v \circ p_e(e) = v(e) = e'$.

(3) Our hypothesis implies that there exists a basis $(e_1, e_2, \ldots, e_n)$ for E whose elements belong to S'. Since the dimension of $\mathcal{T}$ is precisely n (Proposition 12.3.2), it suffices to prove that the projections p_i onto the vectors e_i have independent projections onto $\mathcal{T}$.

Otherwise, there exists a relation $\sum_{i=1}^n \lambda_i \left(p_i - p_0 \circ p_i\right) = 0$ with coefficients λ_i not all zero, for instance, with $\lambda_1 \neq 0$. Applying this relation to a nonzero vector $x \in E$ orthogonal to $e_2, \ldots, e_n$, we obtain the equality

$p_1(x) - p_0(p_1(x)) = 0$, a contradiction, for $p_1(x)$ is nonzero and proportional to $e_1 \notin E_0$, whereas $p_0(p_1(x)) \in E_0$.

(4) It suffices to verify that the projections p_x, $x \in S(\Lambda)$ span $\mathrm{End}^s(E)$, hence that the projections of p_x onto the complementary subspaces $\mathcal{T}$ and $\mathcal{T}'$ of $\mathrm{End}^s(E)$ span each of these subspaces. This is clear for $\mathcal{T}$, by the very definition of relative perfection, and this is proved for $\mathcal{T}'$ by identifying this space with $\mathrm{End}^s(E_0)$.

[(3) is also a consequence of (4) together with Theorem 3.5.3.] □

12.4 A Characterization of Relatively Extreme Lattices

The aim of this section, in which we keep the notation of the previous one, is to prove the following theorem, which reduces to the usual Voronoi theorem (Theorem 3.4.6) when $E_0 = 0$:

Theorem 12.4.1. *A lattice is relatively extreme if and only if it is relatively perfect and relatively eutactic. Moreover, relatively extreme lattices are strictly relatively extreme.*

Proof. The first step consists in characterizing relatively extreme lattices by a condition analogous to that in Theorem 3.4.5. It results from Proposition 12.3.2 that for sufficiently small v, the fact that Λ_0 is stable under the $u \in \mathrm{GL}(E)$ such that $uu = \mathrm{Id} + v$ is equivalent to the condition $v(E_0) \subset E_0^{\perp}$, i.e. $v \in \mathcal{T}$. Using then Lemmas 3.4.3 and 3.4.4, we obtain the following variant of the theorem above (recall that $S' = S(\Lambda) \smallsetminus S(\Lambda_0)$):

Theorem 12.4.2. *A lattice Λ is extreme relative to Λ_0 if and only if*

$$v \in \mathcal{T}\,, \quad \min_{x \in S'} \varphi_x(v) = 0 \quad \textit{and} \quad \mathrm{Tr}(v) \leq 0 \Longrightarrow v = 0\,. \qquad (*)$$

Moreover, every relatively extreme lattice is strictly relatively extreme. □

Proof of 12.4.1. To prove Theorem 12.4.1, it suffices to prove that condition $(*)$ above is equivalent to the fact that Λ is both $\mathcal{T}$-perfect and $\mathcal{T}$-eutactic.

We first show that a lattice which is both $\mathcal{T}$-eutactic and $\mathcal{T}$-perfect is relatively extreme in the same way we proved Theorem 3.4.6: we consider a $v \in \mathcal{T}$ such that $\min_{x \in S} \varphi_x(v) = 0$ and $\mathrm{Tr}(v) \leq 0$; we have $\varphi_x(v) \geq 0$ and $\mathrm{Tr}(v) \leq 0$, hence $\varphi_x(v) = 0$ by the definition of $\mathcal{T}$-eutaxy, and then $v = 0$ by the definition $\mathcal{T}$-perfection.

Consider now a lattice Λ which satisfies condition $(*)$.

We again prove as in Chapter 3 that Λ is $\mathcal{T}$-perfect: if $v \in \mathcal{T}$ is such that $\varphi_x(v) = 0$ for all $x \in S$ (or in S'), we may assume (replacing v by $-v$ if need be) that $\mathrm{Tr}(v) \leq 0$, and Theorem 12.4.2 then implies that $v = 0$.

There remains to show that Λ is $\mathcal{T}$-eutactic, which is proved once more using Stiemke's Theorem 3.3.1. but replacing in the proof given in Chapter 3 the identity map by the endomorphism j which is the identity on E_0 and zero on $E_0^\perp$. Let $v \in \mathcal{T}$. We must verify that the system of inequalities $\varphi_x(v) \geq 0$ and $\mathrm{Tr}(v) \leq 0$ implies the system of equalities $\varphi_x(v) = 0$ and $\mathrm{Tr}(v) = 0$. For $k \in \mathbb{R}$, set $v' = v - kj$. We have $v' \in \mathcal{T}$ by Lemma 12.3.4, $\mathrm{Tr}(v') = \mathrm{Tr}(v) - kn$, and $\varphi_x(v') = \varphi_x(v) - kN(\Lambda)$, hence $\min_{x \in S} \varphi_x(v') = \min_{x \in S} \varphi_x(v) - kN(\Lambda)$. We now choose k in order that this minimum be zero. It is clear that $k \geq 0$, hence that $\mathrm{Tr}(v') \leq \mathrm{Tr}(v) \leq 0$. Condition $(*)$ implies that v' is zero, hence that v is the homothetic transformation of ratio k. From $\varphi_x(v) = -kN(\Lambda) \leq 0$ and $\mathrm{Tr}(v) = kn \geq 0$, we deduce that $k = v = 0$, whence the equalities $\varphi_x(v) = 0$ for all $x \in S$ and $\mathrm{Tr}(v) = 0$. This completes the proof of Theorem 12.4.1. □

12.5 Patchwork Lattices

In this section, we consider the family $\mathcal{F}$ of lattices $\Lambda \subset E$ which are direct sums $\Lambda_1 \bigoplus \Lambda_2 \bigoplus \dots \bigoplus \Lambda_r$, where the Λ_i are relative lattices, isometric to given lattices Λ_i° in Euclidean spaces $E_1^\circ, E_2^\circ, \dots, E_r^\circ$, all having the same norm. Such lattices are called *patchwork lattices*. (The patches are the lattices Λ_i.) We set $m_i = \dim E_i^\circ$; thus $m_1 + m_2 + \dots + m_r = n$. We are chiefly interested in lattices in $\mathcal{F}$ whose Hermite invariant is a local maximum among the lattices in $\mathcal{F}$ (*$\mathcal{F}$-extreme lattices*). The family $\mathcal{F}$ is compact, which guarantees the existence of *critical lattices for $\mathcal{F}$* (on which γ attains its absolute maximum).

Without loss of generality, we may (and shall) assume that E is the orthogonal sum of the E_i°; of course, this does not imply that the lattices in $\mathcal{F}$ are an *orthogonal* sum of lattices isometric to the Λ_i°, but simply that there exists a decomposition $\Lambda = \bigoplus \Lambda_i$ (hence also a decomposition $E = \bigoplus E_i$ with $E_i = \mathbb{R}\Lambda_i$), and for all i an isometry σ_i of Λ_i° onto Λ_i.

There exists a neighbourhood of the identity in $\mathrm{GL}(E)$ on which the transformations u which are isometries on the Λ_i are those which satisfy the conditions $v(E_i) \subset E_i^\perp$ for $1 \leq i \leq r$, where $v = {}^t uu - \mathrm{Id}$, as usual. Thus, after having chosen a particular lattice Λ_0 of $\mathcal{F}$, for instance $\Lambda^\circ = \perp_i \Lambda_i^\circ$, we have the following interpretation for $\mathcal{F}$:

$$\mathcal{F} = \{u(\Lambda_0) \mid \forall i,\, ({}^t uu - \mathrm{Id})(E_i) \subset E_i^\perp \text{ and } N(u(\Lambda_0)) = N(\Lambda_0)\}\,.$$

This description of $\mathcal{F}$ suggests the following definitions, which can be interpreted along the line of Proposition 12.3.2. using a suitable subgroup of $\mathrm{GL}(E)$ and the tangent space at the origin of its symmetrized manifold:

Definition 12.5.1. *We set*

$$\mathcal{T} = \{v \in \mathrm{End}^s(E) \mid \forall i,\, v(E_i^\circ) \subset E_i^\perp\}\,.$$

We say that Λ is $\mathcal{F}$-perfect *(resp.* $\mathcal{F}$-eutactic*) if it is $\mathcal{T}$-perfect (resp. $\mathcal{T}$-eutactic).*

In a basis for E which is the union of bases for the spaces E_i°, the matrices of $\mathcal{T} \subset \mathrm{Sym}_n$ are the matrices of the form

$$\begin{pmatrix} 0 & A_{1,2} & A_{1,3} & \dots & A_{1,r} \\ {}^tA_{1,2} & 0 & A_{2,3} & \dots & A_{2,r} \\ {}^tA_{1,3} & {}^tA_{2,3} & 0 & \dots & A_{3,r} \\ \vdots & \vdots & \vdots & \ddots & \vdots \\ {}^tA_{1,r} & {}^tA_{2,r} & {}^tA_{3,r} & \dots & 0 \end{pmatrix}, \qquad (\mathbf{12.5.2})$$

with $A_{i,j} \in \mathcal{M}_{m_i,m_j}(\mathbb{R})$ and the null matrices of dimensions $m_1, m_2, \dots, m_r$ on the diagonal.

We have for patchwork lattices results analogous to those of Theorems 12.3.5 and 12.4.1, due to Baril. We state some of them below, leaving the proofs to the reader (who can find them in [Bari]).

Theorem 12.5.3

1. *Let $\Lambda = \bigoplus_i \Lambda_i$ be an $\mathcal{F}$-perfect lattice. Then $S(\Lambda) \smallsetminus \cup S(\Lambda_i)$ spans E, and has at least $\sum_{i<j} m_i m_j$ elements.*
2. *If Λ is $\mathcal{F}$-perfect and if all lattices Λ_i° are perfect, Λ itself is perfect.* □

Theorem 12.5.4 *A lattice is $\mathcal{F}$-extreme if and only if it is $\mathcal{F}$-perfect and $\mathcal{F}$-eutactic, and $\mathcal{F}$-extreme lattices are strictly $\mathcal{F}$-extreme.* □

12.6 Exercises for Chapter 12

12.2.1. Let Λ be a norm 1, well-rounded lattice. The aim of this exercise is to prove that if Λ possesses at least $n+1$ pairs $\pm x$ of minimal vectors, then $\det(\Lambda) \leq \frac{3}{4}$. We denote by $\pm e_1, \dots, \pm e_n, \pm e_{n+1}$ the minimal vectors in Λ, and we assume that $(e_1, e_2, \dots, e_n)$ is a basis for E.

1. Let $\Lambda' \subset \Lambda$ be the sublattice generated by $S(\Lambda)$. Show that if Λ contains strictly Λ', then $\det(\Lambda) \leq \frac{1}{4}\det(\Lambda')$.

From now on, we suppose that $S(\Lambda)$ generates Λ.

2. Show that if $e_{n+1} = \sum_{i \leq n} e_i$, we have $\det(\Lambda) \leq \frac{1}{n+1}(1+\frac{1}{n})^n \sim_{n\to\infty} \frac{e}{n}$. [Use Example 9.5.1 (3).]

3. Calculate an upper bound of $\det(\Lambda)$ in case e_{n+1} is a linear combination of $e_1, e_2, \dots, e_n$ with coefficients $0, \pm 1$ and $(m \geq 2)$ nonzero coefficients, and prove in this case the upper bound $\det(\Lambda) \leq \frac{3}{4}$.

4. Suppose that $a_{n+1}e_{n+1} = a_1 e_1 + \dots + a_n e_n$, where the a_i are coprime integers. Show that $\det(\Lambda) \leq \frac{1}{4}$ if there is at least one $a_i \geq 2$.

5. Prove that the bound $\det(\Lambda) \leq \frac{3}{4}$ is general, and that equality holds if and only if $\Lambda \sim \mathbb{A}_2 \perp \mathbb{A}_1^{n-2}$.

12.3.1. Let E_0' be a subspace of E_0 such that $\Lambda_0' = \Lambda_0 \cap E_0$ is a lattice in E_0'. Show that if Λ_0 is perfect relative to Λ_0', a lattice $\Lambda \in \mathcal{F}$ is perfect relative to Λ_0' if and only if it is perfect relative to Λ_0.

12.3.2. Show that a lattice which is perfect relative to a rational lattice is rational.

12.3.3. On the set of lattices having a given section Λ_0, we consider the equivalence relation

$$\Lambda \sim \Lambda' \Longleftrightarrow \exists u \in \mathrm{GL}(E),\ u(\Lambda) = \Lambda',\ u(S(\Lambda)) = S(\Lambda'),\ \text{and } u_{|E_0} = \mathrm{Id}\,,$$

whose classes are called Λ_0*-minimal classes*.

1. Show that there are only finitely many classes.

[Argue as in Section 9.1 and use Exercise 2.2.1.]

2. Deduce from this that modulo isometries preserving E_0, there are only finitely many perfect lattices relative to Λ_0.

12.4.1. Let G be a finite subgroup of $\mathrm{O}(E)$ which stabilizes Λ_0. Let $\mathcal{F}_G$ be the set of lattices in $\mathcal{F}$ whose automorphism group contains G, and let $\mathcal{T}_G$ be the centralizer of G in $\mathcal{T}$. [Note that lattices in $\mathcal{F}_G$ are G-lattices in the sense of Chapter 11.]

1. Show that a lattice is $\mathcal{F}_G$-extreme if and only if it is $\mathcal{T}_G$-perfect and $\mathcal{T}_G$-eutactic.
2. Show that $\mathcal{F}_G$-perfect lattices such that Λ_0 is $G \cap \mathrm{O}(E_0)$-perfect are G-perfect G-lattices.

12.5.1. Show that $\mathbb{D}_n^+$ ($n \geq 10$ even) is extreme relative to its sections isometric to $\mathbb{D}_{n-1}$, but not as a patchwork lattice for the splitting $\mathbb{D}_{n-1} \oplus \mathbb{A}_1$.

12.5.2. Show that if the lattices Λ_i° are all rational, the corresponding patchwork lattices which are perfect as patchwork lattices are themselves rational.

12.5.3. Define the notion of a minimal class for patchwork lattices, and prove in this setting the finiteness result of Exercise 12.3.3.

12.5.4. We consider the generalization of the notion of a patchwork lattice in which we drop the condition $\sum_i m_i = n$. Generalize to this setting the results of Sections 12.3 to 12.5, including those of the exercises.

12.7 Notes on Chapter 12

The results proved in the first two sections are extracted from the joint work [Bt-M] with Batut. [In [Bt-M], we forgot to fill in the row for $\mathbb{A}_{11}^2$ in Table 6.3. This error has been corrected in Table 12.1.10.]

To consider lattices having a given section with the same norm from the point of view of extremality is not something new: the famous error of Chaundy consists in claiming that Λ_9 is extreme after having only proved that is extreme *relative* to $\Lambda_8 \sim \mathbb{E}_8$!

The first general relative results were proved by Barnes in [Bar5], who discovered in particular the characterization of relative perfection with respect to a hyperplane section (see Section 3.7), and gave examples showing that a good notion of a relatively eutactic lattice is not the easy one that could have been expected. (In his language of quadratic forms, Barnes considers *extensions of quadratic forms*.)

Systematic investigations on relatively perfect lattices were carried out only recently. Since they rely on an extension of the Voronoi algorithm, we postpone comments to the notes on Chapter 13.

13 Extensions of the Voronoi Algorithm

Introduction

As usual, E stands for an n-dimensional Euclidean vector space.

Our aim is to classify perfect lattices within certain families by making use of a contiguity algorithm which generalizes the classical Voronoi algorithm of Chapter 7, which only applied to the family $\mathcal{L}$ of all lattices in E. We shall consider some families of lattices defined in Chapters 11 and 12 (G-lattices, lattices with a given section, patchwork lattices).

As in the usual case, the algorithm is described in a natural way in terms of quadratic forms on $\mathbb{R}^n$, or which amounts to the same thing, in terms of positive definite matrices in Sym_n. Nevertheless, we shall put emphasis where possible on the more intuitive point of view of lattices.

The fact that we are able to generalize the Voronoi algorithm to the families above, but not to other important families (e.g., isodual lattices, or pairs (Λ, Λ^*) in $E \times E$) is related to the affine structure of these families. In the case of G-lattices, we consider the orbits of a lattice Λ_0 under the action of the centralizer $\mathcal{G}$ of G. For $u, v \in \mathcal{G}$, if $u + v$ is invertible, this is again an element of $\mathcal{G}$. In the case of lattices with a given section (or of patchwork lattices), we obtain all lattices from a particular one Λ_0 by means of the formula $\Lambda = (\mathrm{Id} + v)(\Lambda_0)$, where v runs through a cone in a vector subspace of $\mathrm{End}^s(E)$.

In each case, we attach to the family we consider a vector subspace of $\mathrm{End}^s(E)$, denoted by $\mathcal{T}$ or $\mathcal{T}_0$, and we make use of the notion of $\mathcal{T}$-perfection as defined in Chapter 3. Using the dictionary lattices–quadratic forms, we attach to $\mathcal{T}$ and to a pair $(\mathcal{B}, \mathcal{B}^*)$ of bases for E a vector subspace of Sym_n, which we shall often still denote by $\mathcal{T}$. The Voronoi algorithm for $\mathcal{T}$ is defined for very general subspaces of Sym_n, but the classification of $\mathcal{T}$-perfect forms is not interesting in general, for only very special spaces $\mathcal{T}$ yield fruitful applications of the notion of $\mathcal{T}$-perfection.

The generalization of the classical theory is carried out in Sections 13.1 and 13.2. Sections 13.3—13.5 are devoted to G-lattices; in particular, we give in Section 13.5 relatively general results on cyclotomic G-lattices. Section 13.6 is devoted to lattices having a given section and to patchwork constructions previously considered in Chapter 12. The chapter ends with a seventh section containing some examples.

13.1 Contiguity Relative to a Space of Symmetric Matrices

In this section, we consider a subspace $\mathcal{T}$ of the vector space Sym_n of $n\times n$ real symmetric matrices, which we identify with the corresponding subspace of the space $\mathcal{Q}_n$ of quadratic forms over $\mathbb{R}^n$. The spaces Sym_n and $\mathcal{Q}_n$ are endowed with the *Voronoi scalar product* defined on Sym_n by $\langle A, B\rangle = \mathrm{Tr}(AB)$, and the space $\mathbb{R}^n$ with its usual scalar product $x \cdot y = \sum x_i y_i$.

To avoid certain difficulties (notably in Proposition 13.1.8), we assume throughout this chapter that $\dim \mathcal{T}$ is at least 2.

Recall some notation and calculations of Chapter 3. In the language of lattices, the notion of $\mathcal{T}$-perfection involves the restriction to $\mathcal{T}$ of the forms φ_x, or (which amounts to the same thing), the orthogonal projection onto $\mathcal{T}$ in $\mathrm{End}^s(E)$ (for the scalar product $\langle u, v\rangle = \mathrm{Tr}(v \circ u)$) of the projections p_x (in E) onto the vectors belonging to the family we consider. With respect to a pair $(\mathcal{B}^*, \mathcal{B})$ of dual bases, these notions can be expressed in the following way: p_x corresponds to the (positive, rank 1) matrix $X\,{}^tX \in \mathrm{Sym}_n$, where $X \in \mathcal{M}_{n,1}(\mathbb{Z})$ is the column of components of x in $\mathcal{B}$, hence

$$\mathrm{Mat}(p_x, \mathcal{B}^*, \mathcal{B}) = \frac{1}{x \cdot x}\, X\,{}^tX\,.$$

[The equalities $\mathcal{B} = (e_1, e_2, \dots, e_n)$, $x = \sum_k x_k\, e_k$, and $p_x(e_i^*) = \sum_k a_{k,i}\, e_k$ imply $a_{j,i} = p_x(e_i^*) \cdot e_j$, hence $(x \cdot x)\, a_{i,j} = (x \cdot x)\, a_{j,i} = (x \cdot e_i^*) \cdot (x \cdot e_j^*) = x_i x_j$.]

In practice, the factor $x \cdot x = N(x)$ does not matter, for it does not depend on the vector x we consider, since x will be a minimal vector in a given lattice. So, we shall forget $N(x)$ in Definition 13.1.1 below.

For $u \in \mathrm{End}^s(E)$, with matrix M with respect to a pair $(\mathcal{B}, \mathcal{B}^*)$ of dual bases, we have

$$\mathrm{Mat}(p_x \circ u, \mathcal{B}, \mathcal{B}) = \mathrm{Mat}(p_x, \mathcal{B}^*, \mathcal{B})\, \mathrm{Mat}(u, \mathcal{B}, \mathcal{B}^*)\,,$$

hence $\langle p_x, u\rangle = \langle X^tX, M\rangle = \langle M, X^tX\rangle$. Consequently, the definition below in terms of matrices interprets the orthogonal projection ω_x of p_x onto a subspace $\mathcal{T}_0$ of $\mathrm{End}^s(E)$ as the projection in Sym_n or in $\mathcal{Q}_n$ onto the subspace $\mathcal{T}$ corresponding to $\mathcal{T}_0$:

Definition 13.1.1. *Let $\mathcal{T}$ be a vector subspace of Sym_n (or of $\mathcal{Q}_n$), let $x \in \mathbb{R}^n$, and let $X \in \mathcal{M}(n, 1)$ be the column of components of x. We set $P_x = P_X = X\,{}^tX$, and we denote by Ω_x or Ω_X the orthogonal projection of p_x onto $\mathcal{T}$ in Sym_n (or in $\mathcal{Q}_n$).*

Translating conditions of $\mathcal{T}$-perfection and $\mathcal{T}$-eutaxy, we obtain the following definition, which reduces to the usual one when $\mathcal{T}$ is the whole space $\mathcal{Q}_n$:

Definition 13.1.2. *The* $\mathcal{T}$-perfection rank *of a form Q (or of its matrix A) is the dimension of the subspace of $\mathcal{T}$ spanned by the Ω_x, $x \in S(Q)$. We* *say that a quadratic form Q, or that its matrix A, is $\mathcal{T}$-*perfect *if the Ω_x, $x \in S(Q)$ span $\mathcal{T}$, and that it is $\mathcal{T}$-*eutactic *if there exists a relation $A^{-1} = \sum_{x\in S(Q)} \rho_x \, \Omega_x$ with strictly positive coefficients.*

We know (Corollary 3.2.6) that the minimal vectors of a perfect form span $\mathbb{R}^n$. Such a property no longer holds for general $\mathcal{T}$-perfect forms. The proposition below permits us to analyse this condition:

Proposition 13.1.3. *Let Q be a positive definite quadratic form. The following conditions are equivalent:*

1. *There exist strictly positive coefficients λ_x, $x \in S(Q)$ such that the quadratic form $\sum_{x\in S} \lambda_x \, P_x$ is positive definite.*
2. *For every system λ_x, $x \in S$ of strictly positive coefficients, the quadratic form $\sum_{x\in S} \lambda_x \, P_x$ is positive definite.*
3. *$S(Q)$ spans $\mathbb{R}^n$.*

Proof. For all $y \in \mathbb{R}^n$ and coefficients $\lambda_x > 0$ ($x \in S$), we have

$$\sum_{x\in S(Q)} \lambda_x \, \langle P_x, P_y \rangle = \sum_{x\in S(Q)} \lambda_x \, (x \cdot y)^2 \geq 0 \,,$$

with equality if and only if y is orthogonal to x for all $x \in S(Q)$. We immediately deduce from this that (1) and (3) on the one hand, and (2) and (3) on the other hand, are equivalent. □

Definition 13.1.4. *Let Q be a positive definite quadratic form. The $\mathcal{T}$-*Voronoi domain *$\mathcal{D}_{Q,\mathcal{T}}$ (or $\mathcal{D}_Q$) of Q is the convex hull in $\mathcal{T}$ of the closed half-lines containing the Ω_x, $x \in S(Q)$. If Q has matrix A, we also write $\mathcal{D}_{A,\mathcal{T}}$ or $\mathcal{D}_A$ instead of $\mathcal{D}_Q$.*

The $\mathcal{T}$-domain of Q is the projection onto $\mathcal{T}$ of the usual domain, which we recover when $\mathcal{T} = \mathcal{Q}_n$. We have

$$\mathcal{D}_{Q,\mathcal{T}} = \mathcal{D}_{A,\mathcal{T}} = \Big\{ \sum_{x\in S(Q)} \lambda_x \, \Omega_x \mid \lambda_x \geq 0 \Big\} \,,$$

and the interior *in* $\mathcal{T}$ of the $\mathcal{T}$-Voronoi domain of Q is

$$\mathrm{Int}_{\mathcal{T}}(\mathcal{D}_{Q,\mathcal{T}}) = \Big\{ \sum_{x\in S(Q)} \lambda_x \Omega_x \mid \lambda_x > 0 \Big\} \,.$$

We define in a similar way the $\mathcal{T}$-Voronoi domain $\mathcal{D}_{\Lambda,\mathcal{T}}$ of a lattice.

To say that Q is $\mathcal{T}$-perfect (or that Λ is $\mathcal{T}$-perfect) amounts to saying that the interior of $\mathcal{D}_{Q,\mathcal{T}}$ (or of $\mathcal{D}_{\Lambda,\mathcal{T}}$) *in* $\mathcal{T}$ in not empty.

Remark 13.1.5. *By its very definition, $\mathcal{D}_{Q,\mathcal{T}}$ is the convex hull of the half-lines $\mathbb{R}_{\geq 0}\,\Omega_x$, $\pm x \in S(Q)$. However, since we are considering a projection of the classical domain, it may happen that some half-lines $\mathbb{R}_{\geq 0}\,\Omega_x$ are not edges: such examples are described in Section 13.4. It may also happen that some lines $\mathbb{R}\,\Omega_x$ coincide even if the corresponding vectors x are non-proportional.*

We are now going to define a notion of contiguity as we did in the classical case (Sections 7.1 and 7.2). We begin by a first definition, compare Definition 7.1.9:

Definition 13.1.6. *Let Q be a quadratic form on $\mathbb{R}^n$, with matrix A, and let $\mathcal{F}$ be a facet (i.e., a face of codimension 1) of its domain in $\mathcal{T}$ (for short, a* facet *of $\mathcal{D}_{Q,\mathcal{T}}$). A* facet vector *for $\mathcal{F}$ is a vector F in $\mathcal{T}$ (hence a quadratic form, that we identify with its matrix), which is orthogonal to $\mathcal{F}$ and moreover oriented towards the interior of $\mathcal{D}$ whenever Q is $\mathcal{T}$-perfect.*

When Q is $\mathcal{T}$-perfect, its facet vectors F are well defined up to multiplication by a positive scalar and are characterized by the following conditions:

1. $\forall\, x \in S(Q)$, $\langle F, \Omega_x \rangle \geq 0$;
2. $\langle F, \Omega_x \rangle = 0 \iff \Omega_x \in \mathcal{F}$.

As in the classical situation, we then consider the family of forms $Q_\theta = Q + \theta F$. If x is one of the vectors which "lie in $\mathcal{F}$" (i.e., such that $P_x = X\,{}^tX \in \mathcal{F}$), we have $Q_\theta(x) = Q(x)$ for every θ. Otherwise, we have $Q_\theta(x) = Q(x) + \theta F(x) \neq Q(x)$. For sufficiently small θ, since the minimal vectors of Q_θ come from those of Q, and the norms of the minimal vectors of Q which do not lie in $\mathcal{F}$ are effectively modified, the minimum of Q_θ satisfies exactly one of the following conditions:

- it is equal to that of Q, and $S(Q_\theta)$ coincides with with the set of vectors of $S(Q)$ which lie in $\mathcal{F}$;
- it is strictly smaller than that of Q, and $S(Q_\theta)$ is then contained in the set of vectors of $S(Q)$ which do not lie in $\mathcal{F}$.

When Q is $\mathcal{T}$-perfect, thanks to the chosen orientation of F, the forms Q_θ and Q have the same minimum for any sufficiently small $\theta \geq 0$.

Definition 13.1.7. *Let Q be a $\mathcal{T}$-perfect form, with minimum m, let $\mathcal{F}$ be a facet of $\mathcal{D}_{Q,\mathcal{T}}$, and let F be a facet vector for $\mathcal{F}$. We denote by $\rho \in\,]0, +\infty]$ the least upper bound of the $\theta \in \mathbb{R}_{\geq 0}$ such that Q_θ is positive definite of minimum m. We say that $\mathcal{F}$ (or that F) is a* dead-end *if $\rho = +\infty$. When $\mathcal{F}$ is not a dead-end, the form Q_ρ is called the $\mathcal{T}$-*contiguous form to *Q* through *$\mathcal{F}$ (or* along *F).*
[When $\dim \mathcal{T} = 1$, we make the convention that the unique facet of $\mathcal{D}_{Q,\mathcal{T}}$ is a dead-end.]

Examples of dead-ends will be given in Section 13.4. Note that dead-ends do not exist in Voronoi's usual situation (see Theorem 7.2.1, and the comment at the end of the proof of Proposition 13.1.8 below).

Proposition 13.1.8. *1. If $\mathcal{F}$ is not a dead-end, the form Q_ρ is $\mathcal{T}$-perfect.*
2. $\mathcal{F}$ is a dead-end if and only if the form F is positive.
3. If $\mathcal{F}$ is a dead-end, the vectors $X \in S(Q)$ such that $X\,{}^tX \in \mathcal{F}$ do not span $\mathbb{R}^n$.

Proof. If F is positive, we have $Q_\theta(x) \geq Q(x) \geq m$ for all nonzero $x \in \mathbb{Z}^n$, hence $\rho = +\infty$. Otherwise, there exists $x \in \mathbb{R}^n$ with $F(x) < 0$. Since there exist points in $\mathbb{Z}^n$ which are arbitrary close to the line $\mathbb{R}x$, there exists $y \in \mathbb{Z}^n$ with $F(y) < 0$. Then $Q_\theta(y)$ is strictly negative for θ sufficiently large, and such a θ is an upper bound for ρ, whence (2).

Moreover, we have $Q_\rho(x) \geq m$ for all $x \neq 0$ in $\mathbb{Z}^n$, but since Q_θ takes for all $\theta > \rho$ values smaller than m on $\mathbb{Z}^n \smallsetminus \{0\}$, there exists $z \in \mathbb{Z}^n$ with $Q_\rho(z) = m$ and $F(z) < 0$. The vector z is then a minimal vector of Q_ρ which is not a minimal vector of Q. Since the rank of the system of vectors $\Omega_x \in \mathcal{F}$ ($x \in S(Q)$) is equal to $\dim \mathcal{T} - 1$, the rank of the system of the Ω_x, $x \in \mathcal{F}$ together with Ω_z is equal to $\dim \mathcal{T}$, whence (1).

To prove (3), we consider a dead-end $\mathcal{F}$ whose elements are n matrices $X_1{}^tX_1, \dots, X_n{}^tX_n$ such that the columns X_i span $\mathbb{R}^n$. Since the X_i are then minimal vectors for all forms Q_θ, Hadamard's inequality shows that the polynomial $\det(A + \theta F)$ is bounded on $[0, +\infty[$, hence constant. This is possible only if $F = 0$, a contradiction. [When $\mathcal{F}$ is a dead-end, we have in particular $\dim \mathcal{F} \leq \frac{n(n-1)}{2}$, which suffices to ensure that no dead-end may exist in the case of the usual Voronoi algorithm, see Theorem 7.2.1.] □

Remark 13.1.9. *In terms of lattices, $\mathcal{T}$-contiguity is defined as in the classical case: a facet form F is the matrix with respect to a pair $(\mathcal{B}, \mathcal{B}^*)$ of a basis $\mathcal{B}$ for a lattice Λ and its dual basis $\mathcal{B}^*$ of a symmetric endomorphism u (the corresponding facet vector), and the contiguous lattice to Λ for u is obtained by considering the images of Λ by $\sqrt{\mathrm{Id} + \theta u}$, $\theta \geq 0$.*

It should be noted that the set of facet vectors for a lattice Λ solely depends on $S(\Lambda)$; consequently, this is the same for Λ and for its sublattice Λ_0 generated by $S(\Lambda)$. The invariants ρ and ρ_0 attached to Λ and Λ_0 through a given facet satisfy the inequality $\rho \leq \rho_0$, and if equality holds, we transform the contiguous lattice to Λ_0 into that of Λ by adjoining to Λ_0 vectors which transformed Λ_0 into Λ. In terms of quadratic forms, this comparison of Λ_0 and Λ is not clear at all, since we must consider bases for Λ_0 and for Λ, which are necessarily distinct.

For the proof of the following statement, which partly generalizes Proposition 7.1.4, we refer the reader to Exercise 13.1.2.

Proposition 13.1.10. *The symmetric positive endomorphisms of $\mathcal{T}$ form a convex half-cone, for which the angle of any two edges is not larger than $\frac{\pi}{2}$, and this value is attained only on pairs (u, v) such that $v \circ u \,(= u \circ v) = 0$.* □

13.2 The Voronoi Algorithm Relative to a Space of Symmetric Matrices

We keep the notation of Section 13.1. In particular, $\mathcal{T}$ stands for a subspace of $\mathcal{Q}_n$, of dimension at least 2, that we identify with a subspace of Sym_n.

We moreover suppose that $\mathcal{T}$-perfect forms satisfy the (WR) hypothesis (*Well–Rounded*; see Section 1.4), which holds in all cases we shall consider, and which will allow us to use Proposition 13.1.3:

The minimal vectors of forms span $\mathbb{R}^n$. **(WR)**

We state the following lemma in terms of matrices.

Lemma 13.2.1. *Let* $A, B \in \mathrm{Sym}_n$ *be two positive definite matrices with the same minimum* m, *such that* $A - B \in \mathcal{T}$.

1. *For every* $C \in \mathcal{D}_{A,\mathcal{T}} \cap \mathcal{D}_{B,\mathcal{T}}$, *we have* $\langle C, A\rangle = \langle C, B\rangle$.
2. *If* $\mathrm{Int}(\mathcal{D}_{A,\mathcal{T}}) \cap \mathcal{D}_{B,\mathcal{T}}$ *is not empty, then* $\mathcal{D}_{A,\mathcal{T}} \subset \mathcal{D}_{B,\mathcal{T}}$.
3. *If* A *is moreover* $\mathcal{T}$*-perfect, then* $B = A$.

Proof. Decompose C as a sum $C = \sum_{x\in S(A)} \lambda_x\, \Omega_x$ with coefficients $\lambda_x \geq 0$. Since $A - B \in \mathcal{T}$, we have

$$\begin{aligned}\langle C, A - B\rangle = \sum_{x\in S(A)} \lambda_x \langle \Omega_x, A - B\rangle &= \sum_{x\in S(A)} \lambda_x \langle P_x, A - B\rangle \\ &= \sum_{x\in S(A)} \lambda_x \left(m - \langle P_x, B\rangle\right) \leq 0\,.\end{aligned}$$

We have similarly $\langle C, B - A\rangle \leq 0$, whence (1).

For $C \in \mathrm{Int}(\mathcal{D}_{A,\mathcal{T}})$, we may assume that the λ_x are strictly positive. Since $m - \langle P_x, B\rangle \leq 0$ for all $x \in S(A)$, the equality $\sum_{x\in S(A)} \lambda_x \left(m - \langle P_x, B\rangle\right) = 0$ implies that $\langle P_x, B\rangle = m$ for all $x \in S(A)$, whence $S(A) \subset S(B)$, i.e. $\mathcal{D}_{A,\mathcal{T}} \subset \mathcal{D}_{B,\mathcal{T}}$.

Since $A - B \in \mathcal{T}$, the condition $\forall\, x \in S(A),\ \langle P_x, A - B\rangle = 0$ is equivalent to $\forall\, x \in S(A),\ \langle \Omega_x, A - B\rangle = 0$, which implies $A - B = 0$ whenever A is $\mathcal{T}$-perfect. □

The set of $\mathcal{T}$-perfect forms is the set of vertices of a (non-directed) graph whose edges are the pairs of contiguous forms (contiguity is clearly a symmetric relation). We want to study the connectedness of this graph. To this end, we shall use the following lemma, for the proof of which we refer the reader to the proof of Theorem 7.3.2:

Lemma 13.2.2. (Voronoi.) *Let* Φ *be a positive definite quadratic form, and let* m *and* K *be two constants. The set of minimal vectors of positive definite forms* Q *of minimum* m *such that* $\langle Q, \Phi\rangle \leq K$ *is a finite subset of* $\mathbb{Z}^n$. □

Theorem 13.2.3. *Let $Q_0 \in \mathcal{Q}_n$ and let $\mathcal{E}$ be the set of quadratic forms Q with the same minimum m as Q_0 and such that $Q - Q_0 \in \mathcal{T}$. Then the graph of $\mathcal{T}$-contiguity for $\mathcal{T}$-perfect forms of $\mathcal{E}$ is connected.*

Proof. Let Q be a $\mathcal{T}$-perfect form in $\mathcal{E}$. Let λ_x, $x \in S(Q)$ be strictly positive real numbers, let $\Phi_0 = \sum_{x \in S(Q)} \lambda_x \, \Omega_x$, and let $\Phi = \sum_{x \in S(Q)} \lambda_x \, P_x$. Since Q is $\mathcal{T}$-perfect, the hypothesis on $\mathcal{T}$ implies that Φ is positive definite. If Φ_0 belongs to $\mathcal{D}_{Q_0,\mathcal{T}}$, Lemma 13.2.1 shows that $Q_0 = Q$. Otherwise, there exists a facet form F of Q_0, such that $\langle \Phi_0, F \rangle < 0$, which implies that there exists an $x \in S(Q)$ with $\langle \Omega_x, F \rangle < 0$. Therefore, there exists a $\mathcal{T}$-contiguous form Q_1 to Q_0 along F, and we have

$$\langle Q_1, \Phi_0 \rangle = \langle Q_0, \Phi_0 \rangle + \rho \langle F, \Phi_0 \rangle < \langle Q_0, \Phi_0 \rangle \,.$$

If $\Phi_0 \notin \mathcal{D}_{Q_1,\mathcal{T}}$, we construct similarly a $\mathcal{T}$-contiguous form Q_2 to Q_1 with $\langle Q_2, \Phi_0 \rangle < \langle Q_1, \Phi_0 \rangle$. We obtain this way a sequence $Q_0, Q_1, \dots, Q_i$ of $\mathcal{T}$-perfect forms having the same minimum m and such that

$$\langle Q_0, \Phi \rangle > \langle Q_1, \Phi \rangle > \cdots > \langle Q_i, \Phi \rangle \,,$$

since because $Q_{i+1} - Q_i$ belongs to $\mathcal{T}$, we have

$$\langle Q_{i+1}, \Phi \rangle - \langle Q_i, \Phi \rangle = \langle Q_{i+1}, \Phi_0 \rangle - \langle Q_i, \Phi_0 \rangle < 0 \,.$$

Since Φ is positive definite, and we have $\langle Q_i, \Phi \rangle < \langle Q_0, \Phi \rangle$ for all i, there exists a finite subset S of $\mathbb{Z}^n$ which contains all $S(Q_i)$. Hence there are only finitely many possible sets $S(Q_i)$.

Let i and j be two indices such that $S(Q_i) = S(Q_j)$. For all $x \in S(Q_i)$, we have $\langle Q_i - Q_j, \Omega_x \rangle = \langle Q_i - Q_j, P_x \rangle$ (for $Q_i - Q_j \in \mathcal{T}$), and this is zero, because $\langle Q_i, P_x \rangle = \langle Q_j, P_x \rangle = m$. Since Q_i is $\mathcal{T}$-perfect, the $\Omega_x, x \in S(Q_i)$ span $\mathcal{T}$, and the conditions $\langle Q_i - Q_j, \Omega_x \rangle = 0$ for all $x \in S(Q_i)$ imply $Q_i - Q_j = 0$. Consequently, the sequence $Q_1, Q_2, \dots$ is finite, there exists an index k such that Φ belongs to $\mathcal{D}_{Q_k,\mathcal{T}}$, and Lemma 13.2.1 implies that $Q_k = Q$. □

In the classical Voronoi theory developed in Chapter 7, the contiguity results are written in the setting of equivalence classes modulo $\mathrm{GL}_n(\mathbb{Z})$ of quadratic forms (with a given minimum). To extend this theory, we must adapt the notion of equivalence in order to take into account the space $\mathcal{T}$. This is the aim of the definition below:

Definition 13.2.4. *We say that two quadratic forms Q and Q' with respective matrices A and A' are* $\mathcal{T}$-equivalent *if there exists $P \in \mathrm{GL}_n(\mathbb{Z})$ such that*

$${}^tPAP = A' \quad \text{and} \quad {}^tP\mathcal{T}P \subset \mathcal{T} \,.$$

[This is indeed an equivalence relation, since ${}^tP\mathcal{T}P \subset \mathcal{T}$ is equivalent to ${}^tP\mathcal{T}P = \mathcal{T}$.]

Proposition 13.2.5. *Contiguity is compatible with $\mathcal{T}$-equivalence.*

[This means that a $\mathcal{T}$-equivalence relating two $\mathcal{T}$-perfect forms induces a bijection between their sets of contiguous forms. In particular, two $\mathcal{T}$-equivalent facets of a same form which is stable under this equivalence are connected to $\mathcal{T}$-equivalent contiguous forms.]

Proof. Recall that if two quadratic forms with matrices F and F' are equivalent under a transformation with matrix P, we have $FX = F'P^{-1}X$ for every column-vector X.

Let F be a facet matrix for the domain D_Q of Q, and let $F' = {}^tPFP$. This is an element of $\mathcal{T}$. Let $X \in S(Q)$ be such that Ω_X belongs to the facet of $\mathcal{D}_{Q,\mathcal{T}}$ which is orthogonal to F. Then $P^{-1}X \in S(Q')$ is such that

$$\langle F', \Omega_{P^{-1}X}\rangle = \langle F', P_{P^{-1}X}\rangle = \langle F, P_X\rangle = \langle F, \Omega_X\rangle = 0\,.$$

Hence F' is a facet matrix for $\mathcal{D}_{Q',\mathcal{T}}$. (Note however that $\mathcal{D}_{Q',\mathcal{T}}$ is in general different from ${}^tP\mathcal{D}_{Q,\mathcal{T}}P$, for an equivalence is not an orthogonal transformation with respect to the scalar product $\langle .\,,.\rangle$.)

For every $\theta \in \mathbb{R}$, let Q_θ be the form with matrix $A_\theta = A + \theta F$. Let $\rho \in \mathbb{R} \cup \infty$ be the least upper bound of the $\theta \geq 0$ such that Q_θ is positive of minimum $m(Q)$ on $\mathbb{Z}^n$, and let Q'_θ be the transformed of Q_θ by P. Its matrix is $A'_\theta = A' + \theta F'$, and the equality $Q'_\theta(P^{-1}X) = Q_\theta(X)$ shows that ρ is the least upper bound of the $\theta \geq 0$ such that Q'_θ is positive of minimum $m(Q) = m(Q')$ on $\mathbb{Z}^n$. □

Remark 13.2.6. *When $\mathcal{T}$ comes from a subspace $\mathcal{T}_0$ of* $\mathrm{End}^s(E)$ *by the map which sends $u \in$* $\mathrm{GL}(E)$ *to its matrix in a pair made of a basis $\mathcal{B}$ for a lattice Λ and its dual basis $\mathcal{B}^*$, $\mathcal{T}$-equivalences with matrix P come from isometries $\sigma : \Lambda \to \Lambda'$ with matrix P* in *$\mathcal{B}$* (not *in the pair $(\mathcal{B}, \mathcal{B}^*)$).*

Proposition 13.2.5 justifies the possibility of searching $\mathcal{T}$-perfect forms modulo $\mathcal{T}$-equivalences, starting from an arbitrary given $\mathcal{T}$-perfect form Q_0, see Section 7.4; it may be possible (and sometimes useful; see Section 13.3) to replace $\mathcal{T}$-equivalence by a finer relation. However, it is not possible to weaken the $\mathcal{T}$-equivalence relation by replacing it by a less fine one which does not respect contiguity, for we should not be able to ensure that the algorithm actually runs through all $\mathcal{T}$-perfect forms.

Besides, one cannot guarantee without extra hypotheses that the graph is finite up to $\mathcal{T}$-equivalence, hence that the algorithm actually stops. Moreover, the notion of $\mathcal{T}$-perfection is purely formal when it does not refer to forms having special properties related to $\mathcal{T}$ (for instance, to consider G-perfect lattices which are not G-lattices does not seem to have any interesting application), and even in this case, $\mathcal{T}$-contiguous forms may no longer belong to the family we consider.

Remark 13.2.7. *In the classical case, Voronoi shows that the half-lines containing the matrices Ω_x (for x running through the set of minimal vectors of* *a perfect form Q) are exactly the edges of $\mathcal{D}_{Q,\mathcal{T}}$, and that no dead-end may occur. This is not general; counter-examples will be given in Section 13.4.*

13.3 Perfect G-Lattices

We keep the notation of the first two sections, and we place ourselves in the situation of Sections 11.1—11.4; G is a finite subgroup of $\mathrm{O}(E)$, hence defining a faithful, rational (over $\mathbb{Q}$) representation of G. We denote by $\mathcal{T}_0$ the centralizer $\mathrm{End}^s_G(E)$ of G in $\mathrm{End}^s(E)$, and by $\mathcal{G}_G$ or $\mathcal{G}$ the centralizer of G in $\mathrm{GL}(E)$. Once chosen a basis $\mathcal{B}$ for E, there corresponds to $\mathcal{T}_0$ a subspace $\mathcal{T}$ of $\mathcal{M}_n(\mathbb{R})$, obtained by mapping $u \in \mathrm{GL}(E)$ onto its matrix with respect to the bases $\mathcal{B}$ and $\mathcal{B}^*$. To $s \in G$ there corresponds the matrix $P_s \in \mathrm{Sym}_n(\mathbb{R})$ of s in $\mathcal{B}$.

The contiguity algorithm can be run, according to one's taste, in terms of quadratic forms or of lattices. We have $\dim \mathcal{T}_0 \geq 2$ if and only if the representation of G is not irreducible over $\mathbb{R}$.

As we saw in Proposition 11.3.2, the projection onto $\mathcal{T}_0$ of the projection p_x onto a nonzero vector $x \in E$ is given by the formula

$$\omega_x = \frac{1}{|G|} \sum_{s \in G} p_{sx} \,.$$

In terms of matrices, we must calculate p_x with respect to the pair of bases $(\mathcal{B}^*, \mathcal{B})$ (*not to the pair* $(\mathcal{B}, \mathcal{B}^*)$), hence, setting $A = \mathrm{Gram}(\Lambda)$, we must replace ω_x by

$$\Omega_X = \frac{1}{|G|} \sum_{s \in G} P_s X\, {}^tX {}^tP_s \in A^{-1} \mathcal{T} A^{-1}.$$

The interpretation of $\mathcal{T}$-equivalence in the language of group theory is given by the following proposition, whose proof is immediate:

Proposition 13.3.1. *Let Λ be a G-lattice in E, let $\mathcal{B}$ be a basis for Λ, let $s \in \mathrm{GL}(E)$, and let P_s be the matrix of s in $\mathcal{B}$. Then P_s is a $\mathcal{T}$-equivalence if and only if s is an isometry which stabilizes the centralizer $\mathcal{T}_0$ of G in $\mathrm{End}^s(E)$.* □

It results from Proposition 13.3.1 that two lattices which are G-isometric in the sense of Definition 11.2.1 have in particular $\mathcal{T}$-equivalent matrix representations, since G-isometries are provided by elements of the centralizer of G. We shall sometimes say that they are *G-equivalent lattices*.

Since the minimal vectors in a G-perfect G-lattice span E (Proposition 11.3.4), we may apply Theorem 13.2.3, which ensures the connectedness of the Voronoi graph of G-perfect G-lattices defining a given integral representation of G. Moreover (Corollary 11.9.5), the number of G-similarity classes of G-perfect G-lattices is finite. We can summarize these properties as a theorem, which justifies that the Voronoi algorithm can be used to find all G-perfect lattices affording a given integral representation:

Theorem 13.3.2. *Let Λ be a G-perfect G-lattice. Then the Voronoi graph of G-similarity classes of G-perfect G-lattices which are integrally equivalent to Λ is finite and connected.* □

Given a finite, rational subgroup of $\mathrm{O}(E)$, the classification of G-perfect G-lattices by means of the (extended) Voronoi algorithm needs the knowledge of the set of integral representations which induce the given rational representation of G. This is a difficult problem, which has been solved only for some particular classes of groups.

If we limit ourselves to a given integral representation for which we know a lattice (or a form) which represents it, another problem arises, namely to construct *one* G-perfect G-lattice which represents it, which we shall then be able to use as a starting-point for the Voronoi algorithm. The possibility of constructing a form contiguous to a non-necessarily $\mathcal{T}$-perfect G-form, using facet vectors in the general sense of Definition 13.1.6, allows us to construct a G-perfect form corresponding to a lattice which is integrally equivalent to the original lattice, as we did in Proposition 7.2.5 for the classical algorithm.

Finally, the search for the G-perfect lattices (or forms) in a given class of integral representations, once we know a G-perfect form, works as usual (see Section 7.4), except that usual equivalence must be replaced by G-equivalence.

Warning. Even if we are interested only in G-lattices up to usual similarity, some lattices could be missed if we do not consider the more restrictive notion of G-equivalence.

A proof of the following proposition, which is an analogue for G-lattices of Proposition 7.1.4, is the subject of Exercise 13.3.1.

Proposition 13.3.3. *The set* End_G^{s+} *is a half-cone (the* "Voronoi cone for *G"), which is the convex hull of the half-cone formed with the half-lines generated (in* $\mathrm{End}^s(E)$*) by the orthogonal projections onto the minimal stable subspaces of E (the* "projections cone for *G"), and these half-lines are the extremal edges of* End_G^{s+}. □

We are now going to give a characterization of dead-ends. Beforehand, we recall some useful formulae about projections, in which x and y denote two nonzero vectors in E:

$$p_x(z) = \frac{x \cdot z}{x \cdot x}\, x, \qquad (p_x \circ p_y)(z) = \frac{(x \cdot y)}{x \cdot x}\frac{(y \cdot z)}{y \cdot y}\, x, \qquad \textbf{(13.3.3 a)}$$

$$\langle \mathrm{Id}, p_x \rangle = \mathrm{Tr}(p_x) = 1, \quad \langle p_x, p_y \rangle = \mathrm{Tr}(p_x \circ p_y) = \frac{(x \cdot y)^2}{(x \cdot x)(y \cdot y)}. \qquad \textbf{(13.3.3 b)}$$

In view of the equality $\langle \omega_x, \omega_y \rangle = \sum_{s,t \in G} \langle p_{sx}, p_{ty} \rangle$, they imply in particular

$$\langle \omega_x, \omega_y \rangle \geq 0 \text{ and } \langle \omega_x, \omega_y \rangle = 0 \Longleftrightarrow \forall\, s,\, t \in G,\ sx \cdot ty = 0\,. \qquad \textbf{(13.3.3 c)}$$

[In other words, the angle of ω_x and ω_y is less than or equal to $\frac{\pi}{2}$, and attains this value if and only if the orbits of x and y span orthogonal subspaces in E.]

Similarly, since the orthogonal projection onto a subspace of E is the sum of the orthogonal projections onto the vectors of an an orthogonal basis for this subspace, we then have for all vectors $x \neq 0$ of E and all subspaces E', E_1, E_2 in E

$$\begin{aligned} &\langle \omega_x, p_{E'} \rangle = 0 \Longleftrightarrow \forall\, s \in G,\ sx \in E'^{\perp}, \\ &\langle p_{E_1}, p_{E_2} \rangle = 0 \Longleftrightarrow E_1 \perp E_2\,. \end{aligned} \qquad \textbf{(13.3.3 d)}$$

[When E' is stable under G, we similarly have $\langle \omega_x, p_{E'} \rangle = 0 \Longleftrightarrow x \in E'^{\perp}$.]

Theorem 13.3.4. *Assume that G is not $\mathbb{R}$-irreducible. Let Λ be a G-perfect G-lattice, let $\mathcal{F}$ be a facet of its Voronoi domain, and let $E_{\mathcal{F}}$ be the span in E of the (nonzero) vectors x such that $\omega_x \in \mathcal{F}$. Then one of the following conditions is satisfied:*

1. *We have $E_{\mathcal{F}} = E$, and $\mathcal{F}$ is not a dead-end.*
2. *$E_{\mathcal{F}}$ is a maximal stable subspace of E, the orthogonal projection $F = p_{E'}$ onto $E' = E_{\mathcal{F}}^{\perp}$ is a facet vector for $\mathcal{F}$, and $\mathcal{F}$ is a dead-end.*

Proof. That $\mathcal{F}$ is not a dead-end in the first case follows from Proposition 13.1.10. Suppose now that $E_{\mathcal{F}}$ does not span E. Its orthogonal is then a non-empty orthogonal sum of irreducible subspaces of E, and the orthogonal projections onto these subspaces constitute together with $\mathcal{F}$ a set of independent vectors in $\mathcal{T}_0 = \mathrm{End}_G^s(E)$. Since $\mathcal{F}$ has codimension 1 in $\mathcal{T}_0$, $E_{\mathcal{F}}^{\perp}$ is an irreducible subspace E' of E. For all $x \in S(\Lambda)$, we have $\langle \omega_x, p_{E'} \rangle \geq 0$ (because the ω_x are positive symmetric endomorphisms), and $\langle \mathcal{F}, p_{E'} \rangle = 0$ (because x is orthogonal to E' when ω_x lies in $\mathcal{F}$). This shows that $p_{E'}$ is a facet vector for $\mathcal{F}$; see Definition 13.1.6. □

When $\mathcal{F}$ is a dead-end, since $E_{\mathcal{F}}$ is rational over $\mathbb{Q}$, E' is also rational over $\mathbb{Q}$, which implies:

Corollary 13.3.5. *A necessary condition for the existence of dead-ends is that one of the $\mathbb{R}$-irreducible components of X be rational over $\mathbb{Q}$. In particular, if the representation afforded by G is irreducible over $\mathbb{Q}$, but not over $\mathbb{R}$, the Voronoi graph for G does not contain any dead-end.* □

Let H be a subgroup of G, and let Λ be a H-perfect G-lattice. A facet of Λ *viewed as a H-lattice*, if it is stable under G, canonically defines a facet of Λ *viewed as a G-lattice*, and the contiguous lattice for H is then also a contiguous lattice for G relative to the G-facet obtained from the original H-facet by projection in $\mathrm{End}^s_H(E)$ onto $\mathrm{End}^s_G(E)$.

As far as a non-stable facet is concerned, the contiguous lattice to Λ through a H-facet has *a priori* nothing to do with that obtained through the corresponding G-facet. For instance, if $n = \ell - 1$, $\ell \geq 5$ a prime, and G is cyclic of order ℓ, the contiguous lattices to $\mathbb{A}_n$ (together with its structure of G-lattice) cannot be isometric to $\mathbb{D}_n$ as in the usual situation where $H = \{\mathrm{Id}\}$, since $\mathbb{D}_n$ has no automorphism of order ℓ. (For $\ell = 5$ and $\ell = 7$, one finds lattices isometric to $\mathbb{A}_5$ and to $\mathbb{A}_7^{(2)}$ respectively; see Section 11.4.)

Remark 13.3.6. *If Λ is a G-perfect G-lattice such that the sublattice Λ_0 generated by its minimal vectors is reducible as a G-lattice, the facet vectors for Λ (or for Λ_0, it amounts to the same thing; see the comment after Remark 13.1.9) are the facet vectors for the G-irreducible components of Λ_0. When $\Lambda = \Lambda_0$, the Voronoi algorithm for G-perfect G-lattices integrally equivalent to Λ can be run for each of the G-irreducible components of Λ.*

13.4 Two-Dimensional Centralizers

We keep the notation of Section 13.3. The uninteresting case where $\mathcal{T} = \mathrm{End}^s_G(E)$ is 1-dimensional is that of the $\mathbb{R}$-irreducible representations, for which there exists (up to G-similarity) only finitely many G-lattices.

Dimension formulae of Theorem 11.3.8 show that according the nature of the centralizer of G, the dimension of $\mathcal{T}$ is bounded from below by 3, 4, or 6 if an $\mathbb{R}$-irreducible character is contained twice in the character χ of G. So, we have $\dim \mathcal{T} = 2$ if and only if χ is a sum of two distinct $\mathbb{R}$-irreducible characters χ_1 and χ_2, which are rational over $\mathbb{Q}$ or conjugate over a quadratic field.

More generally, the fact that χ be a sum of m distinct $\mathbb{R}$-irreducible characters $\chi_1, \chi_2, \ldots, \chi_m$ allows a simple description of the space $\mathcal{T}$: the projections $p_i = p_{E_i}$ onto the irreducible subspaces E_i with characters χ_i form an orthogonal basis for $\mathcal{T}$ (the p_i are indeed nonzero, mutually orthogonal, and their number is equal to $\dim \mathcal{T}$). The edges of the cone of positive endomorphisms of $\mathcal{T}$ are pairwise orthogonal half-lines through the p_i, and facet vectors which are dead-ends lie among the p_i.

Suppose now that $\dim \mathcal{T} = 2$. Let Λ be a G-perfect G-lattice. The Voronoi domain $\mathcal{D}$ of Λ is then an angular domain with angle at most $\frac{\pi}{2}$, whose edges are the two half-lines D_x and D_y containing the endomorphisms ω_x and ω_y associated with minimal vectors x and y in Λ. The Voronoi cone of $\mathcal{T}$ is the angular domain C with angle $\frac{\pi}{2}$, bounded by the two half-lines Δ_1 and Δ_2 containing the projections p_1 and p_2; D_x and D_y lie inside C. Exchanging x

and y if need be, we may assume that the angle $\alpha_1 = \widehat{\Delta_1, D_x}$ is a minimum, and hence also $\alpha_2 = \widehat{D_y, \Delta_2}$.

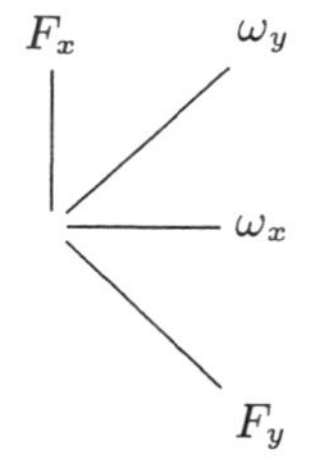

Figure 13.4.0. The Voronoi domain when $\dim \mathcal{L}_G = 2$.
This figure shows a 2-dimensional domain; its facets are the lines spanned by ω_x and ω_y, and F_x and F_y are the corresponding facet vectors.

The search for a pair (x, y) can be done explicitly in the following way: one first looks for a set $\mathcal{R}$ of representatives for the orbits of $\pm G$ acting on $S(\Lambda)$. Since E_1 and E_2 are stable under G, the scalar products $\langle p_{sz}, p_1\rangle$ and $\langle p_{sz}, p_2\rangle$, $z \in S(\Lambda)$, do not depend on the choice of $s \in G$; we thus have $\langle \omega_z, p_i\rangle = \langle p_z, p_i\rangle$. One then chooses $x \in \mathcal{R}$ which maximizes $\langle p_z, p_1\rangle$ and $y \in \mathcal{R}$ which minimizes $\langle p_z, p_2\rangle$.

When D_x and D_y are perpendicular, we have $\{D_x, D_y\} = \{\Delta_x, \Delta_y\}$, and the two facets $\{D_x\}$ and $\{D_y\}$ are dead-ends. We shall discard this case in what follows, which amounts to assuming that $S(\Lambda)$ is not contained in $E_1 \cup E_2$.

If $\alpha_1 = 0$, the facet D_x is a dead-end, since Δ_2 is a facet vector. Suppose now that $\alpha_1 > 0$, and search for a facet vector for D_x of the form $F_x = \lambda_1 p_1 + \lambda_2 p_2$. The real numbers λ_1 and λ_2 must be chosen in such a way that the two conditions

$$\lambda_1 \langle p_x, p_1\rangle + \lambda_2 \langle p_x, p_2\rangle = 0 \quad \text{and} \quad \lambda_1 \langle p_y, p_1\rangle + \lambda_2 \langle p_y, p_2\rangle > 0$$

be satisfied. Moreover, we have for all $z \in E$ the equalities

$$\langle p_z, p_1\rangle + \langle p_z, p_2\rangle = \langle p_z, \mathrm{Id}\,\rangle = 1\,.$$

To find F_x, it now suffices to calculate explicitly the scalar products $\langle p_x, p_1\rangle$ and $\langle p_y, p_1\rangle$.

When $\mathcal{X}$ contains the unit representation, we may choose for E_1 a line. In particular, this is the case whenever the representation of G defined by E is a *permutation representation*, G acting twice transitively on a basis $\mathcal{B} = (e_1, e_2, \dots, e_n)$; see Exercise 13.4.1. We then have $E_1 = \mathbb{R}(e_1 + e_2 + \dots + e_n)$ and $E_2 = \langle e_i - e_j\rangle$, $1 \le i < j \le n$. We give below a few examples of this type, which play an interesting rôle in the construction of some isodual lattices involving the root lattices $\mathbb{A}_n$ and $\mathbb{D}_n$.

The symmetric group $G = S_n$ acts on the canonical basis $\mathcal{B}_0 = (\varepsilon_0, \dots, \varepsilon_n)$ (resp. $\mathcal{B}_0 = (\varepsilon_1, \varepsilon_2, \dots, \varepsilon_n)$) for $\mathbb{R}^{n+1}$ (resp. $\mathbb{R}^n$) by permuting the indices $1, 2, \dots, n$. The lattices $\mathbb{D}_n$ and $\mathbb{A}_n$, and more generally Coxeter's lattices $\mathbb{A}_n^r$ (r an integer dividing $n+1$; see Section 5.2), are endowed with structures of G-lattices. We calculate below a few Voronoi G-graphs for such lattices.

Proposition 13.4.1. *1. The lattices $\mathbb{D}_n$, $n \geq 4$ are S_n- perfect and the orbits of minimal vectors modulo $\pm\mathrm{Id}$ are $\{\varepsilon_i - \varepsilon_j\}$ and $\{\varepsilon_i + \varepsilon_j\}$, both with $\frac{n(n-1)}{2}$ elements.*

2. *The lattices $\mathbb{A}_n^r$ ($n \geq 2$), except for $\mathbb{A}_3^2$ and $\mathbb{A}_5^2$, are S_n-perfect.*
3. *Except if $(n,r) = (3,2), (5,2), (7,2), (8,3)$, the minimal vectors modulo $\pm\mathrm{Id}$ share out among two orbits, which are $\{\varepsilon_0 - \varepsilon_i\}$ $(1 \leq i \leq n)$, n vectors, and $\{\varepsilon_j - \varepsilon_i\}$ $(1 \leq i < j \leq n)$, $\frac{n(n-1)}{2}$ vectors, except if $r = n+1$ or $r = \frac{n+1}{2}$.*
4. *The Voronoi domains of these lattices have two facets, one of which a dead-end, and their Voronoi graphs are of type Λ_1—Λ_2 (two vertices and one edge).*

Proof. Using the description of the minimal vectors in these lattices (Sections 4.2 and 4.3; Section 5.2), we easily prove the assertions which concern orbits; we verify that $A_n^{n+1} = \mathbb{A}_n^*$ is S_n-perfect and has two orbits with respectively 1 and n elements. The other assertions concerning S_n-perfection are evident, since the lattices we consider are perfect in the ordinary sense.

The assertions about dead-ends follow from the positions of minimal vectors with respect to minimal stable subspaces.

In the case of $\mathbb{D}_n$, the stable line is $\mathbb{R}(\varepsilon_1 + \varepsilon_2 + \cdots + \varepsilon_n)$ and the first orbit is contained in E_2; hence it defines a dead-end, whereas the second orbit is not a dead-end, since it is not a subset of E_1.

For $\mathbb{A}_n$, as well as for the lattices $\mathbb{A}_n^r$ whose minimal vectors are those of $\mathbb{A}_n$, since $E_1 = \mathbb{R}e$, with $e = n\,\varepsilon_0 - \varepsilon_1 - \varepsilon_2 \cdots - \varepsilon_n$, and $E_2 = \langle \varepsilon_j - \varepsilon_i \rangle$ $(1 \leq i < j \leq n)$, the orbit with $\frac{n(n-1)}{2}$ elements consisting of the $\varepsilon_j - \varepsilon_i$, $1 \leq i < j \leq n$ is a dead-end, whereas that with n elements is not.

We have analogous results for A_n^* (the dead-end corresponds to the stable vector $\frac{1}{n+1}(n\,\varepsilon_0 - \varepsilon_1 - \varepsilon_2 - \cdots - \varepsilon_n) \in E_1$) and for $A_n^{(n+1)/2}$, in which case we need to calculate explicitly some projections. The vectors

$$e_i' = \frac{2}{n+1}\left(\frac{n-1}{2}(\varepsilon_0 + \varepsilon_i) - \sum_{j \neq 0,\, i} \varepsilon_j\right)$$

constitute one orbit with n elements, and the sum of the projections of the invariant vector e onto these vectors is proportional to

$$\frac{n(n-1)}{2}\varepsilon_0 + \sum_{i \neq 0} \frac{n-1}{2}\varepsilon_i - (n-1)\sum_{j \neq 0} \varepsilon_j = n\,e\,,$$

so that this orbit is a dead-end, whereas that of the $\frac{n(n-1)}{2}$ vectors

$$\frac{2}{n+1}\left(\frac{n-1}{2}(\varepsilon_i+\varepsilon_j)-\sum_{k\neq i,j}\varepsilon_k\right),\ 1\leq i<j\leq n$$

is not. □

Remark 13.4.2. *In the case of $\mathbb{A}_7^2 \simeq \mathbb{E}_7$ (resp. $\mathbb{A}_8^3 \simeq \mathbb{E}_8$), there are three (resp. four) orbits of pairs of minimal vectors. These two lattices provide examples with both a dead-end and edges lying in the interior of the Voronoi domain.*

Remark 13.4.3. *Proposition 13.4.1 shows that the Voronoi graph itself (not only the graph modulo equivalence) reduces to only two vertices and one edge. This is not general; see [B-M6] for various examples.*

As an illustration of the preceding theory, we now give a few explicit determinations of contiguous lattices, beginning with $\mathbb{A}_n$, for which calculations can be avoided.

Example 13.4.4. *The Voronoi S_n-graph of $\mathbb{A}_n$ $(n\geq 2)$ is $\mathbb{A}_n$ —$\mathbb{A}_n^*$.*

Indeed, the map $\varepsilon_0-\varepsilon_i\mapsto\varepsilon_i$ extends to an isomorphism of S_n-modules of $\mathbb{A}_n$ onto $\mathbb{Z}^n$. Since $\mathcal{L}_G$ contains a pair (Λ,Λ^*) (e.g., for $\Lambda=\mathbb{Z}^n$), $\mathbb{A}_n^*$ also lies in $\mathcal{L}_G$. Since it is S_n-perfect, the two S_n-perfect lattices (up to scale) are $\mathbb{A}_n$ and $\mathbb{A}_n^*$.

The proof can also be carried out as follows, using explicit calculations. We verify that $F_1=-\frac{n-1}{n+1}\,p_1+p_2=\mathrm{Id}-\frac{2n}{n+1}\,p_1$ is a facet vector for the orbit which is not a dead-end. Let A be the Gram matrix of the Korkine–Zolotareff basis for $\mathbb{A}_n$ (Section 4.2). We have

$$A+\theta\,F_1(e_i)\cdot e_j=\begin{pmatrix}2&1&1&\dots&1\\1&2&1&\dots&1\\ \vdots&\vdots&\vdots&\ddots&\vdots\\1&1&1&\dots&2\end{pmatrix}+\theta\begin{pmatrix}0&-1&-1&\dots&-1\\-1&0&-1&\dots&-1\\ \vdots&\vdots&\vdots&\ddots&\vdots\\-1&-1&-1&\dots&0\end{pmatrix},$$

and we obtain for $\theta=\frac{n+2}{n}$ a matrix which is proportional to the matrix with entries n on the diagonal and -1 outside, a scaled copy of Gram matrix for $\mathbb{A}_n^*$; see the proof of Proposition 4.2.3. Since the new minimal vector which appears when θ attains the value $\rho=\frac{n+2}{n}$ is invariant under S_n, it comes from e, as we can see by noticing that $p_2(e)=0$. Hence

$$(\mathrm{Id}-\rho F_1)(e)\cdot e=\left(1-\tfrac{n+2}{n}\,\tfrac{n-1}{n+1}\right)(n^2+n)=2.$$

Example 13.4.5. *The Voronoi S_n-graph of $\mathbb{D}_n$ $(n\geq 4)$ is of the form $\mathbb{D}_n$ —$\mathbb{D}'_n$, where $\mathbb{D}'_n$ is isometric to $A_n^{(n+1)/2}$ for odd n and to its hyperplane sections of maximal density for even n.*

[In [B-M6], these lattices are called the *Coxeter lattices* Cox_n for all n.]

For odd n, the calculation is carried out starting from $\mathbb{A}_n^{(n+1)/2}$ and identifying $\mathbb{D}_n$ as a root lattice; for even n, we restrict ourselves to the previous case by considering hyperplane cross-sections.

It is worth noticing that since the dual of $\mathbb{A}_n^{(n+1)/2}$ is perfect for all $n \geq 7$ (it is similar to $\mathbb{A}_n^2$), the orbit of $\mathbb{D}_n$ is not stable under duality, not even merely up to similarity; we deduce from this that an analogous result holds for all even $n \geq 6$, and the case where $n = 4$ is dealt with by a direct calculation. However, it is easy to verify that $\mathbb{D}_5^*$ is similar to a lattice which belongs to the orbit of $\mathbb{D}_5$. This shows that this orbit contains an isodual lattice. This is indeed the densest known 5-dimensional isodual lattice; it was discovered by Conway and Sloane, and is described in Exercise 11.8.4.

13.5 Cyclotomic Lattices

In this section, G denotes a cyclic group of order $q \geq 3$ acting faithfully on E, and such that the corresponding representation comes from a $\mathbb{Q}$-irreducible representation. Its character is a sum of degree 1 complex characters which are conjugates over $\mathbb{Q}$, and G-lattices are then cyclotomic lattices in the sense of Section 11.4. Our goal is to classify those of these lattices which are G-perfect.

We shall not write down explicit calculations, but rather give an account of work done in particular by Sigrist in Neuchâtel, who has implemented the Voronoi algorithm for cyclotomic lattices, following the article [B-M-S], whose methods are more efficient than those of a combinatorial nature that we discussed in Chapter 11, and which look difficult to generalize in relatively large dimensions.

For each group G, we give the largest possible Hermite constant. We often obtain upper bounds which are better than the general ones, either (in dimension $n \leq 8$) because the automorphism groups of critical lattices do not define the representation of G we consider, or (in dimension $n \geq 9$) because the Hermite constant is not known. For instance, to know that the maximum possible density of a 12-dimensional lattice whose automorphism group contains a $\mathbb{Q}$-irreducible cyclic group of order 28 is that of K_{12} is a new information, even if everybody conjectures that K_{12} is the unique critical lattice in dimension 12.

Perfect lattices whose automorphism group contains a subgroup isomorphic to G and defining the cyclotomic representation of G over $\mathbb{Q}$ of course belong to the list of G-perfect G-lattices. Here are a few examples.

• Let p be an odd prime, and let G be cyclic of order p. For $n = p-1$ and $q = p$, the Craig lattices $\mathbb{A}_n^{(r)}$ are G-lattices, which are known to be (G)-perfect when r is not too large relatively to p, see Sections 5.3 and 5.4. This applies in particular to the cases $n = 4$, $r = 1$; $n = 6$, $r = 1, 2$; $n = 10$, $r = 1, 2, 3, 4$; $n = 12$, $r = 1, 2, 3$.

• The cycle $(1, 2, \dots, n)$ followed by the change of sign of ε_1 defines an automorphism σ of order $2n$ of $\mathbb{Z}^n$, hence also of $\mathbb{D}_n$. For $n = 2^m, m \geq 2$, σ is of order $q = 2^{m+1}$, and we have $\varphi(q) = n$. This applies in particular to pairs $(n, q) = (4, 8)$ and $(8, 16)$.

• For $n = 6$ and $q = 9$, we find the perfect lattices $\mathbb{E}_6$ and $\mathbb{E}_6^*$; see Theorem 11.4.6.

• Using some of its descriptions of simple groups in terms of lattices, we can find in [ATLAS] the orders and the nature of the corresponding representation of various cyclic subgroups of $\mathrm{Aut}(\Lambda)$ for certain important lattices Λ. For instance, the table for the group $O_8^+(2)$ (page 86) shows that $\mathbb{E}_8$ is q-cyclotomic for $q = 15, 20, 24$, but not for $q = 16$. Similarly, a close look at the tables relative to $U_4(3)$ shows that K_{12} is q-cyclotomic for $q = 21$ and $q = 28$, but not for $q = 13$ nor $q = 36$.

In the table below, we give for a dimension n and an integer $q \not\equiv 2 \mod 4$ with $\varphi(q) = n$, the number N (resp. P) of similarity classes of G-perfect (resp. perfect) G-lattices, the maximum γ of the Hermite invariant on the set of G-lattices (which is known to be attained on a G-perfect G-lattice), followed by its numerical value rounded down to three decimal places, the analogous approximation for γ_n (only conjectural beyond $n = 8$, i.e. that of Λ_{10} or of K_{12}), and finally the G-critical lattice (which is unique in all cases we consider) when this lattice is known from a previous construction.

Table 13.5.1. G-perfect cyclotomic lattices of dimension $n \leq 12$.

n	4	4	6	6	8	8	8	10	12	12	12
q	5	8, 12	7	9	16	15, 20	24	11	13	21, 28	36
N	1	1, 1	2	2	3	1, 1	2	5	25	8, 13	23
P	1	1, 1	2	2	1	1, 1	1	5	23	8, 7	8
γ	$2/5^{1/4}$	$2^{1/2}$	$4/7^{1/2}$	$2/3^{1/6}$	$*$	2	2	$4/11^{3/10}$	$4/13^{1/4}$	$4/3^{1/2}$	$**$
γ	1.337	1.414	1.511	1.665	1.707	2	2	1.948	2.106	2.309	2.106
γ_n	1.414	1.414	1.665	1.665	2	2	2	2.058	2.309	2.309	2.309
Λ	$\mathbb{A}_4$	$\mathbb{D}_4$	$\mathbb{A}_6^{(2)}$	$\mathbb{E}_6$		$\mathbb{E}_8$	$\mathbb{E}_8$	$\mathbb{A}_{11}^{(2)}$		K_{12}	

$(*) : \gamma = 8/(2 \cdot 241)^{1/4}$; $(**) : \gamma = 36 \cdot (4463 \cdot 5581)^{-1/6}$.

The only isometries among lattices in Table 13.5.1 concern $\mathbb{D}_4$ and K_{12}, which occur twice, and $\mathbb{E}_8$, which occurs three times. Thus Table 13.5.1 contains 5 (resp. 45) G-perfect lattices of dimension 10 (resp. 12). Inspection of the lists of lattices found by Sigrist shows that the minimal norm which makes integral a perfect 12-dimensional lattice takes every even value in the range 2—28, plus the two extra values 32 and 42. (On a lattice which is only G-perfect, one even attains the value 78, with $q = 13$.)

Cyclotomic G-perfect lattices have been completely classified up to dimension 12. Since the equation $\varphi(q) = 14$ has no solution, the next dimension to consider is $n = 16$, for which cyclotomic lattices exist with $q = 17$, 32, 40, 48, and 60, but the classification was completed only for $q = 17$ (and very recently for $q = 32$; see the notes). These results are due to Sigrist ([Sig3]). For $n = 16$ and $q = 17$, the Hermite constant attains its maximum on a lattice which is integral of minimum 6, with determinant $17 \cdot 2^{16}$; among the 1344 G-perfect lattices, an integral and primitive perfect lattice with minimum 278 shows up!

13.6 Lattices with a Fixed Section and Patchwork Lattices

The notation of this section is either that of Sections 13.3 and 13.4, or that of Section 12.5, depending on the question we consider. We first consider lattices having a given section, and then sketch the corresponding results for patchwork lattices.

Let E_0 be a subspace of E, of dimension $m < n$, and let Λ_0 be a lattice in E_0. Denote by $\mathcal{F}$ the family of lattices $\Lambda \subset E$ which have the same norm as Λ_0, and intersect E_0 along Λ_0. In a basis $\mathcal{B}$ for E which extends a basis $\mathcal{B}_0$ for E_0, these lattices are represented by Gram matrices with upper left $m \times m$ corner equal to $\mathrm{Gram}(\mathcal{B}_0)$. We denote by the same letter $\mathcal{F}$ the set of quadratic forms corresponding to these matrices; they are extensions of the form Q_0 with matrix $\mathrm{Gram}(\mathcal{B}_0)$.

The subspace $\mathcal{T}$ of $\mathrm{End}^s(E)$ attached to these data and its representation by matrices in the basis $\mathcal{B}$ have been defined in Section 12.3. The notion of perfection for $\mathcal{F}$ is defined using the subspace

$$\mathcal{T} = \left(\begin{smallmatrix} 0 & * \\ {}^t* & * \end{smallmatrix}\right)$$

of Sym_n, in which 0 denotes the null matrix of Sym_m and the diagonal term $*$ is symmetric.

Given a $\mathcal{T}$-perfect matrix A and a facet vector F for the Voronoi $\mathcal{T}$-domain of A, the contiguity algorithm relative to A and F consists in considering matrices of the form $A_\theta = A + \theta F$, and determining the largest value ρ of θ (if any) for which $A + \theta F$ keeps the same minimum as A. The form we obtain in this way is then the *$\mathcal{T}$-contiguous* (or *$\mathcal{F}$-contiguous*) *form to A along F.*

Due to the nature of the space $\mathcal{T}$, all matrices $A + \theta F$ all have an upper left corner equal to $\mathrm{Gram}(\mathcal{B}_0)$, and as long as θ is sufficiently small to ensure that their minimum over $\mathbb{Z}^n$ is equal to that of A, their corresponding lattices lie in $\mathcal{F}$. This implies that the matrix $A + \rho F$ of the contiguous form to that with matrix A (which is $\mathcal{T}$-perfect by its very construction) is again the Gram matrix of a lattice in $\mathcal{F}$.

Incidentally, the proof of Theorem 7.2.1 easily adapts to the situation we consider here. As a consequence, the contiguous form always exists; in other words, there are no dead-ends in the sense of Definition 13.1.7.

Since the minimal vectors of $\mathcal{T}$-perfect lattices span E (Theorem 12.3.5), Theorem 13.2.3 shows that the Voronoi graph relative to $\mathcal{T}$ is connected, hence that all relatively perfect extensions of Q_0 can be obtained from one of them by a chain contiguous forms. However, to obtain an efficient algorithm, we must be able to test $\mathcal{F}$-equivalence (i.e., $\mathcal{T}$-equivalence in the sense of Definition 13.2.4) of relatively perfect forms. The matrices of $\mathcal{T}$-equivalence are the matrices $P \in \mathrm{GL}_n(\mathbb{Z})$ such that ${}^tP\mathcal{T}P = \mathcal{T}$. An easy calculation shows that they are those matrices whose block decomposition is of the form

$$\left(\begin{smallmatrix} X & Y \\ 0 & Z \end{smallmatrix}\right),$$

i.e. those which preserve E_0.

The proof of Theorem 9.1.5 adapts to our case; see Exercise 12.3.3, and implies that the number of classes of $\mathcal{T}$-perfect forms which extend Q_0 is finite modulo $\mathcal{T}$-equivalence.

We summarize the preceding results as a theorem, which we state in terms of quadratic forms:

Theorem 13.6.1. *1. Forms which are $\mathcal{F}$-contiguous to a given $\mathcal{F}$-perfect form Q exist through every facet of the $\mathcal{T}$-domain of Q.*
2. The contiguity graph of $\mathcal{F}$-perfect forms is connected, and finite modulo the group of $\mathcal{F}$-equivalences. □

Remark 13.6.2. *If a facet vector in the usual sense is orthogonal in* $\mathrm{End}^s(E)$ *to all facets of a section, it also defines a facet vector for the Voronoi algorithm relative to this section. Conversely, given a* perfect *section, the contiguity paths associated with the various facet vectors relative to this section define classes of perfection corank* 1, *hence contiguity paths in the usual sense.*

The situation is somewhat similar for patchwork lattices. The notation is now that of Section 12.5. The space $\mathcal{T}$ associated with a family $\mathcal{F}$ of patchwork lattices is that of Definition 12.5.1, or in matrix form, that of Formula 12.5.2, made with matrices of the form

$$\mathcal{T} = \begin{pmatrix} 0 & A_{1,2} & A_{1,3} & \dots & A_{1,r} \\ {}^tA_{1,2} & 0 & A_{2,3} & \dots & A_{2,r} \\ {}^tA_{1,3} & {}^tA_{2,3} & 0 & \dots & A_{3,r} \\ \vdots & \vdots & \vdots & \ddots & \vdots \\ {}^tA_{1,r} & {}^tA_{2,r} & {}^tA_{3,r} & \dots & 0 \end{pmatrix},$$

where $A_{i,j} \in \mathcal{M}_{m_i,m_j}(\mathbb{R})$ and diagonal terms are the null square matrices of dimensions $m_1, m_2, \dots, m_r$.

In fact, contiguity is dealt with in a similar way, and the finiteness assertion is even easier to prove, since we are working with a compact family (modulo proportionality). However, a new difficulty arises for the connectedness assertion, which does not follow directly from Theorem 13.2.3, because it might happen that minimal vectors in the lattices we consider do not span E. Since this difficulty does not occur in most of the applications, where we work with direct sums of perfect lattices having the same minimum, we shall not enter into details, for which we refer to Baril's thesis [Bari]. We consequently state without a proof the last theorem of this section:

Theorem 13.6.3. *Let $L_1^\circ, \dots, L_r^\circ$ be lattices with the same norm, whose dimensions add to n, and let $\mathcal{F}$ be the set of lattices in E which are a direct sum of lattices $\Lambda_1, \dots, \Lambda_r$ isometric to the preceding lattices.*

1. *For every $\mathcal{F}$-perfect matrix A and every facet vector F for its Voronoi $\mathcal{T}$-domain, A possesses a $\mathcal{T}$-contiguous form along F.*
2. *The $\mathcal{T}$-contiguity graph is connected, and finite modulo $\mathcal{F}$-equivalences.*

□

For both families $\mathcal{F}$ we considered in this section, it is theoretically possible to find all $\mathcal{F}$-perfect lattices from one of them, by constructing the $\mathcal{F}$-contiguity graph modulo $\mathcal{F}$-equivalences, and we are left with the problem of finding *one* particular $\mathcal{F}$-perfect lattice.

A first method consists in considering an arbitrary form in $\mathcal{F}$, and if it is not $\mathcal{T}$-perfect, in increasing its perfection rank by contiguity. (Note that in Definition 13.1.4, the form was not assumed to be $\mathcal{T}$-perfect; see Proposition 7.2.5.) This algorithm quickly finds $\mathcal{T}$-perfect form after a small number of steps.

Another, very simple method, proves quicker for families of lattices with a fixed section having a large codimension: it consists in completing a Gram matrix for Λ_0 by adding extra vectors having small components. This method is the one that was used by Laïhem in his search of 8-dimensional perfect lattices having a perfect 7-dimensional section with the same norm, and also by Baril when passing from dimension 6 to 8 by direct sum with a hexagonal lattice.

Let us give a few details in the case where Λ_0 has codimension 1 and norm, say, an integer a, and possesses a basis of minimal vectors, with Gram matrix A_0. We try to add an n-th column (and row) to A_0 in order to obtain a matrix $A = (a_{i,j})$, choosing $a_{n,n} = a$. Since we must preserve the minimum, we must have $a_{i,n} = a_{n,i} \leq \frac{a}{2}$. We try systems of $n-1$ entries $a_{i,n}$ which satisfy this inequality and have "small" denominators (2 if 1 does not work, then 3 if 2 does not work, etc. In practice, a is sufficiently small to make the method efficient.

Remark 13.6.4. *Since we know a priori ([Cs2]) that minimal vectors of a perfect lattice of dimension $n \leq 7$ span E, we could replace in these dimen-*

sions the usual algorithm by its patchwork variant associated with the direct sum of n copies of $\mathbb{Z}$. However (remark of Baril), this is no real progress, for this does not modify the difference between the number of minimal vectors and the dimension of the Voronoi space, and after Jaquet, it is precisely the size of this difference which makes the algorithm a highly time-consuming one.

13.7 Examples

In this section we describe a few results which have been obtained by applying the Voronoi algorithm to families we considered in Sections 13.4—13.6. Since they heavily rely on computer programs, it is not possible to give full proofs here.

Example 13.7.1. *Laïhem lattices are the 8-dimensional lattices having a perfect 7-dimensional hyperplane section with the same norm (including those which lie above one of the three root lattices). The number of lattices which occur above a given 7-dimensional Λ_0 varies from 1 (when $\Lambda_0 = P_7^1 \sim \mathbb{E}_7^*$) to 171 (when $\Lambda_0 \sim P_7^{24}$). Altogether, 1462 perfect lattices occur, whereas only 1175 perfect lattices (up to similarity) are mentioned in Theorem 6.6.5. There are two reasons for this excess of 287 lattices:*

1. *Certain lattices possess several perfect hyperplane sections with the same norm; for instance, $\mathbb{E}_8$ occurs three times, above $\mathbb{E}_7, \mathbb{D}_7, \mathbb{A}_7$; other example: the lattices denoted by $P(8,5)$ and $P(8,6)$ in [Lah] contain the two lattices P_7^5 and P_7^8.*
2. *Certain lattices occur several times above the same section Λ_0, because, they are not isometric under a transformation* which preserves *Λ_0; the case of a lattice lying above P_7^{30} is quoted in [Lah].*

[Experimental data indicate that the smaller the group $|\operatorname{Aut}(\Lambda_0)|$, the larger the number of lattices lying above Λ_0.]

Following Baril, we display below the contiguity graphs for the root lattices above; as in Chapter 7, the notation $\boxed{\Lambda}$ means that Λ is contiguous to itself:

$$\mathbb{E}_7 \;:\; \mathbb{A}_8^2 \text{———} \boxed{\mathbb{E}_8}$$

$$\mathbb{D}_7 \;:\; \mathbb{D}_8 \text{———} \boxed{\mathbb{E}_8}$$

$$\mathbb{A}_7 \;:\; \mathbb{A}_8 \text{———} \mathbb{D}_8 \text{———} \mathbb{E}_8 \text{———} \boxed{\mathbb{A}_8^2}\,.$$

Example 13.7.2. *Still following Baril, we consider certain lattices which are relatively perfect with respect to various known lattices in dimensions 8* *and 9. The table below gives the number and the set of minimal norms (for their canonical scale) of perfect lattices relatively to nine lattices defined in Chapters 4, 5, or 8.*

Table 13.7.3. Relatively perfect lattices in dimensions 8, 9, 10.

Lattice	$\mathbb{E}_8$	$\mathbb{A}_8^2$	$\mathbb{D}_8$	K_8	K_8'	Λ_9	$\mathbb{A}_9^3$	$\mathbb{D}_9$	Λ_{10}
Number	2	5	3	5	3	5	16	9	6
Norms	4,6	4,12	2,4,12	4,8	4	4,12	6,12,18,24,30	2,4,8,12,16,20	4,12

These lattices are all perfect, which is not *a priori* evident for lattices lying above the non-perfect lattice K_8. They all possess a basis of minimal vectors, except for $\mathbb{D}_{10}^+$ (which lies above $\mathbb{D}_9$).

13.8 Exercises for Chapter 13

12.1.1. Show that if u and v are two symmetric endomorphisms, the equalities $v \circ u = 0$ and $u \circ v = 0$ are equivalent, and that they hold true if and only if for every nonzero eigenvalue of u, the corresponding eigenvectors lie in the kernel of v.

12.1.2. This exercise is intended to prove Proposition 13.1.10.

1. Show that the positive symmetric endomorphisms in $\mathcal{T}$ form a convex half-cone. [Use the proof of Theorem 7.1.4.]
2. Show that the angle between two positive endomorphisms is less than or equal to $\frac{\pi}{2}$.
3. Let u and v be two positive symmetric endomorphisms such that $\langle u, v\rangle = 0$. Show that the matrices A and B of u and v in an orthogonal basis for E in which A is diagonal are of the form $\begin{pmatrix} * & 0 \\ 0 & 0 \end{pmatrix}$ and $\begin{pmatrix} 0 & 0 \\ 0 & * \end{pmatrix}$.

12.3.1. The aim of this exercise is to prove Proposition 13.3.3. Let $u \in \mathrm{End}_G^s(E)$ be a nonzero positive (semi-definite) endomorphism.

1. Let λ be a nonzero eigenvalue of u. Show that the eigenspace E_λ associated with λ is stable under G.
2. Show that the restriction of u to E_λ is of the form $\lambda \sum_{F\in\mathcal{E}} p_F$ for some family $\mathcal{E}$ of irreducible stable subspaces of E_λ.
3. Deduce from this that u is a barycentre with positive coefficients of projections onto suitable irreducible subspaces of E.
4. Let $F, F_1, \dots, F_m$ be irreducible subspaces of E. Show that if there exists a relation with strictly positive coefficients $p_F = \sum_i \lambda_i p_{F_i}$, then $F_i = F$ for all i. [Consider the endomorphisms $p_{F_i} \circ p_F$.]
5. Prove Proposition 13.3.3.

12.4.1. We say that a group G is k-fold transitive on a set $\mathcal{E}$ if for all k-tuples $\{x_1, x_2, \dots, x_k\}$ and $\{y_1, y_2, \dots, y_k\}$ of distinct elements of $\mathcal{E}$, there exists an element of G which maps x_i onto y_i for $1 \le i \le k$. We consider a group G acting on a basis $\mathcal{B} = (e_1, e_2, \dots, e_n)$ for E.

1. Show that if G is (once) transitive on $\mathcal{B}$, the representation of G defined by E is induced by the unit representation of the stabilizer of e_1.

2. Deduce from this that the character of the representation of G defined by E contains the unit character a number of times equal to the number of orbits of G acting on $\mathcal{B}$. [Reduce to the case of a single orbit, and use the formula $\langle \operatorname{Ind}_H^G(\chi), \phi \rangle = \langle \chi, \operatorname{Res}_G^H(\phi) \rangle$.]

3. Suppose that G acts transitively on $\mathcal{B}$. Show that the *augmentation representation of* G (whose character is $\chi_E - \mathbf{1}_E$) is irreducible if and only if G acts twice transitively.

4. Show that S_n, $n \geq 1$ (resp. A_n, $n \geq 3$) is n-fold (resp. $(n-2)$-fold) transitive on $\{1, 2, \dots, n\}$.

12.4.2. Verify the assertions of Remark 13.4.2 about $\mathbb{E}_7$ and $\mathbb{E}_8$, and prove that the two extremal edges are those which come from $\mathbb{A}_n$, $n = 7, 8$.

13.9 Notes on Chapter 13

The possibility of developing a Voronoi algorithm for other families of lattices than that of all lattices occurs for the first time in the article [B-M-S], which proves on the one hand the general results of Sections 13.1 and 13.2, and makes explicit on the other hand the practice of the algorithm in the case of G-perfect G-lattices. The results of [B-M-S] are returned to in Section 13.3, and enlarged with the characterization of dead-ends. We have added a fourth section in view of the possible application of 2-dimensional Voronoi spaces for the construction of isodual lattices; it would be interesting to study other examples. Section 13.5 accounts for work by Sigrist. (Recent progress for dimension 16: there are 4624 C_{32}-perfect forms, among which only 1314 are perfect; details can be read at http://www.unine.ch/math/personnel/equipes/sigrist/)

The description of the Voronoi algorithm for lattices with a given section is only sketched in Section 5 of [B-M-S]. The only works on this question (or on the closely related question of patchwork lattices) are the unpublished theses of Laïhem and of Baril ([Lah], [Bari]).

The main problem in this theory is that of extending the Voronoi algorithm to families of lattices which cannot be viewed as an affine subspace of $\operatorname{End}^s(E)$. Such an extension would perhaps allow the calculation of some important constants in low dimensions, such as the Bergé–Martinet constant γ'_5, the Rankin constant $\gamma_{5,2}$, or the exact value of the Hermite invariant γ on 6-dimensional symplectic lattices. For a first example, see Bavard's recent paper [Bav3], in which the author considers families of symplectic lattices which are parametrized by the upper half-plane.

14 Numerical Data

Introduction

In this chapter, we display (or give references to) some numerical data: Gram matrices for a few lattices which appeared previously in this book, and tables of known large values of the Hermite and Bergé–Martinet invariants γ and γ', including their restrictions to isodual lattices. Data can be found in particular on the following two web pages:

- Nebe and Sloane ([Ne-Sl]): http://www.research.att.com/~njas/lattices/;
- Batut and Martinet ([Bt-M']): http://www.math.u-bordeaux.fr/~martinet/.

14.1 Low-Dimensional Perfect Lattices

We display below Gram matrices for the 15 perfect lattices of dimension $n \leq 6$. These matrices are LLL-reduced (for the definition of LLL (=Lenstra–Lenstra–Lovász) reduction, see the original paper [LLL], or Cohen's book [Coh], Chapter 2, Section 2.6.) The use of reduced matrices often makes the running time of numerical computations much quicker; it also puts in evidence the existence of a basis of minimal vectors for perfect lattices up to dimension 7. Since LLL reduction is a fast procedure, and since tables of Gram matrices are available for all known perfect lattices of dimension $n \leq 8$ in [Bt-M'], we have not reproduced here the 33 matrices for 7-dimensional perfect lattices which were printed in the French edition. Moreover, the reference above also contains tables for the usual invariants of these lattices. Another important reference is Conway and Sloane's [C-S5], whose displayed matrices allow the description of both the roots and the automorphism groups.

Reduced Gram matrices for perfect lattices in dimension ≤ 6.

$$P_1^1 = (1)\,;\ P_2^1 = \begin{pmatrix} 2&1\\1&2 \end{pmatrix};\ P_3^1 = \begin{pmatrix} 2&1&1\\1&2&1\\1&1&2 \end{pmatrix};\ P_4^1 = \begin{pmatrix} 2&1&1&1\\1&2&1&0\\1&1&2&1\\1&0&1&2 \end{pmatrix};\ P_4^2 = \begin{pmatrix} 2&1&1&1\\1&2&1&1\\1&1&2&1\\1&1&1&2 \end{pmatrix}.$$

$$P_5^1 = \begin{pmatrix} 2&1&1&1&1 \\ 1&2&1&0&1 \\ 1&1&2&1&1 \\ 1&0&1&2&1 \\ 1&1&1&1&2 \end{pmatrix}; \quad P_5^2 = \begin{pmatrix} 4&-2&-2&1&1 \\ -2&4&1&-2&1 \\ -2&1&4&1&-2 \\ 1&-2&1&4&-2 \\ 1&1&-2&-2&4 \end{pmatrix}; \quad P_5^3 = \begin{pmatrix} 2&1&1&1&1 \\ 1&2&1&1&1 \\ 1&1&2&1&1 \\ 1&1&1&2&1 \\ 1&1&1&1&2 \end{pmatrix}.$$

$$P_6^1 = \begin{pmatrix} 2&1&1&1&1&1 \\ 1&2&1&0&1&1 \\ 1&1&2&1&1&1 \\ 1&0&1&2&1&0 \\ 1&1&1&1&2&1 \\ 1&1&1&0&1&2 \end{pmatrix}; \quad P_6^2 = \begin{pmatrix} 4&-2&1&-2&-2&-1 \\ -2&4&-2&1&1&-1 \\ 1&-2&4&-2&-2&2 \\ -2&1&-2&4&1&-1 \\ -2&1&-2&1&4&-1 \\ -1&-1&2&-1&-1&4 \end{pmatrix}; \quad P_6^3 = \begin{pmatrix} 2&1&1&1&1&1 \\ 1&2&1&0&1&1 \\ 1&1&2&1&1&1 \\ 1&0&1&2&1&1 \\ 1&1&1&1&2&1 \\ 1&1&1&1&1&2 \end{pmatrix};$$

$$P_6^4 = \begin{pmatrix} 4&2&1&2&2&2 \\ 2&4&2&1&1&2 \\ 1&2&4&2&2&2 \\ 2&1&2&4&1&2 \\ 2&1&2&1&4&2 \\ 2&2&2&2&2&4 \end{pmatrix} \quad P_6^5 = \begin{pmatrix} 4&2&2&1&2&2 \\ 2&4&2&2&1&2 \\ 2&2&4&0&2&1 \\ 1&2&0&4&1&2 \\ 2&1&2&1&4&0 \\ 2&2&1&2&0&4 \end{pmatrix}; P_6^6 = \begin{pmatrix} 4&2&1&2&2&2 \\ 2&4&2&1&2&2 \\ 1&2&4&2&2&2 \\ 2&1&2&4&2&2 \\ 2&2&2&2&4&1 \\ 2&2&2&2&1&4 \end{pmatrix}; P_6^7 = \begin{pmatrix} 2&1&1&1&1&1 \\ 1&2&1&1&1&1 \\ 1&1&2&1&1&1 \\ 1&1&1&2&1&1 \\ 1&1&1&1&2&1 \\ 1&1&1&1&1&2 \end{pmatrix}.$$

14.2 Root Lattices

Up to a permutation of the basis vectors, Korkine–Zolotareff bases for root lattices (see Chapter 4) are those with entries 2 on the diagonal and 1 or 0 outside, with a number of symmetric pairs of zeros as small as possible, namely 0, 1, and 2 for $\mathbb{A}_n$, $\mathbb{D}_n$, and $\mathbb{E}_n$ respectively. Such a basis is *LLL*-reduced in the case of $\mathbb{A}_n$, and an appropriate ordering of the basis vectors makes it *LLL*-reduced in the other two cases. For $\mathbb{D}_n$, a possible choice consists in setting the zeros at the positions $(2,4)$ and $(4,2)$. This was our choice in Section 14.1.

The two matrices below represent $\mathbb{E}_8$; the second one is *LLL* reduced.

$$\mathbb{E}_8 \;:\; \begin{pmatrix} 2&0&1&1&1&1&1&1 \\ 0&2&0&1&1&1&1&1 \\ 1&0&2&1&1&1&1&1 \\ 1&1&1&2&1&1&1&1 \\ 1&1&1&1&2&1&1&1 \\ 1&1&1&1&1&2&1&1 \\ 1&1&1&1&1&1&2&1 \\ 1&1&1&1&1&1&1&2 \end{pmatrix}, \quad \begin{pmatrix} 2&1&1&1&1&1&1&1 \\ 1&2&1&0&1&1&1&1 \\ 1&1&2&1&1&1&1&1 \\ 1&0&1&2&0&1&1&1 \\ 1&1&1&0&2&1&1&1 \\ 1&1&1&1&1&2&1&1 \\ 1&1&1&1&1&1&2&1 \\ 1&1&1&1&1&1&1&2 \end{pmatrix}.$$

Suppressing successively the last row and the last column in the matrices above, we obtain the two sequences of cross-sections displayed below:

$$\mathbb{E}_8 \supset \mathbb{E}_7 \supset \mathbb{E}_6 \supset \mathbb{E}_5 \simeq \mathbb{D}_5 \supset \mathbb{E}_4 \simeq \mathbb{A}_4 \supset \mathbb{E}_3 \simeq \mathbb{A}_2 \perp \mathbb{A}_1 \supset \mathbb{A}_1 \perp \mathbb{A}_1 \supset \mathbb{A}_1$$

and

$$\mathbb{E}_8 \supset \mathbb{E}_7 \supset \mathbb{E}_6 \supset \mathbb{D}_5 \supset \mathbb{D}_4 \supset \mathbb{A}_3 \supset \mathbb{A}_2 \supset \mathbb{A}_1 \,.$$

The successive cross-sections of the second matrix are *LLL*-reduced. We have thus obtained *LLL*-reduced Korkine–Zolotareff bases for all irreducible root lattices.

14.3 Eutactic Lattices up to Dimension 4

Each of the 25 well-rounded minimal classes of dimension $n = 2$, 3, or 4 (Sections 9.2 and 9.3) contains a unique weakly eutactic lattice, to within the exception of the 4-dimensional class a_6, for which the least upper bound of the Hermite invariant is attained on the class b_7, the only one for which the weakly eutactic lattice is not eutactic (it is only semi-eutactic). These eutactic lattices are rational over $\mathbb{Q}$, except those of the classes a_7 and a_8, which are, however, rational over a quadratic field ($\mathbb{Q}(\sqrt{13})$ and $\mathbb{Q}(\sqrt{3})$ respectively).

Since the rings $\mathbb{Z}$, $\mathbb{Z}[\sqrt{3}]$ and $\mathbb{Z}[\frac{1+\sqrt{13}}{2}]$ are principal ideal domains, all eutactic lattices can be defined by an integral and primitive Gram matrix A over one of these domains. In what follows, we denote by B an integral, primitive matrix which is proportional to A^{-1}.

Each class is represented by a lattice endowed with a basis $(e_1, \dots, e_n)$ of minimal vectors; we denote by $e_{n+1}, \dots, e_s$ the other minimal vectors (one for each pair $\pm x$). To each of the pairs $\{\pm e_i\}$, there corresponds the rank 1 matrix $E_i \in \mathrm{Sym}_n$, equal to $X_i\,{}^tX_i$, where X_i denotes the column of the components of e_i in the basis above. The eutaxy relations can be written in the form

$$B = \sum_{i=1}^{s} \lambda_i E_i \,,$$

indeed uniquely except in the case of the configuration $a_{10} = S(\mathbb{D}_4)$ (because s is then strictly larger than the perfection rank).

We list below matrices A and B for every weakly eutactic lattice the determinant of A, representatives for $S/\{\pm 1\}$, a sequence of eutaxy coefficients $(\lambda_1, \lambda_2, \dots, \lambda_n)$ (which is unique except in the case of a_{10}), and the order of the group $\mathrm{Aut}^+(L)$ of special automorphisms of the lattices L corresponding to A.

$\boxed{n = 2}$

a_2 : $S = \{e_1, e_2\}$

$$A = \left(\begin{smallmatrix} 1 & 0 \\ 0 & 1 \end{smallmatrix}\right) \qquad B = A^{-1} = \left(\begin{smallmatrix} 1 & 0 \\ 0 & 1 \end{smallmatrix}\right)$$

$\det(A) = 1\,,\quad |\,\mathrm{Aut}^+(L)| = 4\,,\quad$ eutaxy : $(1, 1)$.

a_3 : $S = \{e_1, e_2, e_1 + e_2\}$

$$A = \left(\begin{smallmatrix} 2 & -1 \\ -1 & 2 \end{smallmatrix}\right) \qquad B = 3A^{-1} = \left(\begin{smallmatrix} 2 & 1 \\ 1 & 2 \end{smallmatrix}\right)$$

$\det(A) = 3\,,\quad |\,\mathrm{Aut}^+(L)| = 6\,,\quad$ eutaxy : $(1, 1, 1)$.

$\boxed{n = 3}$

a_3 : $S = \{e_1, e_2, e_3\}$

$$A = \begin{pmatrix} 1 & 0 & 0 \\ 0 & 1 & 0 \\ 0 & 0 & 1 \end{pmatrix} \qquad B = A^{-1} = \begin{pmatrix} 1 & 0 & 0 \\ 0 & 1 & 0 \\ 0 & 0 & 1 \end{pmatrix}$$

$\det(A) = 1\,,\quad |\,\mathrm{Aut}^+(L)| = 24\,,\quad$ eutaxy : $(1, 1, 1)$.

a_4 : $S = \{e_1, e_2, e_3, e_1 + e_2 + e_3\}$

$$A = \begin{pmatrix} 3 & -1 & -1 \\ -1 & 3 & -1 \\ -1 & -1 & 3 \end{pmatrix} \qquad B = 4A^{-1} = \begin{pmatrix} 2 & 1 & 1 \\ 1 & 2 & 1 \\ 1 & 1 & 2 \end{pmatrix}$$

$\det(A) = 16$, $|\operatorname{Aut}^+(L)| = 24$, eutaxy : $(1,1,1,1)$.

b_4 : $S = \{e_1, e_2, e_3, e_1 + e_2\}$

$$A = \begin{pmatrix} 2 & -1 & 0 \\ -1 & 2 & 0 \\ 0 & 0 & 2 \end{pmatrix} \qquad B = 6A^{-1} = \begin{pmatrix} 4 & 2 & 0 \\ 2 & 4 & 0 \\ 0 & 0 & 3 \end{pmatrix}$$

$\det(A) = 6$, $|\operatorname{Aut}^+(L)| = 12$, eutaxy : $(2,2,3,2)$.

a_5 : $S = \{e_1, e_2, e_3, e_1 + e_2, e_1 + e_3\}$

$$A = \begin{pmatrix} 4 & -2 & -2 \\ -2 & 4 & 1 \\ -2 & 1 & 4 \end{pmatrix} \qquad B = 12A^{-1} = \begin{pmatrix} 5 & 2 & 2 \\ 2 & 4 & 0 \\ 2 & 0 & 4 \end{pmatrix}$$

$\det(A) = 36$, $|\operatorname{Aut}^+(L)| = 8$, eutaxy : $(1,2,2,2,2)$.

a_6 : $S = \{e_1, e_2, e_3, e_1 + e_2, e_1 + e_3, e_2 - e_3\}$

$$A = \begin{pmatrix} 2 & -1 & -1 \\ -1 & 2 & 1 \\ -1 & 1 & 2 \end{pmatrix} \qquad B = 4A^{-1} = \begin{pmatrix} 3 & 1 & 1 \\ 1 & 3 & -1 \\ 1 & -1 & 3 \end{pmatrix}$$

$\det(A) = 4$, $|\operatorname{Aut}^+(L)| = 24$, eutaxy : $(1,1,1,1,1,1)$.

$n = 4$

a_4 : $S = \{e_1, e_2, e_3, e_4\}$

$$A = \begin{pmatrix} 1 & 0 & 0 & 0 \\ 0 & 1 & 0 & 0 \\ 0 & 0 & 1 & 0 \\ 0 & 0 & 0 & 1 \end{pmatrix} \qquad B = A^{-1} = \begin{pmatrix} 1 & 0 & 0 & 0 \\ 0 & 1 & 0 & 0 \\ 0 & 0 & 1 & 0 \\ 0 & 0 & 0 & 1 \end{pmatrix}$$

$\det(A) = 1$, $|\operatorname{Aut}^+(L)| = 192$, eutaxy : $(1,1,1,1)$.

a_5 : $S = \{e_1, e_2, e_3, e_4, e_1 + e_2 + e_3 + e_4\}$

$$A = \begin{pmatrix} 4 & -1 & -1 & -1 \\ -1 & 4 & -1 & -1 \\ -1 & -1 & 4 & -1 \\ -1 & -1 & -1 & 4 \end{pmatrix} \qquad B = 5A^{-1} = \begin{pmatrix} 2 & 1 & 1 & 1 \\ 1 & 2 & 1 & 1 \\ 1 & 1 & 2 & 1 \\ 1 & 1 & 1 & 2 \end{pmatrix}$$

$\det(A) = 125$, $|\operatorname{Aut}^+(L)| = 120$, eutaxy : $(1,1,1,1,1)$.

b_5 : $S = \{e_1, e_2, e_3, e_4, e_1 + e_2 + e_3\}$

$$A = \begin{pmatrix} 3 & -1 & -1 & 0 \\ -1 & 3 & -1 & 0 \\ -1 & -1 & 3 & 0 \\ 0 & 0 & 0 & 3 \end{pmatrix} \qquad B = 12A^{-1} = \begin{pmatrix} 6 & 3 & 3 & 0 \\ 3 & 6 & 3 & 0 \\ 3 & 3 & 6 & 0 \\ 0 & 0 & 0 & 4 \end{pmatrix}$$

$\det(A) = 48$, $|\operatorname{Aut}^+(L)| = 48$, eutaxy : $(3,3,3,4,3)$.

c_5 : $S = \{e_1, e_2, e_3, e_4, e_1 + e_2\}$

$$A = \begin{pmatrix} 2 & -1 & 0 & 0 \\ -1 & 2 & 0 & 0 \\ 0 & 0 & 2 & 0 \\ 0 & 0 & 0 & 2 \end{pmatrix} \qquad B = 6A^{-1} = \begin{pmatrix} 4 & 2 & 0 & 0 \\ 2 & 4 & 0 & 0 \\ 0 & 0 & 3 & 0 \\ 0 & 0 & 0 & 3 \end{pmatrix}$$

$\det(A) = 12$, $|\operatorname{Aut}^+(L)| = 48$, eutaxy : $(2,2,3,3,2)$.

a_6 : $S = \{e_1, e_2, e_3, e_4, e_1 + e_2 + e_3, e_1 + e_2 + e_4\}$

No weakly eutactic lattice

b_6 : $S = \{e_1, e_2, e_3, e_4, e_1 + e_2, e_1 + e_3 + e_4\}$

$$A = \begin{pmatrix} 6 & -3 & -2 & -2 \\ -3 & 6 & 1 & 1 \\ -2 & 1 & 6 & -2 \\ -2 & 1 & -2 & 6 \end{pmatrix} \qquad B = 72A^{-1} = \begin{pmatrix} 22 & 8 & 9 & 9 \\ 8 & 16 & 0 & 0 \\ 9 & 0 & 18 & 9 \\ 9 & 0 & 9 & 18 \end{pmatrix}$$

$\det(A) = 432\,, \quad |\operatorname{Aut}^+(L)| = 12\,, \quad \text{eutaxy} : (5, 8, 9, 9, 8, 9)\,.$

c_6 : $S = \{e_1, e_2, e_3, e_4, e_1 + e_2, e_1 + e_3\}$

$$A = \begin{pmatrix} 4 & -2 & -2 & 0 \\ -2 & 4 & 1 & 0 \\ -2 & 1 & 4 & 0 \\ 0 & 0 & 0 & 4 \end{pmatrix} \qquad B = 12A^{-1} = \begin{pmatrix} 5 & 2 & 2 & 0 \\ 2 & 4 & 0 & 0 \\ 2 & 0 & 4 & 0 \\ 0 & 0 & 0 & 3 \end{pmatrix}$$

$\det(A) = 144\,, \quad |\operatorname{Aut}^+(L)| = 16\,, \quad \text{eutaxy} : (1, 2, 2, 3, 2, 2)\,.$

d_6 : $S = \{e_1, e_2, e_3, e_4, e_1 + e_2, e_3 + e_4\}$

$$A = \begin{pmatrix} 2 & -1 & 0 & 0 \\ -1 & 2 & 0 & 0 \\ 0 & 0 & 2 & -1 \\ 0 & 0 & -1 & 2 \end{pmatrix} \qquad B = 12A^{-1} = \begin{pmatrix} 2 & 1 & 0 & 0 \\ 1 & 2 & 0 & 0 \\ 0 & 0 & 2 & 1 \\ 0 & 0 & 1 & 2 \end{pmatrix}$$

$\det(A) = 9\,, \quad |\operatorname{Aut}^+(L)| = 144\,, \quad \text{eutaxy} : (1, 1, 1, 1, 1, 1)\,.$

a_7 : $S = \{e_1, e_2, e_3, e_4, e_1 + e_2, e_1 + e_3, -e_2 + e_3 + e_4\}\,, \quad \omega = \frac{1+\sqrt{13}}{2}$

$$A = \begin{pmatrix} 2\omega & -\omega & -\omega & 0 \\ -\omega & 2\omega & 2 & \omega-1 \\ -\omega & 2 & 2\omega & 1-\omega \\ 0 & \omega-1 & 1-\omega & 2\omega \end{pmatrix} \qquad B = 6(\omega - 1)A^{-1} = \begin{pmatrix} 2 & 3-\omega & 3-\omega & 0 \\ 3-\omega & \omega & 6-3\omega & 6-3\omega \\ 3-\omega & 6-3\omega & \omega & 3\omega-6 \\ 0 & 6-3\omega & 3\omega-6 & 6\omega-12 \end{pmatrix}$$

$\det(A) = 60 + 48\omega\,, \quad |\operatorname{Aut}^+(L)| = 8\,,$
$\text{eutaxy} : (2\omega - 4, 3 - \omega, 3 - \omega, 3\omega - 6, 3 - \omega, 3 - \omega, 3\omega - 6)\,.$

b_7 : $S = \{e_1, e_2, e_3, e_4, e_1 + e_2, e_1 + e_3, e_1 + e_4\}$
Semi-eutactic lattice

$$A = \begin{pmatrix} 4 & -2 & -2 & -2 \\ -2 & 4 & 1 & 1 \\ -2 & 1 & 4 & 1 \\ -2 & 1 & 1 & 4 \end{pmatrix} \qquad B = 6A^{-1} = \begin{pmatrix} 3 & 1 & 1 & 1 \\ 1 & 2 & 0 & 0 \\ 1 & 0 & 2 & 0 \\ 1 & 0 & 0 & 2 \end{pmatrix}$$

$\det(A) = 108\,, \quad |\operatorname{Aut}^+(L)| = 48\,, \quad \text{eutaxy} : (0, 1, 1, 1, 1, 1, 1)\,.$

c_7 : $S = \{e_1, e_2, e_3, e_4, e_1 + e_2, e_1 + e_3, e_2 + e_4\}$

$$A = \begin{pmatrix} 8 & -4 & -4 & 2 \\ -4 & 8 & 2 & -4 \\ -4 & 2 & 8 & -1 \\ 2 & -4 & -1 & 8 \end{pmatrix} \qquad B = 24A^{-1} = \begin{pmatrix} 5 & 2 & 2 & 0 \\ 2 & 5 & 0 & 2 \\ 2 & 0 & 4 & 0 \\ 0 & 2 & 0 & 4 \end{pmatrix}$$

$\det(A) = 1728\,, \quad |\operatorname{Aut}^+(L)| = 8\,, \quad \text{eutaxy} : (1, 1, 2, 2, 2, 2, 2)\,.$

d_7 : $S = \{e_1, e_2, e_3, e_4, e_1 + e_2, e_1 + e_3, e_2 - e_3\}$

$$A = \begin{pmatrix} 2 & -1 & -1 & 0 \\ -1 & 2 & 1 & 0 \\ -1 & 1 & 2 & 0 \\ 0 & 0 & 0 & 2 \end{pmatrix} \qquad B = 4A^{-1} = \begin{pmatrix} 3 & 1 & 1 & 0 \\ 1 & 3 & -1 & 0 \\ 1 & -1 & 3 & 0 \\ 0 & 0 & 0 & 2 \end{pmatrix}$$

$\det(A) = 8\,, \quad |\operatorname{Aut}^+(L)| = 48\,, \quad \text{eutaxy} : (1, 1, 1, 2, 1, 1, 1)\,.$

a_8 : $\quad S = \{e_1, e_2, e_3, e_4, e_1 + e_2, e_1 + e_3, e_2 + e_4, e_3 + e_4\}\,,\ \theta = \sqrt{3}$

$$A = \begin{pmatrix} 2 & -1 & -1 & \theta-1 \\ -1 & 2 & \theta-1 & -1 \\ -1 & \theta-1 & 2 & -1 \\ \theta-1 & -1 & -1 & 2 \end{pmatrix} \qquad B = 2\theta A^{-1} = \begin{pmatrix} 1+\theta & 1 & 1 & 0 \\ 1 & 1+\theta & 0 & 1 \\ 1 & 0 & 1+\theta & 1 \\ 0 & 1 & 1 & 1+\theta \end{pmatrix}$$

$\det(A) = 12\theta(2-\theta)\,, \quad |\operatorname{Aut}^+(L)| = 8\,,$
eutaxy : $(\theta-1, \theta-1, \theta-1, \theta-1, 1, 1, 1, 1)\,.$

b_8 : $\quad S = \{e_1, e_2, e_3, e_4, e_1 + e_2, e_1 + e_3, e_1 + e_4, e_2 - e_3\}$

$$A = \begin{pmatrix} 4 & -2 & -2 & -2 \\ -2 & 4 & 2 & 1 \\ -2 & 2 & 4 & 1 \\ -2 & 1 & 1 & 4 \end{pmatrix} \qquad B = 24A^{-1} = \begin{pmatrix} 11 & 3 & 3 & 4 \\ 3 & 9 & -3 & 0 \\ 3 & -3 & 9 & 0 \\ 4 & 0 & 0 & 8 \end{pmatrix}$$

$\det(A) = 96\,, \quad |\operatorname{Aut}^+(L)| = 8\,, \quad$ eutaxy : $(1, 3, 3, 4, 3, 3, 4, 3)\,.$

a_9 : $\quad S = \{e_1, e_2, e_3, e_4, e_1 + e_2, e_1 + e_3, e_2 + e_4, e_3 + e_4, e_1 + e_2 + e_3 + e_4\}$

$$A = \begin{pmatrix} 4 & -2 & -2 & 1 \\ -2 & 4 & 1 & -2 \\ -2 & 1 & 4 & -2 \\ 1 & -2 & -2 & 4 \end{pmatrix} \qquad B = 24A^{-1} = \begin{pmatrix} 4 & 2 & 2 & 1 \\ 2 & 4 & 1 & 2 \\ 2 & 1 & 4 & 2 \\ 1 & 2 & 2 & 4 \end{pmatrix}$$

$\det(A) = 81\,, \quad |\operatorname{Aut}^+(L)| = 72\,, \quad$ eutaxy : $(1, 1, 1, 1, 1, 1, 1, 1, 1)\,.$

b_9 : $\quad S = \{e_1, e_2, e_3, e_4, e_1 + e_2, e_1 + e_3, e_1 + e_4, e_2 - e_3, e_2 - e_4,\}$

$$A = \begin{pmatrix} 6 & -3 & -3 & -3 \\ -3 & 6 & 3 & 3 \\ -3 & 3 & 6 & 2 \\ -3 & 3 & 2 & 6 \end{pmatrix} \qquad B = 36A^{-1} = \begin{pmatrix} 10 & 2 & 3 & 3 \\ 2 & 10 & -3 & -3 \\ 3 & -3 & 9 & 0 \\ 3 & -3 & 0 & 9 \end{pmatrix}$$

$\det(A) = 432\,, \quad |\operatorname{Aut}^+(L)| = 12\,, \quad$ eutaxy : $(2, 2, 3, 3, 2, 3, 3, 3, 3)\,.$

a_{10} : $\quad S = \{e_1, e_2, e_3, e_4, e_1 - e_3, e_1 + e_2 - e_3, e_2 - e_3, e_1 - e_4, e_1 + e_2 - e_4,$
$e_1 + e_2 - e_3 - e_4, e_2 - e_4, e_3 - e_4\}$

$$A = \begin{pmatrix} 2 & 0 & 1 & 1 \\ 0 & 2 & 1 & 1 \\ 1 & 1 & 2 & 1 \\ 1 & 1 & 1 & 2 \end{pmatrix} \qquad B = 2A^{-1} = \begin{pmatrix} 2 & 1 & -1 & -1 \\ 1 & 2 & -1 & -1 \\ -1 & -1 & 2 & 0 \\ -1 & -1 & 0 & 2 \end{pmatrix}$$

$\det(A) = 4\,, \quad |\operatorname{Aut}^+(L)| = 576\,, \quad$ eutaxy : $\frac{1}{3}(1, 1, 1, 1, 1, 1, 1, 1, 1, 1, 1, 1)\,.$

b_{10} : $\quad S = \{e_1, e_2, e_3, e_4, e_1 - e_2, e_1 - e_3, e_2 - e_3, e_1 - e_4, a_2e_2 - e_4, e_3 - e_4\}$

$$A = \begin{pmatrix} 2 & 1 & 1 & 1 \\ 1 & 2 & 1 & 1 \\ 1 & 1 & 2 & 1 \\ 1 & 1 & 1 & 2 \end{pmatrix} \qquad B = 5A^{-1} = \begin{pmatrix} 4 & -1 & -1 & -1 \\ -1 & 4 & -1 & -1 \\ -1 & -1 & 4 & -1 \\ -1 & -1 & -1 & 4 \end{pmatrix}$$

$\det(A) = 5\,, \quad |\operatorname{Aut}^+(L)| = 120\,, \quad$ eutaxy : $(1, 1, 1, 1, 1, 1, 1, 1, 1, 1)\,.$

14.4 The Hermite Constant

In this section, we have collected the largest known values (to our knowledge) of the Hermite constant for low-dimensional lattices.

Table 14.4.1 contains values of the invariant $\gamma(\Lambda)$ for all dimensions $n \leq 32$. These values are known to be optimal only for $n \leq 8$. Beyond, everything is conjectural, even in the case of lattices which are exceptionally dense with regard to Roger's bounds, such as the Leech lattice in dimension 24. Numerical values are rounded down to three decimal places.

Table 14.4.1. Large values of the Hermite invariant.

n	1	2	3	4	5	6	7	8
γ	1	$\frac{2}{3^{1/2}}$	$2^{1/3}$	$2^{1/2}$	$2^{3/5}$	$\frac{2}{3^{1/6}}$	$2^{6/7}$	2
γ	1	1.154	1.259	1.414	1.515	1.665	1.811	2
n	9	10	11	12	13	14	15	16
γ	2	$\frac{2^{6/5}}{3^{1/10}}$	$\frac{2^{20/11}}{3^{5/11}}$	$\frac{2^2}{3^{1/2}}$	$\frac{2^{24/13}}{3^{5/13}}$	$\frac{2^{10/7}}{3^{1/14}}$	$2^{7/5}$	$2^{3/2}$
γ	2	2.058	2.140	2.309	2.356	2.488	2.639	2.828
n	17	18	19	20	21	22	23	24
γ	$2^{26/17}$	$\frac{2^{5/3}}{3^{1/18}}$	$2^{31/19}$	$2^{17/10}$	$2^{37/21}$	$\frac{2^{21/11}}{3^{1/22}}$	$2^{44/23}$	4
γ	2.886	2.986	3.098	3.249	3.391	3.572	3.766	4
n	25	26	27	28	29	30	31	32
γ	$2^{49/25}$	$\frac{2^2}{3^{1/26}}$	$\frac{2^2}{3^{1/27}}$	$\frac{2^{29/14}}{3^{1/14}}$	$\frac{2^2}{3^{1/29}}$	$2^{8/15}3^{9/10}$	$2^{15/31}3^{30/31}$	$2^{1/2}3$
γ	3.890	3.834	3.840	3.885	3.851	3.890	4.049	4.242

These values are attained on laminated lattices Λ_n for $n = 1, \dots, 10$ or $n = 14, \dots, 26$, on the K_n lattices for $n = 11, 12, 13$, on cross-sections of any of the four "Quebbemann-like" lattices for $n = 30, 31, 32$ (see the notes on Chapter 8; note that these lattices are modular, and possess cross-sections isometric to $\sqrt{3}\,\mathbb{D}_4$), and on lattices recently found by Bacher ([Bc1]) in dimensions $n = 27, 28, 29$. All these lattices are perfect.

The smallest norm which makes these lattices integral is equal to 1, 2, 4, 8, and 6, according to whether $n = 1$, $2 \le n \le 8$, $9 \le n \le 24$, $25 \le n \le 29$, and $30 \le n \le 32$; for $n = 26$, minimum 4 is also possible. Dimensions 25 to 29 deserve a few comments.

• $n = 25$. There are 23 laminated lattices, corresponding to the 23 deep holes of the Leech lattice (Conway and Sloane; see Chapters 6 and 24 of [C-S]).
• $n = 26$—29. Lattices of norm 4 and determinant 3 (resp. 4) if $n = 26$ (resp. $27, 28, 29$) have the same density as the laminated lattices of same dimensions. An example for $n = 26$ was found by Nebe in [Ne2]. For $n = 27, 28, 29$, examples are provided by the even parts of unimodular lattices of norm 3. Bacher and Venkov proved in [Bc-V2] that there exist for $n = 27$ (resp. 28) exactly three (resp. 38) unimodular lattices of minimum 3, producing three

(resp. 33) even lattices of minimum and determinant 4, and that all even lattices of minimum and determinant 4 are obtained in this way.
• Bacher's lattice in dimension 28 (resp. 29) is obtained by a kind of lamination process above unimodular cross-sections of codimension 2 and norm 3 (resp. 4); see above; for $n = 27$, Bacher makes use of a cross-section of his 28-dimensional lattice.

Remark 14.4.2. *If Λ is a dense lattice with odd minimum m, its even sublattice will generally have minimum $m+1$, which implies*

$$\gamma(\Lambda_0) = 4^{-1/n}\left(1+\frac{1}{m}\right)\gamma(\Lambda),$$

whence

$$\gamma(\Lambda) \geq \gamma(\Lambda_0) \Longleftrightarrow n \leq \frac{\log 4}{\log\left(1+\frac{1}{m}\right)}.$$

One expects m to be "small" with respect to n; hence Λ_0 is expected to be denser than Λ (despite the near miss $\sqrt{2}\,\mathbb{E}_7^* - \sqrt{2}\,\mathbb{E}_7$). However, it is somewhat surprising that minimum 6 does not show up in the table above before dimension 30.

Beyond dimension 32, one can use embeddings of the "Quebbemann-like" lattices in unimodular lattices of dimension 64 and norm 6; the orthogonal lattices to lattices of dimension $31, 30, \dots$ are reasonably dense lattices of dimension $33, 34, \dots$ A better example is known in dimension 36 ([Ks-P]; see [C-S'], p. xvi).

Around dimension 48, the best known examples are obtained by making use of norm 6 unimodular lattices of dimension 48; for $40 \leq n \leq 48$, see [C-S], Chapter 6, Section 2, Corollary 8.

14.5 Invariants Related to Duality

This section is devoted to the "dual Hermite constant" γ'_n ("Bergé–Martinet constant") and its analog for isodual lattices, in particular for symplectic lattices.

Recall (Definition 2.8.1) that γ' is the geometrical mean of the Hermite invariant of a lattice and its dual: we have

$$\gamma'(\Lambda) = \big(\gamma(\Lambda)\,\gamma(\Lambda^*)\big)^{1/2} = \big(N(\Lambda)\,N(\Lambda^*)\big)^{1/2}.$$

Note that the invariants γ' and γ coincide on isodual lattices. Thus the data in Table14. 4.1 can be used for γ' in dimensions 1, 2, 4, 8, 12, 16, 24, and 32.

In the table below, we have listed the greatest known values of γ' we know in dimension $n \leq 24$. (Sources: [B-M1] ($n \leq 9$), [C-S9], [Nap1], [Nap2], [Ne2], [Ne-Pl], [Sl1], [Sou2].) Examples become sparse beyond dimension 24;

Table 14.5.1. Large values of the *square* of γ' (rounded down to three decimal places).

n	1	2	3	4	5	6	7	8
${\gamma'_n}^2$	1	$\frac{4}{3}$	$\frac{3}{2}$	2	2	$\frac{8}{3}$	3	4
${\gamma'_n}^2$	1	1.333	1.5	2	2	2.666	3	4
n	9	10	11	12	13	14	15	16
${\gamma'_n}^2$	$\frac{16}{5}$	4	4	$\frac{16}{3}$	$\frac{24}{5}$	$\frac{16}{3}$	6	8
${\gamma'_n}^2$	3.2	4	4	5.333	4.8	5.333	6	8
n	17	18	19	20	21	22	23	24
${\gamma'_n}^2$	6	8	6	$\frac{64}{7}$	9	$\frac{32}{3}$	12	16
${\gamma'_n}^2$	6	8	6	9.142	9	10.666	12	16

see however Table 14.5.2 below. [Note that the value given for $n = 19$ (${\gamma'}^2 = 6$) is attained on Λ_{19} and on K'_{19}, which are not dual-extreme. Hence it is certainly not optimal.]

Beyond dimension 24, the largest known values of the invariant γ' are attained on isodual lattices; we shall return to this question at the end of this section.

Since γ' takes the same value on a lattice and on its dual, we only describe one lattice per pair (Λ, Λ^*), even if Λ and Λ^* are not similar. For $n \le 8$, the displayed values of γ' are attained on root lattices of Table 14.4.1, except that in dimension 5, $P_5^2 \sim \mathbb{A}_5^3$ and $\mathbb{D}_5$ have the same γ'. In dimension 9, our lattice is Coxeter's $\mathbb{A}_9^2$; another 9-dimensional lattice with the same invariant γ' as $\mathbb{A}_9^2$ occurs in [Bari].

In dimensions 10 and 11, we again have several possible choices: $\mathbb{D}_{10}^+$ and Q_{10}, which are isodual, K'_{10}, K_{11}, $\mathbb{A}_{11}^3$ which are not. Isodual lattices again show up in dimensions 12 (the Coxeter–Todd lattice), 14 (the lattice dicovered by Souvignier and denoted by Q_{14} in [C-S9]; Λ_{14}, which is not isodual, can also be used), 16 (the Barnes–Wall lattice), 20 (a lattice described in [ATLAS], and also found by Nebe, related to the group $[2 \cdot M_{22} \cdot 2]$), and 24 (the Leech lattice). The other lattices which occur in the table are the lattice denoted by $Q_{13}(2)$ in [C-S9] (see also [Pl1]), Λ_{15}, K'_{17} (the lattice $L_{17.4}$ of [Pl-P2] is also a suitable one), K'_{18}, K'_{21}, Λ_{22}, Λ_{23}.

Let us now restrict ourselves to isodual lattices, and first consider ℓ-modular lattices Λ (i.e., Λ is integral, and there exists a similarity of ratio $\sqrt{\ell}$, ℓ an integer, of Λ^* onto Λ; see Section 11.5). For such a lattice, we have $\det(\Lambda) = \ell^{n/2}$, hence

$$\gamma(\Lambda) = \gamma'(\Lambda) = \frac{N(\Lambda)}{\ell^{1/2}};$$

in particular, ${\gamma'}^2$ takes rational values on modular lattices. Note that in odd dimensions, ℓ must be a square, so that this notion is then less interesting; indeed, extensive results are then available only for $\ell = 1$.

The data of Table 14.5.1 can be used for dimensions 1, 2, 4, 8, 10, 12, 14, 16, 20, 24, and even extended to dimensions 26, 28, 30, for which Nebe has given in [Ne2] other interesting examples yielding maximal finite subgroups of $\mathrm{GL}_n(\mathbb{Q})$, and of course in dimension 32 where 2-modular lattices of norm 6 occur in Table 14.4.1. All these lattices are ℓ-modular for some $\ell \in \{1, 2, 3, 7, 39\}$. Here are the values of ℓ and $N(\Lambda)$ for certain even lattices of even dimension n (numerical values are as above rounded down to three decimal places):

Table 14.5.2. Large values of ${\gamma'}^2$ on modular lattices.

n	1	2	4	8	10	12	14	16	20	24	26	28	30	32
ℓ	1	3	2	1	4	3	3	2	7	1	3	39	3	2
$N(\Lambda)$	1	2	2	2	4	4	4	4	8	4	6	22	6	6
$\gamma'(\Lambda)^2$	1	1.333	2	4	4	5.333	5.333	8	9.142	16	12	12.410	12	18

[D_{10}^+ and Q_{10} have different Smith invariants, namely $(4, 2^8, 1)$ and $(4^4, 2^2, 1^4)$.]

For other dimensions, [C-S9] is the unique extensive reference; this is the data source for Table 14.5.3 below. Lattices with a reasonably high density are known only in small dimensions. For instance, the comparison with data of Table 14.4.1 for $n = 14$ and 16 suggests that for $n = 15$, the displayed value ${\gamma'}^2 = 4$ (attained on $\mathbb{A}_{15}^4$) could be far from being optimal. Similarly, Baril has verified that the 9-dimensional orthogonal lattice which we use in Table 14.5.3 below is not extreme as an isodual lattice.

[As usual, the numerical values are rounded down to three decimal places.]

14.6 The Kissing Number

The kissing number is twice the invariant we denote by s, and we shall as previously consider s rather than the "true" kissing number $2s$. We also consider its dual version $s + s^*$ (see Exercise 3.8.7).

The maximum $s_{\max}$ of $s(\Lambda)$ in a given dimension n is known for $n \le 9$ (Watson) and $n = 24$ (Bannai, Odlyzko, Sloane); see notes on Chapter 6. We refer to Conway and Sloane's book ([C-S], Chapter 1, Tables 1.2 and 1.3) for

Table 14.5.3. Large values of γ'^2 (other isodual lattices).

n	3	5	6	7	9	11	13	15	23
γ'	$\frac{1+\sqrt{2}}{2}$	$\frac{7}{5}$	$\frac{3+\sqrt{3}}{3}$	$\frac{5}{3}$	$\frac{12}{7}$	$\frac{7}{4}$	2	2	3
γ'^2	1.457	1.96	2.488	2.777	2.938	3.062	4	4	9

[The Smith invariants of the three lattices of dimensions $5, 7, 9$ are $(25, 5^3, 1)$, $(9, 3^5, 1)$, and $(49, 7^7, 1)$; note that those of the lattices above which are rational are scaled copies of modular lattices.]

a list of large values of s in various dimensions, including the whole interval [1,24], and to [E-R-S] for dimensions beyond 32; see also the recent paper [El1] by Elkies. We simply recall here that for $n \le 8$, $s_{\max}$ is attained uniquely on the laminated lattices, which are scaled copies of root lattices for all $n \le 8$. We remark that $s_{\max}$ is very often attained on lattices on which (conjecturally beyond dimension 8) the Hermite invariant attains its maximum.
[Notable exception: $s(\Lambda_{11}^{\max}) = 219 > s(K_{11}) = 216$.]

We now state and prove the unique theorem of this chapter.

Theorem 14.6.1. *For $1 \le n \le 8$, the maximal value of $s + s^*$ and the lattices Λ on which this value is attained are as given in the table below. (For each possible pair (Λ, Λ^*), we only display one lattice Λ, with $s(\Lambda) \ge s(\Lambda^*)$; there are two such lattices in dimension 5.)*

n	1	2	3	4	5	6	7	8
$s + s^*$	2	6	10	24	25	63	91	240
Lattices	$\mathbb{Z}$	$\mathbb{A}_2$	$\mathbb{A}_3$	$\mathbb{D}_4$	$\mathbb{D}_5$, $\mathbb{A}_5^3$	$\mathbb{E}_6$	$\mathbb{E}_7$	$\mathbb{E}_8$

Proof. We shall end with the proof for dimension 5, which relies on some unpublished work.

If $s_{\max}$ is attained on an isodual lattice, then $(s + s^*)_{\max} = 2\, s_{\max}$. This remark solves our problem for $n = 1, 2, 4, 8$.

In general, we choose a lattice Λ_0 for which $\sigma = s + s^*$ is reasonably large, and we then just look for lattices such that $s \ge s^*$ and $s \ge \frac{\sigma}{2}$.

If $n = 3$, we have $s_{\max} = 6$, attained uniquely on $\mathbb{A}_3$, for which $s + s^* = 10$. We may take $\frac{\sigma}{2} = 5$, but $s = 5$ implies $s^* = 2$ (Theorem 9.2.2), which completes the proof in this case.

If $n = 6$, we have $s_{\max} = 36$, attained uniquely on $\mathbb{E}_6$, for which $s + s^* = 63$. We may take $\lceil \frac{\sigma}{2} \rceil = 32$. By the results of Chapter 6, we have $s(\Lambda) \le s(\mathbb{D}_6) = 30 < \frac{\sigma}{2}$ if Λ is any perfect lattice non-similar to $\mathbb{E}_6$; by Barnes's determination of the Voronoi graph, the greatest value of s on a non-perfect lattice is $s = 28$ (attained on one path $\mathbb{E}_6$—$\mathbb{E}_6$), whence the result in this case.

If $n = 7$, we have $s_{\max} = 63$, attained uniquely on $\mathbb{E}_7$, for which $s + s^* = 91$. We may take $\lceil \frac{\sigma}{2} \rceil = 46$. By the results of Chapter 6, we have $s(\Lambda) \leq s(\mathbb{D}_7) = 42 < \frac{\sigma}{2}$ if Λ is any perfect lattice non-similar to $\mathbb{E}_7$; by Jaquet's determination of the Voronoi graph, the greatest value of s on a non-perfect lattice is $s = 46$, attained on a unique Voronoi path $\mathcal{P}$ (which of course connects $\mathbb{E}_7$ to itself, since $s > 42$). For a non-perfect lattice not on $\mathcal{P}$, we certainly have $s + s^* \leq 90$. It is thus sufficient to prove that $s(\Lambda^*) \leq 44$ holds on the whole path $\mathcal{P}$. To this end, we have verified that 36 out of the 46 minimal vectors are orthogonal to a conveniently chosen vector in E, which shows that all lattices lying on $\mathcal{P}$ have an $\mathbb{E}_6$ section. Since $2\,s(\mathbb{E}_6) - s(\mathbb{D}_5) = 52 > 46$, this section is unique. Hence we have $s^* = 1$ on $\mathcal{P}$, which completes the proof for $n = 7$.

Finally, if $n = 5$, a difficulty arises, due to the fact that for the two putative lattices, s or s^* is comparatively small. We saw in Chapters 6 and 7 that for a 5-dimensional lattice Λ, the perfection rank r is equal to s, except if Λ has a $\mathbb{D}_4$ section, in which case either $\Lambda \sim \mathbb{D}_5$, or Λ is not perfect, and then $s(\Lambda) = r + 2 \leq 16$ and $s^*(\Lambda) = 1$. Also, in the case of a perfect lattice, we have $(s, s^*) = (20, 5)$, $(15, 10)$, or $(15, 6)$. We may take $\lceil \frac{\sigma}{2} \rceil = 13$, and we are left with the lattices belonging to a minimal class with $s = r \in \{13, 14\}$. After Batut ([Bt]), there are 11 such classes. In the course of an unpublished work on γ'_5, Bergé has shown that the upper bound $s^* \leq 10$ holds on all these classes. □

Except for nice examples provided by a few remarkable modular lattices, almost nothing is known beyond dimension 8. We have $(s + s^*)_{\max} \geq 137 = s(\Lambda_9) + s(\Lambda_9^*)$ for $n = 9$; the existence of the isodual lattice Q_{10} suggests that we could have $(s + s^*)_{\max} = 260$ in dimension 10, larger than $255 = s(K'_{10}) + s({K'_{10}}^*)$.

15 Appendix 1: Semi-Simple Algebras and Quaternions

Some constructions of lattices that we developed in this book, notably in Chapter 8, heavily rely on arithmetic in semi-simple algebras, in particular in C.M. fields and quaternion division algebras. The algebras we considered are endowed with a canonical involution. The aim of this appendix is to explain some of the properties we previously used.

Here is a short list of references for the proofs.

• The great classic is Deuring's 1935 *Algebren* ([De]), which covers both the general theory of algebras and the theory of orders over a Dedekind domain, but not Eichler's theory, published after the book was written.
• Theory of Dedekind domains: [Se1], [C-F]; connection with quadratic forms: [O'M].
• Theory of algebras: [Bou2], Chapter VIII; [K-M-R-T].
• Theory of orders: [Rei].
• Quaternion algebras : [Vi].

15.1 Semi-Simple Algebras

By an *algebra*, we mean a finite–dimensional algebra L over a field K. An *involution on* L is an isomorphism $x \mapsto \bar{x}$ of L onto the opposite algebra L^{op}, which is simply L, endowed with the multiplication $(x, y) \mapsto yx$. We say that L is *semi-simple* if it does not contain any nonzero nilpotent ideal, and that it is *simple* if $\{0\}$ and L itself are the only two-sided ideals in L. (In general, any algebra contains a maximal (say, left) nilpotent ideal R; this is a two-sided ideal, called the *radical of* L, and L/R is semi-simple.)

Theorem 15.1.1. *1. A semi-simple algebra is isomorphic to a direct product of simple algebras.*

2. A simple algebra L is isomorphic to a matrix algebra $\mathcal{M}_m(D)$, where D is a division algebra over *K, i.e. a skew-field whose centre contains K. The integer m and the isometry class of D only depend on the isomorphism class of L.*

A semi-simple algebra is thus a direct product $L = \prod_{i=1}^{r} L_i$ of simple algebras L_i. Each L_i has a centre C_i, which is a finite extension of K, and L has centre $C = \prod_i C_i$.

The nice notion is that of a separable algebra: L is *separable* (over K) if all C_i are separable extensions of K, a condition which is always verified in our applications, for the fields we consider are either of characteristic zero, or finite. More intrinsically, this amounts to saying that for any extension K' of K (or for an algebraic closure K' of K), the extended algebra $L_{K'} = K' \bigotimes L$ is semi-simple. Actually, this is easily proved once we know that if L is simple with centre K (we then say that L is *central simple*), then $K' \bigotimes L$ is central simple over K'.

We summarize in a theorem various characterizations of separable algebras. As usual, we denote by $\overline{K}$ an algebraic closure of K.

Theorem 15.1.2. *For a finite–dimensional algebra L over K, the following properties are equivalent:*

1. *L is separable.*
2. *For any extension K' of K, $L_{K'} = K' \bigotimes_K L$ is semi-simple.*
3. *For any finite extension K' of K, $L_{K'} = K' \bigotimes_K L$ is semi-simple.*
4. *$L_{\overline{K}}$ is semi-simple.*

Let L be a central simple K-algebra. Since $L_{\overline{K}}$ is isomorphic to a matrix ring $\mathcal{M}_m(\overline{K})$ (which proves that the rank of a central simple algebra is a square), there corresponds to any $x \in L$ a matrix M_x with entries in $\overline{K}$, whose characteristic polynomial $\chi_{\mathrm{red},x}(X)$ is called the *reduced characteristic polynomial of x*. It can be shown that it only depends on x and has coefficients in K. It is related to the usual characteristic polynomial of x (that of $y \mapsto xy$) by $\chi_x(X) = \chi_{\mathrm{red},x}(X)^m$. Write

$$\chi_{\mathrm{red},x}(X) = X^m - a_1 X^{m-1} + \cdots + (-1)^m a_m \,.$$

We define the *reduced trace of x* by $\mathrm{Trd}(x) = a_1$ and the *reduced norm of x* by $\mathrm{Nrd}(x) = a_m$. They are related to the usual trace and usual norm by

$$\mathrm{Tr}(x) = m \ \mathrm{Trd}(x) \quad \text{and} \quad \mathrm{N}(x) = \mathrm{Nrd}(x)^m \,.$$

We now define the reduced trace for general separable algebras by

$$\mathrm{Trd}_{L/K}(x) = \sum_i \mathrm{Tr}_{C_i/K} \circ \mathrm{Trd}_{L/C_i}(x)$$

(and the general reduced norm in a similar way).

When dealing with simple algebras, it is natural to work with the reduced trace rather than with the trace. However, the trace may prove useful when dealing with certain semi-simple algebras; see the example below.

For an algebra with involution, the bilinear form $\mathrm{Trd}(x\bar{y})$ (as well as $\mathrm{Tr}(x\bar{y})$ if $\mathrm{Char}\, K = 0$) is non-degenerate. Over the base field $\mathbb{Q}$, the natural trace form is not in general positive definite (indeed, it is only on algebras which are products of totally real number fields). To consider this twisted form rather than the natural one is important to obtain positive definite forms on algebras over $\mathbb{Q}$.

Simple algebras with involution have been classified ("Landherr's theorem"; see [K-M-R-T]). Two kinds of algebras must be considered; algebras of the first kind, namely those whose centre is invariant under the involution, have many applications to lattice theory (see e.g. Chapter 8). As far as I know, algebras of the second kind (whose involution induces a non-trivial involution on the centre) have never been systematically considered to construct Euclidean lattices.

Example 15.1.3. *Let G be a finite group, of order n. Then the group ring $K[G]$ (with canonical basis $(s, s \in G)$) is a separable algebra if and only if* $\operatorname{Char} K$ *does not divide n. It possesses a canonical involution, characterized by $\bar{s} = s^{-1}$. When $K = \mathbb{Q}$, both the forms* $\operatorname{Tr}(x\bar{y})$ *and* $\operatorname{Trd}(x\bar{y})$ *are positive definite.* [The forms $\operatorname{Tr}(xy)$ and $\operatorname{Tr}(x\bar{y})$ are easy to handle; in particular, the determinant of the Gram matrix $\operatorname{Tr}(st^{-1})$ is simply n^n, whereas calculating with the reduced trace needs some knowledge of the simple components of $K[G]$.]

15.2 Quaternion Algebras

Let K be a field of characteristic not 2.

Definition 15.2.1. *Let $a, b \in K^\times$. The* quaternion algebra of type (a, b) *is the rank-4 algebra $H = K_{a,b}$ with a basis $(1, i, j, k)$, where 1 is the unit element of H and i, j, k satisfy the multiplication rules $i^2 = a$, $j^2 = b$, $ij = -ji = k$.* [If $\operatorname{Char} K = 2$, the relations take the form $i^2 = i + a$, $j^2 = b$, $ij = k$, $ji = k + j$, with $a \in K$ and $b \in K^\times$.]

Assuming associativity, the other products are well-defined by the rules above; in particular, $k^2 = -ab$. Embedding H into $\mathcal{M}_2(K(\sqrt{a}, \sqrt{b}))$ by

$$1 \mapsto \begin{pmatrix} 1 & 0 \\ 0 & 1 \end{pmatrix},\ i \mapsto \begin{pmatrix} \sqrt{a} & 0 \\ 0 & -\sqrt{a} \end{pmatrix},\ j \mapsto \begin{pmatrix} 0 & \sqrt{b} \\ \sqrt{b} & 0 \end{pmatrix},\ k \mapsto \begin{pmatrix} 0 & \sqrt{ab} \\ -\sqrt{ab} & 0 \end{pmatrix}.$$

shows that H is indeed an associative algebra. It is easy to verify that H is simple, with centre K (and it can be proved that all rank 4 central simple algebras are quaternion algebras). The map

$$q = x + yi + zj + tk \ \mapsto \ \bar{q} = x - yi - zj - tk$$

is an involution, and we have

$$\operatorname{Trd}_{H/K}(q) = 2\,x \quad \text{and} \quad \operatorname{Nrd}(q) = x^2 - a\,y^2 - b\,z^2 + ab\,t^2.$$

(Note that $\operatorname{Trd}(q\bar{q}) = 2\operatorname{Nrd}(q)$, a formula which explains why quaternions are suitable to construct even lattices.) It is clear that a quaternion q is invertible if and only if $\mathrm{Nrd}(q) \neq 0$ (if $\mathrm{Nrd}(q) \neq 0$, then $q^{-1} = \mathrm{Nrd}(q)^{-1}\bar{q}$). Hence H is a skew-field if the quadratic form $q \mapsto \mathrm{Nrd}(q\bar{q})$ does not represent 0 on K, and is isomorphic to $\mathcal{M}_2(K)$ otherwise.

Over $\mathbb{R}$, $H_{a,b}$ is a skew-field if and only if both a and b are negative. Replacing i by $\frac{1}{\sqrt{a}}i$ and j by $\frac{1}{\sqrt{b}}j$, we see that there exists a unique skew-field of rank 4 over $\mathbb{R}$, which can be obtained by taking $a = b = -1$, whence $i^2 = j^2 = k^2 = -1$.

Definition 15.2.2. *The skew-field* $\mathbb{H} = \mathbb{R}_{-1,-1}$ *is called the* field of Hamilton quaternions.

Until the end of this section, we assume that K is a number field. Let v be an infinite prime of K, i.e. an embedding $K \hookrightarrow \mathbb{C}$. We say that v is *real* if $v(K) \subset \mathbb{R}$, and that v is *complex* otherwise.

Definition 15.2.3. *We say that a* <u>*central simple*</u> *K-algebra L is* definite at v *if* $v(\mathrm{Nrd}(x)) \geq 0$ *for all* $x \in L$.

The completion $\widehat{L}_v$ of L at v (for any norm on the K-vector space L) can be identified with $\widehat{K}_v \bigotimes L$, where $\widehat{K}_v$ stands for the completion of K at v, isomorphic to $\mathbb{R}$ or to $\mathbb{C}$. If v is real (resp. complex), $\widehat{L}_v$ (viewed as an algebra over $\widehat{K}_v$) is isomorphic to an algebra of the form either $\mathcal{M}_r(\mathbb{R})$ or $\mathcal{M}_r(\mathbb{H})$ (resp. $\mathcal{M}_r(\mathbb{C})$). Hence L is definite at v if and only if v is real and $\mathbb{R} \bigotimes L$ is isomorphic to a matrix ring over $\mathbb{H}$.
[If $\widehat{L}$ is isomorphic to $\mathcal{M}_r(\mathbb{R})$ or $\mathcal{M}_r(\mathbb{C})$, the reduced norm, which coincides with the determinant, is not positive. In the other direction, $\mathrm{GL}_r(\mathbb{H})$, as well as $\mathrm{GL}_r(\mathbb{C})$, is arcwise connected, hence mapped into (and indeed onto) $\mathbb{R}^*_{>0}$, since the reduced norm is a continuous map.]

Definition 15.2.4. *We say that a central simple K-algebra L is* totally definite *if it is definite at all infinite primes of K. In particular, K must be* totally real.

15.3 Algebraic Lattices over Dedekind Domains

In this section, we consider an integral domain R, its fraction field K, and a K-vector space V of finite dimension $n > 0$.

Definition 15.3.1. *A* fractional ideal of R *(or* of K*) is a* <u>*nonzero*</u> *sub-R-module I of V such that there exists $d \in R \smallsetminus \{0\}$ with $dI \subset R$. An* (algebraic) lattice in V *is a sub-R-module M of V such that there exist free sub-R-modules L_1 and L_2 of V with $L_1 \subset M \subset L_2$.*

We remark that lattices in K are the fractional ideals of R. Moreover, the following properties hold:

1. If I is a fractional ideal, so is $I' = \{x \in K \mid x\,I \subset R\}$.
2. If I, J are fractional ideals, so is their *product*
$$I\,J = \{x_1 y_1 + \cdots + x_k y_k \mid x_i \in I,\, h_j \in J\}.$$
3. If M, M' are lattices in V, then $M + M'$ and $M \cap M'$ also are.
4. Any finitely generated submodule M in V containing a basis for V is a lattice in V.
5. The converse holds true wherever R is *Noetherian* (i.e., all ideals in R are finitely generated).
6. Suppose that V is endowed with a non-degenerate (symmetric) bilinear form b. Then for any lattice M in V, its dual module
$$M^\sharp = \{x \in V \mid \forall y \in M,\, b(x,y) \in R\}$$
is again a lattice in V. [Proof: utilize dual bases for b.]

The set $\mathcal{I}_R$ (or $\mathcal{I}_K$) of invertible fractional ideals constitutes a group under multiplication (in which $I^{-1} = I'$), which contains its subgroup $\mathcal{P}_R$ (or $\mathcal{P}_K$) of *principal fractional ideals* (those of the form $(x) = Rx$, $x \in K^\times$). The group $\mathcal{I}_R/\mathcal{P}_R$, denoted by Cl_R (or Cl_K), is the *class group of* R (or *of* K).

Definition 15.3.2. *A* Dedekind domain *is an integral domain in which every fractional ideal is invertible.*

Here are two important characterizations of Dedekind domains:

Theorem 15.3.3. *For an integral domain R, the following conditions are equivalent:*

1. *R is a Dedekind domain.*
2. *R is Noetherian, and for every nonzero prime ideal $\mathfrak{p}$ in R, the* local ring $R_\mathfrak{p} = \{xy^{-1} \mid x \in R,\, y \in R \smallsetminus \mathfrak{p}\}$ of R at $\mathfrak{p}$ *is a principal ideal domain (for short, a PID).*
3. *R is Noetherian, integrally closed, and every nonzero prime ideal in R is maximal.* ["R integrally closed" means that every $x \in K$ which is integral over R (i.e., a root of a monic polynomial in R[X]) indeed belongs to R.]

From now on, we assume that R is a Dedekind domain.

Thanks to property (2) above, we can define the *elementary divisors* for lattices M and $N \subset M$, and in particular the *Smith invariant* of an integral lattice (for a bilinear form). Moreover, lattices can be constructed locally:

Proposition 15.3.4. *If M, N are two lattices, then $M_\mathfrak{p} = N_\mathfrak{p}$ for almost all prime ideals in R. Conversely, given an R-lattice M and a family $N(\mathfrak{p})$ of $R_\mathfrak{p}$-lattices such that $N(\mathfrak{p}) = M_\mathfrak{p}$ for almost all $\mathfrak{p}$, there exists a unique lattice N such that $N_\mathfrak{p} = N(\mathfrak{p})$ for all prime ideals $\mathfrak{p} \subset R$; we have $N = \bigcap_\mathfrak{p} N_\mathfrak{p}$.*

Using property (3), one proves the following theorem, which is one of the fundamental results in algebraic number theory. To state it, we first recall that an element x in some ring $R' \supset R$ is *integral over* R if it is a root of a monic polynomial with coefficients in R. If $x, y \in R'$ are integral over R, and if they commute, then both $x+y$ and xy are integral over R. This shows that if R' is commutative, then the set of elements in R' which are integral over R is a ring, called the *integral closure of R in R'*.

Theorem 15.3.5. *Let L be a finite separable extension of K. Then the integral closure of R in L is a lattice in L and a Dedekind domain.*

15.4 Arithmetic in Separable Algebras

In this section, we consider a Dedekind domain R, its fraction field K, and a separable algebra L over K. When L is not commutative, the set of elements of L which are integral over R is not in general a ring.
[A remarkable exception: this is a ring whenever R is a complete local ring and L is a skew-field, see Theorem 15.4.4 below.]

The notion of a maximal order is an ersatz for that of the integral closure.

Definition 15.4.1. *An* order *in L (with respect to R) is a subring of L which is a lattice over R.*

The following proposition shows that the definition above is linked with integrality:

Proposition 15.4.2. *A subring of L containing a basis for L over K is an order if and only if all its elements are integral over R.*

Let $\mathfrak{O}$ be an order. The integrality property above shows that the reduced trace is integral on $\mathfrak{O}$. Hence $\mathfrak{O}$ is contained in its dual $\mathfrak{O}^*$ for the form $\operatorname{Trd}(xy)$ (the *codifferent of $\mathfrak{O}$*). This implies the following generalization of the first assertion in Theorem 15.3.5:

Proposition 15.4.3. *Any order is contained in a maximal order.*

It is easy to prove that the ring generated by an order $\mathfrak{O}$ and an integer e in the centre of L is again an order. Since the simple components of L are of the form $L\,e$, where e is a central idempotent in L, a maximal order is a product of maximal orders in the simple components of L, so that properties of maximal order reduce to the case of simple algebras. Replacing K by the centre C of L and R by its integral closure in C, we may even restrict ourselves to the case of central simple algebras.

We define as in the previous section the notion of a fractional (left, right, or two-sided) ideal for an order $\mathfrak{O}$. Define *prime ideals* to be two-sided ideals $\mathfrak{p}$ which satisfy the following property: if $\mathfrak{a}$, $\mathfrak{b}$ are two-sided ideals, then $\mathfrak{a}\mathfrak{b} \subset \mathfrak{p} \Rightarrow \mathfrak{a} \subset \mathfrak{p}$ or $\mathfrak{b} \subset \mathfrak{p}$. In a maximal order, any two-sided fractional ideal possesses a unique decomposition as a product of prime ideals. We thus have as usual a notion of *ramification* for prime ideals in R. However, it is important in practice to consider more generally left (or right) ideals. The fundamental result is that these ideals are locally free, provided they have a well–defined rank.

To analyse ramification (assuming that we know how to handle ramification in field extensions), we are left with the case of a central simple algebra. Let $\mathfrak{p}$ be a prime in R. To complete L at $\mathfrak{p}$ amounts to considering the extended algebra $\widehat{L}_\mathfrak{p} = \widehat{K}_\mathfrak{p} \bigotimes L$. This is then a matrix algebra over some skew-field $D'_\mathfrak{p}$ with centre $\widehat{K}_\mathfrak{p}$. [**Warning:** the rank of $D'_\mathfrak{p}$ is generally smaller than that of the skew-field with centre K attached to L.] The result is easy to state in the complete local case; thus R is a PID which has a unique nonzero prime ideal $\mathfrak{P}$ (a "discrete valuation ring") and is moreover complete:

Theorem 15.4.4. *Let R be complete with maximal ideal $\mathfrak{p}$, and let D be a skew-field with centre K and rank e^2 over K. Then the set $\mathfrak{M}$ of elements in D which are integral over R is the unique maximal order in D, and we have $\mathfrak{p}\mathfrak{M} = \mathfrak{P}^e$ where $\mathfrak{P}$ is the unique maximal ideal in D; moreover, $\mathfrak{P}$ has residue degree e (i.e., $[\mathfrak{M}/\mathfrak{P} : R/\mathfrak{p}] = e$).*

The theorem above shows that ramification above a prime ideal $\mathfrak{p}$ occurs in a given central simple algebra if and only if its completion is not a pure matrix ring over $\widehat{K}_\mathfrak{p}$. As usual, ramification can be detected via the *reduced discriminant*, defined locally (for any order) as the discriminant for the reduced trace of the lattice $\mathfrak{M}$, where $\mathfrak{M}$ is any maximal order. It does not depend on the choice of $\mathfrak{M}$, and it coincides as usual with the reduced norm of the *reduced different* (the inverse of the codifferent, which always exist for maximal orders).

[It may prove useful (see Example 15.1.3 above) to define the *different* and the *discriminant*, using the trace instead of the reduced trace; the discriminant is the (non-reduced) norm of the different.]

15.5 Number Fields

The fundamental result concerning algebras is a theorem of Hasse, which describes central simple algebras over a number field in terms of the centre itself, using local data. This is proved together with theorems of class field theory (which describe Abelian extensions in terms of the base field).

We first consider the local case, in which the centre is the fraction field K of a complete, discrete valuation ring R with finite residue field; these fields (assuming that their characteristic is zero) are exactly the fields of the form $\widehat{K}_\mathfrak{p}$, where K is a number field. Then skew-fields (always finite dimensional) with centre K are canonically in one-to-one correspondence with the group $\mathbb{Q}/\mathbb{Z}$, in such a way that the invariant corresponding to a skew-field D of rank e^2 has order e in $\mathbb{Q}/Z$.

Let K be a number field, and let L be a central simple K-algebra. For every finite prime of K, $\widehat{L}_{\mathfrak{p}}$ is an algebra of the form $\mathcal{M}_r(D'_{\mathfrak{p}})$, to which we attach an invariant in $\mathbb{Q}/Z$. For an infinite prime v, we also have an invariant in $\mathbb{Q}/\mathbb{Z}$, which is 0 if v is complex, and 0 or $\frac{1}{2}$ mod 1 if v is real ($\frac{1}{2}$ if $\widehat{K}_v$ is of the form $\mathcal{M}_r(\mathbb{H})$, and 0 otherwise; in the first case, we say that v is *ramified*; if $L \simeq \mathcal{M}_r(D)$, the local invariants only depend on D).

Theorem 15.5.1. *Let K be a number field. For every prime v of K, let $m_v \in \mathbb{Q}/Z$. Then the collection of the m_v is a system of local invariants for a central simple algebra with centre K if and only if the following conditions are satisfied:*

1. *$m_v = 0$ for almost all v.*
2. *$m_v = 0$ if v is complex.*
3. *$m_v = 0$ or $m_v = \frac{1}{2}$ if v is real.*
4. *$\sum_v m_v = 0$.*

The last condition above is known as *Hasse's law of reciprocity.*

15.6 Quaternions Again

The special case of quaternions, specially useful to construct lattices, deserves some comments.

First, over a local field $K \neq \mathbb{C}$, there exists exactly *one* quaternion field (up to isomorphism) with centre K. Its construction as an algebra $H_{a,b}$ is easy if K has an odd residue characteristic (or if $K = \mathbb{R}$, where this is the field of Hamilton quaternions), but may need some tedious calculations in the case of residue characteristic 2; a global construction may then prove useful.

If K is a number field, Hasse's law of reciprocity reduces to a parity condition: a finite set S of primes is the set of ramified places in a quaternion algebra H if and only if S has even cardinality (S is empty exactly when $H \simeq \mathcal{M}_2(K)$).

Definition 15.6.1. *We say that a quaternion algebra H is* totally definite *if K is totally real, and if H is definite at all real places of K.*

This amounts to the same thing as saying that $\widehat{H}_v \simeq \mathbb{H}$ for all infinite primes of K. These algebras are of a great importance in lattice theory, because the form $\mathrm{Trd}(x\bar{y})$ is then positive definite.

If $[K : \mathbb{Q}]$ is odd, there exists for every ideal $\mathfrak{p}$ a unique totally definite quaternion algebra $H_{\mathfrak{p}}$ (of course, a skew-field) ramified exactly at $\mathfrak{p}$ and at the infinite primes of K. If $K = Q$, the algebra of "usual" quaternions ($i^2 = j^2 = -1$) is H_2; for $p \equiv -1 \mod 4$ (resp. $p \equiv 5 \mod 8$), H_p can be defined by the conditions $i^2 = -1$ (resp. $i^2 = -2$) and $j^2 = -p$.

(If $p \equiv 1 \mod 8$, one can construct H_p using Dirichlet's theorem on primes in an arithmetic progression.)

If $[K : \mathbb{Q}]$ is even, there exists a unique totally definite quaternion algebra which is unramified outside infinity. For $[K : \mathbb{Q}] = 2$, we met such algebras in Chapter 8. Moreover, when 2 does not split in K, this is the algebra of usual quaternions.

15.7 Ideal Class Set

We consider now left fractional ideals for a maximal order $\mathfrak{M}$, first modulo isomorphism, which is equivalent to the condition

$$I \sim J \iff \exists x \in L^*,\ J = I\,x,$$

then modulo stable equivalence, which is

$$I \sim' J \iff I \bigoplus \mathfrak{M} \simeq J \bigoplus \mathfrak{M}.$$

There are only finitely many classes, and the set of ideal classes maps canonically onto the group of ideal classes.

To state the following theorem, essentially due to Eichler, we need to define the reduced norm of a fractional ideal. This is done in the following way: one first proves that fractional ideals are locally free, and then defines the *reduced norm* as the fractional ideal in the centre which coincides everywhere locally with the reduced norm of its local generators.

Theorem 15.7.1. *Let L be a central simple algebra over a number field. Consider on the set of* stable classes *of L the commutative group law such that $I_1 * I_2 = J$, where J is any ideal with $I_1 \bigoplus I_2 \simeq \mathfrak{M} \bigoplus J$ (the* class group *of $\mathfrak{M}$). Denote by $\mathrm{Cl}_{K,L}$ the group of ideal classes of K modulo the equivalence relation $J = \alpha J$ where α is positive at all infinite primes of K which are ramified in L. Then:*

1. *The reduced norm induces an isomorphism of the stable class group of L onto $\mathrm{Cl}_{K,L}$.*
2. *If L is not a totally definite quaternion algebra, stable classes coincide with isomorphism classes.*

Applied with $K = \mathbb{Q}$, this theorem says that a quaternion algebra with centre $\mathbb{Q}$ has only one stable class, and even only one class if it is indefinite. However, the number of classes tends to infinity with the (reduced) discriminant in the totally definite case. It is known that class number one occurs only if $H = H_p$ with $p \in \{2, 3, 5, 7, 13\}$ (and in general that there are altogether only finitely many totally definite quaternion algebras for which each stable class corresponds to a single isomorphism class; in particular, the centres of such algebras have bounded degrees). For more details on quaternion algebras, see [Vi] and the references given there.

In algebraic number theory, it is important to consider systematically unit groups together with class groups. Dirichlet's unit theorem provides a fairly nice description of the unit group. No such result of this kind may be expected for semi-simple algebras; indeed, we have $(\mathcal{M}_2(\mathbb{Z}))^\times = \mathrm{SL}_2(\mathbb{Z})$, a group whose description, even in this particularly simple example, is highly non-trivial. However, something fairly important for lattice theory can be said about elements of finite order in totally definite quaternion algebras. The situation is similar to the case of *C.M. fields* (totally imaginary quadratic extensions L of a totally real number field L_0), for which Dirichlet's unit theorem shows that the unit group of L_0 has finite index in that of L.

Indeed, consider in the field $\mathbb{H}$ of Hamilton quaternions the 3-dimensional sphere $S = \{q \in \mathbb{H} \mid \mathrm{Nrd}(q) = 1\}$ and the 3-dimensional subspace $F = \ker \mathrm{Trd}$ of *pure quaternions* in $\mathbb{H}$. The map

$$S \to \mathrm{SO}(F) \simeq \mathrm{SO}_3(\mathbb{R}) : q \mapsto (x \mapsto qxq^{-1})$$

is a concrete realization of the universal covering $\mathrm{Spin}_3(\mathbb{R})$ of $\mathrm{SO}_3(\mathbb{R})$. Now the finite subgroups of $\mathrm{SO}_3(\mathbb{R})$ (up to conjugacy) are cyclic groups C_m, $m \geq 1$, dihedral groups D_m, $m \geq 2$, or isomorphic to $\mathbb{A}_4$, S_4, or S_5, and described (except for the first one) as the automorphism group of a regular polygon or polyhedron (see e.g. [Zas]). By pull back in S, we obtain groups isomorphic to one of the following groups, which are double covers of the previous ones: C_{2m}, H_{4m}, $m \geq 2$ (the quaternion groups), and three special groups, denoted by $\widehat{A}_4$, $\widehat{S}_4$, $\widehat{S}_5$, which can be realized over the fields $\mathbb{Q}(\zeta_{2m})$, $\mathbb{Q}(\zeta_{2m})$, $\mathbb{Q}$, $\mathbb{Q}(\sqrt{2})$, and $\mathbb{Q}(\sqrt{5})$ respectively.

Given a maximal order $\mathfrak{M}$ in a totally definite quaternion algebra with centre a (totally real) number field L_0, $\mathfrak{M}^\times$ contains a unique maximal finite subgroup U of the type above and contains $\mathbb{Z}_{L_0}^\times$ to finite index. We can thus define a "Hasse index" $[\mathfrak{M}^\times : U\,\mathbb{Z}_{L_0}^\times]$ as in the C.M. case (but this index need no longer be equal to 1 or 2).

16 Appendix 2: Strongly Perfect Lattices

Introduction

This chapter is intended to give a survey of some recent results which have been discovered after the French version of this book was written, and which are of great importance in the theory of extreme lattices: on the one hand, they allow us to prove that some lattices having a comparatively high dimension are extreme without having any precise description of their minimal vectors (example: 80-dimensional even unimodular lattices of minimum 8, for which two lattices are known from [Bac-Ne2]); on the other hand, they connect the theory of extreme lattices to that of spherical designs, which properly belongs to combinatorics. The basic reference is Venkov's [Ven3] (English title: "Lattices and Spherical Designs"). We account for some of the results proved in [Ven3] or in some other articles which appeared in the same issue of "L'Enseignement Mathématique" (Bachoc, Martinet, Venkov), and also for results involving group theory, due to Dummigan, Lempken, Schröder and Tiep ([D-T],[L-S-T]).

This chapter essentially contains no proofs. Unless otherwise stated, they can be found in [Ven3]. As usual, E stands for an n-dimensional Euclidean space. "Sphere" means "sphere centred at the origin of E".

16.1 Spherical Designs

The notion of spherical designs corresponds to a "nice" distribution of a finite set of points on a sphere. We consider a sphere S, of radius $\sqrt{m}$, endowed with the spherical measure dx, normalized so as to give S the mass 1.

Definition 16.1.1. *Let t be a positive integer. We say that a non-empty finite set $X \subset S$ is a* spherical t-design *if the equality*

$$\int_{S^{n-1}} f\, dx = \frac{1}{|X|} \sum_{x \in X} f(x)$$

*holds for all polynomials f on E of degree at most t. We say that a lattice Λ is a t-*design *if $S(\Lambda)$ is. (Despite the existence of* combinatorial designs, *we shall often say for short* "design" *instead of "spherical design".)*

A 1-design is simply a set whose barycentre is the origin of E. Since we want to apply the theory of designs only to layers of lattices, we shall now assume that all designs are *symmetric*. Then any $(2t)$-design is also a $(2t+1)$-design.

In the definition of a t-design, it amounts to the same thing as requiring that the integral should be zero for all *harmonic polynomials* of degree at most t. (We shall denote by Harm_k the space of homogeneous harmonic polynomials of degree k.) This result is used to prove the following characterization of spherical t-designs, the one we shall use in practice, and that we only state for symmetric designs:

Proposition 16.1.2. *Let $t = 2p+1$ be an odd integer. Then, for all $n \geq 2$, a (symmetric) subset X of S is a t-design if and only if there exists a constant c_p such that the equality*

$$\sum_{x \in X} (x \cdot \alpha)^{2p} = c_p\, m^p\, |X|\, (\alpha \cdot \alpha)^p$$

holds for all $\alpha \in E$. Moreover, when this condition is satisfied, we have

$$c_p = \frac{1 \cdot 3 \cdot 5 \ldots (2p-1)}{n(n+2) \cdots (n+2p-2)}\,.$$

[If $n = 1$, S has two elements, which constitute a t-design for all t. If $n > 1$, there is for every X a maximal value of $t \geq 0$ for which X is a t-design.]

It is clear from the definition that any t-design is also a t'-design for all $t' \leq t$, so that the proposition above indeed holds for every $p \leq \frac{t-1}{2}$. This remark is crucial for proving classification results.

Applied with $t = 3$, and taking into account Theorem 3.2.5, the proposition above immediately implies:

Corollary 16.1.3. *A lattice is a 3- (or 2-) spherical design if and only if it is strongly eutactic.*

The classification of strongly eutactic lattices is known only for $n \leq 5$, as a byproduct of the classification of *all* weakly eutactic lattices. The complete list is given in [Mar-V]. Among the 118 eutactic lattices of dimension 5 (Theorem 9.5.2), only eight are strongly eutactic. Thus strong eutaxy looks much more restrictive than ordinary eutaxy. Note also that strongly eutactic lattices are rational.

To test whether a given symmetric set X is a $(2p+1)$-design is easy thanks to the following theorem:

Theorem 16.1.4. *For any finite subset X of a sphere of radius $\sqrt{m}$, we have*

$$\sum_{x,y \in X} (x \cdot y)^{2p} \geq \frac{1 \cdot 3 \cdot 5 \cdots (2p-1)}{n(n+2) \cdots (n+2(p-1))}\, m^{2p}\, |X|^2\,,$$

and equality holds if and only if X is a $2p+1$-design.

16.2 Strong Perfection

The theorem below, due to Venkov, establishes a connection between perfect lattices and designs. For a proof, see Section 6 of [Ven3].

Definition 16.2.1. *We say that a lattice is* strongly perfect *if it is a spherical* 5*-design.*

Theorem 16.2.2. *A strongly perfect lattice is extreme.*

No obvious connection seems to exist with the notion of dual-extremality. However, the invariant γ' plays an important rôle:

Theorem 16.2.3. *If Λ is strongly perfect, then $\gamma'(\Lambda)^2 \geq \frac{n+2}{3}$, and this inequality is strict if $S(\Lambda)$ is a higher design.*

Proof. Applying the formula in Proposition 16.1.2 for $p = 1$ and $p = 2$ with $\alpha \in S(\Lambda^*)$, and taking the difference, we obtain the equality

$$\sum_{x \in S/\{\pm 1\}} (x \cdot \alpha)^2 \big((x \cdot \alpha)^2 - 1\big) = \frac{s}{n}\, \gamma'(\Lambda)^2 \Big(\frac{3}{n+2}\, \gamma'(\Lambda)^2 - 1\Big),$$

in which the left–hand side is non-negative. Applying Proposition 16.1.2 with $p = 3$ proves the last assertion. □

Using the classification of root lattices and of perfect lattices up to dimension 7 (Chapters 4 and 6) together with Proposition 16.1.2, we immediately obtain:

Proposition 16.2.4. *The strongly perfect root lattices are $\mathbb{Z}$, $\mathbb{A}_2$, $\mathbb{D}_4$, $\mathbb{E}_6$, $\mathbb{E}_7$, $\mathbb{E}_8$, and the strongly perfect lattices of dimension $n \leq 7$ (up to scale) are $\mathbb{Z}$, $\mathbb{A}_2$, $\mathbb{D}_4$, $\mathbb{E}_6$, $\mathbb{E}_6^*$, $\mathbb{E}_7$, and $\mathbb{E}_7^*$.*

To obtain further classification results is much more difficult. Results obtained by Venkov and completed by Nebe and Venkov in [Ne-V2] for dimension 10 are summarized in the theorem below; the proofs make use of the notions of a strongly perfect lattice of *minimal type* (if $\gamma'(\Lambda)^2 = \frac{n+2}{3}$) and *general type* (if $\gamma'(\Lambda)^2 > \frac{n+2}{3}$):

Theorem 16.2.5. *1. In dimensions* 8 *to* 11*, the only strongly perfect lattices are $\mathbb{E}_8$, K'_{10}, and ${K'_{10}}^*$.*
2. The strongly perfect lattices which are integral when scaled to minimum 3 *are $\mathrm{O}_1 = \sqrt{3}\,\mathbb{Z}$, $\mathrm{O}_7 = \sqrt{2}\,\mathbb{E}_7^*$, O_{23} (Section 5.7), and the lattices O_{16} and O_{22}, orthogonal to O_7 and O_1 respectively in O_{23}.*

3. *A strongly perfect lattice with* $s = \frac{n(n+1)}{2}$ *is either similar to* $\mathbb{Z}$ *or to* $\mathbb{A}_2$, *or its dimension is of the form* $n = m^2 - 2$ *with* m *odd; moreover, its minimal vectors define an equiangular configuration of lines, and the lattice they generate, scaled to norm* m, *is integral.*

The fact that there exists a unique strongly perfect lattice in dimension 8 whereas nearly 11 000 perfect lattices are known shows that the notion of a strongly perfect lattice is much more restrictive than that of an ordinary perfect lattice.

To classify integral strongly perfect lattices with a given minimum m seems very difficult for $m \geq 4$, despite the existence of the bound $n \leq 3(m^2 - 1)$ that can be derived for $m \geq 2$ from Proposition 16.1.2. For 7-designs, the lattices with $m \leq 5$ have been classified in [Mar7]. Minimum 5 is impossible, and minimum $m \leq 4$ occurs exactly for $\mathbb{Z}$, $\mathbb{E}_8$, Λ_{16}, C_{23}, Λ_{23}, Λ_{24}, and for the not yet classified even, 32-dimensional unimodular lattices of minimum 4.

Remark 16.2.6. *The upper bound above for* n *can also be read in the following way:*

If Λ *is an integral strongly perfect lattice, and if* $n \geq 2$, *then* $m \geq \sqrt{\frac{n+3}{3}}$.
Moreover, this inequality is strict if Λ *is a higher design.*
[These lower bounds, which could have been deduced from Theorem 16.2.3, were announced by Nottebaum and Schröeder (reference [N-S] in [L-S-T]).]

16.3 An Infinite Series

Using the bound $n \leq 3(m^2 - 1)$ together with Voronoi's finiteness Theorem 3.5.4, we see that there are only finitely many integral strongly perfect lattices with a given minimum, so that an infinite series of integral strongly perfect lattices must have unbounded minima. One such series is known, namely the sequences of Barnes–Wall lattices, defined in Section 8.8. This was first proved by Nottebaum ([No]); Venkov then proved in [Ven3] that their sets of minimal vectors are indeed higher designs:

Theorem 16.3.1. *The set of minimal vectors of the Barnes–Wall lattice* BW_n *is a 7-design for every dimension* $n = 2^p \geq 8$.

Both Nottebaum and Venkov used the nice description of the set of minimal vectors in BW_{2^p} provided by the Reed–Muller codes. Whether all layers of these lattices are also 7-designs is an open question. Probably they are, and this should follow from the calculation of the invariants for their automorphism groups. However, these invariants are known only for a twice larger group (see [Ven3], Remark 18.10). It would be interesting to construct other infinite series.

16.4 Modular Lattices

In this section, we only consider *even* ℓ-modular lattices, with $\ell = 1$ or a prime such that $\ell + 1$ divides 24. Quebbemann proved in [Q6] that under these hypotheses, these lattices (i.e., their *theta-series*) are modular for the *Fricke group*, a group twice as large as the usual group $\Gamma_0(\ell)$. Using these results, Quebbemann obtained the following results:

Theorem 16.4.1. *Let ℓ be as above. Then the minimum of an even ℓ-modular n-dimensional lattice is bounded from above by* $2\lfloor\frac{(\ell+1)\,n}{48}\rfloor$.

To find the exact upper bound for the minimum of a lattice as above is an interesting problem in the theory of modular lattices. But the fact that the maximum above is attained on some lattices has important consequences, which justify the following definition:

Definition 16.4.2. *A lattice which attains the bound above is called* extremal.

Using theta-series with harmonic coefficients, Venkov proved in [Ven3] that extremal unimodular lattices are designs of a high level, depending on the dimension modulo 24. These results were later extended to other ℓ-modular lattices in [Bac-V]. We refer the reader to Ogg's book [Ogg] for all that concerns modular forms and theta-series.

Before stating the main results they obtained on this question, we recall that for $\ell = 1$ (resp. $\ell = 2$, resp. $\ell > 2$ as above), the dimension of an even ℓ-modular lattice is divisible by 8 (resp. by 4, resp. by 2).

Theorem 16.4.3. (Venkov, Bachoc and Venkov.) *Let Λ be an even ℓ-modular lattice. Then the layers of Λ are t-designs with t as follows:*

- $t \geq 11$ *if* $\ell = 1$ *and* $n \equiv 0 \mod 24$.
- $t \geq 7$ *if* $\ell = 1$ *and* $n \equiv 8 \mod 24$, *or* $\ell = 2$ *and* $n \equiv 0 \mod 16$.
- $t \geq 5$ *if* $\ell = 2$ *and* $n \equiv 4 \mod 16$, *or* $\ell = 3$ *and* $n \equiv 0, 2 \mod 12$, *or* $\ell = 5$ *and* $n = 16$.
- $t \geq 3$ *if* $\ell = 1$ *and* $n \equiv 16 \mod 24$, *or* $\ell = 2$ *and* $n \equiv 8 \mod 16$, *or* $\ell = 3$ *and* $n \equiv 4, 6 \mod 12$, *or* $\ell = 5$, *and* $n \equiv 0 \mod 8$, *or* $\ell = 7$ *and* $n \equiv 0 \mod 6$.

In all instances where the exact level is known, the value of t given above is optimal, at least for the first layer; this is proved in some cases in [Mar7]. For a recent update of the known pairs (n, ℓ) for which there exist extremal lattices, see [Sch-SP]. The largest dimension for which the theorem above is known to apply is $n = 80$, with $\ell = 1$.

16.5 Group Theory

The Brauer–Coxeter theorem (Theorem 3.6.6) asserts that a lattice Λ whose automorphism group acts irreducibly on E is strongly eutactic (i.e., that $S(\Lambda)$ is a 3-design. The paper [L-S-T] by Lempken, Schröder, and Tiep gives an analogous, useful result for 5-designs, using the symmetric square of the representation afforded by $\mathrm{Aut}(\Lambda)$. To state it, we first recall a few facts from the representation theory of groups.

Given a finite group G and a (complex) representation ρ of G, with character ϕ, the tensor square $\rho\otimes\rho$ has a canonical decomposition

$$\rho\otimes\rho = \mathrm{Sym}^2(\rho) \oplus \mathrm{Alt}^2(\rho)\,,$$

corresponding to the decomposition

$$\phi^2(s) = \frac{\phi(s)^2 + \phi(s^2)}{2} + \frac{\phi(s)^2 - \phi(s^2)}{2}\,,$$

of its character. If ϕ is irreducible, the *Schur indicator*

$$\mathrm{Sch}(\phi) = \frac{1}{|G|}\sum_{s\in G} \phi(s)$$

has value $+1$, 0, or -1 according to whether ϕ is real, complex, or quaternionic, i.e. the centralizer of ρ is a matrix algebra of the form $\mathcal{M}_r(\mathbb{R})$, $\mathcal{M}_r(\mathbb{C})$, or $\mathcal{M}_r(\mathbb{H})$ respectively. Characters which can be realized over $\mathbb{R}$ are sums of characters of the form ϕ (ϕ real), $\phi+\bar{\phi}$ (ϕ complex), or 2ϕ (ϕ quaternionic). (Proofs can be found in [Se2].) An easy calculation with characters shows that the symmetric square of an $\mathbb{R}$-irreducible character contains the unit character exactly once.

Definition 16.5.1. *We say that a representation is* almost irreducible *if it is the sum of the unit character* $\mathbf{1}_G$ *and of an irreducible character distinct from* $\mathbf{1}_G$.

Theorem 16.5.2. (Lempken–Schröder–Tiep.) *If the symmetric square of the representation of* $\mathrm{Aut}(\Lambda)$ *is almost irreducible, all layers of* Λ *are* 5-*designs; in particular,* Λ *is strongly perfect.*

Combined with Remark 16.2.6, the theorem above has important applications for evaluating the minimum of certain lattices, a generally difficult problem. Various examples can be found in [L-S-T]. One of the most striking results concerns the Thompson–Smith lattice (an even unimodular lattice of dimension 248, used by Thompson to construct the finite sporadic simple group which bears his name):

Proposition 16.5.3. *The Thompson–Smith lattice has minimum* $m \geq 10$. [It has has been proved by Schulz ([Schu]; see also Nebe–Sloane's catalogue [Ne-Sl]) that the Thompson-Smith lattice actually contains norm 12 vectors.]

16.6 Designs on Grassmannian Varieties

Symmetric subsets of the sphere $S^{n-1} \subset \mathbb{R}^n$ are in one-to-one correspondence with sets of lines in $\mathbb{R}^n$, so that they can be viewed as points in the projective space $P^{n-1}\mathbb{R}$. In this respect, it is natural to consider more generally Grassmannians $\mathcal{G}_{m,n}$, which are sets of of m-dimensional subspaces of $\mathbb{R}^n$, where we assume w.l.o.g. that $1 \leq m \leq \frac{n}{2}$. After analogues to the packing problem were considered in [C-H-S], Bachoc, Coulangeon, and Nebe introduced in [Bac-C-N] the notion of a t-design on $\mathcal{G}_{m,n}$, and developed for lattices a theory which plays with respect to Rankin invariants (Definition 2.8.3 and Section 10.6) the rôle that played Venkov's theory with respect to the Hermite invariant.

The authors introduce spaces $H^{\mu}_{m,n}$ which play for Grassmannian designs the rôle that played the spaces Harm_k for spherical designs; here μ is a sequence $\mu_1 \geq \mu_2 \geq \cdots \geq \mu_n \geq 0$ with at most m parts, which are assumed to be even; $\sum_i \mu_i$ is the *degree of* μ, denoted by $\deg(\mu)$. They then define a *t-design ($t \geq 2$ even)* to be a finite set $\mathcal{D} \subset \mathcal{G}_{m,n}$ such that

$$\sum_{x \in \mathcal{D}} f(x) = 0 \quad \text{for all } f \in H^{\mu}_{m,n} \text{ and all } \mu \text{ with } 2 \leq \deg(\mu) \leq t\,.$$

If $m = 1$, μ is simply an even integer $2k$, and $H^{2k}_{1,n}$ coincides with Harm_{2k}, so that we recover the usual notion of a spherical t-design.

We have defined (Definitions 10.6.1 and 10.6.3) the notions of m-perfect, m-eutactic, and m-extreme lattices, and quoted a Voronoi-like theorem of Coulangeon (Theorem 10.6.4), which characterizes m-extreme lattices by the properties of m-perfection and m-eutaxy. In [Bac-C-N], the authors define a *strongly m-perfect lattice* as a lattice for which the set of minimal m-dimensional sublattices is a 4-design (in the sense of designs in $\mathcal{G}_{m,n}$), and prove an analogue of Venkov's theorem, namely that a strongly m-perfect lattice is in particular m-extreme. They also prove for their designs a group theoretical theorem analogous to the Lempken–Schröder–Tiep theorem 16.5.2. Using this device, they are able to prove that $\mathbb{D}_4$, $\mathbb{E}_6$, $\mathbb{E}_7$, $\mathbb{E}_8$, Λ_{16}, and Λ_{24} are strongly m-perfect for all $m \leq \frac{n}{2}$. This gives a quick proof of the local properties of the Rankin invariants for the six lattices above.

Of course, as for the usual notion, strong m-perfection is a much more restrictive property than the mere m-extremality. Various lattices are known to be m-extreme, though not strongly so. A striking example is provided by the Coxeter–Todd lattice K_{12}, which is strongly m-perfect for $m = 1$, but not for $m = 2$.

References

[Ash1] A. Ash, *On eutactic forms*, Can. J. Math. **29** (1977), 1040–1054.

[Ash2] A. Ash, *On the existence of eutactic forms*, Bull. London Math. Soc. **12** (1980), 192–196.

[A-MC] A. Ash, M. Mc Connel, *Cohomology at infinity and the well-rounded retract for general linear groups*, Duke Math. J. **90** (1997), 549–576.

[ATLAS] J.H. Conway, R.T. Curtis, S.P. Norton, R.A. Parker, R.A. Wilson, ATLAS of Finite Groups, Oxford Univ. Press, Oxford (1985).

[Bc1] R. Bacher, *Dense lattices in dimensions* 27, 28 *and* 29, Invent. Math. **130** (1997), 153–158.

[Bc2] R. Bacher, *Tables de réseaux entiers unimodulaires construits comme k-voisins de* $\mathbb{Z}^n$, J. Th. Nombres de Bordeaux **9** (1997), 479–497.

[Bc-V1] R. Bacher, B. Venkov, *Lattices and association schemes: a unimodular example without roots in dimension* 28, Ann. Inst. Fourier **45** (1995), 1163–1176.

[Bc-V2] R. Bacher, B. Venkov, *Réseaux entiers unimodulaires sans racines en dimension* 27 *et* 28, in Réseaux euclidiens, designs sphériques et groupes, J. Martinet, ed., L'Ens. Math., Monographie **37**, Genève (2001), 212–267.

[Bac1] C. Bachoc, *Voisinages au sens de Kneser pour les réseaux quaternioniens*, Comm. Math. Helvet. **70** (1995), 350–374.

[Bac2] C. Bachoc, *Applications of coding theory to the construction of modular lattices*, J. Comb. Th., A **78-1** (1997), 92–119.

[Bac-Bt] C. Bachoc, C. Batut, *Étude algorithmique de réseaux construits avec la forme trace*, J. Exp. Math. **1** (1992), 183–190.

[Bac-C-N] C. Bachoc, R. Coulangeon, G. Nebe, *Designs in Grassmannian spaces and lattices*, J. Algebraic Combinatorics, to appear.

[Bac-Ne1] C. Bachoc, G. Nebe, *Classification of Two Genera of* 32*-Dimensional Lattices of Rank* 8 *over the Hurwitz Order*, Exp. Math. **6** (1997), 151–162.

[Bac-Ne2] C. Bachoc, G. Nebe, *Extremal lattices of minimum* 8 *related to the Mathieu group* M_{22}, J. reine angew. Math. **494** (1998), 155–171.

[Bac-V] C. Bachoc, B. Venkov, *Modular forms, lattices and spherical designs*, in Réseaux euclidiens, designs sphériques et groupes, L'Ens. Math., Monographie **37**, J. Martinet, ed., Genève (2001), 87–111.

[B-I] R. Baeza, M.I. Icaza, *On Humbert-Minkowski constant for a number field*, Proc. Amer. Math. Soc. **125** (1997), 3195–3202.

[B-C-I-O] R. Baeza, R. Coulangeon, M.I. Icaza, M. O'Ryan, *Hermite's constant for quadratic number fields*, Exp. Math., to appear.

E. Bannai. See [C-S], Chapter 14.

[Br-R] E.P. Baranovskii, S.S. Ryshkov, *Derivation of perfect lattices from admissible centerings*, Russian Math. Survey **40, 4** (1985), 155–156. (Russian original: 1985.)

[Bari] J-L. Baril, *Autour de l'algorithme de Voronoï : construction de réseaux euclidiens*, Thèse, Bordeaux (1996).

[Bar1] E.S. Barnes, *Note on extreme forms*, Canad. J. Math **7** (1955), 150–154.

[Bar2] E.S. Barnes, *On a theorem of Voronoï*, Proc. Cambridge Phil. Soc. **53** (1957), 537–539.

[Bar3] E.S. Barnes, *The perfect and extreme senary forms*, Canad. J. Math. **9** (1957), 235–242.

[Bar4] E.S. Barnes, *The complete enumeration of extreme senary forms*, Phil. Trans. Roy. Soc. London **(A), 249** (1957), 461–506.

[Bar5] E.S. Barnes, *The construction of perfect and extreme forms I, II*, Acta Arith. **5** (1959), 57–79, 461–506.

[Bar6] E.S. Barnes, *Criteria for extreme forms*, J. Austral. Math. Soc. **1** (1959), 17–20.

[B-W] E.S. Barnes, G.E. Wall, *Some extreme forms defined in terms of abelian groups*, J. Austral. Math. Soc. **1** (1959), 47–63.

[Bt] C. Batut, *Classification of quintic eutactic forms*, Math. Comp. **70** (2001), 395–417.

[Bt-M] C. Batut, J. Martinet, *Radiographie des réseaux parfaits*, Exp. Math. **3** (1994), 39–49.

[Bt-M'] C. Batut, J. Martinet, *A Catalogue of Perfect Lattices*, http://www.math.u-bordeaux.fr/~martinet .

[Bt-Q-S] C. Batut, H-G. Quebbemann, R. Scharlau, *Computations of cyclotomic lattices*, Exp. Math. **4** (1995), 175–179.

C. Batut. See also C. Bachoc and the appendix to [Mar5].

[Bav1] C. Bavard, *Systole et invariant d'Hermite*, J. reine angew. Math. **482** (1997), 93–120.

[Bav2] C. Bavard, *Classes minimales des réseaux et rétractions géométriques équivariantes dans les espaces symétriques*, J. London Mat. Soc. **64** (2001), 275–286.

[Bav3] C. Bavard, *Familles hyperboliques de réseaux symplectiques*, Math. Ann. **320** (2001), 799–833.

[Bav4] C. Bavard, *Propriétés de finitude pour les familles de réseaux et analogues*, preprint (52 pages).

[Bav5] C. Bavard, *Invariant d'Hermite isotrope et densité des réseaux orthogonaux lorenziens*, in preparation.

[Bay1] E. Bayer-Fluckiger, *Definite unimodular lattices having an automorphism of given characteristic polynomial*, Comm. Math. Helvet. **59** (1984), 509–538.

[Bay2] E. Bayer-Fluckiger, *Réseaux unimodulaires*, Sém. Th. Nombres Bordeaux (=J. Th. Nombres de Bordeaux) **1** (1989), 189–196.

[Bay3] E. Bayer-Fluckiger, *Lattices and number fields*, Contemporary Math. **241** (1999), 69–84.

[Bay4] E. Bayer-Fluckiger, *Cyclotomic modular lattices*, J. Th. Nombres de Bordeaux **12** (2000), 273–280.

[Bay5] E. Bayer-Fluckiger, *Ideal lattices*, preprint (17 pages), Cambridge Univ. Press, to appear.

[Bay-M] E. Bayer-Fluckiger, J. Martinet, *Formes quadratiques liées aux algèbres semi-simples*, J. reine angew. Math. **451** (1994), 51–69.

[Be-B] E.F. Beckenbach, R. Bellman, *Inequalities*, Springer-Verlag, Ergebnisse **30**, Heidelberg (1961).

R. Bellman. See E.F. Beckenbach.

C.T. Benson. See L.C. Grove.

[Ber1] A-M. Bergé, *Minimal vectors of pairs of dual lattices*, J. Number Theory **52** (1995), 284–298.

[Ber2] A-M. Bergé, *Classification of positive forms having prescribed automorphisms*, Contemporary Math. **249** (1999), 199–204.

[Ber3] A-M. Bergé, *Symplectic Lattices*, Contemporary Math. **272** (2000), 9–22.

[Ber4] A-M. Bergé, *Une forme duale-extrême irrationnelle*, J. Th. Nombres de Bordeaux **12** (2000), 281–291.

[B-M1] A-M. Bergé, J. Martinet, *Sur un problème de dualité lié aux sphères en géométrie des nombres*, J. Number Theory **32** (1989), 14–42.

[B-M2] A-M. Bergé, J. Martinet, *Sur les minorations géométriques des régulateurs*, in Sém. Th. Nombres de Paris 1987–88, Prog. Math. **81**, Birkhäuser, Basel (1989), 23–49.

[B-M3] A-M. Bergé, J. Martinet, *Réseaux extrêmes pour un groupe d'automorphismes*, Astérisque **198–200** (1992), 41–66.

[B-M4] A-M. Bergé, J. Martinet, *Densité dans des familles de réseaux. Application aux réseaux isoduaux*, L'Ens. Math. **41** (1995), 335–365.

[B-M5] A-M. Bergé, J. Martinet, *Sur la classification des réseaux eutactiques*, J. London Math. Soc. **53** (1996), 417–432.

[B-M6] A-M. Bergé, J. Martinet, *Representations of symmetric groups and lattices*, to appear.

[B-M-S] A-M. Bergé, J. Martinet, F. Sigrist, *Une généralisation de l'algorithme de Voronoï*, Astérisque **209** (1992), 137–158.

[Be-S] N. Bernstein, N.J.A. Sloane, *Some Lattices Obtained from Riemann Surfaces*, Contemporary Math. **201** (2002), 29–32.

[Bl1] H.F. Blichfeldt, *The minimum value of quadratic forms and the closest packing of spheres*, Math. Ann. **101** (1929), 605–608.

[Bl2] H.F. Blichfeldt, *The minimum value of positive quadratic forms in six, seven and eight variables*, Math. Z. **39** (1935), 1–15.

[Bor1] R. Borcherds, *The Leech lattice and other lattices*, Ph. D. Thesis, Cambridge, U.K. (1984).

[Bor2] R. Borcherds, *Classification of positive definite lattices*, Duke Math J. 105 (2000), 525-567.

[Borel] A. Borel, *Introduction aux groupes arithmétiques*, Hermann, Paris (1969).

[Bou1] N. Bourbaki, *Topologie Generale II, Chap. 5–10*, Hermann, Paris (1974).

[Bou2] N. Bourbaki, *Algèbre, chapitres I–X*, Masson, Paris (1980).

[Bou3] N. Bourbaki, *Groupes et algèbres de Lie*, Masson (now Dunod), Paris (1981).

[B-C] R. Brauer, H.S.M. Coxeter, *A generalization of theorems of Schönhardt and Mehmke on polytopes*, Trans. Roy. Soc. Canada III **(3), 34** (1940), 29–34. [= Richard Brauer, Coll. papers III, The M.I.T. Press, Cambridge, Massachusetts (1980), 653–658.]

[B-E] M. Broué, M. Enguehard, *Une famille infinie de formes quadratiques entières ; leurs groupes d'automorphismes*, Ann. Sc. E.N.S. **6** (1973), 17–53.

[Bu-Sa] P. Buser, P. Sarnak, *On the period matrix of a Riemann surface of large genus (with an Appendix by J.H. Conway and N.J.A. Sloane)*, Invent. Math. **117** (1994), 27–56.

[Cas1] J.W.S. Cassels, *An Introduction to Diophantine Approximation*, Cambridge University Press, Tract **45**, Cambridge, U.K. (1965; first edition: 1957).

[Cas2] J.W.S. Cassels, *An Introduction to the Geometry of Numbers*, Springer-Verlag, Grundlehren **99**, Heidelberg (1959).

[Cas3] J.W.S. Cassels, *Rational Quadratic Forms*, Academic Press, London (1978).

[C-F] J.W.S. Cassels, A. Fröhlich, ed., *Algebraic Number Theory*, Academic Press, New–York (1967).

[Chn] T.W. Chaundy, *The arithmetic minima of positive quadratic forms (1)*, Quart. J. Math. Oxford **17** (1946), 166–192.

[Coh] H. Cohen, *A Course in Computational Algebraic Number Theory*, Graduate Texts in Math. **138**, Springer-Verlag, Heidelberg (1993).

[C-S] J.H. Conway, N.J.A. Sloane, *Sphere Packings, Lattices and Groups*, Grundlehren **290**, Springer-Verlag, Heidelberg (1988).

[C-S'] J.H. Conway, N.J.A. Sloane, second edition of [C-S] (1993).

[C-S"] J.H. Conway, N.J.A. Sloane, third edition of [C-S] (1999).

[C-S1] J.H. Conway, N.J.A. Sloane, *Complex and integral laminated lattices*, Trans. Amer. Math. Soc. **280** (1983), 463–490.

[C-S2] J.H. Conway, N.J.A. Sloane, *The Coxeter–Todd lattice, the Mitchell group, and related sphere packings*, Math. Proc. Cambridge Phil. Soc. **93** (1983), 421–440.

[C-S3] J.H. Conway, N.J.A. Sloane, *Low-Dimensional Lattices. I. Quadratic Forms of Small Determinant*, Proc. Royal Soc. London **A 418** (1988), 17–41.

[C-S4] J.H. Conway, N.J.A. Sloane, *Low-Dimensional Lattices. II. Subgroups of $GL(n,\mathbb{Z})$*, Proc. Royal Soc. London **A 419** (1988), 29–68.

[C-S5] J.H. Conway, N.J.A. Sloane, *Low-Dimensional Lattices. III Perfect Forms*, Proc. Royal Soc. London **A 418** (1988), 43–80.

[C-S6] J.H. Conway, N.J.A. Sloane, *Low-Dimensional Lattices. IV. The Mass Formula*, Proc. Royal Soc. London **A 419** (1988), 259–286.

[C-S7] J.H. Conway, N.J.A. Sloane, *Low-Dimensional Lattices V. Integral Coordinates for Integral Lattices*, Proc. Royal Soc. London **A 426** (1989), 211–232.

[C-S8] J.H. Conway, N.J.A. Sloane, *Low-Dimensional Lattices. VI. Voronoi Reduction of Three-Dimensional Lattices*, Proc. Royal Soc. London **A 436** (1992), 55–68.

[C-S9] J.H. Conway, N.J.A. Sloane, *On Lattices Equivalent to Their Duals*, J. Number Theory **48** (1994), 373–382.

[C-S10] J.H. Conway, N.J.A. Sloane, *A Lattice Without a Basis of Minimal Vectors*, Mathematika **42** (1995), 175–177.

[C-S11] J.H. Conway, N.J.A. Sloane, *Low-Dimensional Lattices. VII. Coordination Sequences*, Proc. Royal Soc. London **A 453** (1997), 2369–2389.

[C-H-S] J.H. Conway, R. H. Hardin, N. J. A. Sloane, *Packing Lines, Planes, etc.: Packings in Grassmannian Space*, Experimental Math. **5** (1996), 139–159.

J.H. Conway. See also [ATLAS] and Appendix 2 to [Bu-Sa].

[Cou1] R. Coulangeon, *Réseaux quaternioniens et invariant de Venkov* Manuscripta Math. **82** (1994), 44–50.

[Cou2] R. Coulangeon, *Réseaux unimodulaires quaternioniens en dimension ≤ 32*, Acta Arith. **70** (1995), 9–24.

[Cou3] R. Coulangeon, *Réseaux k-extrêmes*, Proc. London Math. Soc **73** (1996), 555–574.

[Cou4] R. Coulangeon, *Minimal vectors in the second exterior power of a lattice*, J. Algebra **194** (1997), 467–476.

[Cou5] R. Coulangeon, *Tensor products of Hermitian lattices*, Acta Arith. **92** (2000), 115–130.

[Cou6] R. Coulangeon, *Voronoi theory over algebraic number fields*, in Réseaux euclidiens, designs sphériques et groupes, L'Ens. Math., Monographie **37**, J. Martinet, ed., Genève (2001), 147–162.

R. Coulangeon. See also Bachoc, Baeza, and the appendix to [Mar-V].

[Cox1] H.S.M. Coxeter, *Integral Cayley Numbers*, Duke Math. J. **13** (1946), 561–578.

[Cox2] H.S.M. Coxeter, *Extreme forms*, Canad. J. Math. **3** (1951), 391–441.

[C-T] H.S.M. Coxeter, J.A. Todd, *An extreme duodenary form*, Canad. J. Math. **5** (1953), 384–392.

H.S.M. Coxeter. See also R. Brauer.

[Cra1] M. Craig, *A characterization of certain extreme forms*, Illinois J. Math. **20** (1976), 706–717.

[Cra2] M. Craig, *Extreme forms and cyclotomy*, Mathematika **25** (1978), 44–56.

[Cra3] M. Craig, *A cyclotomic construction for Leech's lattice*, Mathematika **25** (1978), 236–241.

[Cs1] G. Csóka, *Three theorems on the minimal vectors of a perfect lattice*, Ann. Univ. Sc. Budapest Eőtvős, Sect. Math. **23** (1986), 283–291.

[Cs2] G. Csóka, *There exists a basis of minimal vectors in every $n \leq 7$ dimensional perfect lattice*, Ann. Univ. Sc. Budapest Eőtvős, Sect. Math. **30** (1987), 245–258.

R.T. Curtis. See [ATLAS].

[De] M. Deuring, *Algebren*, Chelsea Publishing Company, New York (1948); Springer-Verlag, Heidelberg (1968). [Original edition: Julius Springer, Berlin (1935).]

[Die] J. Dieudonné, *Sur les groupes classiques* (third edition), Hermann, Paris (1967).

[D-T] N. Dummigan, P.H. Tiep, *Lower bounds for the minima of certain symplectic and unitary group lattices*, Amer. J. Math. **121** (1999), 889–918.

[Eb] W. Ebeling, *Lattices and codes*, Vieweg, Wiesbaden (1994).

[E-R-S] Y. Edel, E.M. Rains, N.J.A. Sloane, *On kissing numbers in dimensions 32 to 128*, Elect. J. Combinatorics **5** (1998), paper R22.

[El1] N. Elkies, *Mordell-Weil lattices in characteristic 2, I: Construction and first properties*, Internat. Math. Res. Notice **8** (1994), 343–361.

[El2] N. Elkies, *A characterization of the $\mathbb{Z}^n$ lattice*, Math. Res. Lett. **2** (1995), 321–326.

[El3] N. Elkies, *Lattices and codes with long shadows*, Math. Res. Lett. **2** (1995), 643–651.

[El4] N. Elkies, *Mordell-Weil lattices in characteristic 2, II: The Leech lattice as a Mordell-Weil lattice*, Invent. Math. **128** (1997), 1–8.

[El5] N. Elkies, *The Klein quartic in number theory*, in The Eightfold Way, S. Levy ed., Math. Sc. Research Inst. Pub. **35**, Cambridge University Press, Cambridge, U.K. (1999).

[El6] N. Elkies, *Lattices, linear codes, and invariants, Part I* and *Part II*, Notices of the A.M.S. **47** (2000), 1238–1245 *and* 1382–1391.

[El7] N. Elkies, *Mordell-Weil lattices in characteristic 2, III: A Mordell-Weil lattice of rank* 128, Exp. Math. **10** (2001), 467–473.

M. Enguehard. See M. Broué.

[Fe1] W. Feit, *On integral representations of finite groups*, J. London Math. Soc. **29** (1974), 633–683.

[Fe2] W. Feit, *Some lattices over* $\mathbb{Q}(\sqrt{-3})$, J. Algebra **52** (1978), 248–263.

[Gi] M. Gindraux, after a talk at the Bordeaux Number Theory Seminar, December 1st, 2002 (in preparation).

[Gro] B. Gross, *Group representation and lattices*, J. Amer. Math. Soc **3** (1990), 929–960.

[G-B] L.C. Grove, C.T. Benson, *Finite Reflection Groups*, Graduate Texts in Math. **99**, Springer-Verlag, Heidelberg (1971, 1985).

[Ha] A.J. Hahn, *The coset lattices of E.S. Barnes and G.E. Wall*, J. Austral. Math. Soc. **49** (1990), 418–433.

[Hl] T.C. Hales, *The Kepler Conjecture*, http://www.math.lsa.umich.edu/~hales/.

R.H. Hardin. See J.H. Conway.

[H-W] G.H. Hardy, E.M. Wright, *An Introduction to the Theory of Numbers*, Clarendon Press (Oxford Science Publication), Oxford (1979). [Fifth edition; first edition: 1938.]

[Hecke] E. Hecke, *Mathematische Werke*, VandenHoeck & Ruprecht, Göttingen (1970).

[He1] Ch. Hermite, *Sur la théorie des formes quadratiques ternaires*, J. reine angew. Math **40** (1850), 173–177. [= [He3], tome I, pp. 94–99.]

[He2] Ch. Hermite, *Lettres à Jacobi*, J. reine angew. Math. **40** (1850), 261–278. [= [He3], tome I, pp. 100–121.]

[He3] Ch. Hermite, *Œuvres*, Gauthier-Villars, Paris (1905).

[Hof] N. Hofreiter, *Über Extremformen*, Monatshefte Math. Phys. **40** (1933), 129–152.

[Hmb1] P. Humbert, *Théorie de la réduction des formes quadratiques définies positives dans un corps algébrique K fini*, Comment. Math. Helv. **12** (1940), 263-306.

[Hmb2] P. Humbert, *Réduction de formes quadratiques dans un corps algébrique fini*, Comment. Math. Helv. **23** (1949), 50–63.

[Hmp1] J.E. Humphreys, *Introduction to Lie algebras and representation theory*, Springer-Verlag, Heidelberg (1980).

[Hmp2] J.E. Humphreys, *Reflection groups and Coxeter groups*, Cambridge University Press, Cambridge, U.K. (1992).

D. Husemoller. See J. Milnor.

[I] M.I. Icaza, *Hermite constant and extreme forms for algebraic number fields*, J. London Math. Soc. **55** (1977), 11–22.

M.I. Icaza. See also R. Baeza.

[Ja1] D-O. Jaquet, *Domaines de Voronoï et algorithme de réduction des formes quadratiques définies positives*, Sém. Th. Nombres de Bordeaux (= J. Th. Nombres de Bordeaux) **2** (1990), 163–215.

[Ja2] D-O. Jaquet, *Énumération complète des classes de formes parfaites en dimension* 7, Thèse, Neuchâtel (1991).

[Ja3] D-O. Jaquet, *Classification des réseaux dans* $\mathbb{R}^7$ *(à l'aide de la notion de formes parfaites)*, Astérisque **198–199–200** (1992), 177–185.

[Ja4] D-O. Jaquet-Chiffelle, *Description des voisines de* E_7, D_7, D_8 *et* D_9, Sém. Th. Nombres de Bordeaux (= J. Th. Nombres de Bordeaux) **4** (1992), 273–374.

[Ja5] D-O. Jaquet-Chiffelle, *Énumération complète des classes de formes parfaites en dimension* 7, Ann. Inst. Fourier **43,1** (1993), 21–55.

[Ja6] D-O. Jaquet-Chiffelle, *Trois théorèmes de finitude pour les G-formes*, J. Th. Nombres de Bordeaux **7** (1995), 165–176.

[Ja7] D-O. Jaquet-Chiffelle, Mail to J. Martinet, June 14, 1993.

[Ja-S1] D-O. Jaquet, F. Sigrist , *Formes quadratiques contiguës à D_7*, C. R. Acad. Sc. Paris **A 309** (1989), 641–644.

[Ja-S2] D-O. Jaquet-Chiffelle, F. Sigrist, *Classification des formes quadratiques réelles : un contre-exemple à la finitude*, Acta Arith. **68** (1994), 291–294.

[Ka-L] G.A. Kabatiansky, V.I. Levenshtein, *Bounds for packings on a sphere and in space*, Problems of Information Transmission **14** (1978), 1–17.

[Kan] R. Kane, Reflection Groups and Invariant Theory, CMS Books in Mathematics, Springer-Verlag, Heidelberg (2001).

[Ker] M. Kervaire, *Unimodular lattices with a complete root system*, L'Ens. Math. **40** (1994), 59–104.

[Kg] O. King, *A mass formula for unimodular lattices with no roots*, preprint; see also: *Table of masses of 32-dimensional even unimodular lattices having any given root system* (2 megabytes). Available at http://www.math.berkeley.edu/~king/.

[Ki] Y. Kitaoka, *Arithmetic of quadratic forms*, Cambridge University Press, Tract **106**, Cambridge, U.K. (1993).

[Kn1] M. Kneser, *Two remarks on extreme forms*, Canad. J. Math. **7** (1955), 145–149.

[Kn2] M. Kneser, *Klassenzahlen definiter quadratischen Formen*, Archiv der Math. **8** (1957), 241–250.

[K-M-R-T] M.-A. Knus, A. Merkurjev, M. Rost, J.-P. Tignol, The Book of Involutions, American Mathematical Society (1998).

[K-V] H. Koch, B. Venkov, *Über ganzzahlige unimodulare euklidische Gitter*, J. reine angew. Math. **398** (1989), 144–168.

[K-Z1] A. Korkine, G. Zolotareff, *Sur les formes quadratiques positives quaternaires*, Math. Ann. **5** (1872), 581–583.

[K-Z2] A. Korkine, G. Zolotareff, *Sur les formes quadratiques*, Math. Ann. **6** (1873), 366–389.

[K-Z3] A. Korkine, G. Zolotareff, *Sur les formes quadratiques positives*, Math. Ann. **11** (1877), 242–292.

[Ks-P] F.R. Kschischang, S. Pasupathy, *Some ternary and quaternary codes and associated sphere packings*, Trans. Information Theory **38** (1992), 227–246.

[Lah] M. Laïhem, *Construction algorithmique de réseaux parfaits*, Thèse, Bordeaux (1992).

[Lar] J. Larmouth, *The enumeration of perfect forms*, in Computers in Number Theory, A.O.L. Atkin and J. Birch ed., Academic Press, New–York, 237–239.

[Le-Sl] J. Leech, N.J.A. Sloane, *Sphere packing and error-correcting codes*, Canad. J. Math. **23** (1971), 718–745. [= Chap. 5 of [C-S].]

[Le-Se] P.W.H. Lemmens and J.J. Seidel, *Equiangular lines*, J. Algebra **24** (1973), 494–512.

[L-S-T] W. Lempken, B. Schröder, P.H. Tiep, *Symmetric squares, spherical designs, and lattice minima*, J. Algebra **240** (2001), 185–208.

[LLL] A.K. Lenstra, H.W. Lenstra, L. Lovász, *Factoring polynomials with rational coefficients*, Math. Ann. **261** (1982), 515–534.

V.I. Levenshtein. See G.A. Kabatiansky.

[vL] J.H. van Lint, *Introduction to coding theory*, Graduate Texts in Math. **86**, Springer-Verlag, Heidelberg (1999). [Third edition; first edition: 1982.]

[Lo-So] P. Loyer, P. Solé, *Les réseaux* BW_{32} *et* U_{32} *sont équivalents*, J. Th. Nombres de Bordeaux **6** (1994), 359–362.

M. Mc Connel. See Ash.

[McW-S] F.J. MacWilliams, N.J.A. Sloane, *The Theory of Error Correcting Codes*, North Holland, Amsterdam (2nd printing, 1978).

[Mar] J. Martinet, *Les réseaux parfaits des espaces euclidiens*, Masson (now Dunod), Paris (1996).

[Mar1] J. Martinet, *Structures algébriques sur les réseaux*, in Number Theory, (Sém. Th. Nombres de Paris 1992–93), S. David ed., Cambridge University Press, Cambridge, U.K. (1995), 167–186.

[Mar2] J. Martinet, *Une famille de réseaux dual-extrêmes*, J. Th. Nombres de Bordeaux **9** (1997), 169–181.

[Mar3] J. Martinet, *Algebraic constructions of lattices; isodual lattices*, Acts of the Eger congress (Hungary, 1996), Walter de Gruyter (1998), 349–360.

[Mar4] J. Martinet, *Perfect and eutactic lattices: extensions of Voronoï's theory*, in Voronoï's Impact on Modern Science, P. Engel, H. Syta ed., Proc. Inst. Math. National Acad. Sc. Ukrain **21**, book 1, Kyiv (1998), 186–198.

[Mar5] J. Martinet, *Sur la classification des réseaux parfaits de dimension* 5, J. Th. Nombres de Bordeaux **11** (Actes des Journées Arithmétiques de Limoges de 1997) (1999), 149–159.

[Mar6] J. Martinet, *Sur l'indice d'un sous-réseau* (with an appendix by C. Batut), in Réseaux euclidiens, designs sphériques et groupes, L'Ens. Math., Monographie **37**, Genève (2001), 163–211.

[Mar7] J. Martinet, *Sur certains designs sphériques liés à des réseaux entiers*, in Réseaux euclidiens, designs sphériques et groupes, L'Ens. Math., Monographie **37**, J. Martinet, ed., Genève (2001), 135–146.

[Mar8] J. Martinet, *Reduction Modulo 2 and 3 of Euclidean Lattices*, J. Algebra **251** (2002), 864–887.

[Mar-V] J. Martinet, B. Venkov, *Les réseaux fortement eutactiques* (with an appendix by R. Coulangeon), in Réseaux euclidiens, designs sphériques et groupes, L'Ens. Math., Monographie **37**, J. Martinet, ed., Genève (2001), 112–134.

J. Martinet. See also C. Batut, E. Bayer-Fluckiger, A-M. Bergé.

[Mic] L. Michel, E-mail to the author (September 12, 1997).

[M-H] J. Milnor, D. Husemoller, *Symmetric Bilinear Forms*, Springer-Verlag, Ergebnisse **73**, Heidelberg (1973).

[Min1] H. Minkowski, *Gesammelte Abhandlungen*, Chelsea, New York (1967; first edition: Leipzig, 1911).

[Min2] H. Minkowski, *Geometrie der Zahlen*, Johnson reprint Co., New York (1968; first edition: Leipzig, 1896).

[Mor] L.J. Mordell, *Observation on the minimum of a positive definite quadratic form in eight variables*, J. London Math. Soc. **9** (1944), 3–6.

[Nap1] H. Napias, *Sur quelques réseaux contenus dans les réseaux de Leech et de Quebbemann*, Comptes Rendus Acad. Sc. Paris **319**, série I (1994), 653–658.

[Nap2] H. Napias, *Étude expérimentale et algorithmique des réseaux euclidiens*, Thèse, Bordeaux (1996).

[Ne1] G. Nebe, *Endliche rationale Matrixgruppen vom Grad* 24, Aachener Beiträge zur Mathematik, Verlag der Augustinus Buchhandlung (1995).

[Ne2] G. Nebe, *Finite subgroups of* $\mathrm{GL}_n(\mathbb{Q})$ *for* $25 \leq n \leq 31$, Comm. Alg. **24** (1996), 2341–2397.

[Ne3] G. Nebe, *Finite subgroups of* $\mathrm{GL}_{24}(\mathbb{Q})$, Exp. Math. **5** (1996), 163–195.

[Ne4] G. Nebe, *The normalizer action and strongly modular lattices*, L'Ens. Math. **43** (1997), 67–76.

[Ne5] G. Nebe, *Some Cyclo-quaternionic Lattices*, J. Algebra **199** (1998), 472–498.

[Ne-Pl] G. Nebe, W. Plesken, *Finite rational matrix groups*, Memoirs A.M.S. **116, no. 556** (1995), 1–144.

[Ne-R-S] G. Nebe, E. Rains, N.J.A. Sloane, *A Simple Construction for the Barnes–Wall Lattices*, in The Forney Festschrift, R. Blahut ed., (2002, to appear).

[Ne-Sl] G. Nebe, N.J.A. Sloane, *A Catalogue of Lattices*, http://www.research.att.com/~njas/lattices/index.html.

[Ne-V1] G. Nebe, B. Venkov, *Nonexistence of extremal lattices in certain genera of modular lattices*, J. Number Theory **60** (1996), 310–317.

[Ne-V2] G. Nebe, B. Venkov, *The strongly perfect lattices in dimension* 10, J. Th. Nombres de Bordeaux **12** (2000), 503–518.

G. Nebe. See also C. Bachoc.

[Nie] H-V. Niemeier, *Definite Quadratischen Formen der Discriminante* 1 *und der Dimension* 24, J. Number Theory **5** (1973), 141–178.

[No] J. Nottebaum, *Sphärische 4-designs in Gittern*, Diplomarbeit, Universität Oldenburg (1995).

S.P. Norton. See [ATLAS].

A. M. Odlyzko. See [C-S], Chapters 13 and 19.

[Oe1] J. Oesterlé, *Le théorème des classes jumelles de Zimmert et les formules explicites de Weil*, in Sém. Th. Nombres de Paris 1983–84, Prog. Math. **59**, Birkhäuser, Basel (1985), 181–197.

[Oe2] J. Oesterlé, *Empilements de sphères*, in Séminaire Bourbaki 1989-1990, Astérisque **189–190** (1990), 375–397.

[Ogg] A. Ogg, *Modular forms and Dirichlet series*, Benjamin, New York (1969).

[O-W] S. Ohno, T. Watanabe, *Estimates of Hermite Constants for Algebraic Number Fields*, Comm. Math. Univ. Sancti Pauli **50** (2001), 53–63.

[O'M] O.T. O'Meara, *Introduction to Quadratic Forms*, Springer-Verlag, Grundlehren **117**, Heidelberg (1963).

M. M. O'Ryan. See R. Baeza.

R.A. Parker. See [ATLAS].

S. Pasupathy. See F.R. Kschischang.

[Pl1] W. Plesken, *Finite modular groups of prime degree and circulants*, J. Algebra **97** (1985), 286–312.

[Pl2] W. Plesken, *Lattices of covariant quadratic forms*, L'Ens. Math. **47** (2001), 21–56.

[Pl-P1] W. Plesken, M. Pohst, *Constructing integral lattices with prescribed minimum. I*, Math. Comp. **45** (1985), 209–221 and S5–S16.

[Pl-P2] W. Plesken, M. Pohst, *Constructing integral alttices with prescribed Minimum. II*, Math. Comp. **60** (1993), 817–825 .

[Pl-S] W. Plesken, B. Souvignier, *Computing isometries of lattices*, J. Symbolic Computation **24** (1997), 327–334.

W. Plesken. See also G. Nebe.

M. Pohst. See Plesken.

[Q1] H.-G. Quebbemann, *Zur Klassification unimodularer Gitter mit Isometrie von Primzahlordnung*, J. reine angew. Math. **326** (1981), 158–170.

[Q2] H.-G. Quebbemann, *An application of Siegel's formula over quaternion orders*, Mathematika **31** (1984), 12–16.

[Q3] H.-G. Quebbemann, *A construction of integral lattices*, Mathematika **31** (1984), 138–141.

[Q4] H.-G. Quebbemann, *Lattices with theta-functions for $G(\sqrt{2})$ and linear codes*, J. Algebra **105** (1987), 443–450.

[Q5] H.-G. Quebbemann, *Unimodular lattices with isometries of large prime order. II*, Math. Nachr. **156** (1992), 219–224.

[Q6] H.-G. Quebbemann, *Modular lattices in Euclidean spaces*, J. Number Theory **54** (1995), 190–202.

[Q7] H.-G. Quebbemann, *Atkin-Lehner eigenforms and strongly modular lattices*, L'Ens. Math **43** (1997), 55–65.

H.-G. Quebbemann. See also Batut.

[Qi-Z] J.R. Quine, P. L. Zhang, *Extremal symplectic lattices*, Israel J Math. **108** (1998), 237–251.

[R-S] E. Rains, N.J.A. Sloane, *Shadow theory of modular and unimodular lattices*, J. Number Theory **73** (1998), 359–389.

[Ran1] R.A. Rankin, *On positive definite quadratic forms*, J. London Math. Soc. **28** (1953), 309–314.

[Ran2] R.A. Rankin, *On the minimal points of a perfect quadratic form*, Math. Z. **84** (1964), 228–232.

[Rei] I. Reiner, *Maximal orders*, London Math. Soc. Monographs **5**, London (1975).

[Rog] C.A. Rogers, *Packing and covering*, Cambridge University Press, Cambridge, U.K. (1964).

S.S, Ryshkov. See E.P. Baranovskii.

[Sch1] R. Scharlau, *Unimodular lattices over real quadratic fields*, Math. Zeitschrift **216** (1994), 437–452.

[Sch-SP] R. Scharlau, R. Schulze-Pillot, *Extremal lattices*, in Algorithmic Algebra and Number Theory, B.H. Matzat, G.-M. Greul, G. Hiss ed., Springer-Verlag, Heidelberg (1999), 139–170.

[Sch-V] R. Scharlau, B.B. Venkov, *The genus of the Barnes–Wall lattice*, Comment. Math. Helvetici **69** (1994), 322–333.

R. Scharlau. See also C. Batut.

[Sch2] W. Scharlau, *Quadratic and Hermitian Forms*, Grundlehren **270**, Springer-Verlag, Heidelberg (1985).

[Scho] B. Schoeneberg, *Das Verhalten von mehrfachen Thetareihen bei Modulsubstitutionen*, Math. Ann. **116** (1939), 511–523.

[S-W] R. Schoof, L. Washington, *Quintic polynomials and real cyclotomic fields with large class numbers*, Math. Comp. **50** (1988), 543–556.

B. Schröder. See W. Lempken.

[Schu] T. Schulz, *Konstruktion rationaler Darstellungen endlicher Gruppen*, Thesis, Aachen, Lehrstuhl B (2002).

R. Schulze-Pillot. See R. Scharlau.

[Sco1] P.R. Scott, *On perfect and extreme forms*, J. Austral. Math. Soc. **4** (1964), 56–77.

[Sco2] P.R. Scott, *The construction of perfect and extreme forms*, Canad. J. Math. **18** (1966), 147–158.

J.J. Seidel. See P.W.H. Lemmens.

[Se1] J-P. Serre, *Corps locaux*, second edition, Hermann, Paris (1968), translated into English as:

[Se'1] J-P. Serre, *Local Fields*, Springer-Verlag, Heidelberg (1995).

[Se2] J-P. Serre, *Cours d'arithmétique*, P.U.F., Paris (1970), translated into English as:

[Se'2] J-P. Serre, *A Course in Arithmetic*, Graduate Texts in Math. **7**, Springer-Verlag, Heidelberg (1973).

[Se3] J-P. Serre, *Représentations linéaires des groupes finis*, second edition, Hermann, Paris (1971); translated into English as:

[Se'3] J-P. Serre, *Linear Representations of Finite groups*, Graduate Texts in Math. **42**, Springer-Verlag, Heidelberg (1977).

[Se4] J-P. Serre, *Œuvres*, Springer-Verlag, Heidelberg (1986, 2000).

[Sh1] T. Shioda, *Some remarks on elliptic curves over function fields*, Astérisque **209** (1992), 99–114.

[Sh2] T. Shioda, *A uniform construction of the root lattices E_6, E_7, E_8 and their dual lattices* , Proc. Japan Academy **71(A)** (1995), 140–143.

[Sig1] F. Sigrist, E-mail to J. Martinet (September 11, 1990).

[Sig2] F. Sigrist, *Formes Quadratiques encapsulées*, J. Th. Nombres de Bordeaux **2** (1990), 425–429.

[Sig3] F. Sigrist, *Cyclotomic quadratic forms*, J. Th. Nombres de Bordeaux **12** (2000), 519–530.

F. Sigrist. See also A-M. Bergé, D-O. Jaquet.

[Sl1] N.J.A. Sloane, Letter to J. Martinet (May 11, 1992); now expanded in [Ne-Sl] as *A table of the highest kissing numbers presently known.*

[Sl2] N.J.A. Sloane, *The Sphere packing problem*, in Proceedings Internat. Congress Math. Berlin 1998, Documenta Mathematika, III (1998), pp. 387–396.

N.J.A. Sloane. See also N. Bernstein, J.H. Conway, Y. Edel, F.J. MacWilliams, G. Nebe, E. Rains, and Appendix 2 to [Bu-Sa].

P. Solé. See P. Loyer.

[Sou1] B. Souvignier, *Die irreduziblen Bravaisgruppen in $GL_8(\mathbb{Z})$*, Diplomarbeit, RWTH Aachen (1991).

[Sou2] B. Souvignier, *Irreducible finite integral matrix groups of degree 8 and 10*, Math. Comp. **63** (1994), 335–350.

B. Souvignier. See also W. Plesken.

[Sta1] K. C. Stacey, *The enumeration of perfect septenary forms*, J. London Math. Soc. **10** (1975), 97–104.

[Sta2] K. C. Stacey, *The perfect septenary forms with $D_4 = 2$*, J. Austral. Math. Soc. **22** (1976), 144–164.

[Sti] E. Stiemke, *Über positiven Lösungen homogener linearer Gleichungen*, Math. Ann. **76** (1915), 340–342.

[St] M.I. Štogrin, *Locally quasi-densest lattice packings of spheres*, Soviet Math. Dokl. **15** (1974), 1288–1292.

P.H. Tiep. See N. Dummigan, W. Lempken.

[Ti] J. Tits, *Quaternions over $\mathbb{Q}(\sqrt{5})$, Leech's lattice and the sporadic group of Hall-Janko*, J. Algebra **63** (1980), 56–75.

J.A. Todd. See H.S.M. Coxeter.

[Var] A. Vardy, *A new sphere packing in* 20 *dimensions*, Invent. Math. **121** (1995), 119–133.

[Ven1] B.B. Venkov, *On even unimodular Euclidean lattices of dimension 32*, J. Soviet Math. **26** (1984), 1860–1867. (Russian original: 1982.)

[Ven2] B.B. Venkov, *Even unimodular extremal lattices*, Proc. Steklov Inst. Math. **165** (1984), 47–52. (Russian original: 1984.)

[Ven3] B.B. Venkov, *Réseaux et "designs" sphériques* (Notes by J. Martinet), in Réseaux euclidiens, designs sphériques et groupes, L'Ens. Math., Monographie **37**, J. Martinet, ed., Genève (2001), 10–86.
(Referred to erroneously as a future [Bc-V3] in the French edition.)

[Ven4] B.B. Venkov, *On even unimodular Euclidean lattices of dimension 32, II*, J. Soviet Math. **36** (1987), 21–38. (Russian original: 1984.)

[Ven5] B.B. Venkov, *Unimodular lattices and strongly regular graphs*, J Soviet Math. **29** (1985), 1121–1127. (Russian original: 1983.)

B. Venkov. See also R. Bacher, C. Bachoc, J. Martinet, G. Nebe, H. Koch, R. Scharlau.

[Vet] N.M. Vetchinkin, *Uniqueness of the classes of positive quadratic forms on which the values of Hermite constants are attained for* $6 \leq n \leq 8$, Proc. Steklov Inst. Math. **152**, 34–86.

[Vi] M-F. Vignéras, *Arithmétique des Algèbres de Quaternions*, Lecture Notes **800**, Springer-Verlag, Heidelberg (1980).

[Vo1] G. Voronoï, *Nouvelles applications des paramètres continus à la théorie des formes quadratiques : 1 Sur quelques propriétés des formes quadratiques positives parfaites*, J. reine angew. Math **133** (1908), 97–178.

[Vo2] G. Voronoï, *Nouvelles applications des paramètres continus à la théorie des formes quadratiques : 2 Recherches sur les paralléloèdres primitifs, première partie*, J. reine angew. Math **134** (1908), 198–287.

[Vo3] G. Voronoï, *Nouvelles applications des paramètres continus à la théorie des formes quadratiques : 2 Recherches sur les paralléloèdres primitifs, deuxième partie*, J. reine angew. Math **136** (1909), 67–181.

[vdW] B.L. van der Waerden, *Die Reduktionstheorie der positiven quadratischen Formen*, in Studien zur Theorie der quadratischen Formen, B.L. van der Waerden and H. Gross ed., Birkhäuser, Basel (1968), 17–44.

G.E. Wall. See E.S. Barnes.

[Was] L.C. Washington, *Introduction to Cyclotomic Fields*, Graduate Texts **83**, Springer-Verlag, Heidelberg (1982).

L. Washington. See also R. Schoof.

[Wata1] T. Watanabe, *On an analog of Hermite's constant*, J. Lie Theory **10** (2000), 33–52.

[Wata2] T. Watanabe, *Upper bounds for Hermite's constants for orthogonal groups*, Comm. Math. Univ. Sancti Pauli **48** (1999), 25–33.

T. Watanabe. See also S. Ohno.

[Wat1] G.L. Watson, *Transformations of a quadratic form which do not increase the class-number*, Proc. London Math. Soc. **12** (1962), 577–587.

[Wat2] G.L. Watson, *On the minimum of a positive quadratic form in* $n (\leq 8)$ *variables (verification of Blichfeldt's calculation)*, Proc. Camb. Phil. Soc **62** (1966), 719.

[Wat3] G.L. Watson, *On the minimal points of a perfect septenary quadratic form*, Mathematika **16** (1969), 170–177.

[Wat4] G.L. Watson, *On the minimum points of a positive quadratic form*, Mathematika **18** (1971), 60–70.

[Wat5] G.L. Watson, *The number of minimum points of a positive quadratic form*, Dissertationes Math. **84** (1971), 1–46.

[Wat6] G.L. Watson, *The least common denominator of the coefficient of a perfect quadratic form*, Acta Arith. **18** (1971), 29–36.

[Wat7] G.L. Watson, *The number of minimum points of a positive quadratic form having no perfect binary section with the same minimum*, Proc. London Math. Soc. **24** (1972), 625–646.

R.A. Wilson. See [ATLAS].

[Zah] N.V. Zahareva, *Centerings of 8-dimensional lattices that preserve a frame of successive minima*, Proc. Steklov Inst. Math. **152** (1982), 107–134. (Russian original: 1980.)

[Zas] H. Zassenhaus, *The Theory of Groups* (second edition), Chelsea, New York (1958).

P.L. Zhang. See J.R. Quine.

[Zi] R. Zimmert, *Ideale kleiner Norm in Idealklassen und eine Regulator Abschätzung*, Invent. Math. **62** (1981), 367–380.

G. Zolotareff. See A. Korkine.

List of Symbols

$\lVert x \rVert$	$\sqrt{x \cdot x} = \sqrt{N(x)}$, 4
$\lVert \Lambda \rVert$	the square root of the norm of Λ, 4
$\langle u, v \rangle$	Voronoi scalar product of $u, v \in \mathrm{End}^s(E)$, 70
$*$	notation for duality over $\mathbb{Z}$, 8
$\simeq$	an isometry or an algebraic isomorphism, 12
$\prec$	ordering relation on minimal classes, 321

A

$\mathbb{A}_n$	root lattice A_n, 110
$\boldsymbol{A}_n$	root system of type A_n, 133
$\mathbb{A}_n^{(r)}$	r-th Craig lattice, 164
$\mathbb{A}_n^r$	Coxeter lattices ($r \mid n+1$), 153
$\mathbb{A}_n^r$	Coxeter–Barnes lattices, 147
$\mathbb{A}_{n,r}$	Conway–Sloane unified definition for L_n^r and M_n^r, 209, 284
A_{r_1,r_2}	star body attached to a number field, 53
Arf_Q	Arf invariant of the quadratic form Q, 18
$\mathrm{Aut}(\Lambda)$	automorphism group of Λ, 12
$\mathrm{Aut}^{\#}(\Lambda)$	isoduality group of Λ, 400

B

B_{r_1,r_2}	Minkowski's convex body for number fields, 55
$\boldsymbol{B}_n$	root system of type B_n, 133
Bac_{32}	a 2-modular lattice found by Bachoc, 318
BW_{16}	16-dimensional Barnes–Wall lattice, 274
BW_n	Barnes–Wall lattices ($n = 2^r$), 310

C

$\boldsymbol{C}_n$	root system of type C_n, 133
ccc	central centred-cuboidal lattice, 403
$\mathcal{C}_{\mathfrak{O}}$	codifferent of the order $\mathfrak{O}$, 267
Cox_n	a generalization of some Coxeter $\mathbb{A}_n^r$, 457

CT_n	some lattices over the ring of Eisenstein integers ($n \equiv 0 \mod 6$), 285

D

$\mathbb{D}_n$	root lattice D_n, 112
$\mathbb{D}_n^+$	lattice (or packing) D_n^+, 114
$\boldsymbol{D}_n$	root system of type D_n, 133
$\sharp$	notation for other kinds of duality, 267
$\mathbb{D}_{n,t}$	Barnes's enlargement of $\mathbb{D}_n$, 172
$\mathcal{D}_\Lambda$, $\mathcal{D}_Q$,	Voronoi domain of the lattice Λ, or of the quadratic form Q, 230, 445
$\mathcal{D}_{\mathfrak{O}}$	different of the order $\mathfrak{O}$, 267
$\det(\Lambda)$	determinant of the lattice Λ, 5
$\Delta(\Lambda, \mathcal{S})$	difference lattice of Λ for S, 163
$\Delta(\Lambda)$	discriminant of the lattice Λ, 5
d_Q	discriminant of the quadratic form Q, 18
d_K	discriminant of the number field K, 54

E

$\mathbb{E}_6$, $\mathbb{E}_7$	root lattices E_6, E_7, 117
$\mathbb{E}_8$	root lattice E_8, 115
$\boldsymbol{E}_n$ ($n = 6, 7, 8$)	root system of type E_n, 133
$\mathrm{End}(V)$	endomorphism ring of a vector space V, VII
$\mathrm{End}^a(E)$	space of skew-symmetric endomorphisms of E, 67
$\mathrm{End}^s(E)$	space of symmetric endomorphisms of E, 67
$\mathrm{End}^s_G(E)$	centralizer of G in $\mathrm{End}^s(E)$, 385
$\mathrm{End}^{s+}(E)$	space of positive (sym.) endomorphisms of E, 68
$\mathrm{End}^{s++}(E)$	space of positive def. endomorphisms of E, 68

F

$\boldsymbol{F}_4$	root system of type F_4, 133

G

$G^\#(\Lambda)$	G-isoduality group (of a G-lattice Λ), 401
$\boldsymbol{G}_2$	root system of type G_2, 133
Γ_n	lattice constant of the n-dimensional unit sphere, 46
$\gamma(\Lambda)$	Hermite invariant of Λ, 40
γ_n	Hermite constant for dimension n, 40
$\gamma_{\mathcal{C}}$	Hermite invariant of the minimal class $\mathcal{C}$, 343
$\gamma_{n,k}$	Rankin invariants or constants, 56, 376
$\gamma'(\Lambda)$	dual Hermite invariant, 55
γ'_n	dual Hermite constant for dimension n (alias Bergé–Martinet constant), 55
$\mathcal{G}$	generic notation for a closed subgroup of $\mathrm{GL}(E)$, 363

$\mathcal{G}^\circ$	the connected component of the unit element in a topological group $\mathcal{G}$, 364
$\mathrm{GL}(V)$	general linear group of a vector space V, VII
$\mathrm{GL}(\Lambda)$	general linear group of a lattice Λ, 3
$\mathrm{Gram}(\cdot)$	Gram matrix of a system of vectors, 4

H

$h(R)$	Coxeter number of a simple root system R, 137
HNF	Hermite normal form, 170

J

J_n	some lattices related to the Hurwitz order, 271
J'_n	some lattices over the Hurwitz order (defined only for $n \equiv 0 \mod 4$), 274

K

$K_e(\Lambda)$, Λ^e	Kneser neighbour of Λ for e, 179
K_{12}	Coxeter–Todd lattice, 285
K_n	a sequence of lattices related to the lattice K_{12}, 287, 305, 309
K'_n	a sequence of lattices related to the lattice K_{12}, 289, 305, 309
κ_{r_1,r_2}	lattice constant of A_{r_1,r_2}, 53

L

lattice	a discrete subgroup of maximal rank in E, 1
Λ_e	half-lattice of Λ and e, 178
Λ^e, $K_e(\Lambda)$	Kneser neighbour of Λ for e, 179
Λ_n, $\Lambda_n^{\max}, \ldots$	laminated lattices, 93, 309
$\Lambda_{11}^{\mathrm{mid}}$	a sublattice of $\Lambda_{12}^{\mathrm{mid}}$ (not a laminated lattice), 317
$\mathcal{L}$	space of lattices in E, 1
$\mathcal{L}_G$	space of G-lattices, 385
L_n	see L_n^r, $(n,r) = (2m,m)$ or $(2m+1,m)$, 280
L_n^r	Barnes lattices related to Eisenstein integers, 282

M

$\min(\Lambda)$	minimum (norm) of Λ, 4
M_n^r	Barnes lattices related to Eisenstein integers, 283
$\mathcal{M}_{p,q}(R)$	module of $p \times q$ matrices with entries in a ring R, VII
$\mathcal{M}_n(R)$	algebra of $n \times n$ matrices with entries in a ring R, VII

N

$N(\Lambda)$	norm (minimum) of Λ, 4

$N(x)$	norm of $x \in E$ $(N(x) = x \cdot x)$, 4
N, $\mathrm{N}_{L/K}$	norm in an algebra, 480
Ne_{32}	a 2-modular lattice found by Nebe, 318
Nrd, $\mathrm{Nrd}_{L/K}$	reduced norm in a semi-simple algebra, 480

O

$\mathrm{O}(E)$	orthogonal group of E, 12
O_{23}	shorter Leech lattice, 181
O_{24}	odd Leech lattice, 181
ω_x	a projection inside $\mathrm{End}^s(E)$ of a projection inside E, 75, 388
Ω_x	variant of ω_x for quadratic forms, 444

P

p_F	orthogonal projection in E onto a subspace F, 8
p_x	orthogonal projection in E onto $\mathbb{R}x$ $(x \neq 0)$, 71
φ	Euler function, 393
φ_x	the linear form $u \mapsto u(x) \cdot x$ on $\mathrm{End}^s(E)$, 73
P_n	Barnes's lattices P_n, 157
P_n^i	*i–th* perfect lattice in dimension n, 208
$\wp$	Artin–Schreier map, 19

Q

Q_{10}, Q_{14}	Souvignier's lattices (see [C-S9]), 317
Q_{10}, Q_{14}	Souvignier's lattices, see [C-S9], 475
Qb_{32}, Qb'_{32}	2-modular lattices found by Quebbemann, 318
Q_n	Barnes's lattices Q_n, 162

S

$S(\Lambda)$, S	sphere of Λ, 4
$s(\Lambda)$, s	kissing number of Λ, 4
$\mathrm{SL}(V)$	special linear group of a vector space V, VII
$\mathrm{Smith}(\Lambda)$	Smith invariant of Λ, 27, 483

T

${}^tM$, ${}^tu$	transpose of $M \in \mathcal{M}_{p,q}(R)$, of $u \in \mathrm{End}(E)$, 5, 67
Trd, $\mathrm{Trd}_{L/K}$	reduced trace in a semi-simple algebra, 480

U

U_n, U'_n	(in Chap. 4) quadratic forms for the lattice $\mathbb{A}_n$, 111
U_n	(in Chap. 8) quaternionic lattices, 302

V

V_n, V_n' (in Chap. 4) quadratic forms for the lattice $\mathbb{D}_n$, 113
V_n (in Chap. 8) lattices related to U_n, 304

W

W_6, W_7 (in Chap. 4) quadratic forms for $\mathbb{E}_6$ and $\mathbb{E}_7$, 117
W_8, W_8' (in Chap. 4) quadratic forms for $\mathbb{E}_8$, 115
W_n (in Chap. 10) "Watson's lattices", 378
WR well–ounded (lattice or form), 14
WR well–rounded (lattice or form), 448

Z

$\mathbb{Z}^n$ lattice with integer coordinates, 110

Index

A

absolutely extreme lattice
for a body, 45
for a distance function, 47
absolutely irreducible representation, 383
additive determinant, 19
adjoint
form, 23
matrix, 23
admissible lattice
for a body, 45
admissible subset of $\mathbb{Z}$, 354
alternating, 17
antilamination, 317
Arf invariant, 18
Artin–Schreier map, 19
Ash, 347
augmentation, 165
automorphism
of a configuration, 86
of a distance-function, 48
of a root system, 131
automorphism group
of a lattice, 12
of a minimal class, 355
of a quadratic form, 22
axial centred-cuboidal lattice, 344

B

Bachoc, 167, 182, 318, 493, 495
Baeza, 108
Bannai, 84, 225, 476
Baranovskii, 208, 225
Baril, 217
Barnes, 93, 106, 149, 157, 187, 208, 225, 282
Barnes–Wall lattice BW_{16}, 274
Barnes–Wall lattices BW_{2^k}, 310
Batut, 167, 182, 218, 355
Bavard, 356, 381, 426
Bayer, 176, 394, 426
Bergé, 99, 100, 106, 323, 330, 349, 355
Bergé and Martinet, 97, 380, 425, 457
Bergé–Martinet constant, 64
bilinear form, 8
Borcherds, 176
Bravais group, 387
Buser-Sarnak, 426

C

C.M. field, 264, 487
Cayley order, 127
central centred-cuboidal lattice, 403
centre density, 40
centred cubic lattice, 51
characteristic
determinant, 191
vector, 11
class group, 487
code, 128
codifferent (of an order), 267, 484
cofactor, 23

compact type (for a distance-function), 63
contiguous
form, 235, 446
lattice, 236
$\mathcal{F}$-, $\mathcal{T}$-contiguous form, 460
Conway, 175, 177
Conway and Sloane, 14, 93, 176, 211, 284, 305, 416, 426, 467
Conway–Sloane, 107
Conway-Sloane, 318
Coulangeon, 108, 378, 415, 495
Coxeter, 153
groups, 136
lattices, 149, 457
number, 137
transformation, 137
Coxeter–Barnes lattices, 149
Coxeter–Dynkin diagram, 120, 135
Coxeter–Todd lattice, 285, 298
Craig, 167, 177, 187
Craig lattices, 164
critical lattice, 40, 47, 78
for a body, 45
for a distance function, 47
relatively to a section, 92
k-critical, 377
cross-section, 427
Csóka , 210
cyclotomic lattices, 393

D

dead-end, 446
decomposition into squares, 23
Dedekind domain, 483
deficiency (of a simple root system), 137
definite at v, 482
definite endomorphism, 68
density, 24
design
(Grassmannian t-) design, 495
(spherical t-) design, 489
determinant
of a lattice, 5
of a set of vectors, 2
difference lattice, 163
different (of an order), 267
differential manifold, 364
discriminant
of a lattice, 5
of a number field, 54
of a quadratic form, 18
distance-function, 46
division algebra, 479
dual
(partial), 30
basis, 8
code, 128
Hermite constant, 55
Hermite invariant, 55
lattice, 8
module, 8
$\mathfrak{O}$-dual (of a lattice), 267
dual-critical lattice, 55
dual-eutactic lattice, 96
dual-extreme lattice, 55, 95
G-dual-extreme lattice, 421
dual-minimal class, 321
dual-perfect lattice, 96
G-dual-perfect lattice, 421
Dummigan, 489
Dynkin diagram, 120, 135

E

Eichler, 15
Eisenstein integers, 266
elementary divisors, 2, 483
Elkies, 477
equivalent
quadratic forms, 21, 449
representations, 383
G-equivalent lattices, 451
$\mathcal{T}$-equivalent quadratic forms, 449
Euclidean structure, 4
Euler characteristic of a group, 347
Euler function, 393

eutactic
 family of vectors, 73
 lattice, 75
 lattice (relative notion), 435
 quadratic form, 76
 star, 91
$\mathcal{G}$-eutactic lattice, 368
k-eutactic lattice, 377
σ-eutactic lattice, 412
eutaxy coefficients
 of a family of vectors, 73
 of a lattice, 75
even lattice, 6
even part (of a lattice), 6
exceptional
 root lattices, 133
 root systems, 133
exponential map, 365
extended Hamming code, 128
extension (of a quadratic form), 440
exterior powers
 of a lattice, 32
 of a module, 32
extremal lattice, 493
extreme lattice, 47, 78
 for a body, 45
 for a distance function, 47
 relatively to a section, 92, 434
G-extreme (G-lattice), 388
$\mathcal{G}$-extreme lattice, 368
strictly $\mathcal{G}$-extreme lattice, 368
k-extreme, 377
strictly k-extreme, 377
σ-extreme lattice, 412

F

face, 232
face–centred cubic lattice, 113
facet, 232, 446
 (opposite), 241
 (ordinary), 245
 (special), 245
facet vector, 232, 235, 446
faithful representation, 384
Feit, 170
field of definition of a lattice, 342
form
 (bilinear), 8
 (quadratic), 17
fractional ideal, 482, 483
fragile lattice, 344
fundamental parallelotope, 1

G

Gaussian integers, 266
Gram matrix, 4
Gram–Schmidt orthogonalization, 23
Gross, 318

H

Hadamard inequality, 37
Hadwiger, 90
half-lattice, 178
Hall–Janko simple group J_2, 318
Hamilton quaternions, 482
Hamming code, 128
Hermite
 constant, 40
 inequality, 39, 41
 invariant, 40
 normal form, 170
 reduction, 60
($\mathfrak{O}$-)Hermitian lattice, 264
hexagonal lattice, 40
HKZ reduction, 60
hole, 25
Humbert, 107
Hurwitz order, 266
hyperbolic plane, 19

I

Icaza, 108
inner coefficients, 60
integral
 lattice, 5
 representation, 385

$\mathfrak{O}$-integral lattice, 267
invariant
lattice under $G \Leftrightarrow G$-lattice, 385
matrix, 386
modulo 8, 12
quadratic form, 386
involution (on an algebra), 479
irreducible
lattice, 15
representation, 87, 383
root system, 131
(almost) irred. representation, 494
isodual
lattice, 400
minimal class, 423
$(G^{\#}, G)$-isodual lattice, 401
σ-isodual lattice, 372, 400
isoduality, 400
group, 400
of orthogonal type, 409
of symmetric type, 409
of symplectic type, 409
isolated lattice, 48
G-isometry, 385
isotypic
lattice, 15
representation, 391

J

Jacobi and Bravais, 2
Jaquet, 208, 217, 225, 419, 426

K

Kervaire, 137
kissing number, 4
Kneser, 15, 106, 179
neighbour, 179
neighbouring algorithm, 178
weak — neighbour, 186
Korkine–Zolotareff, 61, 76, 77, 80, 81, 190

L

Laïhem, 217
laminated lattice, 93, 309
(strongly), 93
(weakly), 93
lattice, 1
G-lattice, 385
$\mathfrak{O}$-lattice, 266
constant, 45
of index (at most) d, 202
Leech lattice, 175, 177
Lempken, 494
length
of a code, 128
of a lattice ($\|\Lambda\|$), 4
of a vector ($\|x\| = \sqrt{x \cdot x}$), 4
2-length of a lattice, 203
level (of a lattice), 27, 401
Lie
algebra, 365
group, 364
LLL-reduction, 467
Lorentz form, 120

M

Mahler's compactness theorem, 43
Martinet, 100, 314, 394, 493
maximal order, 265
minimal
class, 321
norm (of a lattice), 4
vector, 4
k-minimal section (of a lattice), 378
minimal-equivalent
G-lattices, 419
$(G^{\#}, G)$-lattices, 420
lattices, 321, 373
quadratic forms, 322
minimal k-class, 379
k-minimal-equivalent lattices, 379
minimum
of a lattice, 4
of a quadratic form, 22

Minkowski, 50, 52, 53, 55, 60
Minkowski's theorem
on convex bodies, 52
on successive minima, 50
modular lattice, 267, 401
$\mathfrak{a}$-, λ-modular lattice, 267
q-modular lattice, 401
Mordell, 379
Mordell inequality, 41

N

Napias, 218
Nebe, 491, 495
neighbouring form, 235
Niemeier lattices, 175
nonsingular (quadratic form), 18
norm, 265, 480
in an algebra, 480
of a lattice, 4
of a similarity, 401
of a vector, 4
normal lattice, 401
Nottebaum, 492

O

odd lattice, 6
odd Leech lattice, 181
Odlyzko, 225, 476
Oesterlé, 107
opposite facet, 241
order, 264, 484
(maximal) order, 265
orthogonal
basis, 4
group, 12
sum, 15
type (isoduality of), 409
orthonormal basis, 4
outer coefficients, 60

P

parity check, 128
parity vector, 11
partial dual, 30
patchwork lattices, 438
perfect
class, 323
family of vectors, 73
lattice, 75
lattice (relative notion), 435
pair, 349
quadratic form, 76
G-perfect (G-lattice), 388
$\mathcal{G}$-perfect lattice, 368
k-perfect lattice, 377
σ-perfect lattice, 412
perfection corank
of a family of vectors, 73
of a lattice, 75
perfection rank
of a family of vectors, 73
of a lattice, 75
of a minimal class, 324
of a quadratic form, 76, 444
Plesken and Pohst, 107, 305, 317
positive definite endomorphism, 68
positive endomorphism, 68
prime ideal, 485
primitive
lattice, 5
vector in a lattice, 7
projections cone, 228, 452
Prouhet and Tarry, 167

Q

quadratic form, 17
quasi-perfect pair, 350
quaternion algebra, 481
Quebbemann, 309, 314, 318, 426, 493

R

Rankin, 57, 210, 376, 495
constants, 56, 376
invariants, 56, 376
rational
lattice, 5
representation, 383

reduced
    characteristic polynomial, 480
    different, 485
    discriminant, 485
    norm, 265, 480, 487
    root system, 131
    trace, 480
reduction, 23
    (HKZ), 60
    (Hermite), 60
    (LLL), 467
    (Minkowski), 60
    (Voronoi), 262
    of quadratic forms, 60
Reed–Muller code, 310
relative lattice, 1
representation, 383
    augmentation —, 465
    permutation —, 455
root
    in a lattice, 132
    in a root system, 131
root system, 131
Ryshkov, 208, 225

S

scalar product, 4
Schoof and Washington, 425
Schröder, 494
Scott, 107, 187
section, 427
semi-eutactic
    family of vectors, 73
    lattice, 75
    quadratic form, 76
semi-simple
    algebra, 479
    endomorphism, 17
shorter Leech lattice, 181
signature
    of a number field, 54
    of a real quadratic form, 21
    of an isoduality, 409
Sigrist, 169, 279, 426
similarity, 9
G-similarity, 385
simple
    algebra, 479
    root system, 134
skew-symmetric endomorphism, 67
Sloane, 84, 225, 476
Smith invariant, 27, 483
Souvignier, 317, 475
spectrum, 12
sphere (of a lattice), 4
spinor norm, 19
square root of a positive symmetric endomorphism, 69
stable equivalence, — classes, 487
Stacey, 208, 261
star body, 46
Steinitz class, 393
Stiemke, 77
Stirling formula, 53
strong arithmetical laminations, 94
strongly m-perfect lattice, 495
strongly eutactic
    family of vectors, 73
    lattice, 75
    quadratic form, 76
strongly laminated lattice, 93
strongly perfect lattice, 491
successive minima, 49
    for a function, 49
Sylvester's law of inertia, 21
symmetric
    endomorphism, 67
    type (isoduality of), 409
symmetrized
    manifold, 434
    space, 365
symplectic
    group, 20
    type (isoduality of), 409

T

tangent space, 364
Tarry and Escott, 167

tensor product
of Euclidean spaces, 31
of modules, 30
tetracode, 129
Tiep, 494
Tits, 318
totally
definite, 482, 486
trace, 480
transpose of the matrix M, of the endomorphism u, 5, 67
twisted (reduced) trace, 265
two-sided modules, 264

U

unimodular
lattice, 5
symmetric bilinear module, 175
unitary group, 20

V

Venkov, 119, 175, 489, 491–493
index formula, 101
Voronoi, 80, 83, 106, 238
algorithm, 240
cell, 25
cone, 228, 452
domain, 230, 445
graph, 239
graph of classes, 239
reduction, 262
scalar product, 70
space, 228

W

van der Waerden, 51
Watanabe, 108
Watson, 106, 202, 210, 219, 223, 225
weak (ℓ)-neighbours, 186
weak arithmetical laminations, 94
weakly eutactic
family of vectors, 73
lattice, 75
quadratic form, 76
weakly laminated lattice, 93
weight (of a code), 128
well–rounded
form, 448
lattice, 14
Weyl group, 109, 131

Z

Zimmert, 107

Grundlehren der mathematischen Wissenschaften

A Series of Comprehensive Studies in Mathematics

A Selection

223. Bergh/Löfström: Interpolation Spaces. An Introduction
224. Gilbarg/Trudinger: Elliptic Partial Differential Equations of Second Order
225. Schütte: Proof Theory
226. Karoubi: K-Theory. An Introduction
227. Grauert/Remmert: Theorie der Steinschen Räume
228. Segal/Kunze: Integrals and Operators
229. Hasse: Number Theory
230. Klingenberg: Lectures on Closed Geodesics
231. Lang: Elliptic Curves. Diophantine Analysis
232. Gihman/Skorohod: The Theory of Stochastic Processes III
233. Stroock/Varadhan: Multidimensional Diffusion Processes
234. Aigner: Combinatorial Theory
235. Dynkin/Yushkevich: Controlled Markov Processes
236. Grauert/Remmert: Theory of Stein Spaces
237. Köthe: Topological Vector Spaces II
238. Graham/McGehee: Essays in Commutative Harmonic Analysis
239. Elliott: Probabilistic Number Theory I
240. Elliott: Probabilistic Number Theory II
241. Rudin: Function Theory in the Unit Ball of C^n
242. Huppert/Blackburn: Finite Groups II
243. Huppert/Blackburn: Finite Groups III
244. Kubert/Lang: Modular Units
245. Cornfeld/Fomin/Sinai: Ergodic Theory
246. Naimark/Stern: Theory of Group Representations
247. Suzuki: Group Theory I
248. Suzuki: Group Theory II
249. Chung: Lectures from Markov Processes to Brownian Motion
250. Arnold: Geometrical Methods in the Theory of Ordinary Differential Equations
251. Chow/Hale: Methods of Bifurcation Theory
252. Aubin: Nonlinear Analysis on Manifolds. Monge-Ampère Equations
253. Dwork: Lectures on ρ-adic Differential Equations
254. Freitag: Siegelsche Modulfunktionen
255. Lang: Complex Multiplication
256. Hörmander: The Analysis of Linear Partial Differential Operators I
257. Hörmander: The Analysis of Linear Partial Differential Operators II
258. Smoller: Shock Waves and Reaction-Diffusion Equations
259. Duren: Univalent Functions
260. Freidlin/Wentzell: Random Perturbations of Dynamical Systems
261. Bosch/Güntzer/Remmert: Non Archimedian Analysis – A System Approach to Rigid Analytic Geometry
262. Doob: Classical Potential Theory and Its Probabilistic Counterpart
263. Krasnosel'skiĭ/Zabreĭko: Geometrical Methods of Nonlinear Analysis
264. Aubin/Cellina: Differential Inclusions
265. Grauert/Remmert: Coherent Analytic Sheaves
266. de Rham: Differentiable Manifolds
267. Arbarello/Cornalba/Griffiths/Harris: Geometry of Algebraic Curves, Vol. I
268. Arbarello/Cornalba/Griffiths/Harris: Geometry of Algebraic Curves, Vol. II
269. Schapira: Microdifferential Systems in the Complex Domain
270. Scharlau: Quadratic and Hermitian Forms
271. Ellis: Entropy, Large Deviations, and Statistical Mechanics
272. Elliott: Arithmetic Functions and Integer Products

273. Nikol'skiĭ: Treatise on the Shift Operator
274. Hörmander: The Analysis of Linear Partial Differential Operators III
275. Hörmander: The Analysis of Linear Partial Differential Operators IV
276. Liggett: Interacting Particle Systems
277. Fulton/Lang: Riemann-Roch Algebra
278. Barr/Wells: Toposes, Triples and Theories
279. Bishop/Bridges: Constructive Analysis
280. Neukirch: Class Field Theory
281. Chandrasekharan: Elliptic Functions
282. Lelong/Gruman: Entire Functions of Several Complex Variables
283. Kodaira: Complex Manifolds and Deformation of Complex Structures
284. Finn: Equilibrium Capillary Surfaces
285. Burago/Zalgaller: Geometric Inequalities
286. Andrianaov: Quadratic Forms and Hecke Operators
287. Maskit: Kleinian Groups
288. Jacod/Shiryaev: Limit Theorems for Stochastic Processes
289. Manin: Gauge Field Theory and Complex Geometry
290. Conway/Sloane: Sphere Packings, Lattices and Groups
291. Hahn/O'Meara: The Classical Groups and K-Theory
292. Kashiwara/Schapira: Sheaves on Manifolds
293. Revuz/Yor: Continuous Martingales and Brownian Motion
294. Knus: Quadratic and Hermitian Forms over Rings
295. Dierkes/Hildebrandt/Küster/Wohlrab: Minimal Surfaces I
296. Dierkes/Hildebrandt/Küster/Wohlrab: Minimal Surfaces II
297. Pastur/Figotin: Spectra of Random and Almost-Periodic Operators
298. Berline/Getzler/Vergne: Heat Kernels and Dirac Operators
299. Pommerenke: Boundary Behaviour of Conformal Maps
300. Orlik/Terao: Arrangements of Hyperplanes
301. Loday: Cyclic Homology
302. Lange/Birkenhake: Complex Abelian Varieties
303. DeVore/Lorentz: Constructive Approximation
304. Lorentz/v. Golitschek/Makovoz: Construcitve Approximation. Advanced Problems
305. Hiriart-Urruty/Lemaréchal: Convex Analysis and Minimization Algorithms I. Fundamentals
306. Hiriart-Urruty/Lemaréchal: Convex Analysis and Minimization Algorithms II. Advanced Theory and Bundle Methods
307. Schwarz: Quantum Field Theory and Topology
308. Schwarz: Topology for Physicists
309. Adem/Milgram: Cohomology of Finite Groups
310. Giaquinta/Hildebrandt: Calculus of Variations I: The Lagrangian Formalism
311. Giaquinta/Hildebrandt: Calculus of Variations II: The Hamiltonian Formalism
312. Chung/Zhao: From Brownian Motion to Schrödinger's Equation
313. Malliavin: Stochastic Analysis
314. Adams/Hedberg: Function Spaces and Potential Theory
315. Bürgisser/Clausen/Shokrollahi: Algebraic Complexity Theory
316. Saff/Totik: Logarithmic Potentials with External Fields
317. Rockafellar/Wets: Variational Analysis
318. Kobayashi: Hyperbolic Complex Spaces
319. Bridson/Haefliger: Metric Spaces of Non-Positive Curvature
320. Kipnis/Landim: Scaling Limits of Interacting Particle Systems
321. Grimmett: Percolation
322. Neukirch: Algebraic Number Theory
323. Neukirch/Schmidt/Wingberg: Cohomology of Number Fields
324. Liggett: Stochastic Interacting Systems: Contact, Voter and Exclusion Processes
325. Dafermos: Hyperbolic Conservation Laws in Continuum Physics
326. Waldschmidt: Diophantine Approximation on Linear Algebraic Groups
327. Martinet: Perfect Lattices in Euclidean Spaces

Druck: Strauss Offsetdruck, Mörlenbach
Verarbeitung: Schäffer, Grünstadt